International Series in Operations Research & Management Science

Volume 335

The book series **International Series in Operations Research and Management Science** encompasses the various areas of operations research and management science. Both theoretical and applied books are included. It describes current advances anywhere in the world that are at the cutting edge of the field. The series is aimed especially at researchers, advanced graduate students, and sophisticated practitioners.

The series features three types of books:

- Advanced expository books that extend and unify our understanding of particular areas.
- Research monographs that make substantial contributions to knowledge.
- Handbooks that define the new state of the art in particular areas. Each handbook will be edited by a leading authority in the area who will organize a team of experts on various aspects of the topic to write individual chapters. A handbook may emphasize expository surveys or completely new advances (either research or applications) or a combination of both.

The series emphasizes the following four areas:

Mathematical Programming: Including linear programming, integer programming, nonlinear programming, interior point methods, game theory, network optimization models, combinatorics, equilibrium programming, complementarity theory, multiobjective optimization, dynamic programming, stochastic programming, complexity theory, etc.

Applied Probability: Including queuing theory, simulation, renewal theory, Brownian motion and diffusion processes, decision analysis, Markov decision processes, reliability theory, forecasting, other stochastic processes motivated by applications, etc.

Production and Operations Management: Including inventory theory, production scheduling, capacity planning, facility location, supply chain management, distribution systems, materials requirements planning, just-in-time systems, flexible manufacturing systems, design of production lines, logistical planning, strategic issues, etc.

Applications of Operations Research and Management Science: Including telecommunications, health care, capital budgeting and finance, economics, marketing, public policy, military operations research, humanitarian relief and disaster mitigation, service operations, transportation systems, etc.

This book series is indexed in Scopus.

Eduardo Souza de Cursi

Uncertainty Quantification using R

Springer

Eduardo Souza de Cursi
Department Mechanics/Civil Engineering
INSA Rouen Normandie
Saint-Etienne du Rouvray, France

ISSN 0884-8289 ISSN 2214-7934 (electronic)
International Series in Operations Research & Management Science
ISBN 978-3-031-17784-2 ISBN 978-3-031-17785-9 (eBook)
https://doi.org/10.1007/978-3-031-17785-9

This Springer imprint is published by the registered company Springer Nature Switzerland AG
The registered company address is: Gewerbestrasse 11, 6330 Cham, Switzerland

Introduction

This book presents a collection of methods of uncertainty quantification (UQ) – *id est*, a collection of methods for the analysis of numerical data, namely when uncertainty or variability is involved.

The general aim of UQ is to characterize the observed variability in a quantity X by using a random variable U. In the ideal situation, the connection between X and U is perfectly known and the random variable U has a known distribution. Unfortunately, such a situation may be unrealistic in practice and we must also consider situations where this knowledge is imperfect or even non-existent – for example, situations where U is simply unknown: variability is observed without precise knowledge of the cause.

UQ tries to use all the available information about (X, U) to construct an explanation of X by U, into a form which will be useful for use in numerical calculations involving X. The information may be, for instance, an equation, a numerical problem involving both the variables, or samples.

The methods of UQ are general and may be applied to a wide range of situations. They generally belong to the large and well-supplied family of methods based on functional representations, id est, on expansions of the unknowns in series of functions – we find these approaches in particular in Fourier analysis, spectral methods, finite elements, Bayesian optimization, quantum algorithms, etc. It is a large family with numerous and very diversified applications.

Our objective is to present the practical use of UQ techniques under R. We assume that you are a mean user of this kind of software: if you are a user of Scilab, Octave, or Matlab®, you will recognize the instructions and the codes presented will appear as familiar. Evidently, if you are an expert in R, you will find a large number of improvements in our codes and programs: do not hesitate in making your own enhancements and, eventually, in sharing them.

R is a GNU project to develop a tool for language and environment for statistical computing and graphics. An IDE is proposed by RStudio. The popularity of R and RStudio is no more to be demonstrated: you will find on the web many sites and information about R. A wide literature can be found about this software. The community of the users of R proposes a wide choice of packages to extend the possibilities of R. You will find repositories containing them, for instance, https://search.r-project.org/R/doc/html/packages.html#lib-1

Contents

Chapter 1
Some Tips to Use R and RStudio

Abstract This chapter presents the essentials of R, with a focus on the manipulation of variables, plotting, and the use of data frames and classes. We present also some useful packages for standard numerical methods, which will be used in the sequel. Examples of use are given.

1.1 How to Install R and RStudio

If you do not have R installed on your computer, you must download and install it. Installers can be found on Internet, namely at https://cran.r-project.org/ and https://www.r-project.org/. These websites also propose manuals for the installation and the use of R. A large part of this chapter was inspired from these manuals, so that you can download and use them to learn the basic uses of R, which are summarized in the sequel. In this book, we use mainly the versions 4.1.0 to 4.1.3, downloaded from https://cran.r-project.org/.

RStudio is a useful IDE (integrated development environment), which will make the use of R easier. RStudio exists in commercial or free versions. You will find both at the website https://www.rstudio.com/. In this book, we use the free version of RStudio Desktop 1.4.1717.

After installing R and RStudio, you will have on your desktop an icon analogous to this one, which allows launching RStudio by a double click. If it is not created, you can create it by generating a shortcut for RStudio. A double click on the icon opens the window of RStudio, shown in Fig. 1.1.

E. Souza de Cursi, *Uncertainty Quantification using R*, International Series in Operations Research & Management Science 335,
https://doi.org/10.1007/978-3-031-17785-9_1

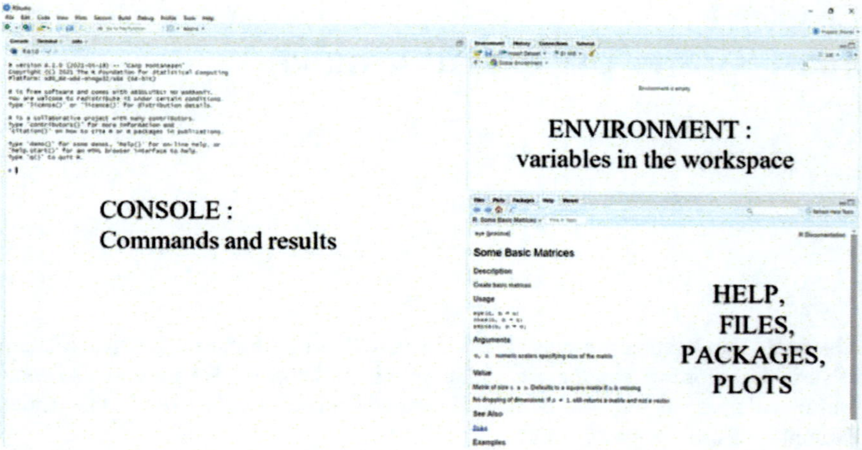

Fig. 1.1 The main window of RStudio

1.2 How to Include a Third-Part Add-In

R can be extended by add-ins usually called packages, created by the community, and offered to the users of R. You will find many packages at the website https://cran.r-project.org/web/packages/. RStudio automatically installs the most used packages. In addition, RStudio possesses a package manager that helps you in adding new packages.

Let us illustrate the use of the package manager by installing the add-in `pracma`, which emulates many MATLAB functions: launch RStudio and click on Packages (at right, in the middle of the window)

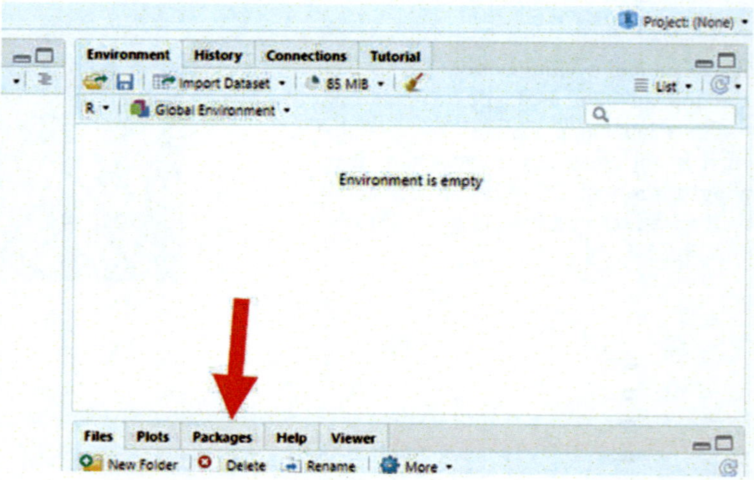

Then, you can see a list of the installed packages. A button *Update* offers the possibility of updating your installed packages. To add a new one, click on *Install*.

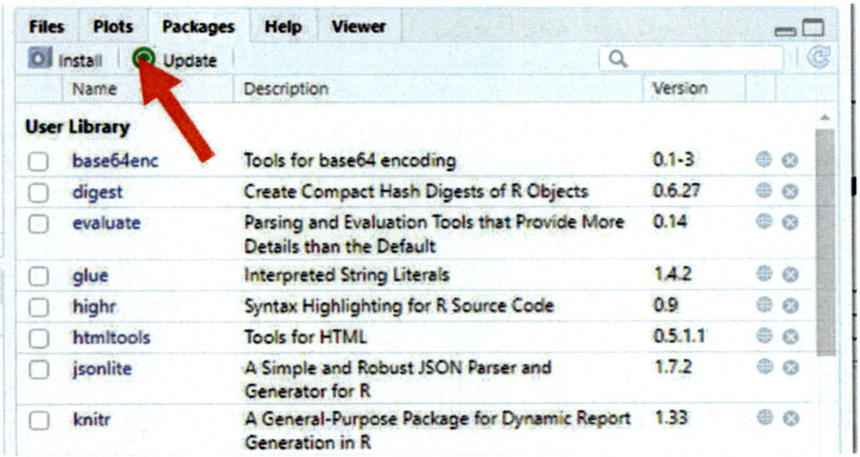

Enter the name of the package (a complete list is available at https://cran.r-project.org/web/packages/). If the name is correct, RStudio will recognize it:

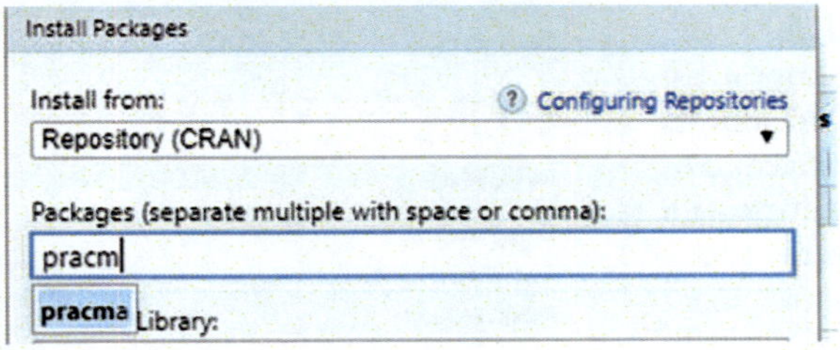

Then, click on the button `Install`.

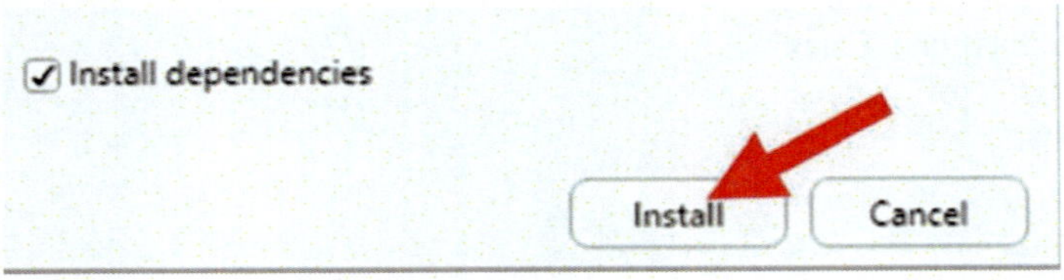

The installation begins. Maybe some warnings appear, namely if you have not development tools installed: do not matter, RStudio will run the install. At the end, you will get a message on the console indicating that the installation is achieved and that the package is available:

```
package 'pracma' successfully unpacked and MD5 sums checked
```

The newly installed package appears in the list:

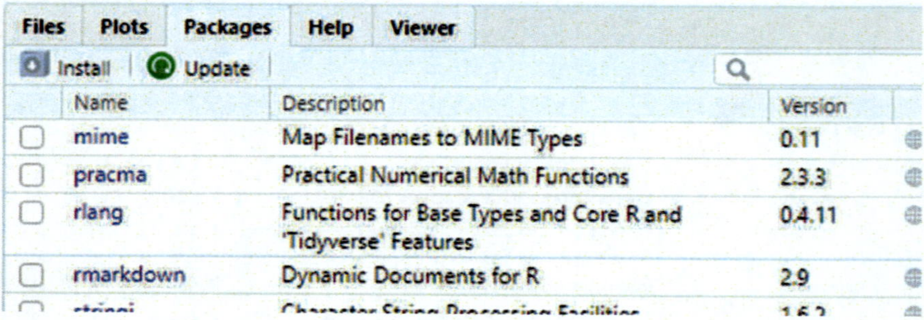

1.3 How to Create a Document with RStudio

RStudio offers the possibility of creating many types of documents: go to *File>New File* and RStudio will show a list of the types of documents that can be created. For instance, you can create a Notebook, where the code is executed in the document itself (analogously to other notebooks, such as Jupyter, Mathematica®, Maple®, and Matlab®). You can integrate Latex in some of these documents.

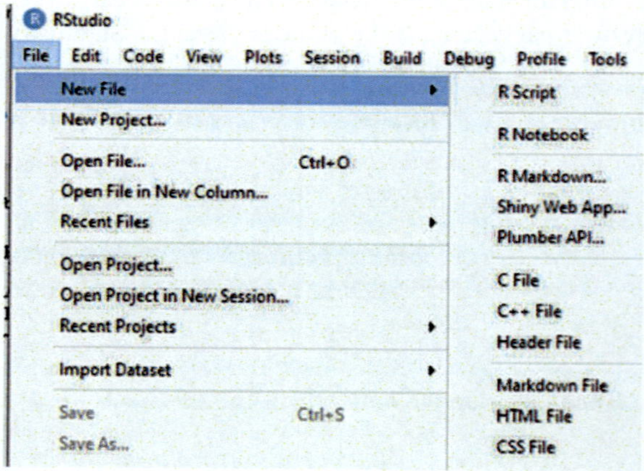

If you choose a R Notebook, you can also choose the type of output: HTML, Microsoft Word, PDF, etc. You may generate the desired output of a notebook by using the button *Preview* and selecting the desired output.

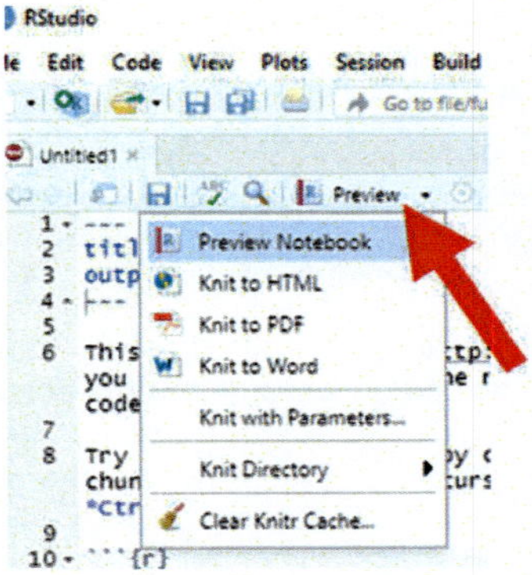

For a R Markdown document, the corresponding button is *Knit*.

In this book, we use mainly R Script files, which are sequences of instructions. They are generated by the first line and correspond to the shortcut CTRL+SHIFT+N.

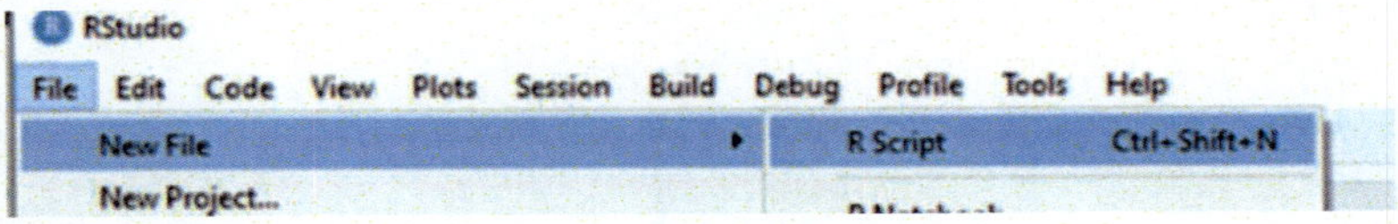

1.4 How to Create a Script with RStudio

A script is a document containing a sequence of instructions. As indicated in the previous section, go to *File>New File>R Script* and open a blank script Enter the instructions $x <- 2$ and $y <- x \wedge 2 + 2$. Select the lines and press the button *Run* (which executes either the current line or the selected lines)

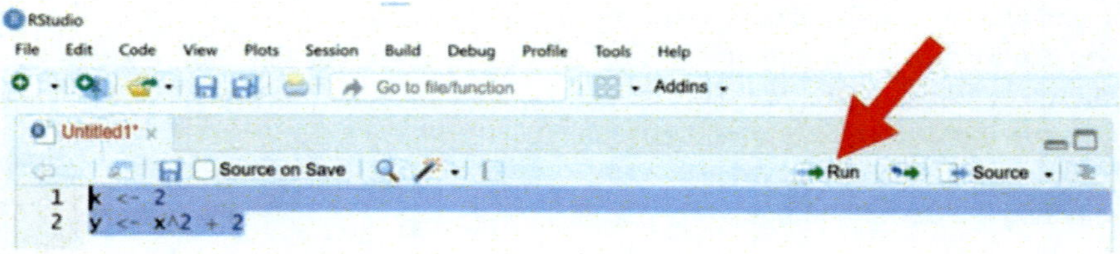

The code is executed in the console:

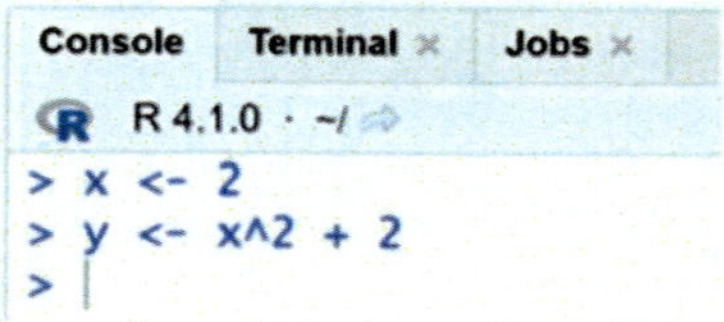

The results appear in the Environment:

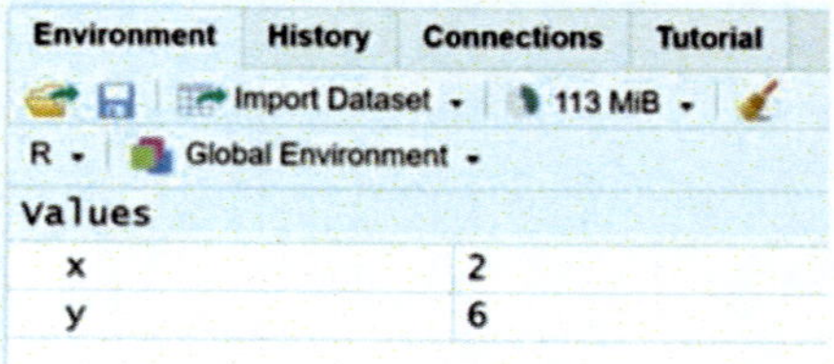

As an alternative, you can use the tab *Code>Run Region*. Then, you can choose the part of code to be executed (eventually, all the code).

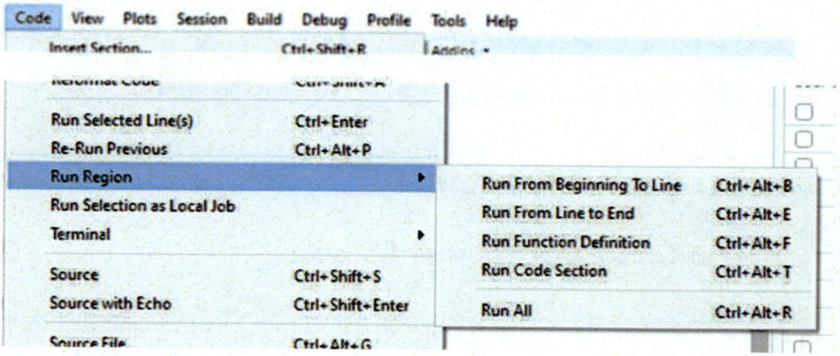

If the script is saved in a file *test.R*, you can execute it by using the command `source`. To use this command, you must furnish the entire path to the file, for instance:

```
source("C:/Users/username/Desktop/Rfiles/test.R")
```

This command uses as source file `test.R`, located in the directory `Rfiles`, at the Desktop of user `username`. The result is similar, except by the fact that the commands in the file are not shown in the console – if you desire to see the commands in the file exhibited in the console, you must use the option "`echo = TRUE`":

```
source("C:/Users/username/Desktop/Rfiles/test.R", echo = TRUE)
```

The command `source` is available in the tab *Code> Source File* (shortcut CTRL+ALT+G). In this case, the File manager opens, and you must navigate to the file to be executed.

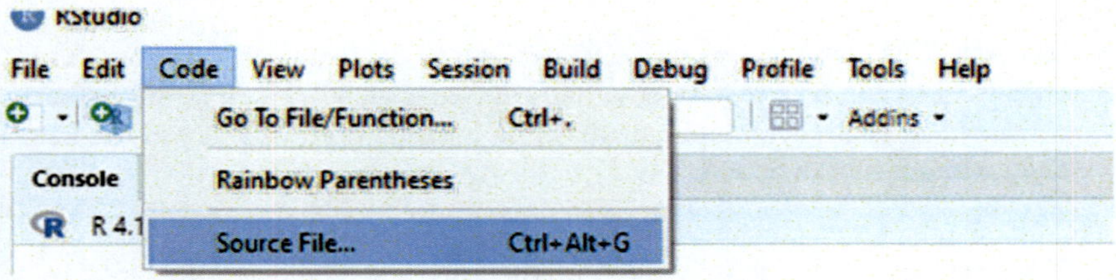

Alternatively, you can define the working directory (by default, it is *username>Documents*) and set it as being the one containing your source file.

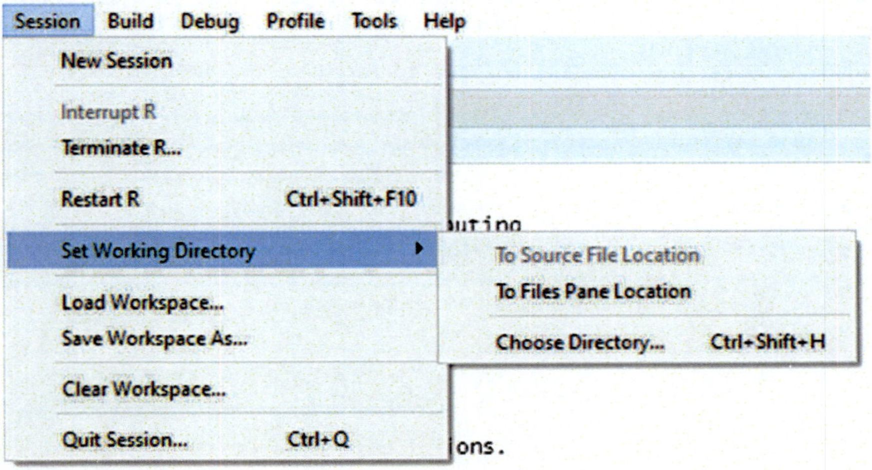

The option Choose Directory opens the file manager and allows you to navigate to the desired directory: you do not need the entire path to run the files in the working directory. Alternatively, you can use the command `setwd` in the console. For instance,

```
setwd("C:/Users/username/Desktop/Rfiles")
```

sets the working directory as `C:/Users/username/Desktop/Rfiles.`. Then, the command

```
source("test.R", echo=TRUE)
```

will produce the result at right:

```
> x <- 2

> y <- x^2 + 2
```

Remark The command `source(file)` brings the variables, functions, and objects created in `file` to your environment (see page 60). Namely, the classes contained in your source file are created in the environment and you can use them without further action. ∎

You can save the output in a file by using the command `sink(file = "filename")`. To stop it, use the command `sink()`. The option `Append=True` allows to append the results to an existing file without erasing the precedent content. For instance, add a line `print(paste("x = ",x, ", y = ",y))` to the script and save it in a file `test.R`. Choose it as source file: you will get the results below:

```
> source("test.R")
[1] "x =  2 , y =  6"
```

Create a second file `test2.R` as follows:

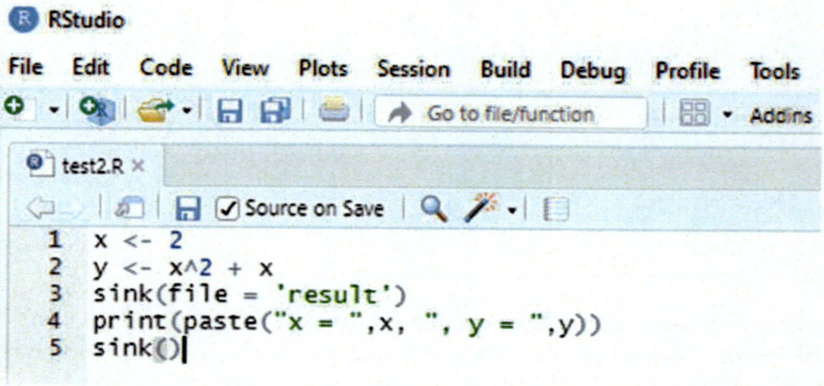

Choose this file as source and execute it. Then, a file `result` is created in your working directory :

```
> source("test2.R")
```

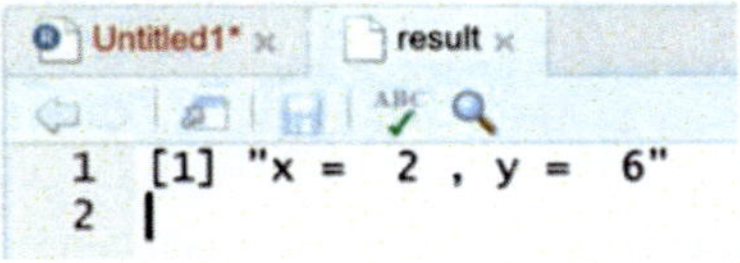

1.5 How to Manipulate Numeric Variables, Vectors, and Factors

The assignment of values to variables is made by the symbols "−>" and "<−" or "=". In many situations, "<−" can be replaced by "=". The instructions

```
2 -> z
z <- 2
z = 2
```

produce both the same result: the value 2 is assigned to the variable `z`. Vectors are created by the instructions `()` (`concatenate`). A vector has no dimension but is treated as a column vector. For instance:

```
> x = c(1,3,6,9,12)
> dim(x)
NULL
> y = t(x)
> dim(y)
[1] 1 5
```

The first instruction creates a vector $x = (1, 3, 6, 9, 12)$. As shown by the second command "`dim(x)`", x has no dimension. The instruction "`y = t(x)`" generates a second variable $y = x^t$, the transpose of x. As we can see from the instruction "`dim(y)`", y is a row vector of dimensions 1×5: for the transposition, x was considered as a column vector of dimensions 5×1.

For instance, let us create a script containing the instructions

```
c(1,3,6,9,12) -> x
y <- c(2,3,4,5,6)
z <- x+y
print(z)
w <- 1/x
print(w)
```

Executing the script produces (in the console) the result below:

```
> c(1,3,6,9,12) -> x
> y <- c(2,3,4,5,6)
> z <- x+y
> print(z)
[1]  3  6 10 14 18
> w <- 1/x
> print(w)
[1] 1.00000000 0.33333333 0.16666667 0.11111111 0.08333333
```

As you can observe, R makes the operation on a vector term by term. For instance, adding to the script the instructions below

```
q<-x^2
print(q)
```

produces in the console the following result:

```
> q <- x^2
> print(q)
[1]   1   9  36  81 144
```

The result is analogous for the instructions

```
> q <- x*x
> print(q)
[1]   1   9  36  81 144
```

To make a matrix product, the operator is `%*%`:

```
> p <- x %*% x
> print(p)
     [,1]
[1,]  271
```

Notice that R automatically transposed the second occurrence of vector x to make the product realizable and that the result is a matrix 1 × 1. You can explicitly calculate the transpose by using the operator `t()`:

```
> t(x) %*% x
     [,1]
[1,]  271
> x %*% t(x)
     [,1] [,2] [,3] [,4] [,5]
[1,]    1    3    6    9   12
[2,]    3    9   18   27   36
[3,]    6   18   36   54   72
[4,]    9   27   54   81  108
[5,]   12   36   72  108  144
```

If you desire to generate a matrix containing the values $x_i x_j$ and ensure that the results have the correct dimensions, you can redefine the vectors as matrices. For instance, redefine the dimension of x as 5×1 and make the product xx^t – as expected, you will get a matrix 5×5 :

```
> dim(x)
NULL
> dim(x) <- c(5,1)
> dim(x)
[1] 5 1
> t(x) %*% x
     [,1]
[1,]  271
> x %*% t(x)
     [,1] [,2] [,3] [,4] [,5]
[1,]    1    3    6    9   12
[2,]    3    9   18   27   36
[3,]    6   18   36   54   72
[4,]    9   27   54   81  108
[5,]   12   36   72  108  144
```

After transforming a vector in a matrix, you must respect the rules for matrix operations, namely multiplication – but you can continue to perform operations term by term. The example below shows an error when we attempt to make the product of x by itself (the dimensions do not agree), but the result is correct when calculating `x*x` or `1/x`, which are evaluated term by term:

```
> x %*% x
Error in x %*% x : non-conformable arguments
> x*x
     [,1]
[1,]    1
[2,]    9
[3,]   36
[4,]   81
[5,]  144
> 1/x
           [,1]
[1,] 1.00000000
[2,] 0.33333333
[3,] 0.16666667
[4,] 0.11111111
[5,] 0.08333333
```

As an alternative, you can evaluate the outer product to generate a matrix containing the values $x_i x_j$:

```
> x = c(1,3,6,9,12)
> o = outer(x,x)
> print(o)
     [,1] [,2] [,3] [,4] [,5]
[1,]    1    3    6    9   12
[2,]    3    9   18   27   36
[3,]    6   18   36   54   72
[4,]    9   27   54   81  108
[5,]   12   36   72  108  144
>
```

If x is transformed in a matrix 5 × 1, the result is an array 5 × 1 × 5 × 1:

```
> dim(x) <- c(5,1)
> o = outer(x,x)
> dim(o)
[1] 5 1 5 1
```

`outer` can be used for other operations than multiplication: for instance, `outer (x,y, "+")` generates a matrix $x_i + y_j$. You can also define a function $f(x, y)$ and generate a matrix of values $f(x_i, y_j)$. For instance:

```
> c(1,3,6,9,12) -> x
> y <- c(2,3)
> f <- function(x,y) x**2 + y**2
```

```
> o = outer(x,y,f)
> print(o)
     [,1] [,2]
[1,]    5   10
[2,]   13   18
[3,]   40   45
[4,]   85   90
[5,]  148  153
```

Again, if x is transformed in a matrix 5 × 1 and y is transformed in a matrix 2 × 1, the result is an array 5 × 1 × 2 × 1:

```
> dim(x) <- c(5,1)
> dim(y) <- c(2,1)
> o = outer(x,y,f)
> dim(o)
[1] 5 1 2 1
```

R performs operations even for vectors of unequal lengths. In such a situation, the vector containing the smaller number of elements is extended by recycling to attain the same number of elements as the other element. For instance:

```
> c(1,3,6,9,12) -> x
> y <- c(2,3)
> z <- x + y
Warning message:
In x + y : longer object length is not a multiple
 of shorter object length
> print(z)
[1]  3  6  8 12 14
> w = x/y
Warning message:
In x/y : longer object length is not a multiple of
 shorter object length
> print(w)
[1] 0.5 1.0 3.0 3.0 6.0
```

Observe that R made the operations extending y to $(2, 3, 2, 3, 2)$.

You can refer to a vector element by its index: x[[2]] refers to the second element of x. For instance:

```
> x[[2]]
[1] 3
```

R uses also the notation x[2] to refer to a new vector formed by the single element x[[2]]. If the result of the operation is a vector, you *must* use single brackets. For instance, using a negative value generates a new vector where the element corresponding to the index was eliminated: x[−2] eliminates the second element of x. A trial to use double brackets produces an error:

```
> x[-2]
[1]  1  6  9 12
> x[[-2]]
Error in x[[-2]] : invalid negative subscript in get1index
 <real>
```

You can use several indexes to generate parts of a vector or vectors formed by elements of another vector – again, you *must* use single brackets for such an operation. As you can see below, using double brackets produces errors:

```
> x[c(4,2)]
[1] 9 3
> x[[c(4,2)]]
Error in x[[c(4, 2)]] :
  attempt to select more than one element in vectorIndex
> x[c(-4,-2)]
[1]  1  6 12
> x[[c(-4,-2)]]
Error in x[[c(-4, -2)]] :
  attempt to select more than one element in vectorIndex
> x[-1:-3]
[1]  9 12
> x[[-1:-3]]
Error in x[[-1:-3]] :
  attempt to select more than one element in vectorIndex
> x[1:3]
[1] 1 3 6
> x[[1:3]]
Error in x[[1:3]] :
  attempt to select more than one element in vectorIndex
```

Under R, you can give names to the elements of a vector:

```
> names(x) = c("first","second","third","fourth","last")
> print(x[c("fourth","second")])
fourth second
     9      3
> print(x[ ! names(x) %in% c("fourth","second")])
first third  last
    1     6    12
```

Symbol “!” is a logical operator “not”. Logical or is “|”, while logical “and” is “&”.R can manipulate logical vectors either on logical or on arithmetic operations (for arithmetic operations, TRUE = 1, FALSE = 0 – R is case sensitive: you *must* use capital letters):

```
> x <- c(TRUE, FALSE, TRUE, FALSE)
> y = c(FALSE, TRUE)
> print(!x)
[1] FALSE  TRUE FALSE  TRUE
> print (x | y)
[1] TRUE TRUE TRUE TRUE
> print (x + y)
[1] 1 1 1 1
> print(x & y)
[1] FALSE FALSE FALSE FALSE
> print(x*y)
[1] 0 0 0 0
> 2*x
[1] 2 0 2 0
```

Notice that y is extended to (*FALSE, TRUE, FALSE, TRUE*) to match the dimensions of x. Notice also that, for numerical operations, TRUE = 1, FALSE = 0.

Logical vectors can be generated by conditions on numerical vectors

```
> x = c(1:5)
> print(x < 3)
[1]  TRUE  TRUE FALSE FALSE FALSE
```

To determine the indexes that verify the condition, you must use the command `which` are shown at right.

```
> which(x < 3)
[1] 1 2
```

R manipulates complex numbers – but you must write both the real and the imaginary part: -1 is a real number, while$-1 + 0i$ is a complex number, as you can see at right.

```
> sqrt(-1)
[1] NaN
Warning message:
In sqrt(-1) : NaNs produced
> sqrt(-1 + 0i)
[1] 0+1i
```

As you can observe, $\sqrt{-1}$ produces an error, while $\sqrt{-1+0i}$ produces the correct result $i = 0 + 1i$. Analogously, “*i*” is not a complex number, but “1*i*” is :

```
> (1-i)*(1+i)
Error: object 'i' not found
> (1 - 1i)*(1 + 1i)
[1] 2+0i
> i*(1 + i)
Error: object 'i' not found
> 1i*(1 + 1i)
[1] -1+1i
```

You can create empty objects and assign them values in a second step. The length and the type of the vector will adapt to your assignments:

```
> x <- numeric()
> x[[2]] <- 1
> print(x)
[1] NA  1
> x[3] = 4
> print(x)
[1] NA  1  4
> x[[4]] = TRUE
> print(x)
[1] NA  1  4  1
> x[[5]] <- "name"
> print(x)
[1] NA     "1"    "4"    "1"    "name"
> x[1] = 1
> print(x)
[1] "1"    "1"    "4"    "1"    "name"
```

NA is an undefined value ("not assigned"). Observe that the logical TRUE was considered as a numeric 1, since the vector is declared as numeric. But the assignment of a string to one of the elements of the vectors transforms all the elements in strings. Trying to assign a numeric value to one of the elements becomes inoperative: the value is transformed in a chain of characters.

```
> x = logical()
> x[[3]]= TRUE
> print(x)
[1]    NA   NA TRUE
> x[[1]] = 0
> print(x)
[1]  0 NA  1
> x[[5]] = "name"
> print(x)
[1] "0"    NA     "1"    NA     "name"
> x[[4]] = TRUE
> print(x)
[1] "0"    NA     "1"    "TRUE" "name"
```

In this case, the assignment of a numeric value to a logical vector transforms all the elements in numeric values. Again, the assignment of a chain of characters to an element transforms all the elements in strings. Then, the assignment of a logical value is inoperative: it is transformed into a string.

You can modify the length of an object: to retain the first two elements of y, you can use

```
> z <- y[1:2]
> length(y) <-2
```

The first instruction creates a new variable z containing the first two elements from y, while the second one modifies y itself by reducing it to its two first elements.

You can reshape a vector into a matrix having the same number of elements – an attempt to reshape to a different number of elements produces an error:

```
> x <- c(1:6)
> print(x)
[1] 1 2 3 4 5 6
> dim(x) <- c(2,2)
Error in dim(x) <- c(2, 2) :
  dims [product 4] do not match the length of object [6]
> dim(x) = c(3,4)
Error in dim(x) = c(3, 4) :
  dims [product 12] do not match the length of object [6]
> dim(x) <- c(2,3)
> print(x)
     [,1] [,2] [,3]
[1,]    1    3    5
[2,]    2    4    6
```

Notice that the object is reshaped by filling the columns. Analogously to arithmetic operations, standard functions are applied term by term:

```
> x <- c(-1,0,1)
> print(exp(x))
[1] 0.3678794 1.0000000 2.7182818
> print(cos(pi*x))
[1] -1  1 -1
> print(sin(pi*x/2))
[1] -1  0  1
```

Finally, you can create vectors of equally spaced elements : if $a < b$, then $a : b$ generates a sequence starting at a, with step 1, until the value of b is attained. If $b < a$, the sequence has a step -1. You can also define the step s using the command `seq(a,b,by = s)` or the number of elements *nel* by `seq(a,b, length = nel)`:

```
> print(1:3.75)
[1] 1 2 3
> print(pi:0)
[1] 3.1415927 2.1415927 1.1415927 0.1415927
> print(1:3)
[1] 1 2 3
> print(6:0)
[1] 6 5 4 3 2 1 0
> print(seq(0,1,by = 0.2))
[1] 0.0 0.2 0.4 0.6 0.8 1.0
> print(seq(0,1, length = 6))
[1] 0.0 0.2 0.4 0.6 0.8 1.0
> print(seq(1,0, by = - 0.2))
[1] 1.0 0.8 0.6 0.4 0.2 0.0
```

R offers many intrinsic functions and add-ins for numerical vectors, namely numerical and statistical functions.

You can also create empty vectors and assign values to their elements in a second step. Analogously to the preceding manipulations, assignments of different types transform the vector: assigning a numeric value to an element of a logical vector transforms it into a numeric one; assigning a string to an element of a vector transforms all the elements in strings:

```
> x = vector()
> x[[3]]= TRUE
> print(x)
[1]    NA    NA TRUE
> x[[1]] = 0
> print(x)
[1]  0 NA  1
> x[[5]] = "name"
> print(x)
[1] "0"    NA     "1"    NA     "name"
> x[[4]] = TRUE
> print(x)
[1] "0"    NA     "1"    "TRUE" "name"
```

R offers also intrinsic commands for set operations: `intersect` evaluates the intersection of two vectors, `union` generates the set union of the vectors, and `setdiff` evaluates the set difference.

```
> x = c(1:4)
> print(x)
[1] 1 2 3 4
> y = c(3:6)
> print(y)
[1] 3 4 5 6
> intersect(x,y)
[1] 3 4
> union(x,y)
[1] 1 2 3 4 5 6
> setdiff(x,y)
[1] 1 2
> setdiff(y,x)
[1] 5 6
```

A common operation on numeric vectors is sorting. Under R, you can use the command sort and choose if you desire to get the indexes of the elements sorted. The command `y <- sort(p)` generates a vector `y` containing the elements of `p` sorted in an ascending order. `z <- sort(p, index.return=TRUE)` generates a list `z` containing two elements: `z$x` contains the sorted vector and `z$ix` contains the corresponding indexes. To sort in a descending order, use the option `decreasing = TRUE`.

```
> p <- c(9,3,6,1,2,8,7,4,5)
> p
[1] 9 3 6 1 2 8 7 4 5
> y = sort(p)
> y
[1] 1 2 3 4 5 6 7 8 9
> z <- sort(p,index.return=TRUE)
> z
$x
[1] 1 2 3 4 5 6 7 8 9

$ix
[1] 4 5 2 8 9 3 7 6 1

> yd <- sort(p,decreasing = TRUE)
> yd
[1] 9 8 7 6 5 4 3 2 1
```

You can sort vectors of strings. Notice that the number of capitals, their position, and the length of the strings influence the result:

```
> q = c("Abcd", "abcd","aBCd","aBcd","abcD","aBCD")
> q
[1] "Abcd" "abcd" "aBCd" "aBcd" "abcD" "aBCD"
> qs = sort(q)
> qs
[1] "abcd" "abcD" "aBcd" "aBCd" "aBCD" "Abcd"
```

```
> q = c("Abc", "abcd","aBC","aBc","abcD","abc","ab")
> q
[1] "Abc"  "abcd" "aBC"  "aBc"  "abcD" "abc"  "ab"
> qs = sort(q)
> qs
[1] "ab"   "abc"  "aBc"  "aBC"  "Abc"  "abcd" "abcD"
```

If you are interested only in the indexes, you can use the command order, which furnishes the vector of indexes. Option `decreasing` can be used with order:

```
> p <- c(9,3,6,1,2,8,7,4,5)
> p
[1] 9 3 6 1 2 8 7 4 5
> iy = order(p)
> iy
[1] 4 5 2 8 9 3 7 6 1
> p[iy]
[1] 1 2 3 4 5 6 7 8 9
```

A `factor` is a vector of text data under the form of strings, corresponding to classes to which individuals belong. It is often used to define categories. For instance, consider the instructions:

```
> aux <- c("A","B","O","AB","O","A","O","O","O")
> x = factor(aux)
> attributes(x)
$levels
[1] "A"  "AB" "B"  "O"

$class
[1] "factor"

> levels(x)
[1] "A"  "AB" "B"  "O"
> nlevels(x)
[1] 4
> summary(x)
 A AB  B  O
 2  1  1  5
```

The command summary gives the number of elements in each class. For numeric vectors, the summary gives statistical properties:

```
> x = c(rep(1,3), rep(2,4), rep(3,3), rep(4,5))
> summary(x)
   Min. 1st Qu.  Median    Mean 3rd Qu.    Max.
  1.000   2.000   3.000   2.667   4.000   4.000
```

Here, the command `rep(value, times)` replicates the value the number of times indicated: we have $x = (1, 1, 1, 2, 2, 2, 2, 3, 3, 3, 4, 4, 4, 4, 4)$. The command `summary` furnishes statistics of the vector.

You can generate factors having a particular order by using the instruction ordered:

```
> aux <- c("A","B","O","AB","O","A","O","O","O")
> y = ordered(aux,levels <- c("A","B","AB","O"))
> summary(y)
 A  B AB  O
 2  1  1  5
```

You can use labels to refer to factors. Below, the levels are coded 1,2,3,4 but have labels A, B, AB, O. Only the labels appear in outputs – they replace the codes of the levels:

```
> y = factor(c(1,2,4,3,4,1,4,4,4),levels = c(1,2,3,4)
 labels = c("A","B","AB","O"))
> y
[1] A  B  O  AB O  A  O  O  O
Levels: A B AB O
```

1.6 How to Manipulate Matrices and Arrays

Under R, a matrix is a particular case of an array. Matrices have two indexes, while arrays may have an arbitrary number of indexes. You can create empty matrices by giving their numbers of rows and columns. Then, you can assign values to particular elements from the matrix created:

```
> a = matrix(nrow = 2, ncol = 3)
> print(a)
     [,1] [,2] [,3]
[1,]   NA   NA   NA
[2,]   NA   NA   NA
> a[[2,2]] = 1
> a[[1,3]] <- 2
> 3 -> a[[1,1]]
> print(a)
     [,1] [,2] [,3]
[1,]    3   NA    2
[2,]   NA    1   NA
```

You can also furnish the data to fill the matrix – if the data are insufficient, R recycles the data, starting at the first value; if the data are excessive, R uses only the first elements:

```
> a = matrix(c(1:4),nrow = 2, ncol = 3)
Warning message:
In matrix(c(1:4), nrow = 2, ncol = 3) :
  data length [4] is not a sub-multiple or multiple of
 the number of columns [3]
> print(a)
     [,1] [,2] [,3]
[1,]    1    3    1
[2,]    2    4    2
> b <- matrix(c(1:8),nrow = 2, ncol = 3)
Warning message:
In matrix(c(1:8), nrow = 2, ncol = 3) :
  data length [8] is not a sub-multiple or multiple of
 the number of columns [3]
> print(b)
     [,1] [,2] [,3]
[1,]    1    3    5
[2,]    2    4    6
```

Again, notice that R creates the matrix by columns. If you desire to fill it by rows, you must specify it:

```
> a = matrix(c(1:4),nrow = 2, ncol = 3,byrow = TRUE)
Warning message:
In matrix(c(1:4), nrow = 2, ncol = 3, byrow = TRUE) :
  data length [4] is not a sub-multiple or multiple of
 the number of columns [3]
> print(a)


     [,1] [,2] [,3]
[1,]    1    2    3
[2,]    4    1    2
> b <- matrix(c(1:8),nrow = 2, ncol = 3, byrow = TRUE)
Warning message:
In matrix(c(1:8), nrow = 2, ncol = 3, byrow = TRUE) :
  data length [8] is not a sub-multiple or multiple of
 the number of columns [3]
> print(b)
     [,1] [,2] [,3]
[1,]    1    2    3
[2,]    4    5    6
```

The procedures are analogous for arrays:

```
> a <- array(dim = c(2,3,2))
> print(a)
, , 1

     [,1] [,2] [,3]
[1,]   NA   NA   NA
[2,]   NA   NA   NA

, , 2

     [,1] [,2] [,3]
[1,]   NA   NA   NA
[2,]   NA   NA   NA
```

```
> a <- array(c(1:12),dim = c(2,3,2))
> print(a)
, , 1

     [,1] [,2] [,3]
[1,]    1    3    5
[2,]    2    4    6

, , 2

     [,1] [,2] [,3]
[1,]    7    9   11
[2,]    8   10   12
```

You can transform a matrix or an array into a vector by destroying its dimensions or concatenating it:

```
> print(c(a))
 [1]  1  2  3  4  5  6  7  8  9 10 11 12
```

```
> dim(a) <- NULL
> print(a)
 [1]  1  2  3  4  5  6  7  8  9 10 11 12
```

You can also transform arrays in matrices by redefining their dimensions:

```
> a = array(1:12,dim = c(2,3,2))
> dim(a) <- c(3,4)
> print(a)
     [,1] [,2] [,3] [,4]
[1,]    1    4    7   10
[2,]    2    5    8   11
[3,]    3    6    9   12
```

You can add rows or columns to a matrix using `rbind` and `cbind`:

```
> x <- c(1:3)
> y = c(4:6)
> z = rbind(x,y)
> print(z)
  [,1] [,2] [,3]
x    1    2    3
y    4    5    6
> s = cbind(x,y)
> print(s)
     x y
[1,] 1 4
[2,] 2 5
[3,] 3 6
> a = matrix(1:8,nrow = 2, ncol = 4)
> b = rbind(a,9:12)
> print(b)
     [,1] [,2] [,3] [,4]
[1,]    1    3    5    7
[2,]    2    4    6    8
[3,]    9   10   11   12
> c = cbind(a,9:10)
> print(c)
     [,1] [,2] [,3] [,4] [,5]
[1,]    1    3    5    7    9
[2,]    2    4    6    8   10
```

Again, recall that the arithmetic operation "*" is performed term by term and the matrix product is "`%*%`":

```
> a = matrix(1:8,nrow = 2, ncol = 4)
> x <- numeric()
> x[1:4] <- 1
> w = a*x
> print(w)
     [,1] [,2] [,3] [,4]
[1,]    1    3    5    7
[2,]    2    4    6    8
> s = a %*% x
> print(s)
     [,1]
[1,]   16
[2,]   20
```

You can manipulate parts of matrices by using sections, which are vectors. For instance, the instructions `y<-x[,1]`, `s<-x[1:3,1:1]`, `p<-x[1:3,1]` generate vectors containing the first column of x. Analogously, `y<-x[1,]`, `s<-x[1:1,1:4]`, `p<-x[1,1:4]` generate vectors containing the first line of x. You can transform them in matrices by redefining their dimensions. Single brackets must be used since the results are vectors or matrices.

You can also define arrays of indexes and take the corresponding part of a matrix. For instance, `y <- x[1:2,3:4]` generates a submatrix y containing $x_{ij} : 1 \leq i \leq 2, 3 \leq j \leq 4$. You can use indexes which are not sequentially defined, as shown at right.

```
> x = matrix(1:12,nrow = 3, ncol = 4)
> print(x)
     [,1] [,2] [,3] [,4]
[1,]    1    4    7   10
[2,]    2    5    8   11
[3,]    3    6    9   12
> i <- c(3,2)
> j = c(4,1,2)
> x[i,j] -> y
> print(y)
     [,1] [,2] [,3]
[1,]   12    3    6
[2,]   11    2    5
```

You can also define arrays of indexes and take the corresponding part of a matrix. For instance, `y <- x[1:2,3:4]` generates a submatrix y containing $x_{ij} : 1 \leq i \leq 2$, $3 \leq j \leq 4$. You can use indexes which are not sequentially defined:


```
> x = matrix(1:12,nrow = 3, ncol = 4)
> print(x)
     [,1] [,2] [,3] [,4]
[1,]    1    4    7   10
[2,]    2    5    8   11
[3,]    3    6    9   12
> i <- c(3,2)
> j = c(4,1,2)

> x[i,j] -> y
> print(y)
     [,1] [,2] [,3]
[1,]   12    3    6
[2,]   11    2    5
```


Analogously to vectors, standard function applies to matrices and arrays element by element. For instance:

```
> x <- matrix(1:6, nrow = 2, ncol = 3)
> sin(pi*x)
              [,1]          [,2]          [,3]
[1,]  1.224606e-16  3.673819e-16  6.123032e-16
[2,] -2.449213e-16 -4.898425e-16 -7.347638e-16

> cos(pi*x)
     [,1] [,2] [,3]
[1,]   -1   -1   -1
[2,]    1    1    1
```

For functions taking vectors as arguments, you can use the instruction `apply (matrix, MARGIN =..., function)`. Using MARGIN = 1 evaluates the function on the rows, while MARGIN = 2 uses the columns as arguments.

For instance, assume that you want to evaluate the sums of the columns of a `x`:

```
> apply(x,MARGIN=2,sum)
[1]  3  7 11
```

Notice that `sum(x)` evaluates the sum of all the elements of `x`:

```
> sum(x)
[1] 21
```

You can also apply a custom function:

```
> f = function(x) x[[1]] + 2*x[[2]]
> apply(x, MARGIN=2, f)
[1]  5 11 17
```

For instance, you can sort the columns of a matrix as shown at right:

```
> A = cbind(c(2,1,4,3), c(4,5,1,2),c(1,3,5,2))
> A
     [,1] [,2] [,3]
[1,]    2    4    1
[2,]    1    5    3
[3,]    4    1    5
[4,]    3    2    2
> B = apply(A,MARGIN=2,sort)
> B
     [,1] [,2] [,3]
[1,]    1    1    1
[2,]    2    2    2
[3,]    3    4    3
[4,]    4    5    5
```

If you desire to sort the same matrix by lines, the result has to be transposed:

```
> B = apply(A,MARGIN=1,sort)
> B
     [,1] [,2] [,3] [,4]
[1,]    1    1    1    2
[2,]    2    3    4    2
[3,]    4    5    5    3
> t(B)
     [,1] [,2] [,3]
[1,]    1    2    4
[2,]    1    3    5
[3,]    1    4    5
[4,]    2    2    3
```

R offers two other functions to apply functions to matrices: `outer(x,y,f)` generates an array $f(x_{ij}, y_{k\ell})$, and `mapply(f,x,y)` generates a matrix $f(x_{ij}, y_{ij})$

```
> f = function(x,y) x + 2*y
> mapply(f,x,x)      # x + 2*x = 3*x
[1]  3  6  9 12 15 18
> outer(x,x,f)
, , 1, 1   # x + 2*x[1,1] = x + 2

     [,1] [,2] [,3]
[1,]    3    5    7
[2,]    4    6    8

, , 2, 1   # x + 2*x[2,1] = x + 4

     [,1] [,2] [,3]
[1,]    5    7    9
[2,]    6    8   10

, , 1, 2   # x + 2*x[1,2] = x + 6

     [,1] [,2] [,3]
[1,]    7    9   11
[2,]    8   10   12

, , 2, 2   # x + 2*x[2,2] = x + 8

     [,1] [,2] [,3]
[1,]    9   11   13
[2,]   10   12   14

, , 1, 3   # x + 2*x[1,3] = x + 10

     [,1] [,2] [,3]
[1,]   11   13   15
[2,]   12   14   16

, , 2, 3   # x + 2*x[2,3] = x + 12

     [,1] [,2] [,3]
[1,]   13   15   17
[2,]   14   16   18
```

R proposes many intrinsic functions and add-ins to deal with matrices, arrays, eigenvalues, eigenvectors, and linear systems. For instance, the command `eigen` produces the eigenvectors and eigenvalues:

```
> A = cbind(c(1,2,3), c(3,4,5),c(4,5,6))
> eigen(A)
eigen() decomposition
$values
[1]  1.176498e+01 -7.649820e-01 -1.161812e-15

$vectors
          [,1]       [,2]       [,3]
[1,] 0.4210993  0.8470672  0.2672612
[2,] 0.5652295  0.1721100 -0.8017837
[3,] 0.7093596 -0.5028471  0.5345225
```

Each column of `$vectors` is an eigenvector. The first instruction generates the matrix

$$A = \begin{pmatrix} 1 & 3 & 4 \\ 2 & 4 & 5 \\ 3 & 5 & 6 \end{pmatrix}$$

The second one evaluates the eigenvalues and eigenvectors of A. If you are interested only in the eigenvalues, you can use the instruction `eigen(A)$values`. Analogously, `eigen(A)$vectors` produces the eigenvectors:

```
> eigen(A)$values
[1]  1.176498e+01 -7.649820e-01 -1.161812e-15
> eigen(A)$vectors
          [,1]       [,2]       [,3]
[1,] 0.4210993  0.8470672  0.2672612
[2,] 0.5652295  0.1721100 -0.8017837
[3,] 0.7093596 -0.5028471  0.5345225
```

If the package `pracma` is installed, you can use the command `eig` to get the eigenvalues:

```
> A = cbind(c(1,2,3), c(3,4,5),c(4,5,6))
> require(pracma)
Loading required package: pracma
> eig(A)
[1]  1.176498e+01 -7.649820e-01 -1.161812e-15
```

A linear system $Ax = b$ can be solved by the commend `solve(A, b)`. For instance, let us consider

$$A = \begin{pmatrix} 1 & -1 & 1 \\ 2 & 2 & -2 \\ 3 & 3 & 1 \end{pmatrix}, b = \begin{pmatrix} 2 \\ 0 \\ 12 \end{pmatrix} \Rightarrow x = \begin{pmatrix} 1 \\ 2 \\ 3 \end{pmatrix}$$

Under R, we have:

```
> A = cbind(c(1,2,3), c(-1,2,3),c(1,-2,1))
> b = c(2,0,12)
> solve(A,b)
[1] 1 2 3
```

If you have installed (and required) the package `pracma`, you can use the equivalent of the Matlab® command "`A\b`" :

```
> mldivide(A,b)
     [,1]
[1,]    1
[2,]    2
[3,]    3
```

Difficulties arise when considering singular systems. For instance:

$$A = \begin{pmatrix} 1 & 3 & 4 \\ 2 & 4 & 5 \\ 3 & 5 & 6 \end{pmatrix}, \; b = \begin{pmatrix} 19 \\ 25 \\ 31 \end{pmatrix} \Rightarrow x = \begin{pmatrix} 1 \\ 2 \\ 3 \end{pmatrix} + \lambda \begin{pmatrix} -1 \\ 3 \\ -2 \end{pmatrix}, \; \lambda \in \mathbb{R}$$

In this case, $det(A) = 0$, so that the system is singular: it admits infinitely many solutions. We have

```
> A = cbind(c(1,2,3),c(3,4,5),c(4,5,6))
> b = c(19,25,31)
> solve(A,b)
 Error in solve.default(A, b) :
   system is computationally singular: reciprocal
  condition number = 4.93432e-18
 > mldivide(A,b)
          [,1]
[1,] 0.9285714
[2,] 2.2142857
[3,] 2.8571429
```

Here, `solve` fails in furnishing a solution, while `mldivide` furnishes an approximated solution. Many packages offer methods for the solution of Linear Systems. For instance, you can use `matlib`, `limSolve`, `OptR`, ...

1.7 How to Use Lists

Under R, a list is a collection of objects. While vectors and arrays contain only values of an unique type, lists can contain any objects of distinct types, including other lists.

For instance, let us create a list containing the strings "January", "February", "March":

```
> list1 = list("January","February","March")
> list1
[[1]]
[1] "January"

[[2]]
[1] "February"

[[3]]
[1] "March"
```

Now, let us create a new list containing a matrix, a vector, and the list `list1`:

```
> A = cbind(c(1,2,3), c(3,4,5), c(4,5,6))
> b = c(1,4,6)
> list2 = list(A,b,list1)
> list2
[[1]]
     [,1] [,2] [,3]
[1,]    1    3    4
[2,]    2    4    5
[3,]    3    5    6

[[2]]
[1] 1 4 6

[[3]]
[[3]][[1]]
[1] "January"

[[3]][[2]]
[1] "February"

[[3]][[3]]
[1] "March"
```

You can refer to an element of a list analogously to vectors and matrices: `list2[[1]]` refers to the first element of the `list2` and `list2[1]` refers to a list formed of the first element of `list2`. Notice that the dimensions of `list2[[1]]` correspond to the dimensions of the matrix, while the dimensions of `list2[1]` is NULL.

```
> list2[[1]]
     [,1] [,2] [,3]
[1,]    1    3    4
[2,]    2    4    5
[3,]    3    5    6
> list2[1]
[[1]]
     [,1] [,2] [,3]
[1,]    1    3    4
[2,]    2    4    5
[3,]    3    5    6

> dim(list2[[1]])
[1] 3 3
> dim(list2[1])
NULL
> is.list(list2[1])
[1] TRUE
> is.matrix(list2[[1]])
[1] TRUE
```

You can concatenate lists to add elements to an existing list.

```
> list3 = c(list1,list("April", "May"))
> View(list3)
```

The command `View(list3)` exhibits `list3` in the same window where scripts are shown.

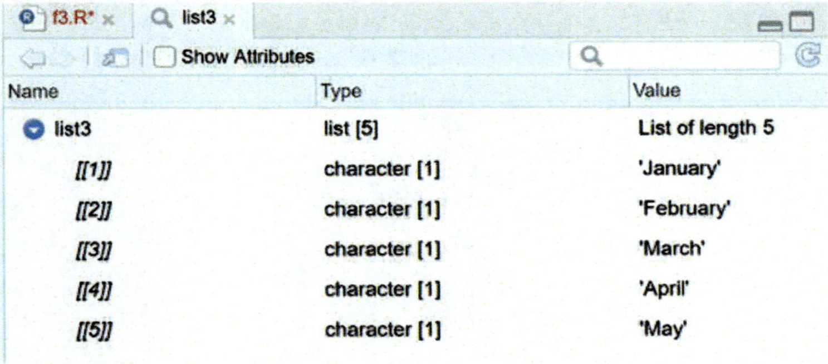

You can also transform lists in vectors by using `unlist` and conversely by `as.list`:

```
> p = unlist(list3)
> p
[1] "January"  "February" "March"    "April"
[5] "May"
> p[[1]]
[1] "January"
> is.list(p)
[1] FALSE
> is.vector(p)
[1] TRUE
> q = as.list(p)
> is.list(q)
[1] TRUE
```

You can define names for the objects in lists by different ways. For instance, you can define an empty list and include elements with their names:

```
> l1 <- list()
> l1$value = 2
> l1$matrix = cbind(c(1,2), c(2,1))
> l1$logic = TRUE
> l1
$value
[1] 2

$matrix
     [,1] [,2]
[1,]    1    2
[2,]    2    1

$logic
[1] TRUE
```

You can also define names when creating the list:

```
> l2 = list(value = 2, logic = TRUE)
> l2$matrix = cbind(c(1,2), c(2,1))
> l2
$value
[1] 2

$logic
[1] TRUE

$matrix
     [,1] [,2]
[1,]    1    2
[2,]    2    1
```

You can access the elements by their names or by their indexes, as shown previously.

```
> l2[[1]]
[1] 2
> l2[[2]]
[1] TRUE
> l2$value
[1] 2
> l2$logic
[1] TRUE
```

Lists can be reshaped by redefining their dimensions and you can add lines and rows, analogously to matrices and arrays:

```
> l3 = list(TRUE,2,3,4,5,"6")
> dim(l3) <- c(2,3)
> l3
     [,1] [,2] [,3]
[1,] TRUE 3    5
[2,] 2    4    "6"
> is.list(l3)
[1] TRUE
> dim(l3)
[1] 2 3
> l3a = cbind(l3, list(7, "8"))
> l3a
     [,1] [,2] [,3] [,4]
[1,] TRUE 3    5    7
[2,] 2    4    "6"  "8"
> is.list(l3a)
[1] TRUE
> dim(l3a)
[1] 2 4
```

```
> l4 = list(TRUE,2,3,4,5,"6")
> dim(l4) <- c(2,1,3)
> is.list(l4)
[1] TRUE
> dim(l4)
[1] 2 1 3
```

You can use set operations on lists:

```
> y <- intersect(l1,l3)
> is.list(y)
[1] TRUE
> dim(y)
NULL
> length(y)
[1] 2
```

```
> z <- union(l1,l3)
> is.list(z)
[1] TRUE
> dim(z)
NULL
> length(z)
[1] 7
```

You can also apply functions on lists with `lapply`:

```
> l5 = list(1,2,3,4,5)
> fun = function(x) cos(pi*x)
> y = lapply(l5,"fun")
> unlist(y)
[1] -1  1 -1  1 -1
```

As an example, let us create a list containing the matrices:

$$A=\begin{pmatrix}1 & 3\\ 2 & 4\end{pmatrix}, B=\begin{pmatrix}2 & 5\\ 3 & 6\\ 4 & 7\end{pmatrix}, C=\begin{pmatrix}3 & 5 & 7\\ 4 & 6 & 8\end{pmatrix}.$$

```
> A = matrix(1:4, nrow = 2, ncol = 2)
> B = matrix(2:7, nrow = 3, ncol = 2)
> C = matrix(3:8, nrow = 2, ncol = 3)
> l6 = list(A,B,C)
```

Then, we create a list containing the second line of each matrix:

```
> A = matrix(1:4, nrow = 2, ncol = 2)
> B = matrix(2:7, nrow = 3, ncol = 2)
> C = matrix(3:8, nrow = 2, ncol = 3)
> l6 = list(A,B,C)
> y = lapply(l6,"[",2,)
> y
[[1]]
[1] 2 4

[[2]]
[1] 3 6

[[3]]
[1] 4 6 8
```

To create a list containing the first column of each matrix, use `lapply (l6,"[", ,1)`.

```
> z = lapply(l6,"[", , 1)
> z
[[1]]
[1] 1 2

[[2]]
[1] 2 3 4

[[3]]
[1] 3 4
```

An operation internal to a list can be made by calling the function within. For instance, let us consider a list formed by two vectors $x = (1,2,3)$, $y = (4,5,6)$ and let us create a third element in the list: $z \leftarrow x + y$.

```
> x = c(1,2,3)
> y = c(4,5,6)
> l1 = list(x,y)
> within(l1,z <- x+y)
[[1]]
[1] 1 2 3

[[2]]
[1] 4 5 6

$z
[1] 5 7 9
```

1.8 Using data.frames

A data.frame is a table of structured data, containing the individual values at rows and the variables at columns. Rows and columns may have names corresponding to individuals and characteristics, respectively. For instance, let us consider the code below:

```
> var1 <- c(1,2,2,3)
> var2 <- c("blue","red", "green", "red")
> var3 <- c(FALSE, TRUE, TRUE, FALSE)
> var4 <- factor(c("A","B","B","A"))
> aux = list(var1,var2,var3,var4)
> d = data.frame(aux)
> colnames(d) <- c("type","color","logic","factor")
> rownames(d) <- c("x1","x2","x3","x4")
> print(d)
   type color logic factor
x1    1  blue FALSE      A
x2    2   red  TRUE      B
x3    2 green  TRUE      B
x4    3   red FALSE      A
```

You can access the names of the variables and individuals

```
> colnames(d)
[1] "type"   "color"  "logic"  "factor"
> rownames(d)
[1] "x1" "x2" "x3" "x4"
```

The command summary gives statistics of the variables:

```
> summary(d)
      type          color               logic          factor
 Min.   :1.00   Length:4           Mode :logical   A:2
 1st Qu.:1.75   Class :character   FALSE:2         B:2
 Median :2.00   Mode  :character   TRUE :2
 Mean   :2.00
 3rd Qu.:2.25
 Max.   :3.00
```

You can refer to the variables by their names or their indexes:

```
> d$color
[1] "blue"  "red"   "green" "red"

> d[,2]
[1] "blue"  "red"   "green" "red"
```

Analogously for the individuals

```
> d["x1",]
   type color logic factor
x1    1  blue FALSE      A
> d[1,]
   type color logic factor
x1    1  blue FALSE      A
```

You can append data to the data.frame using another data.frame. For instance

```
> da
   type color logic factor
x5    2   red  TRUE      A
x6    2  blue  TRUE      B
x7    3 green FALSE      A
> dc = rbind(d,da)
> dc
   type color logic factor
x1    1  blue FALSE      A
x2    2   red  TRUE      B
x3    2 green  TRUE      B
x4    3   red FALSE      A
x5    2   red  TRUE      A
x6    2  blue  TRUE      B
x7    3 green FALSE      A
```

You can also include other variables from other data.frame:

```
> db
   nt val s
x1  A   0 m
x2  B   0 b
x3  A   1 t
x4  B   1 s
> cbind(d,db)
   type color logic factor nt val s
x1    1  blue FALSE      A  A   0 m
x2    2   red  TRUE      B  B   0 b
x3    2 green  TRUE      B  A   1 t
x4    3   red FALSE      A  B   1 s
```

To select the individuals corresponding to given value of a variable, you may use `which`:

```
> ind = which(dc$color == "red")
> dc[ind, ]
   type color logic factor
x2    2   red  TRUE      B
x4    3   red FALSE      A
x5    2   red  TRUE      A
```

You can find in the R repositories packages to deal with `data.frames` and public `data.frames` for tests, for instance, the package `dplyr` (it is part of `tidyverse`, but may be used independently) which proposes functions for filtering the data:

```
> library(dplyr)
> filter(dc, color=="red")
   type color logic factor
x2    2   red  TRUE      B
x4    3   red FALSE      A
x5    2   red  TRUE      A
```

It is interesting to notice that you can create `data.frames` using EXCEL® worksheets and .csv files. For instance, assume that you have a file named *data1.xlsx* containing the data below:

	A	B	C	D
1				
2				
3	type	color	logic	factor
4	1	blue	FALSE	A
5	2	red	TRUE	B
6	2	green	TRUE	B
7	3	red	FALSE	A
8				

The package `xlsx` allows you to read the data to a `data.frame`. For instance, let us consider that the path to the file is

```
C:/Users/username/Desktop/Rfiles/data1.xlxs
```

Then, you can use the commands:

```
setwd("C:/username/Desktop/Rfiles")
library(xlsx)
df1 <- read.xlsx("data1.xlsx", 1,header = TRUE)
```

These commands generate the `data.frame` below:

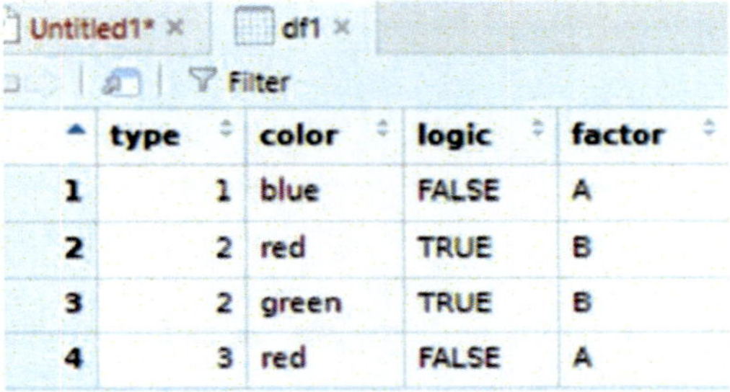

	type	color	logic	factor
1	1	blue	FALSE	A
2	2	red	TRUE	B
3	2	green	TRUE	B
4	3	red	FALSE	A

Other packages can read EXCEL® files, for instance, `readxl`– the reader can refer to the R repositories on the web to find many packages to connect EXCEL® and R. Concerning *.csv* files, R offers a standard instruction `read.csv` which reads a *csv* file into a `data.frame`. As an example, `df2 = read.csv("data1.csv", header = TRUE, sep = ";")` reads the file *data1.csv* using ";" as separator into the `data.frame df2`.

In addition, `RStudio` proposes a tool to import data: go to to the subwindow *Environment* (at right high side) and click on *Import Dataset*

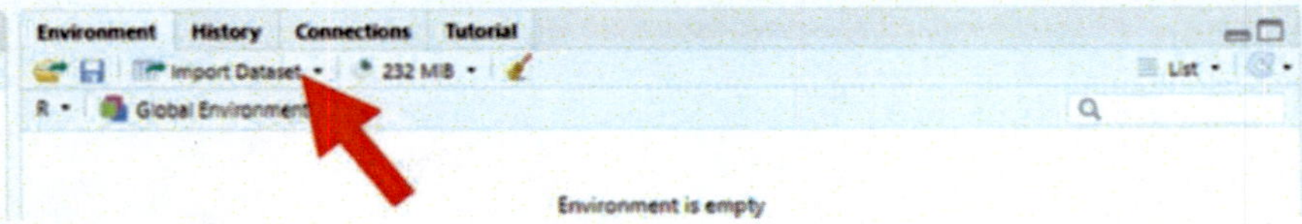

Select *"From Excel"*

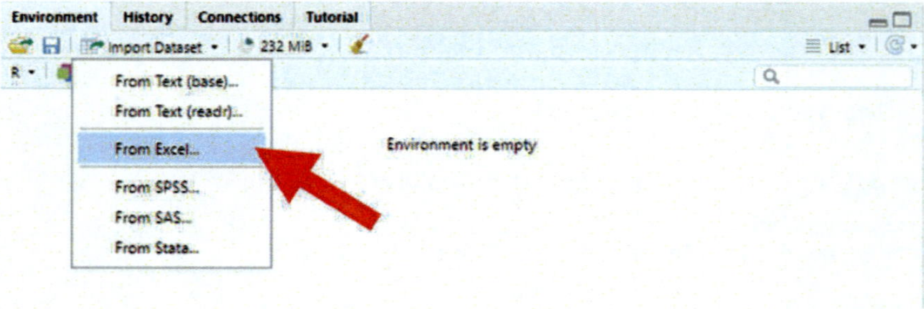

A window *"Import Excel Data"* appears. Click on *"Browse"*:

Notice that you have an option "First row as names" which is automatically checked. Uncheck it if necessary.

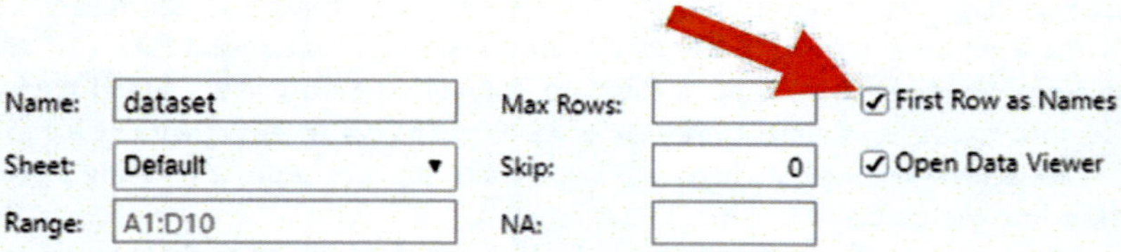

Navigate to the file to be imported: its preview appears. Click on *"Import"*

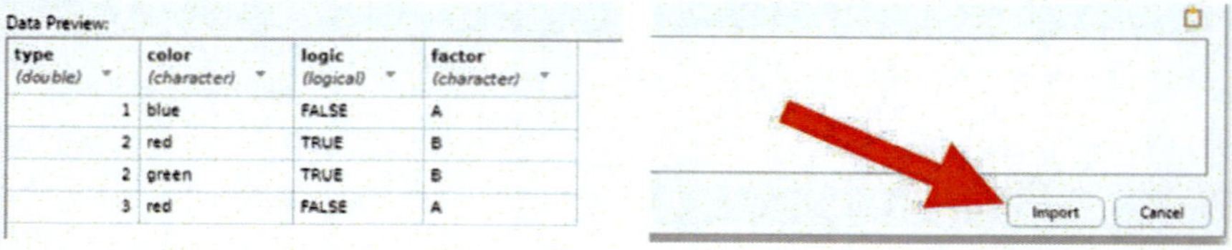

Now, you have a new variable in your environment:

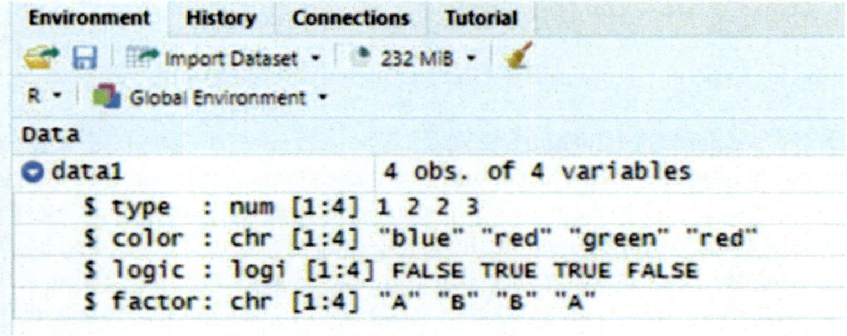

Analogously to lists, operations can be made internally to a `data.frame` by calling the function within. For instance, let us consider a `data.frame` containing two columns $x = (1,2,3)$, $y = (4,5,6)$ and let us create a third column: $z \leftarrow x + y$.

```
> x = c(1,2,3)
> y = c(4,5,6)
> d1 = data.frame(x = c(1,2,3), y = c(4,5,6))
> within(d1,z <- x+y)
  x y z
1 1 4 5
2 2 5 7
3 3 6 9
```

1.9 Plotting with RStudio

Plots appear in the subwindow in the right inferior side of the window:

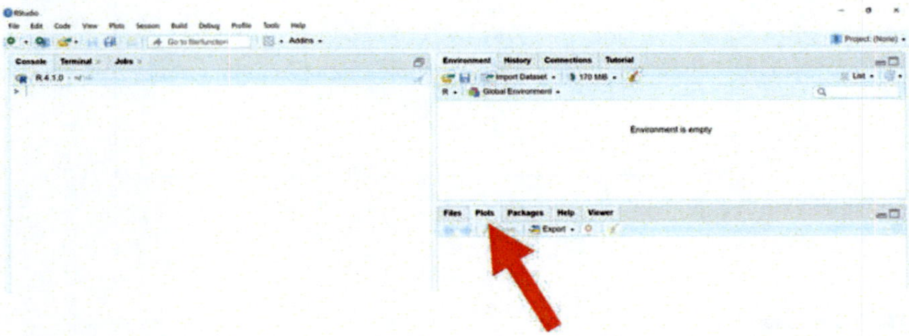

R proposes many plots. We do not explore all the possibilities here, and we limit ourselves to the basic types used in this book. For instance, you can generate a line plot by using "plot". The basic use is

```
plot(x,y, type = " ...", col = "...")
```

Here, type describes the line to be used. For instance: p = points, l = lines, o = points covered by lines, s = steps, and h = histogram. Parameter col defines the color: red, blue, black, etc. R proposes hundreds of predefined colors (enter colors() in the console to see the list). The type of points is controlled by pch = ... : 0= square, 1 = circle, etc. You can add labels to the axes by using xlab = "...", ylab= "...". A title is added by main = "...", A subtitle is added by sub = "...". As an example, let us plot $y = \exp(x)$ on the interval $(-1, 1)$ using 10 subintervals, triangles, and a line:

```
> nint = 10
> a = -1
> b = 1
> x = seq(a,b,length = nint + 1)
> y = exp(x)
> plot(x,y,type = "b", col = "black", pch = 6)
```

The result is shown in Fig. 1.2. To superpose plots or use different colors, you can start by an empty plot and then add the different curves:

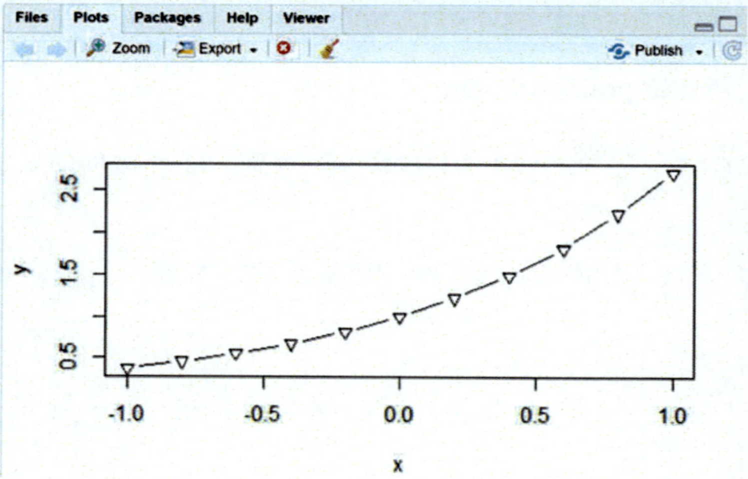

Fig. 1.2 A simple plot

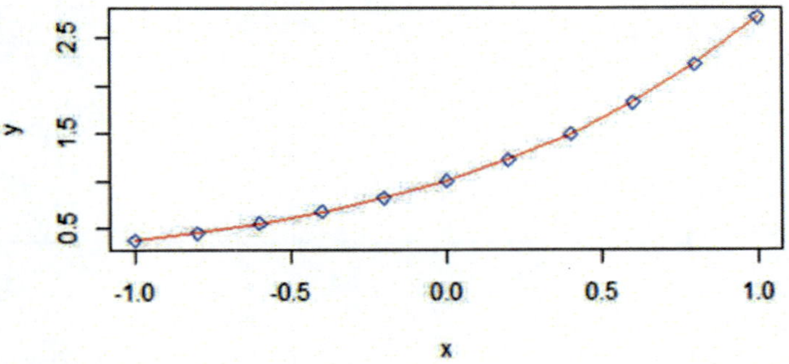

Fig. 1.3 Superposition of plots

```
> nint = 10
> a = -1
> b = 1
> x = seq(a,b,length = nint + 1)
> y = exp(x)
> plot(x,y,type="n")
> lines(x,y,col="red")
> points(x,y, col = "blue", pch = 23)
```

The result is shown in Fig. 1.3:

You can choose the type of line by using `lty` = . . . and the line width by `lwd` = ... For points, you can control their size (`cex` = . . .), their filling color (`bg` = . . .), and the width of their line (`lwd` = . . .). A legend can be added by `legend()`:

```
> nint = 10
> a = -1
> b = 1
> x = seq(a,b,length = nint + 1)
> y = exp(x)
> plot(x,y,type="n")
> lines(x,y,col="red",lty = 3, lwd = 3)
> points(x,y, col = "blue", pch = 23,bg = "green", cex=0.75,
lwd=2)
> legend("topleft",legend = c("line","points"),pch=c(NA,23),
lty = c(3,NA),lwd = c(3,NA), col = c("red","blue"))
```

The result appears as below (Fig. 1.4)

R provides a large panel of graphic controls and types of charts. You can plot pie charts (`pie`), histograms (`hist`), bar plots (`barplot`), curves defined by analytical expressions (`curve`), etc. You can split the graphic window to show several graphics at the same time: `par(mfrow=c(nrow,ncol))` or `layout (matrix(positions), ...)`. The results are shown in Figs. 1.5 and 1.6.

```
> par(mfrow = c(1,2))
> curve(x**2, from = -1, to=1)
> plot(x,y,col="blue",type="l",lty = 4, lwd = 3)
```

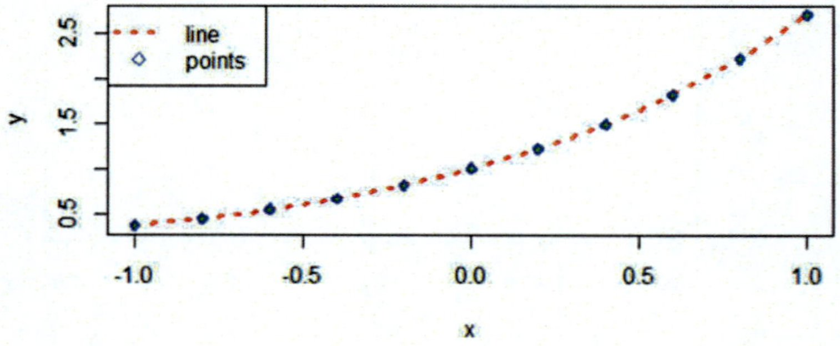

Fig. 1.4 Choosing the type of line

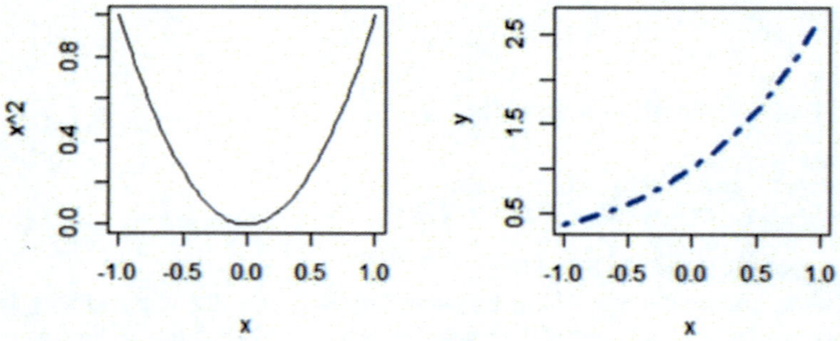

Fig. 1.5 Using mfrow to split the window

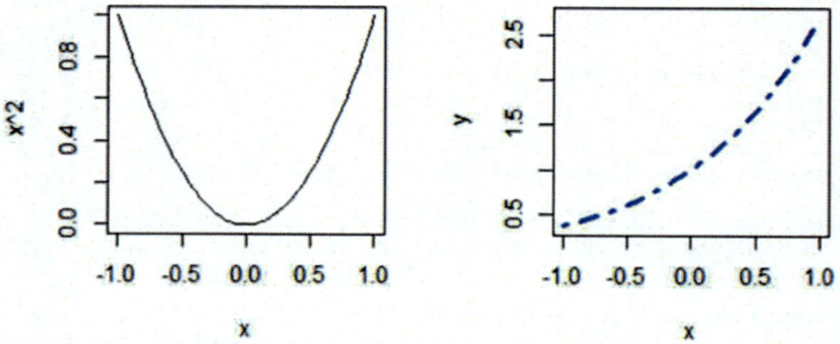

Fig. 1.6 Using layout to split the window

```
> layout(matrix(c(1,2),ncol = 2))
> curve(x**2, from = -1, to=1)
> plot(x,y,col="blue",type="l",lty = 4, lwd = 3)
```

The command layout is more flexible: you can merge positions to exhibit a single chart in a line or a row. Below, we assign the last two positions (last column) to the third chart (recall that R works by columns). The result is shown in Fig. 1.7.

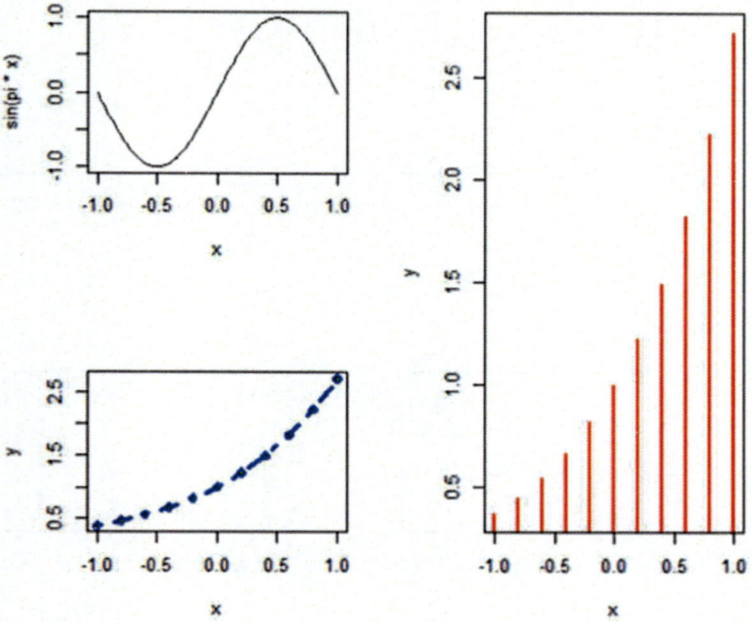

Fig. 1.7 Flexible use of layout

```
> nint = 10
> a = -1
> b = 1
> x = seq(a,b,length = nint + 1)
> y = exp(x)
> layout(matrix(c(1,2,3,3),ncol = 2))
> curve(sin(pi*x), from = a, to=b)
> plot(x,y,col="blue",type="l",lty = 4, lwd = 3)
> points(x,y, col = "blue", pch = 23,bg = "green", cex=0.75,lwd=2)
> plot(x,y,col="red", type="h", lwd=2 )
```

Below, we assign the first line (positions 1 and 3) to the first chart. The result is shown in Fig. 1.8

```
> nint = 10
> a = -1
> b = 1
> x = seq(a,b,length = nint + 1)
> y = exp(x)
> layout(matrix(c(1,2,1,3),nrow = 2))
> curve(sin(pi*x), from = a, to=b)
> plot(x,y,col="blue",type="l",lty = 4, lwd = 3)
> points(x,y, col = "blue", pch = 23,bg = "green", cex=0.75,lwd=2)
> plot(x,y,col="red", type="h", lwd=2 )
```

You can also use `split.screen` to split the plot window. For instance:

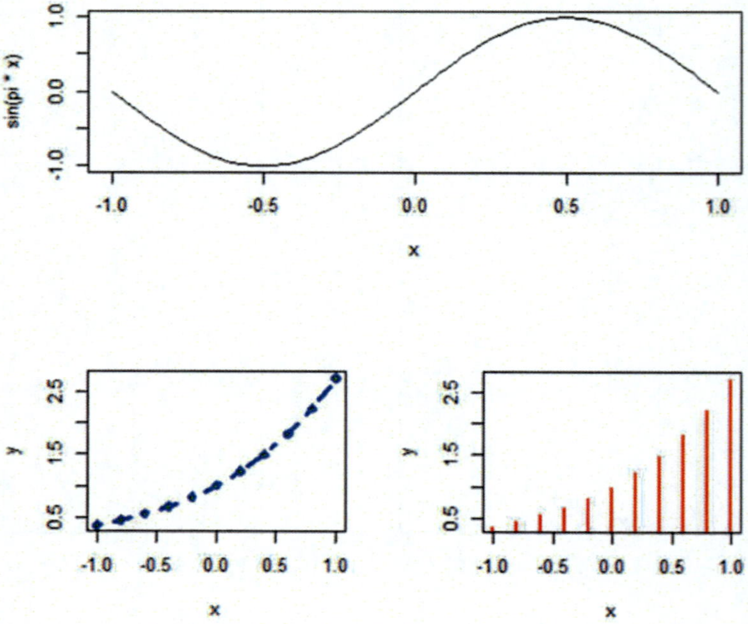

Fig. 1.8 Other example of use of layout

```
nint = 10
a = -1
b = 1
x = seq(a,b,length = nint +1)
y = exp(x)
# divide the region in 1 x 2 (side to side)
split.screen(c(1,2))  # numbered 1 and 2
# divide the first region in 2 x 1 (up and down)
split.screen(c(2,1), screen = 1) # numbered 3 and 4
# plot sinus on screen 3 (left side, up)
screen(3)
curve(sin(pi*x),from=a, to=b)
# plot exp on screen 4 (left side, down)
screen(4)
plot(x,y,col="blue",pch=23,bg="green",cex = 0.75,lwd = 2)
lines(x,y,lty=4,lwd = 3,col="blue")
# plot exp on screen 2 (right side)
screen(2)
plot(x,y,col="red",type="h",lwd = 2)
```

The result is shown in Fig. 1.9.

The R repositories propose many graphic packages destined to enhance the plots, for instance, `ggplot2` (it is also part of `tidyverse`, but may be used independently), `plot3D`, `rgl`, `chplot`, `MASS`, `plotly`, `googleVis`, `highchart`, `ggiraph`, and `gridExtra`. You can find downloads of these packages and their

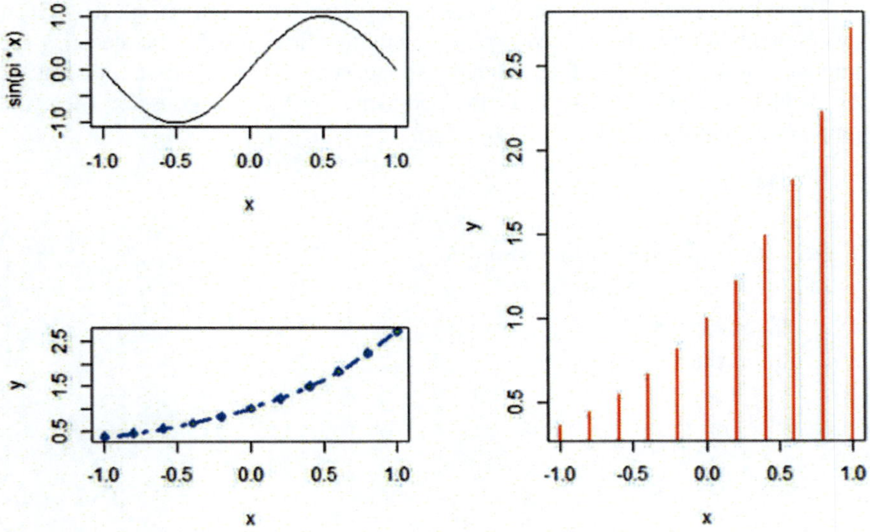

Fig. 1.9 Using `split.screen`

documentation on Internet. They allow you to customize and manipulate your plots. Among them, one of the most popular is `ggplot2`, which conceives a graph as an ordered set of layers to be superposed. `ggplot2` offers a simple command `qplot` (*quick plot*) that is equivalent to the standard `plot` of R. Under `ggplot2`, the type of line, its color, and its size are defined by a parameter named `geom_name`. You add data to the graphic by using `aes`. Titles are added by `labs`. A new layer is added by "+". If you give names to the graphics created, they will not appear in the window, and you can continue to manipulate them. For instance, let us consider the instructions below:

```
nint = 10
a = -1
b = 1
x = seq(a,b,length = nint + 1)
y = exp(x)
library(ggplot2)
g1 <- ggplot()
g2 <- g1 + aes(x,y)
g3 <- g2 + geom_line(colour = 'blue', size = 2)
g4 <- g3 + labs(
  title = "Exponential",
  subtitle = "from -1 to 1",
  x = "x",
  y = "exp(x)"
)
```

These instructions create graphics elements `g1` (an empty graph), `g2` (`g1` with a layer containing the points (x, y)), `g3` (`g2` with blue line of width 2 connecting the points), and `g4` (`g3` with a title, subtitle, and labels on the axes). None is displayed. To display `g4`, you need to enter the command >`g4` in the console. Then, the graphics appear (Fig. 1.10):

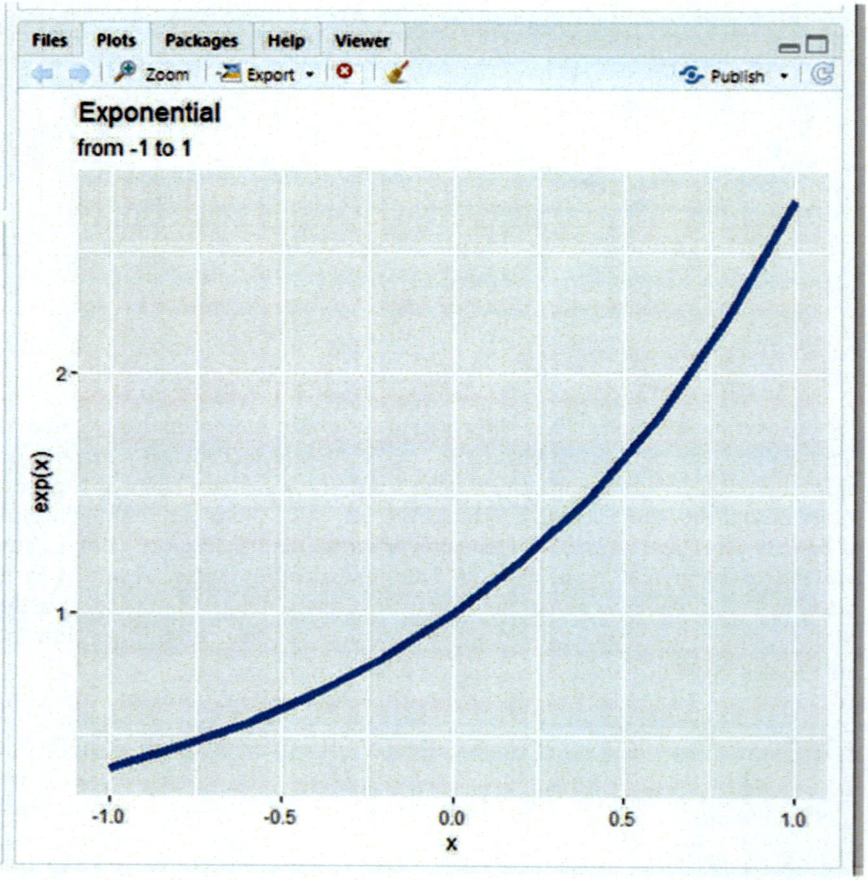

Fig. 1.10 A chart generated by ggplot

You can superpose graphics by adding then to an existing graph. For instance:

```
nint = 10
a = -1
b = 1
x = seq(a,b,length = nint + 1)
y = exp(x)
z = exp(-x)
library(ggplot2)
g1 <- ggplot()
g2 <- g1 + geom_line(aes(x,y),color = 'blue', size = 2)
g3 <- g2 + geom_line(aes(x,z),color = 'red', size = 2)
g4 <- g3 + labs(
  title = "Exponentials",
  subtitle = "from -1 to 1",
  x = "x",
  y = "exp(x), exp(-x)"
)
```

The result appears as below (Fig. 1.11):

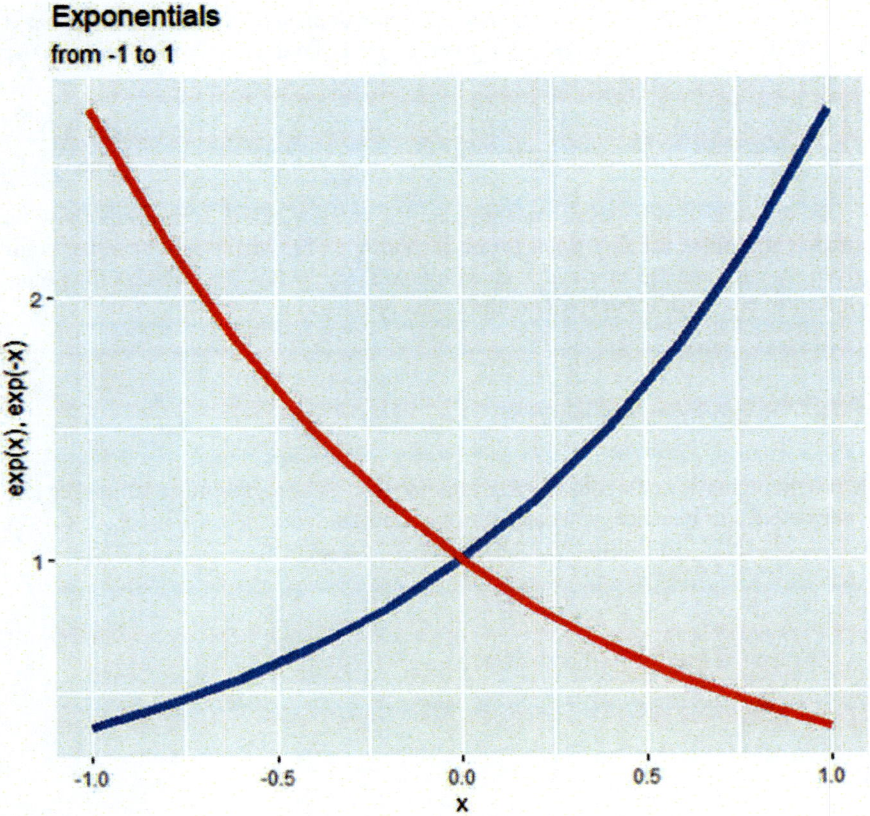

Fig 1.11 Superposition of plots by ggplot – each plot is a layer

To split the window, you can use the packages `patchwork`, `ggpubr`, `gridExtra`, `cowplot`, ... You can find documentation and examples for these packages on Internet easily. For instance, let us illustrate the use of `patchwork`: delimiter "|" separates charts on the same line and delimiter "/" changes to the next line. The charts are automatically adjusted to the line, but it is possible to let a place vacant by using `plot_spacer`, which generates an empty chart (Fig. 1.12).

R proposes an interesting command named `locator`, which captures the positions of points on the graphics window by clicking. This command is useful to include annotations, arrows, etc. Its use is

```
variable = locator(number of points).
```

For instance, `aux = locator(2)` captures the position of 2 points clicked on the graphics window. To write a text at a position clicked, you can use:

```
aux = locator(1)
text(x=aux$x[[1]],y=aux$y[[1]],labels="...",col="...")
```

By clicking on the graphics window, the string defined by `labels` will appear at the position you clicked in the color given by `col`. To draw an arrow, you can use

```
aux = locator(2)
arrows(x0=aux$x[[1]],y0=aux$y[[1]],x1 = aux$x[[2]],y1=aux$y[[2]],
col="...",lwd = ...,xpd = TRUE)
```

By clicking in two points on the graphics window, these commands draw an arrow going from the first point to the second one. The arrow will have the color given by `col` and the line width given by `lwd`. `xpd=TRUE` authorizes the points clicked to lay outside the box containing the plot region.

1.10 Programming with R

R programming is analogous to programming in usual programming languages and interpreters. For instance, you can control a condition with `if ... {}`, `if ... { } ... else {}`, `if ...{} else if ...{} else {}`, `switch` :

```
> x  <- 1
> if (x < 1) {
+   print(paste("x = ",x," < 1"))
+ } else {
+     print(paste("x = ",x," >= 1"))
+   }
[1] "x =  1  >= 1"
```

```
nint = 10
a = -1
b = 1
x = seq(a,b,length = nint + 1)
y = exp(x)
z = exp(-x)
library(ggplot2) 2
c1 = geom_line(aes(x,y),color = 'blue', size = 2)
g1 <- ggplot()
g1 <- g1 + c1
c2 = geom_line(aes(x,z),color = 'red', size = 2)
g2 <- ggplot()
g2 <- g2 + c2
g3 <- ggplot()
g3 <- g3 + c1 + c2
g4 <- g3 + labs(
  title = "Exponentials",
  subtitle = "from -1 to 1",
  x = "x",
  y = "exp(x), exp(-x)"
)
library(patchwork)
(g1 | g2 )/ (g4 | plot_spacer())
```

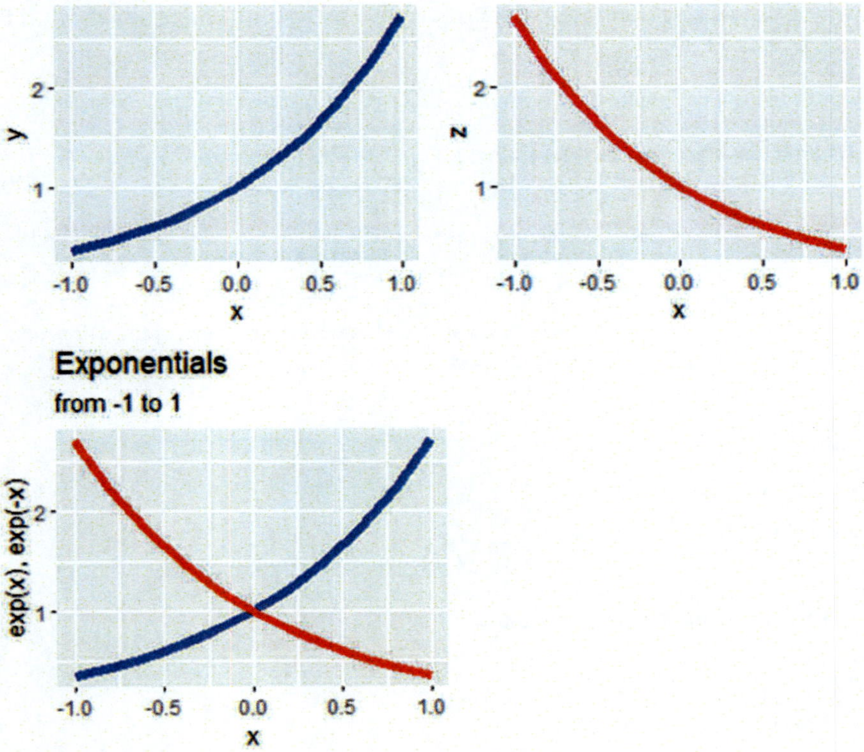

Fig. 1.12 Splitting the window with patchwork

```
> x  <- 1
> if (x < 0) {
+   print(paste("x = ",x," < 0"))
+ } else if (x < 1){
+   print(paste("0 <= x = ",x," < 1"))
+ } else {
+   print(paste("x = ",x," >= 1"))
+ }
[1] "x =  1  >= 1"
```

```
> switch (toString(x < 0),
+         "TRUE" = {print(paste("x = ",x," < 0"))},
+         "FALSE" = {
+           switch( toString(x < 1),
+                   "TRUE" = {print(paste("0 <= x = ",x," < 1"))}
+                   "FALSE" = {paste("x = ",x," >= 1")}
+           )
+         }
+         )
[1] "x =  1  >= 1"
```

You can also generate loops with `for`...{ }, `while`... { } or `repeat` { ... `if` ...{ `break`} }:

```
> s = 0
> for (i in 1:10) {
+   s = s + i
+ }
> cat("s = ",s)
s =  55
```

```
> s = 0
> i = 0
> while(i < 10){
+   i = i + 1
+   s = s + i
+ }
> sprintf("s = %i",s)
[1] "s = 55"
```

```
> s = 0
> i = 0
> repeat{
+   i = i + 1
+   s = s + i
+   if (i >= 10) {break}
+ }
> cat("s = ",s)
s =  55
```

To force to exit a loop, you can use `next` or `break`: `next` jumps the next instruction, while `break` exits the loop.

```
> i = 0
> while (TRUE){
+   i = i + 1
+   if (i >= 10) {break}
+ }
> cat("i = ", i)
i =  10
```

```
> i = 0
> while (TRUE){
+   i = i + 1
+   if (i < 10) {next}
+   break
+ }
> cat("i = ", i)
i =  10
```

You can define functions as follows:

```
name <- function(arguments)
  instructions
  return(result)
}
```

For instance, let us create a function that receives as arguments a vector of numeric values and returns a vector containing its minimum, its maximum, and its mean:

```
> myfun <- function(x){
+    n = length(x)
+    s = sum(x)/n
+    m1 = min(x)
+    m2 = max(x)
+    resu = c(m1,m2,s)
+    names(resu) = c("min","max","mean")
+    return(resu)
+ }
> z = c(1:10)
> aux = myfun(z)
> aux
 min  max mean
 1.0 10.0  5.5
```

Under R, subroutines are just functions that do not return values. For instance, let us write a subroutine that plots the cumulative sums of a vector:

```
> myfun <- function(x){
+    cs = cumsum(x)
+    plot(cs,type='h')
+ }
>  x = c(1:20)
>  myfun(x)
```

Running this code generates the chart (Fig. 1.13):

Under R, ...you do not have the declaration of global and local variables or the possibility of passing arguments by reference to modify them within a function. R works by environments: there is a global environment `.GlobalEnv`, which contains the variables declared outside the function. To modify one of these values, you

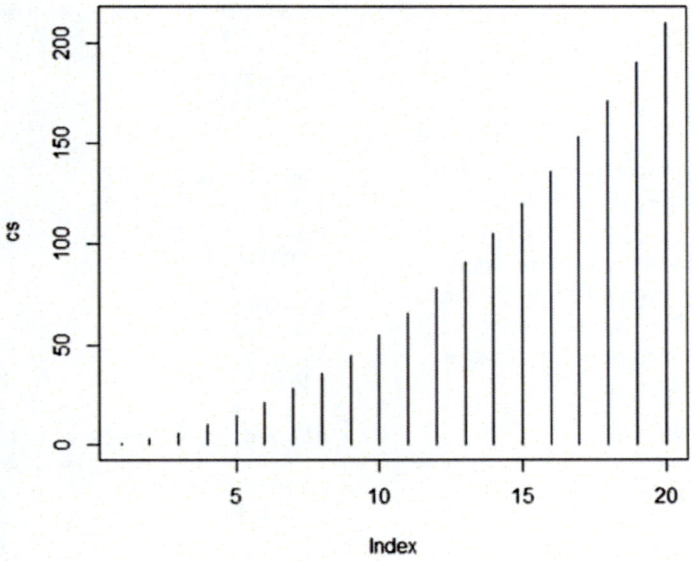

Fig. 1.13 Chart generated by myfun

must use the global assignment operator "<<−" or the instruction `assign`. For instance, let us create a function that receives as arguments a vector and a number and adds to the number the sum of the elements of the vector.

```
> myfun <- function(x,y){
+     aux = sum(x)
+     y <<- y + aux
+ }
>  x = c(1:20)
>  y = 0
>  myfun(x,y)
>  y
[1] 210
```

```
>  myfun <- function(x,y){
+     aux = sum(x)
+     assign("y",y+aux,envir = .GlobalEnv)
+  }
>  x = c(1:20)
>  y = 0
>  myfun(x,y)
>  y
[1] 210
```

Notice that you can create and manipulate environments. As an example, the commands at right create two environments named `env1` and `env2`.

```
> env1 <- new.env()
> env2 <- new.env()
```

You refer to a variable `x` in environment `env` as `env$x`. Variables created in one environment exist only in it. In the example at right, variable `x` exists only in `env2` and variable `y` exists only in `env1`.

```
> env1$y <- "name 1'
> env2$x <- 2
> env1$x
NULL
> env2$x
[1] 2
> env1$y
[1] "name 1"
> env2$y
NULL
```

You can also use the command assign:

```
> assign("z",3,envir=env1)
> assign("z","text",envir=env2)
> env1$z
[1] 3
> env2$z
[1] "text"
```

Finally, you can use data by calling the command `with (d1, instructions)` which executes the operations given by instructions using the data in `d1` – which can be an environment, a `data.frame` or a `list`:

```
> env1 <- new.env()
> env1$x = c(1,2,3)
> env1$y = c(4,5,6)
> with(env1, x+y)
[1] 5 7 9
> x = c(1,2,3)
> y = c(4,5,6)
> l1 = list(x,y)
> with(l1,x+y)
[1] 5 7 9
> d1 = data.frame(x = c(1,2,3),
+                 y = c(4,5,6))
> with(d1, x + y)
[1] 5 7 9
```

Documentation about environments and their use can be found in the literature and Internet

1.11 Classes in R

Classes can be useful to define objects that are destined to be replicated, for which we desire to manipulate different copies simultaneously. R furnishes different methods for the creation of classes, and you can find packages furnishing supplementary methods.

Here, we examine one of these methods, which is close to the usual methods of creation of objects in other languages, based on the package R6. You can install the package R6 from the R repositories on Internet.

R6 classes follow the standard model of objected oriented programming: they contain public properties and methods, and private properties and methods. Public elements may be accessed by external elements, while private elements are accessible only internally. The general scheme of a R6 class is the following one:

```
myclass=R6class("myclass",
     public=list(
        properties,
        initialize=function(pub_attrib,priv_attrib)
        {
           self$properties=pub_attrib
           private$properties = priv_attrib
        },
        finalize = function()
        {
          ...
        }
        pub_method =function(parameters)
        {
           return(result)
        }
     ),
     private=list(
        properties,
        priv_method = function(parameters)
        {
           return(result)
        }
     )
)
```

Methods `initialize` and `finalize` are optional and correspond to the creator and the destructor, respectively – the creator creates an instance (id est, a copy of the object) and the destructor makes the actions prescribed for the destruction of the instance.

Let us consider a simple class, defining a point in $\mathbb{R}^n$: the point is characterized by a vector of coordinates $x = (x_1, \ldots, x_n)$. These coordinates and the dimension n are properties of the point. We can define the class as shown at right:

```
library(R6)
Rn_point=R6Class("Rn_point",
    public=list(
      coords = NA,
      dimension = NA,
      initialize=function(x)
      {
        if (missing(x)==FALSE)
        {
         self$coords=x
         self$dimension = length(x)
        }
      }
    )
)
```

Saving this class in a file `Rnpoint.R`, this class can be used as shown at right:

```
> source("Rnpoint.R")
> x = Rn_point$new(c(1,2,3))
> y = Rn_point$new()
> y$coords = c(1,2,3,4)
> x$dimension
[1] 3
> y$dimension
[1] NA
> x$coords
[1] 1 2 3
> y$coords
[1] 1 2 3 4
```

Notice that the dimension of `y` is undefined at this stage – we have defined only its coordinates. For the class `Rn_point` defined, we must use the instruction `y$dimension = 4` to set this value. As an alternative, we can define the dimension as a public method without argument:

```
library(R6)
Rn_point=R6Class("Rn_point",
    public=list(
      coords = NA,
      initialize=function(x)
      {
        if (missing(x)==FALSE)
        {
         self$coords=x
        }
      },
      dimension = function()
      {return(length(self$coords))}
    )
)
```

This code is saved in the file `Rnpoint2.R`. Then, we obtain the results at right. Notice that `dimension` must be used as a function.

```
> source("Rnpoint2.R")
> x = Rn_point$new(c(1,2,3))
> y = Rn_point$new()
> y$coords = c(1,2,3,4)
> x$dimension()
[1] 3
> y$dimension()
[1] 4
> x$coords
[1] 1 2 3
> y$coords
[1] 1 2 3 4
```

Let us introduce a new class `vec` introducing manipulations of vectors, saved in a file named `vec.R`:

```
library(R6)
#
vec=R6Class("vec",
      public=list(
        dimension = function(x)
        {return(length(x))},
        norm2 = function(x)
        {return(norm(matrix(x,nrow=1),"2"))},
        unitary = function(x)
        {return(x/self$norm2(x))},
        tangent = function(x,y)
        {return(self$unitary(y-x))}
      )
)
```

Modify the file `Rnpoint.R` as follows:

```
library(R6)
source("vec.R")
Rn_point=R6Class("Rn_point",
    inherit = vec,
    public=list(
        coords = NA,
        initialize=function(x)
        {
          if (missing(x)==FALSE)
          {
           self$coords=x
          }
        },
        dimension = function()
        {return(super$dimension(self$coords))},
        dist = function(x,y){
            return(super$norm2(x$coords - y$coords))
        },
        dist2 = function(x){
            return(super$norm2(self$coords - x$coords))
        },
        unity = function(x)
        {return(x$coords/super$norm2(x$coords))},
        tangent = function(x,y)
        {return(super$tangent(x$coords,y$coords))}
    )
)
```

Under this form, the class Rn_point is derived from the class vec and inherits from it. super$name calls the method vec$name from the class vec. The inherited methods evaluate the dimension, the norm, an unitary vector oriented from the origin to a point having the coordinates of the vector, and an unitary vector oriented by the difference of two vectors of coordinates. The inherited methods allow the determination of the dimension of a point, its distance to another point, the distance between two points, an unity vector oriented from the origin to the point, and an unity vector oriented according to the difference of two points. Let us consider a class defining a segment, which is formed by two points (start and end points): create a file Rnsegment.R as follows:

```
library(R6)
source("Rnpoint.R")
#
Rn_segment=R6Class("Rn_segment",
     inherit = Rn_point,
     public=list(
         start = NA,
         end = NA,
         initialize=function(x,y)
         {
           if (missing(x)==FALSE)
           {
            self$start=Rn_point$new(x)
           }
          if (missing(y)==FALSE)
          {
           self$end=Rn_point$new(y)
          }
         },
         dimension = function()
           {return(self$start$dimension())},
         length = function()
           {return(super$dist(self$end, self$start))},
         tangent = function()
           {return(super$tangent(self$start,self$end))}
     )
)
```

This class inherits from Rn_point, so that the methods from Rn_point can be used in Rn_segment. This allows to the determination of the dimension, the length, and the tangent to the segment by using the methods from Rn_point – which uses the methods from vec. An example of use of this class is given below:

```
> source("Rnsegment.R")
> x = c(1:3)
> y = c(4:6)
> s = Rn_segment$new(x,y)
> s$length()
[1] 5.196152
> s$dimension()
[1] 3
> s$tangent()
[1] 0.5773503 0.5773503 0.5773503
```

One of the advantages of using classes is the possibility of easily extend the features and operations of the object, by adding new properties and methods to the original source file. As an example, if we desire to include the possibility of plotting the segment, we can simply add a new public method to the file `Rnsegment.R`, for instance:

```
plot2 = function(){
    x = c(self$start$coords[[1]], self$end$coords[[1]])
    y = c(self$start$coords[[2]], self$end$coords[[2]])
    plot(x,y,type = "n")
    lines(x,y)
},
```

Then, we have:

```
> x = c(1:2)
> y = c(4:5)
> s = Rn_segment$new(x,y)
> s$length()
[1] 4.242641
> s$dimension()
[1] 2
> s$tangent()
[1] 0.7071068 0.7071068
> s$plot2()
```

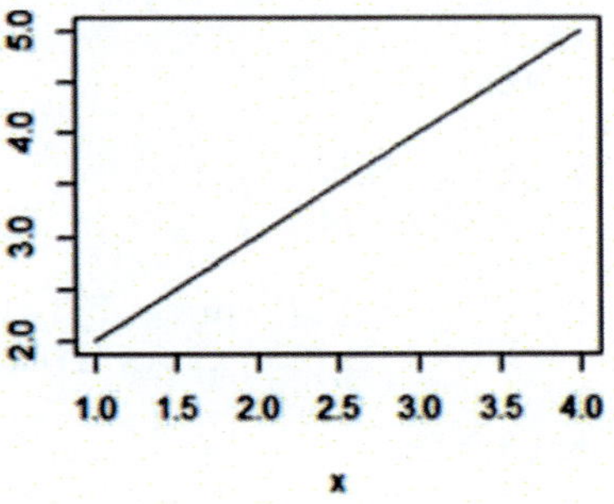

Plots of 3D segments may be obtained by using packages such as `rgl` or `plot3D`.

Let us extend the segments to curves by defining a new class `Rn_curve`, formed by ordered lists of points. The total length of such a curve is evaluated as the sum of the lengths of the segments connecting successive points. The class may be defined in a file `Rncurve.R` containing the code below:

```
library(R6)
source("Rnsegment.R")
#
Rn_curve=R6Class("Rn_curve",
    inherit = Rn_segment,
    public=list(
       point = NA,
       initialize=function(lp)
       {if (missing(lp)==FALSE)
         {aux_list = list()
          for (i in 1:length(lp)){
            aux_point=Rn_point$new(lp[[i]])
            aux_list[[i]] = aux_point
           }
           self$point = aux_list
         }
       },
       dimension = function()
         {return(self$point[[1]]$dimension())},
       segments = function(){
        lp = self$point
        lseg = list()
        for (i in 1:(length(lp)-1)){
         lseg[[i]] = Rn_segment$new(lp[[i]],lp[[i+1]])
         }
         return(lseg)
       },
       length = function(){
        lp = self$point
        s = 0
        for (k in 1:(length(lp)-1)){
         s = s + lp[[k]]$dist2(lp[[k+1]])
        }
        return(s)
       },
       tangent = function(){
        lp = self$point
        lt = list()
        for (i in 1:(length(lp)-1)){
         seg = Rn_segment$new(lp[[i]],lp[[i+1]])
         lt[[i]] = seg.tangent()
        }
        return(lt)
       },
       plot2 = function(){
```

```
lp = self$point
        plot(x,y,type = "n")
        for (i in 1: (length(lp)-1)) {
      x = c(lp[[i]]$coords[[1]], lp[[i+1]]$coords[[1]])
      y = c(lp[[i]]$coords[[2]], lp[[i+1]]$coords[[2]])
      lines(x,y)
      }
    }
  )
)
```

Then, run the code below:

```
> source("Rncurve.R")
> #
> ang = seq(0,2*pi,length=101)
> r = 1
> x = r*cos(ang)
> y = r*sin(ang)
> lp = list()
> for (i in 1: length(ang)){lp[[i]] = c(x[[i]],y[[i]])}
> c1 = Rn_curve$new(lp)
> c1$length()
[1] 6.282152
> c1$plot2()
```

The result appears as follows (Fig. 1.14):

In this book, we shall manipulate some classes, useful to avoid repetitive programming. For instance, we use polynomials, trigonometrical functions, and more generally, families of functions φ_i forming Hilbert basis for expansions representing functions and random variables. A typical finite expansion is $PX = \sum_{i=0}^{k} x_i\varphi_i$. Repetitive operations are performed that use these elements, so that we create a class and call it when necessary. As an example, consider the trigonometric family

Fig. 1.14 Circle generated by class Rn_curve

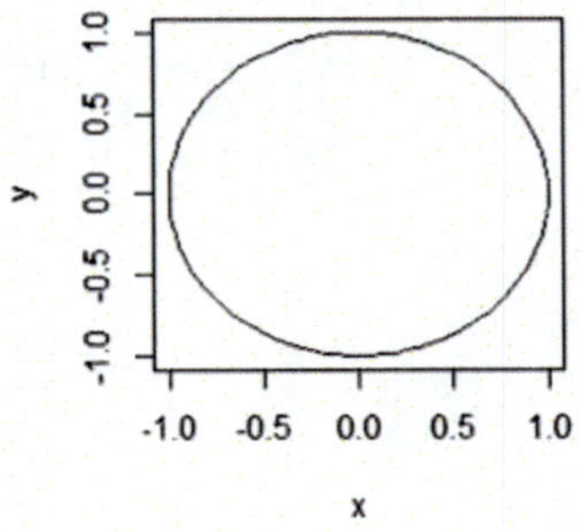

$$\varphi_0(x) = 1;\ \varphi_{2k-1}(x) = \sin\left(k\left(\frac{x-a}{b-a}\right)\right),\ \varphi_{2k}(x) = \cos\left(k\left(\frac{x-a}{b-a}\right)\right)\ (k \geq 1).$$

An example of class for this family is a file `trigo.R` as follows:

```
library(R6)
#
trigo=R6Class("trigo",
    public=list(
        lb = 0,
        ub = 1,
        zero = 1e-3,
        initialize=function(lb1,ub1,zero1)
        {if (missing(lb1)==FALSE)
         {self$lb = lb1}
         if (missing(ub1)==FALSE)
         {self$ub = ub1}
         if (missing(ub1)==FALSE)
         {self$ub = ub1}
        },
        value = function(i, x){
         d = self$ub - self$lb
         z = self$zero
         if (i == 0){v=1}
          else{
           s = (x - self$lb) / d
           if (i == 0){
            c = 0.5 * i
            if (abs(s) < z){v=1}
             else if (abs(s-1) < z){v = cos(c)}
              else {v = cos(c*s)}
           }
           else{
            c = 0.5 * (i + 1)
            if (abs(s) < z) {v = 0}
             else if (abs(s - 1) < z){v = sin(c)}
              else {v = sin(c*s)}
            }
           }
          return(v)
        },
        d1 = function(i, x){
         d = self$ub - self$lb
         if (i == 0){v=0}
```

```
            else{
             s = (x - self$lb) / d
             if (i %% 2 == 0){
              c = 0.5 * i
              if (abs(s) < z){v=1}
               else if (abs(s-1) < z){v = -sin(c)}
               else {v = -c*sin(c*s)}
              }
              else{
               c = 0.5 * (i + 1)
               if (abs(s) < z) {v = c}
                else if (abs(s - 1) < z){v = c*cos(c)}
                 else {v = c*cos(c*s)}
               }
              }
              return(v/d)
        },
        d2 = function(i, x){
         d = self$ub - self$lb
         if (i == 0){v=0}
          else{
           s = (x - self$lb) / d
           if (i == 0){
            c = 0.5 * i
            if (abs(s) < z){v=1}
             else if (abs(s-1) < z){v = -c*c*cos(c)}
              else {v = -c*c*cos(c*s)}
            }
            else{
             c = 0.5 * (i + 1)
             if (abs(s) < z) {v = 0}
              else if (abs(s - 1) < z){v = -c*c*sin(c)}
               else {v = -c*c*sin(c*s)}
             }
            }
            return(v/(d*d))
        }
    )
)
```

An example of use is given below:

```
> source("trigo.R")             > v3**2 + v4**2
> a = -1                        [1] 1
> b = 1                         > v5 = phi$d1(1,0.5)
> phi = trigo$new(a,b)          > v6 = phi$d1(2,0.5)
> v1 = phi$value(1,0.5)         > (v5**2 + v6**2)*(b-a)**2
> v2 = phi$value(2,0.5)         [1] 1
> v3 = phi$value(3,0.5)         > v7 = phi$d2(3,0.5)
> v4 = phi$value(4,0.5)         > v8 = phi$d2(4,0.5)
> v1**2 + v2**2                 > (v7**2 + v8**2)*(b-a)**4/2**4
[1] 1                           [1] 1
```

You can create as many classes as you need and call them in your programs.

A second operation that will be frequent in this book is the evaluation of the expansion $\sum_{i=0}^{k} c_i \varphi_i(u_j), j = 1, \ldots, ns$. We create a class `expansion1D` to handle it:

```
library(R6)
#
expansion1D=R6Class("expansion1D",
    public=list(
      basis = NA,
      initialize=function(phi)
      {if (missing(phi)==FALSE)
        {self$basis = phi}
      },
      proj = function(cc,u) {
       phi = self$basis
       ns = length(u) # points
       k = length(cc) # coefficients
       px = numeric(length = ns)
       for (i in 1:ns){
        s = 0
        for (j in 1:k){
         s = s + cc[[j]] * phi$value(j - 1, u[[i]])}
         px[[i]] = s
       }
       return(px)
      },
      value = function(i,u){
       phi = self$basis
       ns = length(u) # points
       vx = numeric(length = ns)
       for (j in 1:ns){
        vx[[j]] = phi$value(i, u[[j]])
       }
       return(vx)
      },
      d1 = function(i,u){
       phi = self$basis
       ns = length(u) # points
       vdx = numeric(length = ns)
       for (j in 1:ns){
        vdx[[j]] = phi$d1(i, u[[j]])
       }
       return(vdx)
      },
      meanprod = function(i,j,u){
       v1 = self$value(i,u)
       v2 = self$value(j,u)
       return (mean(v1*v2))
      },
      meansp = function(u,k){
          msp = matrix(0,nrow=k,ncol=k)
          for (i in 1:k){
           msp[[i,i]] = self$meanprod(i,i,u)


           for (j in (i+1):k){
            aux = self$meanprod(i,j,u)
            msp[[i,j]] = aux
            msp[[j,i]] = aux
           }
          }
          return(msp)
      }
    )
)
```

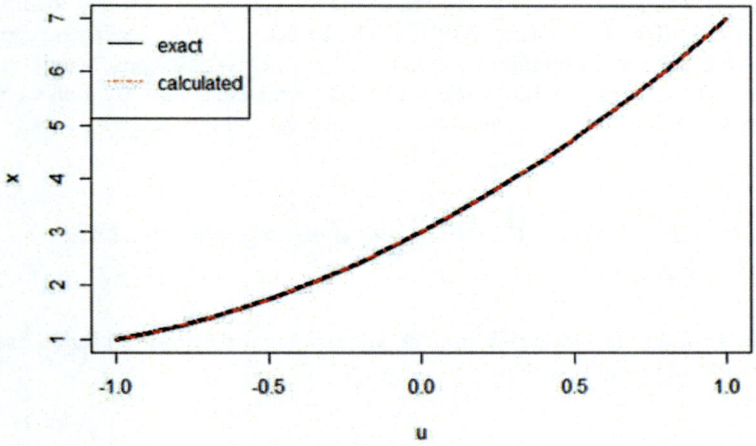

Fig. 1.15 An example of Hilbert expansion

An example of use is given below: we calculate $x = 3 + 3u + u^2$ using a polynomial basis $\varphi_i(x) = \left(\frac{x-a}{b-a}\right)^i$, $i=0,1,2$ and the expression $px = \varphi_0 + 2\varphi_1 + 4\varphi_2$. The result is shown in Fig. 1.15.

```
> source("expansion1D.R")
> source("polynom.R")
> a = -1
> b = 1
> coef = c(1,2,4) # coefficients
> ns = 21
> u = seq(a,b,length = ns)
> phi = polynom$new(a,b)
> expl = expansion1D$new(phi)
> px = expl$proj(coef,u)
> x = numeric(length = ns)
> for (i in 1: ns){
+    s = 3 + 3* u[[i]] + u[[i]]**2
+    x[[i]] = s
+ }
> plot(u,x,type = "n")
> lines(u,x,col="black",lwd = 3)
> lines(u,px,col="red",lty = 4, lwd = 2)
> legend("topleft",legend = c("exact","calculated"),
 col = c("black","red"),lty = c(1,4))
```

1.12 How to Solve Differential Equations with R

The community of R users proposes packages for the solution of differential equations. For instance, `pracma` emulates MATLAB® commands such as *ode23* and *bvp*. In addition, you can find in the R repositories the packages `deSolve` for initial value problems, bvpSolve for boundary value problems, and ReacTran for partial differential equations (PDE). The authors of these packages propose a book with code and examples (Soetaert et al., 2012). In this book, we do not use package ReacTrans – when considered, PDE will be solved by finite element methods. We recall here the basic use of deSolve and bvpSolve.

1.12.1 Initial Value Problems for Ordinary Differential Equations

Let us recall that the general form of an initial value problem (IVP) for an ordinary differential equation (ODE) is

$$\frac{dy}{dt} = f(t, y) \text{ on } (0, T), y(0) = y_0$$

When using deSolve, the basic instruction is ode, as follows:

```
ode(y=..., times = ..., func = ..., parms= ...,method= ...)
```

method is optional: if it is not specified, a default method is used. The first parameter "y=..." gives the initial value y_0; "times=..." gives a numeric vector tc containing discretization of the interval $(0, T)$ in subintervals: for instance, tc = seq(0,T,length.out = ...) – the solution will be calculated in the times contained in the numeric vector tc; "func=..." gives the function f, which uses parameters parms to be evaluated: f = function (t,y,parms){ ... }.

Let us use this instruction to solve the equation

$$y' = Ay, A = \begin{pmatrix} 0 & 1 \\ -1 & 0 \end{pmatrix}, \quad y(0) = \begin{pmatrix} 0 \\ 1 \end{pmatrix},$$

which has an exact explicit solution $y(t) = (\sin(t), \cos(t))^t$. The solution is generated on the interval $(0, 2\pi)$. We present in Fig. 1.16 an example of solution, generated by a standard call to ode. This result was furnished by the code below:

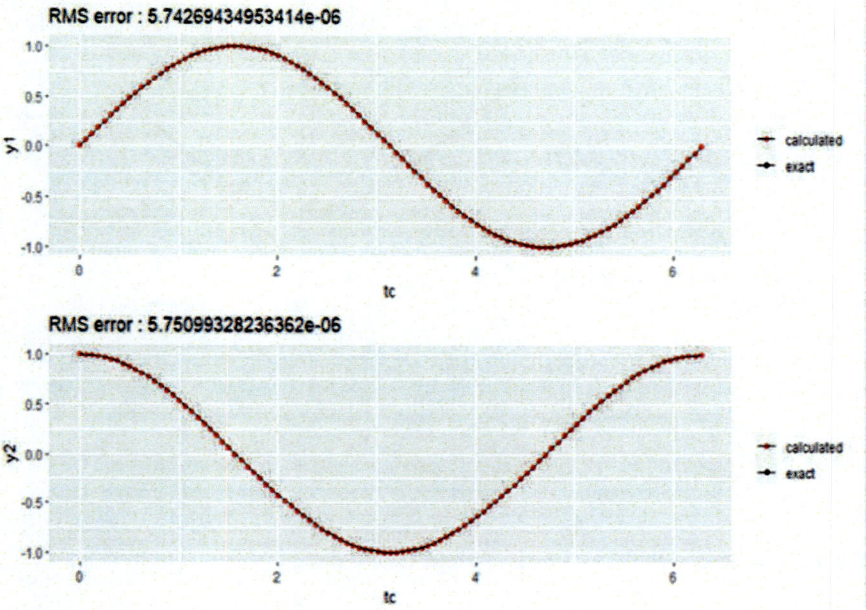

Fig. 1.16 Solution of an IVP

```
f <- function(t, y, parms) {
  with(parms, {dy <- A %*% y
  return(list(c(dy)))})}
  parms <- list(
    A = matrix(c(0, 1, -1, 0),nrow = 2, ncol = 2, byrow = TRUE)
  )
  y0 = c(0,1)
  tc = seq(from= 0, to = 2*pi, length.out=101)
library(deSolve)
resu <- ode(y0, tc, f, parms)
y1 <- subset(resu, select = "1")
y2 <- subset(resu, select = "2")
err1 = norm(y1 - sin(tc))/sqrt(length(tc))
err2 = norm(y2 - cos(tc))/sqrt(length(tc))
library(ggplot2)
g1 <- ggplot()
g2 <- ggplot()
g1 <- g1 + geom_point(aes(tc,y1,color = "calculated"))
g1 <- g1 + labs(title = paste("RMS error :",as.character(err1)))
g1 <- g1 + geom_function(fun = sin, show.legend = TRUE)
g1 <- g1 + scale_color_manual(values = c(
  "calculated" = "red",
  "exact" = "black"))
g1 <- g1 + theme(legend.title = element_blank())
g2 <- g2 + geom_point(aes(tc,y2, color = "calculated"))
g2 <- g2 + labs(title = paste("RMS error :",as.character(err2)))
g2 <- g2 + geom_function(fun = cos, show.legend = TRUE)
g2 <- g2 + scale_color_manual(values = c(
  "calculated" = 'red',
  "exact" = 'black'))
g2 <- g2 + theme(legend.title = element_blank())
library(patchwork)
g1 / g2
```

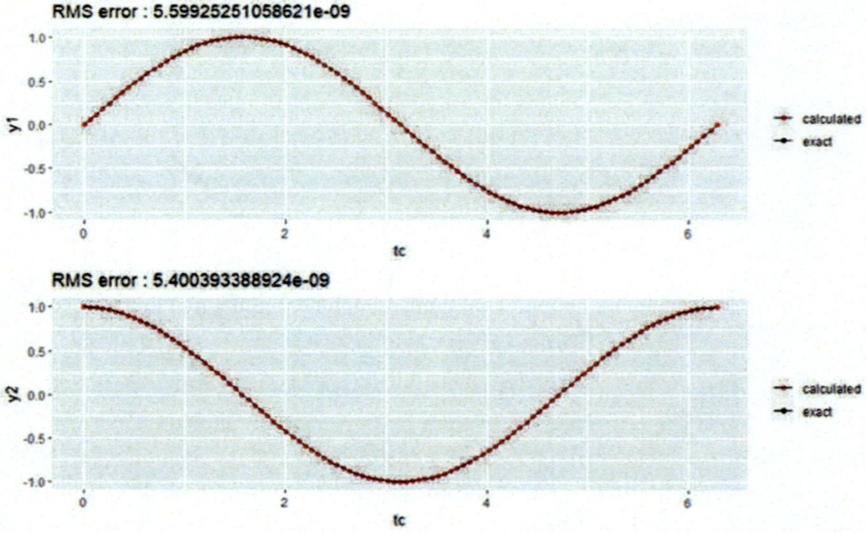

Fig. 1.17 Solution of an IVP by Dormand–Prince method rk45dp7

The results furnished by using `resu <- rk(y0, tc, f, parms, method = rkMethod("rk45dp7"))` (Dormand–Prince method) are shown below (Fig. 1.17)

1.12.2 Boundary Value Problems for Ordinary Differential Equations

The general form of a boundary value problem (IVP) for an ordinary differential equation (ODE) is

$$\frac{dy}{dt} = f(t, y) \text{ on } (0, \mathrm{T}), y_i(0) = y_{i0}(i \in I), y_j(T) = y_{T_j} \ (j \in J)$$

I and J are subsets of indexes $\{1, \ldots, n\}, = \varnothing$, where n is the dimension of $\mathbf{y}$. In general, $I \cup J = \{1, \ldots, n\}, I \cap J$.

Package `bvpSolve` proposes methods for the solution of the boundary value problem (BVP). For instance, let us consider the following BVP:

$$y'''' = \mathrm{y} \ on \ (0, 2\pi), y(0) = y(2\pi) = 0, y'(0) = \mathrm{y}'(2\pi) = 1.$$

The solution is $y(t) = \sin (t)$. This problem is reformulated as

$$Y = \begin{pmatrix} y \\ y' \\ y'' \\ y''' \end{pmatrix}, Y' = \begin{pmatrix} Y_2 \\ Y_3 \\ Y_4 \\ Y_1 \end{pmatrix}, Y_1(0) = Y_1(2\pi) = 0, Y_2(0) = Y_2(2\pi) = 1.$$

We use the code below:

```
f <- function(t, y, parms) {
  with(parms, {dy <- A %*% y + B
  return(list(c(dy)))})}
parms <- list(
  A <- matrix(c(0, 1, 0, 0, 0, 0, 1, 0, 0, 0, 0, 1, 1, 0, 0, 0),
                nrow = 4, ncol = 4, byrow = TRUE),
  B <- matrix(c(0, 0, 0, 0),nrow = 4, ncol = 1, byrow = TRUE)
)
  y0 = c(0,1, NA,NA)
  yend = c(0,1,NA,NA)
  tc = matrix(seq(from= 0, to = 2*pi, length.out=101), ncol = 1)
library(bvpSolve)
resu <- bvpcol(y0, tc, f, yend, parms)
y1 <- subset(resu, select = "1")
y2 <- subset(resu, select = "2")
f1 <- function(t){
  aux <- sin(t)
  return(matrix(aux,ncol = 1))
}
f2 <- function(t){
  aux <- cos(t)
  return(matrix(aux,ncol = 1))
}
y1c <- f1(tc)
y2c <- f2(tc)
err1 = norm(y1 - y1c)/sqrt(length(tc))
err2 = norm(y2 - y2c)/sqrt(length(tc))
library(ggplot2)
g1 <- ggplot() + geom_point(aes(tc,y1,color = "calculated"))
g1 <- g1 + labs(title = paste("RMS error :",as.character(err1)))
g1 <- g1 + geom_line(aes(tc,y1c, color='exact'))
g1 <- g1 + scale_color_manual(values = c(
  "calculated" = "red",
  "exact" = "black"))
g1 <- g1 + theme(legend.title = element_blank())
g2 <- ggplot() + geom_point(aes(tc,y2, color = "calculated"))
g2 <- g2 + labs(title = paste("RMS error :",as.character(err2)))
g2 <- g2 + geom_line(aes(tc,y2c, color='exact'))
g2 <- g2 + scale_color_manual(values = c(
  "calculated" = 'red',
  "exact" = 'black'))
g2 <- g2 + theme(legend.title = element_blank())
library(patchwork)
g1 / g2
```

The results appear in the subwindow "plot" (Fig. 1.18):

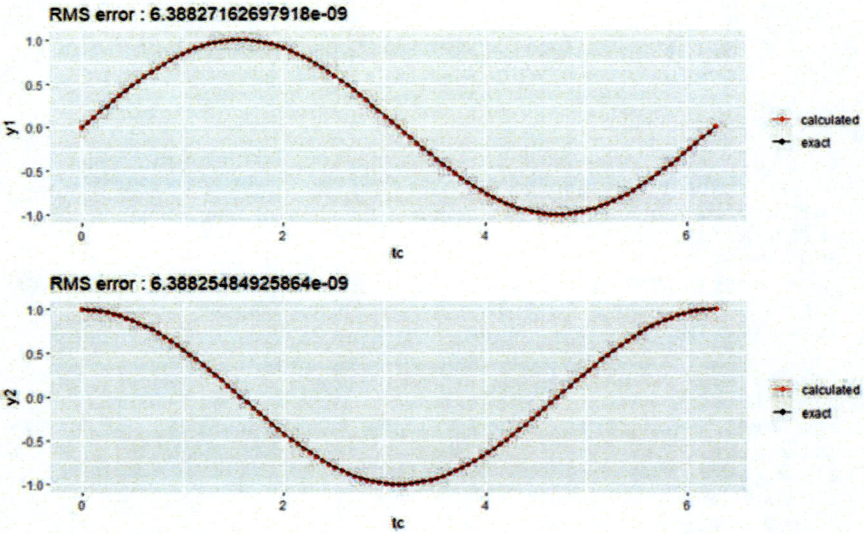

Fig. 1.18 Solution of a BVP

As a second example, let us consider

$$y'''' = 1 \; on \; (0, 1), y(0) = y(1) = 0, y'(0) = y'(1) = 0.$$

The code is analogous, except for

```
f <- function(t, y, parms) {
  with(parms, {dy <- A %*% y + B
  return(list(c(dy)))})}
parms <- list(
  A <- matrix(c(0, 1, 0, 0, 0, 0, 1, 0, 0, 0, 0, 1, 0, 0, 0, 0),
              nrow = 4, ncol = 4, byrow = TRUE),
  B <- matrix(c(0, 0, 0, 1),nrow = 4, ncol = 1, byrow = TRUE)
)
  y0 = c(0,0, NA,NA)
  yend = c(0,0,NA,NA)


f1 <- function(t){
  aux <- t^4/24 - t^3/12 + t^2/24
  return(matrix(aux,ncol = 1))
}
f2 <- function(t){
  aux <- t^3/6 - t^2/4 + t/12
  return(matrix(aux,ncol = 1))
}
```

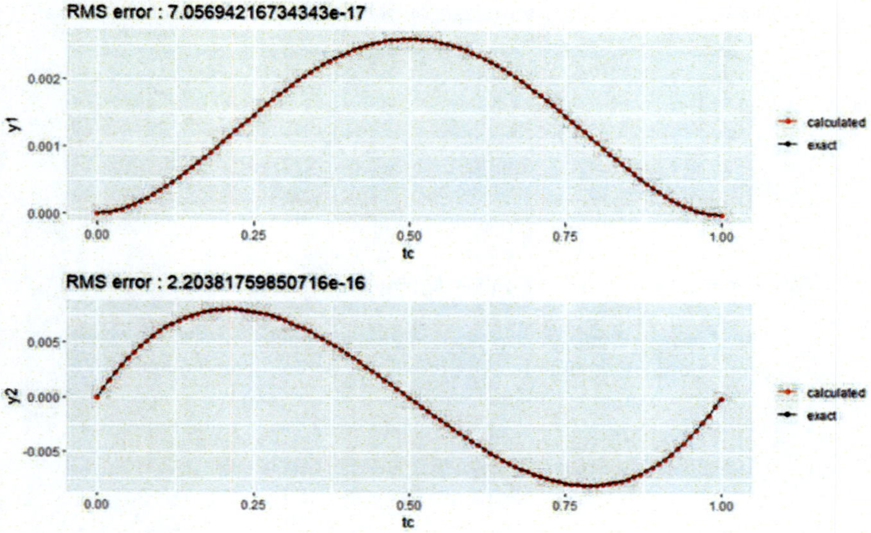

Fig. 1.19 A second example of solution of a BVP

The results are shown below (Fig. 1.19):

If the BVP has multiple solutions, R will evaluate to one among them. For instance, let us consider

$$y''' = -y' \text{ on } (0, 2\pi),\ y(0) = y(2\pi) = 0,\ y'(0) = 1.$$

This problem has infinitely many solutions $(t) = \sin(t) + c(1 - \cos(t)), c \in \mathbb{R}$. Then, `bvpSolve` furnishes a solution corresponding to $c \approx 0.61182225$ (Fig. 1.20)

1.13 Optimization with R

The general form of a mono-objective optimization problem is

$$\boldsymbol{x}^* = \arg\min_C J = \arg\min \{J(\boldsymbol{y}) : \boldsymbol{y} \in C\}.$$

C is the admissible set and J is the objective function. The optimal point is $\boldsymbol{x}^*$ and the optimal value of the objective is $J(\boldsymbol{x}^*)$. In practice, we have often $J : \mathbb{R}^n \to \mathbb{R}$ and either $C = \mathbb{R}^n$ or

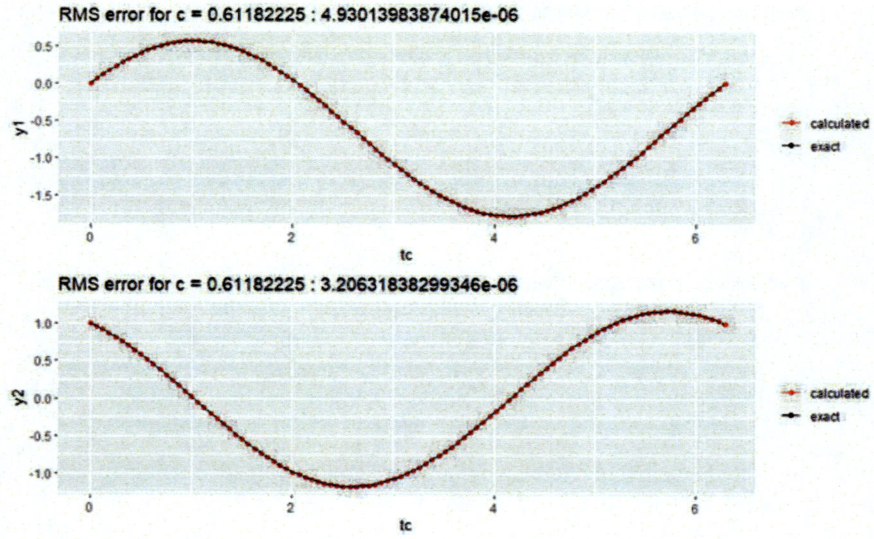

Fig. 1.20 Solution of a BVP having infinitely many solutions

$$C = \{y \in \mathbb{R}^n : \psi_i(y) \leq 0,\ 1 \leq i \leq p; \psi_{i+p}(y) = 0,\ 1 \leq i \leq q\},$$

with ψ_j: $\mathbb{R}^n \rightarrow \mathbb{R}$, $1 \leq j \leq p + q$. When $C = \mathbb{R}^n$, the problem is said to be unconstrained – we refer to it as a problem of unconstrained optimization. Otherwise, it is said to be constrained – we refer to it as a problem of constrained optimization (or optimization under restrictions).

Assuming differentiability of J, the solution of an unconstrained optimization problem verifies the classical equation

$$\nabla J(\boldsymbol{x}^*) = 0 \Leftrightarrow \frac{\partial J}{\partial x_i}(x^*) = 0,\ 1 \leq i \leq n.$$

For constrained optimization problems, the solution involves Lagrange's multipliers $\boldsymbol{\lambda} = (\lambda_1, \ldots, \lambda_{p+q})^t$ and verifies the classical KKT equations

$$\frac{\partial J}{\partial x_i}(\boldsymbol{x}^*) + \sum_{j=1}^{p+q} \lambda_j \frac{\partial \psi_j}{\partial x_i}(\boldsymbol{x}^*) = 0,\ 1 \leq i \leq n;$$

$$\lambda_j \geq 0,\ \psi_j(\boldsymbol{x}^*) \geq 0,\ \lambda_j \psi_j(x^*) = 0\ (1 \leq j \leq p);$$

$$\psi_j(\boldsymbol{x}^*) = 0,\ p + 1 \leq j \leq p + q.$$

Constrained optimization can be reformulated in terms of Lagrange's multipliers by introducing the Lagrangian

$$L(\boldsymbol{x};\boldsymbol{\mu}) = J(\boldsymbol{x}) + \sum_{j=1}^{p+q} \mu_j \psi_j(\boldsymbol{x}).$$

Assuming the existence of $\boldsymbol{x}^*$ and the regularity of the functions J and ψ_j, $1 \leq j \leq p + q$, $(\boldsymbol{x}^*;\boldsymbol{\lambda})$ is a saddle point of L on $\mathbb{R}^n \times \Lambda$, where

$$\Lambda = \{\boldsymbol{\mu} \in \mathbb{R}^{p+q} : \mu_i \geq 0 \, for \; 1 \leq i \leq p\}.$$

Then,

$$\boldsymbol{x}^* \in \mathbb{R}^n, \boldsymbol{\lambda} \in \Lambda \text{ and } L(\boldsymbol{x}^*;\boldsymbol{\mu}) \leq L(\boldsymbol{x}^*;\boldsymbol{\lambda}) \leq L(\boldsymbol{x};\boldsymbol{\lambda}), \forall \boldsymbol{x} \in \mathbb{R}^n, \boldsymbol{\mu} \in \Lambda$$

and

$$J(\boldsymbol{x}^*) = L(\boldsymbol{x}^*;\boldsymbol{\lambda}) = \inf_{\boldsymbol{x}\in\mathbb{R}^n} \sup_{\boldsymbol{\mu}\in\Lambda}\{L(\boldsymbol{x}, \; \boldsymbol{\mu})\} = \sup_{\boldsymbol{\mu}\in\Lambda} \inf_{\boldsymbol{x}\in\mathbb{R}^n}\{L(\boldsymbol{x}, \; \boldsymbol{\mu})\}.$$

This problem can be solved by duality methods, such as Uzawa's or Arrow–Hurwicz ones.

However, in practice, penalty, barrier, or projection approaches are used – furnished by methods such as SQP, interior points, trust regions, and active set. R offers built-in commands for these methods. For instance, `optimize` (for 1D optimization); `nlm` and `optim` (for n-dimensional optimization).

In addition, the community of R users proposes many packages for optimization. A list of available packages can be found in https://cran.r-project.org/web/views/Optimization.html. For instance, you can find `optimx`, which extends `optim`; `lpSolveAPI;` and `lpSolve` for linear programming, `quadprog` for quadratic programming, and `ROI` (R Optimization Infrastructure), which furnishes an interface to other packages and solvers. Among the packages in R repositories, you can find metaheuristics, genetic algorithms, surrogate-based optimization, multiobjective optimization, etc. Package `NlcOptim` clones `fmincon` by `solnl`. Package `pracma` uses `NlcOptim` and clones Matlab functions `linprog` (linear programming) `fmincon` (constrained optimization), `fminunc` (unconstrained optimization), `fminsearch` (Nelder–Meade for unconstrained optimization).

1.13.1 Linear Programming

As an example, let us consider the following optimization problem: determine the minimum of $f(\boldsymbol{x}) = \boldsymbol{c}^t\boldsymbol{x}$, with $\boldsymbol{x} \in \mathbb{R}^{12}$,

$$c = (-34, -28, -5, -24, -12, 3, -32, -20, -5, 7, -5, 10)$$

under the constraints $x_i \geq 0$, $i = 1, \ldots, 12$

$$x_1 + x_2 + x_3 \leq 3500$$
$$x_4 + x_5 + x_6 \leq 2200$$
$$x_7 + x_8 + x_9 \leq 4200$$
$$x_{10} + x_{11} + x_{12} \leq 1800$$
$$0.7x_1 - 0.3x_4 - 0.3x_7 - 0.3x_{10} \leq 0$$
$$-0.5x_1 - 0.5x_4 + 0.5x_7 - 0.5x_{10} \leq 0$$
$$0.7x_2 - 0.3x_5 - 0.3x_8 - 0.3x_{11} \leq 0$$
$$0.3x_3 - 0.7x_6 - 0.7x_9 - 0.7x_{12} \leq 0$$
$$0.1x_2 - 0.9x_5 + 0.1x_8 + 0.1x_{11} \leq 0$$
$$0.4x_1 - 0.6x_4 + 0.4x_7 + 0.4x_{10} \leq 0$$

`lpSolve` assumes that the variables are non-negative. Thus, we can use the code below:

```
library(lpSolve)
objective.in <- c(-34,-28,-5,-24,-12,3,-32,-20,-5,7,-5,10)
const.mat <- matrix(
  c(1, 1, 1, 0, 0, 0, 0, 0, 0, 0, 0, 0,
    0, 0, 0, 1, 1, 1, 0, 0, 0, 0, 0, 0,
    0, 0, 0, 0, 0, 0, 1, 1, 1, 0, 0, 0,
    0, 0, 0, 0, 0, 0, 0, 0, 0, 1, 1, 1,
    0.7, 0, 0, -0.4, 0, 0, -0.3, 0, 0, -0.3, 0, 0,
   -0.5, 0, 0, -0.5, 0, 0, 0.5, 0, 0, -0.5, 0, 0,
    0, 0.7, 0, 0, -0.3, 0, 0, -0.3, 0, 0, -0.3, 0,
    0, 0, 0.3, 0, 0, -0.7, 0, 0, -0.7, 0, 0, 0.7,
    0, 0.1, 0, 0, -0.9, 0, 0, 0.1, 0, 0, 0.1, 0,
    0.4, 0, 0, -0.6, 0, 0, 0.4, 0, 0, 0.4, 0, 0), nrow = 10)
const.dir = c(rep("<=",10))
const.rhs = c(3500,2200,4200,1800,0,0,0,0,0,0)
sol <- lp(direction="min", objective.in, const.mat,
          const.dir, const.rhs)
sol$solution
```

The result appears as follows:

```
> sol$solution
[1] 4.2e+03 0.0e+00 1.0e+30 0.0e+00 0.0e+00 5.0e+03
2.7e+03 4.5e+03 0.0e+00 0.0e+00 0.0e+00 0.0e+00
```

The main parameters are `objective.in` (vector c), `const.mat` (matrix of restrictions), `const.rhs` (second member of restrictions), `const.dir` (direction of the restrictions : "=", "<=", ">="), `direction` ("min" for minimization, "max" for maximization).

`lpSolve` can solve integer, binary, and mixed programming linear problems. For instance, to get integer solutions, you can use the option `all.int = TRUE`. It can also compute sensitivities by the option `compute.sens=1`.

1.13.2 Nonlinear Programming

The package `optimx` offers classical methods of optimization for constrained and unconstrained problems, such as "BFGS" (variable metric), "CG" (Fletcher-Reeves), "spg" (spectral projected gradient), "Nelder-Mead" (Nelder-Meade), and many others. It offers the possibility of running several methods for the same problem.

For instance, let us consider the unconstrained minimization of the following Rosenbrock's function:

$$f(\boldsymbol{x}) = 100\sum_{i=1}^{n-1}\left(x_{i+1} - x_i^2\right)^2 + \sum_{i=1}^{n-1}(x_i - 1)^2.$$

The unconstrained minimization problem looks for

$$\boldsymbol{x}^* = \arg\min\{f(\boldsymbol{x}) : x \in \mathbb{R}^n\} = (1, 1, \ldots, \pm 1), f(\boldsymbol{x}^*) = 0$$

We run the code below, corresponding to $n = 5$ and initial guess $(0, 0, 0, 0, 0)$:

```
f <- function(x) {
  n <- length(x)
  c_rose <- 100
    s1 = 0
    s2 = 0
    for (i in seq(1,n-1)){
        s1 = s1 + (x[i+1]^2 - x[i])^2
        s2 = s2 + (x[i] - 1)^2
    }
    resu <- c_rose*s1 + s2
  return(resu)}
library(optimx)
optimx(par= c(rep(0,5)), fn=f, method =
         c("Nelder-Mead", "CG", "BFGS", "spg",
           "nlm", "Rcgmin", "bobyqa","newuoa",
           "ucminf","nmkb","hjkb"))
```

The results appear as follows ($p = (p_1, \ldots, p_5)$ is the computed solution; value contains (p)):

```
                    p1         p2         p3         p4           p5         value fevals gevals ni
Nelder-Mead 0.33817328 0.57502580 0.76651148 0.87702452 9.590009e-01  8.922511e-01    502     NA
CG          0.01000410 0.01020415 0.01020617 0.01010312 0.000000e+00  3.959794e+00     93     21
BFGS        0.01000408 0.01020417 0.01020618 0.01010312 0.000000e+00  3.959794e+00     31      9
spg                 NA         NA         NA         NA           NA 8.988466e+307     NA     NA
nlm         0.01000358 0.01020364 0.01020566 0.01010262 4.326130e-08  3.959794e+00     NA     NA
Rcgmin      0.00000000 0.00000000 0.00000000 0.00000000 0.000000e+00  4.000000e+00      1      1
bobyqa              NA         NA         NA         NA           NA 8.988466e+307     NA     NA
newuoa              NA         NA         NA         NA           NA 8.988466e+307     NA     NA
ucminf              NA         NA         NA         NA           NA 8.988466e+307     NA     NA
nmkb                NA         NA         NA         NA           NA 8.988466e+307     NA     NA
hjkb                NA         NA         NA         NA           NA 8.988466e+307     NA     NA
            convcode  kkt1  kkt2 xtime
Nelder-Mead        1 FALSE  TRUE     0
CG                 0  TRUE FALSE     0
BFGS               0  TRUE FALSE     0
spg             9999    NA    NA     0
nlm                0  TRUE FALSE     0
Rcgmin             0 FALSE FALSE     0
bobyqa          9999    NA    NA     0
newuoa          9999    NA    NA     0
ucminf          9999    NA    NA     0
nmkb            9999    NA    NA     0
hjkb            9999    NA    NA     0
```

As we observe, the results are incorrect: indeed, due to the non-convexity of the objective function, global optimization is required. For instance, we can use the package GA to run a genetic algorithm. For instance, let us run the code

```
library(GA)
suggestedSol <- matrix(c(rep(0,5)),nrow = 1)
GA1 <- ga(type = "real-valued",
          fitness =  function(x) -f(x),
          lower = c(rep(-100,5)), upper = c(rep(100,5)),
          suggestions = suggestedSol,
          popSize = 50, maxiter = 1000, run = 100)
GA1@solution
```

For comparison purposes, we included the initial guess in the initial population, using the parameter `suggestedSol`. Notice that GA maximizes the fitness: to minimize *f*, we must maximize –*f*. The result is

```
> GA1@solution
     x1 x2 x3 x4 x5
[1,]  0  0  0  0  0
```

The result of a pure genetic algorithm appears again as incorrect. But the package GA offers the possibility of using hybrid algorithms – convergence theorems exist for hybrid algorithms (see, for instance Pogu & Souza de Cursi, 1994; Souza de Cursi et al., 2004; El Mouatasim et al., 2011). For instance, let us combine the genetic algorithm with Fletcher–Reeves:

```
optimArgs = list(method = "CG",
                 poptim = 0.05,
                 pressel = 0.5,
                 control = list(fnscale = -1, maxit = 100))

library(GA)
suggestedSol <- matrix(c(rep(0,5)),nrow = 1)
GA1 <- ga(type = "real-valued",
         fitness =  function(x) -f(x),
         lower = c(rep(-100,5)), upper = c(rep(100,5)),
         suggestions = suggestedSol,
         popSize = 50, maxiter = 1000, run = 100,
         optim = TRUE)
GA1@solution
```

Then, the result is

```
            x1        x2       x3        x4         x5
[1,] 0.9998819 0.9999401 0.999969 0.9999832 -0.9999911
```

For constrained optimization, you can use `pracma` (or `NlcOptim` directly). For instance, let us consider the minimization of the Rosenbrock's function previously introduced under the restrictions

$$\varphi_1(\boldsymbol{x}) = \|\boldsymbol{x}\| - \sqrt{n} \leq 0,\ \varphi_2(\boldsymbol{x}) = \sum_{i=1}^{n} x_i - n = 0.$$

We can use the code below:

```
f <- function(x) {
  n <- length(x)
  c_rose <- 100
    s1 = 0
    s2 = 0
    for (i in seq(1,n-1)){
        s1 = s1 + (x[i+1]^2 - x[i])^2
        s2 = s2 + (x[i] - 1)^2
    }
    resu <- c_rose*s1 + s2
  return(resu)}
hi <- function(x){
  aux = sqrt(sum(x^2))-length(x)
  return(aux)
}
he <- function(x){
  aux = sum(x)-length(x)
  return(aux)
}
library(pracma)
sol = fmincon(x0= c(0.2, 0.5, 0.7, 0.4, 0),
                 fn=f, hin = hi, heq = he)
print(sol$par)
```

This code produces the result

```
[1] 1.0000007 1.0000004 1.0000001 0.9999995 0.999999
```

We can use `NlcOptim` directly as follows:

```
nlc = function(x){
  aux1 = he(x)
  aux2 = hi(x)
  return(list(aux1,aux2))
}
library(NlcOptim)
sol = solnl(X= c(0.2, 0.5, 0.7, 0.4, 0)|,
               objfun=f, confun = nlc)
print(t(sol$par))
```

The result is

```
          [,1]      [,2]      [,3]      [,4]      [,5]
[1,] 0.9999854 0.9999932 0.9999965 0.9999977 0.9999994
```

1.13.3 Duality Methods

As previously indicated, the solution x^* of the problem

$$x^* = \arg\min_C J = \arg\min \{J(y) : y \in C\},$$

$$C = \left\{y \in \mathbb{R}^n : \psi_i(y) \leq 0,\ 1 \leq i \leq p; \psi_{i+p}(y) = 0,\ 1 \leq i \leq q\right\}\#$$

verifies

$$J(x^*) = L(x^*; \lambda) = \inf_{x\in\mathbb{R}^n} \sup_{\mu\in\Lambda} \{L(x,\ \mu)\} = \sup_{\mu\in\Lambda} \inf_{x\in\mathbb{R}^n} \{L(x,\ \mu)\}.$$

where λ is a vector of Lagrange's multipliers such that

$$\lambda \in \Lambda = \{\mu \in \mathbb{R}^{p+q} : \mu_i \geq 0\, for\ 1 \leq i \leq p\}.$$

An alternative method of solution consists in finding

$$\lambda = \sup_{\mu\in\Lambda} g(\mu),\ g(\mu) = \inf_{x\in\mathbb{R}^n} \{L(x,\ \mu)\}$$

Once $\boldsymbol{\lambda}$ is determined, we can find

$$\boldsymbol{x}^* = \arg\min L(\boldsymbol{x}, \boldsymbol{\lambda}).$$

Numerically, we can use Uzawa's algorithm, which reads as

Uzawa's Algorithm
Input: The objective function J, the restrictions ψ_i, $1 \le i \le p + q$, the values of p and q, a real parameter $\mu > 0$
Output: the values of $\boldsymbol{\lambda}$ and $\boldsymbol{x}^*$.
Initialization: set initial values $\boldsymbol{\lambda}^{(0)}$, $\boldsymbol{x}^{(0)}$. Set $n \leftarrow 0$

1. Determine $\boldsymbol{x}^{(n+1)} = \arg\min L(\boldsymbol{x}, \boldsymbol{\lambda}^{(n)})$.
2. Update

$$\lambda_i^{(n+1)} = \max\left\{0, \lambda_i^{(n)} + \mu\varphi_i\left(\boldsymbol{x}^{(n+1)}\right)\right\}, 1 \le i \le p$$

$$\lambda_i^{(n+1)} = \lambda_i^{(n)} + \mu\varphi_i\left(\boldsymbol{x}^{(n+1)}\right), p + 1 \le i \le p + q$$

3. Test for the stopping conditions. If the iterations continue, then go to 1, else estimate $\boldsymbol{x} \approx \boldsymbol{x}^{(r)}$.

To apply this method, we consider:

- `lbdineq` are the $p=$ `nineq` multipliers for the inequalities ($1 \le i \le p$);
- `lbdeq` are the $q=$ `neq` multipliers for the inequalities ($p + 1 \le i \le p + q$);
- `fobj` is the objective function;
- `hineq` are the inequalities;
- `he` are the equalities.

Then,

Start by creating a function that evaluates the Lagrangian L.

```
lagrangian = function(x,lbdineq,lbdeq,fobj,nineq,hineq,neq,heq){
  s = fobj(x)
  if (missing(nineq) != TRUE){
    if (nineq > 0) {
      s = s + sum(lbdineq*hineq(x))
    }
  }
  if (missing(neq) != TRUE){
    if (neq > 0) {
      s = s + sum(lbdeq*heq(x))
    }
  }
  return(s)
}
```

Create a function that updates the Lagrange multipliers:

```
actlbda = function(sol1,nineq,hineq,neq,heq,mu){
  resu = list(sol1[[1]],NA,NA)
  if (missing(nineq) != TRUE){
    if (nineq > 0) {
      aux = sol1[[2]] + mu*hineq(resu[[1]])
      ind = which(aux < 0)
      if (length(ind) > 0){
        aux[ind] = 0
      }
      resu[[2]] = aux
    }
  }
  if (missing(neq) != TRUE){
    if (neq > 0) {
      resu[[3]] = sol1[[3]] + mu*heq(resu[[1]])
    }
  }
  return(resu)
}
```

Create a function that evaluates the norm of the updates

```
distsol = function(sol1,sol2,nineq,neq){
  s = sqrt(sum((sol1[[1]] -sol2[[1]])^2))
  if ( missing(nineq) != TRUE){
    if (nineq > 0) {
      s = s + sqrt(sum((sol1[[2]] -sol2[[2]])^2))
    }
  }
  if (missing(neq) != TRUE){
    if (neq > 0) {
      s = s + sqrt(sum((sol1[[3]] -sol2[[3]])^2))
    }
  }
  return(s)
}
```

and a function to initialize the variables

```
initvars = function(n,nineq,neq){
  resu = list(NA,NA,NA)
  resu[[1]] = c(rep(0,n))
  if (missing(nineq) != TRUE){
    if (nineq > 0) {
      resu[[2]] = c(rep(0,nineq))
    }
  }
  if (missing(neq) != TRUE){
    if (neq > 0) {
      resu[[3]] = c(rep(0,neq))
    }
  }
  return(resu)
}
```

Then, iterate:

```
solac = initvars(n,ni,ne)
solnew = list(NA,NA,NA)
prec = 1 + precmin
nit = 0
continue = TRUE
while(continue) {
  nit = nit + 1
  fuz = function(x){lagrangian(x,solac[[2]],solac[[3]],fobj,ni,hi,ne,he)}
  sol1 = solac
  aux = fminunc(solac[[1]],fuz, maxiter = 1000)
  sol1[[1]] = aux$par
  solnew = actlbda(sol1,ni,hi,ne,he,mu)
  prec = distsol(solnew,solac,ni,ne)
  solac = solnew
  print(paste("nit = ",nit,", prec = ",prec))
  continue = (prec > precmin) & (nit < nitmax)
}
```

For instance, let us consider the minimization of the Rosenbrock's function under the restrictions

$$\varphi_1(x) = \|x\| - \sqrt{n} \leq 0,\ \varphi_2(x) = \sum_{i=1}^{n} x_i - n = 0.$$

In this case,

```
ni = 1
hi = function(x){
  aux = sqrt(sum(x^2)) - sqrt(length(x))
  return(aux)
}
```

```
ne = 1
he = function(x){
  aux = sum(x) - length(x)
  return(aux)
}
```

We use the parameters given at right:

```
n=5
nitmax = 100
precmin = 1e-5
mu = 0.1
fobj = rose
```

The results were

```
[1] "nit =  1 , prec =  2.23606796408196"
[1] "nit =  2 , prec =  2.24536121204785e-08"
> solnew
[[1]]
[1] 1 1 1 1 1

[[2]]
[1] 0

[[3]]
[1] -3.971074e-09
```

`solnew[[1]]` contains the values of x^*, `solnew[[2]]` contains $\lambda_i : 1 \leq i \leq p$, `solnew[[3]]` contains $\lambda_i : p+1 \leq i \leq p+q$,

As a second example, let us consider the determination of the point on the line $x_1 + x_2 = 2$ which is the closest to the circle $x_1^2 + x_2^2 = 1$. We are looking for two points $P_1 = (x_1, x_2)$, $P_2 = (x_3, x_4)$ such that $x_1^2 + x_2^2 = 1$ (id est, P_1 belongs to the circle), $x_3 + x_4 = 2$ (id est, P_2 belongs to the line), and $\|P_1 - P_2\|$ is minimal. Let us set $\boldsymbol{x} = (x_1, x_2, x_3, x_4)$,

$$J(\boldsymbol{x}) = \sqrt{(x_1 - x_3)^2 + (x_2 - x_4)^2},$$

$$\varphi_1(\boldsymbol{x}) = \sqrt{x_1^2 + x_2^2} - 1,$$

$$\varphi_2(\boldsymbol{x}) = x_3 + x_4 - 2,$$

The exact solution is $\left(\sqrt{2}/2,\ \sqrt{2}/2,\ 1,\ 1\right) \approx (0.7071, 0.7071, 1, 1)$. In this case, we look for the minimum of `fdist` :

```
fdist = function(x){
  aux = sqrt((x[[1]]-x[[3]])^2 + (x[[2]]-x[[4]])^2)
  return(aux)
}
```

The restrictions are

```
ni = 0
hi = function(x){
  aux = NA
  return(aux)
}
```

```
ne = 2
he = function(x){
  aux1 = abs(sqrt(x[[1]]^2 + x[[2]]^2) - 1)
  aux2 = abs(x[[3]] + x[[4]] - 2)
  return(c(aux1,aux2))
}
```

We obtain

```
[1] "nit =  100 , prec =  1.09775959962387e-05"
[1] "nit =  100 , prec =  1.09775959962387e-05"
> solnew
[[1]]
[1] 0.7070255 0.7072013 0.9998693 1.0001250

[[2]]
[1] NA

[[3]]
[1] 1.0181879 0.7150872
```

1.13.4 Multiobjective Optimization

R repositories offer packages for multiobjective optimization, such as `rmoo` or `Gpareto`. In this book, the multiobjective problems to be solved are not so difficult as to require sophisticated methods. We consider only biobjective problems of a particular kind and we use a simple variational method introduced in (Souza de Cursi, 2015) and (Zidani, 2013), based on the minimization of the area limited by the Pareto front and the system of axis. Indeed, for a biobjective problem involving $\boldsymbol{f} = (f_1, f_2)$ and an admissible set C, we can consider the individual solutions

$$\boldsymbol{x}_i^* = \arg\min \{f_i(\boldsymbol{x}) : \boldsymbol{x} \in C\}$$

and the curves

$$\boldsymbol{x}(t) = \boldsymbol{x}_1^* + (\boldsymbol{x}_2^* - \boldsymbol{x}_1^*)t + \sum_{i=1}^{k} \boldsymbol{c}_i \varphi_i^0(t),\ t \in (0,1),\ \varphi_i^0(0) = \varphi_i^0(1) = 0.$$

Notice that each $\boldsymbol{c}_i$ is a vector having the same dimension as $\boldsymbol{x}_1^*$ and $\boldsymbol{x}_2^*$. $\boldsymbol{c} = (\boldsymbol{c}_1, \ldots, \boldsymbol{c}_k)$ is a matrix of coefficients (each line corresponds to a component of $\boldsymbol{x}$). Given a curve, we can generate a 2D curve as follows:

$$\boldsymbol{P}(t) = (f_1(\boldsymbol{x}(t)), f_2(\boldsymbol{x}(t))),\ t \in (0,1).$$

Let us consider the utopic point $\boldsymbol{f}_u^* = (f_1(\boldsymbol{x}_1^*), f_2(\boldsymbol{x}_2^*))$ and the axes passing by $\boldsymbol{f}_u^*$ - $H : (f_1(\boldsymbol{x}_1^*), t_2)$ and $V : (t_1, f_2(\boldsymbol{x}_2^*))$, $(t_1, t_2) \in \mathbb{R}^2$. H is a horizontal axis passing by $\boldsymbol{f}_u^*$; V is a vertical axis passing by $\boldsymbol{f}_u^*$. Notice that $\boldsymbol{f}_1^* = (f_1(\boldsymbol{x}_1^*), f_2(\boldsymbol{x}_1^*)) \in H$ and $\boldsymbol{f}_2^* = (f_1(\boldsymbol{x}_2^*), f_2(\boldsymbol{x}_2^*)) \in V$. Since $\boldsymbol{x}(0) = \boldsymbol{x}_1^*$ and $\boldsymbol{x}(1) = \boldsymbol{x}_2^*$, $\boldsymbol{P}$, H and V limit a region. Let $J(\boldsymbol{c})$ be the area of this region. Then, we look for the coefficients $\boldsymbol{c}$ that minimize J (see Fig. 1.21).

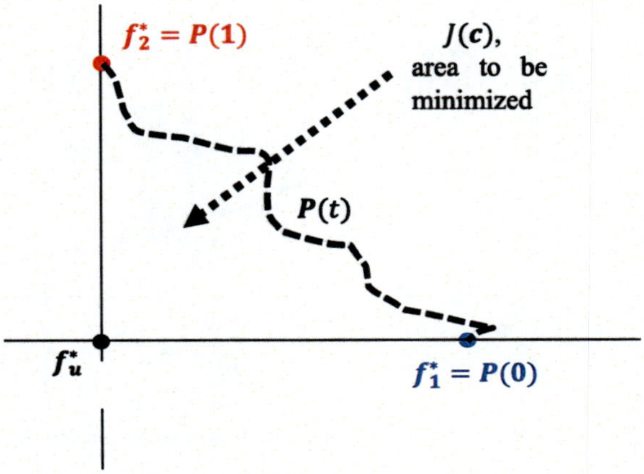

Fig. 1.21 A simple variational method for biobjective optimization

As examples, let us consider the 3 problems below

$$\text{Problem 1}: f_1(x) = \sum_{i=1}^{n} x_i^2, f_2(x) = \sum_{i=1}^{n} (x_i - 1)^2, n = 4$$

$$\text{Problem 2}: f_1(x) = |x_1|, f_2(x) = (1 + 9|x_2|)\left(1 - \sqrt{\frac{|x_1|}{1 + 9|x_2|}}\right).$$

$$\text{Problem 3}: f_1(x) = |x_1|, f_2(x) = (1 + 9|x_2|)\left(1 - \left(\frac{|x_1|}{1 + 9|x_2|}\right)^2\right).$$

In all these situations, the exact Pareto's front is known, so that we can compare the results furnished by the method with the exact solutions. We evaluate the area by a trapezoidal rule involving the points $t_i = i/npt$, $i = 0, \ldots, npt$. The approximation uses $npt = 100$, $k = 3$, and $\varphi_i^0(t) = t^{i+1} - t$. $\boldsymbol{x}_1^*$ and $\boldsymbol{x}_2^*$ are given by the user. The minimization of the area is carried by `fminunc` from package `pracma`. The results appear in Figs. 1.22, 1.23, 1.24: we see that the method furnishes results close to the exact ones. The method is implemented in class `multobj.R`.

The solution in Fig. 1.25 corresponds to the following Binh–Korn problem:

$$f_1(x) = 4\left(x_1^2 + x_2^2\right), f_2(x) = (x_1 - 5)^2 + x_2^2,$$
$$(x_1 - 5)^2 + x_2^2 \leq 25, \ (x_1 - 8)^2 + (x_2 + 3)^2 \geq 7.7,$$

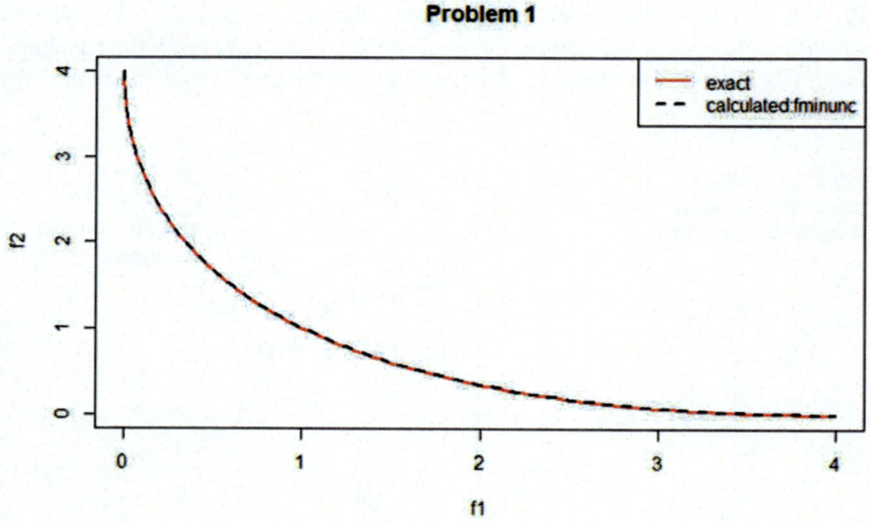

Fig. 1.22 Pareto's front of Problem 1 furnished by the simple variational method

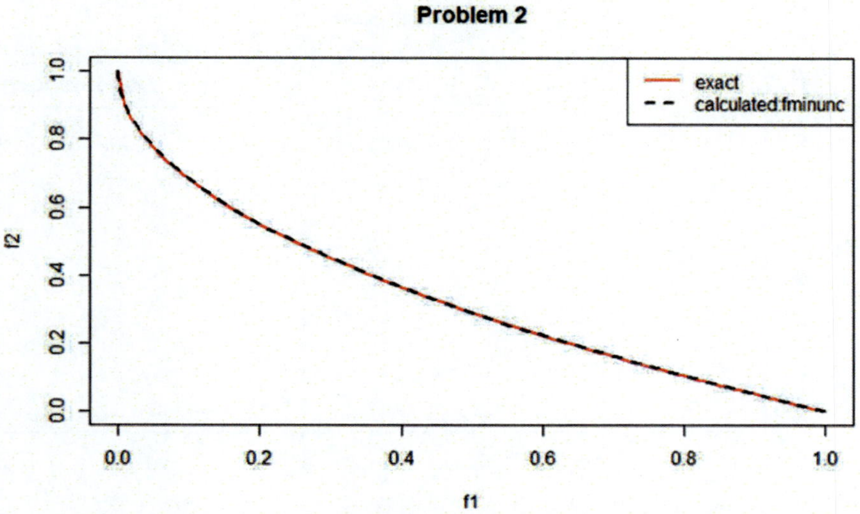

Fig. 1.23 Pareto's front of Problem 2 furnished by the simple variational method

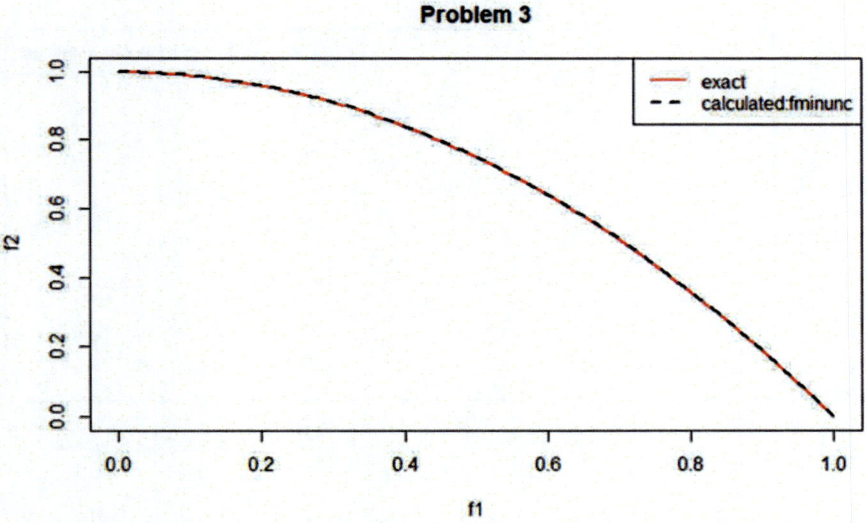

Fig. 1.24 Pareto's front of Problem 3 furnished by the simple variational method

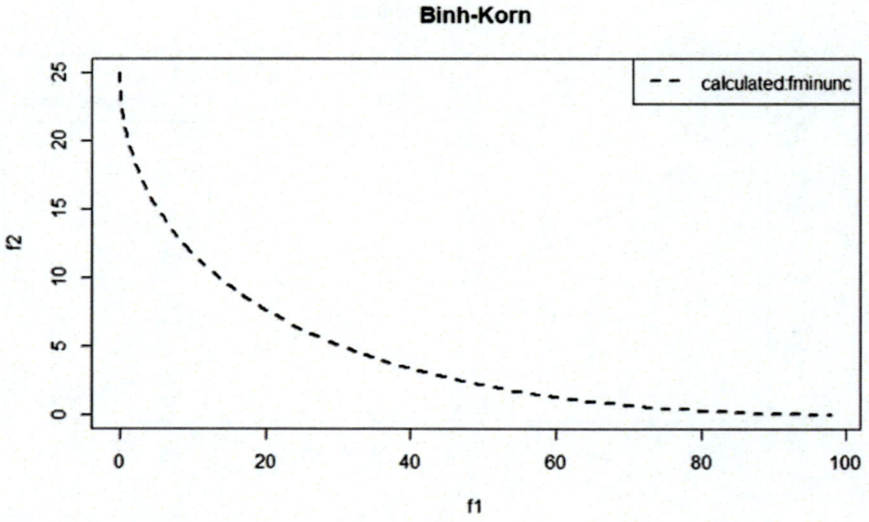

Fig. 1.25 Pareto's front of the Binh–Korn problem by the simple variational method

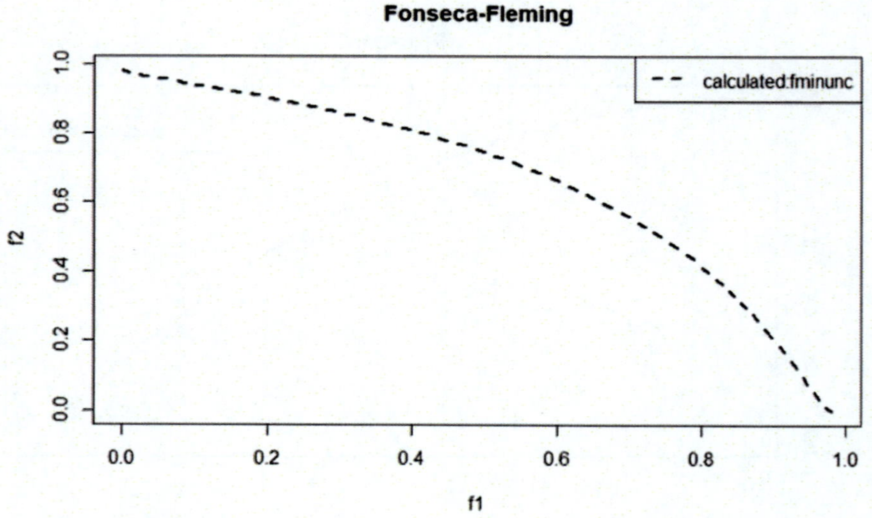

Fig. 1.26 Pareto's front of the Fonseca–Fleming problem by the simple variational method

$$0 \leq x_1 \leq 5, 0 \leq x_2 \leq 3$$

In Fig. 1.26, we exhibit the solution found for the Fonseca–Fleming problem:

$$f_1(x) = 1 - \exp\left(-\sum_{i=1}^{n}\left(x_i - \frac{1}{\sqrt{n}}\right)^2\right), f_2(x) = 1 - \exp\left(-\sum_{i=1}^{n}\left(x_i + \frac{1}{\sqrt{n}}\right)^2\right),$$

$$n = 3,\ -4 \le x_1 \le 4,\ -4 \le x_2 \le 4,\ -4 \le x_3 \le 4$$

1.14 Solving Equations with R

In Sect. 1.6, we introduced methods for the solution of systems of linear equations. Here, we complete the presentation considering overdetermined linear systems and nonlinear equations.

1.14.1 Systems of Linear Equations

Let us consider a linear system $\boldsymbol{Ax} = \boldsymbol{B}$, where $\boldsymbol{A}$ is a $m \times n$ matrix and $\boldsymbol{B}$ is a $m \times 1$ vector. It is usual to distinguish among

- **Undetermined** linear systems, which have infinitely many solutions. Examples of undetermined systems are systems of compatible equations having less equations than unknowns ($m < n$). In this case, the general solution is $\boldsymbol{X} = \boldsymbol{X_p} + N.z$, where N is a basis of the null space of $\boldsymbol{A}$ ($\boldsymbol{A}.N = \boldsymbol{0}$, $rank(N) = n - \ rank\ (\boldsymbol{A})$, $z \in \mathbb{R}^{rank(N)}$)
- **Determined** linear systems, where $m = n$ and $\det(\boldsymbol{A}) \neq 0$.Such a systems have a unique solution.
- **Overdetermined** linear systems, which contain incompatible equations. For instance, linear systems where $m > n$ and the equations are incompatible. Such a systems have only generalized solutions, for instance, least squares ones.

As previously mentioned, R offers the command `solve(A,B)` which determines the solution of **determined** linear systems. For instance, let us consider the linear system with

$$\boldsymbol{A} = \begin{pmatrix} 1 & 4 & 6 & 1 \\ 2 & 1 & 7 & 1 \\ 3 & 2 & 8 & 1 \\ 4 & 3 & 5 & 1 \end{pmatrix},\ \boldsymbol{B} = \begin{pmatrix} 31 \\ 29 \\ 35 \\ 29 \end{pmatrix}.$$

Since $\det(\boldsymbol{A}) = 16$, the linear system is determined: its unique solution is $\boldsymbol{X} = (1, 2, 3, 4)^t$. Under R, we have:

Since det $(A) = 16$, the linear system is determined: its unique solution is $X = (1,2,3,4)^t$. Under R, we have the results at left.

```
> A = matrix(c(1:4,4,1:3,6,7, 8,5, 1,1,1,1),ncol=4)
> B = matrix(c(31,29,35,29),ncol = 1)
> det(A)
[1] 16
> sol = solve(A,B)
> print(sol)
     [,1]
[1,]    1
[2,]    2
[3,]    3
[4,]    4
```

Now, let us consider the situation where the two last lines of A do not exist:

$$A = \begin{pmatrix} 1 & 4 & 6 & 1 \\ 2 & 1 & 7 & 1 \end{pmatrix}, \; B = \begin{pmatrix} 31 \\ 29 \end{pmatrix}.$$

Now, the system is undetermined: the system has infinitely many solutions and solve fails.

```
> A = matrix(c(1,2,4,1,6,7,1,1),ncol=4)
> B = matrix(c(31,29),ncol = 1)
> det(A)
Error in determinant.matrix(x, logarithm =
 TRUE, ...) :
  'x' doit être une matrice carrée
> sol = solve(A,B)
Error in solve.default(A, B) : 'a' (2 x 4)
 must be square
```

For undetermined systems, you can use a package such as `pracma`:

```
> library(pracma)
> rank(A)
[1] 2.5 5.0 6.0 2.5 7.0 8.0 2.5 2.5
> N <- null(A)
> Xp <- mldivide(A,B)
> sol = function(z){Xp + N %*% z}
```

```
> z0 = matrix(c(0,0), ncol = 1
> sol(z0)
          [,1]
[1,] 0.6643234
[2,] 2.0790861
[3,] 3.5729350
[4,] 0.5817223
```

```
> z1 = matrix(c(1,2), ncol = 1)
> sol(z1)
           [,1]
[1,] -0.5198414
[2,]  1.7254379
[3,]  3.6961552
[4,]  2.4411587
```

`null` generates the basis of the null space and `mldivide` generates a particular solution. As an alternative, you can use MASS:

```
> library(MASS)
> N <- Null(t(A))
> Ainv <- ginv(A)
> Xp <- Ainv %*% B
> sol = function(z){Xp + N %*% z}
```

```
> z0 = matrix(c(0,0), ncol = 1)
> sol(z0)
          [,1]
[1,] 0.6643234
[2,] 2.0790861
[3,] 3.5729350
[4,] 0.5817223
```

```
> z1 = matrix(c(1,2), ncol = 1)
> sol(z1)
           [,1]
[1,] -0.5198414
[2,]  1.7254379
[3,]  3.6961552
[4,]  2.4411587
```

Here, `Null` generates the basis of the null space, `ginv` generates a generalized inverse ***Ainv*** of A. A particular solution is generated by the product ***Ainv.B***.

Now, let us consider the situation where the two last lines are repeated:

$$A = \begin{pmatrix} 1 & 4 & 6 & 1 \\ 2 & 1 & 7 & 1 \\ 3 & 2 & 8 & 1 \\ 4 & 3 & 5 & 1 \\ 3 & 2 & 8 & 1 \\ 4 & 3 & 5 & 1 \end{pmatrix}, \; B = \begin{pmatrix} 31 \\ 29 \\ 35 \\ 29 \\ 30 \\ 30 \end{pmatrix}.$$

In this case, there are incompatible equations (lines 3 and 5 and lines 4 and 6). The system has not any classical solution, but only a generalized one, which can be furnished by the same packages.

```
> library(pracma)
> mldivide(A, B)
       [,1]
[1,]  0.500
[2,]  1.375
[3,]  1.625
[4,] 15.250
```

```
> library(MASS)
> Ainv <- ginv(A)
> Ainv %*% B
       [,1]
[1,]  0.500
[2,]  1.375
[3,]  1.625
[4,] 15.250
```

Both the solutions correspond to least squares ones. Of course, you can compute it directly, as shown at left: the least squares solution satisfies $(A^tA)X = A^tB$

```
> AtA = t(A) %*% A
> AtB = t(A) %*% B
> solve(AtA,AtB)
       [,1]
[1,]  0.500
[2,]  1.375
[3,]  1.625
[4,] 15.250
```

1.14.2 Systems of Nonlinear Equations

The community of R users proposes packages for the solution of nonlinear systems of equations.

For instance, `pracma` proposes a command `fsolve` to determine the solutions of systems of nonlinear equations. As an example, let us consider the nonlinear system

$$x_1^2 + x_2^2 + x_3^2 = 1, x_1^2 + 2x_2^2 + 2x_3^2 = 2, x_1 + x_2 + x_3 = 1.$$

This system has two solutions $(0,1,0)$ and $(0,0,1)$. The results furnished by `fsolve` of `pracma` are shown below. Notice the warning concerning the non-invertibility of the Jacobian – a solution is found, nevertheless.

```
> f1 = function(x){sum(x^2)-1}
> f2 = function(x){x[1]^2 + 2*x[2]^2 + 2*x[3]^2 - 2}
> f3 = function(x){sum(x)-1}
> f = function(x){as.matrix(c(f1(x),f2(x),f3(x)),ncol = 1)}
> x0 = as.matrix(c(1,0,0),ncol=1)
> library(pracma)
> sol = fsolve(f,x0)
Warning message:
In inv(A0) : Matrix appears to be singular.
> sol$x
[1]  5.884803e-05  1.000000e+00 -5.884805e-05
> x0 = as.matrix(c(0,0,0),ncol=1)
> sol = fsolve(f,x0)
Warning message:
In inv(A0) : Matrix appears to be singular.
> sol$x
[1]  6.945451e-05 -6.945290e-05  1.000000e+00
```

As an alternative, you can use an optimization approach: `lsqnonlin(f,x0)` minimizes the squared norm of $f(x)$ starting at $x0$. Observe that the method **does not** converge to a solution for the second starting point.

```
> sol = lsqnonlin(f, x0a)
> sol$x
[1]  0.0001077880  0.9999999912 -0.0001077792
> sol = lsqnonlin(f, x0b)
> sol$x
[1] -0.3476993  0.6802836  0.6802836
```

1.14.3 Optimization and Systems of Equations

Let us come back to the optimization problems considered in Sect. 1.13:

$$x^* = \arg\min_C J = \arg\min\left\{J(y) : y \in C \subset \mathbb{R}^n\right\}.$$

The solutions of optimization problems can be characterized also by systems of algebraic equations. For instance, the solution of the unconstrained optimization problem verifies

$$\frac{\partial J}{\partial x_i}(x^*) = 0,\ 1 \le i \le n.$$

If the expression of the gradient $\nabla J(x)$ is available, we can solve this system to determine a solution – keeping in mind that a solution can be also a local minimum or local maximum, except for convex function, so that we must verify if the solution corresponds to a real minimum by other ways. As an example, let us consider the Rosenbrock's function

$$f(x) = 100 \sum_{i=1}^{n-1} \left(x_{i+1} - x_i^2\right)^2 + \sum_{i=1}^{n-1} (x_i - 1)^2.$$

In this case,

$$\frac{\partial f}{\partial x_1} = 400x_1\left(x_1^2 - x_2\right) + 2(x_1 - 1);$$

$$\frac{\partial f}{\partial x_k} = 200\left(x_k - x_{k-1}^2\right) + 400x_k\left(x_k^2 - x_{k+1}\right) + 2(x_k - 1)\ (2 \le k \le n-1);$$

$$\frac{\partial f}{\partial x_n} = 200\left(x_n - x_{n-1}^2\right).$$

We can solve the equations $\nabla f(\boldsymbol{x}) = 0$ using `pracma`:

```
gradrose = function(x){
  n = length(x)
  aux = numeric(length=n)
  aux[[1]] = 2*x[[1]]*(x[[1]]^2 - x[[2]]) + (x[[1]]-1)/100
  for (k in 2:(n-1)){
    aux[[k]] = (x[[k]] - x[[k-1]]^2) + 2*x[[k]]*(x[[k]]^2 - x[[k+1]]) + (x[[k]]-1)/100
  }
  aux[[n]] = x[[n]] - x[[n-1]]^2
  return(aux)
}

n = 5
x0 = c(rep(0,n))

library(pracma)
xsol = fsolve(gradrose,x0,maxiter = 1000)
```

We obtain the exact result

```
> xsol$x
[1] 1 1 1 1 1
```

Now, let us consider constrained optimization. When

$$C = \left\{y \in \mathbb{R}^n : \psi_i(y) \le 0,\ 1 \le i \le p; \psi_{i+p}(y) = 0,\ 1 \le i \le q\right\},$$

the system of nonlinear equations is generated by the KKT conditions

$$\frac{\partial J}{\partial x_i}(\boldsymbol{x}^*) + \sum_{j=1}^{p+q} \lambda_j \frac{\partial \psi_j}{\partial x_i}(\boldsymbol{x}^*) = 0,\ 1 \le i \le n;$$

$$\lambda_j \psi_j(\boldsymbol{x}^*) = 0\ (1 \le j \le p);$$

$$\psi_j(\boldsymbol{x}^*) = 0,\ p+1 \le j \le p+q;$$

$$\lambda_j - |\lambda_j| = 0, 1 \leq j \leq p;$$
$$\psi_j(\boldsymbol{x}^*) + |\psi_j(\boldsymbol{x}^*)| = 0, 1 \leq j \leq p$$

The two last equations are destined to ensure that $\lambda_j \geq 0$ *and* $\psi_j(\boldsymbol{x}) \leq 0, 1 \leq j \leq p$. As an example, let us consider the minimization of the Rosenbrock's function under the restrictions

$$\varphi_1(x) = \|x\| - \sqrt{n} \leq 0, \varphi_2(x) = \sum_{i=1}^{n} x_i - n = 0.$$

Create the restrictions as shown in Sect. 1.13.2:

```
hi = function(x){
  aux = sqrt(sum(x^2)) - sqrt(length(x))
  return(aux)
}

he = function(x){
  aux = sum(x) - length(x)
  return(aux)
}
```

Create the system of nonlinear equations

```
eqs2solve = function(x,lbd1,lbd2){
  n = length(x)
  aux = numeric(length = n + 4)
  aux[1:n] = gradrose(x)
  aux[[n+1]] = lbd1*hi(x)
  aux[[n+2]] = he(x)
  aux[[n+3]] = lbd1 - abs(lbd1)
  aux[[n+4]] = hi(x) + abs(hi(x))
  return(aux)
}
```

Use pracma to find the solution

```
n = 5
x0 = c(rep(0,n+2))

eqs = function(x){
  n = length(x) - 2
  aux = eqs2solve(x[1:n],x[[n+1]],x[[n+2]])
  return(aux)
}

library(pracma)
sol = fsolve(eqs,x0,maxiter = 1000)
```

We obtain the exact result

```
> sol$x
[1] 1 1 1 1 1 0 0
```

If $q = 0$, i.e., if all the restrictions are inequalities, an alternative method consists in defining

$$g(t) = h(\boldsymbol{x}(t),\ t), \boldsymbol{x}(t) = \text{argmin}\{h(\boldsymbol{x},\ t) : x \in \mathbb{R}^n\},$$
$$h(\boldsymbol{x},\ t) = \max\left\{J(\boldsymbol{x}) - t; \psi_j(\boldsymbol{x}),\ 1 \le j \le p\right\}.$$

Then, we look for t^* such that $g(t^*) = 0$: we have $\boldsymbol{x}^* = \boldsymbol{x}(t^*)$. For instance, let us consider the minimization of the Rosenbrock's function under the restrictions

$$\varphi_1(\boldsymbol{x}) = \|\boldsymbol{x}\| - \sqrt{n} \le 0,\ \varphi_2(\boldsymbol{x}) = \sum_{i=1}^{n} x_i - n \le 0.$$

Define the Rosenbrock's function

```
rose = function(x){
  n = length(x)
  s1 = 0
  s2 = 0
  for (k in 1:(n-1)){
    s1 =  s1 + (x[[k+1]] - x[[k]]^2)^2
    s2 = s2 + (x[[k]]-1)^2
  }
  aux = 100*s1 + s2
  return(aux)
}
```

Define $h(\boldsymbol{x}, t)$. Functions `hi` and `he` are those previously defined.

```
h = function(x,t){
  aux = numeric(length = 3)
  aux[[1]] = rose(x)-t
  aux[[2]] = hi(x)
  aux[[3]] = he(x)
  return(max(aux))
}
```

Create a function that will evaluate g – here, we use `pracma` to determine $\boldsymbol{x}(t)$ (it must be charged previously). Notice that `xac` is a global variable: we use the instruction `xac <<- sol$par`

```
gaux = function(t,xini){
  fobj = function(x){h(x,t)}
  sol = fminunc(xini,fobj,maxiter = 1000)
  xac <<- sol$par
  return(sol$value)
```

Iterate to get a correct value of t and $x(t)$.

```
n=5
nitmax = 100
precmin = 1e-5
x0 = c(rep(0,n))
xdeb = matrix(x0, ncol = n)
xac = matrix(x0, ncol = n)
tac = matrix(-1,ncol = 1)
prec = 1 + precmin
nit = 0
continue = TRUE
while(continue) {
  nit = nit + 1
  g = function(t){gaux(t,xdeb)}
  t0 = tac
  sol = uniroot(g,c(-10,10),maxiter = 1000)
  prec = sqrt(sum(xdeb-xac)^2) + abs(sol$root - tac)
  xdeb = xac
  tac = sol$root
  print(paste("nit = ",nit,", prec = ",prec))
  continue = (prec > precmin) & (nit < nitmax)
}
```

We obtain as result

```
[1] "nit =  1 , prec =  5.99999394503208"
[1] "nit =  2 , prec =  2.66356158266576e-10"
> tac
[1] 3.67585e-08
> xac
          [,1]      [,2]      [,3]      [,4]      [,5]
[1,] 0.9999998 0.9999996 0.9999993 0.9999985 0.9999967
```

Equality constraints can be treated approximately by replacing $\psi_j(\boldsymbol{x}) = 0$ by $|\psi_j(\boldsymbol{x})| - \varepsilon_j \leq 0$. Then

$$h(\boldsymbol{x}, t) = \max\left\{J(\boldsymbol{x}) - t; \psi_j(\boldsymbol{x}), 1 \leq j \leq p; |\psi_j(\boldsymbol{x})| - \varepsilon_j, p+1 \leq j \leq p+q\right\}.$$

As an example, let us consider the minimization of the Rosenbrock's function under the restrictions

$$\varphi_1(\boldsymbol{x}) = \|\boldsymbol{x}\| - \sqrt{n} \leq 0, \varphi_2(\boldsymbol{x}) = \sum_{i=1}^{n} x_i - n = 0.$$

Redefine $h(\boldsymbol{x}, t)$. Functions `hi` and `he` are the same as before. We use $\varepsilon_2 = 1E - 3$.

```
h = function(x,t){
  aux = numeric(length = 3)
  aux[[1]] = rose(x)-t
  aux[[2]] = hi(x)
  aux[[3]] = abs(he(x)) - 1e-3
  return(max(aux))
}
```

Then, run the code – we obtain:

```
> tac
[1] 2.067453e-08
> xdeb
          [,1]      [,2]      [,3]      [,4]      [,5]
[1,] 0.9999999 0.9999998 0.9999996 0.9999992 0.9999983
> he(xdeb)
[1] -3.229343e-06
```

As a second example, let us consider again the determination of the point on the line $x_1 + x_2 = 2$ which is the closest to the circle $x_1^2 + x_2^2 = 1$ (see Sect. 1.13.3). Setting $\boldsymbol{x} = (x_1, x_2, x_3, x_4)$, we have

$$J(\boldsymbol{x}) = \sqrt{(x_1 - x_3)^2 + (x_2 - x_4)^2},$$

$$\varphi_1(\boldsymbol{x}) = x_1^2 + x_2^2 - 1,$$

$$\varphi_2(\boldsymbol{x}) = x_3 + x_4 - 2,$$

Then, we are looking for

$$\boldsymbol{x}^* = \arg\min \left\{ J(\boldsymbol{x}) : \boldsymbol{x} \in \mathbb{R}^4,\ \varphi_1(\boldsymbol{x}) = \varphi_2(\boldsymbol{x}) = 0 \right\}.$$

As previously observed, the exact solution is $\left(\sqrt{2}/2, \sqrt{2}/2, 1, 1\right) \approx (0.7071, 0.7071, 1, 1)$. Let us apply the procedure above:

Define the objective function

```
fdist = function(x){
  aux = sqrt((x[[1]]-x[[3]])^2 + (x[[2]]-x[[4]])^2)
  return(aux)
}
```

Define $h(\boldsymbol{x}, t)$.

```
h1 = function(x){
  aux = abs(sqrt(x[[1]]^2 + x[[2]]^2) - 1) - 1e-3
  return(aux)
}

h2 = function(x){
  aux = abs(x[[3]] + x[[4]] - 2) - 1e-3
  return(aux)
}

h = function(x,t){
  aux = numeric(length = 3)
  aux[[1]] = fdist(x)-t
  aux[[2]] = h1(x)
  aux[[3]] = h2(x)
  return(max(aux))
}
```

Then, run the code. We used as starting point:

```
n=4
nitmax = 100
precmin = 1e-5
x0 = c(cos(rnorm(1)),sin(rnorm(1)),1,1)
xdeb = matrix(x0, ncol = n)
```

The result was close to the exact solution.

```
[1] "nit =  1 , prec =  1.97320494353909"
[1] "nit =  2 , prec =  1.40805813791944e-06"
> tac
[1] 0.4125098
> xac
          [,1]      [,2]      [,3]      [,4]
[1,] 0.7078069 0.7078198 0.9994946 0.9995079
> h1(xac)
[1] -7.549343e-07
> h2(xac)
[1] -2.528522e-06
```

1.15 Interpolation and Approximation with R

R provides functions for linear, cubic, and spline interpolation of data. You can also find in the R repositories packages for other kinds of interpolation, namely for spatial bivariate data. The basic interpolation is made by calling functions `approx` and `approxfun`: the last one generates a function making the interpolation, while the first one returns lists of values.

As an example, let us consider the data corresponding to 21 points equally spaced on $(0, 2\pi)$, $x_i = (i-1)\pi/10$, $y_i = \sin(x_i)$, $1 \le i \le 21$. The corresponding data are shown in Table 1.1:

Of course, the data can be generated by R as shown at right:

```
> x = seq(0,2*pi,length.out=21)
> y = sin(x)
```

Table 1.1 Unnoisy data used in the interpolation

x	0.0	0.3	0.6283185	0.9424778	1.256637	1.570796	1.884956
y	0.0	0.309017	0.5877853	0.809017	0.9510565	1.0	0.9510565
x	2.199115	2.513274	2.827433	3.141593	3.455752	3.769911	4.08407
y	0.809017	0.5877853	0.309017	0	−0.309017	−0.5877853	−0.809017
x	4.39823	4.712389	5.026548	5.340708	5.654867	5.969026	6.283185
y	−0.9510565	−1.0	−0.9510565	−0.809017	−0.5877853	−0.309017	0

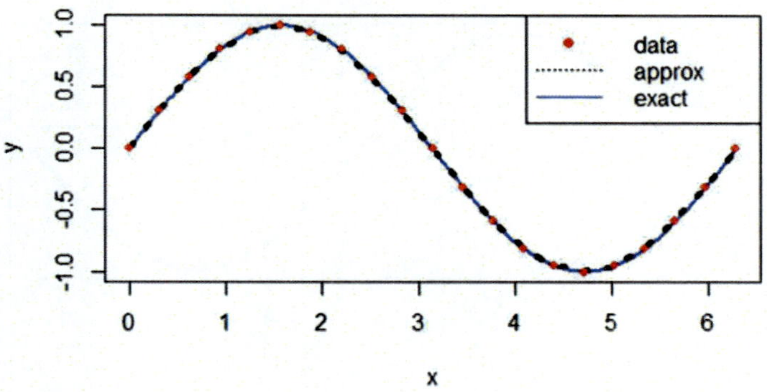

Fig. 1.27 Interpolation of unnoisy data using `approx`

The approximation is generated and plotted as follows:

```
> data_ap = approx(x,y)
> xc = seq(0,2*pi,by=pi/100)
> plot(x,y,type="n")
> lines(xc,sin(xc),col='blue',lty='solid',lwd = 2)
> lines(data_ap$x,data_ap$y,col='black',lty='dotted',lwd=4)
> points(x,y,'p',col='red',pch=19,cex=0.8)
> legend("topright", legend=c("data", "approx","exact"),
+         col=c("red", "black","blue"), lty=c(NA,3,1),pch = c(19,NA,NA))
```

The results are exhibited in Fig. 1.27.

Spline interpolation can be generated as follows:

```
> f <- splinefun(x, y)
> plot(x,y,type="n");
> lines(xc,yc,col="blue",lty="solid",lwd=2)
> par(new=TRUE)
> curve(f(x),0,2*pi,col="black",lty="dotted",lwd=4)
> points(x,y,col="red",pch=19,cex=0.8)
> legend("topright",legend = c("data","approx","exact"),col = c("red","blac
k","blue"),lty = c(NA,3,1),pch = c(19,NA,NA))
```

The results are shown in Fig. 1.28.

For noisy data, the result losses in quality. As an example, let us consider the corrupted data in Table 1.2, corresponding to multiplicative and additive noise added to the data in Table 1.1

The results are shown in Figs. 1.29 and 1.30

When dealing with noisy data, filtering can be considered if you have enough data – R proposes a function `filter` (but you must furnish the coefficients of the filter). However, one of the objectives of UQ is to deal with limited quantities of data – in such a situation, standard filtering techniques can appear as not adapted. We examine two alternative methods in the sequel.

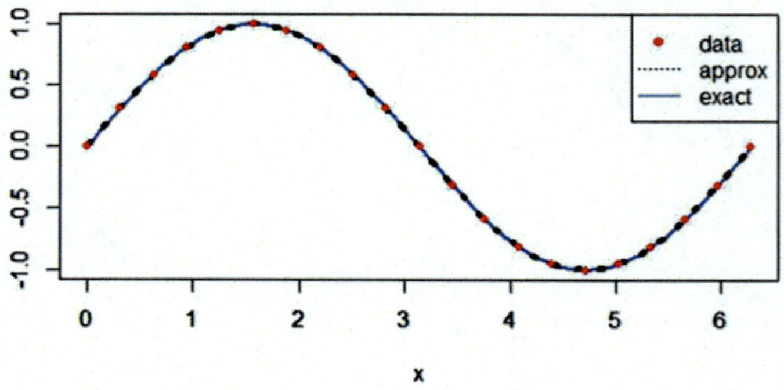

Fig. 1.28 Interpolation of unnoisy data using `spline`

Table 1.2 Noisy data used in the interpolation

x	0.0	0.3	0.6283185	0.9424778	1.256637	1.570796	1.884956
y	0.155613	0.494447	0.403890	0.879648	0.894935	1.051131	0.9679962
x	2.199115	2.513274	2.827433	3.141593	3.455752	3.769911	4.08407
y	1.026136	0.626192	0.156230	−0.081455	−0.405254	−0.440917	−0.935802
x	4.39823	4.712389	5.026548	5.340708	5.654867	5.969026	6.283185
y	− 0.768096	− 0.998765	− 1.152535	− 0.790915	− 0.622155	− 0.050190	− 0.285916

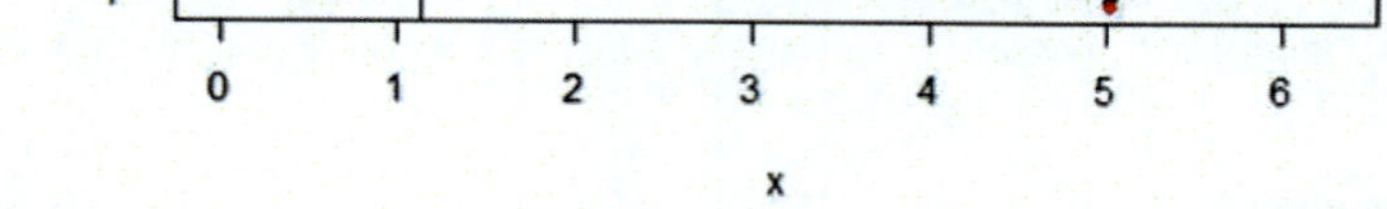

Fig. 1.29 Interpolation of noisy data using `approx`

1.15.1 Variational Approximation

Variational approximations may be interpreted as generalizations of filtering by Fourier expansions – recall that a simple way to denoise a signal consists in taking its Fourier transform and filtering the frequencies which appear to the user as corresponding to noise. A generalization consists in use expansions in a general Hilbert basis, eventually non-trigonometric: for instance, we can consider a Hilbert

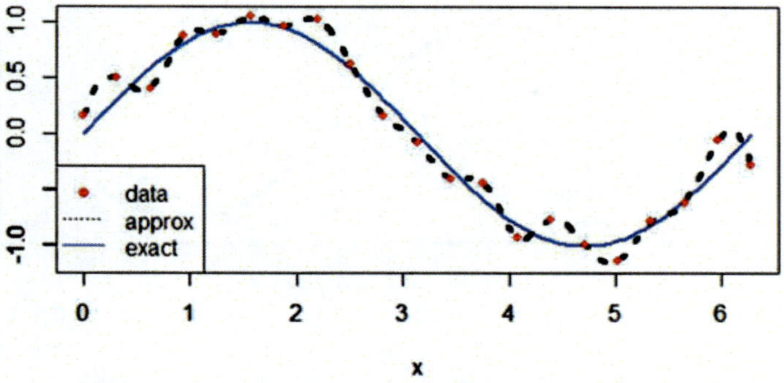

Fig. 1.30 Interpolation of noisy data using `spline`

basis $\{\varphi_i: i \in N\}$ (see Sects. 1.12 and 2.13) and look for an approximation using the k first terms of the basis:

$$y \approx Py = \sum_{i=0}^{k} c_i \varphi_i(x).$$

Typical Hilbert basis are furnished by polynomials and trigonometric functions. A general method for the generation of Hilbert basis consists in the use of eigenfunctions of compact operators (see, for instance (Souza de Cursi & Sampaio, 2010)). If the Hilbert space under consideration has a scalar product $(\blacksquare, \blacksquare)$, the coefficients are determined by the equations defining the orthogonal projection (see Sect. 2.5):

$$\sum_{j=0}^{k} c_j (\varphi_i, \varphi_j) = (y, \varphi_i), \; 0 \le i \le k.$$

These equations correspond to a linear system

$$AC = B, A_{i+1,\, j+1} = (\varphi_i, \varphi_j), \; B_{i+1} = (y, \varphi_i), \; C_{i+1} = c_i.$$

In practice, a common scalar product for functions defined on the interval (a, b) is

$$(u, v) = \int_a^b u(x) v(x) dx.$$

When the expression of $y(x)$ is unknown, but only the data (x_i, y_i), $1 \le i \le n$ are available, we can use an interpolation by `approx.` to generate a function which can be integrated: `f = approxfun(x,y)` generates a function $f(x)$ which is a

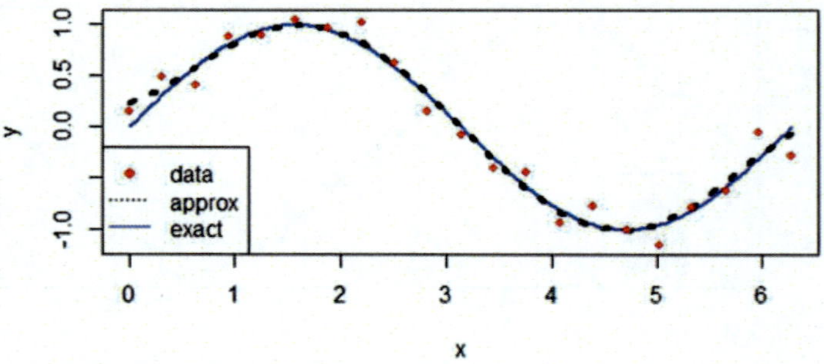

Fig. 1.31 Interpolation of noisy data using the variational approach and `approx`

piecewise linear approximation of $y(x)$ – we use f to evaluate the elements of B. For instance, we can define the function `phi(i,x)` evaluating $\varphi_i(x)$ and determine

```
ff = function(x){f(x)*phi(i,x)}
aux = integrate(ff,0,2*pi)
B[[i+1]] = aux$value
```

Analogously, A_{ij} is evaluated as follows:

```
ff = function(x){phi(i,x)*phi(j,x)}
aux = integrate(ff,0,2*pi)
A[[i+1,j+1]]= aux$value
```

As an example, let us consider the data in Table 1.2 and $\varphi_i(x) = (x/2\pi)^i$. For these data and $k = 5$, we obtain

$$C = (0.2309068, 1.425763, 34.84047, -157.3934, 199.8775, -79.04984)^t$$

The results appear in Fig. 1.31: the approximation is close to the exact curve.

1.15.2 Smoothed Particle Approximation

Smoothed particle approximations are based on the Dirac's delta, *id est*, on the probability distribution of the random variable D_0: $\Omega \rightarrow \mathbb{R}$ such that $D_0(\omega) = 0 \in \mathbb{R}, \forall\, \omega \in \Omega$. Such a variable verifies

$$P(D_0 = 0) = 1;\ P(D_0 = s) = 0,\ if\ s \neq 0.$$

In this case, for $y : \mathbb{R} \rightarrow \mathbb{R}$, $E(y(D)) = y(0)$. $D_x = x + D_0$ is an analogous variable such that $P(D_x = x) = 1$ and $E(y(D_x)) = y(x)$.

Let $h > 0$, Z_h be a random variable such that $E(Z_h) = 0$ and $V(Z_h) \leq h$. We have from Chebychev's inequality (see Sect. 2.4):

$$P(|Z_h| \geq \varepsilon) \leq \frac{h}{\varepsilon^2} \rightarrow 0 \text{ for } h \rightarrow 0+, \forall \varepsilon > 0.$$

Thus, $Z_h \rightarrow D_0$ in probability and in distribution (see Sect. 2.9), so that $E(u(-Z_h)) \rightarrow u(0)$. Analogously, $X_h = x + Z_h \rightarrow D_x$ and

$$E(y(X_h)) \rightarrow E(y(D_x)) = y(x), \textit{for } h \rightarrow 0+.$$

Consequently, we can consider approximations of $y(x)$ by means of $E(y(X_h))$. If Z_h has a CDF G_h, then $P(X_h \leq p) = P(Z_h \leq p - x) = G_h(p - x)$. Analogously, if the PDF of Z_h is g_h, then X_h has a PDF $g_h(p - x)$ and we have

$$y(x) \approx \int_{-\infty}^{+\infty} y(p) g_h(p - x) dp.$$

In practice, the integral is approximated by its truncation to an interval around x, having the form $(x - A_h, x + A_h)$ and is numerically evaluated. Thus, the approximation reads as

$$y(x) \approx \int_{x - A_h}^{x + A_h} y(p) g_h(p - x) dp.$$

For the numerical evaluation, we may consider a partition $-A_h = z_0 < z_1 < \ldots < z_n = A_h$ and an approximation by rectangular sums:

$$y(x) \approx \sum_{i=1}^{n} y(p_i) g_h(p_i - x) V_i,\ V_i = z_{i+1} - z_i,\ p_i = x + \frac{z_i + z_{i-1}}{2}$$

This kind of approximation is usually referred as smoothed particle (SP) approximation. In the language of SP, each p_i is a ***particle*** and V_i is its ***volume***. It is usual to write $V_i = m_i/\rho_i$, where m_i is referred as being the ***mass*** of the particle and ρ_i as its ***volumic mass*** (or ***density***). $k(p, x) = g_\delta(p - x)$ is called the ***kernel***. Thus, the approximation reads as

$$y(x) \approx \sum_{i=1}^{n} \frac{m_i}{\rho_i} y(p_i) k(p_i, x).$$

Taking $u = \rho$, we obtain

$$\rho_j = \rho(p_j) = \sum_{i=1}^{n} \frac{m_i}{\rho_i} \rho(p_i) k(p_i, p_j) = \sum_{i=1}^{n} m_i k(p_i, p_j).$$

Thus, we can choose the masses and determine the densities. Once these quantities are determined, we can use the data $y(p_i)$ on the particles to determine $y(x)$. If the data are (x_i, y_i), $1 \leq i \leq n$, we can use as particles $p_i = x_i$ and as values $y(p_i) = y_i$.

It may be appeared as convenient to use Z_δ such that $g_h(z) = 0$ for $|z| > A_h$, eventually by using a truncation of a PDF f supported by $\mathbb{R}$ on an interval $(-A_h, A_h)$. Then, the PDF of the truncated distribution is

$$g_h(z) = \frac{f(z)}{\int_{-A_h}^{A_h} f(s) ds}, \text{ if } |z| \leq A_h; g_h(z) = 0, \text{ otherwise.}$$

In this case, taking $k(p, x) = f(\rho - x)$, we have

$$y(x) \approx \frac{1}{S(x)} \sum_{i=1}^{n} \frac{m_i}{\rho_i} y(p_i) k(p_i, x),$$

with

$$S(x) = \sum_{j=1}^{n} \frac{m_j}{\rho_j} k(p_j, x)$$

Different kernels may be found in the literature. The most popular approaches use

$$g_h(p, x) = \alpha(h) \psi\left(\frac{\|p - x\|}{h}\right),$$

where $\alpha(h)$ is a normalizing factor such that $\int_{\mathbb{R}^n} g_h(p, x) dp = 1$. For instance:

- Cubic kernel: $\alpha(h) = \frac{2}{3}, \frac{10}{7\pi}, \frac{1}{\pi}$ (dimension 1, 2, or 3, respectively)

$$\psi(r) = \begin{cases} 1 - \frac{3r^2}{2} + \frac{3r^3}{4}, \text{ if } 0 \leq r \leq 1 \\ \frac{1}{4}(2 - r)^3, \text{ if } 1 \leq r \leq 2 \\ 0, \text{ otherwise} \end{cases}$$

- Quartic kernel: $\alpha(h) = \frac{1}{24}, \frac{96}{1199\pi}, \frac{1}{20\pi}$ (dimension 1, 2, or 3, respectively)

$$\psi(r) = \begin{cases} \left(\frac{5}{2} - r\right)^4 - 5\left(\frac{3}{2} - r\right)^4 + 10\left(\frac{1}{2} - r\right)^4, \textit{ if } 0 \le r \le \frac{1}{2} \\ \left(\frac{5}{2} - r\right)^4 - 5\left(\frac{3}{2} - r\right)^4, \textit{ if } \frac{1}{2} \le r \le \frac{3}{2} \\ \left(\frac{5}{2} - r\right)^4, \textit{ if } \frac{3}{2} \le r \le \frac{5}{2}, \\ 0, \text{ otherwise} \end{cases}$$

- Quintic kernel: $\alpha(h) = \frac{1}{120}, \frac{7}{478\pi}, \frac{1}{120\pi}$ (dimension 1, 2, or 3, respectively)

$$\psi(r) = \begin{cases} (3 - r)^5 - 6(2 - r)^5 + 15(1 - r)^5, \textit{ if } 0 \le r \le 1 \\ (3 - r)^5 - 6(2 - r)^5, \textit{ if } 1 \le r \le 2 \\ (3 - r)^5, \textit{ if } 2 \le r \le 3, \\ 0, \text{ otherwise} \end{cases}$$

- Gaussian kernel: $\alpha(h) = \frac{1}{h\sqrt{2\pi}}, \frac{7}{2\pi h^2}, \frac{1}{h^3 (2\pi)^{3/2}}$ (dimension 1, 2, or 3, respectively)

$$\psi(r) = e^{-r^2/2}$$

As an example, let us consider again the data in Table 1.2 and the SP approximation with $h = x_2 - x_1$ and a Gaussian kernel. The results are exhibited in Fig. 1.32: the approximation is close to the exact curve.

Notice that parameter h controls the smoothness of the approximation. For instance, Fig. 1.33 shows the approximations furnished by a quintic kernel with $h = x_2 - x_1$, $h = 2(x_2 - x_1)$, $h = 3(x_2 - x_1)$

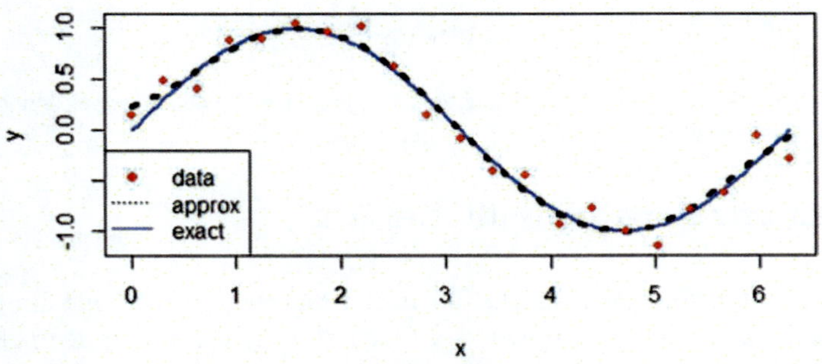

Fig. 1.32 Interpolation of noisy data using the SP approach with Gaussian kernel

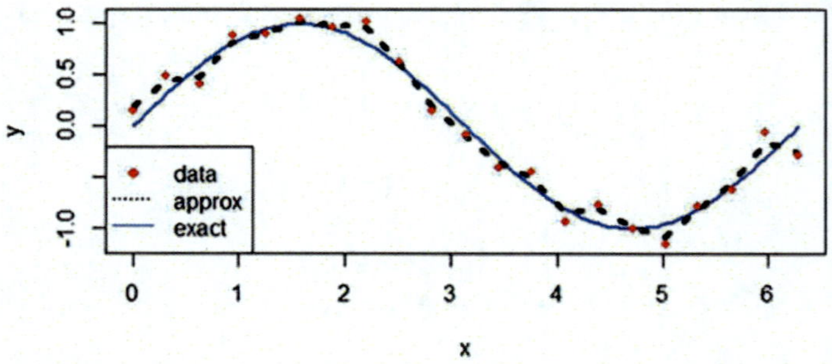

a) $h = x_2 - x_1$

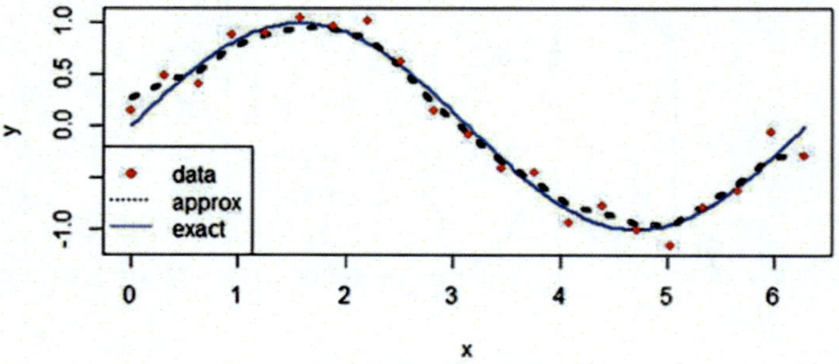

b) $h = 2(x_2 - x_1)$

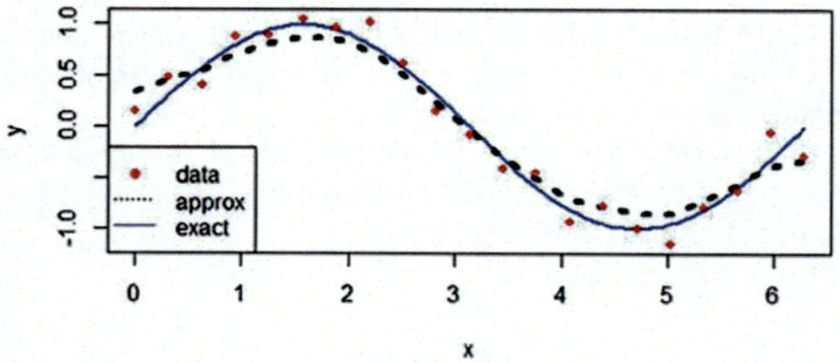

b) $h = 3(x_2 - x_1)$

Fig. 1.33 Interpolation of noisy data using the SP approach with Quintic kernel. Smoothness is controlled by parameter h

1.16 Integrals and Derivatives with R

R proposes a function `integrate` for numerical integration of one-dimensional functions. Package `pracma` proposes a function integral for the same purpose. It may be useful to vectorize the function to be integrated.

As an example, consider
$f(x) = \frac{\sin(x)}{x}$
to be integrated over (0, 1)

```
> f <- function(x){sin(x)/x}
> fi <- Vectorize(f)
> integrate(fi,0,1)
0.9460831 with absolute error < 1.1e-14
> library(pracma)
> integral(fi,0,1,method="Kron")
[1] 0.9460831
> integral(fi,0,1,method="Clen")
[1] 0.9460831
> integral(fi,0,1,method="Simp")
[1] 0.9460831
```

Multiple integrals can be evaluated by `integrate` using functions and nested integration. As an example, let us consider the evaluation of

$$\int_0^1 e^{-y} dy \int_0^{y^2} e^x \sin(x+y) dx$$

We may use the code

```
> cmin = function(y){0}
> cmax = function(y){y^2}
> fxy <- function(x,y) { exp(x)* sin(x+y) }
> fya <- function(y) sapply(y, function (y_i) integrate(fxy,
 lower = cmin(y_i), upper = cmax(y_i), y = y_i)$value)
> fy <- function(y) {exp(-y)*fya(y)}
> integrate(fy, lower = 0, upper = 1)
0.174964 with absolute error < 1.9e-15
```

`pracma` proposes functions `integral2` and `integral3` for 2D and 3D integration. For instance:

```
> library(pracma)
> fyx <- function(y,x) {exp(-y)*fxy(x,y)}
> xmin <- 0; xmax <- cmax
> ymin <- 0; ymax <- 1
> integral2(fyx, ymin, ymax, xmin, xmax)
$Q
[1] 0.174964

$error
[1] 4.862539e-18
```

R repositories propose also packages for the numerical evaluation of derivatives, such as `numDeriv`:

```
> library(numDeriv)
> f = function(x){x*sin(x) + exp(-x)/x}
> df = function(x){sin(x) + x * cos(x) - (1 + 1/x) *
 exp(-x)/x}
> dfn = grad(f,1)
> df(1)
[1] 0.6460144
> dfn
[1] 0.6460144
```

```
> f = function(x){x[1]*sin(x[2])+exp(-x[1])/x[3]}
> df = function(x){c(sin(x[2])-exp(-x[1])/x[3],x[1]*
cos(x[2]),-exp(-x[1])/x[3]^2)}
> p = as.vector(c(1,2,3))
> dfn = grad(f,p)
> df(p)
[1]  0.78667095 -0.41614684 -0.04087549
> dfn
[1]  0.78667095 -0.41614684 -0.04087549
```

Symbolic integrals may be evaluated by using packages `mosaic` and `mosaicCalc`:

```
> f <- function(x){x*sin(x)}
> F <- antiD(f(x)~x)
> F
function (x, C = 0)
{
    numerical_integration(.newf, .wrt, as.list(match.call())
[-1],
        formals(), from, ciName = intC, .tol)
}
<environment: 0x0000016b02599e20>
> F(1)
[1] 0.3011687
> sin(1) - 1*cos(1)
[1] 0.3011687
```

Notice that R was not capable of furnishing the analytic expression $FF(x) = \sin(x) - x \cos(x)$, but generated a function which evaluates numerically the result. Symbolic integrals or derivatives of FF can be calculated, as shown at left – we determine the derivative of FF and we compare its value at $x = 1$ with the value of $f(1)$.

```
> library(mosaic)
> library(mosaicCalc)
> f <- function(x){x*s
> FF <- antiD(f(x)~x)
> FF(1)
[1] 0.3011687
> sin(1) - 1*cos(1)
[1] 0.3011687
> f2 <- D(FF(x)~x)
> f2(1)
[1] 0.841471
> f(1)
[1] 0.841471
```

The manipulation of sequences of antiderivatives requests some precaution, due to the existence of constants: by default, R sets the value at $x = 0$ to 0, as shown at left – to compare the results, we must use an integral taking the value 0 at $x = 0$.

```
> F2 <- antiD(FF(x)~x)
> F2(0)
[1] 0
> F2(1)
[1] 0.0779244
> -2*cos(1) - 1*sin(1) -(-2*cos(0)-0*sin(0))
[1] 0.0779244
```

As shown above, symbolic derivatives are also available with R, using packages `Deriv`, `mosaic`, and `mosaicCalc`:

```
> f <- function(x){x*sin(x) + exp(-x)/x}
> library(mosaic)
> library(mosaicCalc)
> df <- D(f(x)~x)
> df
function (x)
numerical.first.partial(.function, .wrt, .hstep, match.cal
l())
<environment: 0x0000016b04c22608>
> df1 <- D(x*sin(x)~x)
> df1
function (x)
sin(x) + x * cos(x)
> df2 <- D(exp(-x)/x~x)
> df2
function (x)
-(exp(-x)/x + exp(-x)/x^2)
> df(1)
[1] 0.6460144
> df1(1)+df2(1)
[1] 0.6460144
```

As we can observe, R was not able to determine an explicit expression for $f^{'}(x)$, while it furnished an explicit formula for the derivative of each part of the function f. However, the function generated is a function and we can evaluate its derivative symbolically, as shown at left.

```
> d2f = D(df(x)~x)
> d2f1 = D(df1(x)~x)
> d2f2 = D(df2(x)~x)
> d2f(1)
[1] 2.078615
> d2f1(1)+d2f2(1)
[1] 2.078531
```

The results furnished by package `Deriv` are analogous:

```
> f <- function(x){x*sin(x) + exp(-x)/x}
> library(Deriv)
> dfs <- Deriv("f(x)","x", cache.exp=FALSE)
> dfs
[1] "sin(x) + x * cos(x) - (1 + 1/x) * exp(-x)/x
> df = function(x) eval(parse(text=dfs))
> df(1)
[1] 0.6460144
> d2fs = Deriv(dfs,"x", cache.exp=FALSE)
> d2f = function(x) eval(parse(text=d2fs))
> d2f(1)
[1] 2.078531
```

Notice that `Deriv` works on strings, and we must use `eval(parse (text=))` to create a function which can be manipulated.

1.16.1 Variational Approximation of the Derivatives

When dealing with noisy data in limited quantity, the numerical evaluation of the derivatives may be performed by using the variational approach shown in Sect. 1.15. For instance, we can determine an approximation

$$y \approx Py = \sum_{i=0}^{k} c_i \varphi_i(x).$$

and evaluate

$$y' \approx (Py)' = \sum_{i=0}^{k} c_i \varphi_i'(x).$$

Let us apply this approach to the unnoisy data in Table 1.1: for $k = 6$ and $\varphi_i(x) = (x/2\pi)^i$, we obtain the results exhibited in Fig. 1.34.

Now, let us consider the noisy data in Table 1.2 and $\varphi_i(x) = (x/2\pi)^i$. For $k = 6$, we obtain the results in Fig. 1.35

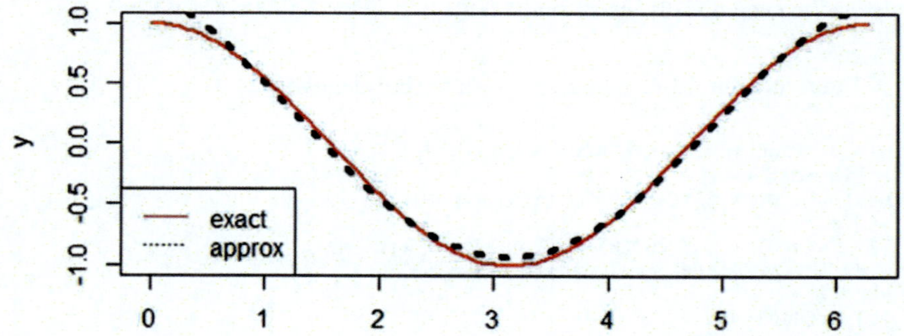

Fig. 1.34 Derivation of unnoisy data using the variational definition of derivative

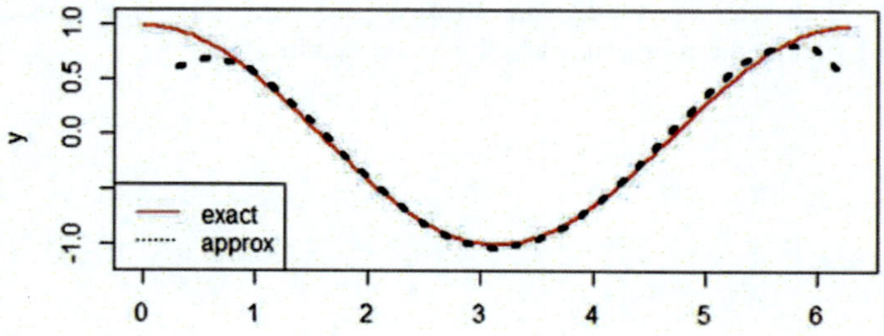

Fig. 1.35 Derivative of noisy data using the variational approximation of the data

As an alternative, we can use the variational definition of derivative – the notion of weak derivative. Indeed, the variational derivative Dy of y verifies

$$\int_a^b v(x)Dy(x)dx = y(b)v(b) - y(a)v(a) - \int y(x)v'(x)dx, \forall v.$$

Let

$$Dy \approx \sum_{i=0}^{k} c_i \varphi_i(x).$$

Then, we can determine the coefficients by solving the linear system $AC = B$, with

$$A_{i+1,j+1} = \int_a^b \varphi_i(x)\varphi_j(x)dx,$$

$$B_{i+1} = y(b)\varphi_i(b) - y(a)\varphi_i(a) - \int_a^b y(x)\varphi_i'(x)dx.$$

Let us apply this approach to the unnoisy data in Table 1.1: for $k = 5$ and $\varphi_i(x) = (x/2\pi)^i$, we obtain the results exhibited in Fig. 1.36.

Applying this approach to the noisy data in Table 1.2, with $\varphi_i(x) = (x/2\pi)^i$ and $k = 4$ produces the results in Fig. 1.37.

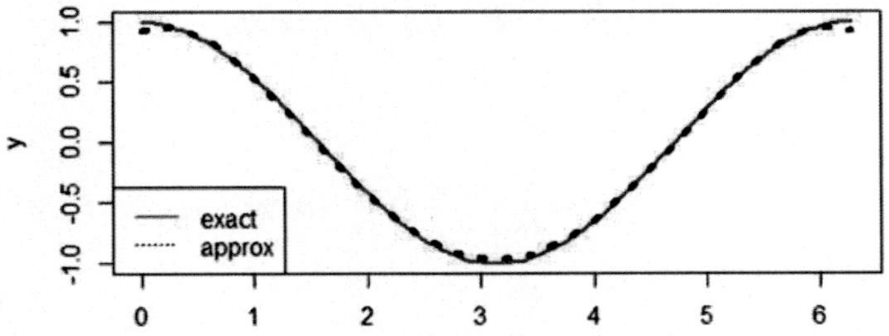

Fig. 1.36 Derivation of unnoisy data using the variational definition of derivative

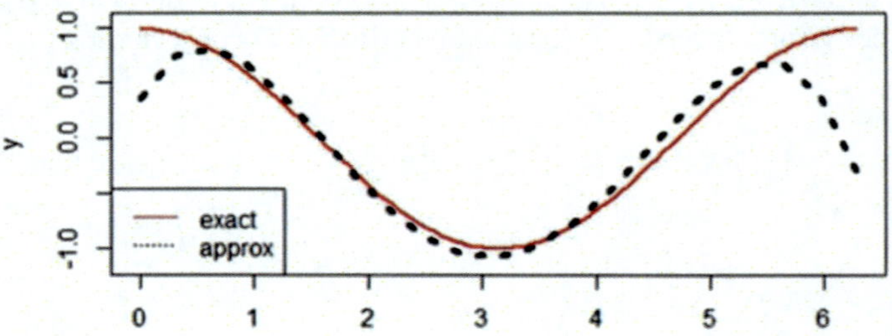

Fig. 1.37 Derivative of noisy data using the variational definition of derivative

1.16.2 Smoothed Particle Approximation of the Derivative

Since

$$y(x) = \int_{-\infty}^{+\infty} y(p)k(p, x)dp.$$

we have

$$y'(x) = \int_{-\infty\_}^{+\infty} y(p)\frac{d}{dx}k(p, x)dp.$$

what leads to the approximation

$$y'(x) = \int_{x-A_\delta}^{x+A_\delta} y(p)\frac{d}{dx}k(p, x)dp.$$

$$y'(x) \approx \sum\nolimits_{j=1}^{n} \frac{m_j}{\rho_j} y_j \frac{d}{dx} k(p_j, x).$$

A second form of approximation is based on the equation

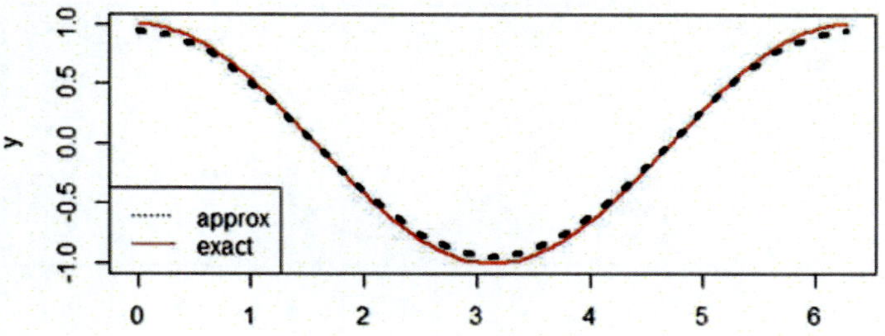

Fig. 1.38 Derivation of unnoisy data using the SPH approach

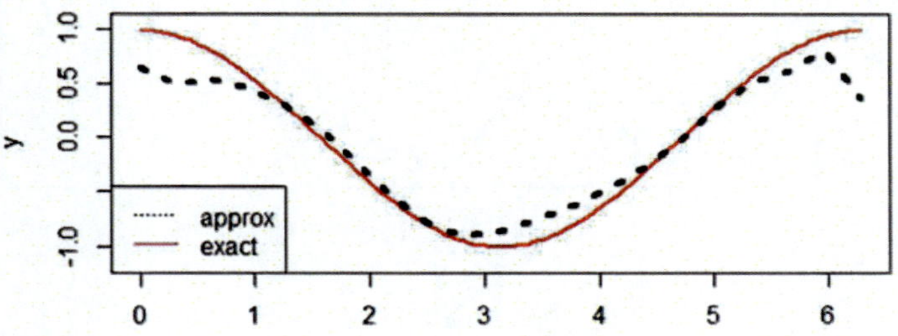

Fig. 1.39 Derivation of noisy data using the SP approach

$$y'(x) = \frac{\int \delta'(p-x)(y(p)-y(x))dp}{\int (p-x)\delta'(p-x)dp}$$

what leads to the approximation

$$y'(x) \approx \frac{\sum_{j=1}^{n} \frac{m_j}{\rho_j} \left(y_j - y(x)\right) \frac{d}{dx} k\left(p_j, x\right)}{\sum_{j=1}^{n} \frac{m_j}{\rho_j} \left(p_j - x\right) \frac{d}{dx} k\left(p_j, x\right)}$$

Let us apply this approach to the unnoisy data in Table 1.1: $h = x_2 - x_1$ and a Gaussian kernel, we obtain the results exhibited in Fig. 1.38.

Now, let us consider the noisy data in Table 1.2 and the SP approximation with $h = 1.75(x_2 - x_1)$ and a Gaussian kernel. The results are exhibited in Fig. 1.39: the approximation is close to the exact curve.

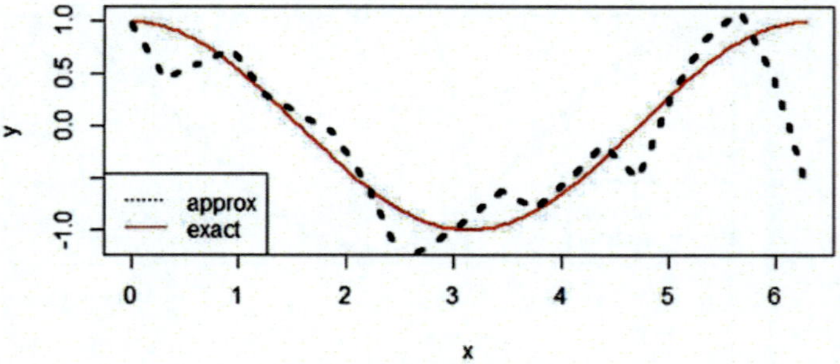

a) $h = x_2 - x_1$

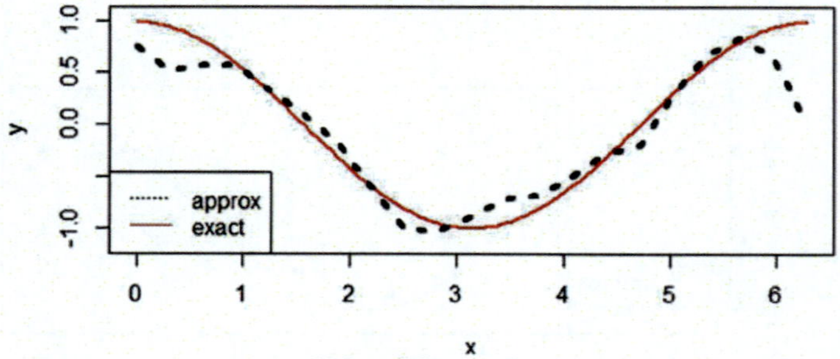

b) $h = 2(x_2 - x_1)$

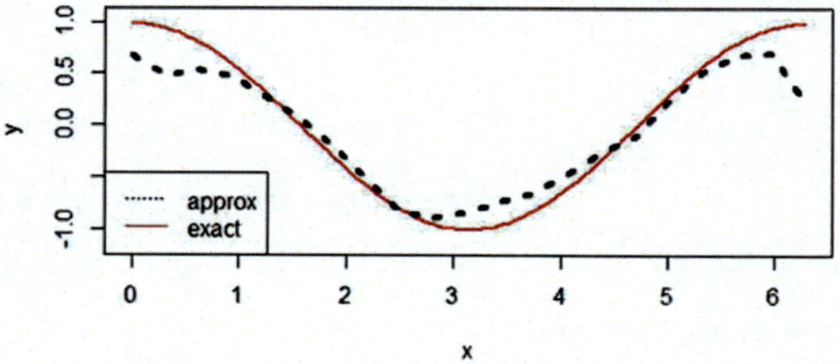

b) $h = 3(x_2 - x_1)$

Fig. 1.40 Derivation of noisy data using the SP approach with Quintic kernel. Smoothness is controlled by parameter h

Analogously to interpolation (Sect. 1.15.2), the smoothness of SP derivatives is controlled by parameter h. Examples are shown in Fig. 1.40.

Chapter 2
Probabilities and Random Variables

Abstract In this chapter, we present the fundamental elements of probability and statistics that are used in the book, namely the elements about random variables and random vectors, with particular attention to the use of R for the calculations. The classical topics of a standard course are considered, such as statistics, regression, correlations, confidence intervals, and hypothesis testing. We consider also random functions and random curves, as well as their generation, evaluation of means, and confidence intervals for such objects.

2.1 Notation

In the sequel, $\mathcal{N}$ denotes the set of the natural numbers; $\mathcal{N}^* = \mathcal{N} - \{0\}$ denotes the set of the strictly positive natural numbers. Analogously, $\mathbb{R}$ denotes the set of the real numbers and $\mathbb{R}^* = \mathbb{R} - \{0\}$ denotes the set of the strictly positive real numbers. $\mathbb{R}_e = \mathbb{R} \cup \{-\infty, +\infty\}$ is the set of the extended real numbers. The notation $(a, b) = \{x \in \mathbb{R}\colon a < x < b\}$ defines an open interval of real numbers.

$\mathbb{R}^n = \{\boldsymbol{x} = (x_1, \ldots, x_n) : x_i \in \mathbb{R}, 1 \leq i \leq n\}$ is the set of the n-tuples formed by real numbers. Analogous notation is used for $\mathbb{R}_e^n$. The standard scalar product of $\boldsymbol{x}$ and $\boldsymbol{y} = (y_1, \ldots, y_n) \in \mathbb{R}^n$ is

$$(\boldsymbol{x}, \boldsymbol{y})_{\mathbb{R}^k} = \sum_{i=1}^{n} x_i y_i. \tag{2.1}$$

If no confusion is possible, we drop the indexing expression and use the lighter notation $(\boldsymbol{x}, \boldsymbol{y})$. In some contexts, we use also the notatiions:

Supplementary Information The online version of this chapter (https://doi.org/10.1007/978-3-031-17785-9_2) contains supplementary material, which is available to authorized users.

E. Souza de Cursi, *Uncertainty Quantification using R*, International Series in Operations Research & Management Science 335,
https://doi.org/10.1007/978-3-031-17785-9_2

$$x^t y = (x, y) \text{ or } x.y = (x, y). \tag{2.2}$$

We denote by $|\bullet|_p$ the standard p-norm of $\mathbb{R}^k$:

$$|x|_p = \left(\sum_{i=1}^{n} |x_i|^p \right)^{\frac{1}{p}}. \tag{2.3}$$

We have

$$|x|_2 = \sqrt{(x, x)}. \tag{2.4}$$

A $m \times n$-matrix $\mathbf{A}$ of real numbers is $\mathbf{A} = (A_{ij}\colon 1 \le i \le m, 1 \le j \le n)$. $M(m, n)$ denotes the space of the $m \times n$-matrices of real numbers. The elements of $\mathbb{R}^n$ may be identified to elements of $M(n, 1)$ (column vectors): $\mathbb{R}^n \sim M(n, 1)$. Thus, we can set

$$y = A x \in M(m, 1) \sim \mathbb{R}^m, \; y_i = \sum_{j=1}^{n} A_{ij} x_j \; (1 \le i \le m) \tag{2.5}$$

If no confusion is possible, we shall alleviate the notation by using the convention of an implicit sum on repeated indexes (also called Einstein's convention) by writing simply $y_i = A_{ij} x_j$.

2.2 Probability

We use the formal definition of probability introduced by (Kolmogorov, 1950):

Definition 2.1 Let Ω be a non-empty set and $\mathcal{P}(\Omega) = \{E : E \subset \Omega\}$ be the set of the parts of Ω. A probability defined on Ω is an application $P\colon \mathcal{P}(\Omega) \longrightarrow \mathbb{R}$ such that

1. $P(E) \ge 0, \forall E \in \mathcal{P}(\Omega)$;
2. $P(\Omega) = 1$;
3. $P\left(\bigcup_{n \in \mathbb{N}} E_n \right) = \sum_{n \in \mathbb{N}} P(E_n), \forall \{E_n : n \in \mathbb{N}\}$ *such that* $E_i \cap E_j = \emptyset$, *if* $i \neq j$.

Ω is the ***universe***. The subsets of $E \subset \Omega$ are usually called ***events***. (Ω, P) is the ***probability space***. This definition leads to many known properties of probabilities. For instance:

Proposition 2.1 Let (Ω, P) be a probability space. Then

1. $P(\varnothing) = 0$;
2. $P(E - F) = P(E) - P(E \cap F)$;
3. $P(E \cup F) = P(E) + P(F) - P(E \cap F)$;
4. $P(E^c = \Omega - E) = 1 - P(E)$;
5. $E \subset F \Rightarrow P(E) \leq P(F)$;
6. $0 \leq P(E) \leq 1, \forall E \subset \Omega$;
7. $P\left(\bigcup_{n \in \mathbb{N}} E_n\right) \leq \sum_{n \in \mathbb{N}} P(E_n), \forall \{E_n : n \in \mathbb{N}\} \subset \mathcal{P}(\Omega)$.
8. Let $\{E_n : n \in \mathbb{N}\} \subset \mathcal{P}(\Omega)$. If $\sum_{n \in \mathbb{N}} P(E_n) < \infty$, then $P\left(\bigcup_{i=n}^{+\infty} E_i\right) \longrightarrow 0$ for $n \longrightarrow +\infty$;
9. Let $\{E_n : n \in \mathbb{N}\} \subset \mathcal{P}(\Omega)$ be such that $E_n \subset E_{n+1}$, $\forall n \in \mathbb{N}$. then, $P\left(\bigcup_{n=0}^{+\infty} E_n\right) = \lim_{n \to +\infty} P(E_n)$;
10. Let $\{E_n : n \in \mathbb{N}\} \subset \mathcal{P}(\Omega)$ be such that $E_{n+1} \subset E_n$, $\forall n \in \mathbb{N}$. then, $P\left(\bigcap_{n=0}^{+\infty} E_n\right) = \lim_{n \to +\infty} P(E_n)$.

Proof We have

1. Let $P(\varnothing) = a \in \mathbb{R}, a \geq 0$. Consider $\{E_n : n \in \mathbb{N}\}$ such that $E_n = \varnothing, \forall n \in \mathbb{N}$. then $E_i \cap E_j = \varnothing$, *if* $i \neq j$. Thus, $P\left(\bigcup_{n \in \mathbb{N}} E_n\right) = \sum_{n \in \mathbb{N}} P(E_n)$. Since $\bigcup_{n \in \mathbb{N}} E_n = \varnothing$, we have $a = \sum_{n \in \mathbb{N}} a$, with $a \geq 0$. This equality is impossible if $a > 0$ (the series becomes divergent), so that $a = 0$.
2. Consider $\{E_n : n \in \mathbb{N}\}$ such that $E_0 = E - F, E_1 = E \cap F, E_n = \varnothing, \forall n \geq 2$. Then, $E_i \cap E_j = \varnothing$, *if* $i \neq j$, so that $P\left(\bigcup_{n \in \mathbb{N}} E_n\right) = \sum_{n \in \mathbb{N}} P(E_n)$. Or, on the one hand, $\bigcup_{n \in \mathbb{N}} E_n = (E - F) \cup (E \cap F) = E$ and, on the other hand, $P(E_0) = P(E - F)$; $P(E_1) = P(E \cap F)$; $p(E_0) = 0, \forall n \geq 2$. Thus, $P(E) = P(E - F) + P(E \cap F)$.
3. $P(E \cup F - F) = P(E \cup F) - P((E \cup F) \cap F) = P(E \cup F) - P(F)$. In addition, $P(E \cup F - F) = P(E - F) = P(E) - P(E \cap F)$. Thus,

$$P(E \cup F) - P(F) = P(E) - P(E \cap F).$$

(continued)

4. $P(E^c) = P(\Omega - E) = P(\Omega) - P(E) = 1 - P(E)$.
5. $P(F) = P(F - E) + P(E \cap F) = P(F - E) + P(E) \geq p(E)$.
6. Since $E \subset \Omega$: $0 \leq P(E) \leq P(\Omega) = 1$.
7. Consider $A_0 = E_0, A_n = E_n - \bigcup_{i=0}^{n-1} A_i$, *for* $n > 0$. We have $A_i \cap A_j = \varnothing$ for $i \neq j$. Consequently, $P\left(\bigcup_{n\in\mathbb{N}} A_n\right) = \sum_{n\in\mathbb{N}} P(A_n)$. In addition, on the one hand, $A_n \subset E_n \Rightarrow P(A_n) \leq P(E_n)$ and, on the other hand, $\bigcup_{n\in\mathbb{N}} A_n = \bigcup_{n\in\mathbb{N}} E_n$, hence the result.
8. We have $P\left(\bigcup_{i=n}^{+\infty} E_i\right) \leq \sum_{i=n}^{+\infty} P(E_i) \longrightarrow 0$, since the series is convergent.
9. Let $F_0 = E_0; F_n = E_n - E_{n-1}, n \geq 1$. Then, $P(F_n) = P(E_n) - P(E_{n-1} \cap E_n) = P(E_n) - P(E_{n-1})$. In addition, on the one hand, $F_i \cap F_j = \varnothing$, if $i \neq j$ and, on the other hand, $\bigcup_{n=0}^{+\infty} F_n = \bigcup_{n=0}^{+\infty} E_n$. Thus, $P\left(\bigcup_{n=0}^{+\infty} E_n\right) = P(E_0) + \sum_{n=0}^{+\infty} \times (P(E_n) - P(E_{n-1})) = \lim_{n\to+\infty} P(E_n)$.
10. Apply the preceding result to $\{E_n^c : n \in \mathbb{N}\} \subset \mathcal{P}(\Omega)$ be such that $E_n^c \subset E_{n+1}^c$: $P\left(\bigcup_{n=0}^{+\infty} E_n^c\right) = \lim_{n\to+\infty} P(E_n^c)$. Thus, $P\left(\bigcap_{n=0}^{+\infty} E_n\right) = 1 - \lim_{n\to+\infty} P(E_n^c)$, hence the result.

There are two events that deserve particular attention:

Definition 2.2 Let (Ω, P) be a probability space

1. $E \subset \Omega$ is ***almost sure*** if and only if $P(E) = 1$;
2. $F \subset \Omega$ is ***negligible*** if and only if $P(F) = 0$.

Notice that "almost sure" is not "certain" and "negligible" is not "impossible" (but may be interpreted as "almost impossible").

Example 2.1 Let us consider $\Omega = (0, 1)$ and the probability $P([a, b]) = b - a$. Let $F = \{x \in \Omega, x(1 - x) = 0\}$: We have $F = \{0, 1\}$, so that $P(F) = P([0, 0]) +$

(continued)

Example 2.1 (continued)

$P([1,1]) = (0 - 0) + (1 - 1) = 0$. Thus, the event F is almost impossible – but it ***is not*** impossible. In fact, the events $E = \{x \in \Omega : x \in \mathbb{I} = \mathbb{R} - \mathbb{Q}\}; F = \{x \in \Omega : x \in \mathbb{Q}\}$ are complementary: $E = F^c$. Since the Lebesgue measure of the set of rational numbers is zero, F is negligible and E is almost sure, but any subinterval will contain both rational and irrational numbers.

We have:

Proposition 2.2 Let (Ω, P) be a probability space. Let $E \subset \Omega$ be almost sure and $F \subset \Omega$ be negligible. Then, for any $A \subset \Omega$, $P(A \cap E) = P(A)$; $P(A \cup F) = P(A)$, $P(A \cup E) = 1$; $P(A \cap F) = 0$.

Proof We have

1. $E \subset A \cup E \Rightarrow P(A \cup E) \geq P(E) = 1$;
2. $A \cap F \subset F \Rightarrow 0 \leq P(A \cap F) \leq P(F) = 0 \Rightarrow P(A \cap F) = 0$;
3. $P(A \cup F) = P(A) + P(F) - P(A \cap F) = P(A) + 0 + 0 = P(A)$;
4. $(A \cap E)^c = A^c \cup E^c$ and $P(E^c) = 1 - P(E) = 0$. Thus, $P(A^c \cup E^c) = P(A^c) \Rightarrow P((A \cap E)^c) = P(A^c) \Rightarrow P(A \cap E) = P(A)$.

Proposition 2.3 A finite or enumerable Reunion of negligible events is negligible; a finite or enumerable intersection of almost sure events is almost sure.

Proof Let $\{F_n : n \in N\}$ be a family of negligible events. Proposition 2.1 (item 7) shows that $P\left(\bigcup_{n \in \mathbb{N}} F_n\right) \leq 0$, so that $\bigcup_{n \in \mathbb{N}} F_n$ is negligible. Let $A_n = \bigcup_{i=0}^{n} F_i$: Since $A_n \subset \bigcup_{n \in \mathbb{N}} F_n$, we have $P(A_n) \leq P\left(\bigcup_{n \in \mathbb{N}} F_n\right) = 0, \forall n \in \mathbb{N}$.

(continued)

Let $\{E_n: n \in N\}$ be a family of almost sure events and $\{F_n = E_n^c: n \in \mathbb{N}\}$. Then, $\left(\bigcap_{n\in\mathbb{N}} E_n\right)^c = \bigcup_{n\in\mathbb{N}} E_n^c = \bigcup_{n\in\mathbb{N}} F_n$ is negligible, so that $\bigcap_{n\in\mathbb{N}} E_n$ is almost sure. In addition, $\bigcap_{n\in\mathbb{N}} E_n \subset \bigcap_{i=0}^{n} E_i \Rightarrow 1 = P\left(\bigcap_{n\in\mathbb{N}} E_n\right) \le P\left(\bigcap_{i=0}^{n} E_i\right)$, so that $\bigcap_{n\in\mathbb{N}} E_n$ is almost sure.

2.2.1 *Mass Functions and Mass Densities*

Mass functions can be used to define probabilities on finite or enumerable universes, while **mass densities** are used on subsets of real numbers:

Definition 2.3

1. Let Ω be finite or enumerable. A **mass function** μ on Ω is an application $\mu: \Omega \longrightarrow \mathbb{R}$ such that $\forall \omega \in \Omega: \mu(\omega) = P(\{\omega\})$. We have $\mu(\omega) \ge 0$ on Ω and $\sum_{\omega \in \Omega} \mu(\omega) = 1$. If useful, μ may be extended by zero for $\omega \notin \Omega$.
2. Let $\Omega \subset \mathbb{R}$. A **mass density** μ on Ω is an application $\mu: \Omega \longrightarrow \mathbb{R}$ such that $\forall (a, b) \subset \Omega: P((a, b)) = \int_a^b \mu(\omega) d\omega$. We have $\mu(\omega) \ge 0$ on Ω and $\int_\Omega \mu(\omega) d\omega = 1$. If useful, μ may be extended by zero for $\omega \notin \Omega$ in both the cases.

R offers intrinsic functions that allows us to manipulate mass functions and mass densities. For instance, `sum` and `integrate` (see examples)

Example 2.2 An usual situation concerns $\Omega = (\alpha, \beta)$ and the mass density

$$\mu(\omega) = \frac{1}{\beta - \alpha}, \alpha < \omega < \beta.$$

In this case, the probability of any subinterval reads as

$$P((a, b)) = \int_a^b \mu(\omega) d\omega = \frac{b - a}{\beta - \alpha}, \alpha < a < b < \beta.$$

(continued)

Example 2.2 (continued)

Consider the code below (notice the capital "V" in "Vectorize"):

```
alfa <<- -1  # give a global value for alpha
betta <<- 1  # give a global value for beta
mu <- function(omega){1/(betta-alfa)}
a <- 0
b <- 1/2
p1 = integrate(Vectorize(mu),a,b)
p1$value
library(pracma)
p2 = quad(mu,a,b)
p2
```

Running the code produces the results below:

```
> p1$value        > p2
[1] 0.25          [1] 0.25
```

Example 2.3 Consider $\Omega = \{\omega_1, \ldots, \omega_n\}$ and the mass function

$$\mu(\omega_i) = p_i \geq 0, \sum_{i=1}^{n} p_i = 1.$$

Then, $P(\{\omega_{i_1}, \ldots, \omega_{i_k}\}) = \sum_{j=1}^{k} \mu(\omega_{i_j})$. For instance, consider an urn containing balls: The quantities are $(5, 3, 2, 4, 6)$, corresponding to colors (R, B, G, Y, M). Their probabilities are

$$\mu(R) = \frac{5}{20}, \mu(B) = \frac{3}{20}, \mu(G) = \frac{2}{20}, \mu(Y) = \frac{4}{20}, \mu(M) = \frac{6}{20}.$$

Thus,

$$P(\{R, G, B\}) = \mu(R) + \mu(G) + \mu(B) = \frac{1}{2}.$$

(continued)

Example 2.3 (continued)

Consider the code below:

```
Omega <- c("R","B","G","Y","M")
qtty <- c(5,3,2,4,6)
event = c("R","G","B")
d = data.frame(row.names = Omega, mu = qtty/sum(qtty))
p = 0
for (i in seq(1,length(event)) )
{
  p = p + d[[event[i],"mu"]]
}
p
```

Running the code produces the result at left.

```
> p
[1] 0.5
```

Example 2.4 Consider $\Omega = \{\omega_1, \ldots, \omega_n\}$ and the mass function

$$\mu(\omega_i) = \alpha \ln(1+i) + \beta.$$

Let us determine the possible values of the parameters α, β. Indeed, we have

$$\mu(\omega_i) \geq 0 \Leftrightarrow \alpha \ln(1+i) + \beta \geq 0.$$

This inequality is satisfied for $i = 1, \ldots n$ if and only if

$$\beta \geq \max\{-\alpha \ln 2, \ -\alpha \ln(1+n)\}$$

In addition,

$$\sum_{i=1}^{n} \mu(\omega_i) = 1 \Leftrightarrow \alpha S + n\beta = 1, S = \sum_{i=1}^{n} \ln(1+i).$$

(continued)

Example 2.4 (continued)

We can visualize the possible values using the code below:

```
n <- 4
s <- sum(log(2+seq(1,n)))
alfa = seq(-3,3, length.out = 180)
bettasol = (1 - alfa*s)/n
test = (bettasol >= -alfa*log(2)) & (bettasol >= -alfa*log(1+n))
ns = as.character(n)
ns2 = toString(n)
ns3 = bquote(n == .(n))
graphics.off()
str2 = substitute(paste("possible values of ( ",alpha,", ",beta,"), for n = ",v),list(v=n))
plot(alfa, bettasol, type="n", xlab=expression(alpha),
     ylab=expression(beta),main = str2);
lines(alfa[test <= 0], bettasol[test <= 0], col = "red", lty = 2, lwd = 3)
lines(alfa[test > 0], bettasol[test > 0], col = "blue", lwd = 5)
legend("topright", legend=c("IMPOSSIBLE", "POSSIBLE"),
       col=c("red", "blue"), lty=2:1, lwd = 3:5, cex=1)
```

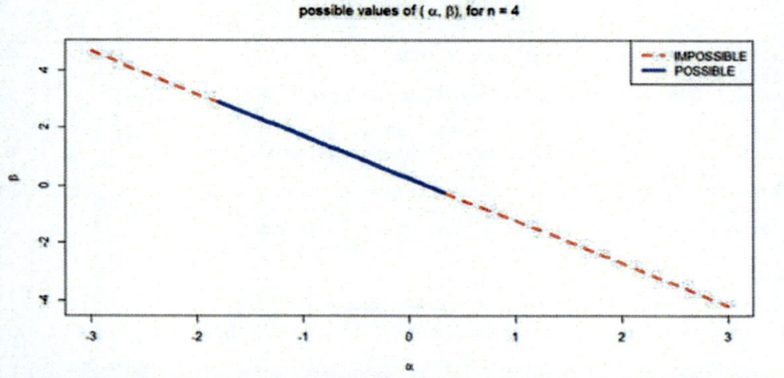

Example 2.5 Consider $\Omega = (0, 1)$ and the mass density

$$\mu(\omega) = \alpha \ln(2 + \omega) + \beta.$$

Let us determine the possible values of the parameters α, β. Indeed, we have

$$\mu(\omega) \geq 0 \Leftrightarrow \alpha \ln(2 + \omega) + \beta \geq 0.$$

(continued)

Example 2.5 (continued)

This inequality is satisfied for $\omega \in (0, 1)$ if and only if

$$\beta \geq \max\{-\alpha \ln 2, \ -\alpha \ln 3\}$$

In addition,

$$\int_0^1 \mu(\omega) d\omega = 1 \Leftrightarrow \alpha S + \beta = 1, S = \log \frac{27}{4} - 1 \approx 0.90954.$$

We can visualize the possible values using the code

```
alfa = seq(-10,10, length.out = 400)
bettasol = 1 - (log(27/4) -1)*alfa
test = (bettasol >= -alfa*log(3)) & (bettasol >= -alfa*log(2))
graphics.off()
str1 = expression(paste("possible values of ( ",alpha,", ",beta,")"))
plot(alfa, bettasol, type="n", xlab=expression(alpha),
     ylab=expression(beta),main = str1);
lines(alfa[test <= 0], bettasol[test <= 0], col = "red", lty = 2, lwd = 3
lines(alfa[test > 0], bettasol[test > 0], col = "blue", lwd = 5)
legend("topright", legend=c("IMPOSSIBLE", "POSSIBLE"),
       col=c("red", "blue"), lty=2:1, lwd = 3:5, cex=1)
```

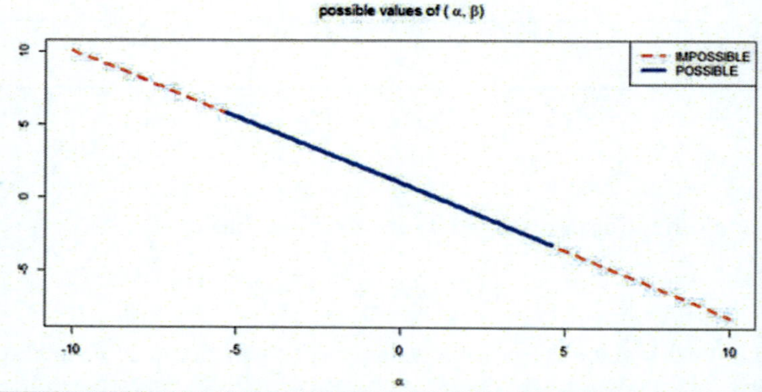

Exercises

1. Consider $\Omega = (0, 1)$ and the mass density $\mu(\omega) = \beta e^{-\alpha\omega}$. What are the possible values of α and β? Plot the admissible region.
2. Consider $\Omega = \{\omega_1, \omega_2, \omega_3\}$ and the mass function $\mu(\omega_i) = \alpha i^2 + \beta$. What are the possible values of α and β? Plot the admissible region.
3. Consider $\Omega = (0, 1)$ and the mass density $\mu(\omega) = \alpha\omega + \beta$. What are the possible values of α and β? Plot the admissible region.
4. Consider $\Omega = \{\omega_1, \omega_2, \omega_3\}$ and the mass function $\mu(\omega_i) = \alpha i + \beta$. What are the possible values of α and β? Plot the admissible region
5. Consider $\Omega = (0, 1)$ and the probability $P((a, b)) = b - a$.(mass density $\mu(\omega) = 1$). Find the probabilities of the following events

   a) $A = \{x \in \Omega : x < \frac{1}{3}\}$;
   b) $B = \{x \in \Omega : x^2 < \frac{1}{2}\}$;
   c) $C = \{x \in \Omega : x^2 > \frac{1}{4}\}$;
   d) $D = \{x \in \Omega : 9x^2 - 5x + 1 \leq 0\}$.
   e) Write *programs* in R to find these probabilities.

6. Consider $\Omega = \{\omega_1, \omega_2, \omega_3\}$ and $\mu(\omega_i) = \frac{i}{6}$.

   a) Find $P(\{\omega_2, \omega_3\})$;
   b) Find $P(\{\omega_1, \omega_3\})$.
   c) Write programs in R to solve these questions.

7. Consider $\Omega = \{\omega_1, \omega_2, \omega_3, \omega_4\}$ and the events $A = \{\omega_1\}$, $B = \{\omega_1, \omega_2\}$, $C = \{\omega_1, \omega_2, \omega_4\}$. Let $P(A) = \frac{1}{4}$, $P(B) = \frac{9}{20}$, $P(C) = \frac{7}{15}$.

   a) Find $P(\{\omega_2\})$, $P(\{\omega_3\})$, $P(\{\omega_4\})$;.
   b) Find $P(\{\omega_1, \omega_2, \omega_3\})$.

8. Consider $\Omega = \{\omega_1, \omega_2, \omega_3\}$ and the events $A = \{\omega_1\}$, $B = \{\omega_1, \omega_2\}$. Let $P(A) = \frac{1}{3}$ and $P(B) = \frac{1}{2}$.

   a) Find $P(\{\omega_2\})$;
   b) Find $P(\{\omega_3\})$
   c) Find $P(\{\omega_1, \omega_3\})$.

2.2.2 *Combinatorial Probabilities*

Combinatorial probabilities arise when considering finite universes with a constant mass function, id est, when Ω is finite and all its elements are equiprobable, id est,$\Omega = \{\omega_1, \ldots, \omega_n\}$ and $p_i = \mu(\omega_i) = P(\{\omega_i\}) = \frac{1}{n}$. Then, probabilities of the events are determined by counting their number of elements: for instance, $P(E = \{\omega_{i_1}, \ldots, \omega_{i_k}\}) = \frac{k}{n}$. Many classical situations involving coins, balls, cards, dices, etc., fall in this category. These problems are mainly combinatorial; id est, they use combinatorics to evaluate the number of elements of E and Ω. R proposes some functions to determine these numbers. For instance, `combn(x,n)` generates all the combinations of n elements from a vector x. In addition, the community offers packages able to deal with combinatorial problems. For instance, the package `arrangements` furnishes functions that evaluate permutations and combinations. For instance,

- `npermutations(n,k)`: number of permutations of k elements among n, without repetition. Returns $\frac{n!}{(n-k)!}$
- `ncombinations(n,k)`: number of combinations of k elements among n, without repetition. Returns $\frac{n!}{k!(n-k)!}$
- `npermutations(n,k,replace = TRUE)`: number of permutations of k elements among n, with repetition. Returns n^k
- `ncombinations(n,k, replace = TRUE)`: number of combinations of k elements among n, with repetition. Returns $\frac{(n+k-1)!}{k!(n-1)!}$

In addition, the function `factorial(n)` returns $n!$.

Example 2.6 Two cards are simultaneously draw at random from a 52-card deck. What is the probability that are number cards? To evaluate the probability, we start by evaluating the global number of possibilities n: There are 52 possibilities for the first card and 51 for the second one, so that $n = 52 \times 51 = npermutations(52, 2)$. The deck has 40 number cards, so that there are 40 possibilities for the first face and 39 for the second: We have $k = 40 \times 39 = npermutations(40, 2)$.

The code for the evaluations is shown at right. The result is

```
> p
[1] 0.5882353
```

```
library(arrangements)
n = npermutations(52,2)
k = npermutations(40,2)
p = k/n
```

Now, assume that you take a card at random, put it back, reshuffle, and take another card at random. In this situation, repetition is possible: On the one hand, $n = 52 \times 52 = npermutations(52, 2, replace = TRUE)$, on the other hand, $k = 12 \times 12 = npermutations(40, 2, replace = true)$.

The code for the evaluations is shown at right. The result is

```
> p
[1] 0.591716
```

```
n = npermutations(52,2, replace=TRUE)
k = npermutations(40,2, replace = TRUE)
p = k/n
```

Example 2.7 In a 52-card deck of playing cards, you draw simultaneously two at random. What is the probability that you get one ace? Here, it is convenient to consider the complementary event: No ace is drawn. For this event, $k = 48 \times 47 = ncombinations(48; 2)$. Thus, its probability is 0.8506787. So, the probability of getting one ace is 0.1493213.

```
n = npermutations(52,2)
k = npermutations(48,2)
p = 1 - k/n
```

```
> p
[1] 0.1493213
```

With repetition, these probabilities are, respectively, `0.852071` and `0,147,929.`

```
n = npermutations(52,2, replace=TRUE)
k = npermutations(48,2, replace = TRUE)
p = 1 - k/n
```

```
> p
[1] 0.147929
```

Exercises

1. In a 52-card deck of playing cards, you draw simultaneously 5 at random. What is the probability of getting
   a) 5 hearts;
   b) A full (5 cards of the same color);
   c) Four aces;
   d) 4 cards of same numbers (or faces);
   e) Evaluate these probabilities using R.
2. In a 52-card deck of playing cards, you draw sequentially 5 at random. At each draw, the card is put back in the deck, which is reshuffled. What is the probability of getting
   a) 5 hearts;
   b) 5 cards of the same color;
   c) Exactly four aces;
   d) At least 4 aces;
   e) Evaluate these probabilities using R.
3. We roll three dice numbered from 1 to 6. All the faces are equiprobable. What is the probability of getting
   a) Three aces
   b) A sum of points equal to 9;
   c) A sum equal to 10;
   d) Evaluate these probabilities using R.

4. An urn contains 20 numbered balls of three colors: 6 red, 12 green, 2 blue. We draw at random 2 balls simultaneously. What is the probability of getting
   a) Two green balls;
   b) Two balls of same color; two balls of different colors
   c) Evaluate these probabilities using R.

2.3 Independent Events

Definition 2.4 The probability of an event E conditionally to an event F such that $P(F) > 0$ is

$$P(E \mid F) = \frac{P(E \cap F)}{P(F)}.$$

The reader can find in the literature an extension of the definition of conditional probability to include negligible events – see, for instance, Coletti & Scozzafava, (2000). We have

Proposition 2.4 (*Bayes' Formulae*) Let (Ω, P) be a probability space, $E, F \subset \Omega$.

1. If $P(E)P(F) > 0$ then $P(E \cap F) = P(F) \times P(E \mid F) = P(E) \times P(F \mid E)$
2. If $P(E)P(F) > 0$ then

$$P(F \mid E) = \frac{P(F) \times P(E \mid F)}{P(E)}$$

3. Let $\{E_1, \ldots, E_n\} \subset \mathcal{P}(\Omega)$ verify $P\left(\bigcap_{i=1}^{n} E_i\right) > 0$. Then $P\left(\bigcap_{i=1}^{n} E_i\right) = P(E_1) \prod_{i=2}^{n} \left(E_i \middle| \bigcap_{j=1}^{i-1} E_j\right)$, id Est,

$$P\left(\bigcap_{i=1}^{n} E_i\right) = P(E_1)P(E_2|E_1)P(E_3|E_1 \cap E_2)\ldots P(E_n|E_1 \cap \ldots \cap E_{n-1})$$

(continued)

Proposition 2.4 (continued)

4. Let $\Omega = \bigcup_{i=1}^{k} E_i$, $P(E_i) > 0, \forall\, i;\ E_i \cap E_j = \varnothing$, *if* $i \neq j$. Then

$$P(F) = \sum_{i=1}^{k} P(E_i)\, P(F \mid E_i);$$

$$P(E_j \mid F) = \frac{P(E_j)\, P(F \mid E_j)}{P(F)} = \frac{P(E_j)\, P(F \mid E_j)}{\sum_{i=1}^{k} P(E_i)\, P(F \mid E_i)}.$$

Proof

1. We have $P(E|F) = P(E \cap F)/P(F)$ and $P(F|E) = P(E \cap F)/P(E)$, hence the result.
2. $P(F) \times P(E|F) = P(E) \times P(F|E)$, hence the result.
3. The proof is made by recurrence on the assumption H_n: $P\left(\bigcup_{i=1}^{k} E_i, \bigcap_{i=1}^{n} E_i\right) = P(E_1) \prod_{i=2}^{n} P\left(E_i \middle| \bigcap_{j=1}^{i-1} E_j\right)$. On the one hand, $P(E_1 \cap E_2) = P(E_1)P(E_2|\ E_1)$ and H_2 is verified. On the other hand, for $n \geq 2$, $H_n \Rightarrow H_{n+1}$. Indeed, $P\left(\bigcap_{j=1}^{n+1} E_j\right) = P\left(E_{n+1} \cap \left(\bigcap_{i=1}^{n} E_i\right)\right) = P\left(\bigcap_{i=1}^{n} E_i\right) P\left(E_{n+1} \middle| \bigcap_{i=1}^{n} E_i\right)$. Thus, using H_n, we obtain H_{n+1}.
4. Since $F = F \cap \Omega = \mathrm{F} \cap \left(\bigcup_{i=1}^{k} E_i\right) = \bigcup_{i=1}^{k} (F \cap E_i)$, we have $P(F) = \sum_{i=1}^{k} P(F \cap E_i) = \sum_{i=1}^{k} P(E_i)\, P(F \mid E_i)$. But $P(E_j|\ F) = P(E_j \cap F)/P(F)$, so that $P(E_j|\ F) = P(E_j)\, P(F \mid E_j)/P(F)$, hence the result.

Definition 2.5 E and F are independent if and only if

$$P(E \cap F) = P(E) \times P(F).$$

Independence means that the events do not affect each other:

Proposition 2.5 Let (Ω, P) be a probability space, E, $F \subset \Omega$, E and F independent. Then:

1. If $P(E)\,P(F) > 0$, then $P(E|\,F) = P(E)$ *and* $P(F|\,E) = P(F)$;
2. E^c and F^c are independent;
3. Let $\{E_n: n \in \mathbb{N}\}$be a family such that E_i, E_j are independent if $i \neq j$. Then $P\left(\bigcap_{n=0}^{+\infty} E_n\right) = \prod_{n=0}^{+\infty} P(E_n)$;
4. Let $\{E_n: n \in \mathbb{N}\}$ be a family such that E_i, E_j are independent if $i \neq j$. If $\sum_{n\in\mathbb{N}} P(E_n) = +\infty$, then $P\left(\bigcup_{i=n}^{+\infty} E_i\right) = 1, \forall n \in \mathbb{N}$.
5. Let $\{E_n: n \in \mathbb{N}\}$ be a family such that E_i, E_j are independent if $i \neq j$. If $\sum_{n\in\mathbb{N}} P(E_n) < +\infty$, then $P\left(\bigcup_{i=n}^{+\infty} E_i\right) \longrightarrow 0$, for $n \longrightarrow +\infty$.

Proof $P(E \cap F) = P(F) \times P(E|\,F) = P(E) \times P(F|\,E) = P(E) \times P(F)$.

Thus, $P(F) \times P(E|\,F) = P(E) \times P(F)$ and $P(E) \times P(F|\,E) = P(E) \times P(F)$, hence the first result.

$P(E^c \cap F^c) = P(E^c) + P(F^c) - P(E^c \cup F^c)$. In addition, $P(E^c \cup F^c) = P((E \cap F)^c) = 1 - P(E \cap F) = 1 - P(E)P(F)$. Thus, $P(E^c \cup F^c) = 1 - (1 - P(E^c))(1 - P(F^c)) = P(E^c) + P(F^c) - P(E^c)P(F^c)$. Hence the second result, since $(E^c \cap F^c) = P(E^c)P(F^c)$.

The third result is obtained by taking $A_n = \bigcap_{i=0}^{n} E_i$. Then, $A_{n+1} \subset A_n$ and $P\left(\bigcap_{n=0}^{+\infty} E_n\right) = P\left(\bigcap_{n=0}^{+\infty} A_n\right) = \lim_{n\to+\infty} P(A_n) = \lim_{n\to+\infty} \prod_{i=0}^{n} P(E_i)$, hence the result.

Consider $F_n = \bigcup_{i=n}^{+\infty} E_i$. We have $P(F_n) \leq \sum_{i=n}^{+\infty} P(E_i)$. Since the series converges, we have $\sum_{i=n}^{+\infty} P(E_i) \longrightarrow 0$ for $n \longrightarrow +\infty$, hence the last result.

Example 2.8 A fair coin has head and tail results equiprobable. It is tossed twice and the result is written. We have the information that one of the results is head. What is the probability that the other is tail? Let us denote H a head and T a tail. Here, $\Omega = \{HH, HT, TH, TT\}$. Let us consider the events

(continued)

Example 2.8 (continued)

$E = \{HH, HT, TH\}$ (one of them is a head) and the event $F = \{HT, TH, TT\}$ (one of them is a tail). We have $E \cap F = \{HT, TH\}$ and

$$P(F|E) = \frac{P(E \cap F)}{P(E)} = \frac{\frac{2}{4}}{\frac{3}{4}} = \frac{2}{3}.$$

Example 2.9 A sample contains 50% of green pins. Globally, 20% of the pins are damaged, but there is a difference between the colors: 30% of the green pins are damaged. What is the probability that a damaged pin is green? Let E_1 = *the pin is green*; E_2 = *the pin is not green*. We have $P(E_1) = 0, 5$, $P(E_2) = 0, 5$. Let F = *pin is damaged*. We have $P(F|\, E_1) = 0, 3$, $P(A) = 0, 2$. Then

$$P(E_1|F) = \frac{P(E_1) \times P(F|E_1)}{P(F)} = \frac{0, 5 \times 0, 3}{0, 2} = 0, 75.$$

Exercises

1. Overall, it is estimated that 1% of the parts manufactured in a factory are defective. The quality control makes mistakes by accepting 1% of non-compliant parts and rejecting 2% of compliant parts. What is the probability that an accepted part is compliant?
2. Two boxes contain the same number of pins, divided between two colors: Red and green. In box 1, 30% are red and 70% are green; in box 2, 40% are red pins and 60% are green. The boxes are equiprobable.
   a) We take a red pin: What is the probability that it came from box 1?
   b) We take a green pin: What is the probability that it came from box 2?
   c) How are these values modified if box 1 is much larger than box 2, so that its probability is $\frac{3}{4}$?
3. An urn contains 20 numbered balls of three colors: 1 to 6 in red, 7 to 18 in green, 19 to 20 in blue. We draw a ball at random.
   a) The number is multiple of 3: What is the probability that the color is green?
   b) The number has two digits: What is the probability that the color is blue?
   c) The number is pair: What is the probability that the color is red?

(continued)

4. Three urns contain 10 numbered balls in red and blue each, but the number of red balls is different in each one: 5 in urn 1, 4 in urn 2, 3 in urn 3. The urns are equiprobable. A red ball is drawn.

   a) What is the probability that it came from urn 1?
   b) What is the probability that it did not come from urn 2?
   c) What is the probability that it came from urn 3??

5. Determine if the events A and B are independent or not, when

   a) $P(A)=\frac{1}{3}, P(B)=\frac{2}{3}, P(A\cap B)=\frac{1}{2}$.
   b) $P(A)=\frac{1}{2}, P(B)=\frac{2}{3}, P(A\cap B)=\frac{1}{3}$.
   c) $P(A)=\frac{3}{5}, P(B)=\frac{5}{6}, P(A\cap B)=\frac{1}{2}$.
   d) $P(A)=\frac{3}{5}, P(B)=\frac{5}{6}, P(A\cap B)=\frac{2}{5}$.

2.4 Numerical Variables on Finite Populations

Let us consider a finite universe $\Omega = \{\omega_1, \ldots, \omega_{np}\}$ – Ω is also referred as ***population***. np is the ***size*** of the population. A member of the population $\omega_i \in \Omega$ is an ***individual***: the population contains np individuals. An event $E=\{\omega_{i_1}, \ldots, \omega_{i_k}\} \subset \Omega$ is a ***subpopulation*** of Ω.

Let X: $\Omega \longrightarrow \mathbb{R}$ be a numerical variable defined on Ω. Since Ω is finite, the image of X is a finite set of m distinct values: $X(\Omega) = \{X_1, \ldots, X_m\}$, $X_i \neq X_j$ for $i \neq j$. We have $0 < m \leq np$. These values can be ranged in a strictly crescent order: in the sequel, we assume that $X_i < X_j$ for $i < j$ without loss of generality.

Let us consider the subpopulation $H_i = X^{-1}(\{X_i\}) = \{\omega \in \Omega\colon X(\omega) = X_i\}$. Let μ be the mass function on Ω. The *probability* of the value X_i is

$$P(X=X_i)=p_i=\sum_{\omega\in H_i}\mu(\omega). \tag{2.6}$$

Assume that X is defined for any $\omega \in \Omega$ (id est, $\Omega \subset$ *dom* (X)). Then, it is immediate that

$$\bigcup_{i=1}^{m} H_i=\Omega;\ \sum_{i=1}^{m} p_i=1; 0\leq p_i\leq 1,\ \forall i. \tag{2.7}$$

More generally, for $I \subset \mathbb{R}$,

$$P(X \in I)=P\left(X^{-1}(I)\right)=\sum_{\omega\in X^{-1}(I)}\mu(\omega). \tag{2.8}$$

Table 2.1 Frequency table of *X*

Value	X_1	...	X_m
Frequency	p_1	...	p_m

Remark 2.1 In many practical situations, the individuals are equiprobable, so that $\mu(\omega) = 1/np, \forall\, \omega \in \Omega$. then, $p_i = \frac{c_i}{np}$, where $c_i =$ *card* (H_i) is the number of elements belonging to H_i. We have

$$\sum_{i=1}^{m} c_i = np;\ 0 \le c_i \le np,\ \forall i.$$

The information about *X* can be summarized in a frequency Table 2.1.

Remark 2.2 In all generality, it is necessary to consider situations where some elements of $X(\Omega)$ may be negligible, i.e., a value p_i can be null: Such a generality leads to the use of the expressions "almost sure" and "almost surely" in most of the results. Of course, the values corresponding to null probabilities can be eliminated from the possible values of *X*, so that $p_i > 0, \forall\, i$—then the expressions "almost sure" and "almost surely" can be eliminated from the results below.

Definition 2.6 Let f: $\mathbb{R} \longrightarrow \mathbb{R}$ be a function. The ***mean*** or ***expectation*** of $f(X)$ is

$$E(f(X)) = \sum_{i=1}^{np} \mu(\omega_i) f(X(\omega_i)).$$

We have

Proposition 2.6 Let X: $\Omega \longrightarrow \mathbb{R}$ be a numerical variable and f, g: $\mathbb{R} \longrightarrow \mathbb{R}$ be two functions. Then

1. $E(\alpha f(X) + \beta g(X)) = \alpha E(f(X)) + \beta E(g(X))$;
2. $E(f(X)) = \sum_{i=1}^{m} p_i f(X_i)$;
3. If $f(X) \ge 0$, then $E(f(X)) \ge 0$;

(continued)

Proposition 2.6 (continued)

4. If $f(X) \geq 0$, then $\lambda P(f(X) \geq \lambda) \leq E(f(X))$, $\forall \lambda \in \mathbb{R}$ (Markov's inequality);
5. Let $1_A(\omega) = 1$, *if* $\omega \in A$; $1_A(\omega) = 0$, *if* $\omega \notin A$. Then $E(1_A) = P(A)$;
6. If $f: \mathbb{R} \longrightarrow \mathbb{R}$ is convex, then $f(E(X)) \leq E(f(X))$ (Jensen's inequality).

Proof We have

$$\sum_{j=1}^{np} \mu(\omega_j)\left(\alpha f(X(\omega_j)) + \beta g(X(\omega_j))\right)$$
$$= \alpha \sum_{j=1}^{np} \mu(\omega_j) f(X(\omega_j)) + \beta \sum_{i=1}^{np} \mu(\omega_j) g(X(\omega_j)),$$

So that $E(\alpha f(X) + \beta g(X)) = \alpha E(f(X)) + \beta E(g(X))$.
Since $\bigcup_{i=1}^{m} H_i = \Omega$, we have

$$\sum_{j=1}^{np} \mu(\omega_j) f(X(\omega_j)) = \sum_{i=1}^{m} \left(\sum_{\omega \in H_i} \mu(\omega) \underbrace{f(X(\omega))}_{= f(X_i) \text{ on } H_i} \right)$$

So that

$$E(f(X)) = \sum_{i=1}^{m} f(X_i) \left(\sum_{\omega \in H_i} \mu(\omega) \right) = \sum_{i=1}^{m} p_i f(X_i).$$

If $f(X) \geq 0$, we have $f(X(\omega)) \geq 0$, $\forall \omega \in \Omega$. thus $E(f(X))$ is a sum of non-negative terms and we have $E(f(X)) \geq 0$.

Let $A = \{\omega \in \Omega: f(X(\omega) \geq \lambda\}$ and $B = \{\omega \in \Omega: f(X(\omega) < \lambda\}$. We have $A \cup B = \Omega$, $A \cap B = \emptyset$. Thus,

$$\sum_{j=1}^{np} \mu(\omega_j) f(X(\omega_j)) = \sum_{\omega \in A} \mu(\omega) f(X(\omega)) + \sum_{\omega \in B} \mu(\omega) f(X(\omega))$$

Since $f \geq 0$,

$$\sum_{j=1}^{np} \mu(\omega_j) f(X(\omega_j)) \geq \sum_{\omega \in A} \mu(\omega) f(X(\omega)) \geq \lambda \sum_{\omega \in A} \mu(\omega) = \lambda P(A),$$

(continued)

Hence the fourth assertion. In addition,

$$E(1_A) = \sum_{j=1}^{np} \mu(\omega_j) 1_A(\omega_j) = \sum_{\omega \in A} \mu(\omega) = P(A).$$

The last assertion results from the convexity condition:

$$f(E(X)) = f\left(\sum_{i=1}^{m} p_i X_i\right) \underbrace{\leq}_{\text{convexity}} \sum_{i=1}^{m} p_i f(X_i) = E(f(X)).$$

Definition 2.7
1. The **moment of order k of X** is $M_k(X) = E(X^k)$.
2. The **central moment of order k of X** is $CM_k(X) = E((X - E(X))^k)$.
3. The **variance of X** is $V(X) = CM_2(X) = E((X - E(X))^2)$.
4. The **standard deviation** of X is $(X) = \sqrt{V(X)}$.
5. A **median** of X is a number $med(X) = m$ such that $(X \leq m) \geq \frac{1}{2}$ *and* $P(X \geq m) \geq \frac{1}{2}$.

We have

Proposition 2.7 Let $X: \Omega \longrightarrow \mathbb{R}$ be a numerical variable. Then

1. $V(X) \geq 0$;
2. $V(X) = E(X^2) - E(X)^2$;
3. $V(X) = 0$ if and only if X is constant almost surely;
4. $E(X^2) = 0$, if and only if $X = 0$ almost surely.

Proof Since $f(X) = (X - E(X))^2 \geq 0$. The preceding proposition yields that $V(X) \geq 0$.

We have $E((X - E(X))^2) = E(X^2 - 2\,E(X)X + E(X)^2)$, so that $V(X) = E(X^2) - 2E(X)E(X) + E(X)^2 = E(X^2) - E(X)^2$.

(continued)

Let $V(X) = 0$. Then $\sum_{i=1}^{m} p_i(X_i - E(X))^2 = 0$. Thus, $X_i \neq E(X) \Rightarrow p_i = 0$. Consider $\Lambda = \{i\colon X_i \neq E(X), 1 \leq i \leq m\}$. Let $F = \bigcup_{i \in \Lambda} X^{-1}(\{X_i\})$. Since $P(X^{-1}(\{X_i\})) = p_i = 0$, F is a finite Reunion of negligible events consequently, $P(F) = 0$ and $F = \{\omega \in \Omega\colon X(\omega) \neq E(X)\}$ is negligible. Thus, $E = \{\omega \in \Omega\colon X(\omega) = E(X)\}$ is almost sure.

Analogously, if $E(X^2) = 0$, then $\sum_{i=1}^{m} p_i X_i^2 = 0$. Thus, $X_i \neq 0 \Rightarrow p_i = 0$. By the same way, we obtain that $E = \{\omega \in \Omega\colon X(\omega) = 0\}$ is almost sure.

Proposition 2.8 (Chebyshev's inequality) let $X\colon \Omega \longrightarrow \mathbb{R}$ be a numerical variable. Then

$$P(|X - E(X)| \geq \varepsilon) \leq \frac{V(X)}{\varepsilon^2}, \forall \varepsilon > 0.$$

Proof Let $f(X) = (X - E(X))^2 \geq 0$. Then, from proposition 2.6:

$$\varepsilon^2 P\big(f(X) \geq \varepsilon^2\big) \leq E\big(f(X)\big) = V(X).$$

Since $P(|X - E(X)| \geq \varepsilon) = P(f(X) \geq \varepsilon^2)$, we obtain the result.

Definition 2.8 The cumulative distribution function (CDF) of X is

$$F(x) = P(X \leq x) = P(X \in]-\infty, x]) = \sum_{\substack{i \text{ such that} \\ X_i \leq x}} p_i = \sum_{\substack{\omega \text{ such that} \\ X(\omega) \leq x}} \mu(\omega).$$

The CDF may be represented by Table 2.2.

Table 2.2 CDF of X

value	X_1	X_2	...	X_{m-1}	X_m
frequency	p_1	$p_1 + p_2$	...	$\sum_{i=1}^{m-1} p_i$	1

Proposition 2.9 Let $X: \Omega \longrightarrow \mathbb{R}$ be a numerical variable and $F: \mathbb{R} \longrightarrow \mathbb{R}$ be its CDF. Then

1. F is monotonically increasing: $x \leq y \Rightarrow F(x) \leq F(y)$;
2. $F(x) = 0$ if $x < X_1$;
3. $F(x) = \sum_{j=1}^{i-1} p_j$, if $X_{i-1} \leq x < X_i$, $1 < i \leq m$.
4. $F(x) = 1$ if $x \geq X_m$;
5. $P(a < X \leq b) = F(b) - F(a)$;
6. $P(X = a) = F(a+) - F(a-)$;
7. F is right continuous: $F(x+) = F(x), \forall x$.
8. m is median of X if and only if $F(m+) \geq \frac{1}{2}$, $F(m-) \leq \frac{1}{2}$.
9. Let F be continuous at m.Then, m is median of X if and only if $F(m) = \frac{1}{2}$.
10. F has a variational derivative: $F' = f = \sum_{i=1}^{m} p_i \delta_{X_i}$, where δ_{X_i} is the Dirac's mass supported by X_i. f is the probability DENSITY (PDF) of X.

Proof

1. If $x \leq y$ then $]-\infty, x] \subset]-\infty, y]$, hence the result
2. If $x < X_1$ then $X^{-1}(]-\infty, x]) = \emptyset$, hence the result.
3. If $X_{i-1} \leq x < X_i$, then $X^{-1}(]-\infty, x]) = \bigcup_{j=1}^{i-1} X^{-1}(\{X_j\})$, hence the result.
4. If $x \geq X_m$ then $X^{-1}(]-\infty, x]) = \Omega$ hence the result.
5. $X^{-1}(]a, b]) = X^{-1}(]-\infty, b]) - X^{-1}(]-\infty, a])$, hence the result.
6. Let $a = X_i \in X(\Omega)$. Then $p_i = P(X = X_i) = F(X_i+) - F(X_i-)$. If $a \notin X(\Omega)$, then $P(X = a) = 0 = F(a+) - F(a-)$.
7. Result 7 follows from results 2,3 and 4.
8. Notice that $F(m+) = F(m)$. Thus, we have $P(X \leq m) \geq 1/2$. Moreover, $P(X \geq m) = 1 - P(X \leq m) + P(X = m)$, so that $P(X \geq m) = 1 - F(m) + F(m+) - F(m-) = 1 - F(m-)$. Consequently, $P(X \geq m) \geq 1/2$ if and only if $F(m-) \leq 1/2$.
9. Result 9 follows from result 8.
10. It yields directly from the definition of variational derivative (see, for instance (Souza de Cursi, 2015)).

You can find R packages and intrinsic commands to create and deal with frequency tables. For instance, consider the dataset *chickwts*, containing the data about the effects of feed supplements on the evolution of the weight of chickens. Enter the commands:

```
> aux <- table(chickwts$feed)
> t1 = prop.table(aux)
> t1

   casein horsebean   linseed  meatmeal   soybean sunflower
0.1690141 0.1408451 0.1690141 0.1549296 0.1971831 0.1690141
> t2 = as.vector(t1)
> t2
[1] 0.1690141 0.1408451 0.1690141 0.1549296 0.1971831 0.1690141
> c2 = cumsum(t2)
> c2
[1] 0.1690141 0.3098592 0.4788732 0.6338028 0.8309859 1.0000000
```

Indeed, *chickwts* contains the measurements of six feed supplements (casein, horsebean, linseed, meat meal, soybean, and sunflower). There are 12, 10, 14, 11, 14, 12 measurements of each, respectively. The first command loads the data in a Table `aux` containing the numbers of occurrences of each feed type. The second one generates a Table `t1` containing the relative frequencies: they are equal to the number of measurements divided by the total number of observations (72). The table `t1` is transformed in a vector `t2` which can be manipulated as usual: for instance, `c2` contains the cumulative sums corresponding to Table 2.2.

Notice that R can determine the CDF of a sample of values directly, using the intrinsic function `ecdf`:

```
> F = ecdf(chickwts$feed)
> plot(F)
> F(1)
[1] 0.1690141
> F(3)
[1] 0.4788732
> F(5)
[1] 0.8309859
>
```

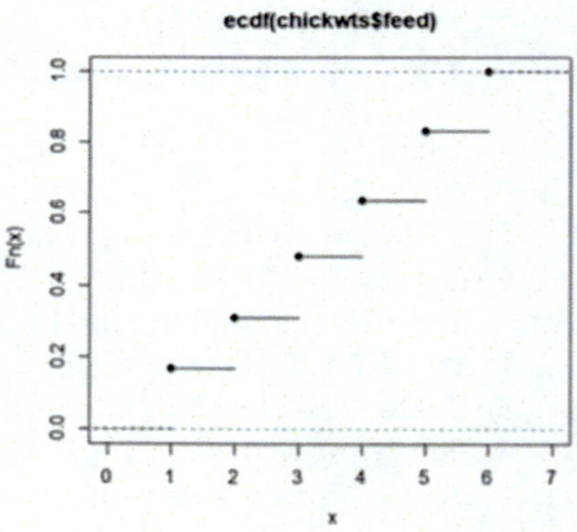

`ecdf` returns a function – in the example above, a function called `F`. Since the feed types (*casein, horsebean, linseed, meat meal, soybean, sunflower*) are not numeric but levels (id est, factors), they are transformed into integers 1, 2, 3, ... by `ecdf`. In this transformation, R uses the alphabetic order of the factors, so that 1 corresponds to casein, 2 to horsebean, 3 to linseed, and so on. If you desire to assign specific numbers to each feed type, you must make it manually (or using a package such as dplyr). For instance, if you want to use the order (*linseed, soybean, sunflower, horsebean, casein, meat meal*), you must start by the code below:

```
> cwtcopy <- chickwts
> ocopy = c("linseed", "soybean", "sunflower",
+           "horsebean", "casein", "meatmeal")
> for (i in seq(1,length(ocopy))){
+   levels(cwtcopy$feed)[levels(cwtcopy$feed) == ocopy[i]] <- i
+ }
> cwtvec1 <- as.vector(cwtcopy$feed)
> prop.table(table(cwtvec1))
cwtvec1
        1         2         3         4         5         6
0.1690141 0.1971831 0.1690141 0.1408451 0.1690141 0.1549296
> F1 <- ecdf(as.numeric(cwtvec1))
> F1(1)
[1] 0.1690141
> F1(3)
[1] 0.5352113
> F1(5)
[1] 0.8450704
```

Here, we create a copy of the data.frame *chickwts* named *cwtcopy* and we replace the levels as indicated. *Cwtopy* can be manipulated as any data.frame: we extract a vector *cwtvec1* containing the feed types, which are transformed in characters. Then, we create a frequency table. As an alternative, you can start by creating a vector of characters containing the levels and replace the values in the vector:

```
> cwtvec2 <- as.vector(chickwts$feed)
> for (i in seq(1,length(ocopy))){
+   cwtvec2[cwtvec2 == ocopy[[i]]] <- i
+ }
> prop.table(table(cwtvec2))
cwtvec2
        1         2         3         4         5         6
0.1690141 0.1971831 0.1690141 0.1408451 0.1690141 0.1549296
> F2 <- ecdf(as.numeric(cwtvec2))
> F2(1)
[1] 0.1690141
> F2(3)
[1] 0.5352113
> F2(5)
[1] 0.8450704
```

In this case, we create a vector *cwtvec2* containing the feed types, which are transformed in characters. Then, we replace them by the desired values. Notice that you can use gsub to make the replacements:

```
> cwtvec3 <- as.vector(chickwts$feed)
> for (i in seq(1,length(ocopy))){
+   cwtvec3 = gsub(ocopy[[i]],i,cwtvec3)
+ }
> prop.table(table(cwtvec3))
cwtvec3
        1         2         3         4         5         6
0.1690141 0.1971831 0.1690141 0.1408451 0.1690141 0.1549296
> F3 <- ecdf(as.numeric(cwtvec3))
> F3(1)
[1] 0.1690141
> F3(3)
[1] 0.5352113
> F3(5)
[1] 0.8450704
```

If your data are contained in a bidimensional table, you must start by converting it to a vector: the instruction `as.vector(t(data))` generates a vector containing the rows of a 2D table `data`. As an example, import the file *colors200.xlsx* to R. This file contains a sample of 200 variates from the colors {*red, green, blue, yellow, black*}, under the form of a 20 × 10 Table. Enter the commands below:

```
> colorvec <- as.vector(t(colors200))
> prop.table(table(colorvec))
colorvec
 black   blue  green    red yellow
 0.210  0.185  0.220  0.215  0.170
> ordcol = c("red", "green", "blue",
+            "yellow", "black")
> for (i in seq(1,length(ordcol))){
+   colorvec = gsub(ordcol[[i]],i,colorvec)
+ }
> prop.table(table(colorvec))
colorvec
    1     2     3     4     5
0.215 0.220 0.185 0.170 0.210
```

As you can observe, the original table `colors200` is transformed in a vector `colorvec`. If R is asked for the evaluation of the frequencies directly on the initial `colorvec`, the colors are ordered in alphabetic order. To keep the desired order, we must perform the replacements: then, the frequencies appear in the order *red, green, blue, yellow, and black* as desired.

If the data are numeric, you can determine statistics such as mean, variance, media, standard deviation by using intrinsic functions of R or by writing simple instructions. Let us illustrate both the methods using the sample data:

$$3, 1, 2, 4, 2, 5, 3, 2, 4, 1, 1, 5, 3, 2, 3, 2, 1, 2, 1, 5, 4, 3, 5, 4, 1, 2, 1$$

The frequency table corresponding to the data is

1	2	3	4	5
0.2592593	0.2592593	0.1851852	0.1481481	0.1481481

It can be generated using the class `fstats.R`, which manipulates frequency tables, defined as lists containing fields "X" (the values of X), "p" (p_i), "Xnames" (labels of the values of X), "cumul" (a numeric vector containing the values $P(X \leq X_i)$.) `$fromTable` creates such an object from a Table, `$mean` evaluates the mean of X, $var. evaluates its variance, `$med` determines its median, `$meanf` evaluates $E(f(X))$, `$mom` evaluates $E(X^k)$, `$centmom` evaluates $E((X - E(X))^k)$, `$cdf` generates a function $F(x) = P(X \leq x)$. Below, we generate a table `t1` as previously shown. Then, we generate a frequency table `ft1` and we evaluate the statistics:

```
source("ftstats.R")
fts = ftstats$new()
x = c(3,1,2,4,2,5,3,2,4,1,1,5,3,2,3,2,1,2,1,5,4,3,5,4,1,2,1)
t1 = prop.table(table(x))
ft1 = fts$fromTable(t1)
p_c = ft1$cumul
mX = fts$mean(ft1)
vX = fts$var(ft1)
sX = fts$sd(ft1)
medX = fts$med(ft1)
mc3 = fts$centmom(ft1,3)
m3 = fts$mom(ft1,3)
m4 = fts$mom(ft1,4)
mc4 = fts$centmom(ft1,4)
f4 = function(x){x^4}
m4b = fts$meanf(ft1,f4)
f3 = function(x){x^3}
m3b = fts$meanf(ft1,f3)
ff = fts$cdf(ft1)
ff15 = ff(1.5)
```

The results appear in the environment:

ff15	0.518518518518518
m3	35.3333333333333
m3b	35.3333333333333
m4	149.925925925926
m4b	149.925925925926
mc3	0.962962962962964
mc4	6.91358024691358
medX	2L
mX	2.66666666666667
p_c	num [1:5] 0.259 0.519 0.704 0.852 1
sX	1.38777733297742
vX	1.92592592592593

You can also evaluate the mean, moments, variance, and standard deviation as follows:

```
> X = as.numeric(names(ft))
> p = as.vector(ft)
> mX = sum(p*X)
> mX2 = sum(p*X^2)
> mX3 = sum(p*X^3)
> mX4 = sum(p*X^4)
> mcX3 = sum(p*(X-mX)^3)
> mcX4 = sum(p*(X-mX)^4)
> vX = mX2 - mX^2
> sX = sqrt(vX)
```

mcX3	0.962962962962964
mcX4	6.91358024691358
mX	2.66666666666667
mX2	9.03703703703704
mX3	35.3333333333333
mX4	149.925925925926
sX	1.38777733297742
vX	1.92592592592593

Exercises

1. Import the file *apples.Xlsx*. Generate a table corresponding to its contents. Modify the table by changing the names into numerical values:

"Pink"	"Gala"	"Canada"	"Delicious"	"Golden"	"Granny"
1	2	3	4	5	6

Generate a frequency table using these numerical values.
Find the mean, the variance, and the CDF of the numerical values.

2. Import the file *continents.Xlsx*. Generate a table corresponding to its contents. Modify the table by changing the names into numerical values:

"Antarctica"	"Europe"	"Oceania"	"Asia"	"America"	"Africa"
1	2	3	4	5	6

Generate a frequency table using these numerical values.
Find the mean, the standard deviation, and the CDF of the numerical values.

3. Import the file *fruits.Xlsx*. Generate a table corresponding to its contents. Modify the table by changing the names into numerical values:

"Strawberry"	"Banana"	"Kiwi"	"Pineapple"	"Orange"	"Apple"
1	2	3	4	5	6

Generate a frequency table using these numerical values.
Find the mean, the standard deviation, and the CDF of the numerical values.

4. When a fair coin is tossed n times, the number of heads X has a mean $\frac{n}{2}$ and a variance $\frac{n}{4}$.

   a) Determine an upper bound for the probability of obtaining a deviation of 10% of the mean. TIP: Consider $P\left(\left|X-\frac{n}{2}\right|>\alpha\frac{n}{2}\right)$.

(continued)

b) Find a lower bound for the probability of obtaining at least 95% of the expected heads. TIP: $P(|X - m| > a) = P(X < m - a) + P(X > m + a)$

c) Determine a number of tosses for which the observed frequency of heads deviates from 1/2 by at most 1% with a probability 99%.

TIP: Find n such that $P\left(\left|X - \frac{n}{2}\right| > 0.01\frac{n}{2}\right) \leq 0.01$.

5. A company has a mean daily production of 1000 units of a given product.

a) Use Markov's inequality to find an upper bound for the probabilities of productions of 1010, 1100, 2000 unities.

b) Assume that the standard deviation of the daily production is 10. Find an upper bound for these probabilities

TIP: Consider $P(|X - 1000| \geq x - 1000)$.

2.4.1 Couples of Numerical Variables

Let Y: $\Omega \longrightarrow \mathbb{R}$ be a second numerical variable defined on Ω, such that $Y(\Omega) = \{Y_1, \ldots, Y_n\}$. The couple (X, Y) forms a 2-dimensional vector of numerical variables defined on Ω. Let us consider the subpopulation $H_{ij} = X^{-1}(\{X_i\}) \cap Y^{-1}(\{Y_j\}) = \{\omega \in \Omega: X(\omega) = X_i \text{ and } Y(\omega) = Y_j\}$. Let μ be the mass function on Ω. The *probability* of the pair (X_i, Y_j) is

$$P(X = X_i, Y = Y_j) = p_{ij} = \sum_{\omega \in H_{ij}} \mu(\omega). \tag{2.9}$$

If both X, Y are defined for any $\omega \in \Omega$ (id est, $\Omega \subset dom(X)$, $\Omega \subset dom(Y)$), then

$$\bigcup_{\substack{i=1,\ldots,m\\ j=1,\ldots,n}} H_{ij} = \Omega; \quad \sum_{\substack{i=1,\ldots,m\\ j=1,\ldots,n}} p_{ij} = 1; \; 0 \leq p_{ij} \leq 1, \forall i, j. \tag{2.10}$$

Remark 2.3 In many practical situations, the individuals are equiprobable, so that $\mu(\omega) = 1/np$, $\forall\, \omega \in \Omega$. then, $p_{ij} = \frac{c_{ij}}{np}$, where $c_{ij} = card\,(H_{ij})$ is the number of elements belonging to H_{ij}. We have

$$\sum_{\substack{i=1,\ldots,m\\ j=1,\ldots,n}} c_{ij} = np; \; 0 \leq c_{ij} \leq np, \forall i, j.$$

Table 2.3 Contingency table of (X, Y)

X \ Y	Y_1	...	Y_n
X_1	p_{11}	...	p_{1n}
...	$\vdots$	$\ddots$	$\vdots$
X_m	p_{m1}	...	p_{mn}

The information about the couple (X, Y) may be synthetized in a Contingency Table 2.3 (or Cross-Tabulation Table):

Definition 2.9 Let f: $\mathbb{R}^2 \longrightarrow \mathbb{R}$ be a function. The ***mean*** or ***expectation*** of $f(X, Y)$ is

$$E(f(X,\ Y)) = \sum_{i=1}^{np} \mu(\omega_i) f(X(\omega_i), Y(\omega_i)).$$

We have

Proposition 2.10 Let X: $\Omega \longrightarrow \mathbb{R}$ be a numerical variable and f, g: $\mathbb{R} \longrightarrow \mathbb{R}$ be two functions. Then

1. $E(\alpha f(X, Y) + \beta g(X, Y)) = \alpha E(f(X, Y)) + \beta E(g(X, Y))$;
2. $E(f(X,\ Y)) = \sum_{\substack{i=1,\ \ldots,\ m \\ j=1,\ \ldots,\ n}} p_{ij} f(X_i, Y_j)$;
3. If $f(X, Y) \geq 0$, then $E(f(X, Y)) \geq 0$.
4. If $f(X, Y) \geq 0$, then $\lambda P(f(X, Y) \geq \lambda) \leq E(f(X, Y))$, $\forall\, \lambda \in \mathbb{R}$.

Proof Analogous to the proof of proposition 2.6.

Definition 2.10

1. The **marginal distribution of *X*** is $P(X = i) = p_{i\blacksquare} = \sum_{j=1}^{n} p_{ij}$.
2. The **marginal distribution of *Y*** is $P(Y = j) = p_{\blacksquare j} = \sum_{i=1}^{m} p_{ij}$.

(continued)

Definition 2.10 (continued)

3. The **covariance of** (X, Y) is $cov(X, Y) = E((X - E(X))(Y - E(Y)))$.
4. The **covariance matrix** of X is $C(X, Y) = \begin{pmatrix} cov(X, X) & cov(X, Y) \\ cov(Y, X) & cov(Y, Y) \end{pmatrix}$.

Proposition 2.11 Let $(X, Y)\colon \Omega \longrightarrow \mathbb{R}^2$ be a couple of numerical variables. Then $cov(X, Y) = E(XY) - E(X)E(Y)$; $V(\alpha X + \beta Y) = \alpha^2 V(X) + \beta^2 V(Y) + 2\alpha\beta\, cov(X, Y)$; $|cov(X, Y)| \leq \sqrt{V(X)}\sqrt{V(Y)}$.

Proof We have

$$(X - E(X))(Y - E(Y)) = XY - XE(Y) - YE(X) - E(X)E(Y),$$

So that

$$cov(X, Y) = E(XY) - E(X)E(Y) - E(Y)E(X) + E(X)E(Y),$$

Hence the first result. Analogously, $E(\alpha X + \beta Y) = \alpha E(X) + \beta E(Y)$ and

$$\begin{aligned}(\alpha X + \beta Y - E(\alpha X + \beta Y))^2 &= (\alpha(X - E(X)) + \beta(Y - E(Y)))^2 \\ &= \alpha^2 E\left((X - E(X))^2\right) + \beta^2\left((Y - E(Y))^2\right) + 2\alpha\beta(X - E(X))(Y - E(Y)),\end{aligned}$$

So that

$$V(\alpha X + \beta Y) = \alpha^2 V(X) + \beta^2 V(Y) + 2\alpha\beta\, cov(X, Y).$$

Let $f(\lambda) = \lambda^2 V(X) + 2\lambda cov(X, Y) + V(Y) = V(\lambda X + Y) \geq 0, \forall \lambda \in \mathbb{R}$. then, the discriminant of the polynomial verifies

$$\Delta = (2\, cov(X, Y))^2 - 4\, V(X)V(Y) \leq 0$$

And we have the last result.

Definition 2.11 The cumulative distribution function (CDF) of (X, Y) is

$$F(x, y) = P(X \leq x, Y \leq y) = P((X, Y) \in]-\infty, x] \times]-\infty, y]) = \sum_{\substack{(i,j) \text{ such that} \\ X_i \leq x \text{ and } Y_j \leq y}} p_{ij} = \sum_{\substack{\omega \text{ such that} \\ X(\omega) \leq x \text{ and } Y(\omega) \leq y}} \mu(\omega).$$

Proposition 2.12 Let (X, Y): $\Omega \longrightarrow \mathbb{R}^2$ be a couple of numerical variables and F: $\mathbb{R}^2 \longrightarrow \mathbb{R}$ be its CDF. Then

1. $x_1 \leq x_2$ *and* $y_1 \leq y_2 \Rightarrow F(x_1, y_1) \leq F(x_2, y_2)$;
2. $F(x, y) = 0$ if $x < X_1$ *or* $y < Y_1$;
3. $F(x, y) = \sum_{r=1}^{i-1} \sum_{s=1}^{j-1} p_{rs}$, if $X_{i-1} \leq x < X_i$ *and* $Y_{j-1} \leq y < Y_j$.
4. $F(x, y) = \sum_{s=1}^{j-1} p_{rs}$ if $x \geq X_m$ *and* $Y_{j-1} \leq y < Y_j$;
5. $F(x, y) = \sum_{r=1}^{i-1} p_{rs}$ if $y \geq Y_n$ *and* $X_{i-1} \leq x < X_i$;
6. $F(x, y) = 1$ if $x \geq X_m$ *and* $y \geq Y_n$;
7. $P(x_1 < X \leq x_2, y_1 \leq Y \leq y_2) = F(x_2, y_2) - F(x_2, y_1) - F(x_1, y_2) + F(x_1, y_1)$;
8. F has a variational derivative: $\partial^2 F / \partial x \partial y = f = \sum_{i=1}^{m} p_{ij} \delta_{X_i Y_j}$, where $\delta_{X_i Y_j}$ is the Dirac's mass supported by (X_i, Y_j). f is the PROBABILITY density (PDF) of (X, Y).

Proof Analogous to the proof of proposition 2.9.

R can deal with contingency tables. To create a contingency table, you may use a `data.frame`. As an example, let us consider the dataset `esoph`, provided by R, containing data about Smoking, Alcohol, and (O)esophageal Cancer. It contains information about age group (agegp), alcohol (alcgp) and tobacco (tobgp) consumption, number of cases (ncases), and number of controls (ncontrols). For instance, enter the commands below to get a contingency table `ct1`:

```
> ct1 = prop.table(table(esoph$alcgp,esoph$tobgp))
> ct1

            0-9g/day      10-19      20-29        30+
  0-39g/day 0.06818182 0.06818182 0.05681818 0.06818182
  40-79     0.06818182 0.06818182 0.06818182 0.05681818
  80-119    0.06818182 0.06818182 0.04545455 0.05681818
  120+      0.06818182 0.06818182 0.05681818 0.04545455
```

Here, variable X is the alcohol consumption, while Y is the tobacco consumption. Let us generate a second contingency table with X = age group, Y = alcohol consumption.

```
> t2 = table(esoph$agegp,esoph$alcgp)
> ct2 = prop.table(t2)
> ct2

         0-39g/day      40-79     80-119       120+
  25-34 0.04545455 0.04545455 0.03409091 0.04545455
  35-44 0.04545455 0.04545455 0.04545455 0.03409091
  45-54 0.04545455 0.04545455 0.04545455 0.04545455
  55-64 0.04545455 0.04545455 0.04545455 0.04545455
  65-74 0.04545455 0.03409091 0.04545455 0.04545455
  75+   0.03409091 0.04545455 0.02272727 0.02272727
```

You can determine the marginal distributions with the instruction `marginSums` and add the marginal distributions to the table using the command `addmargins`:

```
> p1 = marginSums(ct2,1)
> p1

    25-34     35-44     45-54     55-64     65-74       75+
0.1704545 0.1704545 0.1818182 0.1818182 0.1704545 0.1250000
> p2 = marginSums(ct2,2)
> p2

0-39g/day     40-79    80-119      120+
0.2613636 0.2613636 0.2386364 0.2386364
> ct2b <- addmargins(ct2)
> ct2b

         0-39g/day      40-79     80-119       120+        Sum
  25-34 0.04545455 0.04545455 0.03409091 0.04545455 0.17045455
  35-44 0.04545455 0.04545455 0.04545455 0.03409091 0.17045455
  45-54 0.04545455 0.04545455 0.04545455 0.04545455 0.18181818
  55-64 0.04545455 0.04545455 0.04545455 0.04545455 0.18181818
  65-74 0.04545455 0.03409091 0.04545455 0.04545455 0.17045455
  75+   0.03409091 0.04545455 0.02272727 0.02272727 0.12500000
  Sum   0.26136364 0.26136364 0.23863636 0.23863636 1.00000000
```

You can replace the names of the rows and of the columns using `rownames` and `colnames`:

```
> Xnames = rownames(ct1)
> rownames(ct1) <- 1:length(Xnames)
> Ynames = colnames(ct1)
> colnames(ct1) <- 1:length(Ynames)
> print.table(ct1)
```

```
           1          2          3          4
1 0.06818182 0.06818182 0.05681818 0.06818182
2 0.06818182 0.06818182 0.06818182 0.05681818
3 0.06818182 0.06818182 0.04545455 0.05681818
4 0.06818182 0.06818182 0.05681818 0.04545455
```

To evaluate the statistics, you need to transform the labels – which are strings of characters – into numeric variables. Once the transformation is made, you can evaluate the statistics such as covariances or means by using the formulae previously introduced. These expressions are included in the class `ctstats`:

```
source("ctstats.R")
cts = ctstats$new()
ctab1 = cts$fromTable(ct1)
mXY = cts$meanXY(ctab1)
v = cts$varXY(ctab1)
cXY = cts$cov(ctab1)
f <- function(x,y){exp(x)*sin(y)}
mf = cts$meanf(ctab1,f)
mediXY = cts$medXY(ctab1)
```

```
> mXY
$mX
[1] 2.454545

$mY
[1] 2.409091
```

```
> v
$varX
[1] 1.247934

$varY
[1] 1.241736
```

```
> cXY
[1] -0.06095041
> mf
[1] 7.581205
```

```
> mediXY
$medX
[1] 2

$medY
[1] 2
```

Class `ctstats` manipulates contingency tables, defined as a list containing fields named “X” (values of X), “Y” (values of Y), “p” (values of p_{ij}), “Xnames” (names of the values in X), “Ynames” (names of values in Y), “pX” ($p_{i\blacksquare}$), “pY” ($p_{\blacksquare j}$). $fromTable creates such an object from a table. `$meanXY` evaluates the means of X and Y, `$varXY` evaluates their variances, `$sdXY` evaluates their standard deviations, `$cov` calculates their covariance, `$meanf` finds $E(f(X, Y))$, and `$medXY` determines the medians of X and Y. Analogously:

```
ctab2 = cts$fromTable(ct2)
mXY = cts$meanXY(ctab2)
v = cts$varXY(ctab2)
cXY = cts$cov(ctab2)
f <- function(x,y){exp(x)*sin(y)}
mf = cts$meanf(ctab2,f)
mediXY = cts$medXY(ctab2)
```

```
> mXY
$mX
[1] 3.386364

$mY
[1] 2.454545
```

```
> v
$varX
[1] 2.691632

$varY
[1] 1.247934
```

```
> cXY
[1] -0.02789256
> mf
[1] 33.16004
```

```
> mediXY
$medX
[1] 3

$medY
[1] 2
```

As a final example, let us import the data contained in a EXCEL® file *testdata.Xlsx*:

	A	B	C	D
1		type 1	type 2	type 3
2	A	0,015801	0,087542	0,142022
3	B	0,134701	0,001723	0,120232
4	C	0,025932	0,139779	0,054127
5	D	0,081264	0,071766	0,125112

The `data.frame testdata` appears in the environment:

	...1	type 1	type 2	type 3
1	A	0.01580059	0.087541736	0.14202186
2	B	0.13470116	0.001723143	0.12023217
3	C	0.02593187	0.139779003	0.05412673
4	D	0.08126395	0.071765739	0.12511205

We generate a contingency table as follows:

```
> aux = testdata[,1]
> Xnames = aux[[1]]
> aux = names(testdata)
> Ynames = aux[2:length(aux)]
> p = testdata[1:length(Xnames),2:length(aux)]
> source("ctstats.R")
> cst = ctstats$new()
> ct2 = cst$create(p=p,Xnames=Xnames,Ynames=Ynames)
```

`$create` generates a contingency table as previously described. The values of X and Y are missing in the call: default values are provided – integers 1, 2, ... The contingency table `ct2` can be manipulated as in the previous examples.

Exercises

1. Import the file *pets.Xlsx*. Generate a contingency table corresponding to its contents as in the example above (`ct2`). Find the means, the standard deviations and the covariances.
2. Import the file *sport.Xlsx*. Generate a contingency table corresponding to its contents as in the example above (`ct2`). Find the means, the standard deviations and the covariances.
3. Import the file *colors.Xlsx*. Generate a contingency table corresponding to its contents as in the example above (`ct2`). Find the means, the standard deviations and the covariances.

2.4.2 Independent Numerical Variables

Definition 2.12 Let X, Y be two numerical variables defined on (Ω, P). X and Y are independent if and only if

$$P(X \leq x, Y \leq y) = P(X \leq x)P(Y \leq y), \forall (x, y) \in \mathbb{R}^2.$$

We have

Proposition 2.13 Let X, Y be two numerical variables defined on (Ω, P). Then

1. Let F_{XY} be the CDF of the pair (X, Y), F_X be the CDF of X, F_Y be the CDF of Y. X and Y are independent if and only if $F_{XY}(x, y) = F_X(x) F_Y(y)$, $\forall (x, y) \in \mathbb{R}^2$.
2. X and Y are independent if and only if $P(X = X_i, Y = Y_j) = P(X = X_i) P(Y = Y_j)$, $\forall i, j$.
3. X and Y are independent if and only if $p_{ij} = p_{i\bullet} p_{\bullet j}$, $\forall i, j$.
4. Let f_{XY} be the PDF of the pair (X, Y), f_X be the PDF of X, f_Y be the PDF of Y. X and Y are independent if and only if $f_{XY}(x, y) = f_X(x) f_Y(y)$, $\forall (x, y) \in \mathbb{R}^2$.
5. If X and Y are independent, then $cov(X, Y) = 0$.

Proof The first result follows straightly from definition **2.12**. Results 2 to 4 follow from proposition **2.12** (notice that $\delta_{X_i Y_j} = \delta_{X_i}\delta_{Y_j}$). Finally, $p_{ij} = p_{i\blacksquare}p_{\blacksquare j}, \forall\, i, j \Rightarrow E(XY) = E(X)E(Y)$, hence the last result.

Definition 2.13 Let X, Y be two numerical variables defined on (Ω, P). The distribution of X conditional to Y is

$$P(X = X_i | Y = Y_j) = \frac{P(X = X_i, Y = Y_j)}{P(Y = Y_j)}.$$

The distribution of Y conditional to X is

$$P(Y = Y_j | X = X_i) = \frac{P(X = X_i, Y = Y_j)}{P(X = X_i)}.$$

Proposition 2.14 Let X, Y be two numerical variables defined on (Ω, P).

1. X and Y are independent if and only if $P(X = X_i | Y = Y_j) = P(X = X_i), \forall\, i, j$.
2. X and Y are independent if and only if $P(Y = Y_j | X = X_i) = P(Y = Y_j), \forall\, i, j$.

Proof If X and Y are independent, then $P(X = X_i, Y = Y_j) = P(X = X_i)$ $P(Y = Y_j)$, $\forall\, i, j$, *so that*, $\forall\, i, j$: $P(X = X_i | Y = Y_j) = P(X = X_i)$ and P-$(Y = Y_j | X = X_i) = P(Y = Y_j)$. Conversely, let either $P(X = X_i | Y = Y_j) = P$(-$(X = X_i | Y = Y_j) = P(X = X_i)$, $\forall\, i, j$ or $P(Y = Y_j | X = X_i) = P(Y = \mathrm{j})$, $\forall\, i, j$. In both the cases, $P(X = X_i, Y = Y_j) = P(X = X_i)P(Y = Y_j)$, $\forall\, i, j$, so that X and Y are independent.

The class `ctstats` contains a method `condXY` which determines the conditional distributions. It returns a list containing "`X2Y`" (conditional distribution of X to Y) and "`Y2X`" (conditional distribution of Y to X). An example is given below, using the contingency table `ct1`, previously defined.

```
> pc = cts$condXY(ctab1)
> pc$X2Y
     [,1] [,2] [,3] [,4]
[1,] 0.25 0.25 0.25 0.30
[2,] 0.25 0.25 0.30 0.25
[3,] 0.25 0.25 0.20 0.25
[4,] 0.25 0.25 0.25 0.20
> pc$Y2X
          [,1]      [,2]      [,3]      [,4]
[1,] 0.2608696 0.2608696 0.2173913 0.2608696
[2,] 0.2608696 0.2608696 0.2608696 0.2173913
[3,] 0.2857143 0.2857143 0.1904762 0.2380952
[4,] 0.2857143 0.2857143 0.2380952 0.1904762
```

Exercises

1. Consider $\Omega = \{1, 2, \ldots, 12\}$ and the variables $X(\omega) = \omega \ mod \ 2$, $Y(\omega) = \omega \ mod \ 3$. Assume that $\mu(\omega) = 1/12$.

   (a) Verify that $X(\Omega) = \{0, 1\}$ and $P(X = 0) = P(X = 1) = 1/2$.
   (b) Verify that $Y(\Omega) = \{0, 1, 2\}$ and $(Y = 0) = P(Y = 1) = P(Y = 2) = 1/3$.
   (c) Determine $P(X = i, Y = j)$ for all the values of (i, j).
   (d) Are these variables independent?

2. Consider $\Omega = \{1, 2, \ldots, 12\}$ and the variables $X(\omega) = \omega \ mod \ 2$, $Y(\omega) = \omega \ mod \ 4$. Assume that $\mu(\omega) = 1/12$.

   (a) Verify that $X(\Omega) = \{0, 1\}$ and $(X = 0) = P(X = 1) = 1/2$.
   (b) Verify that $Y(\Omega) = \{0, 1, 2, 3\}$ and $(Y = 0) = P(Y = 1) = P(Y = 2) = P(Y = 3) = 1/4$.
   (c) Determine $P(X = i, Y = j)$ for all the values of (i, j).
   (d) Are these variables independent?

3. Consider $\Omega = \{1, 2, \ldots, 12\}$ and the variables $X(\omega) = \omega \ mod \ 3$, $Y(\omega) = |\sin(\omega\pi/2)|$. Assume that $\mu(\omega) = 1/12$.

   (a) Verify that $X(\Omega) = \{0, 1, 2\}$ and $(X = 0) = P(X = 1) = P(X = 2) = 1/3$.
   (b) Verify that $Y(\Omega) = \{0, 1\}$ and $P(Y = 0) = P(Y = 1) = 1/2$.
   (c) Determine $P(X = i, Y = j)$ for all the values of (i, j).
   (d) Are these variables independent?

2.5 Numerical Variables as Elements of Hilbert Spaces

Let us consider the set of numerical variables defined on Ω:

$$\mathcal{H} = \{Z : \Omega \longrightarrow \mathbb{R}, \ \Omega \subset dom(Z)\}. \tag{2.11}$$

For $X, Y \in \mathcal{H}$, we can consider

$$(X, Y)_{\mathcal{H}} = E(XY). \tag{2.12}$$

Proposition 2.15 $(\blacksquare, \blacksquare)_{\mathcal{H}}$ is a symmetric, positive and bilinear on $\mathcal{H}$. in addition, $(X, X)_{\mathcal{H}} = 0$ if and only if $X = 0$ almost surely.

Proof We have $E(XY) = E(YX)$, so that $(\blacksquare, \blacksquare)_{\mathcal{H}}$is symmetric. In addition,

$$E((\alpha X + \beta Y)Z) = E(\alpha XZ + \beta YZ) = \alpha E(XZ) + \beta E(YZ),$$

So that $(\blacksquare, \blacksquare)_{\mathcal{H}}$ is bilinear. We have also $E(XX) = E(X^2) \geq 0$, hence $(\blacksquare, \blacksquare)_{\mathcal{H}}$ is positive. The last assertion follows from proposition 2.7.

This result shows that $(\blacksquare, \blacksquare)_{\mathcal{H}}$ satisfies the conditions to be used as scalar product. Indeed, let us consider the Hilbert space

$$H = L^2(\Omega, P) = \{X \in \mathcal{H} : E(X^2) < \infty\}, \tag{2.13}$$

having scalar product and norm defined by

$$(X, Y)_H = E(XY), \ \|X\|_H = \sqrt{(X, Y)_H}. \tag{2.14}$$

To alleviate the notation, we drop the index H in the sequel, except in the situations where it is necessary to clear some ambiguity.

Remark 2.4 As previously observed, in all generality, it is necessary to consider situations where some elements of $X(\Omega)$ may be negligible, i.e., a value p_i can be null. In such a situation, we must consider a binary equivalence relation

$$W \sim Z \Leftrightarrow W - Z = 0 \text{ almost surely}$$

And the classes

$$[Z]_{\mathcal{H}} = \{W \in \mathcal{H} : W \sim Z\}.$$

$$\mathbb{H} = \{[Z]_{\mathcal{H}} : Z \in \mathcal{H}\}, \ ([X]_{\mathcal{H}}, [Y]_{\mathcal{H}})_{\mathbb{H}} = E(XY).$$

(continued)

Remark 2.4 (continued)

$(\blacksquare, \blacksquare)_{\mathbb{H}}$ is a scalar product on $\mathbb{H}$, which can be identified to $\mathcal{H}$. The completion H of $\mathbb{H}$ generates a Hilbert space referred as $H = L^2(\Omega, P)$, having as scalar product $(X, Y)_H = E(XY)$. Such a formal construction is detailed in (Souza de Cursi & Sampaio, 2010).

The Hilbertian structure allows the use of the standard results concerning Hilbert spaces, namely those concerning orthogonal projections:

Definition 2.14 Let S be a non-empty linear subspace of $L^2(\Omega, P)$ and $X \in L^2(\Omega, P)$. The ***orthogonal projection*** of X onto S is the element PX that minimizes the distance between X and S, id est,

$$PX \in S \text{ and } \|X - PX\| = \min\{\|X - v\| : v \in S\}.$$

We have

Theorem 2.1 Let S be a closed linear subspace of $L^2(\Omega, P)$ and $X \in L^2(\Omega, P)$. Then, the orthogonal projection PX of X onto S exists and is uniquely determined.

Proof See (Souza de Cursi & Sampaio, 2010).

Theorem 2.2 Let $S \subset L^2(\Omega, P)$ be a linear subspace and $X \in L^2(\Omega, P)$. PX is the orthogonal projection of X onto S if and only if

$$PX \in S \text{ and } (X - PX, v) = 0, \forall v \in S.$$

Proof See (Souza de Cursi & Sampaio, 2010).

Theorem 2.3 Let $S \subset L^2(\Omega, P)$ be a linear subspace and $X \in L^2(\Omega, P)$. Then $X = PX$ if and only if $X \in S$.

Proof It is immediate that $X = PX \in S \Rightarrow X \in S$. If $X \in S$, $\|X - PX\| \leq 0$, so that $X = PX$.

Remark 2.5 The results extend straightly to $n - \textit{dimensional}$ vectors $X = (X_1, \ldots, X_n)$, $Y = (Y_1, \ldots, Y_n)$, $(X, Y) = E(X.Y)$.

2.5.1 Conditional Probabilities as Orthogonal Projections

Conditional probabilities can be interpreted as orthogonal projections:

Definition 2.15 Let $A \subset \Omega$. the **indicator function** of A is

$$1_A(\omega) = 1, \text{ if } \omega \in A; 1_A(\omega) = 0, \text{ if } \omega \notin A.$$

Proposition 2.16 Let $A, B \in \mathcal{P}(\Omega)$ be such that $P(A > 0)$. Let $X = 1_B$ and $S = \{\alpha 1_A, \alpha \in \mathbb{R}\}$. Then, the orthogonal projection PX of X onto S is $PX = P(B|A)1_A$.

Proof Indeed, $P(B|A)1_A \in S$. In addition,

$$(1_B - P(B \mid A)1_A, 1_A) = (1_B, 1_A) - P(B \mid A)\,(1_A, 1_A).$$

Since $1_B 1_A = 1_{A \cap B}$ and, as shown in proposition 2.6, $E(1_C) = P(C)$ for $C \subset \Omega$, we have

$$(1_B - P(B \mid A)1_A, 1_A) = P(A \cap B) - P(B \mid A)\,P(A) = 0.$$

Thus, $PX = P(B|A)1_A$

2.5.2 Means as Orthogonal Projections

Means also can be interpreted as orthogonal projections:

Proposition 2.17 Let $X \in L^2(\Omega, P)$. Let $S = \{Z \in L^2(\Omega, P)$: *Z is constant*$\}$. Then, the orthogonal projection PX of X onto S is $PX = E(X)$. In addition, $\|X - PX\|^2 = V(X)$, $\|X - PX\| = \sigma(X)$.

Proof Let $Z \in S$. Then $Z(\omega) = \alpha \in \mathbb{R}$, $\forall \omega \in \Omega$, so that $Z = \alpha 1_\Omega \in S$. In addition, $E(Z) = E(Z 1_\Omega) = (Z, 1_\Omega)$. Then

$$(X - E(X)1_\Omega, 1_\Omega) = (X, 1_\Omega) - E(X)(1_\Omega, 1_\Omega) = E(X) - E(X) = 0.$$

Since $E(X)1_\Omega \in S$, we have $PX = E(X)$. Then,

$$\|X - PX\|^2 = E\left((X - E(X))^2\right) = V(X) \text{ and } \|X - PX\| = \sqrt{V(X)} = \sigma(X).$$

Remark 2.6 The medians can be interpreted as orthogonal projections for the norm $\|X\|_1 = E(|X|)$. Indeed, $m = med(X)$ if and only if m minimizes $\|X - v\|_1$ on the subspace S of the constants, Id Est, if and only if

$$\|X - m\|_1 = \min \left\{\|X - v\|_1 : v \in S\right\}.$$

2.5.3 Affine Approximations and Correlations

In this section, we are interested in the approximation of one of the elements of a couple of numerical variables by an affine function of the other one, id est, in the determination of coefficients $(a, b) \in \mathbb{R}^2$ such that $PX = aU + b$ is the best approximation of X by an affine function of U.

Definition 2.16 Let X, $U \in L^2(\Omega, P)$. The coefficient of linear correlation between X and U is

$$\rho(X, U) = \frac{cov(X, U)}{\sqrt{V(X)}\sqrt{V(U)}}.$$

Then

Proposition 2.18 Let X, $U \in L^2(\Omega, \mathrm{P})$. Let $S = \{Z \in L^2(\Omega, P)\colon\ Z = \alpha U + \beta;\ \alpha, \beta \in \mathbb{R}\}$. Then, the orthogonal projection PX of X onto S is $PX = aU + b$, with

$$a = \frac{cov(X,\ U)}{V(U)},\ b = E(X) - aE(U).$$

In addition,

$$\|X - PX\|^2 = V(X)\left(1 - |\rho(X,\ \ U)|^2\right).$$

Proof Let $PX = aU + b$. Then $(X - PX, v) = 0$, $\forall\ v \in S$, id Est,

$$(X,\ v\ ) = a(U,\ v) + b(1_\Omega,\ v),\ \forall v \in S$$

Taking successively $v = 1_\Omega$, $v = U$, we obtain

$$a\underbrace{(U,\ 1_\Omega)}_{E(U)} + b\underbrace{(1_\Omega,\ 1_\Omega)}_{P(\Omega) = 1} = \underbrace{(X,\ 1_\Omega)}_{E(X)};\ a\underbrace{(U,\ U)}_{E\left(U^2\right)} + b\underbrace{(1_\Omega,\ U)}_{E(U)} = \underbrace{(X,\ U)}_{E(XU)}.$$

These equations form a linear system for the unknowns (a, b), having as solution

$$a = \frac{cov(X,\ U)}{V(U)},\ b = E(X) - aE(U).$$

Since

$$\|X - aU - b\|^2 = \|(X - E(X)) - a(U - E(U))\|^2,$$

We have

$$\|X - aU - b\|^2 = V(X) + a^2 V(U) - 2a\ cov(X,\ U) = V(X) - \frac{(cov(X,\ \ U))^2}{V(U)}.$$

Hence,

$$\|X - PX\|^2 = V(X)\left(1 - |\rho(X,\ \ U)|^2\right).$$

Proposition 2.19 We have $|\rho(X, U)| \leq 1$ and $|\rho(X, U)| = 1 \Leftrightarrow X$ is an affine function of U almost surely.

Proof Proposition 2.18 shows that $|\rho(X, U)| \leq 1$. We have $|\rho(X, U)| = 1 \Leftrightarrow \|X - PX\| = 0$. Theorem 2.3 shows that $\|X - PX\| = 0 \Leftrightarrow X \in S$. Hence the result.

Remark 2.7 For multidimensional vectors $\boldsymbol{X} = (X_1, \ldots, X_n)$, $\boldsymbol{U} = (U_1, \ldots, U_m)$, we consider $S = \{\boldsymbol{Z} \in [L^2(\Omega, P)]^n : \boldsymbol{Z} = \boldsymbol{\alpha U} + \boldsymbol{\beta}; \boldsymbol{\alpha} \in M(n, m), \boldsymbol{\beta} \in M(n, 1)\}$. Then, $\boldsymbol{PX} = \boldsymbol{AU} + \boldsymbol{B}$, $\boldsymbol{A} \in M(n, m)$, $\boldsymbol{B} \in M(n, 1)$. Matrix $\boldsymbol{A} = (A_{ij}: 1 \leq i \leq n, 1 \leq j \leq m)$ verifies

$$\sum_{j=1}^{m} A_{ij}\, cov\left(U_j, U_k\right) = cov(X_i, U_k).$$

These equations show that each line of $\boldsymbol{A}$ is the solution of a linear system: Line i is the solution $\boldsymbol{R}^i$ of $\boldsymbol{MR}^i = \boldsymbol{N}$, $M_{kj} = cov\,(U_j, U_k)$, $N_k = cov\,(X_i, U_k)$. $\boldsymbol{B}$ is given by $\boldsymbol{B} = \boldsymbol{X} - \boldsymbol{AE}(\boldsymbol{U})$.

The class `ctstats` contains methods `corr` and `linmod` for the determination of $\rho(U, X)$ and the approximation $X \sim aU + b$. It is assumed that the contingency table presents X as second variable and U as first variable. To get an approximation $U \sim aX + b$, you must use the method `transpose` to inverse the places between U and X.

As an example, let us consider $\Omega = \{1, 2, 3\}$, $\mu(1) = \mu(3) = 1/4$, $\mu(2) = \frac{1}{2}$, $X(\omega) = \omega + 1$, $U(\omega) = 2\omega$. We have $X(\Omega) = \{2, 3, 4\}$, $U(\Omega) = \{2, 4, 6\}$, $X = U/2 + 1$, $U = 2X - 2$. As shown at right, we obtain the exact results.

```
> Omega = c(1,2,3)
> po = c(0.25, 0.5, 0.25)
> X = Omega+1
> U = 2*Omega
> p = matrix(0,nrow=3,ncol=3)
> for (i in 1:3){
+     p[i,i]=po[i]
+ }
> source("ctstats.R")
> cst = ctstats$new()
> ct1 = cst$create(X=U,Y=X,p=p)
> linap1 = cst$linmod(ct1)
> paste("a = ",linap1$a,", b = ",linap1$b)
[1] "a =  0.5 , b =  1"
> ct2 = cst$transpose(ct1)
> linap2 = cst$linmod(ct2)
> paste("a = ",linap2$a,", b = ",linap2$b)
[1] "a =  2 , b =  -2"
```

```
> linap1$fitted
[1] 2 3 4
> linap1$residual
     [,1] [,2] [,3]
[1,]    0   -1   -2
[2,]    1    0   -1
[3,]    2    1    0
> linap1$error
[1] 0
> linap1$eta
[1] 1
```

`Linmod` returns also a vector of the fitted data $F_i = aX_i + b$ (`"fitted"`), a table of residuals $R_{ij} = F_i - Y_j$ (`"residuals"`), the norm $\|X - PX\|$ of the error (`"error"`) and a coefficient $\eta(X| U) = \exp(-\|X - PX\|^2)$ (`"eta"`). Notice that $\|X - PX\|^2 = E((X - PX)^2)$, so that the values having null probability have a null contribution. $\eta(X| U)$ is an indicator of the distance between X and the function of U determined. We have $0 \leq \eta \leq 1$, with $\eta = 1 \Leftrightarrow X = PX$ (see definition 2.17)

R proposes an intrinsic function `lm` (linear model) for the determination of the coefficients when the values are equiprobable. The instruction `lm(X ~ U)` generates a list containing the results of the approximation of X by a linear model $aU + b$. Coefficient b is referred as Intercept. The list contains the coefficients, the residuals, the fitted values, and other information. The results can be plotted.

As an example, let us consider the universe $\Omega = \{1, 2, 3, 4, 5, 6\}$ and the variables $X(\omega) = 3\omega + 1$, $U(\omega) = 2\omega - 1$. We have $X = \frac{3U+5}{2}$. We can determine the coefficients as shown at right. The results are $a = 1.5$, $b = 2.5$, as expected.

```
> Omega = c(1:6)
> X = 3*Omega + 1
> U = 2*Omega - 1
> model <- lm(X~U)
> coefficients(model)
(Intercept)           U
        2.5         1.5
```

You can plot the data and the results as follows:

```
> plot(U,X)
> abline(model,lty="dotted",col="red")
> legend(("topleft"),legend = c("data","model"),
+        lty = c(NA,"dotted"),pch = c(1,NA),
+        col= c("black","red"))
```

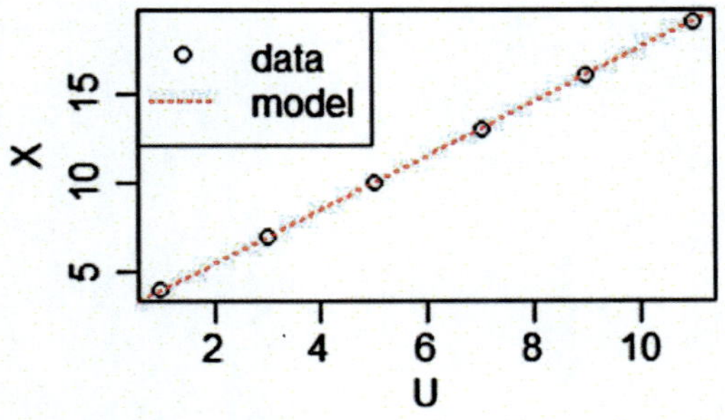

Multilinear approximation can be generated by `lm` (for equiprobable data). For instance, the instruction `lm(X ~ U1 + U2)` generates a list containing the results of the approximation of *X* by a linear model $a_1U_1 + a_2U_2 + b$.

As an example, let us consider the variables $X(\omega) = 3U_1(\omega) + U_2(\omega)$, $U_1(\omega) = 2\omega - 1$, $U_2(\omega) = \omega^2 + 1$. We can determine the coefficients as shown at right. The results are close to the exact values, as expected.

```
> Omega = c(1:6)
> U1 = 2*Omega - 1
> U2 = Omega^2 + 1
> X = 3*U1 + U2
> mo1 <- lm(X~U1+U2)
> mo1$coefficients
 (Intercept)           U1           U2
2.900779e-15 3.000000e+00 1.000000e+00
```

When dealing with noisy data, the command `confint` can also furnish confidence intervals for the coefficients. For instance:

```
> X = 3*omega + 1
> U = 2*omega - 1
> Xn = (1 + 0.1*runif(length(X),-1,1))*X
> Un = (1 + 0.1*runif(length(U),-1,1))*U
> m1 = lm(X~U)
> m1$coefficients
(Intercept)           U
        2.5         1.5
> confint(m1)
            2.5 % 97.5 %
(Intercept)   2.5    2.5
U             1.5    1.5
Warning message:
In summary.lm(object, ...) :
  essentially perfect fit: summary may be unreliable
```

Here, the fitting is evaluated by R as “essentially perfect”, so that the confidence interval for a risk $\alpha = 0.05$ contains a single value. Indeed, m1 contains more information than the coefficients:

```
> summary(m1)

Call:
lm(formula = X ~ U)

Residuals:
         1          2          3          4          5          6 
 1.757e-15 -2.112e-15 -2.601e-16  2.687e-16 -9.078e-17  4.380e-16 

Coefficients:
             Estimate Std. Error   t value Pr(>|t|)    
(Intercept) 2.500e+00  1.159e-15 2.157e+15   <2e-16 ***
U           1.500e+00  1.678e-16 8.937e+15   <2e-16 ***
---
Signif. codes:  0 '***' 0.001 '**' 0.01 '*' 0.05 '.' 0.1 ' ' 1

Residual standard error: 1.404e-15 on 4 degrees of freedom
Multiple R-squared:      1,	Adjusted R-squared:      1 
F-statistic: 7.986e+31 on 1 and 4 DF,  p-value: < 2.2e-16

Warning message:
In summary.lm(m1) : essentially perfect fit: summary may be unreliable
```

Thus, R furnishes also the residuals, the standard errors, the t-values (which are equal to the coefficient divided by the standard errors), and the probability associated to the t-value: a small t-value means a significant probability of error – in such a situation, the results must be considered as not significant. R also furnishes the coefficient of linear correlation and the BFS statistics.

You will find in the R repositories packages that complement the basic information and methods provided by `lm`. For instance, `MASS`, `relaimpo`, `bootstrap`, `DAAG`, `leps`.

2.5.4 Conditional Mean

It is interesting to notice that the problem of the approximation of a numerical variable by a general function of another one has a solution – id est, there is a solution $PX = f(U)$ is the best approximation of X by a generic function of U. We have

Proposition 2.20 Let $X, U \in L^2(\Omega, \mathrm{P})$. Let $S = \{Z \in L^2(\Omega, P): Z = g(U), g: \mathbb{R} \longrightarrow \mathbb{R}\}$. Then, the orthogonal projection PX of X onto S is $PX = E(X|\, U)$, with

(continued)

Proposition 2.20 (continued)

$$E(X|U=U_i)=\frac{1}{p_{i\blacksquare}}\sum_{j=1}^{m}p_{ij}X_j=\sum_{j=1}^{m}P\big(X=X_j\mid U=U_i\big)X_j.$$

Proof Let $Z\in S$ and $Z_k=g(U_k)$. The orthogonal projection minimizes

$$\|X-Z\|^2=\sum_{j=1}^{m}\sum_{k=1}^{n}p_{jk}\big(X_j-Z_k\big)^2.$$

Thus, Z_i nullifies the derivative of this sum of squares. Since

$$\frac{\partial}{\partial Z_i}\|X-Z\|^2=2\sum_{j=1}^{m}p_{ji}\big(X_j-Z_i\big)=2\left(\sum_{j=1}^{m}p_{ji}X_j-Z_ip_{\blacksquare i}\right),$$

We have

$$\frac{\partial}{\partial Z_i}\|X-Z\|^2=0\Leftrightarrow Z_i=\frac{1}{p_{\blacksquare i}}\sum_{j=1}^{m}p_{ji}X_j,$$

Hence the result.

Definition 2.17 Let $PX=E(X|\,U)$. Then,

$$\eta(X|U)=\exp\left(-\|X-PX\|^2\right)$$

measures the distance between X and the subspace S. We can use

$$R(X,U)=\frac{1}{2}\big(\eta(X|U)+\eta(U|X)\big)$$

As ***generalized coefficient of correlation*** between X and U.

Proposition 2.21 Let X, $U \in L^2(\Omega, \mathrm{P})$. Let $S_U = \{Z \in L^2(\Omega, P): Z = g(U), g: \mathbb{R} \longrightarrow \mathbb{R}\}$, $S_X = \{Z \in L^2(\Omega, P): Z = h(X), g: \mathbb{R} \longrightarrow \mathbb{R}\}$. Let us denote $PX = E(X|\,U)$ and $PU = E(U|\,X)$. Then

1. $\eta(X|\,U) = 1$ if and only if $X = PX \in S_U$;
2. $\eta(U|\,X) = 1$ if and only if $U = PU \in S_X$;
3. $X = PX \in S_U \Leftrightarrow U = PU \in S_X$;
4. $\eta(X|\,U) = 1 \Leftrightarrow \eta(U|\,X) = 1$;
5. $R(X, U) = 1$ if and only if $X = PX \in S_U$ *and* $U = PU \in S_X$.

Proof $\eta(X|\,U) = 1$ if and only if $\|X - PX\| = 0$, hence the first result. Analogously, $\eta(U|\,X) = 1$ if and only if $\|U - PU\| = 0$, hence the second result.

Let $X = PX \in S_U$. Then, $X = g(U)$, so that, for each i, there is $j(i)$ such that $X_i = g(U_{j(i)})$ and we have $p_{ik} = 0$, for $k \neq j(i)$. Indeed,

$$\|X - g(U)\|^2 = \sum_{i=1}^{m} \sum_{j=1}^{n} p_{ij} \left(X_i - g(U_j)\right)^2 = 0.$$

Thus, $X_i - g(U_j) \neq 0 \Longrightarrow p_{ij} = 0$. Thus,

$$\sum_{i=1}^{m} \sum_{j=1}^{n} p_{ij} \left(A_i - B_j\right)^2 = \sum_{i=1}^{n} p_{i,\,j(i)} \left(A_i - B_{j(i)}\right)\big)^2$$

Let $h(X_i) = U_{j(i)}$. Taking $A_i = h(X_i)$, $B_j = U_j$, we have

$$\|h(X) - U\|^2 = \sum_{i=1}^{n} p_{i,\,j(i)} \left(h(X_i) - U_{j(i)}\right)\big)^2 = 0,$$

So that $U \in S_X$ and we have $U = PU \in S_X$. The converse is obtained by the same way, hence the third result.

The fourth result follows from the third one. Finally, notice that $0 \leq R(X, U) \leq 1$ and $R(X, U) = 1 \Leftrightarrow \eta(X|\,U) = \eta(U|\,X) = 1$, hence the last result.

The class `ctstats` contains methods `condmean` and `gencorr` for the determination of the conditional means and of the generalized coefficient of correlation. Both return lists containing elements `"X2Y"` and `"Y2X"` corresponding to $E(X|\,Y)$ and $E(Y|\,X)$. `gencorr` returns also the value of R (`"R"`).

As an example, let us consider the universe $\Omega = \{0, 1, 2, 3\}$, $\mu(0) = 0.25$, $\mu(1) = 0.4$, $\mu(2) = 0.1$, $\mu(3) = 0.25$ and the variables $X(\omega) = 2\omega - 1$, $U(\omega) = 1 + \omega^2$. We have $X(\Omega) = \{-1, 1, 3, 5\}$, $U(\Omega) = \{1, 2, 5, 10\}$. As a second example, let us consider the universe $\Omega = \{-2, -1, 1, 2\}\}$, $\mu(-2) = 0.25$, $\mu(-1) = 0.4$, $\mu(1) = 0.1$, $\mu(2) = 0.25$ and the variables $X(\omega) = 2\omega - 1$, $U(\omega) = 1 + \omega^2$. We have $X(\Omega) = \{-5, -3, 1, 3\}$, $U(\Omega) = \{5, 2, 2, 5\} = \{2, 5\}$.

```
> Omega = c(0,1,2,3)
> po = c(0.25, 0.4, 0.1,0.25)
> X = 2*Omega-1
> U = Omega^2 + 1
> p = matrix(0,nrow=4,ncol=4)
> for (i in 1:4){
+     p[i,i]=po[i]
+ }
> source("ctstats.R")
> cst = ctstats$new()
> ct1 = cst$create(X=U,Y=X,p=p)
>
> cst$condmean(ct1)
$X2Y
[1]  1  2  5 10

$Y2X
[1] -1  1  3  5

> cst$gencorr(ct1)
$X2Y
[1] 1

$Y2X
[1] 1

$R
[1] 1
```

```
> Omega = c(-2,-1,1,2)
> po = c(0.25, 0.4, 0.1,0.25)
> X = 2*Omega-1
> U = Omega^2 + 1
> p = matrix(0,nrow=4,ncol=4)
> for (i in 1:4){
+     p[i,i]=po[i]
+ }
> source("ctstats.R")
> cst = ctstats$new()
> ct1 = cst$create(X=U,Y=X,p=p)
>
> cst$condmean(ct1)
$X2Y
[1] 5 2 2 5

$Y2X
[1] -5 -3  1  3

> cst$gencorr(ct1)
$X2Y
[1] 1

$Y2X
[1] 1

$R
[1] 1
```

Exercises

1. Consider the data below:

X	4.9	7.2	9.4	11.2	12.4	14.4	17.7	19.3	20.5	23.7
U	1.0	2.0	3.1	3.9	5.0	5.8	7.1	7.9	8.6	9.7

Assume that the universe is equiprobable: $\mu = constant = 0.1$

(a) Determine the covariance between X and U.
(b) Determine the best linear approximation of X as a function of U.
(c) Plot the approximation and the data
(d) Determine the generalized correlation between X and U.

(continued)

2. Consider the data below:

X	1.0	1.4	1.8	2.0	2.3	2.4	2.6	2.8	3.0	3.2
U	1.0	1.9	3.1	4.0	5.2	5.8	7.0	7.9	9.2	10.3

Assume that the universe is equiprobable: $\mu = constant = 0.1$

(a) Determine the covariance between *X* and *U*.
(b) Determine the best linear approximation of *X* as a function of *U*.
(c) Plot the approximation and the data
(d) Determine the generalized correlation between *X* and *U*.
(e) Determine the best approximation of *X* as a function of *U*.
(f) Plot the approximation and the data.

3. Consider the data below:

X	0.9	0.8	0.7	0.7	0.6	0.6	0.5	0.5	0.4	0.4
U	1.0	1.9	2.9	4.1	5.0	5.9	7.0	7.8	8.9	9.7

Assume that the universe is equiprobable: $\mu = constant = 0.1$

(a) Determine the covariance between *X* and *U*.
(b) Determine the best linear approximation of *X* as a function of *U*.
(c) Plot the approximation and the data
(d) Determine the generalized correlation between *X* and *U*.
(e) Determine the best approximation of *X* as a function of *U*?
(f) Plot the approximation and the data.

2.6 Random Variables

Random variables generalize the notion of numerical variables. Now, we consider a general probability space (Ω, P), where the universe Ω can be uncountable, for instance, a real interval or a subset of $\mathbb{R}^n$.

Analogously to a numerical variable on a finite population, a random variable X is an application $X: \Omega \to \mathbb{R}$. For a given a subset $A \subset \mathbb{R}$, we may consider $X^{-1}(A) = \{\omega \in \Omega: X(\omega) \in A\}$. Then, $P(X \in A) = P(X^{-1}(A))$. Analogously to the case of numerical variables,

Definition 2.18 Let

$$I_X(x) = X^{-1}(]-\infty, x]) = \{\omega \in \Omega: X(\omega) \le x\}$$

(continued)

Definition 2.18 (continued)

The **cumulative distribution function** (CDF) of X is

$$F(x) = P(X \leq x) = P(X \in]-\infty, x]) = P(I_X(x)).$$

F is usually called the **distribution** of X.

Remark 2.8 If the probability P on Ω is defined by a mass function μ, then

$$F(x) = \sum_{\omega \in I_X(x)} \mu(\omega) = \sum_{\omega \in \Omega} \mu(\omega) 1_{I_X(x)}(\omega). \tag{2.15}$$

If the probability P on Ω is defined by a mass density μ, then

$$F(x) = \int_{I_X(x)} \mu(\omega) \mathrm{d}\omega = \int_{\Omega} \mu(\omega) 1_{I_X(x)}(\omega) \mathrm{d}\omega. \tag{2.16}$$

Remark 2.9 Different variables can have the same distribution. For instance, consider $\Omega = (0, 1)$, $\mu(\omega) = 1$, $X(\omega) = \omega$, $Y(\omega) = 1 - \omega$. We have

$$P(X \leq x) = P([0,\ x]) = x; P(Y \leq y) = P([1 - y,\ 1]) = y.$$

Thus, $F_X(s) = F_Y(s)$, $\forall\ s$, but $X \neq Y$.

We have

Proposition 2.22 Let X: $\Omega \longrightarrow \mathbb{R}$ be a random variable and F: $\mathbb{R} \longrightarrow \mathbb{R}$ be its CDF. Then

1. F is monotonically increasing: $x \leq y \Rightarrow F(x) \leq F(y)$;
2. $F(x) \longrightarrow 0$ if $x \longrightarrow -\infty$;
3. $F(x) \longrightarrow 1$ if $x \longrightarrow +\infty$;
4. $P(a < X \leq b) = F(b) - F(a)$;
5. $P(X = a) = F(a+) - F(a-)$.

(continued)

Proposition 2.22 (continued)

6. $P(X > x) = 1 - F(x)$.
7. If the probability P on Ω is defined by a mass function μ, then

$$F(x) = \sum_{\omega \in I_X(x)} \mu(\omega) = \sum_{\omega \in \Omega} \mu(\omega) 1_{I_X(x)}(\omega).$$

8. If the probability P on Ω is defined by a mass density μ, then

$$F(x) = \int_{I_X(x)} \mu(\omega) d\omega = \int_{\Omega} \mu(\omega) 1_{I_X(x)}(\omega) d\omega.$$

Proof The proof of assertions 1–6 is analogous to the proof of proposition 2.9. Details can be found at (Souza de Cursi, 1992) or (Souza de Cursi & Sampaio, 2015).

Results 7 and 8 follow from the definitions of mass functions and mass densities (Sect. 2.2.1).

Definition 2.19 Let $X: \Omega \longrightarrow \mathbb{R}$ be a random variable on the probability space (Ω, P). Let F_X be the CDF of X and $\mu_X((a, b)) = F_X(b) - F_X(a)$. The ***probability density FUNCTION*** (PDF) f_X of X is the density associated to μ_X, if it exists. Then

$$P(X \in A) = \mu_X(A) = \int_A \mu_X(dx) = \int_A f_X dx.$$

Notice that $\mu_X(dx) = dF_X(x)$ and

$$F_X(x) - F_X(a) = \mu_X((a, x)) = \int_a^x f_X dx \Rightarrow F'_X = f_X. \tag{2.17}$$

If P is defined by a mass density μ, then

$$\mu_X((a, x)) = \int_{I_X(x) - I_X(a)} \mu(\omega) d\omega. \tag{2.18}$$

If P is defined by a mass function μ, then

$$\mu_X((a,\ x)) = \sum_{\omega \in I_X(x) - I_X(a)} \mu(\omega). \tag{2.19}$$

Definition 2.20 Let f: $\mathbb{R} \longrightarrow \mathbb{R}$ be a function. The ***mean*** or ***expectation*** of $f(X)$ is

$$E(f(X)) = \int_{\mathbb{R}} f(x)\mu_X(dx) = \int_{\mathbb{R}} f(x) f_X(x) dx.$$

We have

Proposition 2.23 Let X: $\Omega \longrightarrow \mathbb{R}$ be a random variable and f, g: $\mathbb{R} \longrightarrow \mathbb{R}$ be two functions. Then

1. $E(\alpha f(X) + \beta g(X)) = \alpha E(f(X)) + \beta E(g(X))$;
2. $E(f(X)) = \sum_{i=1}^{m} p_i f(X_i)$;
3. If $f(X) \geq 0$, then $E(f(X)) \geq 0$;
4. If $f(X) \geq 0$, then $\lambda P(f(X) \geq \lambda) \leq E(f(X))$, $\forall\ \lambda \in \mathbb{R}$;
5. Let $1_A(\omega) = 1$, *if* $\omega \in A$; $1_A(\omega) = 0$, *if* $\omega \notin A$. Then $E(1_A) = P(A)$;
6. Let F be the CDF of X. Then $F(x) = E\left(1_{I_{X(x)}}\right)$.

Proof The proof of assertions 1–5 is analogous to the proof of proposition 2.6. Result 6 follows from result 5. Details can be found at (Souza de Cursi & Sampaio, 2015).

If P is defined by a mass density μ, then

$$E(f(X)) = \int_{\Omega} \mu(\omega) f(X(\omega)) \mathrm{d}\omega; \tag{2.20}$$

If P is defined by a mass function μ, then

$$E(f(X)) = \sum_{\omega \in \Omega} \mu(\omega) f(X(\omega)). \tag{2.21}$$

Definition 2.21

1. The **moment of order k of X** is $M_k(X) = E(X^k)$.
2. The **central moment of order k of X** is $CM_k(X) = E((X - E(X))^k)$.
3. The **variance of X** is $V(X) = CM_2(X) = E((X - E(X))^2)$.
4. The **standard deviation** of X is $(X) = \sqrt{V(X)}$.
5. The **characteristic function** of X is $\varphi_X(t) = E(e^{itX})$.

We have

Proposition 2.24 Let X: $\Omega \longrightarrow \mathbb{R}$ be a random variable. Then

1. $V(X) \geq 0$;
2. $V(X) = E(X^2) - E(X)^2$;
3. $V(X) = 0$ if and only if X is constant almost surely;
4. $E(X^2) = 0$, if and only if $X = 0$ almost surely;
5. $P(|X - E(X)| \geq \varepsilon) \leq V(X)/\varepsilon^2, \forall\, \varepsilon > 0$;
6. If $M_k(X) < \infty$, *then* $\varphi_X^{(k)}(0) = i^k M_k(X)$.
7. If $M_k(X) < \infty, \forall k \in \mathbb{N}$ *then* $\varphi_X(t) = \sum\limits_{n\in\mathbb{N}} \frac{(it)^n}{n!} M_n(X)$;
8. If $Y = aX + b$ then $\varphi_Y(t) = e^{ibt}\varphi_X(at)$.

Proof For assertions 1–5, the proof is analogous to those of propositions **2.20.6** and **2.20.7**. For the others, $\varphi_X(t) = E(e^{itX}) \Longrightarrow \varphi_X^{(k)}(t) = E\left(i^k X^k e^{itX}\right)$, hence assertion 6. Result 7 follows from the Taylor's series of $\varphi_X(t)$. Result 8 is immediate. For details, see (Souza de Cursi & Sampaio, 2015).

Example 2.10 Let us consider $\Omega = (0, 1)$ and the probability $P((a, b)) = b - a$ $(\mu(\omega) = 1)$. Let $X(\omega) = \log(1 + \omega)$. Then $X(\omega) < x \Leftrightarrow \omega < e^x - 1$. Thus

$$I_X(x) = \begin{cases} \emptyset, & \text{if } x < 0; \\ (0, \; e^x - 1), & \text{if } 0 \leq x < \log(2); \\ \Omega, & \text{if } x \geq \log(2). \end{cases}$$

Thus,

$$F_X(x) = \begin{cases} 0, & \text{if } x < 0; \\ e^x - 1, & \text{if } 0 \leq x < \log(2); \\ 1, & \text{if } x \geq \log(2). \end{cases}$$

(continued)

Example 2.10 (continued)

The PDF is

$$f_X(x) = \begin{cases} 0, & \textit{if } x < 0; \\ e^x, & \textit{if } 0 < x < \log(2); \\ 0, & \textit{if } x > \log(2). \end{cases}$$

The mean of X is

$$E(X) = \int_0^{\log(2)} x e^x dx = 2\log(2) - 1$$

In addition,

$$E\left(X^2\right) = \int_0^{\log(2)} x^2 e^x dx = 2\log^2(2) - 4\log(2) + 2,$$

So that $V(X) = 1 - 2\log^2(2)$. We have also

$$E(X) = \int_0^1 \log(1 + \omega)\mu(\omega)d\omega = 2\log(2) - 1.$$

$$E\left(X^2\right) = \int_0^1 \log^2(1 + \omega)\mu(\omega)d\omega = 2\log^2(2) - 4\log(2) + 2.$$

And, again, $V(X) = 1 - 2\log^2(2)$. The median m verifies

$$F(m) = \frac{1}{2} \Longrightarrow m = \log\left(\frac{3}{2}\right)$$

We can find a numerical approximation of the distribution by two ways. The first one consists in discretizing Ω and generating a sample from X:

(continued)

Example 2.10 (continued)

```
> mu = function(omega){1}
> X = function(omega){log(1+omega)}
> a = 0
> b = 1
> nint = 10000
> omd = seq(from = a, to = b, length = nint + 1)
> X1 = Vectorize(X)
> Xd = X1(omd)
> F1 = ecdf(Xd)
> Xmin = 0
> Xmax = log(2)
> nint = 25
> s = (Xmax - Xmin)/nint
> xabs <<- seq(from = Xmin - 2*s, to = Xmax + 2*s, by = s)
> f2 = (exp(xabs)-1)*((xabs<=log(2)) & (xabs >= 0)) + (xabs>log(2))
> f1 = F1(xabs)
> plot(xabs,f2,type = "n",main = "method 1(sample)",xlab="x",ylab="")
> lines(xabs,f2,lty="solid",lwd="2",col="blue")
> points(xabs,f1,pch=19,cex = 1, col = "red")
> legend("topleft", legend=c("exact", "calculated"),
+        col=c("blue", "red"), lty=c("solid",NA),
+        pch = c(NA,19), lwd = c("2",NA))
```

The second one consists in integrating the mass density:

```
> f3 = numeric(length=length(xabs))
> library(pracma)
> for (i in 1:length(xabs)){
+   fun_a = function(omega){mu(omega)*(X(omega)<=xabs[i])}
+   f3[i] = integral(Vectorize(fun_a),a,b,no_intervals=100 )
+ }
> plot(xabs,f2,type = "n",main = "method 2(integral)",xlab="x",ylab="")
> lines(xabs,f2,lty="solid",lwd="2",col="blue")
> points(xabs,f3,pch=19,cex = 1, col = "red")
> legend("topleft", legend=c("exact", "calculated"),
+        col=c("blue", "red"), lty=c("solid",NA),
+        pch = c(NA,19), lwd = c("2",NA))
```

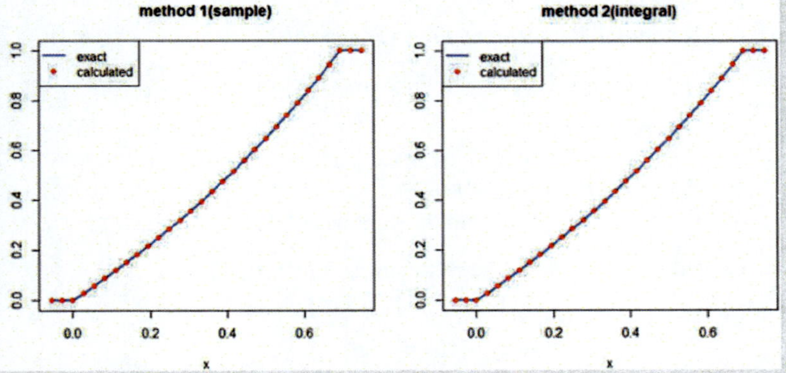

Exercises

1. Consider $\Omega = (0, 1)$ and the mass density $\mu(\omega) = \alpha\omega$.
   a) Show that $\alpha = 2$.
   b) Find the distribution of $X(\omega) = e^{\omega}$.
   c) Find the expression of $M_k(X)$.
   d) Determine the mean and the variance of X.
2. Consider $\Omega = (0, 1)$ and the mass density $\mu(\omega) = \alpha$.
   a) Show that $\alpha = 1$.
   b) Find the distribution of $X(\omega) = e^{\omega}$.
   c) Find the expression of $M_k(X)$.
   d) Determine the mean and the variance of X.
3. Consider $\Omega = (0, 1)$ and the mass density $\mu(\omega) = \alpha\omega^2$.
   a) Show that $\alpha = 3$.
   b) Find the distribution of $X(\omega) = \omega$.
   c) Find the expression of $M_k(X)$.
   d) Determine the mean and the variance of X.
4. Consider $\Omega = (0, 1)$ and the mass density $\mu(\omega) = 1$.
   a) Find the distribution of $X(\omega) = \omega$.
   b) Find the expression of $M_k(X)$.
   c) Determine the mean and the variance of X.
5. Consider $\Omega = \{-1, 0, 1\}$ and the mass function $\mu(1) = \mu(-1) = \frac{1}{3}$.
   a) Determine $\mu(0)$.
   b) Find the distribution of $X(\omega) = \omega^2$.
   c) Find the expression of $M_k(X)$.
   d) Determine the mean and the variance of X.
6. Consider $\Omega = \{-1, 0, 1\}$ and the mass function $\mu(1) = \mu(-1) = \frac{1}{6}$.
   a) Determine $\mu(0)$.
   b) Find the distribution of $X(\omega) = \omega^2$.
   c) Find the expression of $M_k(X)$.
   d) Determine the mean and the variance of X.
7. Consider $\Omega = \{-1, 0, 1\}$ and the mass function $\mu(1) = \mu(-1) = \frac{1}{6}$.
   a) Determine $\mu(0)$.
   b) Find the distribution of $X(\omega) = \omega^2$.

(continued)

c) Find the expression of $M_k(X)$.
d) Determine the mean and the variance of X.

8. Let $\Omega = \{0, 1, \ldots, n\}$ and $X(\omega) = \omega \; mod \; 2$ (remainder after division by 2). Assume the elements of Ω as equiprobable.

   a) Show that $X(\Omega) = \{0, 1\}$.
   b) Determine $p = P(X = 0)$ when $n = 2k$.
   c) Determine $p = P(X = 0)$ when $n = 2k + 1$.
   d) Determine the CDF F as a function of p.

9. Let X be a random variable having the CDF $F_X(x) = x^3$ for $0 < x < 1$.

   a) Show that $P(X \leq 0) = 0$.
   b) Show that $P(X > 1) = 1$.
   c) Find its PDF f_X.
   d) Determine the mean and the variance of X.

10. Let $\Omega = \{1, 2, 3, 4, 5\}$ and $X(\omega) = \frac{1}{\omega+1}$. Assume the elements of Ω as equiprobable.

    a) Determine $X(\Omega)$.
    b) Determine the distribution of X.
    c) Determine the mean and the variance of X.

11. Let $\Omega = (-1, 1)$ and $X(\omega) = \omega^3$, with $P((a, \; b)) = \frac{b-a}{2} \; \left(\mu(\omega) = \frac{1}{2}\right)$.

    a) Determine $X(\Omega)$.
    b) Determine the distribution of X.
    c) Determine the mean and the variance of X.

12. Determine the characteristic functions of the following random variables

    a) X is such that $P(X = 0) = 1$.
    b) X is such that $P(X = 1) = p$, $P(X = 0) = 1 - p$.
    c) X is Poisson distributed with parameter λ: $P(X = k) = \frac{e^{-\lambda}\lambda^k}{k!}$.
    d) X is uniformly distributed on (a, b) – *its PDF is*: $f(x) = \frac{1}{b-a}$ on (a, b)
    e) X is exponentially distributed –*its PDF is*: $f(x) = \lambda e^{-\lambda x}$ on $(0, +\infty)$.

2.6.1 Numerical Evaluation of Statistics

In Sect. 2.10, we consider samples from random variables and their use for the evaluation of statistics. Here, we consider the numerical evaluation of statistics using numerical integration. The use of this approach requests the knowledge of the distribution of the variable or of the mass density. In such a situation, the mean, the moments, and the variance can be obtained by numerical integration. A first approximation consists in evaluating

$$E(\phi(X)) \approx \int_{-A}^{A} \phi(x)dF(x) \tag{2.22}$$

and using numerical integration to evaluate the left-hand side. For instance, we can consider $-A = x_0 < x_1 < \ldots < x_n = A$ and

$$E(\phi(X)) \approx \sum_{i=0}^{n} \phi\left(x_{i+\frac{1}{2}}\right)(F(x_{i+1}) - F(x_i)),\ x_{i+\frac{1}{2}} = \frac{x_{i+1} + x_i}{2}. \tag{2.23}$$

If the PDF f is known, we can use the approximation

$$E(\phi(X)) \approx \int_{-A}^{A} \phi(x)f(x)dx. \tag{2.24}$$

Again, the left-hand side can be evaluated by numerical integration. In this case, the methods presented in Sect. 1.16 can be used. Of course, a simple estimate may be generated by considering a partition $-A = x_0 < x_1 < \ldots < x_n = A$ and

$$E(\phi(X)) \approx \sum_{i=0}^{n} p_{i+\frac{1}{2}}\phi\left(x_{i+\frac{1}{2}}\right),\ p_{i+\frac{1}{2}} = f\left(x_{i+\frac{1}{2}}\right)(x_{i+1} - x_i). \tag{2.25}$$

A mass density μ can be used into an analogous way with a partition of the domain of the values of ω:

$$E(\phi(X)) \approx \sum_{i=0}^{n} p_{i+\frac{1}{2}}\phi\left(X\left(\omega_{i+\frac{1}{2}}\right)\right),\ p_{i+\frac{1}{2}} = \mu\left(\omega_{i+\frac{1}{2}}\right)(\omega_{i+1} - \omega_i). \tag{2.26}$$

The evaluation of these expressions reduces to sums to be calculated by using loops. For instance,

```
evalmean1=function (phi,f,xmin,xmax,n){
  h = (xmax-xmin)/n
  x = seq(xmin,xmax,by=h)
  s=0
  for (i in 1:n){
    xm = 0.5*(x[i]+x[i+1])
    p = f(xm)*h
    s = s + p*phi(xm)
  }
  return(s)
}
```

```
evalmean2=function (phi,mu,X,ommin,ommax,n){
  h = (ommax-ommin)/n
  om = seq(ommin,ommax,by=h)
  s=0
  for (i in 1:n){
    omm = 0.5*(om[i]+om[i+1])
    p = mu(omm)*h
    s = s + p*phi(X(omm))
  }
  return(s)
}
```

Example 2.11 Let us consider $\Omega = (0, 1)$ and the density $\mu(\omega) = 1$. Let $X(\omega) = \omega^2$. Consider a partition of Ω in n equally spaced subintervals: $\omega_i = \frac{i}{n}$. The results furnished by Eq. (2.26) for three values of n are shown in Table 2.4.

Table 2.4 Numerical evaluation of the mean and variance using the mass function

n	100	500	1000	*exact*
$E(X)$	0,333,325	0,333,333	0,33,333,325	$\frac{1}{3}$
$E(X^2)$	0,1,999,833	0,1,999,993	0,1,999,998	$\frac{1}{5}$
$V(X)$	0,088877778	0,088888444	0,088888778	$\frac{4}{45}$

We can also use the PDF $f(x) = \frac{1}{2\sqrt{x}}$ and a partition of (0,1) to evaluate $\int_0^1 x^k f(x)dx$. The results furnished by Eq. (2.25) are shown in Table 2.5.

Table 2.5 Numerical evaluation of the mean and variance using the PDF

n	100	500	1000	*exact*
$E(X)$	0,333,362,736	0,333,336,015	0,333,334,286	$\frac{1}{3}$
$E(X^2)$	0,199,996,957	0,199,999,876	0,199,999,969	$\frac{1}{5}$
$V(X)$	0,088866244	0,088886978	0,088888223	$\frac{4}{45}$

Of course, we can use integrate to determine these values:

```
> mu <- function(om){1}
> X <- function(om){om^2}
> g <- function(om){mu(om)*X(om)}
> integrate(g,0,1)
0.3333333 with absolute error < 3.7e-15.
> f <- function(x){1/(2*sqrt(x))}
> g <- function(x){x*f(x)}
> integrate(g,0,1)
0.3333334 with absolute error < 0.00011
```

Exercises

1. Let X be a continuous random variable having the CDF F. Assume that $F(x) = x^3$ for $0 < x < 1$.
   a) Determine the PDF $f(x)$.
   b) Determine the exact values of $E(X)$, $E(X^2)$, $E(X^3)$, $V(X)$.

(continued)

c) Consider a partition of $(0, 1)$ in $n = 100$ equally spaced subintervals: $x_i = \frac{i}{n}$. Use this partition and $f(x)$ to estimate $E(X)$, $E(X^2)$, $E(X^3)$, $V(X)$.
d) Determine the same values using `integrate`.
e) Determine the mean and the variance of X.

2. Let $\Omega = (0, 1)$, $X(\omega) = \frac{1}{\omega+1}$, $\mu(\omega) = 1$.

a) Determine its CDF $F(x)$.
b) Determine its PDF $f(x)$.
c) Determine the exact values of $E(X)$, $E(X^2)$, $E(X^3)$, $V(X)$.
d) Consider a partition of $(0, 1)$ in $n = 250$ equally spaced subintervals: $x_i = \frac{i}{n}$. Use this partition and $f(x)$ to estimate $E(X)$, $E(X^2)$, $E(X^3)$, $V(X)$.
e) Determine the same values using `integrate`.
f) Determine the mean and the variance of X.

3. Let $\Omega = (-1, 1)$, $X(\omega) = \omega^3$, $\mu(\omega) = 1/2$.

a) Determine its CDF $F(x)$.
b) Determine its PDF $f(x)$.
c) Determine the exact values of $E(X)$, $E(X^2)$, $E(X^3)$, $V(X)$.
d) Consider a partition of $(0, 1)$ in $n = 1000$ equally spaced subintervals: $x_i = \frac{i}{n}$. Use this partition and $f(x)$ to estimate $E(X)$, $E(X^2)$, $E(X^3)$, $V(X)$.
e) Determine the same values using `integrate`.
f) Determine the mean and the variance of X.

2.7 Random Vectors

A random vector $\boldsymbol{X} = (X_1, \ldots, X_n)$ of dimension n is an application $\boldsymbol{X}\colon \Omega \rightarrow \mathbb{R}^n$; id est, $\boldsymbol{X}$ is a vector of numerical characteristics of the elements of Ω. Analogously to random variables, probabilities on the values of $\boldsymbol{X}$ are generated by the probabilities on Ω: for given a subset $\boldsymbol{A} \subset \mathbb{R}^n$, we may consider $\boldsymbol{X}^{-1}(\boldsymbol{A}) = \{\omega \in \Omega\colon \boldsymbol{X}(\omega) \in \boldsymbol{A}\}$. Then, $P(\boldsymbol{X} \in \boldsymbol{A}) = P(\boldsymbol{X}^{-1}(\boldsymbol{A}))$. The cumulative function is defined analogously to the one-dimensional situation: for $\boldsymbol{X} = (X_1, \ldots, X_n)$ and $\boldsymbol{x} = (x_1, \ldots, x_n)$, we denote "$\boldsymbol{X} \leq \boldsymbol{x}$" the region "$X_1 \leq x_1, X_2 \leq x_2, \ldots, X_n \leq x_n$", id est, "$\boldsymbol{X} \leq \boldsymbol{x}$" $= \boldsymbol{A}(\boldsymbol{x})$, where $\boldsymbol{A}(\boldsymbol{x}) = \prod_{i=1}^{n} A_i(x_i)$, $A_i(x_i) = (-\infty, x_i)$. Let $I_i(x_i) = X_i^{-1}(A_i(x_i))$. Then, $\boldsymbol{X}^{-1}(\boldsymbol{A}(\boldsymbol{x})) = \boldsymbol{I}_{\boldsymbol{X}}(\boldsymbol{x}) = \bigcap_{i=1}^{n} I_i(x_i)$ and the CDF of $\boldsymbol{X}$ is $F(\boldsymbol{x}) = P(\boldsymbol{I}(\boldsymbol{x}))$. The PDF is defined as $f = \frac{\partial^n F}{\partial x_1 \partial x_2 \ldots \partial x_n}$. For a couple $\boldsymbol{X} = (X_1, X_2)$ of random variables, we have

Definition 2.22 Let $X = (X_1, X_2)$ be a couple of random variables defined on a probability space (Ω, P).

The **cumulative distribution function (*CDF*)** of X is:

$$F(x_1, x_2) = P(X_1 \leq x_1, X_2 \leq x_2)$$

F is the **distribution** of X.

The **probability density function (*PDF*)** is $f = \frac{\partial^2 F}{\partial x_1 \partial x_2}$ (if it exists).

The **marginal distribution** of X_i is $F_i(x_i) = P(X_i \leq x_i)$. We have

$$F_1(x_1) = F(x_1, +\infty), F_2(x_2) = F(+\infty, x_2)$$

The **marginal density** of X_i is $f_i = \frac{\partial F_i}{\partial x_i} = \int_{-\infty}^{+\infty} f(x_1, x_2) dx_i$.

The **copula** associated to $X = (X_1, X_2)$ is, for $0 < c_1, c_2 < 1$:

$$C(c_1, c_2) = P(F_1 \leq c_1, F_2 \leq c_2)$$

The **covariance** between X_1 and X_2 is

$$cov(X_1, X_2) = E((X_1 - E(X_1))(X_2 - E(X_2))) = E(X_1 X_2) - E(X_1)E(X_2).$$

The **linear correlation coefficient** between X_1 and X_2 is

$$\rho(X_1, X_2) = \frac{cov(X_1, X_2)}{\sqrt{V(X_1)V(X_2)}}.$$

The best approximation of X_i by an affine function of X_j in the least squares sense is $\ell(X_j) = \alpha X_j + \beta$, with

$$\alpha = \frac{cov(X_i, X_j)}{V(X_j)}, \beta = E(X_i) - \alpha E(X_j).$$

We have $E((X_i - \ell(X_j))^2) = V(X_i)(1 - |\rho(X_i, X_j)|^2)$. Thus,

$$X_i = \ell(X_j) \ a.s. \Leftrightarrow |\rho(X_i, X_j)| = 1.$$

The covariance matrix $C(X)$ and the correlation matrix $\rho(X)$ of X are given by

$$C_{ij}(X) = Cov(X_i, X_j), \rho_{ij}(X) = Cov(X_i, X_j).$$

(continued)

Definition 2.22 (continued)

The **conditional mean** of X_i with respect to X_j is the best approximation of X_i by a function $\psi_j(X_j)$ in the least squares' sense. We have

$$E(X_2|X_1) = \psi_1(x_1) = \int x_2 f(x_2|x_1)\, dx_2, f(x_2|x_1) = \frac{f(x_1, x_2)}{f_1(x_1)}.$$

$$E(X_1|X_2) = \psi_2(x_2) = \int x_1 f(x_1|x_2)\, dx_1, f(x_1|x_2) = \frac{f(x_1, x_2)}{f_2(x_2)}.$$

The generalized correlation coefficient is

$$R(X_1, X_2) = \frac{1}{2}(\eta(X_1|X_2) + \eta(X_2|X_1)),$$

$$\eta(X_1|X_2) = \exp\left(-\|X_1 - E(X_1|X_2)\|^2\right),$$

$$\eta(X_2|X_1) = \exp\left(-\|X_2 - E(X_2|X_1)\|^2\right).$$

Remark 2.10 If the probability P on Ω is defined by a mass function μ, then

$$F(x) = \sum_{\omega \in I_X(x)} \mu(\omega) = \sum_{\omega \in \Omega} \mu(\omega) 1_{I_X(x)}(\omega), \tag{2.27}$$

$$1_{I_X(x)}(\omega) = \prod_{i=1}^{n} 1_{I_i(x_i)}(\omega). \tag{2.28}$$

If the probability P on Ω is defined by a mass density μ, then

$$F(x) = \int_{I_X(x)} \mu(\omega) \mathrm{d}\omega = \int_{\Omega} \mu(\omega) 1_{I_X(x)}(\omega) \mathrm{d}\omega. \tag{2.29}$$

Remark 2.11 If $X = (X_1, X_2)$ has a PDF f, then

$$E(\phi(X_1, X_2)) = \iint_{-\infty}^{+\infty} \phi(x_1, x_2) f(x_1, x_2) dx_1 dx_2.$$

(continued)

Remark 2.11 (continued)

If Ω has a mass function μ, then

$$E(\phi(X_1,\ X_2))=\sum_{\omega\in\Omega}\mu(\omega)\phi(X_1(\omega),X_2(\omega)).$$

If Ω has a mass density μ, then

$$E(\phi(X_1,\ X_2))=\int_{\Omega}\mu(\omega)\phi(X_1(\omega),X_2(\omega))d\omega.$$

If the probability P on Ω is defined by a mass density μ, then

$$F(x)=\int_{I_x(x)}\mu(\omega)\mathrm{d}\omega=\int_{\Omega}\mu(\omega)1_{I_x(x)}(\omega)\mathrm{d}\omega.$$

Example 2.12 Let us consider $\Omega = (0, 1)$ and the density $\mu(\omega) = 1$. Let $X_1(\omega) = \omega^2$; $X_2(\omega) = \omega$. We have, for $0 < x_1, x_2 < 1$:

$$X_1<x_1,X_2<x_2\Leftrightarrow\omega<\min\{\sqrt{x_1},x_2\}$$

Thus, for $0 < x_1, x_2 < 1$:

$$F(x_1,x_2)=\min\{\sqrt{x_1},x_2\}=\begin{cases}\sqrt{x_1}, & \text{if } x_1^2<x_2\\ x_2, & \text{if } x_1^2>x_2\end{cases}$$

This function cannot be derived in the usual sense, due to the change of definition across the curve $x_2=x_1^2$. The density is null in the subregions $x_2<x_1^2$ and $x_2>x_1^2$, but corresponds to a Dirac mass – See, for instance (Souza de Cursi, 2015) – On the curve separating these regions. The marginal distributions are, for $0 < x_1, x_2 < 1$:

$$F_1(x_1)=\sqrt{x_1},\quad F_2(x_2)=x_2.$$

Although we cannot derive F, marginal densities may be determined:

$$f_1(x_1)=\frac{1}{2\sqrt{x_1}},\quad f_2(x_2)=1.$$

(continued)

Example 2.12 (continued)

The copula associated to $\boldsymbol{X}$ is $C(c_1, c_2) = \min\{c_1, c_2\}$. We have

$$E\left(X_1^k\right) = \frac{1}{2}\int_0^1 x_1^{k-\frac{1}{2}} dx_1 = \frac{1}{2k+1},$$

$$E\left(X_2^k\right) = \int_0^1 x_2^k dx_2 = \frac{1}{k+1}.$$

Thus,

$$E(X_1) = \frac{1}{3}, V(X_1) = \frac{4}{45};\ E(X_2) = \frac{1}{2}, V(X_2) = \frac{1}{12}.$$

In addition,

$$E(X_1X_2) = \frac{1}{\int_0^1 d\omega}\int_0^1 X_1(\omega)X_2(\omega)d\omega = \int_0^1 \omega^3 d\omega = \frac{1}{4},$$

$$cov(X_1, X_2) = \frac{1}{12},\quad \rho(X_1, X_2) = \frac{\sqrt{15}}{4}.$$

The best least squares approximation by linear functions are

$$X_1 \approx X_2 - \frac{1}{6};\ X_2 \approx \frac{15}{16}X_1 + \frac{1}{16}.$$

The conditional mean $E(X_i|X_j)$ is determined by finding $\psi(X_j)$ that minimizes

$$J(\psi) = E\left(\left(X_i - \psi(X_j)\right)^2\right).$$

Then

$$E(X_1|X_2) = X_2^2, E(X_2|X_1) = \sqrt{X_1}.$$

Thus,

$$\|X_1 - E(X_1|X_2)\| = \|X_2 - E(X_2|X_1)\| = 0,$$

So that

$$\eta(X_1|X_2) = \eta(X_2|X_1) = 1 \Rightarrow R(X_1, X_2) = 1.$$

Example 2.13 Let us consider $\Omega = \{-2, -1, 0, 1, 2\}$. Assume that all the elements are equiprobable. Let $X_1(\omega) = \omega^2$; $X_2(\omega) = |\omega|$. We have

$$\boldsymbol{X}(\Omega) = \{(4,\ 2), (1,\ 1), (0,\ 0), (1,\ 1), (4,\ 2)\}$$

Thus,

$$P(\boldsymbol{X} = (4,\ 2)) = P(\boldsymbol{X} = (1,\ 1)) = \frac{2}{5}; P(\boldsymbol{X} = (0,\ 0)) = \frac{1}{5}.$$

The marginal distributions are

$$P(X_1 = 4) = P(X_1 = 1) = \frac{2}{5};\ P(X_1 = 0) = \frac{1}{5}.$$

$$P(X_2 = 2) = P(X_2 = 1) = \frac{2}{5};\ P(X_2 = 0) = \frac{1}{5}.$$

We have

$$E(X_1) = 2,\ V(X_1) = \frac{14}{5};\ E(X_2) = \frac{6}{5},\ V(X_2) = \frac{14}{5}$$

In addition, $E(X_1 X_2) = \frac{18}{5}$, so that

$$cov(X_1, X_2) = \frac{6}{5},\quad \rho(X_1, X_2) = \frac{3\sqrt{5}}{7}.$$

The best least squares approximation by linear functions are

$$X_1 \approx \frac{15}{7} X_2 - \frac{4}{7};\ X_2 \approx \frac{3}{7} X_1 + \frac{12}{35}.$$

The conditional mean $E(X_i | X_j)$ is determined by finding $\psi(X_j)$ that minimizes

$$J(\psi) = E\left((X_i - \psi(X_j))^2 \right),$$

For instance, we evaluate $E(X_1 | X_2)$ by minimizing

$$J(\psi) = \frac{2}{5}(4 - \psi(2))^2 + \frac{2}{5}(1 - \psi(1))^2 + \frac{1}{5}(0 - \psi(0))^2,$$

(continued)

Example 2.13 (continued)

So that $\psi(2) = 4$, $\psi(1) = 1$, $\psi(0) = 0$: We may take $\psi(X_2) = X_2^2$. In addition

$$P(X_1 = i | X_2 = j) = \frac{P(X = (i,\ j))}{P(X_2 = j)}.$$

Then

$$P(X_1 = 4 | X_2 = 2) = P(X_1 = 1 | X_2 = 1) = P(X_1 = 0 | X_2 = 0) = 1.$$

The other probabilities are null. To evaluate $E(X_2 | X_1)$, we minimize

$$J(\psi) = \frac{2}{5}(2 - \psi(4))^2 + \frac{2}{5}(1 - \psi(1))^2 + \frac{1}{5}(0 - \psi(0))^2,$$

So that $\psi(4) = 2$, $\psi(1) = 1$, $\psi(0) = 0$: We may take $\psi(X_1) = \sqrt{X_1}$. Here

$$P(X_2 = 2 | X_1 = 4) = P(X_2 = 1 | X_1 = 1) = P(X_2 = 0 | X_1 = 0) = 1$$

And the other probabilities are null. Again, we have

$$\|X_1 - E(X_1 | X_2)\| = \|X_2 - E(X_2 | X_1)\| = 0,$$
$$\eta(X_1 | X_2) = \eta(X_2 | X_1) = 1 \Longrightarrow R(X_1,\ X_2) = 1.$$

Let us introduce the independence between variables:

Definition 2.23 Let $X = (X_1, X_2)$ be a couple of random variables defined on a probability space (Ω, P). We say that $X = (X_1, X_2)$ is a couple of **independent** variables if and only if ITS CDF is the product of the marginal CDFs of the components:

$$F(x_1, x_2) = F_1(x_1) F_2(x_2)$$

In such a case, we say that X_1 and X_2 are **independent**.

Proposition 2.25 Let $X = (X_1, X_2)$ be a couple of random variables defined on a probability space (Ω, P). The following assertions are equivalent:

1. X_1 and X_2 are independent;
2. The density of the pair is the product of the marginal densities of the components: $f(x_1, x_2) = f_1(x_1) f_2(x_2)$;

(continued)

Proposition 2.25 (continued)

3. The copula associated to $X = (X_1, X_2)$ is, for $0 < c_1, c_2 < 1$:
4. $C(c_1, c_2) = P(F_1 < c_1)P(F_2 < c_2) = c_1c_2$;
5. The characteristic function of the pair is the product of the of the characteristic functions of the components: $\phi(t_1, t_2) = E(e^{it_1X_1+it_2X_2}) = \phi_1(t_1)\phi_2(t_2)$, $\phi_i(s) = E(e^{isX_i})$.

Proof We have

1. It is immediate that assertions 1 and 2 are equivalent.
2. We have

$$P(F_i < c_i) = P\left(X_i \le F_i^{-1}(c_i)\right) = F_i\left(F_i^{-1}(c_i)\right) = c_i.$$

If the variables are independent, then

$$P(F_1 \le c_1, F_2 \le c_2) = P\left(X_1 \le F_1^{-1}(c_1)\right)P\left(X_2 \le F_2^{-1}(c_2)\right) = c_1c_2.$$

Conversely, assume that $C(c_1, c_2) = c_1c_2$. Then,

$$P\left(X_1 \le F_1^{-1}(c_1), X_2 \le F_2^{-1}(c_2)\right) = P\left(X_1 \le F_1^{-1}(c_1)\right)P\left(X_2 \le F_2^{-1}(c_2)\right).$$

Taking $c_1 = F_1(x_1)$, $c_2 = F_2(x_2)$, we have

$$P(X_1 \le x_1, X_2 \le x_2) = P(X_1 \le x_1)P(X_2 \le x_2),$$

So that the variables are independent. Thus, independence of (X_1, X_2) is equivalent to $C(c_1, c_2) = c_1c_2$.
3. If X_1 and X_2 are independent, then $E(e^{it_1X_1+it_2X_2}) = E(e^{it_1X_1})E(e^{it_2X_2})$, so that $\phi(t_1, t_2) = \phi_1(t_1)\phi_2(t_2)$. Conversely, if the characteristic function of the pair is the product of the characteristic function of the components, then an inverse Fourier transform shows that $f(x_1, x_2) = f_1(x_1)f_2(x_2)$, so that the variables are independent.

Proposition 2.26 If X_1 and X_2 are independent, then $v(X_1, X_2) = \rho(X_1, X_2) = 0$, $E(X_1 | X_2) = E(X_1)$, $E(X_2 | X_1) = E(X_2)$.

Proof Le f be density of the couple and f_i be the marginal density of X_i. We have $f(x_1, x_2) = f_1(x_1)f_2(x_2)$, so that $E(X_1X_2) = E(X_1)E(X_2)$. Thus, $cov(X_1, X_2) = 0$. Consequently, we have also $\rho(X_1, X_2) = 0$. In addition, $f(x_1|x_2) = f_1(x_1)$ and $f(x_2|x_1) = f_2(x_2)$, so that $E(X_1|X_2) = E(X_1)$ and $E(X_2|X_1) = E(X_2)$.

Remark 2.12 The converse of proposition 2.26 is generally false, except for gaussian vectors.

Example 2.14 Let us consider a couple $\boldsymbol{X} = (X_1, X_2)$ having AS PDF

$$f(x_1, x_2) = \begin{cases} 1, & \text{if } 0 < x_1,\ x_2 < 1 \\ 0, & \text{otherwise.} \end{cases}$$

The marginal densities verify

$$f_i(x_i) = \begin{cases} 1, & \text{if } 0 < x_i < 1 \\ 0, & \text{otherwise.} \end{cases}$$

Thus, $f(x_1, x_2) = f_1(x_1)f_2(x_2)$ and the variables are independent. Notice that, for $0 < x_i < 1$: $F_i(x_i) = x_i$ and $F(x_1, x_2) = x_1x_2 = F_1(x_1)F_2(x_2)$.

Analogously, $P(F_1 < c_1, F_2 < c_2) = P(X_1 < c_1, X_2 < c_2) = c_1c_2$. Notice that $f(x_1|x_2) = f(x_2|x_1) = 1$, for $0 < x_1,\ x_2 < 1$. Thus, $E(X_1|X_2) = E(X_2|X_1) = \frac{1}{2} = E(X_1) = E(X_2)$.

Example 2.15 Let us consider a couple $\boldsymbol{X} = (X_1, X_2)$ having AS PDF

$$P(\boldsymbol{X} = (0,\ 0)) = P(\boldsymbol{X} = (0,\ 1)) = P(\boldsymbol{X} = (1,\ 0)) = P(\boldsymbol{X} = (1,\ 1)) = \frac{1}{4}$$

The marginal distributions are

$$P(X_i = 0) = P(X_i = 1) = \frac{1}{2}.$$

(continued)

Example 2.15 (continued)

Thus,

$$P(\boldsymbol{X}=(i,\ j))=P(X_1=i,X_2=j)=P(X_1=i)P(X_2=j)$$

And the variables are independent. Notice that $P(X_1=i|X_2=j)=P(X_1=i)$ and $P(X_2=j|X_1=i)=P(X_2=j)$, so that $E(X_1|X_2)=E(X_2|X_1)=\frac{1}{2}=E(X_1)=E(X_2)$.

Exercises

1. Let $\Omega=(0,1)$, with $P((a,b))=b-a$ ($\mu(\omega)=1$) and $\boldsymbol{X}=(X_1,X_2)$, $X_1(\omega)=\omega^2$, $X_2(\omega)=\omega^4$.
   a) Find the CDF F of $\boldsymbol{X}$.
   b) Determine the PDF f of $\boldsymbol{X}$.
   c) Determine the marginal distributions and densities.
   d) Are the variables independent?
   e) Find the distributions of $E(X_2|X_1)$ and $E(X_1|X_2)$.
2. Let $\boldsymbol{X}=(X_1,X_2)$ be a pair of random variables having as density

$$f(x_1,x_2)=\begin{cases} x_1+x_2, & \text{if } 0<x_1,\ x_2<1; \\ 0, & \text{otherwise.}\end{cases}$$

   a) Determine the marginal distributions and densities.
   b) Are the variables independent?
   c) Find the distributions of $E(X_2|X_1)$ and $E(X_1|X_2)$.
3. Let $\boldsymbol{X}=(X_1,X_2)$ be a pair of random variables having as density

$$f(x_1,x_2)=\begin{cases} \dfrac{2}{(x_2+1)^3}, & \text{if } 0<x_1<x_2; \\ 0, & \text{otherwise.}\end{cases}$$

   a) Determine the marginal distributions and densities.
   b) Are the variables independent?
   c) Find the distributions of $E(X_2|X_1)$ and $E(X_1|X_2)$.

(continued)

4. Let $X = (X_1, X_2)$ be a pair of non-negative independent random variables such that

$$f_1(x_1) = \lambda e^{-\lambda x_1}, f_2(x_2) = \mu e^{-\mu x_2}.$$

Let $Y = X_1 + X_2$.

a) Determine the distribution of Y.
b) Determine the distribution of the couple (X_1, Y).
c) Find the distribution of $E(X_1 | Y)$.

2.8 Discrete and Continuous Random Variables

It is usual to consider **discrete** and **continuous** random variables, according to the range of X: if the image $X(\Omega)$ is a finite or enumerable set, then we say that X is discrete. If the image $X(\Omega)$ is an interval of real numbers (eventually infinite or semi-infinite), we say that X is continuous.

2.8.1 Discrete Variables

As previously remarked, we say that X is discrete if and only if

$$X(\Omega) = \{X_1, \ldots, X_n\} \text{ (finite) } or\ X(\Omega) = \{X_i : i \in \mathbb{N}\} \text{(enumerable)}.$$

Assume that $X_1 < X_2 < \ldots < X_i < X_{i+1} < \ldots$. Let $p_i = P(X = X_i)$. The CDF of X is

$$F(x) = \begin{cases} 0, & \text{if } x \le \min\{X_i\}; \\ \sum_{k=1}^{i} p_k, & \text{if } X_i < x \le X_{i+1}; \\ 1, & \text{if } x > \sup\{X_i\}. \end{cases} \tag{2.30}$$

In addition,

$$E(\phi(X)) = \sum_i p_i \phi(X_i). \tag{2.31}$$

Thus,

$$M_k(X) = \sum_i p_i X_i^k,\ E(X) = \sum_i p_i X_i,\ E(X^2) = \sum_i p_i X_i^2,\ \varphi(t) = \sum_i p_i e^{itX_i}.$$

Examples of discrete variables are:

- The uniform distribution on $\{1,\ldots,n\}$:, $P(X=i)=\frac{1}{n}$, for $1\le i\le n$. We have

$$E(X)=\frac{n+1}{2},\ V(X)=\frac{n^2-1}{12},\ \sigma(X)=\sqrt{\frac{n^2-1}{12}}.$$

You can generate a sample of k variates from the uniform distribution on $\{1,\ldots,n\}$ using the instruction `sample(1:n, k, replace=TRUE)`. Taking `replace=FALSE` generates a sample without repetition – it can be used to generate random permutations. For instance:

```
> n = 5
> k = 8
> sample(1:n,8,replace=TRUE) # a sample of 8 variates
[1] 5 3 1 3 2 3 4 5
> sample(1:n,n,replace=FALSE) # a random permutation
[1] 3 1 4 5 2
```

- Bernoulli $\mathcal{B}(p)$: describes success (value 1) or failure (value 0) in a trial for a probability of success p. $X(\Omega)=\{0,1\}$, $P(X=1)=p$, $P(X=0)=q=1-p$. We have

$$E(X)=p,\ V(X)=pq,\quad \sigma(X)=\sqrt{pq}.$$

- Binomial $\mathcal{B}(n,p)$: gives the number of successful trials among n when the probability of success is p. It may be interpreted as the law of a sum of n independent Bernoulli laws. $X(\Omega)=\{0,1,\ldots,n\}$, $P(X=k)=\binom{n}{k}p^kq^{n-k}$, $q=1-p$. We have

$$E(X)=np,\ V(X)=npq,\quad \sigma(X)=\sqrt{npq}.$$

Notice that $\mathcal{B}(1,p)=\mathcal{B}(p)$.

R offers built-in functions to deal with Bernoulli and Binomial, namely `dbinom`, `pbinom`, `qbinom`, and `rbinom`. For instance, `dbinom(k,n,p)` furnishes the value of $P(X=k)$ for $\mathcal{B}(n,p)$ and `pbinom(k,n,p)` returns the value of $P(X\le k)$ for $\mathcal{B}(n,p)$ – setting the flag `lower.tail` to `FALSE` returns $P(X>k)$. `qbinom(p,n,a)` returns the smallest value of k such that $P(X\le k)\ge a$ for $\mathcal{B}(n,p)$ – the flag `lower.tail=FALSE` furnishes the

(continued)

smallest value of k such that $P(X > k) \geq a$. rbinom(n,k,p) generates a sample of k variates from $\mathcal{B}(n, p)$. For instance, let us consider $n = 5$, $p = 2/10$, $q = 8/10$. We have

```
> n = 5
> p = 0.2
> dbinom(0,n,p) # P(X=0)
[1] 0.32768
> dbinom(1,n,p) # P(X=1)
[1] 0.4096
> dbinom(2,n,p) # P(X=2)
[1] 0.2048
> pbinom(2,n,p) # P(X<=2)
[1] 0.94208
> pbinom(2,n,p,lower.tail=FALSE) # P(X>2)
[1] 0.05792
> qbinom(0.8,n,p) # k s.t. P(X<=k)>= 0.8
[1] 2
> dbinom(0,n,p)+dbinom(1,n,p)+dbinom(2,n,p) # P(X<= 2)
[1] 0.94208
> pbinom(2,n,p) # P(X<= 2)
[1] 0.94208
> pbinom(3,n,p) # P(X<= 3)
[1] 0.99328
> qbinom(0.2,n,p,lower.tail=FALSE) # k s.t. P(X>k)>= 0.2
[1] 2
> dbinom(5,n,p)+dbinom(4,n,p)+dbinom(3,n,p)+dbinom(2,n,p) # P(X>1)
[1] 0.26272
> 1 - pbinom(1,n,p) # P(X>1)
[1] 0.26272
> 1 - pbinom(2,n,p) # P(X>2)
[1] 0.05792
> rbinom(8,n,p)
[1] 1 1 0 0 0 0 1 2
> |
```

As observed above, Bernoulli distributions correspond to $\mathcal{B}(1, p)$. As n alternative, you can use the package Rlab, which proposes analogous instructions dbern, pbern, qbern, and rbern.

- Negative binomial $\mathcal{NB}(n, p)$: gives the number of failures necessary to get n successes when the probability of success is p. $X(\Omega) = \{0, 1, \ldots\}$, $P(X = k) = \binom{n+k-1}{n-1} p^n q^k$, $q = 1 - p$. We have

$$E(X) = n\frac{q}{p},\ V(X) = n\frac{q}{p^2},\quad \sigma(X) = \frac{1}{p}\sqrt{nq}.$$

Under R, the instructions `dnbinom`, `pnbinom`, `qnbinom`, and `rnbinom` deal with the negative binomial distribution. Analogously to the binomial, `dnbinom(k,n,p)` corresponds to $P(X = k)$ for $\mathcal{NB}(n, p)$ and `pnbinom (k,n,p)` to $P(X \leq k)$ for $\mathcal{B}(n, p)$. `qnbinom(p,n,a)` returns the smallest value of k such that $P(X \leq k) \geq a$ for $\mathcal{NB}(n, p)$. The flag `lower.tail` has the same function as in the preceding. `rnbinom(n,k,p)` generates a sample of k variates from $\mathcal{NB}(n, p)$. For instance, let us consider $n = 5, p = 2/10, q = 8/10$. We have

```
> n = 5
> p = 0.2
> dnbinom(0,n,p) # P(X=0)
[1] 0.00032
> dnbinom(5,n,p) # P(X=5)
[1] 0.01321206
> dnbinom(10,n,p) # P(X=10)
[1] 0.0343941
> pnbinom(4,n,p) # P(X<=4)
[1] 0.01958144
> pnbinom(4,n,p,lower.tail=FALSE) # P(X>4)
[1] 0.9804186
> qnbinom(0.8,n,p) # k s.t. P(X<=k)>= 0.8
[1] 28
> qnbinom(0.2,n,p,lower.tail=FALSE) # k s.t. P(X>k)>= 0.2
[1] 28
> rnbinom(8,n,p)
[1] 37 37 14 26 26 12 25 26
```

- Poisson $\mathcal{P}(\lambda)$: $X(\Omega) = \mathbb{N} = \{0, 1, \ldots, n, \ldots\}$, $P(X = k) = \frac{e^{-\lambda}\lambda^k}{k!}$. We have

$$E(X) = \lambda,\ V(X) = \lambda,\ \sigma(X) = \sqrt{\lambda}.$$

A Poisson's law approximates the sum of a large number of Bernoulli laws having small parameter p ($\lambda = np$).

R built-in functions to deal with Poisson's variables are `dpois` and `ppois`, which return the values of $P(X = k)$ and $P(X \leq k)$, respectively. `qpois` returns the smallest value of k such that $P(X \leq k) \geq a$ and `rpois` generates a sample from the Poisson's distribution. The flag `lower.tail` produces the same effects as in the preceding. For instance:

```
> lbda = 2
> dpois(0,lbda) # P(X=0)
[1] 0.1353353
> dpois(5,lbda) # P(X=5)
[1] 0.03608941
> dpois(10,lbda) # P(X=10)
[1] 3.818985e-05
> ppois(4,lbda) # P(X<=4)
[1] 0.947347
> ppois(4,lbda,lower.tail=FALSE) # P(X>4)
[1] 0.05265302
> qpois(0.8,lbda) # k s.t. P(X<=k)>= 0.8
[1] 3
> qpois(0.2,lbda,lower.tail=FALSE) # k s.t. P(X>k)>= 0.2
[1] 3
> rpois(8,lbda)
[1] 2 2 2 3 4 1 2 1
```

- Multinomial distribution $\mathcal{M}(n, p)$: it is a generalization of the binomial distribution to the situation where we consider a population which is subdivided in k distinct subpopulations of probabilities $p_1, \ldots, p_k$, respectively – we have $p_1, + \ldots + p_k = 1$. A trial takes simultaneously n elements from the population and we count the number elements X_i of each subpopulation i. For $n_1 + \ldots + n_k = n$, we have

$$P(X_1 = n_1, \; \ldots, X_k = n_k) = \frac{n!}{n_1! \ldots n_k!} p_1^{n_1} p_2^{n_2} \ldots p_k^{n_k} = \frac{\left(\sum_{i=1}^{k} n_i\right)!}{\prod_{i=1}^{k} (n_i!)} \prod_{i=1}^{k} p_i^{n_i}.$$

Notice that each X_i is a binomial variable $\mathcal{B}(n, p_i)$. The variables $X_1, \ldots, X_k$ are not independent: $cov\,(X_i, X_j) = -np_ip_j, for\ i \neq j;\ cov\,(X_i, X_i) = np_iq_i$.

The function $dmultinom(c(n_1, \ldots, n_k), c(p_1, \ldots, p_k))$ furnishes $P(X_1 = n_1, \ldots, X_k = n_k)$. *rmultinom* generates a variate from the multinomial distribution. For instance,

```
> n = c(3, 4, 5)  # (n1, n2, n3)
> p = c(10,20,20) # will be normalized by R
>                 # to (0.2,0.4,0.4)
> dmultinom(n,sum(n),p) # P(X1=n1, X2=n2,X3=n3)
[1] 0.05813305
> dmultinom(n,sum(n),c(0.2,0.4,0.4)) # verification
[1] 0.05813305
> rmultinom(1,10,p)    # a variate from the multinomial
     [,1]
[1,]    3
[2,]    2
[3,]    5
```

2.8.2 Continuous Variables Having a PDF

Let X be a continuous random variable having f as PDF. Then,

$$E(\phi(X)) = \int_{-\infty}^{+\infty} \phi(x) f(x) dx. \tag{2.32}$$

Thus

$$M_k(X) = \int_{-\infty}^{+\infty} x^k f(x) dx, \; E(X) = \int_{-\infty}^{+\infty} x f(x) dx, \tag{2.33}$$

$$E\left(X^2\right) = \int_{-\infty}^{+\infty} x f(x) dx, \; \varphi(t) = \int_{-\infty}^{+\infty} e^{itx} f(x) dx. \tag{2.34}$$

Examples of continuous variables are as follows:

- Uniform distribution on (a, b):

$$f(x) = \begin{cases} \dfrac{1}{b-a}, & \text{if } x \in (a,b) \\ 0, & \text{if } x \notin (a,b) \end{cases}$$

We have

$$E(X) = \frac{a+b}{2}, \; V(X) = \frac{(b-a)^2}{12}, \; \sigma(X) = \frac{b-a}{\sqrt{12}}.$$

The functions for the uniform distribution are `dunif`, `punif`, `qunif`, and `runif`, illustrated below:

(continued)

```
> a = -1; b = 3;
> dunif(0,a,b) # density f(0)=1/4
[1] 0.25
> dunif(-2,a,b) # density f(-2) = 0
[1] 0
> dunif(4,a,b) # density f(4) = 0
[1] 0
> punif(0,a,b) # P(X<=0)=1/4
[1] 0.25
> punif(0,a,b,lower.tail = FALSE) # P(X>0)=3/4
[1] 0.75
> punif(-2,a,b) # P(X<=-2) = 0
[1] 0
> punif(-2,a,b,lower.tail = FALSE) # P(X>-2)=1
[1] 1
> punif(3,a,b) # P(X<=3) = 1
[1] 1
> punif(3,a,b,lower.tail = FALSE) # P(X>3)=0
[1] 0
> qunif(0.5,a,b) # a s.t. P(X<=a) = 1/2 : a = 1
[1] 1
> qunif(0.75,a,b, lower.tail=FALSE) # a s.t. P(X>a) = 3/4 : a = 0
[1] 0
> runif(5,a,b) # 5 variates
[1] -0.2683086  0.9780069  0.7150712 -0.5944543  2.8579115
```

- Standard Gaussian $N(0, 1)$

$$f(x) = \frac{1}{\sqrt{2\pi}} e^{-\frac{x^2}{2}}$$

We have

$$E(X) = 0,\ V(X) = 1,\ \sigma(X) = 1.$$

Normal $N(m, \sigma)$

$$f(x) = \frac{1}{\sigma\sqrt{2\pi}} e^{-\frac{1}{2}\left(\frac{x-m}{\sigma}\right)^2}$$

We have

$$E(X) = m,\ V(X) = \sigma^2,\ \sigma(X) = \sigma.$$

The built-in functions for normal variables are `dnorm`, `pnorm`, `qnorm`, and `rnorm`:

```
> m = 1;   # mean
> sig = 2; # standard deviation
> dnorm(0,m,sig) # Normal density at x= 0
[1] 0.1760327
> exp(-0.5*((0-m)/sig)^2)/(sig*sqrt(2*pi)) # verification
[1] 0.1760327
> pnorm(0,m,sig) # P(X<=0)
[1] 0.3085375
> pnorm(0,m,sig,lower.tail = FALSE) # P(X > 0)
[1] 0.6914625
> qnorm(0.025,m,sig) # a s. t. P(X <= a)=0.025
[1] -2.919928
> m + sig*qnorm(0.025,0,1) # verification
[1] -2.919928
> m - sig*qnorm(0.975,0,1) # second verification
[1] -2.919928
> qnorm(0.975,m,sig) # a s. t. P(X > a)=0.975
[1] 4.919928
> m + sig*qnorm(0.975,0,1) # verification
[1] 4.919928
> m - sig*qnorm(0.025,0,1) # second verification
[1] 4.919928
> rnorm(5,m,sig) # 5 variates
[1]  1.30517947 -0.05224129  2.48376585 -1.06855284  1.01996856
```

- Chi-squared (Pearson's distribution) χ^2: the distribution of Z^2, where Z is standard Gaussian.

$$f(x) = \frac{1}{\sqrt{2\pi x}} e^{-\frac{x}{2}}.$$

  We have $E(X) = 1$, $V(X) = 2$.

- Chi-squared with n degrees of freedom $\chi^2(n)$: the distribution of $Z_1^2 + \ldots Z_n^2$, where $Z_1, \ldots Z_n$ are independent standard Gaussian variables.

$$f(x) = \frac{1}{2^{\frac{n}{2}}\Gamma\left(\frac{n}{2}\right)} e^{-\frac{x}{2}} x^{\frac{n}{2}-1}.$$

  We have $E(X) = n$, $V(X) = 2n$.

The functions for the chi-squared distribution are `dchisq`, `pchisq`, `qchisq`, and `rchisq`:

(continued)

```
> dof = 4;  # degrees of freedom
> x = 2;
> dchisq(x,dof) #density at x
[1] 0.1839397
> exp(-x/2)*(x^((dof/2)-1))/((2^(dof/2))*gamma(dof/2))
[1] 0.1839397
> pchisq(x,dof) #P(X<=x)
[1] 0.2642411
> pchisq(x,dof,lower.tail = FALSE) #P(X>x)
[1] 0.7357589
> qchisq(0.95,dof) # a s.t. P(X<=a) = 0.95
[1] 9.487729
> qchisq(0.05,dof, lower.tail = FALSE) # a s.t. P(X>a) = 0.05
[1] 9.487729
> rchisq(3,dof) # 3 variates
[1] 0.4135705 4.8750361 2.7646584
```

Student–Fisher with n degrees of freedom *SF*(n): the distribution of $\frac{Y}{\sqrt{\frac{Q}{n}}}$, where Y, Q are independent, Y is standard Gaussian, Q is $\chi^2(n)$.

$$f(x) = \frac{\Gamma\left(\frac{n+1}{2}\right)}{\sqrt{n\pi}\Gamma\left(\frac{n}{2}\right)}\left(1+\frac{x^2}{n}\right)^{\frac{n+1}{2}}.$$

We have $E(X) = 0$, $V(X) = \frac{n}{n-2}$ (for $n > 2$).

The built-in functions for Student–Fisher variables are `dt`, `pt`, `qt`, and `rt`:

```
> dof = 15; # degrees of freedom
> x = 2;
> dt(x,dof) # density at x
[1] 0.05920773
> (gamma((dof+1)/2)/ (sqrt(dof*pi)*gamma(dof/2)))*(1 + x^2/dof)
^(-((dof+1)/2))
[1] 0.05920773
> pt(x,dof) # P(X <= x)
[1] 0.9680275
> pt(x,dof, lower.tail = FALSE) # P(X > x)
[1] 0.0319725
> qt(0.975,dof) # a s.t. P(X <= a) = 0.975
[1] 2.13145
> qt(0.025,dof, lower.tail = FALSE) # a s.t. P(X > a) = 0.025
[1] 2.13145
> rt(3,dof) # 3 variates
[1]  1.4413464 -1.1809253 -0.4272363
```

- Behrens–Fisher–Snedecor (BFS, or simply Fisher) with (n_1, n_2) degrees of freedom: the distribution of $\frac{\frac{Q_1}{n_1}}{\frac{Q_2}{n_2}}$, where Q_i is $\chi^2(n_i)$, Q_1 and Q_2 independent. We have $(x > 0)$

$$f(x) = \left(\frac{n_1}{n_2}\right)^{\frac{n_1}{2}} \frac{\Gamma\left(\frac{n_1+n_2}{2}\right)x^{\frac{n_1}{2}-1}}{\Gamma\left(\frac{n_1}{2}\right)\Gamma\left(\frac{n_2}{2}\right)\left(1+\frac{n_1}{n_2}x\right)^{-\left(\frac{n_1+n_2}{2}\right)}}.$$

The built-in functions for BFS variables are he built-in functions for Student–Fisher variables are `df`, `pf`, `qf`, and `rf` :

```
> dof1 = 10; # degrees of freedom 1
> dof2 = 15; # degrees of freedom 2
> x = 2;
> df(x,dof1,dof2) # density at x for X1=(Q1/n1)/(Q2/n2)
[1] 0.161374
> ((dof1/dof2)^(dof1/2))*((gamma((dof1+dof2)/2)*(x^(dof1/2 -
1)))/(gamma(dof1/2)*gamma(dof2/2)*((1+(dof1/dof2)*x)^((dof1+dof
2)/2))))
[1] 0.161374
> df(x,dof2,dof1) # density at x for X2=(Q2/n2)/(Q1/n1)
[1] 0.1719924
> ((dof2/dof1)^(dof2/2))*((gamma((dof1+dof2)/2)*(x^(dof2/2 -
1)))/(gamma(dof1/2)*gamma(dof2/2)*((1+(dof2/dof1)*x)^((dof1+dof
2)/2))))
[1] 0.1719924
> pf(x,dof1,dof2) # P(X1>=x)
[1] 0.8908599
> pf(1/x,dof2,dof1,lower.tail = FALSE) #P(X2 > 1/x)
[1] 0.8908599
> qf(0.975,dof1,dof2) # a s.t. P(X1 <= a) = 0.975
[1] 3.060197
> qf(0.975,dof2,dof1,lower.tail=FALSE)# a s.t. P(X2 > a) = 0.975
[1] 0.3267764
> 1/qf(0.975,dof1,dof2)
[1] 0.3267764
> rf(3,dof1,dof2) # 3 variates from X1
[1] 2.370917 1.663006 2.075515
> rf(3,dof2,dof1) # 3 variates from X2
[1] 0.7732564 0.5791612 3.6066118
>
```

- Log-normal: the distribution of X such that $\log(X)$ is Normal $N(m, \sigma)$. We have

$$f(x) = \frac{1}{\sigma x\sqrt{2\pi}} e^{-\frac{1}{2}\left(\frac{\log(x)-m}{\sigma}\right)^2}.$$

The built-in functions for log-normal variables are `dlnorm`, `plnorm`, `qlnorm`, and `rlnorm`:

```
> mlog = 1; # mean of log(X)
> slog = 1; # standard deviation of log(X)
> dlnorm(1,mlog,slog) # density at log(1)
[1] 0.2419707
> exp(-0.5*((log(1)-mlog)/slog)^2)/(1*slog*sqrt(2*pi))
[1] 0.2419707
> dlnorm(exp(2),mlog,slog) #density at log(x)=2
[1] 0.03274718
> exp(-0.5*((2-mlog)/slog)^2)/(exp(2)*slog*sqrt(2*pi))
[1] 0.03274718
> plnorm(1,mlog,slog) # P(log(X)<=log(1))
[1] 0.1586553
> plnorm(exp(2),mlog,slog,lower.tail=FALSE) # P(log(X)>2)
[1] 0.1586553
> qlnorm(0.025,mlog,slog) # a s.t. P(log(X)<=log(a))=0.025
[1] 0.3829067
> qlnorm(0.975,mlog,slog,lower.tail=FALSE) # a s.t. P(log
(X)>log(a))=0.975
[1] 0.3829067
> rlnorm(4,mlog,slog) # 4 variates
[1]  5.319203  3.241512  4.026116 10.842443
```

Exercises

1. Let X_1 and X_2 be two independent variables having Bernoulli distributions of same parameter $p > 0$. Determine the range and the probabilities of $Z_1 = X_1 - X_2$, $Z_2 = X_1X_2$.
2. Let X be a continuous random variable uniformly distributed on $(-1, 1)$. Use R to evaluate

   a) $P\left(-\frac{1}{2} < X < \frac{1}{3}\right)$;
   b) $P\left(X > \frac{1}{3}\right)$;
   c) $M_3(X)$; .

3. An urn contains 10 red balls and 30 blue balls. 5 balls are drawn sequentially, with replacement (the ball is put back in the urn after drawing). Use R to determine the probabilities of getting

   a) Exactly three red balls.
   b) At least two red balls.
   c) Less than three red balls.

(continued)

4. An urn contains 10 red balls and 30 blue balls. A ball is drawn. If it is blue, it is put back in the urn and we make another draw. The game stops when a red ball is drawn. Let n be the number of draws before stopping. Use R to determine the probabilities of

   a) $n = 5$;
   b) $n \geq 5$;
   c) $n \leq 10$.

5. Let X be a random variable which is normally distributed $N(1,2)$. Use R to find the probabilities

   a) $P(-1 < X < 1)$;
   b) $P(X < 3)$;
   c) $P(X > 1)$.

6. Let X be a random variable which is normally distributed $N(0,2)$ and $Y = e^{-X+1}$. Use R to find the probabilities

   a) $P(1 < Y < 2)$;
   b) $P(Y < 3)$;
   c) $P(Y > 2)$.

7. Let X be a random variable which is normally distributed $N(0,2)$ and $Y = X^2$. Use R to find the probabilities

   a) $P(1 < Y < 2)$;
   b) $P(Y < 3)$;
   c) $P(Y > 2)$.

8. Let X_1, X_2 be two independent random variables normally distributed $N(0,2)$ and $Y = \frac{X_1}{\sqrt{X_2^2}}$. Find the probabilities

   a) $P(1 < Y < 2)$;
   b) $P(Y < 3)$;
   c) $P(Y > 2)$.

9. Let X_1, X_2 be two independent random variables normally distributed $N(0,2)$ and $Y = \frac{X_1^2}{X_2^2}$. Use R to find the probabilities

   a) $P(1 < Y < 2)$;
   b) $P(Y < 3)$;
   c) $P(Y > 2)$.

Supplementary Exercises

1. The statistics of car crashes show the following connection with the local maximum speed regulation (source NHTSA, data for 2017):

Max speed (mph)	30	40	50	55	≥60	No limit
Accidents (thousands)	418	311	248	327	233	60

   Given a crash, what is the probability of that the max speed was 40 mph? Inferior or equal to 40? At least 55?
2. A candidate for a game must choose between 10 boxes containing prizes whose value is respectively 0, 10, 20, 25, 50, 100, 200, 500, 1000, 10,000 dollars. What is the probability of him making more than a thousand dollars? Less than a hundred dollars? What is the mean value of the winnings?
3. A sphynx waits for travelers to pass by and asks them a question which they answer with a probability of hitting the right answer equal to $p > 0$. In case of error, the traveler is devoured by the sphynx. The number of travelers passing each week near the sphynx follows a Poisson's law of parameter $\lambda > 0$. What is the probability that the sphynx will not eat anyone in a week?
4. A model is destined to predict an event. When the prediction is positive, the success rate is 85%. When the answer is negative, the success rate is of 60%. The global probability of the event is 10%.

   a) The model furnishes a positive answer. What is the probability that the event arises in reality?
   b) The model furnishes a negative answer. What is the probability that the event does not occur in reality?

5. You have in your pocket two dices. One of the dices is regular, numbered from 1 to 6, but the other one is unfair: The face numbered one was replaced by a second face numbered 6. In both the dices, all the faces are equiprobable. You take a dice at random, equiprobably and roll it. The face up is 6.

   a) What is the probability of this result?
   b) What is the probability that the dice is not the unfair one?

6. Two models M_1 and M_2 are destined to predict an event. When the prediction is positive, the success rate is 85% for M_1 and 70% for M_2. When the answer is negative, the success rate is of 60% for M_1 and 80% for M_2. The global probability of the event is 10%.

(continued)

a) Both the models furnish a positive answer. What is the probability that the event arises in reality?
b) Both the models furnish a negative answer. What is the probability that the event does not occur in reality?
c) M_1 furnishes a positive answer and M_2 furnishes a negative answer. What is the probability that the event arises in reality?
d) M_1 furnishes a negative answer and M_2 furnishes a positive answer. What is the probability that the event does not occur in reality?

7. Let $\boldsymbol{X} = (X_1, \ldots, X_n)^t$ be a vector of independent variables. Let $\boldsymbol{A}$ be a $m \times n$ matrix of real numbers and $\boldsymbol{Y} = \boldsymbol{AX}$.

a) Verify that $\boldsymbol{Y}^t\boldsymbol{Y} = \boldsymbol{X}^t\boldsymbol{A}^t\boldsymbol{AX}$.
b) Verify that $E(\boldsymbol{Y}) = \boldsymbol{A}E(\boldsymbol{X})$.
c) Verify that $\mathbf{C}(\boldsymbol{Y}) = E(\boldsymbol{X}^t\boldsymbol{A}^t\boldsymbol{AX}) - E(\boldsymbol{X})^t\boldsymbol{A}^t\boldsymbol{A}E(\boldsymbol{X})$.

8. Let $\boldsymbol{X} = (X_1, \ldots, X_n)$ be a vector of **independent standard gaussian variables**. Let $\boldsymbol{A}$ be a $m \times n$ matrix of real numbers and $\boldsymbol{Y} = \boldsymbol{AX}$.

a) Show that $E(\boldsymbol{Y}) = \mathbf{0}$.
b) Both the models furnish a negative answer. What is the probability that the event does not occur in reality?
c) M_1 furnishes a positive answer and M_2 furnishes a negative answer. What is the probability that the event arises in reality?
d) M_1 furnishes a negative answer and M_2 furnishes a positive answer. What is the probability that the event does not occur in reality?

9. Two models M_1 and M_2 are destined to predict an event. When the prediction is positive, the success rate is 85% for M_1 and 70% for M_2. When the answer is negative, the success rate is of 60% for M_1 and 80% for M_2. The global probability of the event is 10%.

a) Both the models furnish a positive answer. What is the probability that the event arises in reality?
b) Both the models furnish a negative answer. What is the probability that the event does not occur in reality?
c) M_1 furnishes a positive answer and M_2 furnishes a negative answer. What is the probability that the event arises in reality?
d) M_1 furnishes a negative answer and M_2 furnishes a positive answer. What is the probability that the event does not occur in reality?

2.9 Sequences of Random Variables

One of the central points in uncertainty quantification is the approximation of random variables: We look for representations of the observed variability of a system as a function of the known or assumed variability of certain parameters and inputs. Such a representations are generally approximations based on the construction of a sequence $\{X_n: n \geq 0\}$ of random variables that converge to the exact variable X: $X_n \longrightarrow X$. In probability, it is usual to manipulate different definitions of convergence:

1. Convergence in the quadratic mean: in quadratic mean if and only if $E((X_n - X)^2) \longrightarrow 0$, for $n \longrightarrow +\infty$;
2. Almost sure convergence: the event $E = \{X_n \longrightarrow X\}$ is almost sure, id est, $P(E) = 1$;
3. Convergence in probability: for any $\varepsilon > 0$, the event $E_n(\varepsilon) = \{\|X_n - X\| \geq \varepsilon\}$ verifies $P(E_n(\varepsilon)) \longrightarrow 0$, for $n \longrightarrow +\infty$;
4. Convergence in distribution: let F_n be the CDF of X_n and F be the CDF of X. $X_n \longrightarrow X$ in distribution if and only if $F_n(x) \longrightarrow F(x)$ at any point x where F is continuous.

One of the fundamental results concerning the convergence of sequences of random variables is **Lévy's theorem** (Lévy, 1922):

Theorem 2.4 (Lévy) Let $\{X_n: n \in \mathbb{N}\}$ be a sequence of random variables such that the characteristic function of X_n is φ_n. Let X be a random variable having as characteristic function φ. Then $X_n \longrightarrow X$ in distribution if and only if $\varphi_n(t) \longrightarrow \varphi(t)$ almost everywhere.

Levy's theorem furnishes a practical method for the approximation of a random variable X by approximating its characteristic function φ – as previously observed, $\varphi(t) = E(e^{itX})$,expands as

$$\varphi(t) = \sum_{n=0}^{+\infty} \frac{(it)^k}{k!} M_k(X) = 1 + itM_1(X) + \frac{(it)^2}{2} M_2(X) + \frac{(it)^3}{2} M_3(X) + \ldots$$

We may consider X_n as the variable having as characteristic function

$$\varphi_n(t) = \sum_{n=0}^{n} \frac{(it)^k}{k!} M_k(X) \approx E\left(e^{itX_n}\right).$$

Then, from Levy's theorem, $X_n \longrightarrow X$ in distribution. In practice, we look for a variable verifying the equations:

$$M_k(X_n) = M_k(X), for\ 1 \leq k \leq n. \tag{2.35}$$

A second theorem concerns gaussian variables

Theorem 2.5 Let $\{X_n: n \in N\}$ be a sequence of normally distributed random variables. If $X_n \longrightarrow X$ in distribution, then X is normally distributed. In addition, $X_n \longrightarrow X$ in distribution if and only if $E(X_n) \longrightarrow E(X)$, $V(X_n) \longrightarrow V(X)$.

Example 2.16 Let us consider a gaussian variable $W \sim N(0, 1)$. Since

$$M_1(W) = 0, M_2(W) = 1, M_3(W) = 0,$$

We may consider a variable W_3 such that

$$M_k(W_3) = M_k(W),\ 1 \leq k \leq 3.$$

For instance, a discrete variable such that

$$P(W_3 = -1) = P(W_3 = 1) = \frac{1}{2}.$$

Such an approximation is useful to generate random walks, such as approximated Brownian motions (see Sect. 4.8.3): For instance, we may consider $X_0 = (X_3)_0 = 0$, $X_{n+1} = X_n + W_n \approx (X_3)_{n+1} = (X_3)_n + (W_3)_n$. Assume that we are interested in the statistics of X_{100}: We generate $1e5$ variates from both the exact and the approximated values and we analyze the statistics of the samples:

(continued)

Example 2.16 (continued)

```
nr = 1e5;
ns = 100;
x = numeric(length=nr)
x3 = numeric(length=nr)
for (n in 1:nr)
{
  w3=sample(c(-1,1),ns,replace=TRUE);
  w = rnorm(ns);
  a = 0;
  for (i in 1:ns){a=a+w[[i]]}
  x[[n]]=a
  a = 0;
  for (i in 1:ns){a=a+w3[[i]]}
  x3[[n]]=a;
}
> summary(x)
     Min.   1st Qu.    Median      Mean   3rd Qu.      Max.
-44.92428  -6.76575  -0.04387  -0.02865   6.69440  43.89544
> summary(x3)
     Min.   1st Qu.    Median      Mean   3rd Qu.      Max.
-44.00000  -6.00000   0.00000  -0.00618   6.00000  42.00000
> var(x)
[1] 99.17653
> var(x3)
[1] 100.2748
```

As we can observe, the results are close, namely for the mean and the variance: The approximation furnishes good estimates of the statistics. For $1e6$ variates, we obtain

```
> summary(x)
     Min.   1st Qu.    Median      Mean   3rd Qu.      Max.
-47.50099  -6.75998  -0.01202  -0.00667   6.74037  47.02197
> summary(x3)
    Min.  1st Qu.   Median     Mean  3rd Qu.     Max.
-46.0000  -6.0000   0.0000  -0.0013   6.0000  50.0000
> var(x)
[1] 100.0246
> var(x3)
[1] 99.785
```

Exercises

1. Let W be a gaussian variable W~$N(0, 1)$. Find a discrete variable W_5 such that.

$$M_k(W_5) = M_k(W),\ 1 \le k \le 5.$$

(continued)

Analogously to Example 2.16, consider $X_0 = (X_5)_0 = 0$, $X_{n+1} = X_n + W_n \approx (X_5)_{n+1} = (X_5)_n + (W_5)_n$. Generate a sample of X_{100} and of the approximated $(X_5)_{100}$ and compare their statistics.

TIP: $M_4(W) = 3$, $M_5(W) = 0$. Look for a discrete variable taking the values $-\sqrt{a}, -1, 0, 1, \sqrt{a}$.

2. Let W be a gaussian variable W~$N(0, 1)$. Find a discrete variable W_7 such that.

$$M_k(W_7) = M_k(W),\ 1 \leq k \leq 7.$$

Analogously to Example 2.16, consider $X_0 = (X_5)_0 = 0$, $X_{n+1} = X_n + W_n \approx (X_7)_{n+1} = (X_7)_n + (W_7)_n$. Generate a sample of X_{100} and of the approximated $(X_7)_{100}$ and compare their statistics.

TIP: $M_6(W) = 15$, $M_7(W) = 0$. Look for a discrete variable taking the values $-\sqrt{b}, -\sqrt{a}, -1, 0, 1, \sqrt{a}, \sqrt{b}$.

3. Consider $\Omega = (0, 1)$ and the mass density $\mu(\omega) = 1$. Let $X_n, X: \Omega \to \mathbb{R}$ be random variables such that $X_n(\omega) = \frac{\omega}{n}$ and $X(\omega) = 0$.

a) Determine the CDF F_n of X_n.
b) Find the CDF F of X.
c) Show that $X_n \longrightarrow X$ in distribution.
d) Show that $X_n \longrightarrow X$ in probability.
e) Show that $X_n \to X$ almost surely.
f) Show that $X_n \to X$ in the quadratic mean.

4. Let $S_n = \frac{X_n}{n}$, where $\{X_n: n \in N\}$ is a sequence of independent discrete random variables such that $(0 < p < 1)$:

$$P(X_n = k) = \frac{p}{n}\left(1 - \frac{p}{n}\right)^k,\ k \geq 0.$$

a) Let $t \in \mathbb{R}$ and $n(t) \in N$ verify $n(t) \leq nt < n(t) + 1$. Verify that.

$$P(X_n \leq nt) = P(X_n \leq n(t)) = 1 - \left(1 - \frac{p}{n}\right)^{n(t)+1}$$

b) Verify that.

$$\left(1 - \frac{p}{n}\right)^{nt+2} \leq \left(1 - \frac{p}{n}\right)^{n(t)+1} \leq \left(1 - \frac{p}{n}\right)^{nt+1}$$

c) Conclude that $P(S_n \leq t) \longrightarrow 1 - e^{-pt}$.
d) Show that S_n converges in distribution to an exponential law.

2.10 Samples

A **sample** from X is a set of independent observations of X, id est, a set $\mathcal{X}=\{X_1, \ldots, X_n\}$, where each X_i has the same distribution as X and is independent from X_j, $\forall j \neq i$. The **empirical mean** of the sample is

$$\overline{X_n}=\frac{1}{n}\sum_{i=1}^{n} X_i. \tag{2.36}$$

A **median** of the sample is a middle value that cuts the sample in two equal parts. If the sample is ordered in an increasing order, the median is $X_{\frac{n}{2}}$ for n pair; while for n odd, the median is the arithmetic mean of $X_{\frac{n-1}{2}}, X_{\frac{n+1}{2}}$.

Population's variance of the sample is

$$V_p=\frac{1}{n}\sum_{i=1}^{n}\left(X_i-\overline{X_n}\right)^2=\frac{1}{n}\sum_{i=1}^{n}X_i^2-\left(\overline{X_n}\right)^2. \tag{2.37}$$

We have

$$V_p=\frac{D_n^2}{n}, D_n^2=\sum_{i=1}^{n}\left(X_i-\overline{X_n}\right)^2. \tag{2.38}$$

Population's standard deviation is $s_p=\sqrt{V_p}$. This Pearson's approach considers the values $\mathcal{X}$ as being a population of equiprobable individuals and apply the formulae.

Since the elements of the sample are variates from X, all these quantities are random variables – thus, we may evaluate their means and variances. For instance,

$$E\left(\overline{X_n}\right)=E(X), V\left(\overline{X_n}\right)=\frac{1}{n}V(X), \sigma\left(\overline{X_n}\right)=\frac{\sigma(X)}{\sqrt{n}}. \tag{2.39}$$

For normal variables, $E(V_p) \neq V(X)$, so that it may be preferable to use the **sample's variance** and the associated **sample's standard deviation**:

$$V_n=\frac{D_n^2}{n-1}, s_n=\sqrt{V_n}. \tag{2.40}$$

Observe that, from Chebyshev's inequality (Proposition 2.8):

$$P(|\bar{X}_n-E(X)|\geq\varepsilon)\leq\frac{V(X)}{n\varepsilon^2},$$

so that $\bar{X}_n \to E(X)$ in probability – this result is the **weak law of large numbers**. The reader may find in the literature extensions of this basic result, such as the **strong law of large numbers** : for X regular enough, $\overline{X_n} \to E(X)$ almost surely.

In addition, the values of these quantities change with the sample: a **confidence interval** must be associated to the values of the mean and the variance of the sample, to take into account the error margins. To define a confidence interval to the estimation $\widehat{\xi}$ of quantity ξ, we must choose a **confidence level** $1 - \alpha$ (α is the **risk**, id est, the probability of rejection of the real value) and determine an interval (ξ_{min}, ξ_{max})that $P\left(\widehat{\xi} \in (\xi_{min}, \xi_{max})\right) = 1 - \alpha$. Confidence intervals are often based on the **Central Limit Theorem**

Theorem 2.6 (Central Limit) Let $\{X_n: n \in \mathbb{N}\}$ be a sequence of random variables having the same distribution of finite mean m and variance σ^2. Let

$$Z_n = \frac{\overline{X_n} - m}{\frac{\sigma}{\sqrt{n}}}.$$

Then Z_n converges in distribution to a standard Gaussian $N(0, 1)$.

Indeed, we use $\widehat{m} = \overline{X_n}$ as estimation of the mean: by determining from $N(0, 1)$ a number z_α such that $P(|Z| \le z_\alpha) = 1 - \alpha$, we evaluate an error margin $\Delta_\alpha = \frac{z_\alpha \sigma}{\sqrt{n}}$ and we generate a confidence interval $\left(\overline{X_n} - \Delta_\alpha, \bar{X}_n + \Delta_\alpha\right)$. When σ is unknown, we may use the **Cochran's theorem.**

Theorem 2.7 (Cochran) Let $(X_1, \ldots, X_n)$ be a vector of independent random variables having the same distribution $N(m, \sigma)$. Let

$$Q_n^2 = \frac{D_n^2}{\sigma^2}, T_n = \frac{\bar{X}_n - m}{\frac{s_n}{\sqrt{n}}}.$$

Then Q_n^2 is chi-squared $\chi^2(n - 1)$ and T_n is student-fisher $SF(n - 1)$.

Using this result, we may determine from $SF(n - 1)$ a number t_α such that $P(|T_n| \le t_\alpha) = 1 - \alpha$ and evaluate an error margin $\Delta_\alpha = \frac{t_\alpha s_n}{\sqrt{n}}$, what generates a confidence interval $\left(\overline{X_n} - \Delta_\alpha, \bar{X}_n + \Delta_\alpha\right)$. Cochran's theorem furnishes also a confidence interval for the variance: we determine from the distribution $\chi^2(n - 1)$ two numbers A_1, A_2 such that $P\left(Q_n^2 \le A_1\right) = P\left(Q_n^2 \ge A_2\right) = \frac{\alpha}{2}$. Then, $\left(\frac{D_n^2}{A_2}, \frac{D_n^2}{A_1}\right)$ is a confidence interval for $\widehat{\sigma^2}$.

Table 2.6 Non-rejection intervals for tests on the mean

H_0	Non-rejection interval	t_α
$m = m_0$	$\bar{X}_n \in (m_0 - \Delta_\alpha, m_0 + \Delta_\alpha)$	$P(\lvert T_n \rvert \le t_\alpha) = 1 - \alpha$
$m < m_0$	$\bar{X}_n \in (-\infty, m_0 + \Delta_\alpha)$	$P(T_n \le t_\alpha) = 1 - \alpha$
$m > m_0$	$\bar{X}_n \in (m_0 - \Delta_\alpha, +\infty)$	$P(T_n \le t_\alpha) = \alpha$

Table 2.7 Non-rejection intervals for tests on the variance σ^2

H_0	Non-rejection interval	A_1, A_2
$\sigma^2 = \sigma_0^2$	$D_n^2 \in (A_1\sigma_0^2, A_2\sigma_0^2)$ $s_n^2 \in \left(\frac{A_1\sigma_0^2}{n-1}, \frac{A_2\sigma_0^2}{n-1}\right)$	$P(\chi_{n-1}^2 \le A_1) = \frac{\alpha}{2}, P(\chi_{n-1}^2 \ge A_2) = \frac{\alpha}{2}$
$\sigma^2 < \sigma_0^2$	$D_n^2 \in (0, A_2\sigma_0^2)$ $s_n^2 \in \left(0, \frac{A_2\sigma_0^2}{n-1}\right)$	$P(\chi_{n-1}^2 \ge A_2) = \alpha$
$\sigma^2 > \sigma_0^2$	$D_n^2 \in (A_1\sigma_0^2, +\infty)$ $s_n^2 \in \left(\frac{A_1\sigma_0^2}{n-1}, +\infty\right)$	$P(\chi_{n-1}^2 \le A_1) = \alpha$

The data may also be used to test hypothesis made about the mean, such as H_0: $m = m_0$, H_0: $m > m_0$, or H_0: $m < m_0$. A test verifies if the data are incompatible with the hypothesis at a risk level α. If there is an incompatibility, the hypothesis must be rejected. Otherwise, the data cannot reject the hypothesis: the user may consider that the hypothesis remains tentatively valid. Tests are based on non-rejection intervals, analogous to confidence intervals: for instance, we may use the non-rejection regions given in Table 2.6.

Analogously, hypothesis on the variance may be tested, such as H_0: $\sigma = \sigma_0$, H_0: $\sigma > \sigma_0$, or H_0: $\sigma < \sigma_0$. Examples of non-rejection regions are given in Table 2.7.

In practice, you may also be interested in the comparison of statistics of two samples from variables X and Y. For instance, two samples $\mathcal{X} = \{X_1, \ldots, X_{n_X}\}$ and $\mathcal{Y} = \{Y_1, \ldots, Y_{n_Y}\}$ are available and these data are intended to be used to verify the relations between the means m_X and m_Y of X and Y, or between their standard deviations σ_X and σ_Y: we may test the hypothesis H_0: $m_X = m_Y$, H_0: $m_X < m_Y$, H_0: $m_X > m_Y$, $H_0: \sigma_X^2 = \lambda\sigma_Y^2$, $H_0: \sigma_X^2 < \lambda\sigma_Y^2$, $H_0: \sigma_X^2 > \lambda\sigma_Y^2$.

These tests are based in the same principles as the preceding ones: a region of non-rejection is defined for a risk α and the hypothesis is rejected if the observed values are not in the region. Otherwise, the hypothesis is considered as compatible with the data.

The tests for the comparison of the means are reduced to those for a single variable $\bar{Z} = \bar{X}_{n_X} - \bar{Y}_{n_Y}$, which is generally assumed to be normally distributed $N(m_Z, \sigma_{\bar{Z}})$. We have $m_Z = m_X - m_Y$, but a supplementary assumption is needed for the normality of the variable $\bar{Z}$ – for instance, one among the following:

A large amount of data in each sample: sizes n_X, n_Y of the samples large enough to use gaussian approximations for each empirical mean. Let s_X, s_Y be the in this case is approximately $\sigma_{\bar{Z}} \approx s_{\bar{Z}} = \sqrt{\frac{s_X^2}{n_X} + \frac{s_Y^2}{n_Y}}$;

- Coupled observations: the data are formed by couples (X_i, Y_i): in this situation, $n_X = n_Y$ and we have a sample $Z_i = X_i - Y_i$ from $Z = X - Y$, which has as mean $m_Z = m_X - m_Y$ and tests are performed on m_Z using $\sigma_{\bar{Z}} \approx s_Z/\sqrt{n_X}$, where s_Z is evaluated by the formula for the sample's standard deviation;
- The variance is the same for both the samples: $\sigma_X = \sigma_Y = \sigma$. In this case, $\sigma_{\bar{Z}} = \sigma\sqrt{\frac{1}{n_X} + \frac{1}{n_Y}}$ and $\frac{D^2}{\sigma^2} = \frac{(n_X-1)s_X^2 + (n_Y-1)s_Y^2}{\sigma^2}$ is approximately $\chi^2(n_X + n_Y - 2)$. Then, the tests are performed using these variables.

The first approach is often used even for small datasets, but such a use cannot be rigorously justified in such a situation. Under one of these assumptions,

$$m_X = m_Y + a \Leftrightarrow m_Z = a,\ m_X < m_Y + a \Leftrightarrow m_Z < a,\ m_X > m_Y + a \Leftrightarrow m_Z > a.$$

Thus, the comparison reduces to a test on the value of $\bar{Z}$. Examples of non-rejection intervals for the first assumption are given in Table 2.8.

For the comparison of variances, we observe that

$$F = \frac{\sigma_Y^2}{\sigma_X^2}\frac{s_X^2}{s_Y^2} = \frac{s_X^2/\sigma_X^2}{s_Y^2/\sigma_Y^2} \sim BFS(n_X - 1, n_Y - 1).$$

The tests are based on this parameter. Examples of non-rejection regions are given in Table 2.9.

Table 2.8 Non-rejection intervals for comparisons of means, assuming normality of $\bar{Z} = \overline{X}_{n_X} - \overline{Y}_{n_Y}$. $\mathbb{Z}$ is a standard gaussian variable $N(0, 1)$

H_0	Non-rejection interval	$\mathbf{z}_\alpha$
$m_X = m_Y + a$	$\bar{Z} \in (a - \mathbf{z}_\alpha\sigma_{\bar{Z}}, a + \mathbf{z}_\alpha\sigma_{\bar{Z}})$	$P(\mathbb{Z} \le \mathbf{z}_\alpha) = 1 - \frac{\alpha}{2}$
$m_X < m_Y + a$	$\bar{Z} \in (-\infty, a + \mathbf{z}_\alpha\sigma_{\bar{Z}})$	$P(\mathbb{Z} \le \mathbf{z}_\alpha) = 1 - \alpha$
$m_X > m_Y + a$	$\bar{Z} \in (a + \mathbf{z}_\alpha\sigma_{\bar{Z}}, +\infty)$	$P(\mathbb{Z} \le \mathbf{z}_\alpha) = \alpha$

Table 2.9 Non-rejection intervals for comparisons of variances

H_0	Non-rejection interval	F_1, F_2
$\sigma_X^2 = \lambda\sigma_Y^2$	$\frac{s_X^2}{s_Y^2} \in (\lambda F_1, \lambda F_2)$	$P(F \le F_1) = \frac{\alpha}{2}, P(F \ge F_2) = \frac{\alpha}{2}$
$\sigma_X^2 < \lambda\sigma_Y^2$	$\frac{s_X^2}{s_Y^2} \in (0, \lambda F_2)$	$P(F \ge F_2) = \alpha$
$\sigma_X^2 > \lambda\sigma_Y^2$	$\frac{s_X^2}{s_Y^2} \in (\lambda F_1, +\infty)$	$P(F \le F_1) = \alpha$

R has built-in functions to evaluate means, variances, standard deviations, and medians and to perform tests: `mean` evaluates the mean, `var.` determines the variance, `sd` calculates the standard deviation, `median` determines the median, `t.test` compares the mean of the data to a given value or the means of two samples, and `var.test` compares the variances of two samples. To compare the variance of a sample to a given value, you must write some code or use a package such as `OneTwoSamples`. By default, `t.test` and `var.test` use $\alpha = 5\%$. A different value may be defined by setting parameter `conf.level` to the value of $1 - \alpha$.

Example 2.17 Let us generate a sample of 10 variates from $N(2, 1)$:

```
> x = rnorm(10,2,1);
> mx = mean(x); sx = sd(x);
> cat('Xbar = ', mx,', sx = ',sx)
Xbar =  2.198559 , sx =  1.000074
```

The data is

```
> x
 [1] 1.3293631 2.0839996 2.6097986 0.5833949 1.5055909
 [6] 2.9044445 3.3334163 2.1018317 1.6414581 3.8922949
```

Let us generate a confidence interval with $\alpha = 5\%$ for the mean:

```
> alfa = 0.05;
> za = pt(1-alfa/2,length(x)-1);
> dmx = za*sx/sqrt(length(x));
> mx1 = mx - dmx;
> mx2 = mx + dmx;
> cat('mx in ( ',mx1,', ',mx2,' )')
mx in (  1.938448 ,  2.45867  )
```

A confidence interval with $\alpha = 5\%$ for the variance is determined as follows:

```
> a1 = qchisq(alfa/2,length(x)-1);
> a2 = qchisq(1-alfa/2,length(x)-1);
> s1 = sx/sqrt(a2/(length(x)-1));
> s2 = sx/sqrt(a1/(length(x)-1));
> cat('sx in ( ',s1,', ',s2,' )')
sx in (  0.6878863 ,  1.825746  )
```

(continued)

Example 2.17 (continued)

Let us test the hypothesis H_0: $m_X = 2.2$ with $\alpha = 5\%$:

```
> aux = t.test(x,alternative="two.sided",mu=2.2)
> if ((aux$conf.int[[1]] <= aux$estimate) & (aux$conf.int[[2]] >
= aux$estimate))
+ {print("H0 accepted")} else
+ {print("H0 rejected")}
[1] "H0 accepted"
```

`t.test` generates a list containing many fields: `Conf.Int` contains the interval of non-rejection, and `estimate` contains the mean of the sample. If `estimate` belongs to `conf.Int`, the hypothesis is accepted. The default value of the confidence level is $1 - \alpha = 0.95$ – It can be modified by setting `conf.Level` $=1 - \alpha$. For instance, let us test the hypothesis H_0: $m_X < 1.9$ with $\alpha = 10\%$:

```
> aux = t.test(x,alternative="less",mu=1.9, conf.level = 0.9)
> if ((aux$conf.int[[1]] <= aux$estimate) & (aux$conf.int[[2]] >
= aux$estimate))
+ {print("H0 accepted")} else
+ {print("H0 rejected")}
[1] "H0 accepted"
```

Let us test the hypothesis H_0: $m_X > 2.3$ with $\alpha = 20\%$:

```
> aux = t.test(x,alternative="greater",mu=2.3, conf.level = 0.8)
> if ((aux$conf.int[[1]] <= aux$estimate) & (aux$conf.int[[2]] >
= aux$estimate))
+ {print("H0 accepted")} else
+ {print("H0 rejected")}
[1] "H0 accepted"
```

Let us test the hypothesis H_0: $\sigma_X = 1.2$ with $\alpha = 5\%$:

```
> a1 = qchisq(alfa/2,length(x)-1);
> a2 = qchisq(1-alfa/2,length(x)-1);
> s0 = 1.2;
> smin = s0*sqrt(a1/(length(x)-1));
> smax = s0*sqrt(a2/(length(x)-1));
> if ((smin <= sx) & (smax >= sx))
+ {print("H0 accepted")} else
+ {print("H0 rejected")}
[1] "H0 accepted"
```

(continued)

Example 2.17 (continued)

The hypothesis H_0: $\sigma_X < 0.8$ with $\alpha = 5\%$ is tested as follows:

```
> a2 = qchisq(1-alfa/2,length(x)-1);
> s0 = 0.8;
> smax = s0*sqrt(a2/(length(x)-1));
> if (smax >= sx)
+ {print("H0 accepted")} else
+ {print("H0 rejected")}
[1] "H0 accepted"
```

The hypothesis H_0: $\sigma_X > 1.3$ with $\alpha = 5\%$ is tested as follows:

```
> a1 = qchisq(alfa/2,length(x)-1);
> s0 = 1.3;
> smin = s0*sqrt(a1/(length(x)-1));
> if (smin <= sx)
+ {print("H0 accepted")} else
+ {print("H0 rejected")}
[1] "H0 accepted"
```

Notice that the data are consistent with many hypotheses, which may be in contradiction between themselves.

Example 2.18 Let us generate a supplementary sample of 10 variates from $N(1.5, 1.5)$:

```
> y = rnorm(10,1.5,1.5);
> my = mean(y); sy = sd(y);
> cat('Ybar = ', my,', sx = ',sy)
Ybar =  1.363959 , sx =  1.491558> alfa = 0.05;
> za = pt(1-alfa/2,length(y)-1);
> dmy = za*sy/sqrt(length(y));
> my1 = my - dmy;
> my2 = my + dmy;
> cat('my in ( ',my1,', ',my2,' )')
my in (  0.9760172 ,  1.751901  )> a1 = qchisq(alfa/2,length(y)-
1);
> a2 = qchisq(1-alfa/2,length(y)-1);
> s1 = sy/sqrt(a2/(length(y)-1));
> s2 = sy/sqrt(a1/(length(y)-1));
> cat('sy in ( ',s1,', ',s2,' )')
sy in (  1.025946 ,  2.723004  )
```

(continued)

Example 2.18 (continued)

The data are

```
> y
 [1]  3.6586518  0.6540401 -0.2812232  1.4297956  2.2282177
 [6]  1.1312535 -0.3738780 -0.2015139  3.6233055  1.7709413
```

Let us test the hypothesis H_0: $m_X = m_Y$ with $\alpha = 5\%$:

```
> aux = t.test(x,y,alternative="two.sided");
> d = aux$estimate[[1]]-aux$estimate[[2]];
> if ((aux$conf.int[[1]] <= d) & (aux$conf.int[[2]] >= d))
+ {print("H0 accepted")} else
+ {print("H0 rejected")}
[1] "H0 accepted"
```

The result is independent of the order between x and y:

```
> aux = t.test(y,x,alternative="two.sided");
> d = aux$estimate[[1]]-aux$estimate[[2]];
> if ((aux$conf.int[[1]] <= d) & (aux$conf.int[[2]] >= d))
+ {print("H0 accepted")} else
+ {print("H0 rejected")}
[1] "H0 accepted"
```

Let us test the hypothesis H_0: $m_X < m_Y$ with $\alpha = 5\%$:

```
> aux = t.test(x,y,alternative="less");
> d = aux$estimate[[1]]-aux$estimate[[2]];
> if ((aux$conf.int[[1]] <= d) & (aux$conf.int[[2]] >= d))
+ {print("H0 accepted")} else
+ {print("H0 rejected")}
[1] "H0 accepted"
```

An alternative is:

```
> aux = t.test(y,x,alternative="greater");
> d = aux$estimate[[1]]-aux$estimate[[2]];
> if ((aux$conf.int[[1]] <= d) & (aux$conf.int[[2]] >= d))
+ {print("H0 accepted")} else
+ {print("H0 rejected")}
[1] "H0 accepted"
```

Let us test the hypothesis H_0: $\sigma_X = \sigma_Y$

```
> aux = var.test(x,y,ratio=1,alternative="two.sided")
> d = aux$estimate;
> if ((aux$conf.int[[1]] <= d) & (aux$conf.int[[2]] >= d))
+ {print("H0 accepted")} else
+ {print("H0 rejected")}
[1] "H0 accepted"
```

(continued)

Example 2.18 (continued)

Let us test the hypothesis H_0: $\sigma_X > 2\sigma_Y$

```
> aux = var.test(x,y,ratio=2,alternative="greater")
> d = aux$estimate;
> if ((aux$conf.int[[1]] <= d) & (aux$conf.int[[2]] >= d))
+ {print("H0 accepted")} else
+ {print("H0 rejected")}
[1] "H0 accepted"
```

Let us test the hypothesis H_0: $\sigma_X < 0.8\sigma_Y$

```
> aux = var.test(x,y,ratio=0.8,alternative="less")
> d = aux$estimate;
> if ((aux$conf.int[[1]] <= d) & (aux$conf.int[[2]] >= d))
+ {print("H0 accepted")} else
+ {print("H0 rejected")}
[1] "H0 accepted"
```

Again, the data is consistent with various hypotheses, which may be in contradiction among themselves.

Remark 2.13 When dealing with tests, it is usual to refer to *t-values* and *p-values*.

A *t-value* is the absolute value of the variable used in the test and a *p-value* is the probability of getting an absolute value higher than the *t-value*.

A *p-value* gives the odds of observing the value found in the test: It can be used to estimate the reliability of the result.

The tests available in R return more information than *t-values*, namely they return *p-values*. For instance, let us test H_0: $m = 2.2$, using a sample of 10 values from $N(2, 1)$:

```
> x = rnorm(10,2,1);
> res = t.test(x,alternative="two.sided",mu=2.2)
> res

        One Sample t-test

data:  x
t = -0.5649, df = 9, p-value = 0.5859
alternative hypothesis: true mean is not equal to 2.2
95 percent confidence interval:
 1.256341 2.766534
sample estimates:
mean of x
 2.011438
```

(continued)

Remark 2.13 (continued)

As we can see, R returns the values of t, degrees of freedom (df), p-value, a confidence interval and $\overline{X_n}$. Analogously, let us make a chi-squared test:

```
> x = c(60,40)
> pth =c(0.5,0.5)
> resu=chisq.test(x,p=pth)
> resu

        Chi-squared test for given probabilities

data:  x
X-squared = 4, df = 1, p-value = 0.0455
```

Again, we obtain the values of Q_n^2, degrees of freedom and the associated p-value

Exercises

1. In a poll, a sample of 1000 voters gives 52 percent to a candidate. Give a confidence interval for his or her score, with a 5% risk. What is the interval for a 1% risk? What is the size of the sample to get a confidence interval of length $\pm 1\%$?
2. At the end of each day, an agency compiles statistic on the average waiting time (AWT) of customers. Over 30 days, we have the following results (in minutes):

AWT	0–5	5–10	10–15	15–20	20–25	25–30
Number of days	2	7	11	6	3	1

Use these data to estimate the mean and the variance of the mean of the AWT. Give confidence intervals for these quantities with $\alpha = 5\%$.

3. A quality department must check whether the impurity content of a product complies with the legislation, which requires a mean of 5 and a maximum variance of 2, assuming that the distribution of X is Normal,
   a) A sample of 20 elements is analyzed and leads to an empirical mean $\bar{X_{20}} = 3$ and an empirical standard deviation $s_{20} = 1.1$. Determine confidence intervals with risk 5% for the men and the variance.
   b) A new sample of 400 elements is analyzed and leads to an empirical mean $\bar{X_{400}} = 3.2$ and an empirical standard deviation $s_{400} = 1.3$. Determine confidence intervals with risk 5% for the mean and the variance.

(continued)

c) Can the quality department confirm that the variance is inferior to 2 with a risk of 5%?
d) Can it confirm that the mean is inferior to 5, with a risk of 5%?

4. Two methods of measurement of a quantity furnish the results below:

Method 1	44,45,45,48,48,46,46,46,46,47,47,47,47,47,47,47,49,49,49,49,50,50,51,51,52
Method 2	45,45,48,46,46,46,46,46,46,47,47,47, 47,47,47, 47,47,47,49,49,49,50,50,50

a) Determine the empirical mean and the standard variance of each method. Give confidence intervals for these quantities, with a risk of 5%.
b) May we consider the means of the samples as equal, with a risk of 5%?
c) May we consider that the variances of both the methods are equal, with a risk of 5%?

5. The results of the students of two groups are given below:

Group 1	11,8,4,12,7,8,6,6,3,5
Group 2	8,12,9,14,10,11,3,5,9,9

a) Determine the empirical mean and the standard variance of each method. Give confidence intervals for these quantities, with a risk of 5%.
b) May we consider group 2 as better, with a risk of 5%?
c) May we consider that the variances of both the methods are equal, with a risk of 5%?

6. The results of the students of two groups are given below:

Group 1	13,14,19,7,9,13,11,1,12,7,17,17,11,6,16,7,11,20,13
Group 2	7,17,4,12,15,9,7,12,8,14,14,9,9,9,10,8,11,17,14

a) Determine the empirical mean and the standard variance of each method. Give confidence intervals for these quantities, with a risk of 5%.
b) May we consider group 2 as better, with a risk of 5%?
c) May we consider that the variances of both the methods are equal, with a risk of 5%?

2.10.1 *Maximum-Likelihood Estimators*

Samples are often used to estimate parameters from distributions. For instance, to make a prediction about the results of a poll, we need to estimate the parameter $\boldsymbol{p} = (p_1, \ldots, p_k)$ of a multinomial distribution. Analogously, to determine if the proportion of non-compliant elements in a product satisfies a given condition, we must estimate the parameter p of a Bernoulli distribution.

In general, we have a model distribution $f(x, \boldsymbol{\theta})$, where $\boldsymbol{\theta} = (\theta_1, \ldots, \theta_r)$ is a vector of unknown parameters – to be determined from the sample. For a discrete distribution, f is a probability: $f(x, \boldsymbol{\theta}) = P(X = x)$. For a continuous distribution, f is the density of probability.

Example 2.19 Consider a binomial distribution $\mathcal{B}(n, p)$: The distribution is driven by the parameter $\theta = p$ and the model is $f(k, p) = \binom{n}{k} p^k (1-p)^{n-k}$.

For a Poisson's distribution, $\theta = \lambda$ and $f(k, \lambda) = \lambda^k e^{-\lambda} / k!$.

For an uniform distribution on $(-a, a)$, the parameter is $\theta = a$ and the model is $f(x, a) = 1/2a$, *if* $|x| < a; f(x, a) = 0$, *otherwise*.

For a Normal distribution, the parameters are $\theta = (m, \sigma)$ and the model is $f(x, \theta) = \exp\left(-\left((x-m)/\sigma\right)^2\right) / \left(\sigma\sqrt{2\pi}\right)$.

The objective is to find an estimator $\widehat{\boldsymbol{\theta}}$ of $\boldsymbol{\theta}$. A classical method to estimate the values of $\boldsymbol{\theta}$ is the maximum of the likelihood: let $\mathcal{X} = \{X_1, \ldots, X_{ns}\}$ be a sample from X. The likelihood is defined as

$$L(\mathcal{X}, \boldsymbol{\theta}) = \prod_{i=1}^{ns} f(X_i, \boldsymbol{\theta}). \tag{2.41}$$

It is often useful to consider the log-likelihood:

$$\log L(\mathcal{X}, \boldsymbol{\theta}) = \sum_{i=1}^{ns} \log\left(f(X_i, \boldsymbol{\theta})\right) \tag{2.42}$$

A popular way to generate an estimator is the use of the maximum of likelihood:

Definition 2.24 The maximum-likelihood estimator (MLE) of $\boldsymbol{\theta}$ is the element $\widehat{\boldsymbol{\theta}}$ which maximizes L or, equivalently, $\log L$.

Example 2.20 Consider a binomial distribution $\mathcal{B}(n, p)$: The model is $f(x, p) = \binom{n}{x} p^x (1-p)^{n-x}$, so that

$$\log\left(f(x, p)\right) = x \log p + (n-x)\log(1-p) + \log\binom{n}{x}.$$

Thus,

(continued)

Example 2.20 (continued)

$$\log L(X,p) = (\log p)\sum_{i=1}^{ns} X_i + \left(n \times ns - \sum_{i=1}^{ns} X_i\right)\log(1-p) + \sum_{i=1}^{ns}\log\binom{n}{X_i}.$$

To determine the maximum of $\log L(X, p)$, we find the solutions of

$$\frac{\partial}{\partial p}\log L(X,\ p)\bigg|_{\widehat{p}} = 0,$$

id est,

$$\frac{1}{\widehat{p}}\sum_{i=1}^{ns} X_i - \left(n \times ns - \sum_{i=1}^{ns} X_i\right)\frac{1}{1-\widehat{p}} = 0,$$

so that

$$\widehat{p} = \frac{1}{n \times ns}\sum_{i=1}^{ns} X_i.$$

Example 2.21 Consider an uniform distribution on $(-a, a)$,

$$f(x, a) = \frac{1}{2a},\ if\ |x| < a; f(x, a) = 0, \text{ otherwise}.$$

Thus,

$$L(X, a) = \frac{1}{2^n a^n},\ a \geq \max\{|X_i| : 1 \leq i \leq n\}.$$

The maximum of $L(X, a)$ is attained at the minimal value of a. Thus,

$$\widehat{a} = \max\{|X_i| : 1 \leq i \leq n\}.$$

The estimators can be numerically determined by optimization: we can look for the minimum of the negative likelihood $-L$ or the negative log-likelihood $-\log L$. We can use the methods introduced in Sect. 1.13 or use a package, for instance, `EstimationTools` – this package proposes maximum-likelihood estimation for the distributions predefined in R.

Example 2.22 Consider a sample from a binomial distribution $\mathcal{B}(10, p)$: $\mathcal{X} = \{1, 1, 1, 3, 3\}$. As shown above, the MLE estimation for p is $\widehat{p} = 0.18$. We can determine the estimation with `EstimationTools` as follows:

```
> X = c(1,1,1,3,3);
> n = 10;
> library(EstimationTools);
> sol =  maxlogL(X, dist = 'dbinom', fixed = list(size = n),link
 = list(over = "prob", fun = "logit_link"));
> cat("p estimated = ",sol$fit$par)
p estimated =  0.18
```

As an alternative, we can define a function `nlogl` evaluating $-\log L$ and use the function `optimize`:

```
> X = c(1,1,1,3,3);
> n = 10;
> nlogl = function(p){
+    aux = dbinom(X,n,p,log=TRUE)
+    s = -sum(aux)
+    return(s)
+ }
> sol <- optimize(f=nlogl, lower=0, upper=1)
> cat("p estimated = ",sol$minimum)
p estimated =  0.1800048
```

We can also use a method from Sect. 1.13:

```
> library(GA)
> GA1 <- ga(type = "real-valued",
+           fitness =  function(x) -nlogl(x),
+           lower = 0.01, upper = 0.99,
+           popSize = 50, maxiter = 1000, run = 100,
+           optim = TRUE, monitor = FALSE)
> GA1@solution
          x1
[1,] 0.1800014
```

Exercises

1. Let X be a discrete variable Poisson distributed: $P(X = k) = \lambda^k e^{-\lambda}/k!$. Let $\mathcal{X} = \{X_1, \ldots, X_{ns}\}$ be a sample from X.

   a) Determine the likelihood function.
   b) Find the log-likelihood.
   c) Determine the MLE for λ.
   d) Generate a sample of 10 variates of X for $\lambda = 3$. Use R and the sample to determine the MLE of λ.

(continued)

2. Let X be a discrete variable, geometrically distributed: $P(X = k) = p(1 - p)^k$. Let $\mathcal{X} = \{X_1, \ldots, X_{ns}\}$ be a sample from X.

   a) Determine the likelihood function.
   b) Find the log-likelihood.
   c) Determine the MLE for p.
   d) Generate a sample of 10 variates of X for $p = 0.3$. Use R and the sample to determine the MLE of p.

3. Let X be a continuous variable exponentially distributed: Its density is $f(x) = \lambda e^{-\lambda x}$ $(x > 0)$. Let $\mathcal{X} = \{X_1, \ldots, X_{ns}\}$ be a sample from X.

   a) Determine the likelihood function.
   b) Find the log-likelihood.
   c) Determine the MLE for λ.
   d) Generate a sample of 10 variates of X for $\lambda = 2$. Use R and the sample to determine the MLE of λ.

4. Let X be a continuous variable normally distributed: Its density is $f(x) = \frac{1}{\sigma\sqrt{2\pi}} e^{-\frac{1}{2}\left(\frac{x-m}{\sigma}\right)^2}$. Let $\mathcal{X} = \{X_1, \ldots, X_{ns}\}$ be a sample from X.

   a) Determine the likelihood function.
   b) Find the log-likelihood.
   c) Determine the MLE estimators for $\boldsymbol{\theta} = (m, \sigma)$.
   d) Generate a sample of 20 variates of X for $m = 2, \sigma = 1$. Use R and the sample to determine the MLE of (m, σ).

2.10.2 Samples from Random Vectors

When $\boldsymbol{X} = (X_1, \ldots, X_k)$ is a random vector of dimension k, a sample is a set of vectors. Then, the empirical mean and the variance may be determined for each component X_i, $1 \leq i \leq k$. Analogously, tests on the mean and on the variance may be performed for each component. However, other analyses can be performed, such as the calculation of covariances, correlations, affine approximations, and conditional expectations. For instance, we evaluate the population's covariance as

$$cov_P(X_{k_n}, X_{\ell_n}) = \frac{1}{n}\sum_{i=1}^{n}\left(X_{ki} - \overline{X_{k_n}}\right)\left(X_{\ell i} - \overline{X_{\ell_n}}\right) = \overline{X_{kn}X_{\ell n}} - \overline{X_{kn}} \times \overline{X_{\ell n}}.$$

where

$$\overline{X_{kn}X_{\ell n}} = \frac{1}{n}\sum_{i=1}^{n} X_{ki}X_{\ell i}.$$

The standard covariance of the sample is

$$cov_n(X_{k_n}, X_{\ell_n}) = \frac{1}{n-1}\sum_{i=1}^{n}\left(X_{ki} - \overline{X_{k_n}}\right)\left(X_{\ell i} - \overline{X_{\ell_n}}\right).$$

The empirical correlation is

$$r(X_{k_n}, X_{\ell_n}) = \frac{cov_P(X_{k_n}, X_{\ell_n})}{\sqrt{V_P(X_{k_n})V_P(X_{\ell_n})}} = \frac{cov_n(X_{k_n}, X_{\ell_n})}{\sqrt{V_n(X_{k_n})V_n(X_{\ell_n})}}.$$

r is independent of the choice between population's and standard's definitions.

Analogously to Sect. 2.5, we can determine linear approximations and conditional means.

R proposes built-in functions for the analysis of samples of vectors. `lm(Y~X)` determines the linear approximation $Y \approx aX + b$. `cov` evaluates the covariances, `cor` furnishes their correlation. In addition, the class `ctstats` can be used to evaluate approximations and correlations.

Example 2.23 Consider the situation where $X = (X_1, X_2)$ and the available data are given in Table 2.10.

Table 2.10 An example of data

X_1	1	2	3	4	5
X_2	4.5	6.8	9.6	10.8	12.5

We can evaluate the covariance, coefficient of linear correlation, and a linear approximation $X_2 \approx aX_1 + b$ as follows (Fig. 2.1):

```
> x1 = c(1:5);
> x2 = c(4.5, 6.8, 9.6, 10.8, 12.5);
> c1 = cov(x1,x2);
> rho1 = cor(x1,x2);
> cat('covariance = ',c1,', linear correlation = ',rho1);
covariance =  5 , linear correlation =  0.990002>
> m1 <- lm(x2~x1);
> a = m1$coefficients[[2]];
> b = m1$coefficients[[1]]
> cat('a = ',a,', b = ',b)
a =  2 , b =  2.84>
> yap = a*x1+b;
> plot(x1,x2,type="n")
> points(x1,x2,pch=19,col="red")
> points(x1,yap,pch=19)
> lines(x1,yap,lty=2)
> legend('topleft',legend=c('data','approx','approx'), pch=c(19,19,NA),
lty=c(NA,NA,2),col=c('red','black','black'))
```

(continued)

Example 2.23 (continued)

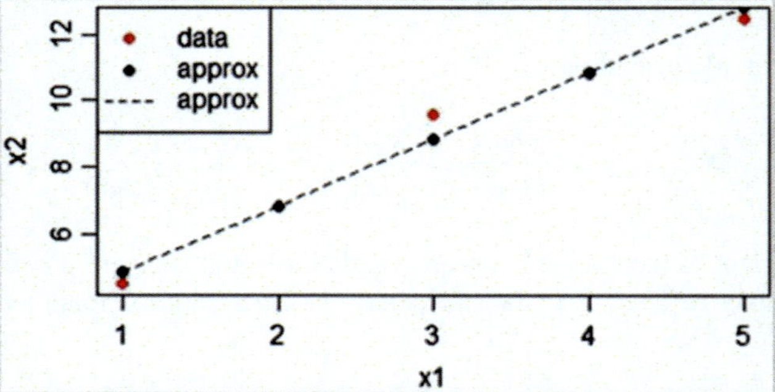

Fig. 2.1 Linear approximation of the data

As indicated, we can use class `ctstats`:

```
> source("ctstats.R");
> cst = ctstats$new()
> ct1 = cst$fromPairs(x1,x2)
> coef = cst$linmod(ct1)
> paste("a = ",coef$a,", b = ",coef$b)
[1] "a =  2 , b =  2.84"
```

2.10.3 Empirical CDF and Empirical PDF

If the sample contains a large number of elements, we may determine approximations of the cumulative function and the probability density of the variable.

To fix the ideas, let us consider a sample $\mathcal{X} = \{X_1, \ldots, X_n\}$ of a random variable X, with n large enough.

To determine the CDF, we can use the function `ecdf`, which generates a function F associating to an abscissa x the value of $F(x) = P(X \leq x)$. Of course, it is also possible to use an user-defined function, by approximating the CDF at a subset of evaluation points $x_1, \ldots, x_{ne}$. For instance, we can fix a number of subintervals n_{sub} and determine a step $\delta x = \frac{\max \mathcal{X} - \min \mathcal{X}}{n_{sub}}$. Then, we set $x_i = \min \mathcal{X} + (i-1)\delta x$. Once the evaluation points are determined, we approximate the CDF as

$$F(x_i) \approx F_e(x_i) = \frac{\text{card}(\mathcal{X}_i)}{n}, \mathcal{X}_i = \{X_j \in \mathcal{X} : X_j \leq x_i\},$$

id est,

$$F_e(x_i) = \frac{\text{number of elements from } \mathcal{X} \text{ that are inferior or equal to } x_i}{n}.$$

This approach is implemented in the method `empcdf` in the class `sampleX`.
Once the CDF was generated, the PDF is approximated as its numerical derivative, generated by one of the methods introduced in Sect. 1.16.

Example 2.24 Consider $X\sim\chi^2(9)$. We generate a sample of 1e4 variates from X and determine the CDF by using `ecdf` and `empcdf` as follows.

```
> X =rchisq(1e4,9);
> F1 = ecdf(X)
> source("sampleX.R")
> s1 = sampleX$new()
> sX1 = s1$create(X)
> sX1=s1$empcdf(sX1,50)
> x = sX1$cdf$x
> F = sX1$cdf$F
> plot(x,F,type="n");
> lines(x,F,lty=2,lwd=3,col='red')
> lines(x,F1(x),lty=1,col='black')
> legend('bottomright',legend=(c('ecdf','empcdf')),col=
c('red','black'),lty=(c(2,1)))
```

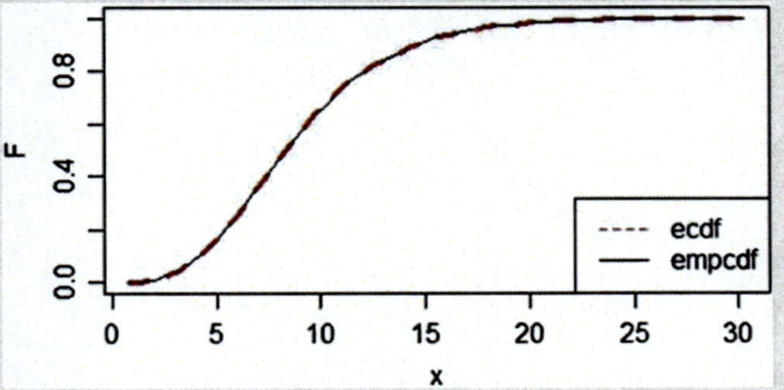

Fig. 2.2 Empirical CDF from a sample of 1e4 variates from $\chi^2(9)$. Both the methods furnish the same result

The CDF is shown in Fig. 2.2. The PDF is determined by a particle derivative using the Gaussian kernel with $h = x_2 - x_1$. The result is shown in Fig. 2.3: The estimated PDF is close to the exact one.

(continued)

Example 2.24 (continued)

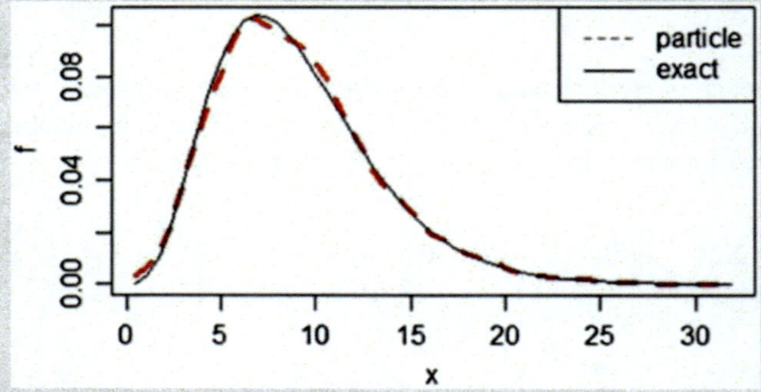

Fig. 2.3 Empirical PDF furnished by the particle derivative of the empirical CDF (Gaussian kernel with $h = x_2 - x_1$)

Notice that you can generate the histogram of the data using the function `hist`, which will furnish an approximation of the PDF. For instance, you can use the code:

```
plot.new()
hx=hist(X,freq = FALSE)
plot(hx,freq = FALSE,xlab="x", ylab="f(x)")
lines(x,dchisq(x,9),lty=1,lwd=3,col='red')
legend('topright',legend=(c('data','exact')),
        col=c('black','red'),lty=(c(1,1)))
```

The result is shown in Fig. 2.4.

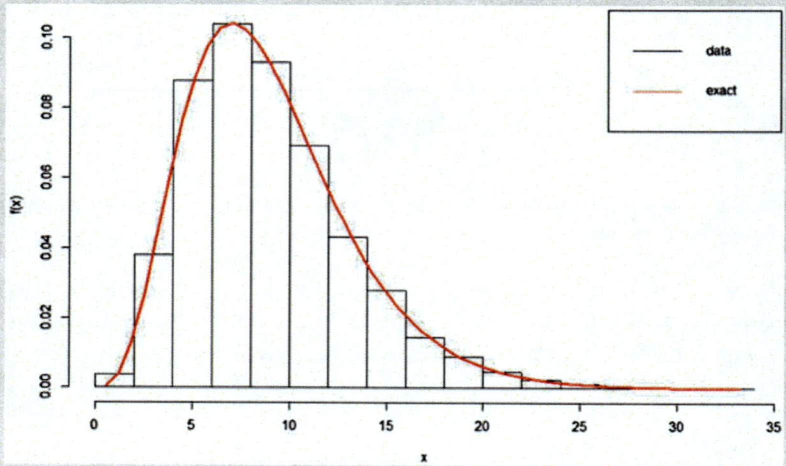

Fig. 2.4 Empirical PDF furnished by `hist`

Table 2.11 A Frequency Table generated from a sample $\{X_1,\ldots,X_n\}$: n_i is the number of occurrences of elements from class number i and $p_i = n_i/n$

Class	$X \le x_1$	$X \in (x_1, x_2)$	...	$X \in (x_{k-1}, x_k)$	$X > x_k$
Observations	n_1	n_2	...	n_k	n_{k+1}
Frequency	p_1	p_2	...	p_k	p_{k+1}

2.10.4 Testing Adequacy of a Sample to a Distribution

Samples can be used to generate frequency tables analogous to Tables of relative frequencies, analogous to Table 2.1. For instance, a sample $\mathcal{X} = \{X_1, \ldots, X_n\}$ can be used to generate a Frequency Table analogous to Table 2.11.

The data in Table 2.11 can be used to verify if prior assumptions on the distribution are coherent with the observations. To do this, one of the popular tests is the χ^2 – *test*, based on the following theorem:

Theorem 2.8 (Fisher-Pearson) Let $\boldsymbol{Y} \in \mathbb{R}^s$ be a variable having a multinomial distribution $\mathcal{M}(N, \boldsymbol{p})$. Let $\boldsymbol{f}: \mathbb{R}^r \to \mathbb{R}^s$ and $\boldsymbol{\theta} \in \mathbb{R}^r$ be such that $\boldsymbol{p} = \boldsymbol{f}(\boldsymbol{\theta})$. Let $\widehat{\boldsymbol{\theta}}$ be the ***maximum-likelihood estimation*** of $\boldsymbol{\theta}$ and $\widehat{\boldsymbol{p}} = \boldsymbol{f}\left(\widehat{\boldsymbol{\theta}}\right)$. Let us define

$$Z_i = \frac{Y_i - n\widehat{p}_i}{\sqrt{n\widehat{p}_i}}, \ 1 \le i \le s.$$

Then $\|\mathbf{Z}\|^2$ is approximatively $\chi^2(s - r - 1)$ for large values of n:

$$\|\mathbf{Z}\|^2 \longrightarrow \chi^2(\mathrm{s} - \mathrm{r} - 1) \text{ in distribution } for \ \mathrm{n} \longrightarrow +\infty.$$

This theorem allows us to test the hypothesis H_0: *X has a density* $f(x, \boldsymbol{\theta})$. This hypothesis is tested as follows: use the data in Table 2.11 to estimate the r parameters θ and estimate the value $\widehat{p}_i = P(Class\ i)$. Then, evaluate ($s = k + 1$, for Table 2.11)

$$Q^2 = \sum_{i=1}^{s} \frac{(n_i - T_i)^2}{T_i} \quad (T_i = n\widehat{p}_i). \tag{2.43}$$

T_i is the expected value if the hypothesis is true. According to the theorem, Q^2 is approximatively $\chi^2(s - r - 1)$(i.e., $\chi^2(k - r)$, for Table 2.11): we may determine a critical value χ^2_α such that $P\left(Q^2 \ge \chi^2_\alpha\right) = \alpha$. The non-rejection region is $\left(0, \chi^2_\alpha\right)$.

R offers a built-in function `chisq.test` to perform a test for the comparison between the observed frequencies and a given probability vector p. In the general situation, `pchisq` can be used to evaluate $a = P(\chi^2 \ge Q^2)$, H_0 is rejected if $a < \alpha$

(This was the original point of view adopted y Pearson – see (Pearson, 1900). An alternative consists in using `qchisq` to determine χ_α^2 and comparing its value to Q^2: the hypothesis is rejected if $Q^2 \geq \chi_\alpha^2$.

Example 2.25 A coin is tossed 100 times. We observe 60 heads (H) and 40 tails (T). To test the hypothesis that the coin is fair, we consider the hypothesis H_0: $P(T) = P(H) = 1/2$. In such a situation, the number of parameters estimated is $r = 0$, the number of classes is $r = 2$ (H and T). The expected number of observations are $T_H = T_T = 100 \times \frac{1}{2} = 50$, corresponding to a fair coin. Then:

$$Q^2 = \frac{(40-50)^2}{50} + \frac{(60-50)^2}{50} = 4.$$

Since $n = 2$, we look for χ_α^2 at the distribution $\chi^2(1)$: For $\alpha = 0.05$, the value is 3.84. Thus, the hypothesis is rejected with a risk of 5%. Analogously, $P(\chi^2 > Q^2) = 0.0455 < 0.05$, so that the hypothesis is rejected by this way too. We can use R as follows:

```
> x = c(60,40); # observations
> pth =c(0.5,0.5); # theoretical probabilities
> xth = pth*sum(x); # theoretical observations under pth
> Q2 = sum((x-xth)^2/xth); # Q2
> s = length(x); # number of classes
> r = 0; # number of parameters estimated
> dof = s - r - 1; # degrees of freedom
> p.cr = pchisq(Q2,dof,lower.tail=FALSE); # critical probability
> alfa = 0.05; # risk
> if (p.cr <= alfa){
+   print(paste0("probability ", p.cr," <= alpha = ",alfa))
+   print(paste0("**************** H0 rejected"))} else{
+     print(paste0("probability ", p.cr," > alpha = ",alfa))
+     print(paste0("**************** H0 accepted"))
+   }
[1] "probability 0.0455002638963585 <= alpha = 0.05"
[1] "**************** H0 rejected"
```

The alternative is implemented as follows:

```
> Q2.cr = qchisq(alfa,dof,lower.tail=FALSE); # critical chi squared
> if (Q2.cr <= Q2){
+   print(paste0("critical chi sq. ", Q2.cr," <= Q2 = ",Q2))
+   print(paste0("**************** H0 rejected"))} else{
+     print(paste0("critical chi sq. ", Q2.cr," > Q2 = ",Q2))
+     print(paste0("**************** H0 accepted"))
+   }
[1] "critical chi sq. 3.84145882069413 <= Q2 = 4"
[1] "**************** H0 rejected"
```

(continued)

Example 2.25 (continued)

In this situation, we can use `chisq.Test`:

```
> resu=chisq.test(x,p=pth)
> if (resu$p.value <= alfa){
+   print(paste0("probability ", resu$p.value," <= alpha = ",alfa))
+   print(paste0("*************** H0 rejected"))} else{
+     print(paste0("probability ", resu$p.value," > alpha = ",alf
a))
+           print(paste0("*************** H0 accepted"))
+   }
[1] "probability 0.0455002638963585 <= alpha = 0.05"
[1] "*************** H0 rejected"
```

Example 2.26 Three products are proposed to 100 consumers. 32 choose product 1, 40 choose product 2, 28 choose product 3. To test the hypothesis that the distribution is uniform, we evaluate

$$Q^2 = \frac{\left(32 - \frac{100}{3}\right)^2}{\frac{100}{3}} + \frac{\left(40 - \frac{100}{3}\right)^2}{\frac{100}{3}} + \frac{\left(60 - \frac{100}{3}\right)^2}{\frac{100}{3}} = 2.24$$

As in the preceding example, no parameter was estimated, so that $r = 0$. Here, $s = 3$ and we look for χ^2_α at the distribution $\chi^2(2)$: The critical value is 5.99. Thus, the hypothesis is not rejected with a risk of 5%. Analogously, the critical probability is $0.33 > 0.05$ and the hypothesis is not rejected with a risk 5%. We can use R as follows:

```
> x = c(32,40,28); # observations
> pth =c(1/3,1/3,1/3); # theoretical probabilities
> xth = pth*sum(x); # theoretical observations under pth
> Q2 = sum((x-xth)^2/xth); # Q2
> s = length(x); # number of classes
> r = 0; # number of parameters estimated
> dof = s - r - 1; # degrees of freedom
> p.cr = pchisq(Q2,dof,lower.tail=FALSE); # critical probability
> alfa = 0.05; # risk
> if (p.cr <= alfa){
+   print(paste0("probability ", p.cr," <= alpha = ",alfa))
+   print(paste0("*************** H0 rejected"))} else{
+     print(paste0("probability ", p.cr," > alpha = ",alfa))
+     print(paste0("*************** H0 accepted"))
+   }
[1] "probability 0.326279794623039 > alpha = 0.05"
[1] "*************** H0 accepted"
```

(continued)

Example 2.26 (continued)

The alternative is implemented as follows:

```
> Q2.cr = qchisq(alfa,dof,lower.tail=FALSE); # critical chi squared
> if (Q2.cr <= Q2){
+   print(paste0("critical chi sq. ", Q2.cr," <= Q2 = ",Q2))
+   print(paste0("*************** H0 rejected"))} else{
+     print(paste0("critical chi sq. ", Q2.cr," > Q2 = ",Q2))
+     print(paste0("*************** H0 accepted"))
+   }
[1] "critical chi sq. 5.99146454710798 > Q2 = 2.24"
[1] "*************** H0 accepted"
```

In this situation also, `chisq.test` can be used:

```
> resu=chisq.test(x,p=pth)
> if (resu$p.value <= alfa){
+   print(paste0("probability ", resu$p.value," <= alpha = ",alfa))
+   print(paste0("*************** H0 rejected"))} else{
+     print(paste0("probability ", resu$p.value," > alpha = ",alf
a))
+           print(paste0("*************** H0 accepted"))
+   }
[1] "probability 0.326279794623039 > alpha = 0.05"
[1] "*************** H0 accepted"
```

Example 2.27 The number of non-compliant parts in 100 samples of 5 elements is the following.

Number	0	1	2	3	4	5
Occurrences	81	10	6	2	1	0

To test the hypothesis that the distribution of the non-compliant parts is binomial $\mathcal{B}(5, p)$ we must start by estimating p. The MLE estimator for p is (see Example 2.26):

$$\hat{p} = \frac{81 \times 0 + 10 \times 1 + 6 \times 2 + 3 \times 2 + 4 \times 1}{5 \times 100}$$

Thus, $\hat{p} = 0.064$. Then, $Q^2 \approx 151$. One parameter was estimated: $r = 1$. We look for χ^2_α at the distribution $\chi^2(4)$: The critical value is 9.5. Thus, the hypothesis is rejected with a risk of 5%.

We can use R as follows:

(continued)

Example 2.27 (continued)

```
> x = c(81,10,6,2,1,0); # observations
> n = length(x)-1; # number of trials (binomial distrib)
> vals = c(0:n); # number of defectuous
> ns = sum(x); # number of samples
> p.est = sum(x*vals)/(ns*n)
> xth = ns*dbinom(vals,n,p.est); # theoretical observations under p
th
> Q2 = sum((x-xth)^2/xth); # Q2
> s = length(x); # number of classes
> r = 1; # number of parameters estimated
> dof = s - r - 1; # degrees of freedom
> p.cr = pchisq(Q2,dof,lower.tail=FALSE); # critical probability
> alfa = 0.05; # risk
> if (p.cr <= alfa){
+    print(paste0("probability ", p.cr," <= alpha = ",alfa))
+    print(paste0("*************** H0 rejected"))} else{
+      print(paste0("probability ", p.cr," > alpha = ",alfa))
+      print(paste0("*************** H0 accepted"))
+    }
[1] "probability 1.31101204693816e-31 <= alpha = 0.05"
[1] "*************** H0 rejected"
```

The alternative is implemented as follows:

```
> Q2.cr = qchisq(alfa,dof,lower.tail=FALSE); # critical chi squared
> if (Q2.cr <= Q2){
+    print(paste0("critical chi sq. ", Q2.cr," <= Q2 = ",Q2))
+    print(paste0("*************** H0 rejected"))} else{
+      print(paste0("critical chi sq. ", Q2.cr," > Q2 = ",Q2))
+      print(paste0("*************** H0 accepted"))
+    }
[1] "critical chi sq. 9.48772903678115 <= Q2 = 150.891844157771"
[1] "*************** H0 rejected"
```

Example 2.28 Let us consider the observations of a variable X:

class	(0,5)	(5,9)	(9,14)	(14,18)	(18,23)
occurrences	61	28	5	4	1

We consider the hypothesis that the distribution of X is exponential $f(x, \lambda) = \lambda e^{-\lambda x}$ $(x > 0)$. The first step consists in estimating λ: For a sample $\mathcal{X} = \{X_1, \ldots, X_{ns}\}$

$$ns \times \log \lambda - \lambda \sum_{i=1}^{ns} X_i,$$

so that

(continued)

Example 2.28 (continued)

$$\frac{1}{\widehat{\lambda}} = \frac{1}{ns}\sum_{i=1}^{ns} X_i.$$

We use the centers of the classes to evaluate $\widehat{\lambda}$:

$$\widehat{\lambda} = \frac{61 + 28 + 5 + 4 + 1}{61 \times 2.5 + 28 \times 7 + 5 \times 11.5 + 4 \times 16 + 1 \times 20.5} \approx 0.21$$

Then, we use this value to determine the probabilities of the classes: For an interval (a, b): $P(X \in (a, b)) = F(b) - F(a)$, where F is the CDF of X. We have

class	(0,5)	(5,9)	(9,14)	(14,18)	(18,23)
$\widehat{p}$	0.61	0.24	0.09	0.04	0.01

Then, $Q^2 \approx 6.4$. One parameter was estimated: $r = 1$. We look for χ^2_α at the distribution $\chi^2(3)$: The critical value is 7.8. Thus, the hypothesis is accepted with a risk of 5%.

We can use R as follows:

```
> xlims = c(0,5,9,14,18,23);
> xcounts= c(61,28,5,4,1);
> xm = vector(length=length(xcounts))
> for (i in 1: (length(xlims)-1)){
+   xm[[i]]=(xlims[[i+1]]+xlims[[i]])/2
+   }
> m.est = sum(xm*xcounts)/sum(xcounts);
> lbda.est = 1/m.est
> pth = vector(length=length(xcounts));
> for (i in 1:length(xcounts)){
+   pth[[i]]=pexp(xlims[[i+1]],lbda.est)-pexp(xlims[[i]],lbda.est)
+ }
> ns = sum(xcounts); # number of samples
> xth = ns*pth; # theoretical observations
> Q2 = sum((xcounts-xth)^2/xth); # Q2
> s = length(xcounts); # number of classes
> r = 1; # number of parameters estimated
> dof = s - r - 1; # degrees of freedom
> p.cr = pchisq(Q2,dof,lower.tail=FALSE); # critical probability
> alfa = 0.05; # risk
> if (p.cr <= alfa){
+   print(paste0("probability ", p.cr," <= alpha = ",alfa))
+   print(paste0("*************** H0 rejected"))} else{
+     print(paste0("probability ", p.cr," > alpha = ",alfa))
+     print(paste0("*************** H0 accepted"))
+   }
[1] "probability 0.0946300383580203 > alpha = 0.05"
[1] "*************** H0 accepted"
```

(continued)

Example 2.28 (continued)

The alternative is implemented as follows:

```
> Q2.cr = qchisq(alfa,dof,lower.tail=FALSE); # critical chi squared
> if (Q2.cr <= Q2){
+   print(paste0("critical chi sq. ", Q2.cr," <= Q2 = ",Q2))
+   print(paste0("*************** H0 rejected"))} else{
+     print(paste0("critical chi sq. ", Q2.cr," > Q2 = ",Q2))
+     print(paste0("*************** H0 accepted"))
+   }
[1] "critical chi sq. 7.81472790325118 > Q2 = 6.37727835169594"
[1] "*************** H0 accepted"
```

Example 2.29 Let us consider the observations of a variable X:

class	(−1.1,0.5)	(0.5,2.1)	(2.1,3.6)	(3.6,5.2)	(5.2,6.8)
occurrences	9	20	38	24	8

We consider the hypothesis that the distribution of X is normal $f(x, m, \sigma) = \exp\left(-\left((x-m)/_{\sigma}\right)^2\right)$. The first step consists in estimating m and σ: For a sample $X = \{X_1, \ldots, X_{ns}\}$, the MLE estimators are

$$\widehat{m} = \frac{1}{ns}\sum_{i=1}^{ns} X_i, \widehat{\sigma}^2 = \frac{1}{ns}\sum_{i=1}^{ns} (X_i - \widehat{m})^2$$

We use the centers of the classes to evaluate $\widehat{m}$ and $\widehat{\sigma}$:

$$\widehat{m} \approx 2.9, \widehat{\sigma} \approx 1.7.$$

As in the previous example, we use this value to determine the probabilities of the classes: For an interval (a, b): $P(X \in (a, b)) = F(b) - F(a)$, where F is the CDF of X. We have

class	(−1.1,0.5)	(0.5,2.1)	(2.1,3.6)	(3.6,5.2)	(5.2,6.8)
$\widehat{p}$	0.07	0.24	0.35	0.25	0.07

Thus, $Q^2 \approx 1.96$. Two parameters were estimated: $r = 2$. We look for χ^2_α at the distribution $\chi^2(3)$: The critical value is 6.0. Thus, the hypothesis is accepted with a risk of 5%.

(continued)

Example 2.29 (continued)

We can use R as follows:

```
> xlims = c(-1.1,0.5,2.1,3.6,5.2,6.8);
> xcounts= c(9,20,38,24,8);
> xm = vector(length=length(xcounts))
> for (i in 1: (length(xlims)-1)){
+   xm[[i]]=(xlims[[i+1]]+xlims[[i]])/2
+   }
> m.est = sum(xm*xcounts)/sum(xcounts);
> sig.est = sqrt(sum((xm-m.est)^2*xcounts)/sum(xcounts))
> pth = vector(length=length(xcounts));
> for (i in 1:length(xcounts)){
+   pth[[i]]=pnorm(xlims[[i+1]],m.est,sig.est)-pnorm(xlims[[i]],
m.est,sig.est)
+ }
> ns = sum(xcounts); # number of samples
> xth = ns*pth; # theoretical observations
> Q2 = sum((xcounts-xth)^2/xth); # Q2
> s = length(xcounts); # number of classes
> r = 2; # number of parameters estimated
> dof = s - r - 1; # degrees of freedom
> p.cr = pchisq(Q2,dof,lower.tail=FALSE); # critical probability
> alfa = 0.05; # risk
> if (p.cr <= alfa){
+   print(paste0("probability ", p.cr," <= alpha = ",alfa))
+   print(paste0("*************** H0 rejected"))} else{
+     print(paste0("probability ", p.cr," > alpha = ",alfa))
+     print(paste0("*************** H0 accepted"))
+   }
[1] "probability 0.375888377389174 > alpha = 0.05"
[1] "*************** H0 accepted"
```

The alternative is implemented as follows:

```
> Q2.cr = qchisq(alfa,dof,lower.tail=FALSE); # critical chi squared
> if (Q2.cr <= Q2){
+   print(paste0("critical chi sq. ", Q2.cr," <= Q2 = ",Q2))
+   print(paste0("*************** H0 rejected"))} else{
+     print(paste0("critical chi sq. ", Q2.cr," > Q2 = ",Q2))
+     print(paste0("*************** H0 accepted"))
+   }
[1] "critical chi sq. 5.99146454710798 > Q2 = 1.95692609662455"
[1] "*************** H0 accepted"
```

Exercises

1. A dice rolls 100 times. The results are the following:

Face	1	2	3	4	5	6
Observations	17	20	25	14	21	13

Can we assume that the dice is fair?

(continued)

2. A sample of a variable furnishes the results below:

Value	0	1	2	3	4
Observations	25	10	30	30	5

Can we assume that X is uniformly distributed?

3. A sample of a variable furnishes the results below:

Value	0	1	2	3	4	5	6
Observations	11	9	7	9	5	10	8

Can we assume that X is uniformly distributed?

4. A sample of a variable furnishes the results below:

Value	0	1	2	3
Observations	40	42	13	5

Can we assume that X is Poisson distributed?

5. A sample of a variable furnishes the results below:

Value	0	1	2	3	4	5	6
Observations	5	15	42	35	30	10	8

Can we assume that X is Poisson distributed?

6. A sample of a variable furnishes the results below:

Value	0	1	2	3	4
Observations	16	50	120	90	42

Can we assume that X is binomially distributed $\mathcal{B}(4, p)$?

7. A sample of a variable furnishes the results below:

Value	0	1	2	3
Observations	200	100	10	5

Can we assume that X is binomially distributed $\mathcal{B}(3, p)$?

8. A sample of a variable furnishes the results below:

Value	8–10	10–12	12–14	14–16	16–18
Observations	5	67	510	370	60

Can we assume that X is normally distributed $N(m, \sigma)$?

9. A sample of a variable furnishes the results below:

Value	0–10	10–20	20–30	30–40	40–50	50–60
Observations	10	210	560	180	20	5

Can we assume that X is normally distributed $N(m, \sigma)$?

2.10.5 Testing the Independence a Couple of Variables

In the preceding section, we used samples from a variable X to generate Frequency Tables and verify if the data are consistent with a theoretical distribution. Analogously, observations of a couple (X, Y) can be used to generate a Contingency Table as Table 2.12.

Such a Table can be transformed in probabilities by using the transformation $p_{ij} = \frac{o_{ij}}{ns}$. The marginal probabilities are $p_{i\blacksquare} = \frac{o_{i\blacksquare}}{ns}, p_{\blacksquare j} = \frac{o_{\blacksquare j}}{ns}, o_{i\blacksquare} = \sum_{j=1}^{n+1} o_{ij}, o_{\blacksquare j} = \sum_{i=1}^{m+1} o_{ij}$.

Table 2.13 can be used to verify if the hypothesis of independence of the variables is consistent with the observations – we can test the hypothesis H_0: X and Y are independent by using the χ^2 test: if the variables are independent, then $P(X = X_i, Y = Y_j) = P(X = X_i)P(Y = Y_j)$. We start by estimating $P(X = X_i) \approx p_{i\blacksquare}$, $P(Y = Y_j) \approx p_{\blacksquare j}$, so that $\widehat{p_{ij}} = p_{i\blacksquare} p_{\blacksquare j}$. Then, we evaluate

Table 2.12 A Contingency Table generated from a sample $\{(X_1, Y_1), \ldots, (X_{ns}, Y_{ns})\}$: o_{ij} is the number of occurrences of elements from class (i, j)

Classes	$Y < y_1$	$Y \in (y_1, y_2)$	...	$Y \in (y_{n-1}, y_n)$	$Y > y_n$
$X < x_1$	o_{11}	o_{12}	...	$o_{1,n}$	$o_{1,n+1}$
$X \in (x_1, x_2)$	o_{21}	o_{22}	...	$o_{2,n}$	$o_{2,n+1}$
$\vdots$	$\vdots$	$\vdots$	$\ddots$	$\vdots$	$\vdots$
$X \in (x_{m-1}, x_m)$	o_{m1}	o_{m2}	...	$o_{m,n}$	$o_{m,n+1}$
$X > x_m$	$o_{m+1,1}$	$o_{m+1,2}$	...	$o_{m+1,n}$	$o_{m+1,n+1}$

$$Q^2 = \sum_{i=1}^{m} \sum_{j=1}^{n} \frac{(o_{ij} - T_{ij})^2}{T_{ij}} \quad \left(T_{ij} = ns \times \widehat{p_{ij}}\right). \tag{2.44}$$

Q^2 is approximatively $\chi^2((m-1)(n-1))$: we may determine a critical value χ^2_α such that $P\left(Q^2 \geq \chi^2_\alpha\right) = \alpha$. The non-rejection region is $\left(0, \chi^2_\alpha\right)$.

The function `chisq.test` determines Q^2 and $P(\chi^2 \geq Q^2)$.

Example 2.30 Let us consider 100 observations of a couple (X, Y) leading to the contingency table at below left. To test the independence, we must determine p_{ij} (Table 2.26.5 at right).

Table 2.13 A contingency table generated by a sample of 100 observations of (X,Y)

X \ Y	-1	0	1
-1	20	0	10
0	0	40	0
1	10	0	20

X \ Y	-1	0	1
-1	0.2	0	0.1
0	0	0.4	0
1	0.1	0	0.2

(continued)

Example 2.30 (continued)

Thus, $p_{i\blacksquare} = (0.3, 0.4, 0.3)$, $p_{\blacksquare j} = (0.3, 0.4, 0.3)$. Assuming independence, the expected values are (Table 2.14)

Table 2.14 Expected values for the test of independence. At left, the values of T_{ij}. At right, the probabilities $\widehat{p_{ij}}$.

X \ Y	-1	0	1
-1	9	12	9
0	12	16	12
1	9	12	9

X \ Y	-1	0	1
-1	0.09	0.12	0.09
0	0.12	0.16	0.12
1	0.09	0.12	0.09

Thus, $Q^2 \approx 111$. The number of degrees of freedom is $(3-1)(3-1) = 4$: For $\alpha = 0.05$, $\chi^2_\alpha = 9.5$. Since $Q^2 > \chi^2_\alpha$, the hypothesis is rejected.

We can use R as follows:

```
> x = matrix(c(20,0,10,0,40,0,10,0,20),nrow=3,ncol=3);
> resu = chisq.test(x)
> x = matrix(c(20,0,10,0,40,0,10,0,20),nrow=3,ncol=3);
> resu = chisq.test(x);
> p.cr = resu$p.value; # critical probability
> alfa = 0.05; # risk
> if (p.cr <= alfa){
+   print(paste0("probability ", p.cr," <= alpha = ",alfa))
+   print(paste0("*************** H0 rejected"))} else{
+     print(paste0("probability ", p.cr," > alpha = ",alfa))
+     print(paste0("*************** H0 accepted"))
+   }
[1] "probability 4.21700441839518e-23 <= alpha = 0.05"
[1] "*************** H0 rejected"
>
> Q2 = resu$statistic
> d = dim(x);
> dof = (d[[1]]-1)*(d[[2]]-1) ; # degrees of freedom
> Q2.cr = qchisq(alfa,dof,lower.tail=FALSE); # critical chi squared
> if (Q2.cr <= Q2){
+   print(paste0("critical chi sq. ", Q2.cr," <= Q2 = ",Q2))
+   print(paste0("*************** H0 rejected"))} else{
+     print(paste0("critical chi sq. ", Q2.cr," > Q2 = ",Q2))
+     print(paste0("*************** H0 accepted"))
+   }
[1] "critical chi sq. 9.48772903678115 <= Q2 = 111.111111111111"
[1] "*************** H0 rejected"
```

Example 2.31 Let us consider 100 observations of a couple (X, Y) leading to the contingency table at below left. To test the independence, we must determine p_{ij} (Table 2.15 at right).

(continued)

Example 2.31 (continued)

Table 2.15 A contingency table generated by a sample of 100 observations of (X,Y). At left, the numbers of observations. At right, the probabilities

X \ Y	0	1
0	28	22
1	21	29

X \ Y	0	1
0	0.28	0.22
1	0.21	0.29

Thus, $p_{i\blacksquare} = (0.5, 0.5)$, $p_{\blacksquare j} = (0.49, 0.51)$. Assuming independence, the expected values are (Table 2.16)

Table 2.16 Expected values for the test of independence. At left, the values of T_{ij}. At right, the probabilities $\widehat{p_{ij}}$

X \ Y	0	1
0	28.5	26.5
1	34.2	31.8

X \ Y	0	1
0	0.285	0.265
1	0.342	0.318

Thus, $Q^2 \approx 1.58$ the number of degrees of freedom is $(2-1)(2-1) = 1$: For $\alpha = 0.05$, $\chi^2_\alpha = 3.8$. Since $Q^2 < \chi^2_\alpha$, the hypothesis is accepted.

We can use R as follows:

```
> x = matrix(c(28,22,21,29),nrow=2,ncol=2);
> resu = chisq.test(x);
> p.cr = resu$p.value; # critical probability
> alfa = 0.05; # risk
> if (p.cr <= alfa){
+   print(paste0("probability ", p.cr," <= alpha = ",alfa))
+   print(paste0("**************** H0 rejected"))} else{
+     print(paste0("probability ", p.cr," > alpha = ",alfa))
+     print(paste0("**************** H0 accepted"))
+   }
[1] "probability 0.230046116594712 > alpha = 0.05"
[1] "**************** H0 accepted"
>
> Q2 = resu$statistic
> d = dim(x);
> dof = (d[[1]]-1)*(d[[2]]-1) ; # degrees of freedom
> Q2.cr = qchisq(alfa,dof,lower.tail=FALSE); # critical chi squared
> if (Q2.cr <= Q2){
+   print(paste0("critical chi sq. ", Q2.cr," <= Q2 = ",Q2))
+   print(paste0("**************** H0 rejected"))} else{
+     print(paste0("critical chi sq. ", Q2.cr," > Q2 = ",Q2))
+     print(paste0("**************** H0 accepted"))
+   }
[1] "critical chi sq. 3.84145882069413 > Q2 = 1.4405762304922"
[1] "**************** H0 accepted"
```

Exercises

1. A sample of two variables furnishes the results below:

X \ Y	1	2	3	4
1	50	15	6	50
2	47	13	9	35
3	35	33	12	34

Can we assume that *X* and Y are independent?

2. A sample of two variables furnishes the results below:

X \ Y	1	2	3
1	200	190	110
2	300	150	50

Can we assume that *X* and Y are independent?

3. A sample of two variables furnishes the results below:

X \ Y	1	2	3	4
1	150	250	80	10
2	75	110	40	10
3	40	100	130	10

Can we assume that *X* and Y are independent?

2.11 Generating Triangular Random Numbers

Triangular distributions are mainly used when the information available on the random variables is limited to a minimal value x_{min}, a maximal value x_{max}, and a probable value x_p. A triangular distribution with these parameters is referred as $T(x_{min}, x_p, x_{max})$. The PDF associated to a triangular distribution forms a triangle and its equation is

$$f(x) = \begin{cases} \dfrac{2(x - x_{min})}{(x_{max} - x_{min})(x_p - x_{min})}, & \text{if } x_{min} \le x \le x_p; \\ \dfrac{2(x_{max} - x)}{(x_{max} - x_{min})(x_{max} - x_p)}, & \text{if } x_p \le x \le x_{max}; \\ 0, & \text{otherwise.} \end{cases}$$

Its CDF is

$$F(x) = \begin{cases} 0, & \text{if } x \le x_{\min}; \\ \dfrac{(x - x_{\min})^2}{(x_{\max} - x_{\min})(x_p - x_{\min})}, & \text{if } x_{\min} \le x \le x_p; \\ 1 - \dfrac{(x_{\max} - x)^2}{(x_{\max} - x_{\min})(x_{\max} - x_p)}, & \text{if } x_p \le x \le x_{\max}; \\ 1, & \text{otherwise.} \end{cases}$$

The generation of samples from a triangular distribution uses the uniform generator. Indeed, let U be uniformly distributed on (0,1) and consider

$$X = \begin{cases} x_{\min} + \sqrt{U(x_{\max} - x_{\min})(x_p - x_{\min})}, & \text{if } U < \dfrac{x_p - x_{\min}}{x_{\max} - x_{\min}} \\ x_{\max} - \sqrt{(1-U)(x_{\max} - x_{\min})(x_{\max} - x_p)}, & \text{otherwise.} \end{cases}$$

Then, X is $T(x_{\min}, x_{\max}, x_p)$.

As an example, let us generate a sample formed by 1E4 variates from $T(-1, 0, 3)$ and determine the empirical CDF and PDF (the PDF is generated by SP numerical derivation of the CDF with a gaussian kernel and $h = (x_2 - x_1)/2$. The preceding equations were implemented in `triang.R`, which contains `dtriang`, `ptriang`, `qtriang`, and `rtriang` – according to the standard names in R, these functions furnish, respectively, the CFD, the PDF, the quantiles, and samples. The results are shown in Fig. 2.5.

2.12 Generating Random Numbers by Inversion

Let X be a random variable having as cumulative function F. Let $U = F(X)$. Then, for $u \in (0, 1)$

$$P(U < u) = P(F(X) < u) = P(X < F^{-1}(u)) = F(F^{-1}(u)) = u.$$

Consequently, U is uniformly distributed on (0, 1). Thus, samples from X may be obtained from samples from the uniform distribution, if F^{-1} is available: we generate $u_1, \ldots, u_{ns}$ from the uniform distribution and we set $x_i = F^{-1}(u_i)$.

This approach is very general and allows the generation of samples from arbitrary distributions for which F^{-1} can be determined. It can be easily implemented in R. For instance, if F^{-1} is evaluated by the member `F$inv.` of a list `F`, we can use the code below (available in file `inversion.R`):

```
rinversion = function(F,n)
{
  u = runif(n);
  x = numeric(length=n)
```

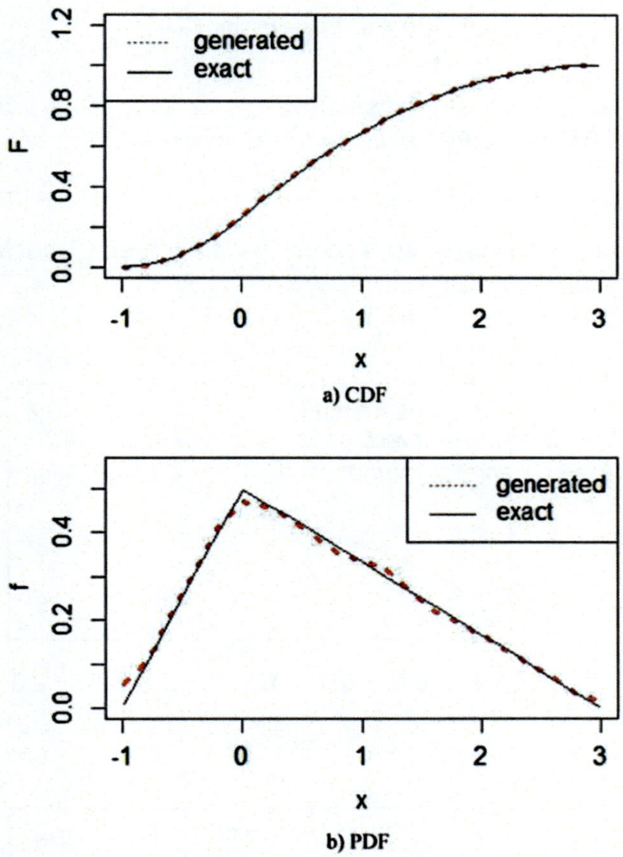

Fig. 2.5 Results furnished by the `triang.R`. The empirical sample is formed by 1E4 variates from $T(-1,0,3)$ furnished by `rtriang`. (**a**) CDF. (**b**) PDF

```
    for (i in 1:n){
      x[[i]] = F$inv(u[[i]])
    }
    return(x)
}
```

`Rinversion(F,n)` generates a sample of n variates by the method of inversion. As an example, let us consider the arcsine distribution, for which

$$F(x) = \frac{2}{\pi} \arcsin \sqrt{\frac{x-a}{b-a}}, \, a < x < b.$$

Here,

$$F^{-1}(u) = a + (b - a)\sin^2\left(\frac{\pi}{2}u\right)$$

Let us generate a sample of 1E4 variates using $a = -1$, $b = 2$ and use it to estimate the CDF and the PDF of X. The results are shown in Fig. 2.6.

2.13 Generating Random Vectors with a Given Covariance Matrix

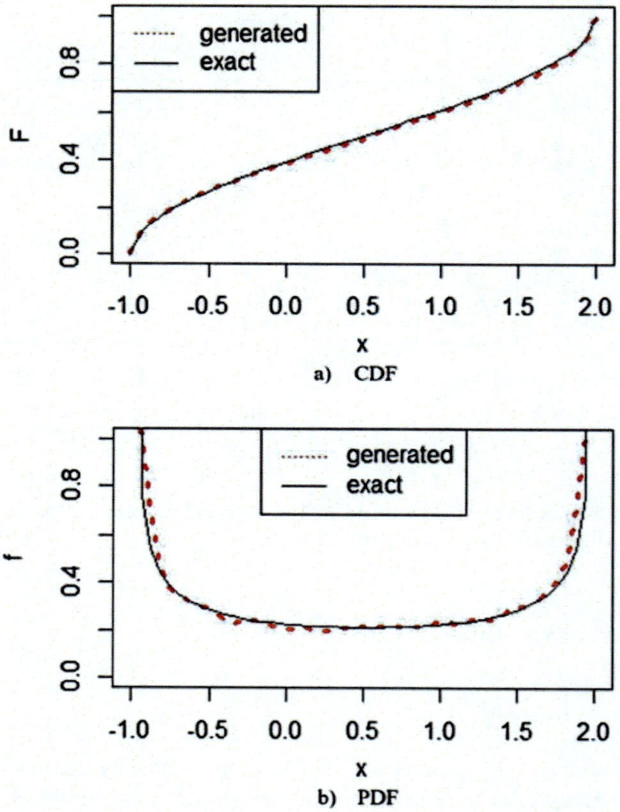

Fig. 2.6. Results furnished by the method of inversion. The empirical sample is formed by 1E4 variates from the *arcsine* distribution, generated by `inversion`. **(a)** CDF. **(b)** PDF

A situation often found in practice is the simulation of a random vector $\boldsymbol{X} = (X_1, \ldots, X_n)$ having a prescribed covariance matrix $\boldsymbol{C} = (C_{ij}, 1 \leq i, j \leq n)$.

Assume initially that $E(\boldsymbol{X}) = \boldsymbol{0}$. A simple solution consists in using a vector of m known independent random variables $\boldsymbol{Z} = (Z_1, .., Z_n)$ such that each Z_i has a mean equal to zero and a variance equal to one:

$$E(Z_i) = 0,\ V(Z_i) = 1\ for\ 1 \leq i \leq m; E(Z_i Z_j) = 0, for\ i \neq j.$$

Then,

$$X_i = \sum_{j=1}^{m} A_{ij} Z_j \Leftrightarrow \boldsymbol{X} = \boldsymbol{Z}\boldsymbol{A}, \boldsymbol{A} = (A_{ij},\ 1 \leq i \leq n,\ 1 \leq j \leq m).$$

In this case (recall that $\boldsymbol{Z}$ is a vector of independent variables with zero mean):

$$E(X_i X_j) = \sum_{k,\ \ell=1}^{m} A_{ik} A_{j\ell} E(Z_k Z\_\ell) = \sum_{k,\ \ell=1}^{m} A_{ik} A_{jk} E(Z_k^2) = (\boldsymbol{A}\boldsymbol{A}^t)_{ij},$$

so that

$$\boldsymbol{C} = \boldsymbol{A}\boldsymbol{A}^t.$$

Taking $m = n$, $\boldsymbol{A}$ may be determined as a Choleski's decomposition of matrix $\boldsymbol{C}$ (recall that $\boldsymbol{C}$ is symmetric positive). Such a decomposition can be determined by the R command `chol(C)`. If $E(\boldsymbol{X}) = \boldsymbol{m}$, we can consider $\boldsymbol{X} = \boldsymbol{Z}\boldsymbol{A} + \boldsymbol{m}$.

Notice that his approach cannot furnish a specific distribution except for special situations, for instance, gaussian vector. In this case, if $\boldsymbol{Z}$ is a gaussian vector, then, $\boldsymbol{X}$ is gaussian too.

Example 2.32 Let us consider a random vector $\boldsymbol{X} = (X_1, X_2, X_3)$ such that

$$\boldsymbol{C} = \begin{pmatrix} 3 & 2 & 3 \\ 2 & 5 & 4 \\ 3 & 4 & 8 \end{pmatrix},\ \boldsymbol{m} = (1\ \ 2\ \ 3).$$

We start by determining $\boldsymbol{A}$:

(continued)

Example 2.32 (continued)

```
> matC = matrix(c(3,2,3,2,5,4,3,4,8),
+               nrow = 3,byrow=TRUE)
> A = chol(matC)
> A
         [,1]     [,2]     [,3]
[1,] 1.732051 1.154701 1.732051
[2,] 0.000000 1.914854 1.044466
[3,] 0.000000 0.000000 1.977142
```

Let us consider Z_i gaussian $N(0, 1)$. We generate a sample of 1E4 variates from X as follows:

```
> ns = 1e4
> Z = matrix(nrow=ns,ncol=3)
> for (j in 1:3) {Z[,j]=rnorm(ns)}
> m = matrix(c(1,2,3),nrow=1)
> X = Z %*% A + matrix(m,nrow=ns, ncol = 3,byrow=TRUE)
> mX = numeric(3)
> for (j in 1:3) {mX[[j]]=mean(X[,j])}
> CX = cov(X)
> mX
[1] 1.034043 1.980979 3.014619
> CX
         [,1]     [,2]     [,3]
[1,] 3.000725 1.991555 3.031528
[2,] 1.991555 5.019668 3.982137
[3,] 3.031528 3.982137 7.931980
```

As we can observe, the empirical means and covariance matrix of the sample are close to the prescribed values. We can use other distributions. For instance, using a uniform distribution:

```
> ns = 1e4
> a = 1
> Z = matrix(nrow=ns,ncol=3)
> for (j in 1:3) {Z[,j]=runif(ns,-a,a)/sqrt(a*a/3)}
> m = matrix(c(1,2,3),nrow=1)
> X = Z %*% A + matrix(m,nrow=ns, ncol = 3,byrow=TRUE)
> mX = numeric(3)
> for (j in 1:3) {mX[[j]]=mean(X[,j])}
> CX = cov(X)
> mX
[1] 1.000233 2.005757 2.969868
> CX
         [,1]     [,2]     [,3]
[1,] 2.985872 1.988777 2.969614
[2,] 1.988777 5.029767 3.944957
[3,] 2.969614 3.944957 7.876151
> |
```

2.14 Generating Regular Random Functions

The main difference between random numbers and random functions is a matter of dimensionality. For instance, let us consider a function y: $(0, 1) \rightarrow \mathbb{R}$: to define such a function, we must give the value $y(t)$ for any $t \in (0, 1)$, so that the definition of y requests the definition of infinitely many values.

Indeed, functions are infinite dimensional objects; id est, functions are elements from infinite dimensional vector spaces: contrarily to vectors from $\mathbb{R}^n$ – which are fully characterized by a finite number of coefficients (n real values) – functions need an infinite number of parameters to be completely defined.

Some classes of functions can be characterized using Hilbert basis – fortunately, these classes are general enough for practical purposes. Recall that a Hilbert basis for a real function space V is a family $\{\varphi_i: i \geq 0, i \in N\} \subset V$ such that

$$\forall y \in V: y = \sum_{i=0}^{+\infty} y_i \varphi_i, \, y_i \in \mathbb{R}, \forall i \geq 0. \tag{2.45}$$

Thus, we may generate an approximation by considering a finite sum, for instance

$$\forall y \in V: y \approx P_k y = \sum_{i=0}^{k} y_i \varphi_i, \, y_i \in \mathbb{R}, 0 \leq i \leq k. \tag{2.46}$$

More generally, we may consider

$$\forall y \in V: y \approx P_k y = \sum_{i=0}^{k} y_{j_i} \varphi_{j_i}, \, y_i \in \mathbb{R}, j_i \geq 0, j_i \in \mathbb{N}. \tag{2.47}$$

These expressions generalize to the case where $\boldsymbol{y} = (y_0, \ldots, y_d)$ is a vector of real functions. Typical Hilbert basis for functions defined on an interval (a, b) are furnished by polynomials:

$$\varphi_i^{poly}(t) = \left(\frac{t-a}{b-a}\right)^i, a < b; \tag{2.48}$$

or trigonometrical functions:

$$\varphi_0^{trigo}(t) = 1; \; \varphi_{2i-1}^{trigo}(t) = \sin\left(i\left(\frac{t-a}{b-a}\right)\right), \tag{2.49}$$

$$\varphi_{2i}^{trigo}(t) = \cos\left(i\left(\frac{t-a}{b-a}\right)\right) \; (i \geq 1). \tag{2.50}$$

We may also consider periodical functions

$$\varphi_0^{per}(t) = 1;\ \varphi_{2i-1}^{per}(t) = \sin\left(2\pi i\left(\frac{t-a}{b-a}\right)\right), \tag{2.51}$$

$$\varphi_{2i}^{per}(t) = \cos\left(2\pi i\left(\frac{t-a}{b-a}\right)\right)\ (i \geq 1). \tag{2.52}$$

To generate approximations of a given y, we may use the standard scalar product of square summable functions

$$(y, z) = \int_a^b y(t)z(t)dt$$

with an orthogonal projection approach. For instance, when using the approximation (2.46), we have

$$(y - P_k y, \varphi_i) = 0,\ 0 \leq i \leq k.$$

Thus, the coefficients $\boldsymbol{y} = (y_0, \ldots, y_k)^t$ are the solutions of a linear system

$$A\boldsymbol{y} = B,\ A_{i+1,\, j+1} = (\varphi_i, \varphi_j),\ B_{i+1} = (y, \varphi_i).$$

If the family $\{\varphi_i: i \geq 0, i \in N\}$ is formed by orthogonal elements, then $(\varphi_i, \varphi_j) = 0$, if $i \neq j$. In this case,

$$y_i = \frac{B_{i+1}}{A_{i+1,\, i+1}}.$$

For an orthonormal family, in addition, $A_{i+1,\, i+1} = 1$, so that $y_i = B_{i+1}$. The reader will find in the literature examples of orthonormal families, for instance, Legendre polynomials. An arbitrary family $\{\phi_i: i \geq 0, i \in N\}$ may be transformed into an orthonormal family $\{\varphi_i: i \geq 0, i \in N\}$ by the Gram–Schmidt procedure:

$$\varphi_i = \frac{r_i}{\sqrt{(r_i, r_i)}};\ r_0 = \phi_0,\ r_i = \phi_i - \sum_{j=0}^{i-1} (\phi_i, \varphi_j)\varphi_j\ (i \geq 1).$$

Polynomials are implemented in the class `polynom.R`, trigonometrical functions in `trigo.R`, and periodical functions are implemented in `trigoper.R`. Class `expansion1d` proposes methods `spmatrix` (determination of A) and `smmatrix` (determination of B). `expansion1d` proposes also a function `coeff` that evaluates the coefficients $\boldsymbol{y}$ and `proj` that evaluates $P_k y$.

Example 2.33 Let us consider the function y: $(-1,1) \to \mathbb{R}$ given by $y(t) = e^t$. Let us look for a polynomial approximation of order $k = 4$. We have

$$y_0 = \frac{5(-539 + 73\mathbf{e}^2)}{2\mathbf{e}} \approx 0.368887,$$

$$y_1 = -\frac{45(-569 + 77\mathbf{e}^2)}{\mathbf{e}} \approx 0.706556,$$

$$y_2 = \frac{840(-133 + 18\mathbf{e}^2)}{\mathbf{e}} \approx 0.930079,$$

$$y_3 = -\frac{70(-2431 + 329\mathbf{e}^2)}{\mathbf{e}} \approx 0.013995,$$

$$y_4 = \frac{630(-133 + 18\mathbf{e}^2)}{\mathbf{e}} \approx 0.697559.$$

Approximated values may be obtained by the methods presented in Sect. 1.16. For instance, we can use

We can use R as follows:

```
> source("expansion1D.R")
> a = -1
> b = 1
> z = 1e-3
> y = function(t){exp(t)}
>
> source("polynom.R")
> k = 4
> phi = polynom$new(a,b,z)
> hel = expansion1D$new(phi)
> c = hel$coeff(k+1,y)
> xc = seq(a,b,length.out=26)
> yc = y(xc)
> yap = hel$proj(c,xc)
> plot(xc,yc,type='n',main = paste('polynomial family, k = ',k))
> lines(xc,yc,lty=1,col='black')
> lines(xc,yap,lty=3,col='red',lwd=3)
> legend("top",legend = c("exact","approx"),col = c("black","red"),lty =
c(1,3))
```

The results for different families are shown in Fig. 2.7. Notice that the periodical family produces a periodic approximation.

(continued)

Example 2.33 (continued)

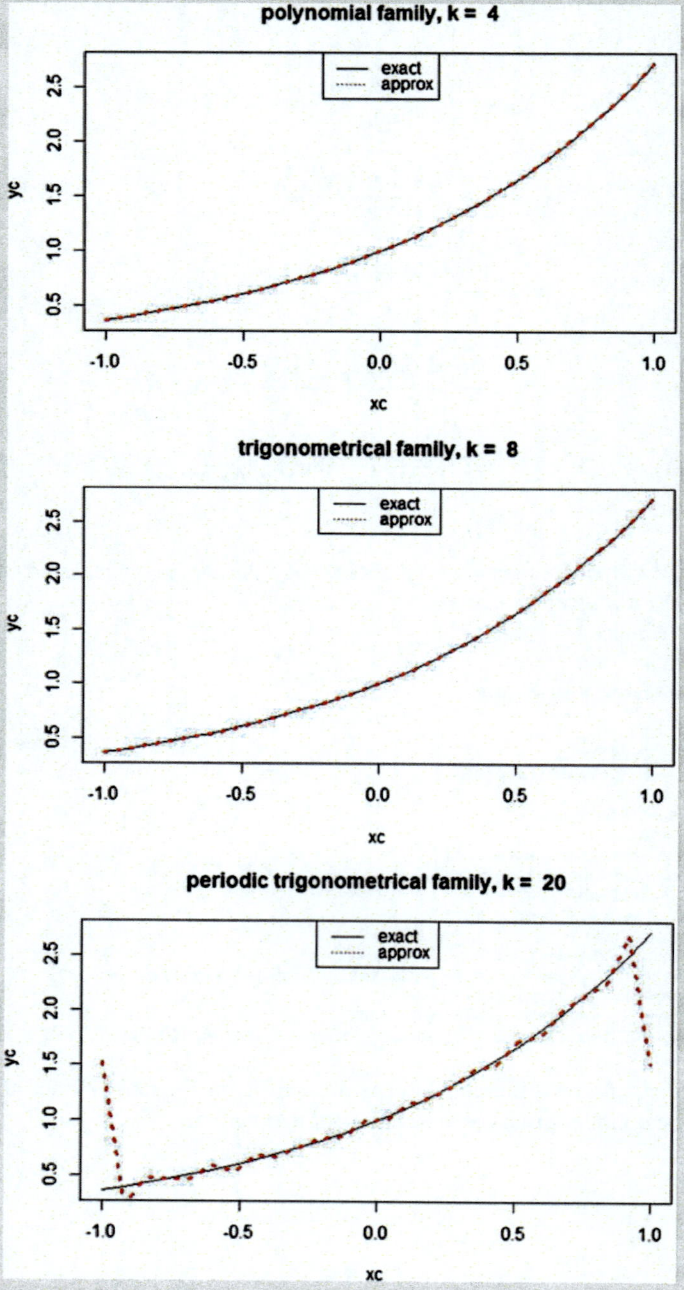

Fig. 2.7 Approximation by different Hilbert basis

Hilbert expansions may be used to generate regular random functions – the functions generated will have the same regularity as the Hilbert basis considered. Different methods may be used to achieve the generation. Whichever you choose, you must previously define the Hilbert basis $\{\varphi_i: i \geq 0, i \in N\}$

1. Method 1: use Eq. (2.46) with a fixed k. If you choose this option, you must select k before starting the generation. Now, let the value of k be given: you must generate the coefficients $y_0, \ldots, y_k$. To do this, you must choose a distribution taking values on $\mathbb{R}^k$: for instance, you can generate each coefficient as a variate from an independent gaussian distribution $N(0, \sigma)$.
2. Method 2: use Eq. (2.46) with a variable k having as minimal value k_0. If you choose this option, you must determine k at each generation. For instance, generate a random integer k_1 from a discrete distribution taking values from 0 to $+\infty$ (for instance Poisson's distribution). Then, take $k = k_0 + k_1$ and apply method 1. As an alternative, you can use `sample(k0:kmax,1)` to generate an integer k such that k0≤k≤kmax.
3. Method 3: use Eq. (2.47) with a fixed k. Again, you must select k before starting the generation. Now, let the value of k be given: you must generate the indexes j_1, . . j_k. For instance, take $j_\ell = 1 + \delta_\ell$, where δ_ℓ is a variable having an independent Poisson distribution. Once the indexes are defined, you must generate the coefficients $y_1, \ldots, y_k$. Analogously to the preceding methods, a distribution taking values on $\mathbb{R}^k$ is needed: for instance, you can use an independent gaussian distributions $N(0, \sigma)$.
4. Method 4: use Eq. (2.47) with a variable k having as minimal value k_0. In this case, select k as in method 2. Then use method 3.

Example 2.34 Let us apply the method 1. The code below

```
# Method 1
k = 5
sig = 5;
c1 = rnorm(k+1,0,sig)
print("Method 1 - values of c :",quote=FALSE)
print(c1, digits=3)
```

Produces the result (due to the `rnorm` generation of random normal numbers, the results differ for each run).

(continued)

Example 2.34 (continued)

```
[1] Method 1 - values of c :
[1] -8.769 8.072 -1.787 0.153 -0.151 -8.036
```

Let us apply method 2. The code:

```
# Method 2
kmin = 4
kmax = 10
k = sample(kmin:kmax,1)
sig = 5;
c2 = rnorm(k+1,0,sig)
print(paste("Method 2 - k = ",k,", values of c :"),quote=FALSE)
print(c2, digits=3)
```

produces

```
[1] Method 2 - k = 6, values of c :
[1] 2.365 8.666 -7.080 3.086 -0.396 -4.548 -6.020
```

Now, use method 3:

```
# Method 3
k = 5
sig = 5;
c3 = rnorm(k+1,0,sig)
ind = rpois(k+1,2*k)
print("Method 3 - values of c :",quote=FALSE)
print(c3, digits=3)
print("Indexes :",quote=FALSE)
print(ind)
```

```
[1] Method 3 - values of c :
[1] 8.844 0.997 1.915 5.546 -9.377 9.654
[1] Indexes :
[1] 9 7 16 10 6 9
```

(continued)

Example 2.34 (continued)

As an alternative, the indexes can be determined by a sample from `0: kmax`, as follows:

```
ind = sample(0:(2*k), k+1)
```

```
[1] Method 3 bis - values of c :
[1]  1.59 4.03 -7.79 6.08 2.97 -4.06
[1] Indexes :
[1]  1  7  5  2 10  4
```

Let us apply method 4: It is analogous to method 3, excepted by the choice of k, which is random. For instance:

```
kmin = 4
kmax = 10
k = sample(kmin:kmax,1)
```

```
[1] Method 4 - k= 6, values of c :
[1] -1.815 4.553 -8.517 -2.827 4.297 1.794 0.767
[1] Indexes :
[1] 10 20  9 12 16 11 17
[1] Method 4 bis - k= 8, values of c :
[1] -1.815 4.553 -8.517 -2.827 4.297 1.794 0.767
[1] Indexes :
[1]  5 13 10  8 16  6 15 11 14
```

The curves generated by these coefficients and different Hilbert basis are exhibited in Fig. 2.8. To get the legend outside the graph's region, we use the commands

```
> par(mar=c(6.1, 4.1, 4.1, 8.1),xpd=TRUE)
> legend("topright", inset=c(-0.2,0),legend = c
("1","2","3","3bis","4","4bis"),col = c("red","black","r
ed","black","red","black"),lty = c(1:6))
```

(continued)

Example 2.34 (continued)

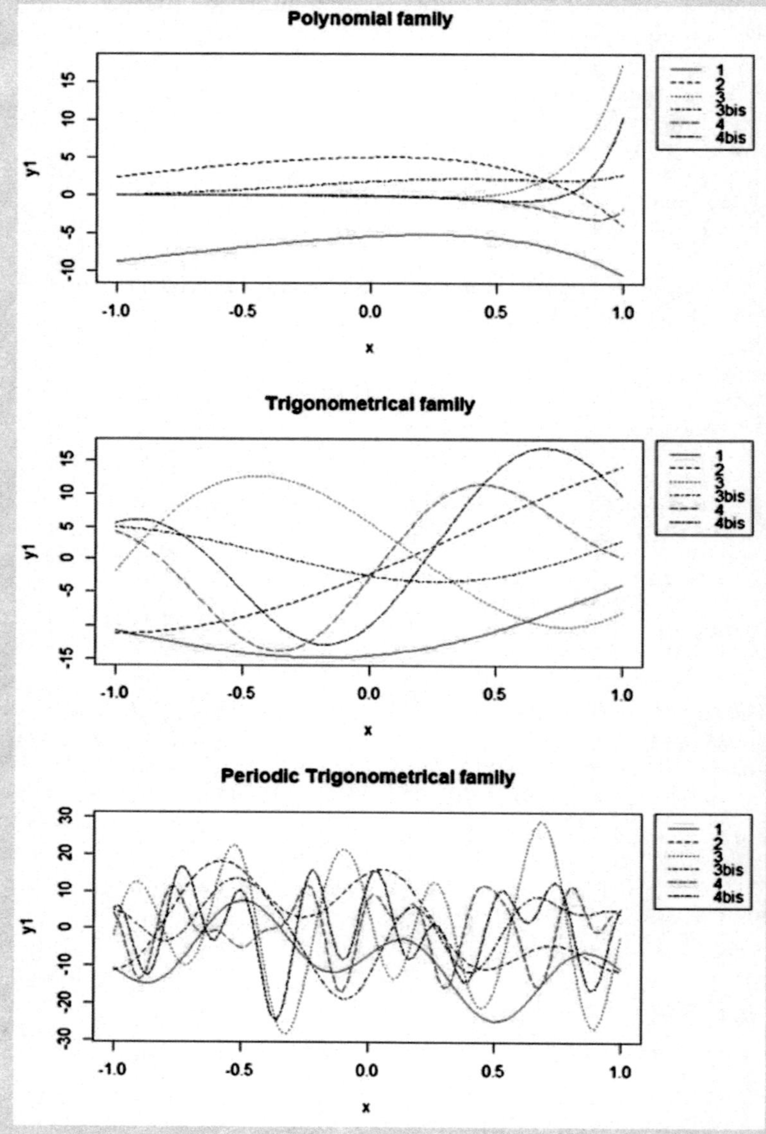

Fig. 2.8 Random regular functions generated by different families and different methods. The coefficients are those given in the example

2.15 Generating Regular Random Curves

A random regular curve in $\mathbb{R}^2$ may be generated by using its parametric form: $\boldsymbol{x} = (x_1(t), x_2(t))$, $t \in (0, T)$. The initial and the final values may be prescribed by considering families $\{\varphi_i: i \geq 0, i \in N\}$ having properties of periodicity or such that their values at the boundary points are equal to zero. For instance, to generate a curve such that $\boldsymbol{x}(0) = \boldsymbol{x_0}, \boldsymbol{x}(T) = \boldsymbol{x_T}$, we may consider

$$x_i(t) = x_i(0) + \frac{x_i(T) - x_i(0)}{T} t + \sum_{j=0}^{k} x_{i,\,j} \varphi_j^0(t), \tag{2.53}$$

where $\varphi_j^0(0) = \varphi_j^0(T) = 0$. For instance,

$$\varphi_j^0(t) = \varphi_j(t) - \left(\varphi_j(T) - \varphi_j(0)\right) \frac{t}{T} - \varphi_j(0). \tag{2.54}$$

To generate periodic curves, we may consider the periodic family (2.51) and (2.52)

Example 2.35 Let us generate 3 random curves such that $x(0) = (1, 0)$ and $x(1) = (0, 2)$. Here, $T = 1$. We apply the method 1 to generate 6 random coefficients, corresponding to $k = 5$:

```
> k = 5
> sig = 5;
> cx1 = list(NA,NA,NA)
> cx2 = list(NA,NA,NA)
> for (i in 1: 3){
+   cx1[[i]] = rnorm(k+1,0,sig)
+   cx2[[i]] = rnorm(k+1,0,sig)
+   print(paste("************ Curve ",i,":"),quote=FALSE)
+   print("Coefficients for x1 :",quote=FALSE)
+   print(cx1[[i]], digits=3)
+   print("Coefficients for x2 :",quote=FALSE)
+   print(cx2[[i]], digits=3)
+ }
[1] ************ Curve  1 :
[1] Coefficients for x1 :
[1]  0.656 -6.914  3.683  3.497 -0.838  3.870
[1] Coefficients for x2 :
[1]  3.73 -4.07 -7.95 -9.09  1.70 -6.21
[1] ************ Curve  2 :
[1] Coefficients for x1 :
[1]  6.208  0.690  0.909  1.803  2.413 -3.830
[1] Coefficients for x2 :
[1] -2.13  7.56 -1.07 -5.35  4.50 -3.55
[1] ************ Curve  3 :
[1] Coefficients for x1 :
[1]   1.211   4.522  -2.396   1.401   0.548 -14.392
[1] Coefficients for x2 :
[1]  3.425  0.659  3.158  0.137 -1.788  1.270
```

(continued)

Example 2.35 (continued)

The results appear in Fig. 2.9.

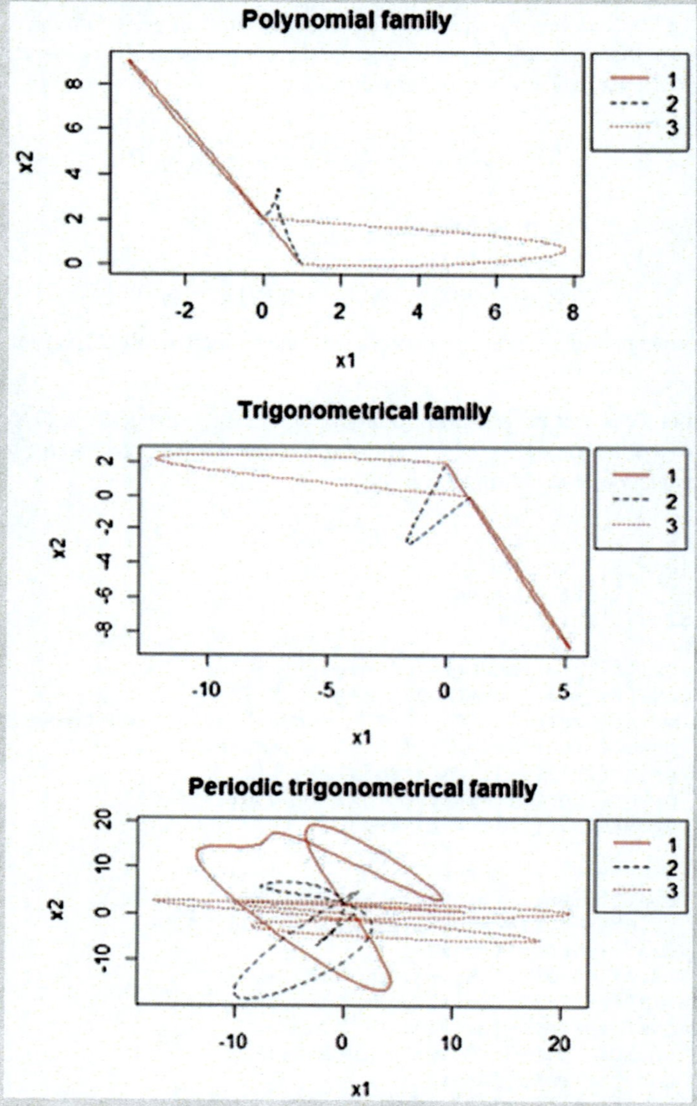

Fig. 2.9 Random regular curves generated by different Hilbert basis, using method 1 to generate the coefficients. All the curves connect (1,0) to (0,2). The coefficients are those given in the example

Example 2.36 Let us generate 3 random curves such that (0) = (1, 0) and $x(1) = (0, 2)$. Here, $T = 1$. We apply the method 3 to generate 6 random coefficients, corresponding to $k = 5$:

```
[1] ************ Curve 1 :
[1] Coefficients for x1 :
[1] -1.883   2.610   0.237  -5.719   3.537  -3.682
[1] indexes for x1 :
[1]  5  4  9  8  7 10
[1] Coefficients for x2 :
[1] -0.761   5.935   1.654   2.210  10.855   3.298
[1] Indexes for x2 :
[1]  6  0 10  4  1  3
[1] ************ Curve 2 :
[1] Coefficients for x1 :
[1] -7.50    6.85   -4.45   -2.15   -5.36    4.31
[1] indexes for x1 :
[1] 4 2 7 3 0 8
[1] Coefficients for x2 :
[1]  1.334   0.952  -5.163  11.603   3.819   9.004
[1] Indexes for x2 :
[1] 10  2  3  7  6  1
[1] ************ Curve 3 :
[1] Coefficients for x1 :
[1]  9.440  -0.958  -4.551  -4.489   0.152   3.616
[1] indexes for x1 :
[1] 6 4 0 2 7 9
[1] Coefficients for x2 :
[1]  2.039  -0.075  -3.252   6.524   9.015   1.640
[1] Indexes for x2 :
[1] 5 9 4 0 6 2
```

(continued)

Example 2.36 (continued)

The results appear in Fig. 2.10.

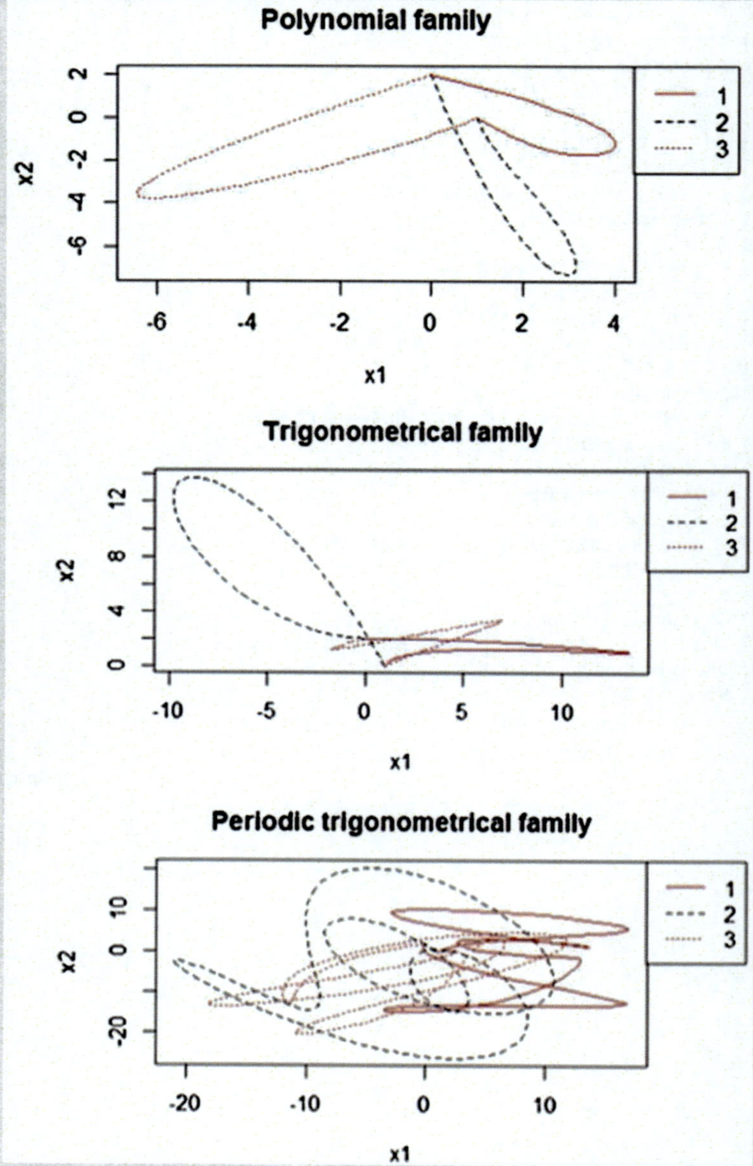

Fig. 2.10 Random regular curves generated by different Hilbert basis, using method 3bis to generate the coefficients. All the curves connect (1,0) to (0,2). The coefficients are those given in the example

Example 2.37 Let us generate 3 periodic random curves such that $x(0) = (1, 0) = (1, 0)$. Here, $T = 1$.

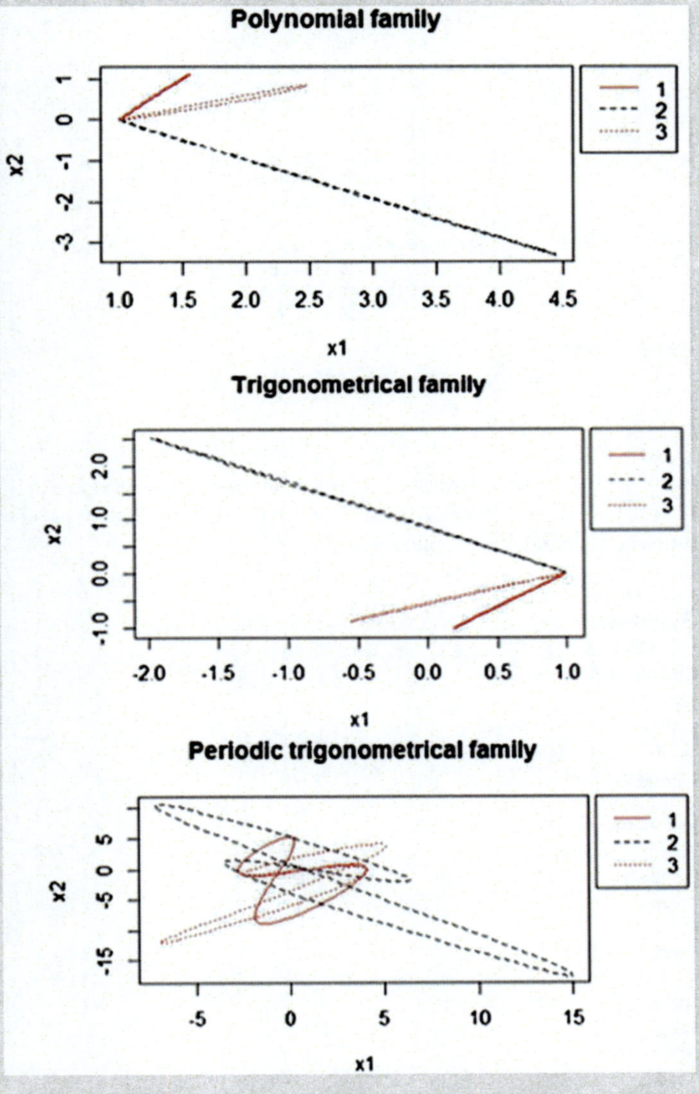

Fig. 2.11 Random regular periodic curves generated by different Hilbert basis, using method 1 to generate the coefficients. The coefficients are those given in the example

We apply the method 1 to generate 4 random coefficients, corresponding to $k = 3$:

(continued)

Example 2.37 (continued)

```
[1] ************* Curve  1 :
[1] Coefficients for x1 :
[1] -1.49 -1.66  1.03 -2.12
[1] Coefficients for x2 :
[1]  1.86  4.45  1.14 -3.62
[1] ************* Curve  2 :
[1] Coefficients for x1 :
[1] -4.90  4.35 -1.76 -7.82
[1] Coefficients for x2 :
[1]  2.37 -8.89  2.22  7.22
[1] ************* Curve  3 :
[1] Coefficients for x1 :
[1] -4.86  1.85  1.18 -4.60
[1] Coefficients for x2 :
[1]  4.71  5.38  2.56 -3.79
```

The results appear in Fig. 2.11.

Example 2.38 Let us generate 3 periodic random curves such that $x(0) = (1,0) = (1,0)$. Here, $T = 1$. We apply the method 3 to generate 4 random coefficients, corresponding to $k = 3$:

```
[1] ************* Curve 1 :
[1] Coefficients for x1 :
[1] -5.85   -3.31    2.61   -5.63
[1] indexes for x1 :
[1] 4 2 1 0
[1] Coefficients for x2 :
[1]  1.839  -0.566  15.882  -7.713
[1] Indexes for x2 :
[1] 4 0 3 1
[1] ************* Curve 2 :
[1] Coefficients for x1 :
[1] 11.281   0.335   0.895   1.076
[1] indexes for x1 :
[1] 6 5 1 2
[1] Coefficients for x2 :
[1] 1.903   3.622   2.711   0.986
[1] Indexes for x2 :
[1] 6 4 2 5
[1] ************* Curve 3 :
[1] Coefficients for x1 :
[1]  1.90   -1.05    6.86   -3.85
[1] indexes for x1 :
[1] 3 6 4 0
[1] Coefficients for x2 :
[1]  3.95   -4.76   -7.30    5.71
[1] Indexes for x2 :
[1] 2 6 4 0
```

(continued)

Example 2.38 (continued)

The results appear in Fig. 2.12.

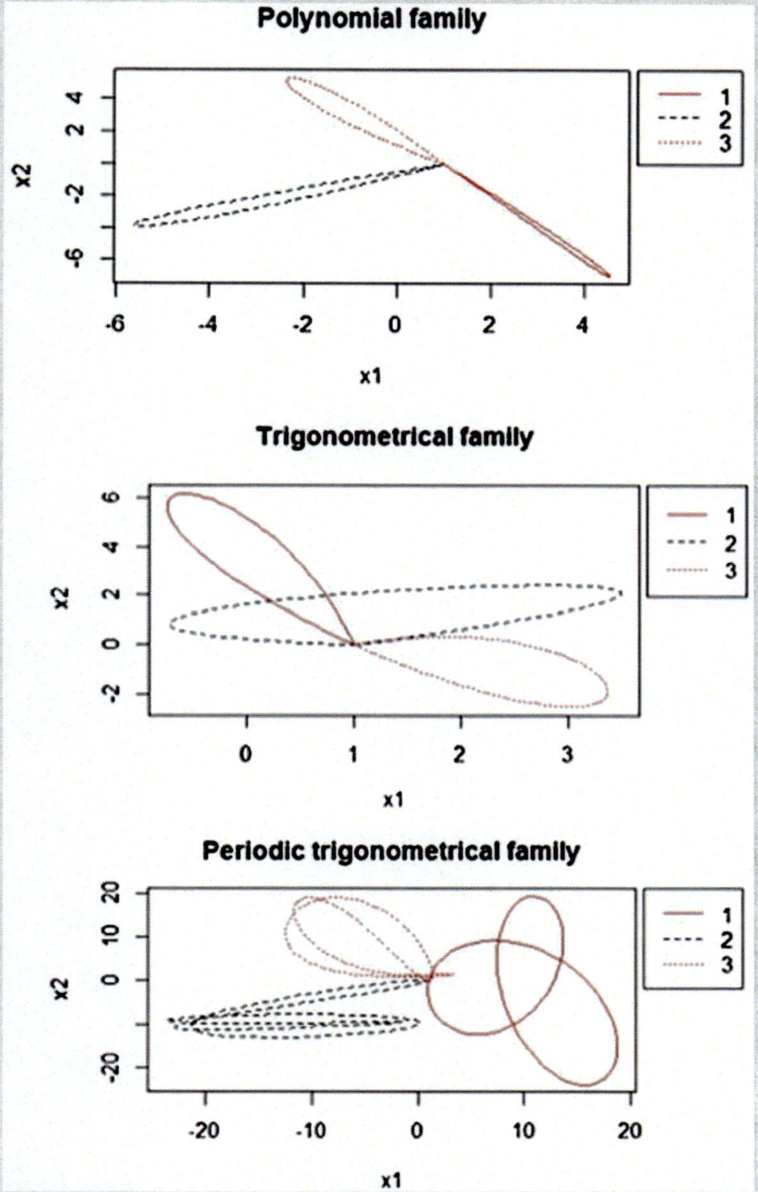

Fig. 2.12 Random regular periodic curves generated by different Hilbert basis, using method 3bis to generate the coefficients. The coefficients are those given in the example

Chapter 3
Representation of Random Variables

Abstract In this chapter, we present some methods to determine the distribution of a random variable from limited-sized samples. The methods are based on the representation of the random variable under consideration as a function of another random variable – which may be unknown and replaced by an artificial variable. The methods extend to the representation of random functions, curves, surfaces, etc. Examples and programs in R are given.

In this chapter, we present some UQ methods for the representation of variability.

3.1 The UQ Approach for the Representation of Random Variables

Let us consider the following situation: assume that U is a random variable and that $X = X(U)$ is other random variable which depends upon U – for instance, X is the solution of a linear system or an optimization problem which depends on U but the expression of the variable X as a function of U is not known: only a few data are available – a reduced sample of the couple (U, X) is given, with a number of elements insufficient to produce a statistical distribution. In other words, we are interested in finding the solution of the model problem below:

Supplementary Information The online version of this chapter (https://doi.org/10.1007/978-3-031-17785-9_3) contains supplementary material, which is available to authorized users.

E. Souza de Cursi, *Uncertainty Quantification using R*, International Series in Operations Research & Management Science 335,
https://doi.org/10.1007/978-3-031-17785-9_3

Problem 3.1
Let (U, X) be a couple of random variables on a probability space (Ω, P). Let be given a sample of ns observations of the couple:

$$\mathcal{S} = \{(U_1, X_1), \ldots, (U_{ns}, X_{ns})\} \tag{3.1}$$

Assuming that $X = X(U)$, Find the probability distribution of X – *id est*, its Cumulative Distribution Function (CDF) and its Probability Density Function (PDF).

Remark 3.1
In some situations, $X_1, \ldots, X_{ns}$ are not given explicitly, but implicitly, as solutions of an equation $\psi(X, U) = 0$ – the equation can be algebraical, differential, or functional, in general.

The UQ approach can be condensed as follows: consider a family of functions (for instance, a subset from a Hilbert basis – see Sect. 2.13).

$$\mathcal{F} = \{\varphi_j : 0 \le j \le k\} \tag{3.2}$$

and approximate

$$X \approx PX(U) = \sum_{j=0}^{k} x_j \varphi_j(U). \tag{3.3}$$

Then, use the sample (3.1) to determine the coefficients

$$\boldsymbol{x} = (x_0, x_1, \ldots, x_k)^t \tag{3.4}$$

Once $\boldsymbol{x}$ is determined, (3.3) can be used to determine approximations of the CDF and the PDF of X, to evaluate probabilities of events connected to X or furnish evaluations of statistics of X.

As shown by this summary, the UQ approach for the representation is based on the determination of the coefficients $\boldsymbol{x}$ of the expansion (3.3). We shall present in the sequel some methods for the determination of $\boldsymbol{x}$, including complex situations, where the distribution of U is unknown or the values of U are missing. Notice that Eq. (3.3) can be written under matrix form as

$$X \approx PX(U) = \boldsymbol{\Phi}(U)\boldsymbol{x}, \ \boldsymbol{\Phi}(\mathrm{U}) = (\varphi_0(U), \ldots, \varphi_k(U)). \tag{3.5}$$

In practice, classical basis given by polynomial or trigonometrical families can be used. For instance, $\varphi_i(U) = ((U - a)/(b - a))^i$, $a = \min U$, $b = \max U$.

If the joint distribution of the couple (U, X) is known, the coefficients $\boldsymbol{x}$ can be determined by solving a linear system furnished by the equations characterizing the orthogonal projection (see Theorem 2.2):

$$E\big(\boldsymbol{\Phi}(U)^t\boldsymbol{\Phi}(U)\big)\boldsymbol{x} = E\big(\boldsymbol{\Phi}(U)^t X\big). \tag{3.6}$$

For instance, let us consider the CDF

$$F(u,x) = \min\left\{\frac{1+\log(x)}{2}, \frac{1+u}{2}\right\}, \textit{if} -1 \le u,\ \log(x) \le 1;\quad 0, \text{otherwise}.$$

This CDF corresponds to a PDF

$$f(u,x) = \frac{1}{2}\delta_\Sigma(u,x),$$

defined by a Dirac's delta δ_Σ supported by the curve

$$\Sigma = \{(u,\ x)\colon u = \log(x),\ -1 \le x \le 1\}.$$

Let $\varphi_i(U) = ((U+1)/2)^i$. Then,

$$E\big(\varphi_i(U)\varphi_j(U)\big) = \frac{1}{2}\int_{-1}^{1}\left(\frac{1+u}{2}\right)^{i+j}du = \int_0^1 s^{i+j}ds = \frac{1}{i+j+1};$$

$$E(X\varphi_i(U)) = \frac{1}{2}\int_{-1}^{1}\left(\frac{1+u}{2}\right)^{i}e^u du = \frac{1}{e}\int_0^1 s^i e^{2s}ds = \frac{\Gamma(i+1) - \Gamma(i+1,\ -2)}{(-2)^{i+1}e};$$

An example of program in R to determine the coefficients for $k = 6$ is given below:

```
source("polynom.R")
source("expansion1D.R")
a = -1
b = 1
z = 1e-3
x = function(u){exp(u)}
nd = 6
phi = polynom$new(a,b,z)
he1 = expansion1D$new(phi)
xpx = he1$coeff(nd,x)
uc = seq(a,b,length.out=41)
xc = x(uc)
px = he1$proj(xpx,uc)
plot(uc,xc,type='n')
lines(uc,xc,lty=1,col='black')
lines(uc,px,lty=3,col='red',lwd=3)
legend("top",legend = c("X","PX"),col = c("black","red"),lty = c(1,3))
```

The method `coeff` in the class `expansion1D` reads as:

```
coeff = function(k,x){
  A = matrix(nrow = k+1, ncol = k+1)
  B = matrix(nrow = k+1, ncol = 1)
  for (i in 1:(k+1)){
    phi_i = function(u){self$basis$value(i-1,u)}
    vphi_i = Vectorize(phi_i)
    vx = Vectorize(x)
    f = function(u){vx(u)*vphi_i(u)}
    aux = integrate(f,lower=self$basis$lb,upper=self$basis$ub)
    B[[i]] = aux$value
    for (j in i:(k+1)){
      phi_j = function(u){self$basis$value(j-1,u)}
      vphi_j = Vectorize(phi_j)
      f = function(u){vphi_i(u)*vphi_j(u)}
      aux = integrate(f,lower=self$basis$lb,upper=self$basis$ub)
       A[[i,j]] = aux$value
      A[[j,i]] = aux$value
    }
  }
  cc = solve(A,B)
  return(cc)
}
```

Running the code produces the result below (Fig. 3.1):

Unfortunately, in practice, such a situation is exceptional. In most of real situations, the available information reduces to limited samples of observations of the couple. As previously indicated, we are mainly interested in such a situation.

For instance, assume that the available data are Table 3.1.

Using these data, we have $\bar{X}_n = 1.511121$, $s_n(X) = 0.7385565$, while the exact values are

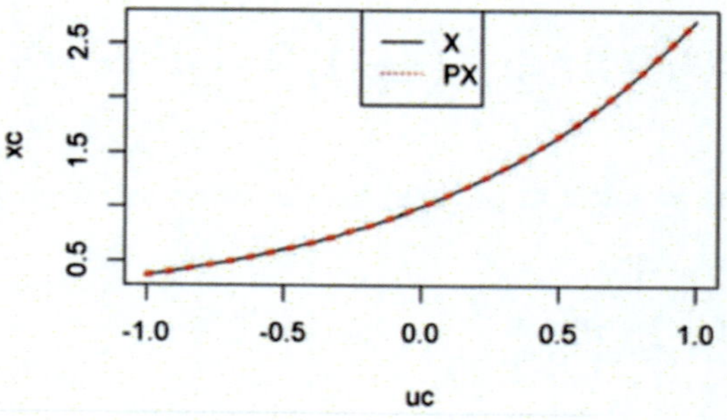

Fig. 3.1 An example of UQ approximation for a couple (U, X) of known distribution. Unfortunately, such a situation is rarely found in practice

Table 3.1 An example of data of limited size

U	0.110263	0.818171	0.218055	−0.428266	−0.521897
X	1.11657	2.26635	1.24366	0.65164	0.59339
U	0.972777	0.498947	−0.063814	0.439192	0.899228
X	2.64528	1.64699	0.93818	1.55145	2.45770

$$E(X) = \frac{1}{2}\left(e + \frac{1}{e}\right) \approx 1.175201, \quad \sigma(X) = \sqrt{\frac{1}{2}\left(1 - \frac{1}{e^2}\right)} \approx 0.6575199.$$

In addition, if we use Table 3.1 to evaluate $P(X > 2.4)$ and $P(X < 0.7)$, we get the values 0.2 for both, while the exact values are

$$P(X < 0.7) = \frac{1}{2}(1 + \log(0.7)) \approx 0.3216625;$$

$$P(X > 2.4) = \frac{1}{2}(1 - \log(2.4)) \approx 0.06226563.$$

Finally, these data appear as insufficient to determine the distribution of X, as shown in Fig. 3.2.

Of course, the available data do not contain the information that the variable U is distributed on $(-1, 1)$, since we have only data on $(-0.521897, 0.972777)$. Thus, we may consider that the comparisons must be made with the restriction of the variables to the interval of the observations. In this case, setting $= -\ 0.521897$, $b = 0.972777$, $U_R = U1_{(a,b)}$, $X_R = e^{U_R}$, we have

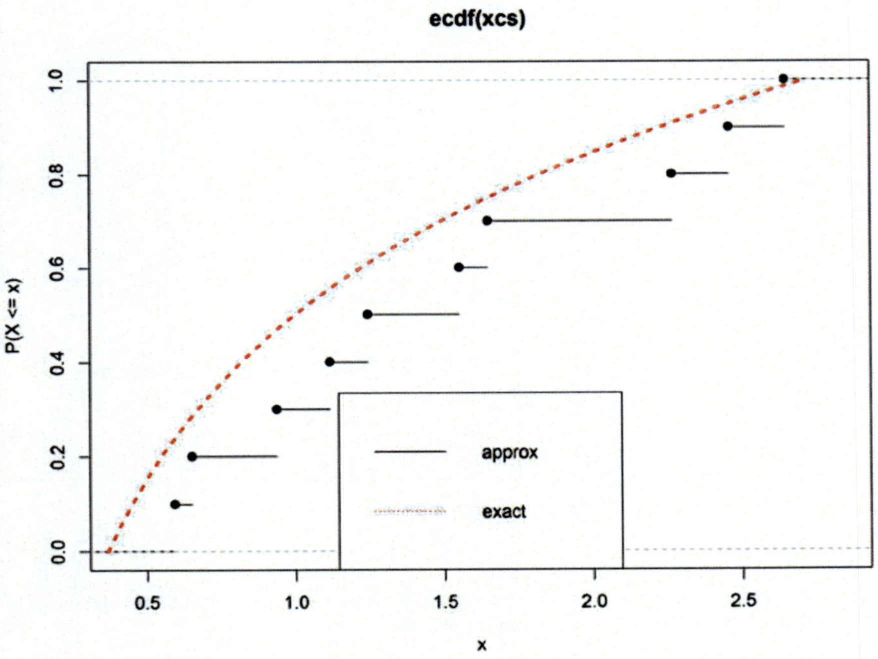

Fig. 3.2 Comparison between the CDF generated by the data in Table 3.1 (in black) and the exact CDF of X. The approximation has poor quality

$$E(X_R) = \frac{e^b - e^a}{\mathrm{b} - \mathrm{a}} \approx 1.372799,$$

$$\sigma(X_R) = \sqrt{\frac{(e^a - e^b)((-2 + a - b)e^a + (2 + a - b)e^b)}{2(a - b)^2}} \approx 0.581762.$$

$$P(X_R < 0.7) = \frac{\log(0.7) - a}{b - a} \approx 0.1105405;$$

$$P(X_R > 2.4) = \frac{b - \log(2.4)}{b - a} \approx 0.06510334.$$

Thus, even in this case, the sample does not furnish good estimates for the probabilities. In addition, as shown in Fig. 3.3, the approximation of the CDF is poor.

We shall examine in the sequel some methods to determine the coefficients when the information reduces to a sample $\mathcal{S}$ (Eq. (3.1)) of limited size.

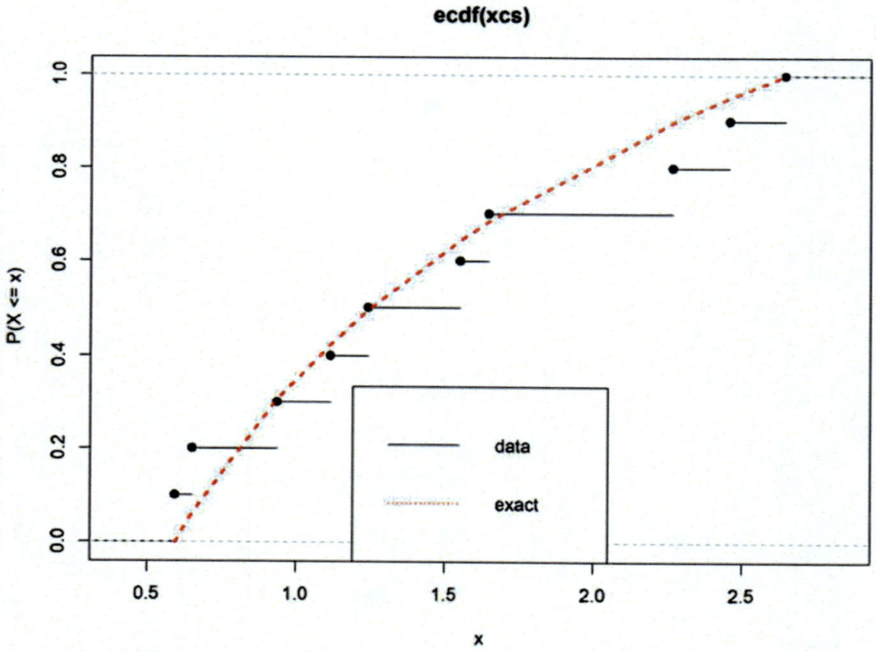

Fig. 3.3 Comparison between the CDF generated by the data in Table 3.1 (in black) and the CDF of the restriction X_R of X to the interval of the observed data

3.2 Collocation

Collocation is the simplest way to determine a representation of random variables – but this approach confronts obstacles in the situations where U is unknown or the data contain significant errors.

For explicitly given data (Problem 3.1.1), collocation can be reduced to the solution of a linear system. Indeed, collocation imposes that

$$X_i = X(U_i) \approx PX(U_i),\ 1 \leq i \leq ns \tag{3.7}$$

id est, that the approximation coincides with the observed value at each point of the sample. Thus,

$$\sum_{j=0}^{k} x_j \varphi_j(u_i) = X_i \tag{3.8}$$

Let us consider the vectors $\boldsymbol{X} = (X_1, \ldots, X_{ns})^t$, $\boldsymbol{U} = (U_1, \ldots, U_{ns})^t$. Then,

$$\boldsymbol{Ax} = \boldsymbol{B}, \tag{3.9}$$

with $\boldsymbol{A} = (\boldsymbol{\Phi}(U_1), \ldots, \boldsymbol{\Phi}(U_{ns}))^t$ (line i of $\boldsymbol{A}$ contains $\boldsymbol{\Phi}(U_i)$) and $\boldsymbol{B} = \boldsymbol{X}$, id est,

$$\boldsymbol{A} = \left(A_{ij}: 1 \leq i \leq ns,\ 1 \leq j \leq k\right),\quad A_{ij} = \varphi_j(U_i); \tag{3.10}$$

$$\boldsymbol{B} = (B_i: 1 \leq i \leq ns),\ B_i = X_i \tag{3.11}$$

Thus, the determination of $\boldsymbol{x}$ reduces to the solution of the linear system (3.9, 3.10, and 3.11). Matrix $\boldsymbol{A}$ is formed by ns rows and $k + 1$ columns: in general, $ns > k + 1$, so that the system is overdetermined. We have

$$(\boldsymbol{A}^t\boldsymbol{A})\boldsymbol{x} = \boldsymbol{A}^t\boldsymbol{B}. \tag{3.12}$$

To obtain a more stable result, we can consider the Levenberg–Marquardt approximation

$$\boldsymbol{x} \approx \boldsymbol{x}_\varepsilon \quad \text{with } (\boldsymbol{A}^t\boldsymbol{A} + \varepsilon \boldsymbol{Id})\boldsymbol{x}_\varepsilon = \boldsymbol{A}^t\boldsymbol{B}. \tag{3.13}$$

Here, $\varepsilon > 0$ is a parameter of regularization – Tikhonov's regularization – to be chosen as small enough (often inferior to $1E - 3$). Collocation is implemented in class `expansion1D` as follows:

```
colloc = function(k,eps,u,x){
  A = matrix(nrow = length(u), ncol = k+1)
  B = matrix(nrow = length(x), ncol = 1)
  for (i in 1:length(u)){
    B[[i]] = x[[i]]
    uu = u[[i]]
    for (j in 1:(k+1)){
      A[[i,j]] = self$basis$value(j-1,uu)
    }
  }
  cc = solve(t(A) %*% A + eps*diag(k+1),t(A) %*% B)
  return(cc)
}
```

To apply this method to the data in Table 3.1, enter the data as follows:

```
source("polynom.R")
source("expansion1D.R")
uc = c(0.110263, 0.818171, 0.218055, -0.428266, -0.521897,
       0.972777, 0.498947, -0.063814, 0.439192, 0.899228)
xc = c(1.11657, 2.26635, 1.24366, 0.65164, 0.59339,
       2.64528, 1.64699, 0.93818, 1.55145, 2.45770)
aux = sort(uc,index.return = TRUE)
ucs = aux$x
xcs = xc[aux$ix]
```

Then, run the code below:

```
a = min(ucs)
b = max(ucs)
z = 1e-3
nd = 6
phi = polynom$new(a,b,z)
hel = expansion1D$new(phi)
eps = 0
xpx = hel$colloc(nd,eps,ucs,xcs)
px = hel$proj(xpx,ucs)
plot(ucs,xcs,type='n')
lines(ucs,xcs,lty=1,col='black')
lines(ucs,px,lty=3,col='red',lwd=3)
legend("top",legend = c("X","PX"),col = c("black","red"),lty = c(1,3))
```

Here, xpx contains the coefficients $\boldsymbol{x}$. The code furnishes

$$\boldsymbol{x} = (0.5933901, 0.8870419, 0.662488, 0.3289636, 0.1313289, 0.02278473, \\ 0.01928178)$$

The result is shown in Fig. 3.4.

Once the coefficients are determined, we can use them to determine the distribution of X, statistics of this variable and probabilities of events connected to it. The simplest way consists in generating a large sample from X, by using a large sample from U : we generate 10^5 variates from $U(-0.521897, 0.972777)$ – notice that no information is available outside this interval – and use them to generate a sample from X by using the approximation – the computational cost is negligible. Since the

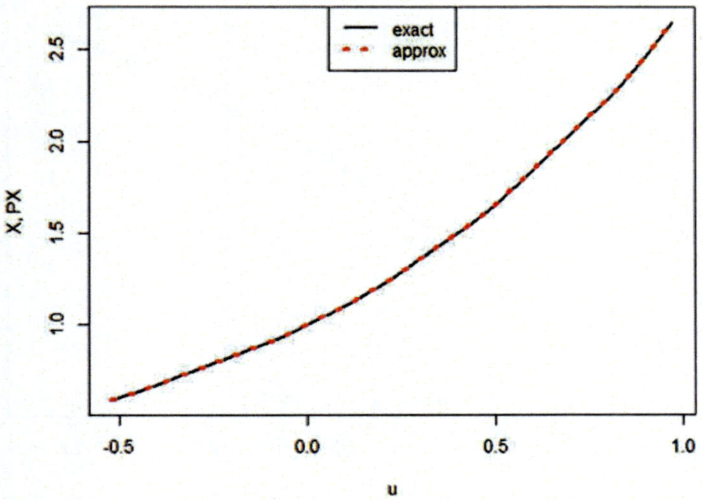

Fig. 3.4 Collocation results for the data in Table 3.1 (polynomial family, $k = 6$)

distribution is uniform, we can use equally spaced values. Then, we use these simulated data to approximate the CDF of X by using the empirical CDF from the simulated data. The results are shown in Figs. 3.5 and 3.6: we see that the accordance is excellent. The PDF is determined by evaluating the numerical derivative of the CDF (a method is available in class `sampleX.R`). The result is exhibited in Fig. 3.7: again, the accordance is good.

In this case, the large sample with 1E5 variates furnishes the estimations $\bar{PX} = 1.3728$, $s(PX) = 0.5818$, $P(X > 2.4) = 0.06511$ and $P(X < 0.7) = 0.1105$ – to be compared with the exact values 1.3728, 0.5818, 0.06510, 0.1105. We observe that these values generated are in good accordance with the real values. If we do not desire to use a large sample, we can use alternative approaches. Indeed, many alternatives to this simple approach exist.

For instance, let f_U be the PDF of U and f_X be the PDF of X. Assuming that X is injective, we have (see, for instance (Souza de Cursi E. , 1992)):

$$f_X(x) = \frac{f_U\left(X^{-1}(x)\right)}{|J(X,\ U)|},\quad J(X,\ U) = \text{jacobien of } X(U).$$

In one-dimensional situations, $J(X, U) = X'(U) = dX/dU$. Thus, the PDF of X can be determined as

$$f_X(\boldsymbol{\Phi}(u)\boldsymbol{x}) \approx f_X(X(u)) = \frac{f_U(u)}{|X'(u)|} \approx \frac{f_U(u)}{|\boldsymbol{\Phi}'(u)\boldsymbol{x}|}. \tag{3.14}$$

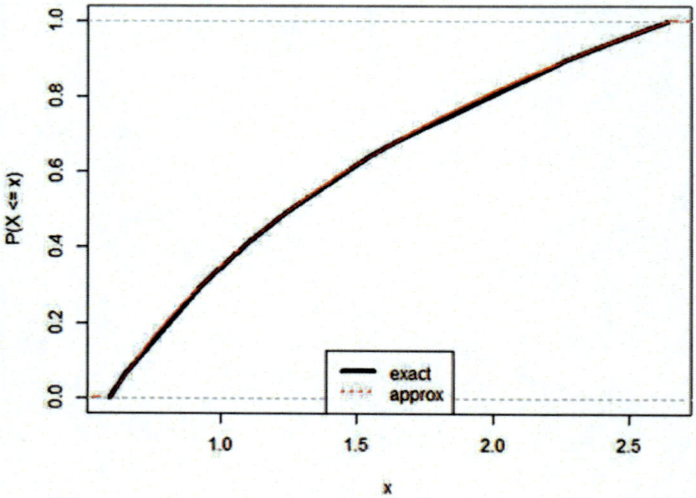

Fig. 3.5 Comparison between the CDF generated by collocation (polynomial family, $k = 6$) and the CDF of the restriction X_R of X to the interval of the observed data. Here, the CDF of PX is generated by `ecdf`

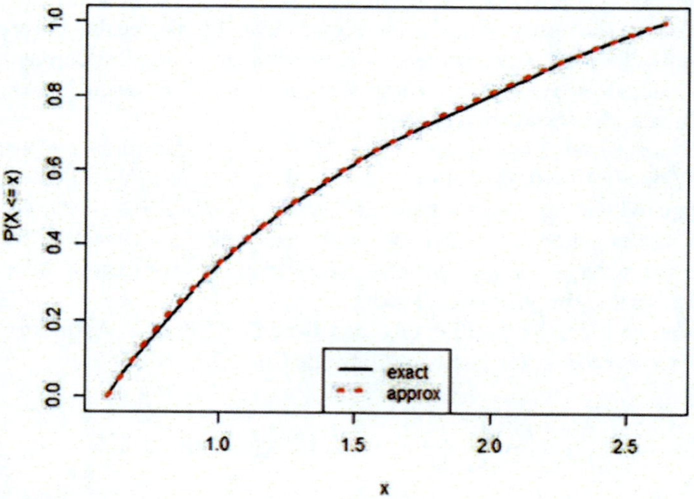

Fig. 3.6 Comparison between the CDF generated by collocation (polynomial family, $k = 6$) and the CDF of the restriction X_R of X to the interval of the observed data. Here, the CDF of PX is generated by `sampleX`

Using (3.14), we obtain the results shown in Fig. 3.8 – they are better than those generated by the simple method, but this approach cannot be used if the distribution of U is unknown and supposes that X is injective: to use it effectively, it is necessary to verify that $\boldsymbol{\Phi}'(u)\boldsymbol{x}$ is injective on the interval under consideration.

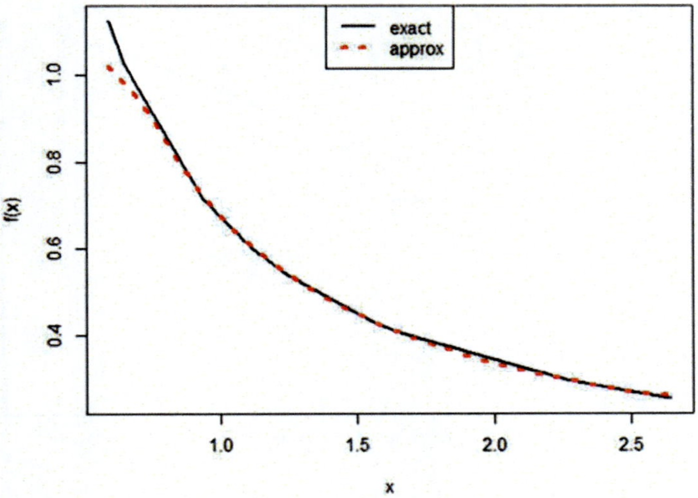

Fig. 3.7 Comparison between the PDF generated by collocation (polynomial family, $k = 6$) and the PDF of the restriction X_R of X to the interval of the observed data. Here, the PDF of PX is generated by `sampleX`

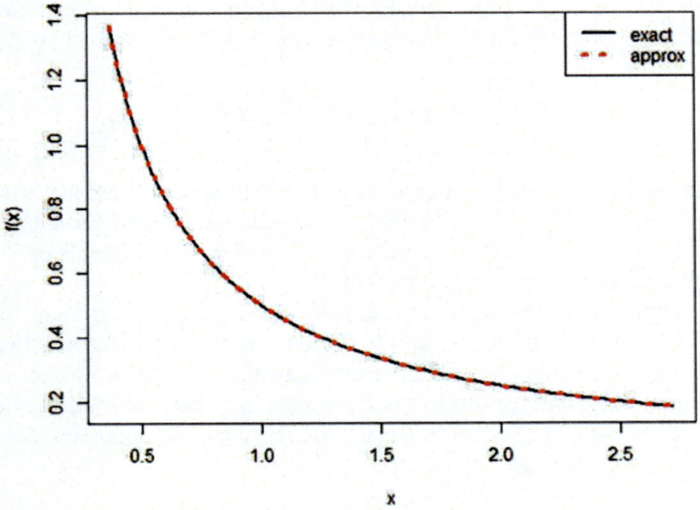

Fig. 3.8 Comparison between the PDF generated by Eq. (3.14) (polynomial family, $k = 6$) and the PDF of the restriction X_R of X to the interval of the observed data

Under these conditions, this approximation can be used to generate the CDF of X by numerical integration of f_X given by (3.14). As an example, a trapezoidal rule involving 51 points generates the results in Fig. 3.9.

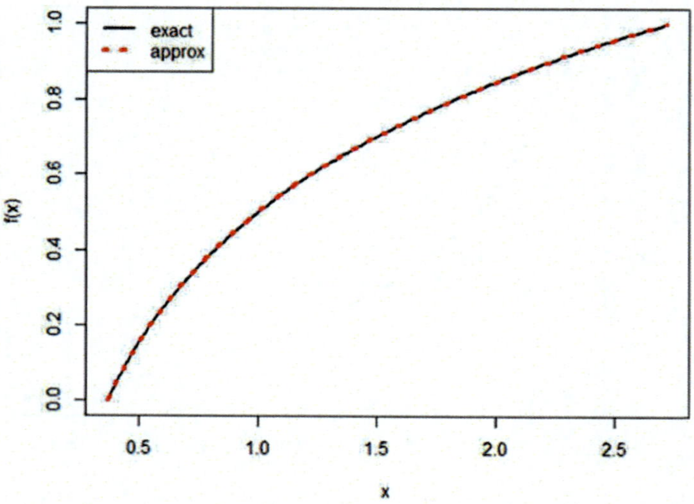

Fig. 3.9 Comparison between the CDF generated by integration of the PDF furnished by Eq. (3.14) (polynomial family, $k = 6$) and the CDF of the restriction X_R of X to the interval of the observed data

As in the preceding, the mean and the standard deviation can be evaluated can be evaluated using a large sample generated by the representation found, but we can also evaluate them by numerical integration using the PDF furnished by Eq. (3.14):

$$E(X^p) = \int x^p f_X(x)dx = \int (X(u))^p f_U(u)du. \tag{3.15}$$

For instance, a trapezoidal integration with 50 subintervals and the first integral produces $\bar{PX} = 1.3729, s(PX) = 0.5820$. The second integral produces $\overline{PX} = 1.3728, s(PX) = 0.5821$. In this case, the empirical mean of the sample of 50 variates from X has $\overline{PX} = 1.3777, s(PX) = 0.6009$.

The probabilities are evaluated directly from the CDF found: we have $P(X > 2.4) = 0.0651$ and $P(X < 0.7) = 0.1105$ – very close to the exact values.

A second alternative consists in determining the CDF of X by the numerical evaluation of the Lebesgue integral of the probability measure $P(dU)$: a numerical method is exposed in (Souza de Cursi, 2015). A simple approximation can be generated by considering

$$I(x) = \{u: X(u) \leq x\}.$$

An approximation of this set can be generated by considering an interval (u_{min}, u_{max}) containing the range of U and a partition

$$u_{\min} = u_0 < u_1 < u_2 < \ldots < u_n = u_{\max}.$$

Such a partition defines n subintervals $I_i = (u_{i-1}, u_i)$, $i = 1, \ldots, n$. Let us introduce

$$\mathcal{K}(I_i, x) = I_i, \text{ if} \max\{\boldsymbol{\Phi}(u_{i-1})\boldsymbol{x}, \boldsymbol{\Phi}(u_i)\boldsymbol{x}\} \le x; \ \mathcal{K}(I_i, x) = \varnothing, \text{ otherwise.}$$

We can approximate

$$I(x) \approx \bigcup_{i=1}^{n} \mathcal{K}(I_i, x),$$

so that

$$P(X \le x) = P(U \in I(X)) \approx \sum_{i=1}^{n} P(U \in \mathcal{K}(I_i, x)),$$

id est,

$$P(X \le x) \approx \sum_{\substack{1 \le i \le n, \\ \max\{\boldsymbol{\Phi}(u_{i-1})\boldsymbol{x}, \boldsymbol{\Phi}(u_i)\boldsymbol{x}\} \le x}} P(I_i), \tag{3.16}$$

In practice, we can consider a second interval $(x_{\min}, x_{\max})$ containing the range of X and a partition $x_{\min} = x_0 < x_1 < x_2 < \ldots < x_m = x_{\max}$: the CDF of X is evaluated at these points.

For instance, assume that the interval $(u_{\min}, u_{\max}) = (-0.521897, 0.972777)$ is discretized in $n = 250$ subintervals, while the interval $(x_{\min}, x_{\max})$ is discretized in $m = 25$ subintervals. This approach furnishes the results shown in Fig. 3.10. The PDF generated by SP derivative of the CDF is shown in Fig. 3.11.

The results can be improved by using smaller steps. For instance, using $n = 2500$, $m = 250$, we obtain the results shown in Fig. 3.12.

In this case, $P(X > 2.4)$ and $P(X < 0.7)$ are evaluated from the calculated CDF. The approximation with $n = 2500$, $m = 250$ furnishes $P(X > 2.4) = 0.06532$ and $P(X < 0.7) = 0.1103$. $\overline{PX}$ and $s(PX)$ can be evaluated either by using the empirical mean of the sample, by Eq. (3.15) or Stieltjes integration using the CDF:

$$E(X^p) = \int x^p dF_X(x) = \int X(u)^p dF_U(u). \tag{3.17}$$

The approach by Riemann integration (3.15) produced $\overline{PX} = 1.3704$ and $s(PX) = 0.5821$; the second approach by (3.17) furnished $\overline{PX} = 1.3701$ and $s(PX) = 0.5822$ – both the values are close to the exact ones.

As previously observed, another Hilbert basis can be considered. For instance, we can consider a trigonometrical family:

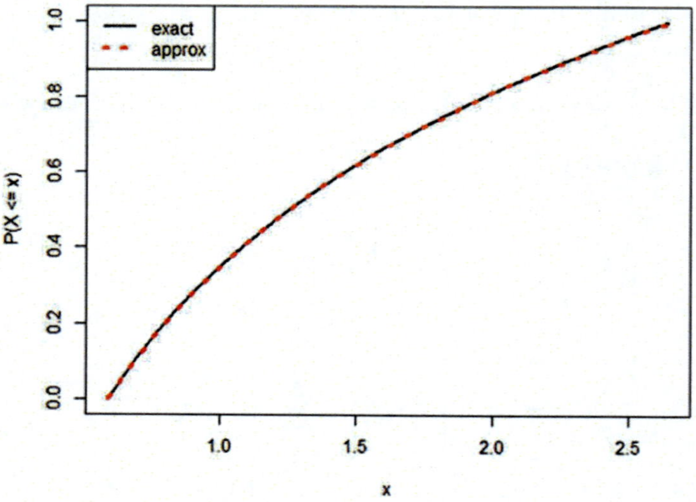

Fig. 3.10 Comparison between the CDF generated by Eq. (3.16) (polynomial family, $k = 6$) and the CDF of the restriction X_R of X to the interval of the observed data

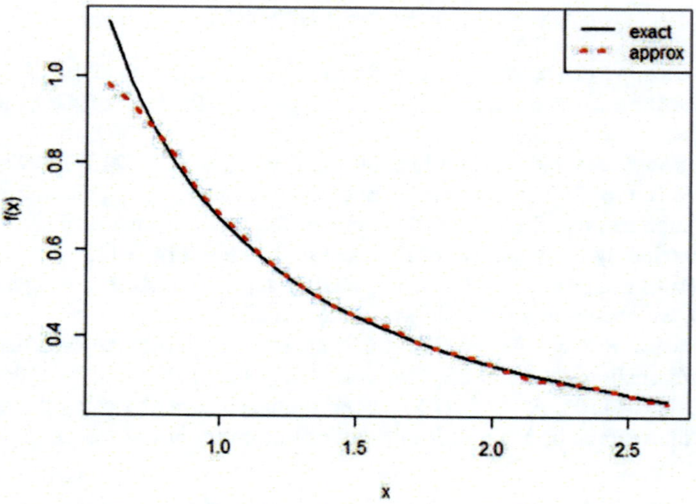

Fig. 3.11 Comparison between the PDF generated by SP derivative of the CDF furnished by Eq. (3.16) (polynomial family, $k = 6$, kernel gaussian) and the PDF of the restriction X_R of X to the interval of the observed data

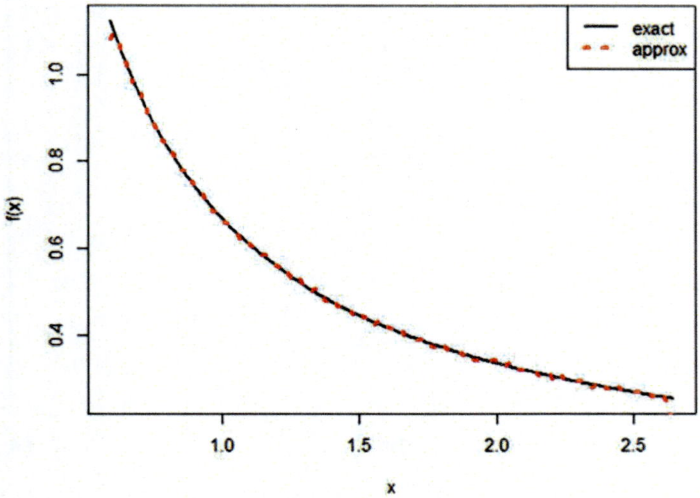

Fig. 3.12 Improvement in the evaluation of the PDF. Using smaller steps, both the CDF and the PDF are better evaluated. Here, the PDF generated by SP derivative of the CDF furnished by Eq. (3.16) (polynomial family, $k = 6$, kernel gaussian) is compared to the PDF of the restriction X_R of X to the interval of the observed data

```
source("trigo.R")
z = 1e-3
nd = 12
phi = trigo$new(a,b,z)
he1 = expansion1D$new(phi)
eps = 1e-5
xpx = he1$colloc(nd,eps,ucs,xcs)
```

In this case, we obtain the results exhibited in Figs. 3.13, 3.14, and 3.15. For this expansion, $\overline{PX} = 1.372671$, $s(PX) = 0.5842$, $P(X > 2.4) = 0.06532$, and $P(X < 0.7) = 0.1100$. With $k = 14$, we obtain $\overline{PX} = 1.3728$, $s(PX) = 0.5814$, $P(X > 2.4) = 0.06530$, and $P(X < 0.7) = 0.1098$. With $k = 22$, $\overline{PX} = 1.3726$, $s(PX) = 0.5819$, $P(X > 2.4) = 0.06521$, and $P(X < 0.7) = 0.1111$.

Again, we can choose the alternative approach furnished by Eq. (3.14). Then, we obtain the results shown in Figs. 3.16 and 3.17. We use $k = 12$ – for this approximation, a trapezoidal integration with 2E3 subintervals produces $\overline{PX} = 1.3744$, $s(PX) = 0.5809$, $P(X > 2.4) = 0.06475$, and $P(X < 0.7) = 0.1106$. With $k = 14$, we obtain $\overline{PX} = 1.3731$, $s(PX) = 0.5810$, $P(X > 2.4) = 0.06495$, and $P(X < 0.7) = 0.1101$. For $k = 22$, we obtain $\overline{PX} = 1.3758$, $s(PX) = 0.5815$, $P(X > 2.4) = 0.06512$, and $P(X < 0.7) = 0.1112$.

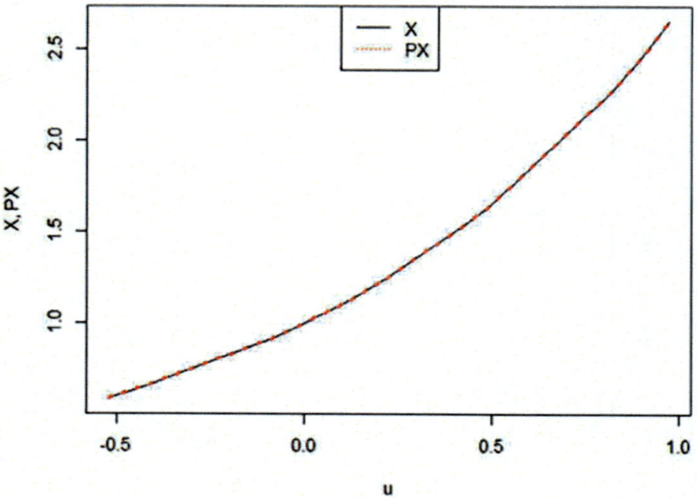

Fig. 3.13 Collocation results for the data in Table 3.1 (trigonometrical family, $k = 12$)

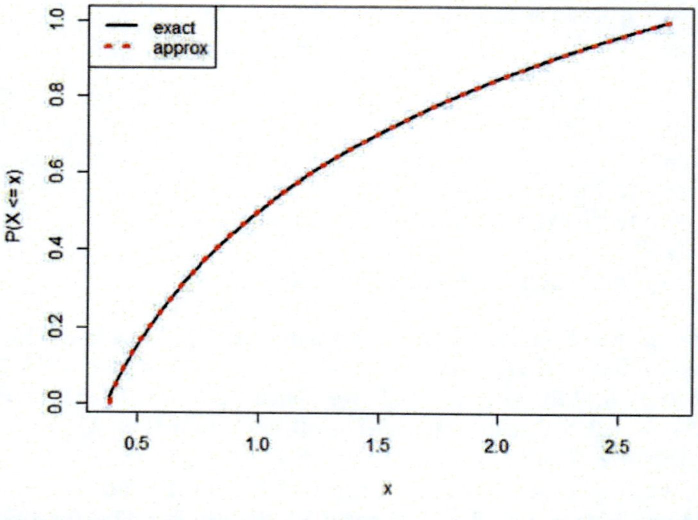

Fig. 3.14 Comparison between the CDF generated by collocation (trigonometrical family, $k = 12$) and the CDF of the restriction X_R of X to the interval of the observed data. Here, the CDF of PX is generated by `ecdf`

The approach furnished by Eq. (3.16) can be used also: with $k = 12$, we obtain the PDF shown in Fig. 3.18 and $P(X > 2.4) = 0.06554$, $P(X < 0.7) = 0.1098$. The approach (3.15) furnishes $\overline{PX} = 1.3710$ and $s(PX) = 0.5820$; using (3.17) gives $\overline{PX} = 1.3699$ and $s(PX) = 0.5821$ – again, both the values are close to the exact ones.

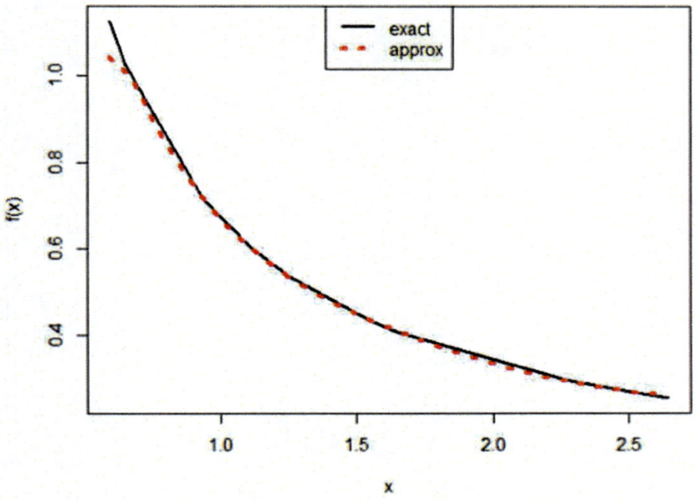

Fig. 3.15 Comparison between the PDF generated by collocation (trigonometrical family, $k = 12$) and the PDF of the restriction X_R of X to the interval of the observed cdata. Here, the PDF of PX is generated by `sampleX`

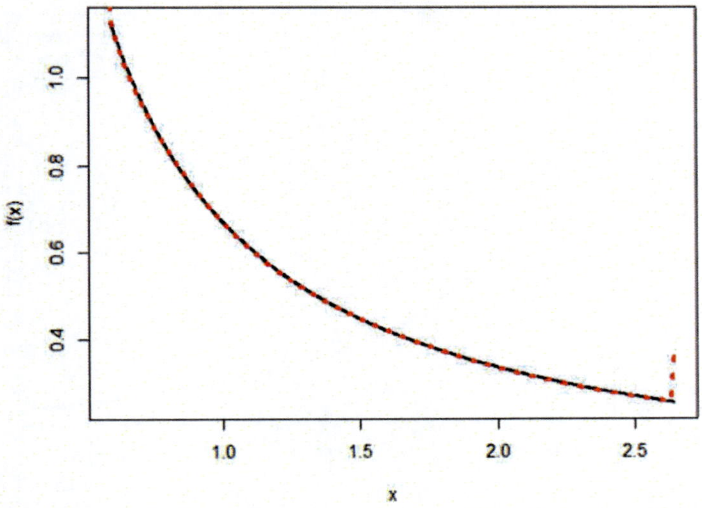

Fig. 3.16 Comparison between the PDF generated by Eq. (3.14) (trigonometrical family, $k = 12$) and the PDF of the restriction X_R of X to the interval of the observed data

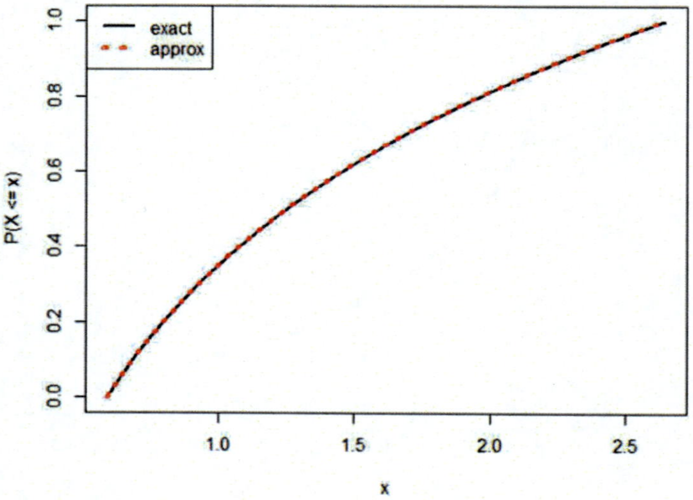

Fig. 3.17 Comparison between the CDF generated by integration of the PDF furnished by Eq. (3.14) (trigonometrical family, $k = 12$) and the CDF of the restriction X_R of X to the interval of the observed data

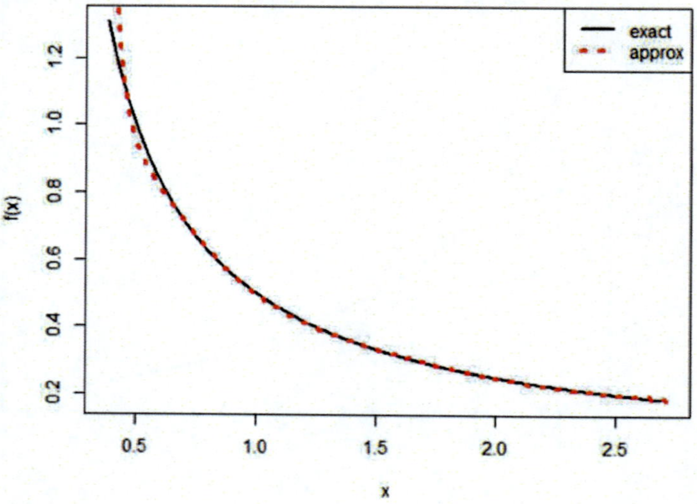

Fig. 3.18 Comparison between the PDF generated by SP derivative of the CDF furnished by Eq. (3.16) (trigonometrical family, $k = 12$, kernel gaussian) and the PDF of the restriction X_R of X to the interval of the observed data

If the information that U is uniformly distributed on $(-1, 1)$ is obtained somewhere, we can use the coefficients to generate an analogous sample but using points on $(-1, 1)$, to get the complete distribution of X. The results are shown in Figs. 3.19 and 3.20. Once the CDF is determined, the PDF is found by the numerical derivative of the CDF, using the method of class `sampleX.R`. The result is exhibited in Fig. 3.21: again, the accordance is good.

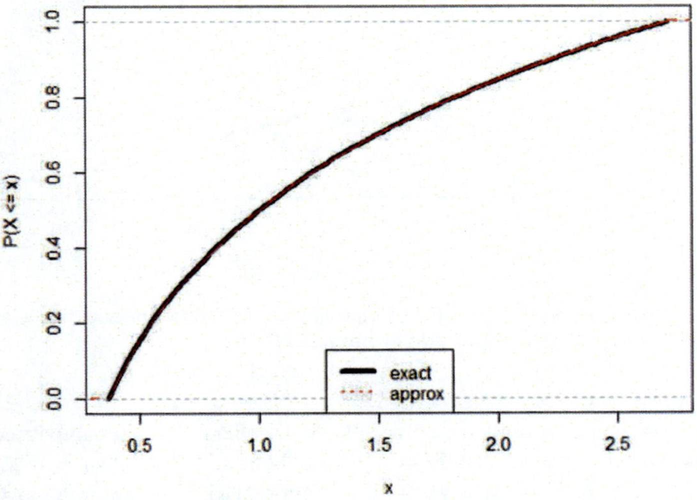

Fig. 3.19 Comparison between the CDF generated by collocation (polynomial family, $k = 6$) and the CDF of X. Here, the CDF is generated by `ecdf`

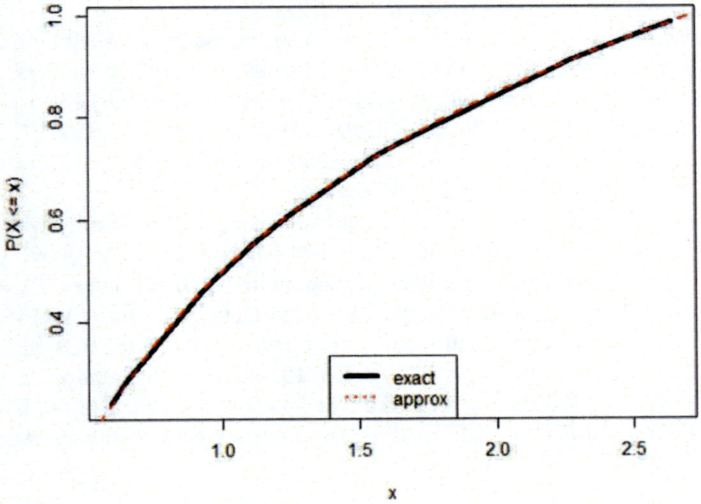

Fig. 3.20 Comparison between the CDF generated by collocation (polynomial family, $k = 6$) and the CDF of X. Here, the CDF is generated by `sampleX`

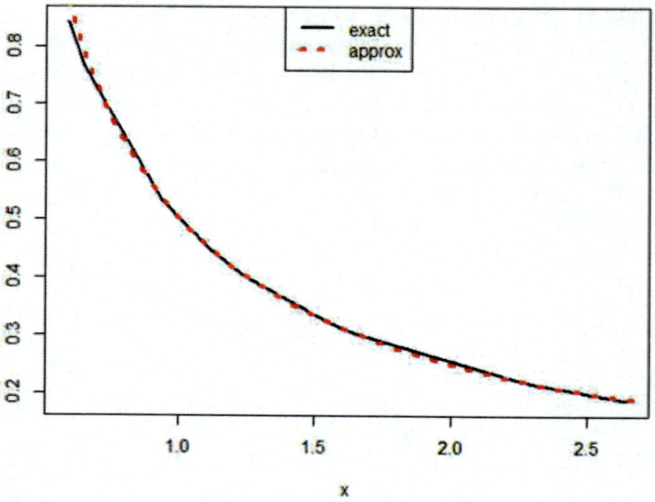

Fig. 3.21 Comparison between the PDF generated by collocation (polynomial family, $k = 6$) and the PDF of X. Here, the PDF is generated by `sampleX`

Here, $\overline{PX} = 1.1752$, $s(PX) = 0.6575$, $P(X > 2.4) = 0.06227$ and $P(X < 0.7) = 0.3217$. Again, the accordance is excellent. The trigonometrical family with $k = 12$ furnishes $\overline{PX} = 1.1748$, $s(PX) = 0.6576$, $P(X > 2.4) = 0.06243$ and $P(X < 0.7) = 0.32126$. For $k = 14$, the trigonometrical approximation furnishes $\overline{PX} = 1.1739$, $s(PX) = 0.6591$, $P(X > 2.4) = 0.06232$ and $P(X < 0.7) = 0.3210$; for $k = 14$, it furnishes $\overline{PX} = 1.1861$, $s(PX) = 0.6451$, $P(X > 2.4) = 0.06241$ and $P(X < 0.7) = 0.3211$; for $k = 22$, it furnishes $\overline{PX} = 1.1795$, $s(PX) = 0.6524$, $P(X > 2.4) = 0.06235$ and $P(X < 0.7) = 0.3221$.

In this case too, we can choose the alternative approach furnished by Eq. (3.14). Using a polynomial family, with $k = 6$, we obtain the results shown in Figs. 3.22 and 3.23. In this case, a trapezoidal rule with 1E3 subintervals furnishes $\overline{PX} = 1.1752$, $s(PX) = 0.6575$, $P(X > 2.4) = 0.06226$ and $P(X < 0.7) = 0.3217$. Using a trigonometrical family with $k = 12$, we obtain $\overline{PX} = 1.1773$, $s(PX) = 0.6571$, $P(X > 2.4) = 0.061284$ and $P(X < 0.7) = 0.3224$.

Choosing the approach by Eq. (3.16) and a trigonometrical family with $k = 12$, we obtain the PDF shown in Fig. 3.24 and $P(X > 2.4) = 0.06248$, $P(X < 0.7) = 0.3212$. Riemann integration of the PDF furnishes $\overline{PX} = 1.1740$ and $s(PX) = 0.6586$; Stieltjes integration of the CDF gives $\overline{PX} = 1.1745$ and $s(PX) = 0.6576$. For a polynomial family with $k = 6$, we obtain the PDF in Fig. 3.25, $P(X > 2.4) = 0.06231$, $P(X < 0.7) = 0.3215$. Riemann integration of the PDF furnishes $\overline{PX} = 1.1750$ and $s(PX) = 0.6575$; Stieltjes integration of the CDF gives $\overline{PX} = 1.1745$ and $s(PX) = 0.6575$. All these values are close to the exact ones.

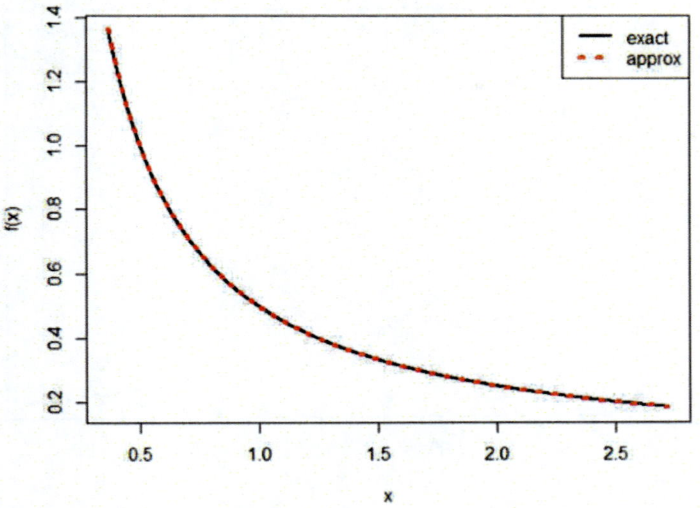

Fig. 3.22 Comparison between the PDF generated by Eq. (3.14) (polynomial family, $k = 6$) and the PDF of X

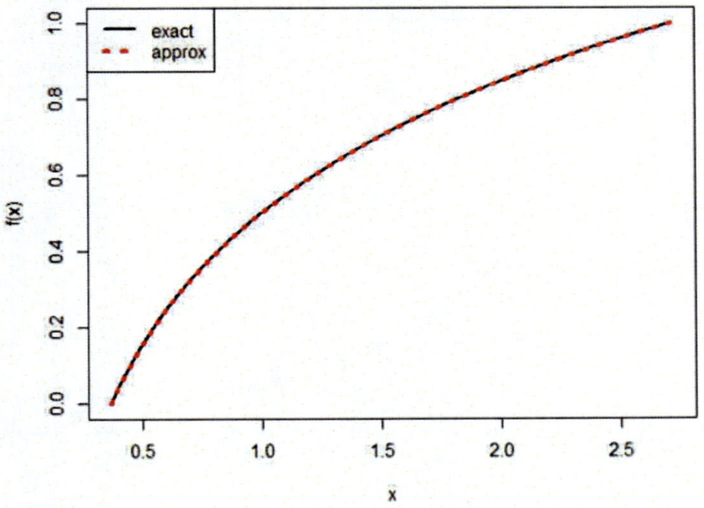

Fig. 3.23 Comparison between the CDF generated by integration of the PDF furnished by Eq. (3.14) (polynomial family, $k = 6$) and the CDF of X

If the distribution of U is unknown, we can choose between two alternatives: either determining the distribution of U firstly and then determining the distribution of X by one of the preceding methods: we obtain a representation of X as a function of U and then, we generate the CDF and PDF of X.

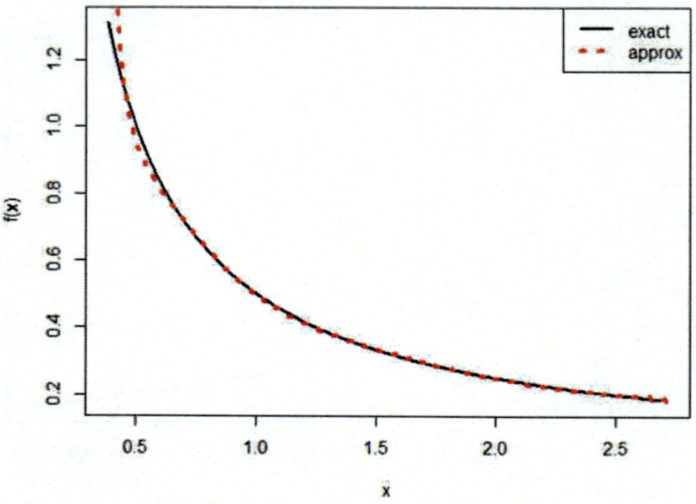

Fig. 3.24 Comparison between the PDF generated by SP derivative of the CDF furnished by Eq. (3.16) (trigonometrical family, $k = 12$, kernel gaussian) and the PDF of X

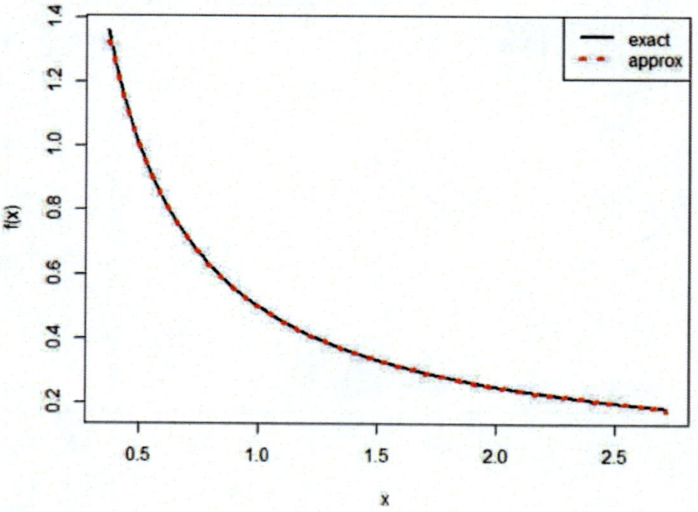

Fig. 3.25 Comparison between the PDF generated by SP derivative of the CDF furnished by Eq. (3.16) (polynomial family, $k = 6$, kernel gaussian) and the PDF of X

- Or directly determining the distribution of X, without determining the distribution of U. In this case, we do not obtain a representation of X as a function of U. – only the CDF and the PDF of X are generated.

Table 3.2 An example of data including an artificial Gaussian variable

A	−1.8596342	−0.8097976	−0.5309607	−0.4362206	−0.2881361
U	−0.521897	−0.428266	−0.063814	0.110263	0.218055
A	0.3361204	0.5652095	1.0876513	1.2547856	2.3231017
U	0.439192	0.498947	0.818171	0.899228	0.972777

In both the cases, it is necessary to introduce an artificial variable A, having a known distribution: for instance, let us consider an artificial variable $A \sim N(0,1)$ – notice that the distribution of A is different from the distribution of U.

Let us examine the first approach: we look for a representation $PU = \mathbf{\Phi}^t(A)\mathbf{u}$ of U as a function of A. Since the distribution of A is known, this representation determines the distribution of U. Then, we determine the distribution of X by one of the preceding approaches.

As an example, let us generate a sample$(A_1, \ldots, A_{ns})$ of ns variates from $N(0, 1)$ and use this sample to find a representation $PU = \sum_{i=0}^{k} u_i \varphi_i(A)$ of U as a function of A. In a first step, it is necessary to create an artificial link between A and U – for instance, we can reorder the data into a way that both A and U are increasingly ordered: $U_1 < U_2 < \ldots < U_{ns}$ and $A_1 < A_2 < \ldots < A_{ns}$ – this generates a positive correlation between U and A. For instance, consider the data in Table 3.2. The values of A were generated by `rnorm`.

We consider an artificial sample formed by the pairs (A_i, U_i) and collocation can furnish coefficients $\mathbf{u} = (u_1, \ldots, u_k)$ defining an expansion of U as a function of A:

$$PU(A) = \sum_{i=0}^{k} u_i \varphi_i(A). \tag{3.18}$$

As an example, using a polynomial family with $k = 3$, we obtain the results exhibited in Fig. 3.26. An alternative to this approach is the *Moments Matching Method* (M3, see Sect. 3.4) – of course, both can be combined, and the coefficients furnished by collocation can be used as starting point for M3.

The approximated distribution was generated using a sample of $1E5$ variates from $N(0, 1)$, truncated to the interval $(-1.8596342, 2.3231017)$, corresponding to the minimal and maximal values of the sample from A. The final sample contained 95797 variates from A, which were used to generate the same number of variates from U. As we can see, the approximated distribution is close to the real one.

Once the sample from U is generated, we can use the coefficients $\mathbf{x}$ and Eq. (3.3) to generate a sample from X: we obtain a sample of 95797 variates from X, which is used to determine the CDF of X. The result is exhibited in Figs. 3.27, 3.28, and 3.29. In this case, we obtain $\overline{PX} = 1.4319$, $s(PX) = 0.6266$, $P(X > 2.4) = 0.1037$ and $P(X < 0.7) = 0.1063$. There is a degradation when compared to the situation where the distribution of U is known, but the approximations can be considered as good,

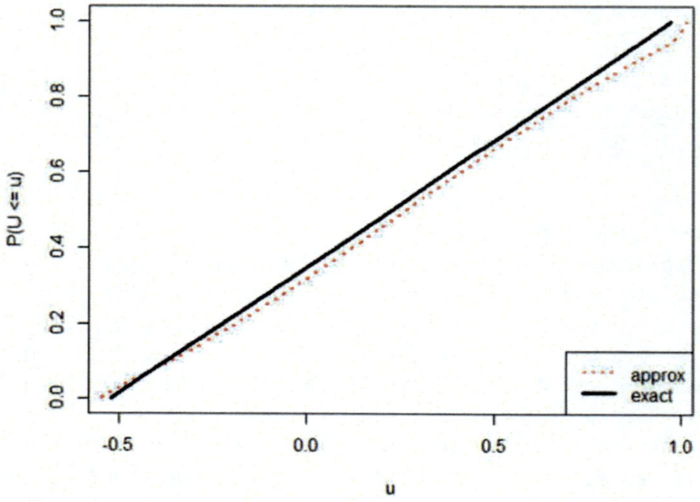

Fig. 3.26 Distribution of U generated by collocation, with an artificial variable $A \sim N(0, 1)$ and a polynomial family with $k = 3$

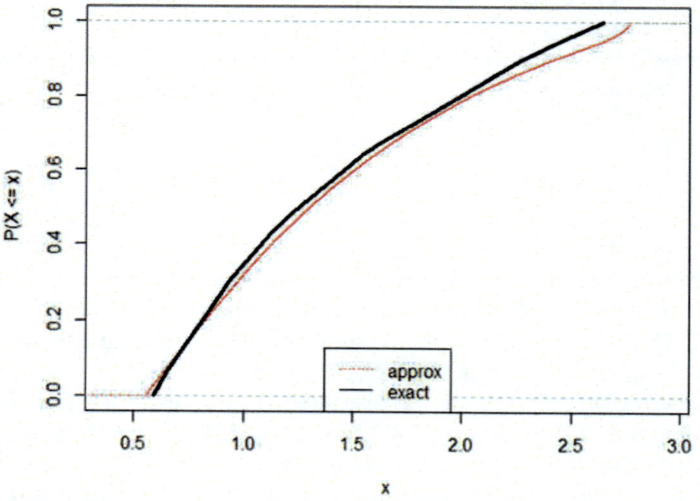

Fig. 3.27 Distribution of X_R generated by collocation, using the sample of U generated by the artificial variable $A \sim N(0, 1)$ and coefficients x determined by collocation (polynomial family, $k = 6$). Here, the CDF was generated by `ecdf`

since the distribution of U is unknown and we used a wrong artificial variable (gaussian instead of uniform).

Analogously to the preceding situations, we can adopt alternative approaches based on Eqs. (3.14) and (3.16). For instance, let us consider (3.14) under the same

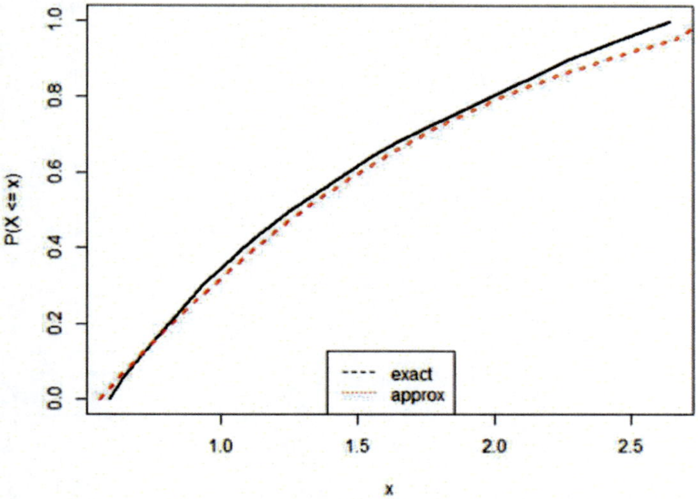

Fig. 3.28 Distribution of X_R generated by collocation, using the sample of U generated by the artificial variable $A \sim N(0,1)$ and coefficients x determined by collocation (polynomial family, $k = 6$). Here, the CDF was generated by `sampleX`

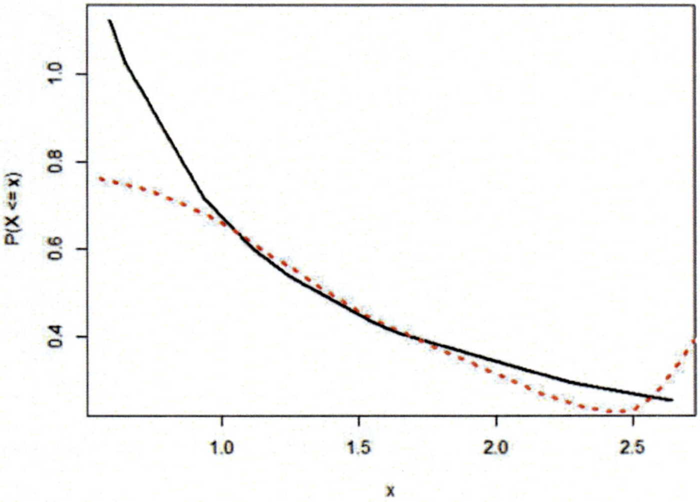

Fig. 3.29 PDF of X_R generated by collocation, using the sample of U generated by the artificial variable $A \sim N(0,1)$ and coefficients x determined by collocation (polynomial family, $k = 6$). Here, the PDF was generated by `sampleX`

conditions: U is approximated by a polynomial of degree 3 in A, then X is approximated by a polynomial of degree 6 in U. The result is exhibited in Figs. 3.30, 3.31, 3.32, and 3.33. In this case, we obtain $\overline{PX} = 1.3889$, $s(PX) = 0.6569$,

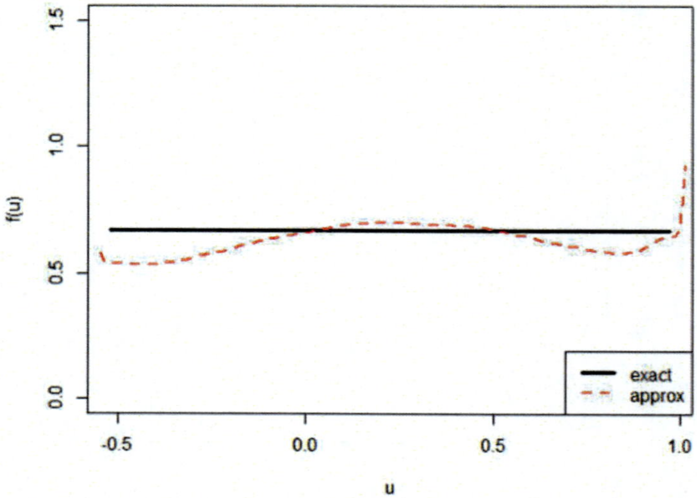

Fig. 3.30 PDF of U generated by Eq. (3.14), with an artificial variable $A \sim N(0, 1)$ and a polynomial family with $k = 3$

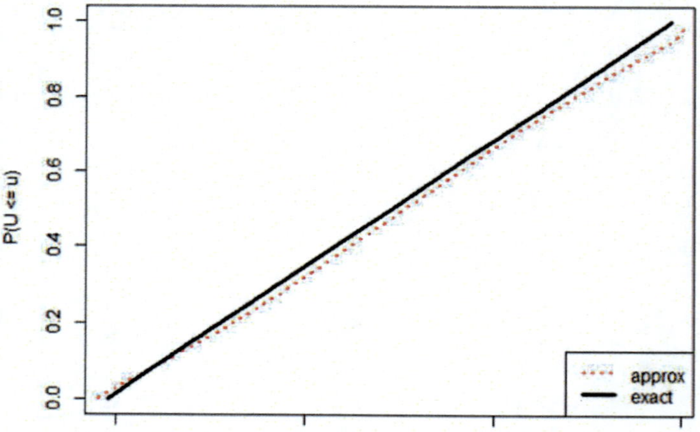

Fig. 3.31 CDF of U generated by integration of the PDF furnished by Eq. (3.14), with an artificial variable $A \sim N(0, 1)$ and a polynomial family with $k = 3$

$P(X > 2.4) = 0.1048$ and $P(X < 0.7) = 0.1674$. Again, we observe a degradation when compared to the situation where the distribution of U is known – we used a wrong artificial variable (gaussian instead of uniform).

Now, let us consider (3.16) under the same conditions: U is approximated by a polynomial of degree 3 in A; then, X is approximated by a polynomial of degree 6 in U. The result is exhibited in Figs. 3.34, 3.35, 3.36, and 3.37. In this case, we obtain

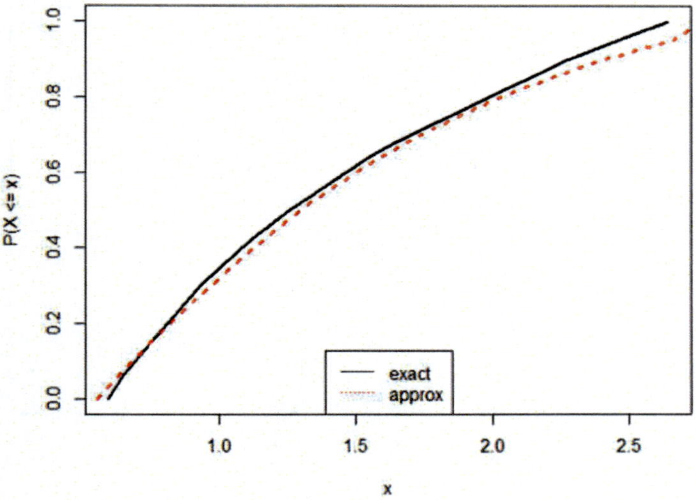

Fig. 3.32 Distribution of X_R generated by collocation, using the sample of U generated by the artificial variable $A \sim N(0, 1)$ and coefficients x determined by collocation (polynomial family, $k = 6$)

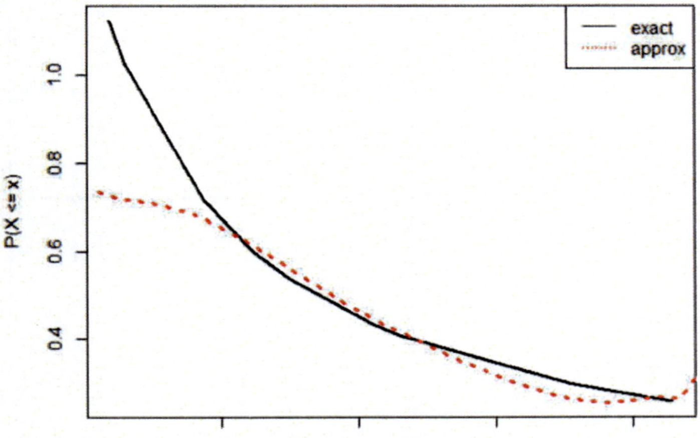

Fig. 3.33 PDF of X_R generated by collocation, using the sample of U generated by the artificial variable $A \sim N(0, 1)$ and coefficients x determined by collocation (polynomial family, $k = 6$)

$\overline{PX} = 1.3744$, $s(PX) = 0.5836$, $P(X > 2.4) = 0.06513$, and $P(X < 0.7) = 0.1103$. Even with a wrong artificial variable, this approach furnishes results not so far from the exact ones.

The results can be improved by using A uniform – for instance, a look to the CDF of U furnished by the methods suggests the use of an uniform variable. Let us

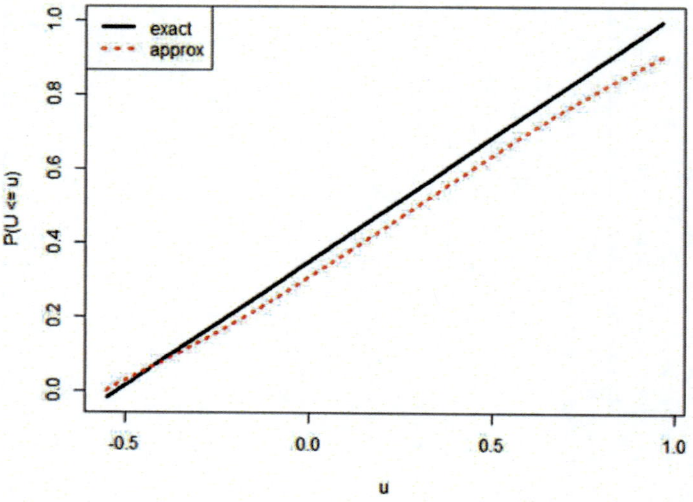

Fig. 3.34 CDF of U generated by Eq. (3.16), with an artificial variable $A \sim N(0, 1)$ and a polynomial family with $k = 3$

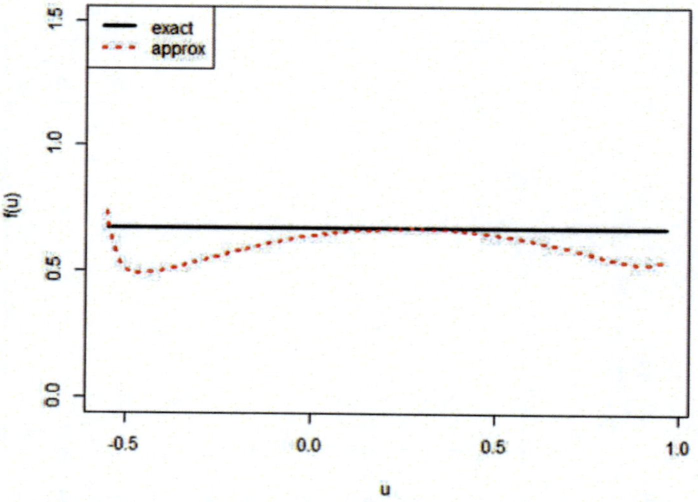

Fig. 3.35 PDF of U generated by SP derivative of the CDF furnished by Eq. (3.16), with an artificial variable $A \sim N(0, 1)$ and a polynomial family with $k = 3$

consider an uniform artificial variable corresponding to Table 3.3. The first approach furnishes the results shown in Figs. 3.38, 3.39, and 3.40. Then, $\overline{PX} = 1.4068$, $s(PX) = 0.6244$, $P(X > 2.4) = 0.0941$, and $P(X < 0.7) = 0.1231$, which are close to the exact values.

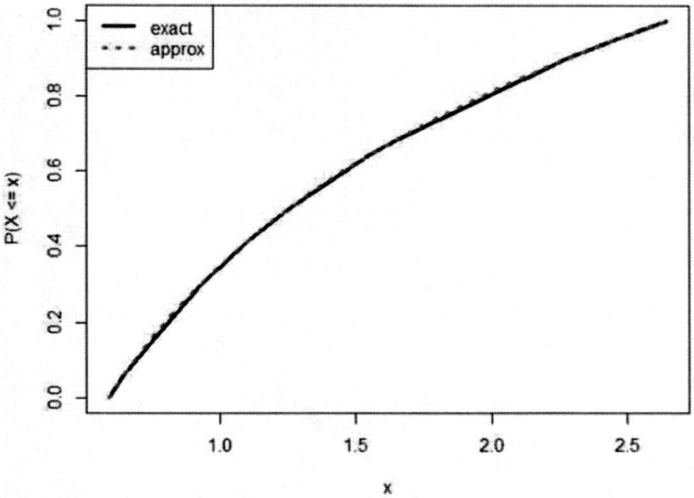

Fig. 3.36 Distribution of X_R generated by collocation, using the sample of U generated by the artificial variable $A \sim N(0,1)$ and coefficients x determined by collocation (polynomial family, $k = 6$)

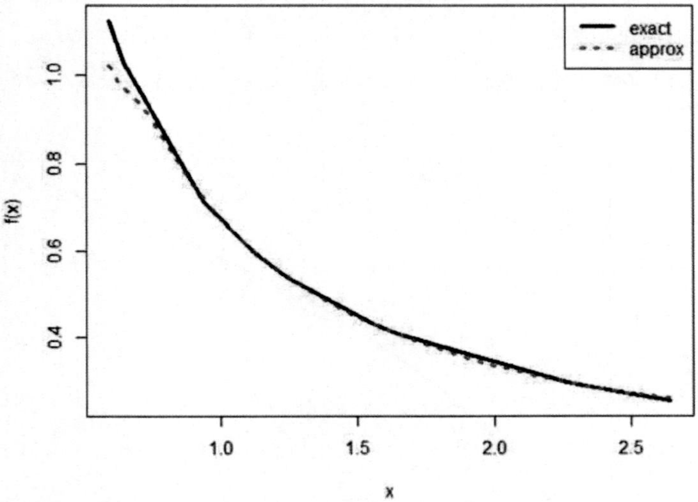

Fig. 3.37 PDF of X_R generated by collocation, using the sample of U generated by the artificial variable $A \sim N(0,1)$ and coefficients x determined by collocation (polynomial family, $k = 6$). Here, the PDF was generated by `sampleX`

Table 3.3 An example of data including an artificial Uniform variable

A	0.002062061	0.222307650	0.242540573	0.270328112	0.534471773
U	−0.521897	−0.428266	−0.063814	0.110263	0.218055
A	0.547722789	0.694275970	0.698030683	0.826904916	0.927155268
U	0.439192	0.498947	0.818171	0.899228	0.972777

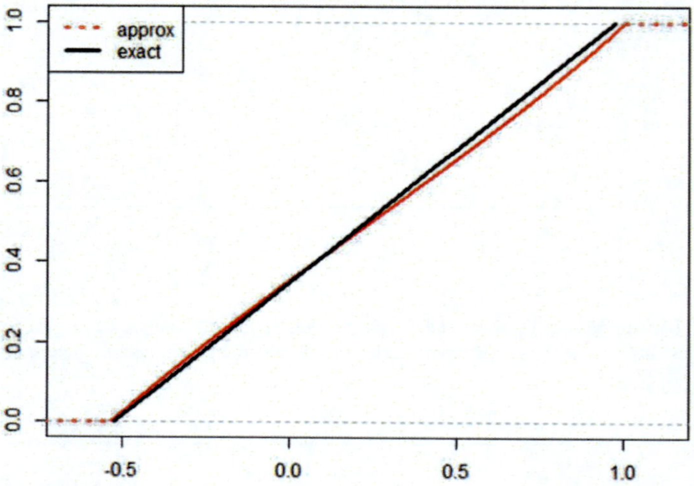

Fig. 3.38 Distribution of U generated by collocation, with an artificial variable $A \sim U(0, 1)$ and a polynomial family with $k = 3$

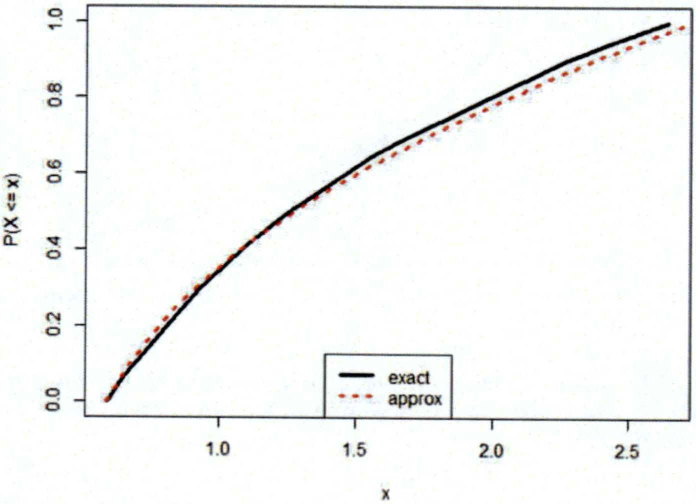

Fig. 3.39 Distribution of X_R generated by collocation, using the sample of U generated by the artificial variable $A \sim U(0, 1)$ and coefficients x determined by collocation (polynomial family, $k = 6$)

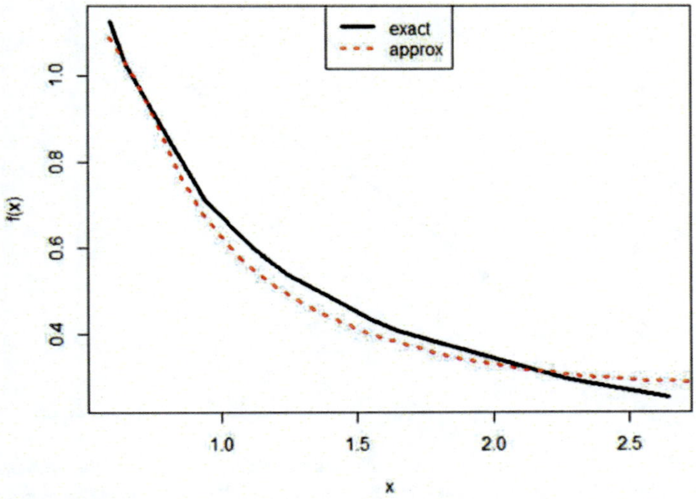

Fig. 3.40 PDF of X_R generated by collocation, using the sample of U generated by the artificial variable $A \sim U(0, 1)$ and coefficients x determined by collocation (polynomial family, $k = 6$)

Table 3.4 A second example of data including an artificial Gaussian variable

A	−1.8596342	−0.8097976	−0.5309607	−0.4362206	−0.2881361
X	0.59339	0.65164	0.93818	1.11657	1.24366
A	0.3361204	0.5652095	1.0876513	1.2547856	2.3231017
X	1.55145	1.64699	2.26635	2.45770	2.64528

If we are not interested in the distribution of U, we can determine the distribution of X directly. For instance, let us consider the artificial sample in Table 3.4:

We can apply the preceding approach to this data. For instance, collocation produces the results in Figs. 3.41 and 3.42. In this case, we obtain $\overline{PX} = 1.3996$, $s(PX) = 0.5672$, $P(X > 2.4) = 0.06145$, and $P(X < 0.7) = 0.09696$,

When dealing with implicit data (Remark 3.1.2), collocation leads to the solution of algebraic equations. Indeed, if (X, U) verifies $\psi(X, U) = 0$, we can consider the system of equations

$$\psi(PX(U_i), U_i) = 0, \; i = 1, \ldots, ns. \tag{3.19}$$

This system of equations can be solved to furnish the coefficients of the expansion (3.3).

As an example, let us consider that the line corresponding to the values of X in Table 3.1 is missing, but we have the information that $\psi(X, U) = U - \log X = 0$. Then, we can consider the system of nonlinear equations

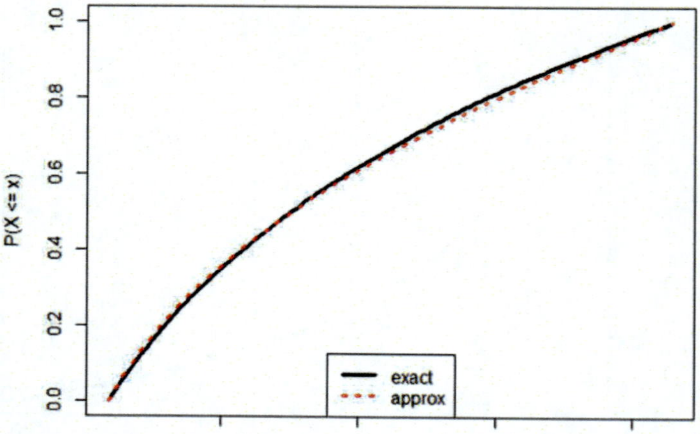

Fig. 3.41 Distribution of X_R generated by collocation, using the sample of the artificial variable $A \sim N(0,1)$ and coefficients x determined by collocation (polynomial family, $k = 4$). Here, the distribution of U is not determined

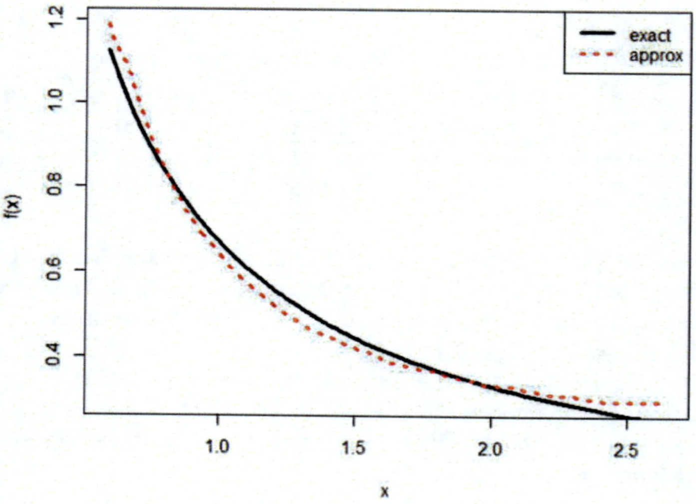

Fig. 3.42 PDF of $X_R$generated by the SP derivative of the CDF obtained with the sample of the artificial variable $A \sim N(0,1)$ and coefficients x determined by collocation (polynomial family, $k = 4$). Here, the distribution of U is not determined

$$\log\left(PX(U_i)\right) - U_i = 0,\ i = 1, \ldots, ns.$$

This system can be solved by R (see Sect. 1.14.2). For instance, we can use the package `pracma` and `fsolve` as follows:

```
rm(list=ls())
cat("\f")
source("polynom.R")
source("expansion1D.R")
psi = function(x,u){log(x)-u}
uc = c(0.110263, 0.818171, 0.218055, -0.428266, -0.521897,
       0.972777, 0.498947, -0.063814, 0.439192, 0.899228)
ucs = sort(uc)

a = min(ucs)
b = max(ucs)
z = 1e-3
phi = polynom$new(a,b,z)
he1 = expansion1D$new(phi)
eqs = function(cc){
  px = he1$proj(cc,ucs)
  v = vector(length = length(ucs))
  for (i in 1:length(ucs)){
    v[[i]]=psi(px[[i]],ucs[[i]])
  }
  aux = as.matrix(v)
  return(aux)
}
k = 6
x0a = as.matrix(rep(1,k+1),ncol=1)
library(pracma)
sol = fsolve(eqs,x0a)
xpx = sol$x
```

Then, variable xpx contains the coefficients $\boldsymbol{x} = (x_0, \ldots, x_k)$ of the expansion. The code furnishes

$$\boldsymbol{x} = (0.59339371, 0.88693384, 0.66292699, 0.32883757, 0.12948471, 0.02573000, 0.01797236)$$

The RMS difference with the result y collocation is 1E-3. The comparison with the exact solution is shown in Figs. 3.43, 3.44, and 3.45. Now, $\overline{PX} = 1.3728$, $s(PX) = 0.5818$, $P(X > 2.4) = 0.06511$ and $P(X < 0.7) = 0.1105$. The exact values are 1.3728, 0.5818, 0.06510, 0.1105.

As a second example, let us consider again that the line corresponding to the values of X in Table 3.1 is missing, but we have the information that

$$\frac{dX}{dU} - X = 0, X(0) - 1 = 0$$

Then, we can consider the system of linear equations

$$\frac{d}{dU}PX(U_i) - PX(U_i) = 0, i = 1, \ldots, ns; \ \ \mathrm{PX}(0) - 1 = 0.$$

Again, we can use the package pracma and fsolve as follows:

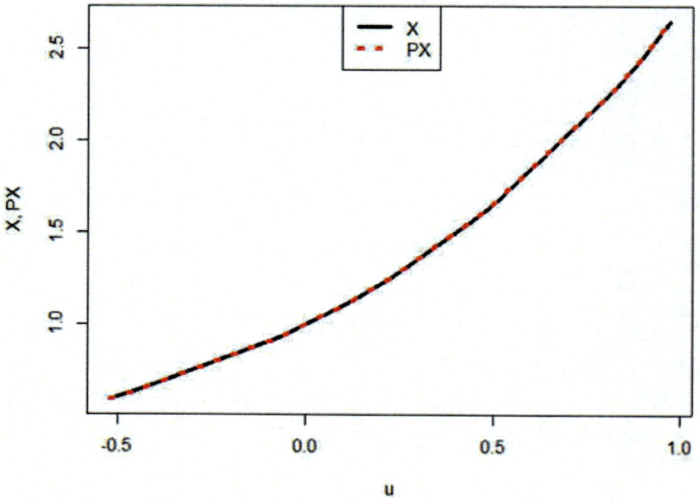

Fig. 3.43 Collocation results for the implicit data (Table 3.1 without the values of X, polynomial family, $k = 6$). The curves almost coincide

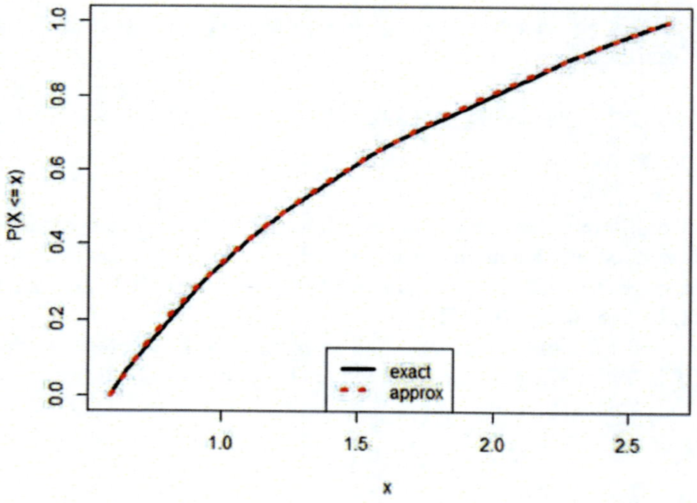

Fig. 3.44 Distribution of X_R generated by collocation, using the implicit data (Table 3.1 without the values of X, polynomial family, $k = 6$)

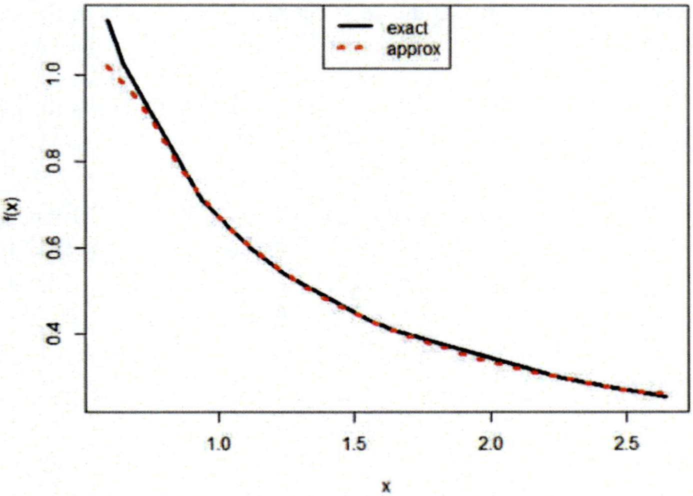

Fig. 3.45 PDF of X_R generated by collocation, using the using the implicit data (Table 3.1 without the values of X, polynomial family, $k = 6$)

```
rm(list=ls())
cat("\f")
source("polynom.R")
source("expansion1D.R")
psi = function(x,dx,u){dx-x}
x0 = 1
u0 = 0
uc = c(0.110263, 0.818171, 0.218055, -0.428266, -0.521897,
       0.972777, 0.498947, -0.063814, 0.439192, 0.899228)
ucs = sort(uc)
a = min(ucs)
b = max(ucs)
z = 1e-3
phi = polynom$new(a,b,z)
he1 = expansion1D$new(phi)
eqs = function(cc){
  px = he1$proj(cc,ucs)
  dpx = he1$dproj(cc,ucs)
  v = vector(length = length(ucs)+1)
  for (i in 1:length(ucs)){
    v[[i]]=psi(px[[i]],dpx[[i]],ucs[[i]])
  }
  v[[length(ucs)+1]] =he1$proj(cc,u0) - x0
  aux = as.matrix(v)
  return(aux)
}
k = 6
x0a = as.matrix(rep(1,k+1),ncol=1)
library(pracma)
sol = fsolve(eqs,x0a)
xpx = sol$x
```

Again, variable xpx contains the coefficients $x = (x_0, \ldots, x_k)$ of the expansion. The code furnishes

$$x = (0.59339613, 0.88692806, 0.66322705, 0.32669316, 0.13487430, 0.02010552, 0.02005645)$$

The RMS difference with the result y collocation is 7E-4. The comparison with the exact solution is shown in Figs. 3.46, 3.47, and 3.48. In this case, $\overline{PX} = 1.3728$,

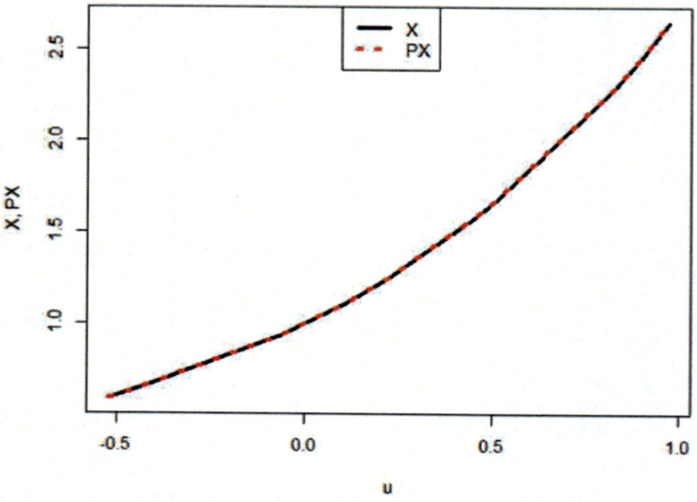

Fig. 3.46. Collocation results for the implicit data (Table 3.1 without the values of X, polynomial family, $k = 6$)

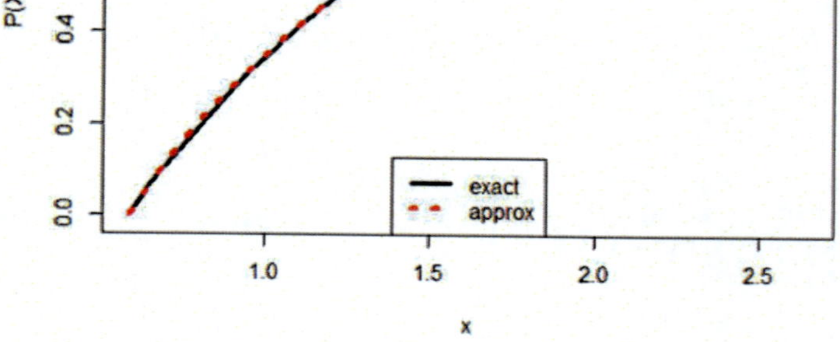

Fig. 3.47 Distribution of X_R generated by collocation, using the implicit data (Table 3.1 without the values of X, polynomial family, $k = 6$). Here, the CDF as generated by `ecdf`

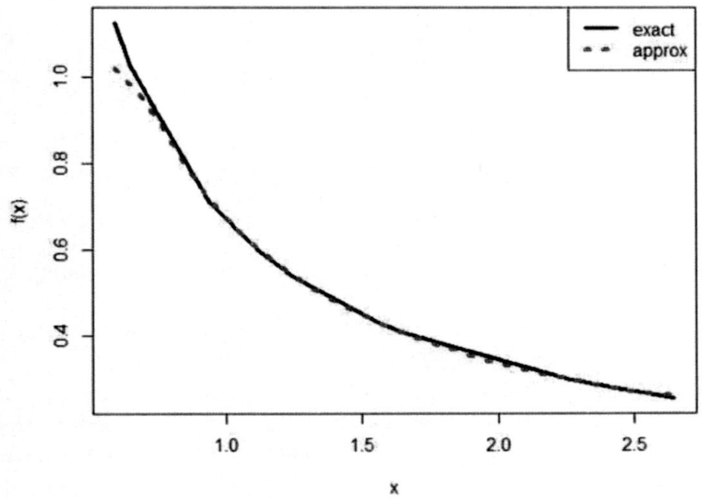

Fig. 3.48 PDF of X_R generated by collocation, using the using the implicit data (Table 3.1 without the values of X, polynomial family, $k = 6$). Here, the PDF was generated by `sampleX`

$s(PX) = 0.5818$, $P(X > 2.4) = 0.06511$ and $P(X < 0.7) = 0.1105$. The exact values are 1.3728, 0.5818, 0.06510, 0.1105.

Exercises

1. The following data were observed:

U	1.000	0.819	0.670	0.549	0.449	0.368
X	−1.033	−0.578	−0.296	−0.169	−0.097	−0.049

U	0.301	0.247	0.202	0.165	0.135	0.111
X	−0.029	−0.014	−0.009	−0.004	−0.002	−0.001

(a) Use the data above to determine by collocation a representation of X as a function of U.
(b) Assume that $U \sim N(0.4, 0.3)$. Use the representation found to determine the CDF and the PDF of X.
(c) Assume that the distribution of U is unknown. Introduce an artificial uniformly distributed variable A and use it to determine a representation of U as function of A by collocation. Use the representation found to determine the CDF and the PDF of X.

(continued)

2. The following data were observed:

U	0.0580	0.0609	0.2197	0.3883	0.4677	0.5439
X	0.0000	0.0021	0.0060	0.0078	0.0123	0.0126
U	0.7550	0.8709	0.8761	0.9045	0.9133	0.9778
X	0.0050	0.0022	0.0023	0.0019	0.0020	0.0001

(a) Use the data above to determine by collocation a representation of X as a function of U.
(b) Assume that $U \sim N(0.5, 0.1)$. Use the representation found to determine the CDF and the PDF of X.
(c) Assume that the distribution of U is unknown. Introduce an artificial uniformly distributed variable A and use it to determine a representation of U as function of A by collocation. Use the representation found to determine the CDF and the PDF of X.

3. The following data were observed:

U	0.2138	0.2483	0.2947	0.2951	0.3354	0.3425
X	8.7482	8.7470	7.0831	6.1580	6.8340	5.7317
U	0.3627	0.3654	0.4223	0.5067	0.5422	0.5549
X	5.1731	5.1117	4.3880	4.6680	4.9407	4.6589

(a) Use the data above to determine by collocation a representation of X as a function of U.
(b) Assume that $U \sim N(0.5, 0.1)$. Use the representation found to determine the CDF and the PDF of X.
(c) Assume that the distribution of U is unknown. Introduce an artificial uniformly distributed variable A and use it to determine a representation of U as function of A by collocation. Use the representation found to determine the CDF and the PDF of X.

4. The following data were observed:

U	0.0866	0.4522	0.7285	0.7902	1.1223	1.3086
X	1.1770	2.0239	2.8004	2.9942	4.1943	5.0098
U	1.3672	1.4038	1.5366	1.5799	1.6650	1.6894
X	5.2917	5.4745	6.1855	6.4346	6.9506	7.1057

(continued)

(a) Use the data above to determine by collocation a representation of X as a function of U.
(b) Assume that $U \sim N(1, 0.5)$. Use the representation found to determine the CDF and the PDF of X.
(c) Assume that the distribution of U is unknown. Introduce an artificial uniformly distributed variable A and use it to determine a representation of U as function of A by collocation. Use the representation found to determine the CDF and the PDF of X.

5. The following data were observed:

U	0.0866	0.4522	0.7285	0.7902	1.1223	1.3086

U	1.3672	1.4038	1.5366	1.5799	1.6650	1.6894

Consider a variable X such that

$$\frac{dX}{dU} = 1 + X, X(0) = 1$$

(a) Use the data above to determine by collocation a representation of X as a function of U.
(b) Assume that $U \sim N(1, 0.5)$. Use the representation found to determine the CDF and the PDF of X.

6. The following data were observed:

U	0.3160	0.3572	0.4611	0.4716	0.5060	0.5162

U	0.5336	0.5379	0.5565	0.5846	0.6349	0.6519

Consider a variable X such that

$$\frac{dX}{dU} = U * X, X(0) = 1$$

(a) Use the data above to determine by collocation a representation of X as a function of U.
(b) Assume that $U \sim N(0.5, 0.1)$. Use the representation found to determine the CDF and the PDF of X.

3.3 Variational Approximation

The variational approach is based on the notion of orthogonal projection, introduced in Sects. 1.15 and 2.5: given a random variable $X \in V = L^2(\Omega, P)$ and a linear subspace, PX – the orthogonal projection of X onto S is the element of S which minimizes the distance to X and it is characterized by

$$PX \in S \;\; and \;\; E(wPX) = E(wX), \forall w \in S \tag{3.20}$$

In the context of UQ, we consider

$$\mathcal{F} = \{\varphi_0, \varphi_1, \ldots, \varphi_k\} \tag{3.21}$$

and

$$S = [\mathcal{F}] = \left\{ \sum_{j=0}^{k} a_j \varphi_j(U) : \boldsymbol{a} = (a_0, \; a_1, \; \ldots, \; a_k) \in \mathbb{R}^{k+1} \right\} \tag{3.22}$$

Then, the condition $PX \in S$ is equivalent to Eq. (3.3). We have

$$E\left(w \sum_{j=0}^{k} x_j \varphi_j(U) \right) = E(wX), \forall w \in S. \tag{3.23}$$

Taking $w = \varphi_i$, we have

$$\sum_{j=0}^{k} x_j E\big(\varphi_j(U)\varphi_j(U)\big) = E(X\varphi_i(U)), 0 \le i \le k. \tag{3.24}$$

Thus, the coefficients $\boldsymbol{x} = (x_0, \ldots, x_k)$ solve the linear system

$$\boldsymbol{Ax} = \boldsymbol{B}, \tag{3.25}$$

with $\boldsymbol{A} = E(\boldsymbol{\Phi}(\mathrm{U})^t\boldsymbol{\Phi}(\mathrm{U}))$ and $\boldsymbol{B} = E(\boldsymbol{\Phi}(\mathrm{U})^t X)$, id est,

$$\boldsymbol{A} = \big(A_{ij} : 1 \le i \le k+1, 1 \le j \le k+1\big), \;\; A_{ij} = E\big(\varphi_{i-1}(U)\varphi_{j-1}(U)\big); \tag{3.26}$$

$$\boldsymbol{B} = (B_i : 1 \le i \le k+1), \; B_i = E(X\varphi_{i-1}(U)) \tag{3.27}$$

Again, $\boldsymbol{x}$ is the solution of a linear system, now given by (3.26) and (3.27). Here, the matrix $\boldsymbol{A}$ is formed by $k+1$ rows and $k+1$ columns, so that the system is, general, determined. If we use a sample analogous to Problem (3.1) to generate the linear system, then we approximate

$$A_{ij} \approx \frac{1}{ns}\sum_{r=1}^{ns} \varphi_{i-1}(U_r)\varphi_{j-1}(U_r); B_i \approx \frac{1}{ns}\sum_{r=1}^{ns} X_r \varphi_{i-1}(U_r) \tag{3.28}$$

In this case, the variational approximation corresponds to the least-squares of the linear system generated by collocation, with $\varepsilon = 0$ (see Eqs. (3.12) and (3.13)).

In the situation where X is defined implicitly by an equation $\psi(X, U) = 0$, we can look for an expansion such that

$$PX \in S \text{ and } E(w\,\psi(PX, \ U)) = 0, \forall w \in S. \tag{3.29}$$

This variational equation leads to the equations

$$E(\varphi_i(U)\,\psi(PX(U), \ U)) = 0, i = 0, \ldots, k. \tag{3.30}$$

This system of equations can be solved to furnish the coefficients $\boldsymbol{x}$ of the expansion (3.3). If we use a sample – as in Problem (3.1) – to generate the equations, we have

$$\frac{1}{ns}\sum_{r=1}^{ns} \varphi_i(U_r)\psi(PX(U_r), U_r) = 0, i = 0, \ldots, k. \tag{3.31}$$

As an example, let us consider again the situation where the line corresponding to the values of X in Table 3.1 is missing, but we have the information that $\psi(X, U) = U - \log X = 0$. Then, we can consider the system of nonlinear equations

$$\frac{1}{ns}\sum_{r=1}^{ns} \varphi_i(U_r)(\log(PX(U_r)) - U_r) = 0, i = 0, \ldots, k.$$

As in the preceding, this system can be solved by R (see Sect. 1.14.2). For instance, we can use the package `pracma` and `lsqnonlin` as follows:

```
rm(list=ls())
cat("\f")
source("polynom.R")
source("expansion1D.R")
psi = function(x,u){log(abs(x)+1e-5) - u}
uc = c(0.110263, 0.818171, 0.218055, -0.428266, -0.521897,
       0.972777, 0.498947, -0.063814, 0.439192, 0.899228)
ucs = sort(uc)
|
k = 6
a = min(ucs)
b = max(ucs)
z = 1e-3
phi = polynom$new(a,b,z)
he1 = expansion1D$new(phi)
eqs = function(cc){
  px = he1$proj(cc,ucs)
  vpsi = vector(length = length(ucs))
  for (r in 1:length(ucs)){
    vpsi[[r]]=psi(px[[r]],ucs[[r]])
  }
  eq = vector(length = k+1)
  for (i in 1:k+1){
    vphi = he1$value(i-1,ucs)
    eq[[i]]=mean(vphi*vpsi)
    }
  aux = as.matrix(eq)
  return(aux)
}
x0a = as.matrix(rep(1,k+1),ncol=1)
library(pracma)
sol = lsqnonlin(eqs,x0a)
xpx = sol$x
```

Again, variable xpx contains the coefficients $\boldsymbol{x} = (x_0, \ldots, x_k)$ of the expansion. The code furnishes

$$\boldsymbol{x} = (0.59552418, 0.89171359, 0.61551074, 0.35995845, 0.17374861, 0.04619873, - 0.03890627)$$

The RMS difference with the result y collocation is 7E-4. The comparison with the exact solution is shown in Figs. 3.49, 3.50, and 3.51. In this case, $\overline{PX} = 1.373436$, $s(PX) = 0.5815821$, $P(X > 2.4) = 0.06539935$ and $P(X < 0.7) = 0.1084989$ – the exact values are 1.372799, 0.581762, 0.06510334, 0.1105405.

Let us consider also that the line corresponding to the values of X in Table 3.1 is missing, but we have the information that

$$\frac{dX}{dU} - X = 0, X(0) - 1 = 0$$

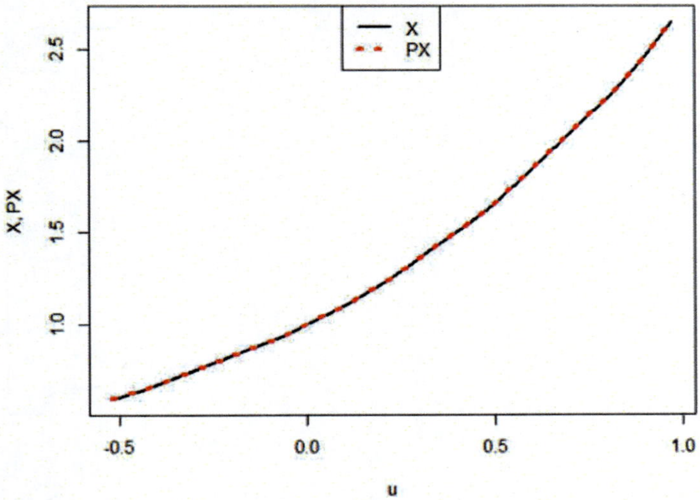

Fig. 3.49 Variational results for the implicit data (Table 3.1 without the values of X, polynomial family, $k = 6$)

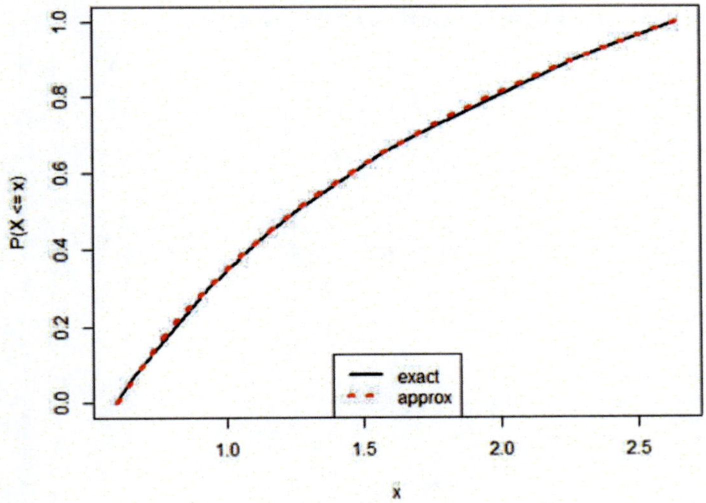

Fig. 3.50 Distribution of X_R generated by the variational approach, using the implicit data (Table 3.1 without the values of X, polynomial family, $k = 6$)

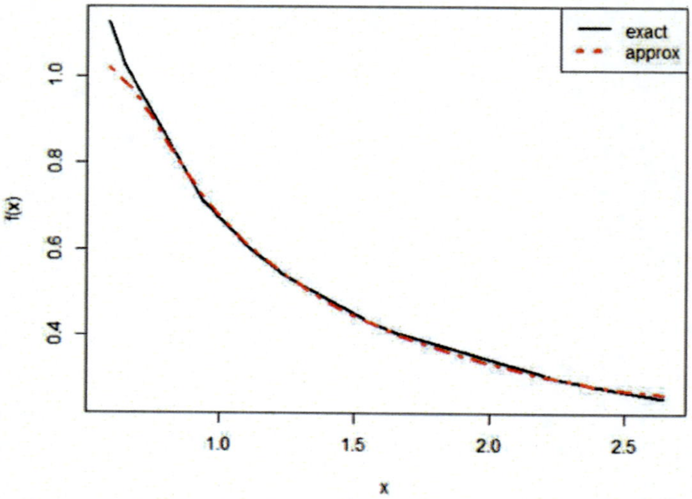

Fig. 3.51 PDF of X_R generated by the variational approach, using the using the implicit data (Table 3.1 without the values of X, polynomial family, $k = 6$). Here, the PDF was generated by `sampleX`

We can use the package `pracma` and `fsolve` as follows:

```
rm(list=ls())
cat("\f")
source("polynom.R")
source("expansion1D.R")
psi = function(x,dx,u){dx-x}
x0 = 1
u0 = 0
uc = c(0.110263, 0.818171, 0.218055, -0.428266, -0.521897,
       0.972777, 0.498947, -0.063814, 0.439192, 0.899228)
ucs = sort(uc)
k = 6
a = min(ucs)
b = max(ucs)
z = 1e-3
phi = polynom$new(a,b,z)
he1 = expansion1D$new(phi)
eqs = function(cc){
  px = he1$proj(cc,ucs)
  dpx = he1$dproj(cc,ucs)
  vpsi = vector(length = length(ucs))
  for (r in 1:length(ucs)){
    vpsi[[r]]=psi(px[[r]],dpx[[r]],ucs[[r]])
  }
  eq = vector(length = k+2)
  for (i in 1:k+1){
    vphi = he1$value(i-1,ucs)
    eq[[i]]=mean(vphi*vpsi)
  }
  eq[[k+2]] =he1$proj(cc,u0) - x0
  aux = as.matrix(eq)
  return(aux)
}
x0a = as.matrix(rep(1,k+1),ncol=1)
library(pracma)
sol = fsolve(eqs, x0a)
xpx = sol$x
```

The code furnishes

$$x = (0.59336931, 0.88790104, 0.65760592, 0.33901535, 0.12324390, 0.02410768, 0.02003295)$$

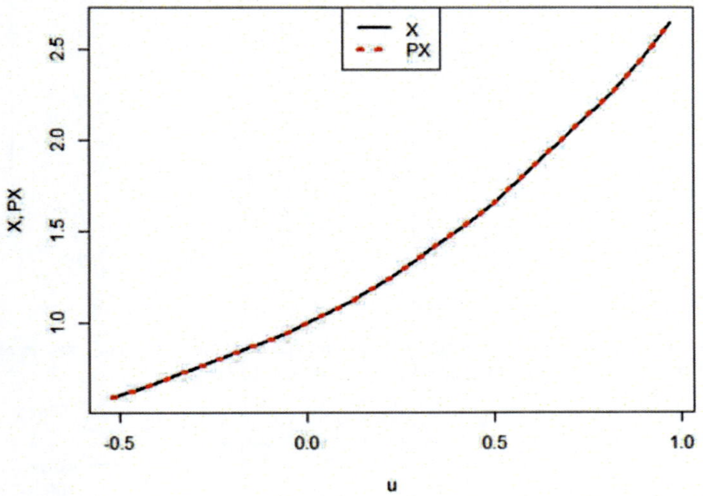

Fig. 3.52 Variational results for the implicit data (Table 3.1 without the values of *X*, polynomial family, $k = 6$)

The comparison with the exact solution is shown in Figs. 3.52, 3.53, and 3.54. Now, $\overline{PX} = 1.372808$, $s(PX) = 0.5817713$, $P(X > 2.4) = 0.06510935$ and $P(X < 0.7) = 0.1105189$ – exact values: 1.372799, 0.581762, 0.06510334, 0.1105405.

Exercises

1. The following data were observed:

U	1.000	0.819	0.670	0.549	0.449	0.368
U	0.301	0.247	0.202	0.165	0.135	0.111

Consider a variable *X* such that

$$X^2 - 3X + U = 0$$

(continued)

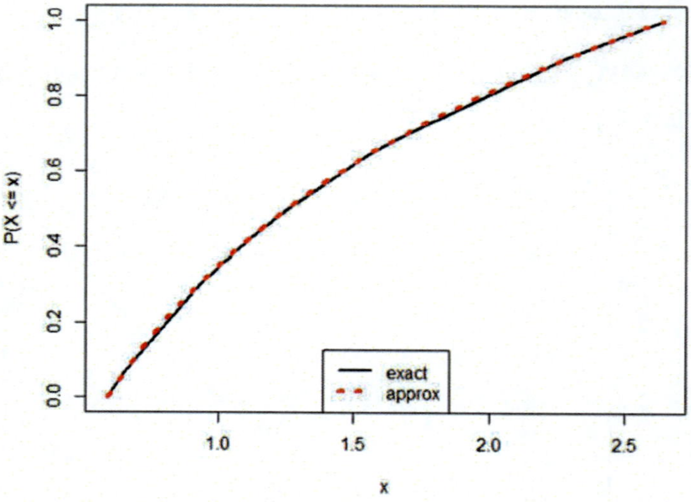

Fig. 3.53 Distribution of X_R generated by the variational approach, using the implicit data (Table 3.1 without the values of X, polynomial family, $k = 6$)

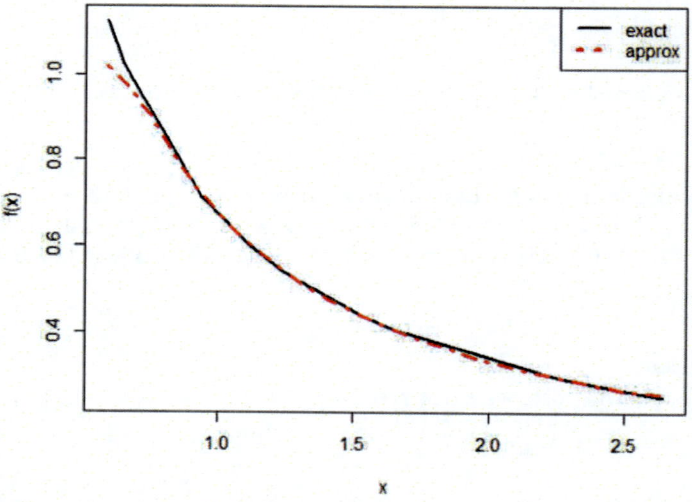

Fig. 3.54 PDF of X_R generated by the variational approach, using the using the implicit data (Table 3.1 without the values of X, polynomial family, $k = 6$). Here, the PDF was generated by `sampleX`

(a) Use the data above to determine a variational approximation of X as a function of U.
(b) Assume that $U \sim N(0.4, 0.3)$. Use the representation found to determine the CDF and the PDF of X.

2. The following data were observed:

U	0.0580	0.0609	0.2197	0.3883	0.4677	0.5439

U	0.7550	0.8709	0.8761	0.9045	0.9133	0.9778

Consider a variable X such that

$$\frac{dX}{dU} + U^2 X = 0, X(0) = 1$$

(a) Use the data above to determine a variational approximation of X as a function of U.
(b) Assume that $U \sim N(0.5, 0.1)$. Use the representation found to determine the CDF and the PDF of X.

3. The following data were observed:

U	0.2138	0.2483	0.2947	0.2951	0.3354	0.3425

U	0.3627	0.3654	0.4223	0.5067	0.5422	0.5549

Consider a variable X such that

$$\frac{dX}{dU} + U^2 + X = 0, X(0) = 1$$

(a) Use the data above to determine a variational approximation of X as a function of U.
(b) Assume that $U \sim N(0.5, 0.1)$. Use the representation found to determine the CDF and the PDF of X.

4. The following data were observed:

U	0.0866	0.4522	0.7285	0.7902	1.1223	1.3086

U	1.3672	1.4038	1.5366	1.5799	1.6650	1.6894

(continued)

Consider a variable X such that

$$\frac{dX}{dU} = 1 + X, X(0) = 1$$

(a) Use the data above to determine a variational approximation of X as a function of U.
(b) Assume that $U \sim N(1, 0.5)$. Use the representation found to determine the CDF and the PDF of X.

5. The following data were observed:

U	0.3160	0.3572	0.4611	0.4716	0.5060	0.5162
U	0.5336	0.5379	0.5565	0.5846	0.6349	0.6519

Consider a variable X such that

$$\frac{dX}{dU} = U * X, X(0) = 1$$

(a) Use the data above to determine a variational approximation of X as a function of U.
(b) Assume that $U \sim N(0.5, 0.1)$. Use the representation found to determine the CDF and the PDF of X.

3.4 Moments Matching Method

In the preceding situations, the connection between the variables X and U was empirically contained in the data or given by an equation.

We may also be interested in situations where such an information is missing. For instance, when we have only information about the values of X: a variability of the response of a system is observed, but we do not know the explaining variable U.

If the information about U is missing, the methods above cannot be used directly: as seen in the preceding sections, we can introduce an artificial variable A and use it to characterize the variability observed. To do this, we must generate an artificial connection between A and X. In the approach by collocation, we ranged both the series in increasing order, to generate a positive correlation between them.

An alternative approach is the *Moments Matching Method* (*M3*). This method usually requests more data (id est, more values of X) than the preceding ones. It is also considered as difficulty to use, since it involves global optimization problems. In addition, *M3* starts by breaking the connection between the observations of X and the values of PX, since it concentrates on the connection of their probability distributions: in principle, PX is not an approximation of X: their distributions may be close, but the variables may differ significantly. Nevertheless, when *M3* is correctly used, it may solve situations where U is unknown. In addition, M3 can be used as a complement of collocation, tending to improve the results.

From the mathematical standpoint, *M3* is based on a consequence of Levy's theorem (see Theorem 2.9.1), which shows that is possible to approach the distribution of a random variable X by approximating its moments. Thus, we may find an approximation of the distribution of X by using the distribution of Z such that

$$M_i(Z) = M_i(X),\ 1 \leq i \leq m. \tag{3.32}$$

To find such a distribution, an arbitrary regular random variable Z may be considered – the unique limitation is the existence of m moments. Taking $Z = PX$ defined as in the preceding (Eq. 3.3) furnishes a nonlinear system of equations

$$F_p(\boldsymbol{x}) = \mu_p(\boldsymbol{x}) - M_p(X) = 0,\ 1 \leq i \leq m, \tag{3.33}$$

where

$$\mu_p(\boldsymbol{x}) = M_p(PX) = E((PX)^p) = E\left(\left(\sum_{j=0}^{k} x_j\varphi_j(U)\right)^p\right) = E((\boldsymbol{\Phi}(U)\boldsymbol{x})^p) \tag{3.34}$$

One of the interesting features of this approach is the following one: here, U appears as arbitrary – the unique limitation remains the one formulated above: the existence of the first m moments of $Z = \boldsymbol{\Phi}(U)\boldsymbol{x} = \sum_{j=0}^{k} x_i\varphi_j(U)$. By this approach, the connection between U and X is generated by Eqs. (3.29) and Levy's theorem, so that – in principle – we can use a variable that has no connection with X. In practice, the use of a variable which is unconnected to X (so, independent from X) often requests more information on X – *id est*, a larger sample – to obtain good results. Of course, if the information about the couple (U, X) is available, the data for U may be used to the fitting, with best results with smaller sets of data.

Errors in the determination of $\boldsymbol{x}$ may be interpreted as the introduction of noise: indeed, the characteristic function ϕ_X of X may be interpreted as the Fourier transform of f_X – the probability density of X. An error in the evaluation of the coefficients $\boldsymbol{x}$ generates an error in the approximation of ϕ_X by ϕ_{PX}, what is equivalent to adding a noise to f_X – observe that a small error on the PDF f_X may correspond to a large error on the CDF $F_X(x)$, which is the integral of f_X on $(-\infty, x)$. This is one of the main difficulties when using the approach by *M3*: on the one hand, an excellent precision on $\boldsymbol{x}$ is necessary; on the other hand, we must solve a hard numerical problem, which requests adapted methods. Nevertheless, as previously remarked, when these problems are correctly solved, the approach *M3* furnishes results for situations where the other approaches cannot be applied, such as, for instance, situations where only observations of X are given and U is unknown.

Table 3.5 Data for the method M3. To situations are considered: U known or missing

U	0.110263	0.818171	0.218055	-0.428266	-0.521897
X	1.11657	2.26635	1.24366	0.65164	0.59339
U	0.972777	0.498947	−0.063814	0.439192	0.899228
X	2.64528	1.64699	0.93818	1.55145	2.45770
U	0.948851	0.825434	−0.158727	−0.874752	0.472945
X	2.58274	2.282872	0.853229	0.416965	1.60471
U	−0.752068	0.217750	−0.823237	0.435408	0.028596
X	0.471391	1.24328	0.439008	1.54559	1.02901
U	0.812861	−0.2440565	0.077486	−0.453719	−0.817560
X	2.25435	0.783443	1.08057	0.635261	0.441507

The approach by Moments Matching may be implemented according to different formulations, which are exposed in the sequel. We shall illustrate the method by considering the situation where U is a random variable having a uniform distribution on the interval $(-1, 1)$ and $X = e^U$. We shall consider three situations: in the first one, the available information is a sample of 25 values of the couple (U, X) (Table 3.5) and the distribution of U is known. In the second one, we have the information in Table 3.5, but the distribution of U is unknown. In the last one, the available information consists only of the values of X – U is missing.

3.4.1 The Standard Formulation of M3

If $m = k$, Eqs. (3.29) and (3.30) can be solved by methods of resolution of nonlinear equations, such as, for instance, Newton-Raphson or fixed-point methods. When $m > k$, we must look for a generalized solution, such as, for instance, the minimum of the distance between the vectors of moments

$$J(\boldsymbol{x}) = d(\boldsymbol{\mu}(\boldsymbol{x}), \boldsymbol{M}(X)), \tag{3.35}$$

where

$$\boldsymbol{\mu}(\boldsymbol{x}) = (\mu_1(\boldsymbol{x}), \ \ldots, \mu_m(\boldsymbol{x})), \ \boldsymbol{M}(X) = (M_1(X), \ \ldots, M_m(X)) \tag{3.36}$$

For instance, we may use a distance generated by a norm:

$$d_p(\boldsymbol{A}, \boldsymbol{B}) = \|\boldsymbol{A} - \boldsymbol{B}\|_p. \tag{3.37}$$

In practice, it may be interesting to use a pseudo-distance generated by using a vector of relative errors:

$$d_{p,rel}(\boldsymbol{A},\boldsymbol{B}) = \left\|\frac{\boldsymbol{A}-\boldsymbol{B}}{\boldsymbol{B}}\right\|_p, \quad \frac{\boldsymbol{A}-\boldsymbol{B}}{\boldsymbol{B}} = \left(\frac{A_1-B_1}{B_1}, \ldots, \frac{A_m-B_m}{B_m}\right). \tag{3.38}$$

Classical norms can be used, such as the mean sum of absolute values (MSA), the root mean square (RMS) or the maximal absolute value (MaxA):

$$\|\boldsymbol{V}\|_1 = \frac{1}{m}\sum_{i=1}^{m}|V_i| \quad (MSA). \tag{3.39}$$

$$\|\boldsymbol{V}\|_2 = \sqrt{\frac{1}{m}\sum_{p=1}^{m}V_i^2} \quad (RMS). \tag{3.40}$$

$$\|\boldsymbol{V}\|_\infty = \max\{|V_i| : 1 \le i \le m\} \quad (MaxA). \tag{3.41}$$

Although these norms are equivalent in theory, the numerical results may change significantly according to the starting point used and the norm chosen. Indeed, the main difficulty in this approach is to obtain a good approximation of the minimizers of J defined by (3.31). Indeed, this function is characterized by its nonconvexity, so that it does not satisfy one of the main assumptions in the convergence results of the standard methods of optimization – the convexity of the objective function – here, of J. In the lack of convexity, the standard methods of deterministic optimization often fail unless for initial points carefully chosen – in fact, for initial points near the solution.

To overcome the difficulty of convergence, global optimization methods were proposed by many authors. R proposes a package `GA` which implements genetic algorithms – see Sect. 1.13), which will be used here. Otherwise, the reader can find in the literature many algorithms to solve global optimization problems, including for constrained optimization (see, for instance, (Pogu & Souza de Cursi, 1994; Autrique & Souza de Cursi, 1997; El Mouatasim et al., 2006; Bouhadi et al., 2004).

Let us consider the first situation: The whole data in Table 3.5 is known and we look for the solution corresponding to $k = 6$, $m = 10$, $d = d_{\infty,rel}$. Many strategies can be used to carry the optimization, namely concerning the choice of the initial point – a good candidate is the result generated by collocation. To illustrate the method, we start by a different procedure: we generate a starting point by a first optimization using $d = d_{1,\ rel}$ and a starting point (`suggestedSol`) randomly generated

$$x^{(0,1)} = \begin{pmatrix} 0.4277091, 0.2848874, 0.9316106, 0.3233701, \ldots \\ \ldots, 0.5220148, 0.1644089, 0.6758046 \end{pmatrix}.$$

The optimization by `GA` using as objective $d_{1,\ rel}$ produces as result:

$$x^{(0,\infty)} = \begin{pmatrix} -0.6806936, 3.402445, 3.595097, -2.45974, \ldots \\ \ldots, 4.659208, -3.720075, -5.146975 \end{pmatrix}.$$

Then, the optimization by GA using $d_{\infty,rel}$ as objective and this point as starting point (`suggestedSol`) produces the result

$$x^{(\infty)} = \begin{pmatrix} 2.673211, -4.507739, -2.057077, 4.546916, \ldots \\ \ldots, 0.4474495, 4.04641, -3.065383 \end{pmatrix}.$$

If the procedure is iterated, using this point to start an optimization by GA using $d = d_{1,\ rel}$ and using the result to an optimization by GA with $d_{\infty,rel}$, we obtain at convergence

$$x^{(\infty)} = \begin{pmatrix} 2.650768, -3.436297, -2.372657, 0.9982764, \ldots \\ \ldots, 3.07271, 2.5276, -1.442318 \end{pmatrix}.$$

The CDF and the PDF corresponding to this point are shown in Figs. 3.55 and 3.56.

Notice that GA includes a stochastic part, so that the results may differ from a run to another one. The starting point influences the result too. For instance, the same procedure with the starting point

$$x^{(0,1)} = \begin{pmatrix} -0.8070135, -0.7510242, 0.7712421, -0.6448800, \ldots \\ \ldots, -0.8149873, 0.5284796, 0.7886524 \end{pmatrix}.$$

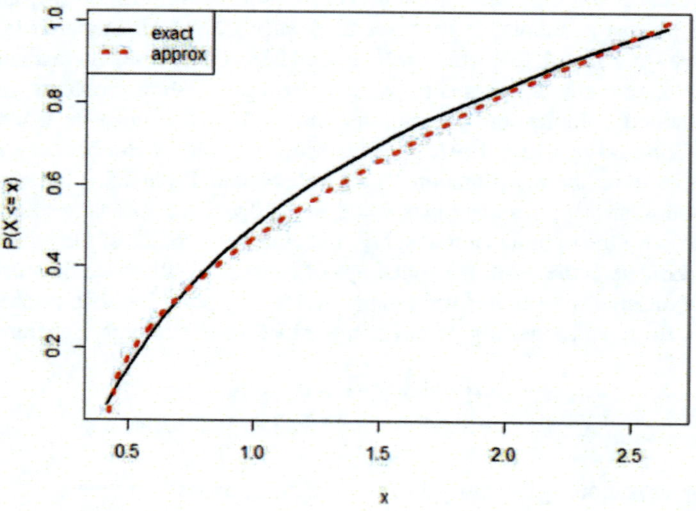

Fig. 3.55 CDF of X generated by M3 (Table 3.5, polynomial family, $k = 6$, $m = 10$, U and its distribution known)

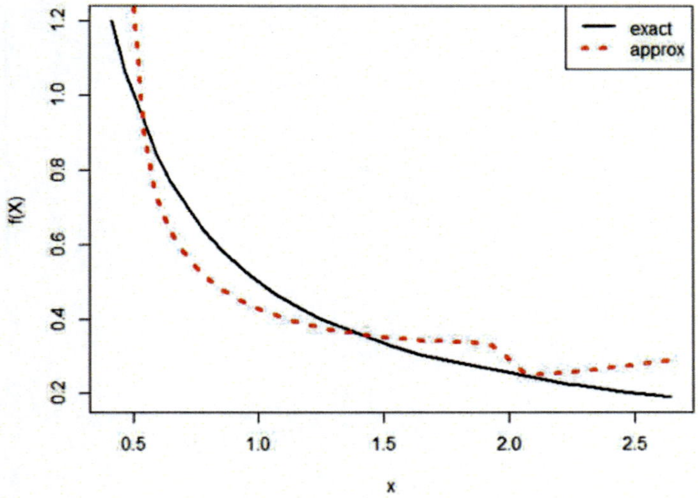

Fig. 3.56 PDF of X generated by M3 (Table 3.5, polynomial family, $k = 6$, $m = 10$, U and its distribution known)

generated

$$x^{(\infty)} = \begin{pmatrix} 2.650768, -3.436297, -2.372657, 0.9982764, \ldots \\ \ldots, 3.07271, 2.5276, -1.442318 \end{pmatrix}.$$

The corresponding results are shown in Figs. 3.57 and 3.58.

As previously observed, the result can be improved by using as starting point the solution generated by collocation:

$$x^{(colloc)} = \begin{pmatrix} 0.4169667, 0.7702149, 0.7138546, 0.4247128, \ldots \\ \ldots,, 0.2411957\ 0.01980785, 0.0585251 \end{pmatrix}.$$

In this case, the starting point corresponds to $d_{\infty,rel} = 5E - 7$, so that GA does not improve the result. The corresponding results are shown in Figs. 3.59 and 3.60.

Let us consider the second situation: the distribution of U is unknown. In this case, we can choose the first approach that starts by determining the distribution of U. Analogously to Sect. 3.2, we consider an artificial variable $A \sim N(0, 1)$. We use `rnorm` to generate the sample in Table 3.6

We look for an approximation $U \approx \boldsymbol{\Phi}(A)\boldsymbol{u}$. For $k = 5$ and a polynomial family, collocation furnishes

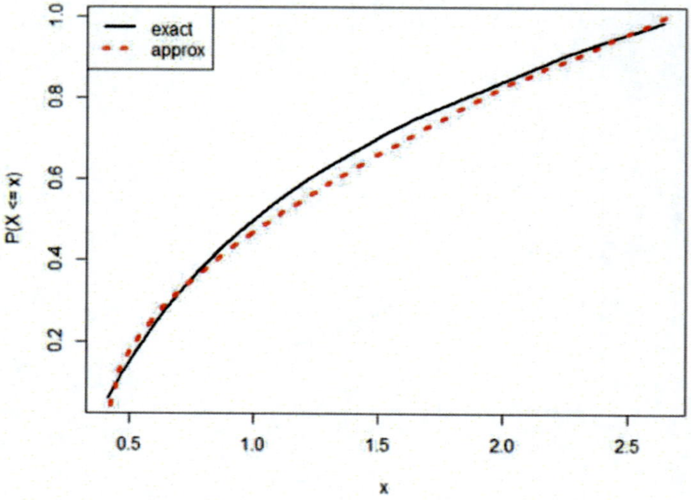

Fig. 3.57 CDF of X generated by M3 (Table 3.5, polynomial family, $k = 6$, $m = 10$, U and its distribution known)

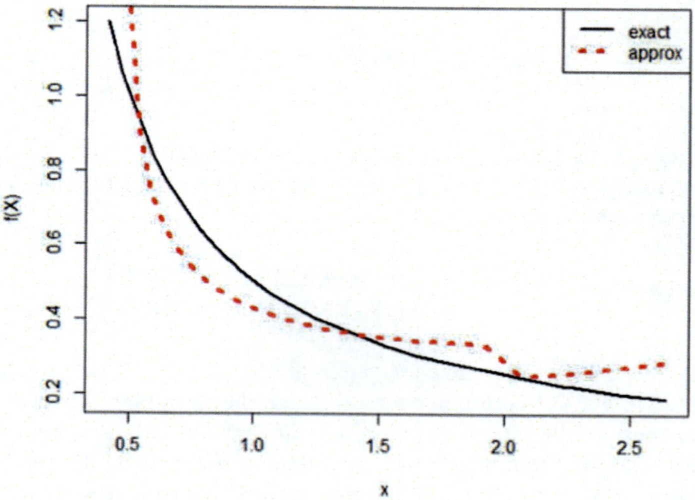

Fig. 3.58 PDF of X generated by M3 (Table 3.5, polynomial family, $k = 6$, $m = 10$, U and its distribution known)

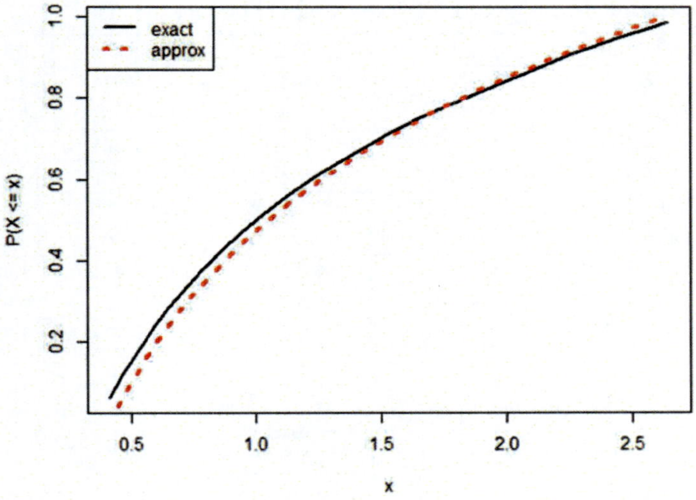

Fig. 3.59 CDF of X generated by M3 (Table 3.5, polynomial family, $k = 6$, $m = 10$, U and its distribution known)

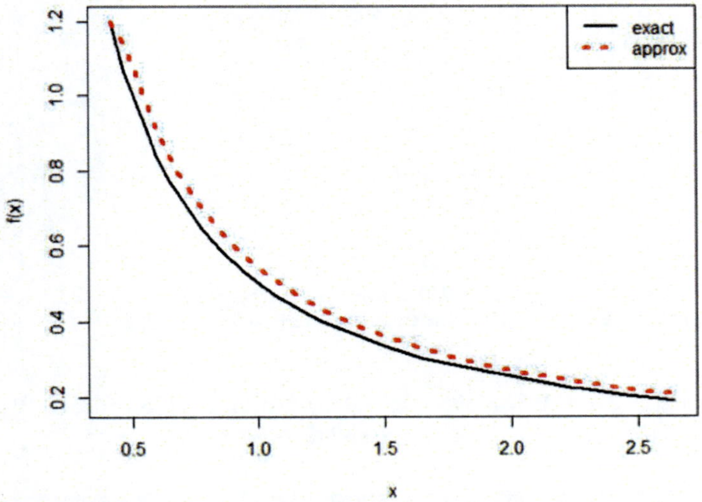

Fig. 3.60 PDF of X generated by M3 (Table 3.5, polynomial family, $k = 6$, $m = 10$, U and its distribution known)

Table 3.6 Data for the determination of the distribution of U by method M3

U	−0.8747520	−0.8232370	−0.8175600	−0.7520680	−0.5218970
A	−1.69282675	−1.36532620	−1.13764893	−0.90435314	−0.85557539
U	−0.4537190	−0.4282660	−0.2440565	−0.1587270	−0.0638140
A	−0.62413905	−0.55531186	−0.55029124	−0.48780361	−0.29600649
U	0.0285960	0.0774860	0.1102630	0.2177500	0.2180550
A	−0.17379275	−0.07466342	0.27389469	0.27389469	0.64491201
U	0.4354080	0.4391920	0.4729450	0.4989470	0.8128610
A	0.68163795	0.73221107	0.90818361	0.93777431	1.00175775
U	0.8181710	0.8254340	0.8992280	0.9488510	0.9727770
A	1.00923872	1.11946666	1.29903777	1.34054714	1.53373080

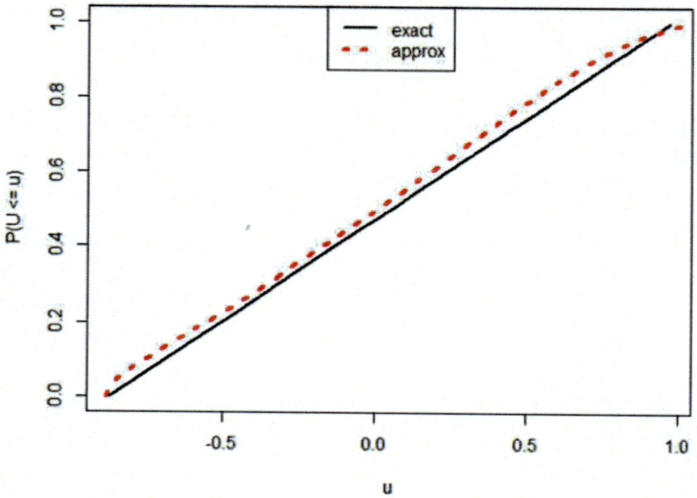

Fig. 3.61 CDF of U generated by M3 (Table 3.6, polynomial family, $k = 5$, $m = 10$, the distribution of U is unknown and we use an artificial variable $A \sim N(0, 1)$)

$$\boldsymbol{u}^{(\text{colloc})} = \begin{pmatrix} -0.8277595, -4.76959, 43.16617, -105.6562, \ldots \\ \ldots, 110.3948, -41.31127 \end{pmatrix}.$$

We use this value as starting point (`suggestedSol`) and the distance $d_{2,\,rel}$, we obtain (Fig. 3.61)

$$\boldsymbol{u}^{(2)} = \begin{pmatrix} -0.8861719, -0.2191236, 5.858474, -3.138678, \ldots \\ \ldots, -3.08195, 2.50002 \end{pmatrix}.$$

Then, we determine an approximation $X \approx \Phi(U)x$. For $k = 6$ and a polynomial family, collocation furnishes

$$x^{(colloc)} = \begin{pmatrix} 0.4169667, 0.7702149, 0.7138546, 0.4247128, \ldots \\ \ldots, 0.2411957, 0.01980785, 0.0585251 \end{pmatrix}.$$

To determine the distribution of X, we can use the approximation $U \approx \Phi(A)u$ to generate a large sample of U: `rnorm` generates $1E5$ variates from A, from which we exclude the values outside the extremes of A in Table 3.6 : only the values in $(-0.62413905, 1.53373080)$ are retained – we keep 89241 values). These values generate a sample from U, which is used to generate a sample of 89241 variates from X. The results are shown in Figs. 3.62 and 3.63.

Analogously to Sect. 3.2, we can choose the direct determination of the distribution of X, without determining the distribution of U. In this case, we consider the data in Table 3.7.

We use the package GA to minimize the distance $d_{2,\ rel}$. The solution obtained by collocation is used as initial point (`suggestedSol`). The results are exposed in Figs. 3.64 and 3.65. For this approximation, $\overline{PX} = 1.2040, s(PX) = 0.7472$, $P(X > 2.4) = 0.09269$, and $P(X < 0.7) = 0.3577$. The values corresponding to collocation were $\overline{PX} = 1.1418, s(PX) = 0.7208$, $P(X > 2.4) = 0.12$, and $P(X < 0.7) = 0.24$. Notice that these results are obtained without any other information about X than the data in Table 3.7 – in addition, we use a Gaussian variable instead of an uniform one.

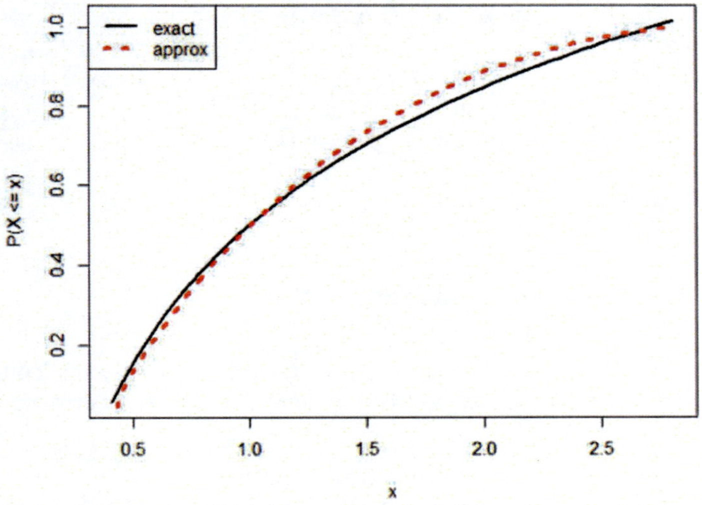

Fig. 3.62 CDF of X generated by collocation, using the data in Table 3.5 and the approximation $U \approx \Phi(A)u$

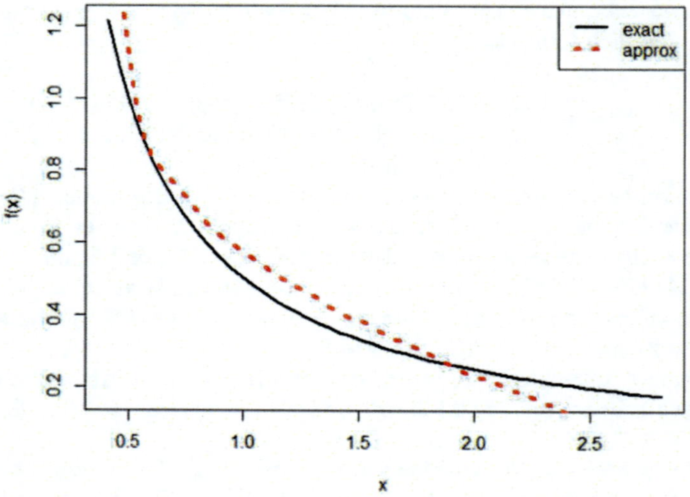

Fig. 3.63 PDF of X generated by collocation, using the data in Table 3.5 and the approximation $U \approx \Phi(A)u$

Table 3.7 Data for the direct determination of the distribution of X by method M3

X	0.416965	0.439008	0.441507	0.471391	0.593390
A	−1.69282675	−1.36532620	−1.13764893	−0.90435314	−0.85557539
X	0.635261	0.651640	0.783443	0.853229	0.938180
A	−0.62413905	−0.55531186	−0.55029124	−0.48780361	−0.29600649
X	1.029010	1.080570	1.116570	1.243280	1.243660
A	−0.17379275	−0.07466342	0.27389469	0.27389469	0.64491201
X	1.545590	1.551450	1.604710	1.646990	2.254350
A	0.68163795	0.73221107	0.90818361	0.93777431	1.00175775
X	2.266350	2.282872	2.457700	2.582740	2.645280
A	1.00923872	1.11946666	1.29903777	1.34054714	1.53373080

3.4.2 *Alternative Formulations of M3*

M3 can also be formulated under other forms, generally less efficient, but often easy to implement and less expensive in terms of computational cost. Alternative forms are

- A constrained optimization problem (CO formulation)
- A bi-objective optimization problem. (BO formulation)

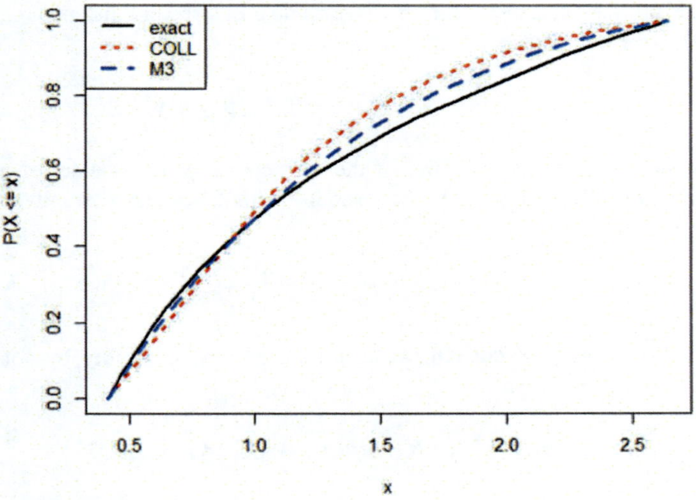

Fig. 3.64 CDF of X generated by M3 (Table 3.7, polynomial family, $k = 4$, $m = 30$. We use an artificial variable $A \sim N(0, 1)$). The starting point is furnished by collocation, which corresponds to the red discontinuous line

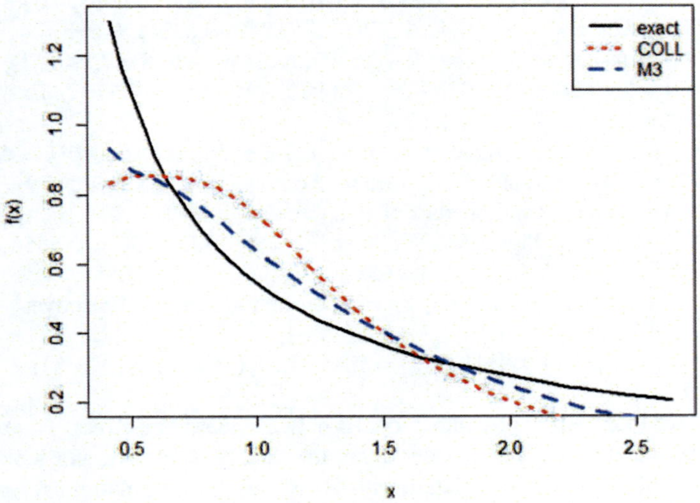

Fig. 3.65 PDF of X obtained by SP (kernel gaussian) derivation of the CDF generated by M3 (Table 3.7, polynomial family, $k = 4$, $m = 30$, artificial variable $A \sim N(0, 1)$). The continuous line is the solution by collocation

For instance, we can consider Eqs. (3.31) as constraints for the determination of the coefficients $\boldsymbol{x}$: let $\boldsymbol{A}$ and $\boldsymbol{B}$ be the collocation matrices given by Eqs. (3.10) and (3.11): a good approximation minimizes the distance between the approximated values $\boldsymbol{Ax}$ and the observed values $\boldsymbol{B}$. By introducing a second distance (or pseudo-distance) δ, we may look for

$$\boldsymbol{x} = \arg\min \left\{ \delta(\boldsymbol{Ay},\ \boldsymbol{B}) : F_p(\boldsymbol{y}) = 0,\ 1 \le p \le m,\ \boldsymbol{y} \in \mathbb{R}^{k+1} \right\}$$

In this formulation, the equality of the moments appears as a set of nonlinear constraints of optimization. As in the standard formulation, an alternative formulation is

$$\boldsymbol{x} = \arg\min \left\{ \delta(\boldsymbol{Ay},\ \boldsymbol{B}) : d(\boldsymbol{\mu}(\boldsymbol{y}),\ \boldsymbol{M}(X)) = 0,\ \boldsymbol{y} \in \mathbb{R}^{k+1} \right\}$$

In practice, it may be convenient to introduce some flexibility by considering $\varepsilon > 0$ and

$$\boldsymbol{x} = \arg\min \left\{ \delta(\boldsymbol{Ay},\ \boldsymbol{B}) : d(\boldsymbol{\mu}(\boldsymbol{y}),\ \boldsymbol{M}(X)) \le \varepsilon,\ \boldsymbol{y} \in \mathbb{R}^{k+1} \right\}$$

Such a problem can be solved, for instance, by `fmincon`, from package `pracma`. We use this function in the examples below.

As an example, let us consider the data in Table 3.6. We look for the distribution of U, using a formulation with $\delta = d = d_2$ and $\varepsilon = 1E-2$. With a polynomial family, $k = 3$, $m = 30$, collocation furnishes $\boldsymbol{u} = (-0.9706, 1.4852, 0.9683, -0.4645)$, and M3-CO furnishes $\boldsymbol{u} = (-0.9584,\ \ 1.3791,\ \ 1.2194, -0.6338)$. For $k = 4$, $m = 30$, collocation furnishes $\boldsymbol{u} = (-0.8957, -0.5307, 10.3349, -14.8767, 7.0651)$; M3-CO furnishes $\boldsymbol{u} = (-0.9066,\ \ -0.06879,\ \ 7.6982, -10.0856,\ \ 4.3702)$. Examples of results appear in Figs. 3.66 and 3.67.

Let us perform a supplementary step: we use the representation $PU = \boldsymbol{\Phi}(A)\boldsymbol{u}$ of U to generate a large sample of $1E5$ variates from U. Using the data in Table 3.5, we find by collocation the coefficients $\boldsymbol{x}$ of an approximation $PX = \boldsymbol{\Phi}(U)\boldsymbol{x}$. An example of result is shown in Fig. 3.68. We have $\overline{PX} = 1.1381$, $s(PX) = 0.5656$, $P(X > 2.4) = 0.0352$, and $P(X < 0.7) = 0.2665$.

In the bi-objective formulation, we can consider a vector of objectives

$$\boldsymbol{f}(\boldsymbol{y}) = (f_1(\boldsymbol{y}), f_2(\boldsymbol{y})), f_1(\boldsymbol{y}) = \delta(\boldsymbol{Ay}, \boldsymbol{B}), f_2(\boldsymbol{y}) = d(\boldsymbol{\mu}(\boldsymbol{y}), \boldsymbol{M}(X)).$$

Then, we look for the possible tradeoffs between these objectives. To exemplify the BO formulation, let us consider again the data in Table 3.6 and look for the distribution of U, using a formulation with $\delta = d = d_2$. With a polynomial family and $k = 4$, the variational approach presented in Sect. 1.13.3 furnishes the result in Fig. 3.69 – the minimization of the area is carried by `fminunc` from package `pracma`. From the figure, we see that the values – this may be interpreted as the fact that all the points furnish similar results. Indeed, as shown in Figs. 3.70, 3.71, 3.72, and 3.73, the results for $t = 0.25$ and $t = 0.5$ are similar and close to those furnished by collocation.

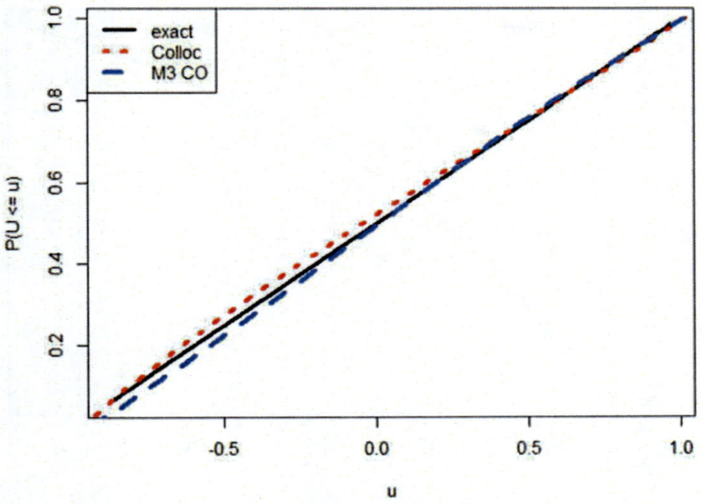

Fig. 3.66 CDF of U generated by M3 in the CO formulation (Table 3.6, polynomial family, $k = 3$, $m = 20$. We use an artificial variable $A \sim N(0, 1)$). The starting point is furnished by collocation

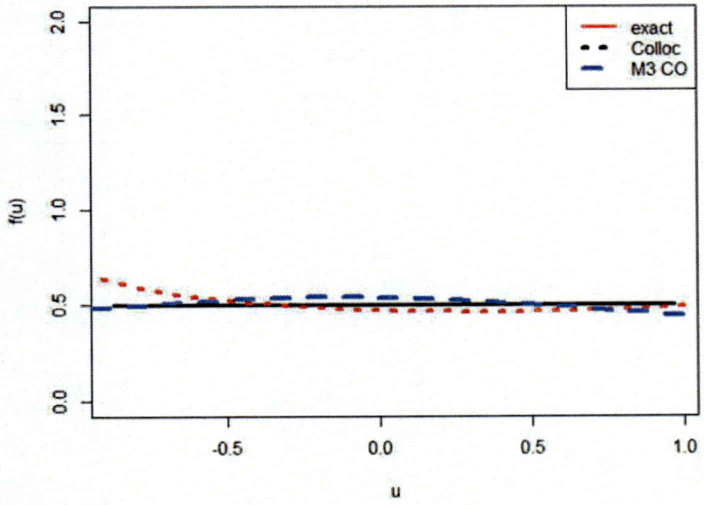

Fig. 3.67 PDF of U obtained by SP (gaussian kernel) derivation of the CDF furnished by the CO formulation of M3 (Table 3.6, polynomial family, $k = 3$, $m = 20$, artificial variable $A \sim N(0, 1)$)

Analogously to the preceding, we can determine the distribution of X using the approximations found for U: we can use them to generate a large sample from X and, then, use this sample to approximate de CDF of X. The CDF found for $t = 0.5$ is shown in Fig. 3.74. The result for $t = 0.25$ appears in Fig. 3.75.

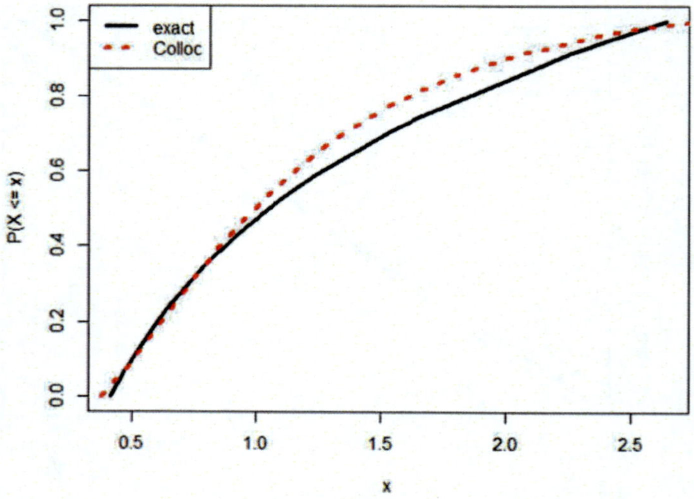

Fig. 3.68 CDF of X generated by the second step: the coefficients $\boldsymbol{x}$ of $PX = \boldsymbol{\Phi}(U)\boldsymbol{x}$ are generated by collocation (data on Table 3.5) and the representation $PU = \boldsymbol{\Phi}(A)\boldsymbol{u}$ previously obtained is used to generate a large sample from $\boldsymbol{U}$, which is used to determine the CDF

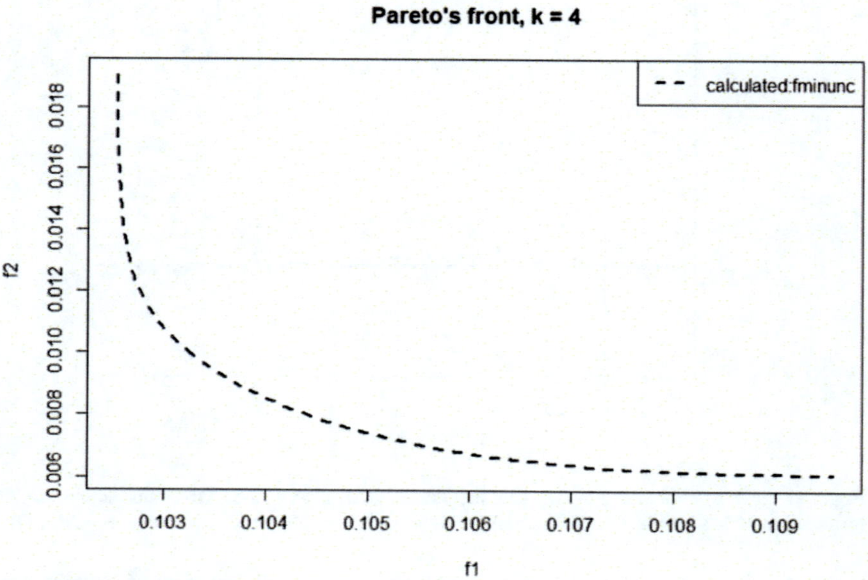

Fig. 3.69 Pareto's front obtained using $\delta = d = d_2$ as distance

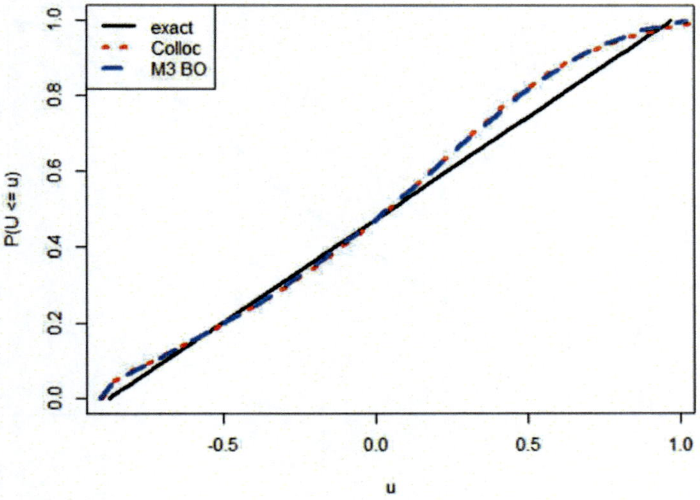

Fig. 3.70 CDF corresponding to $t = 0.25$

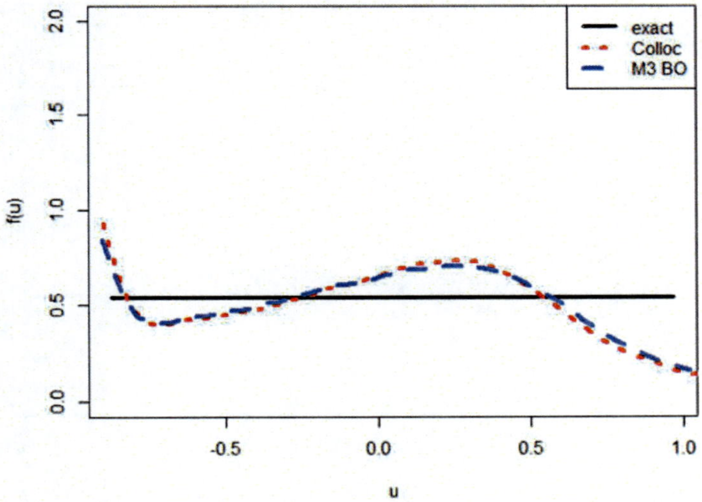

Fig. 3.71 PDF corresponding to $t = 0.25$

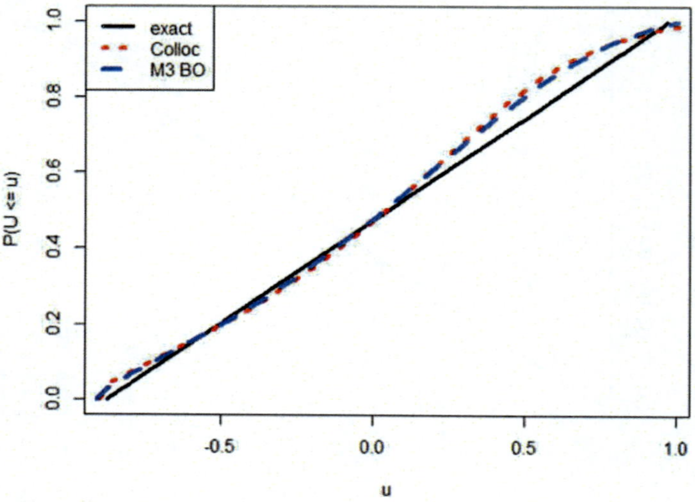

Fig. 3.72 CDF corresponding to $t = 0.5$

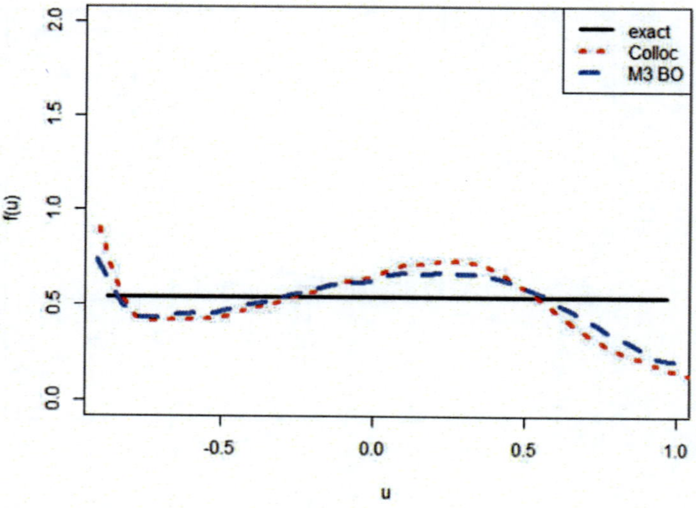

Fig. 3.73 PDF corresponding to $t = 0.5$

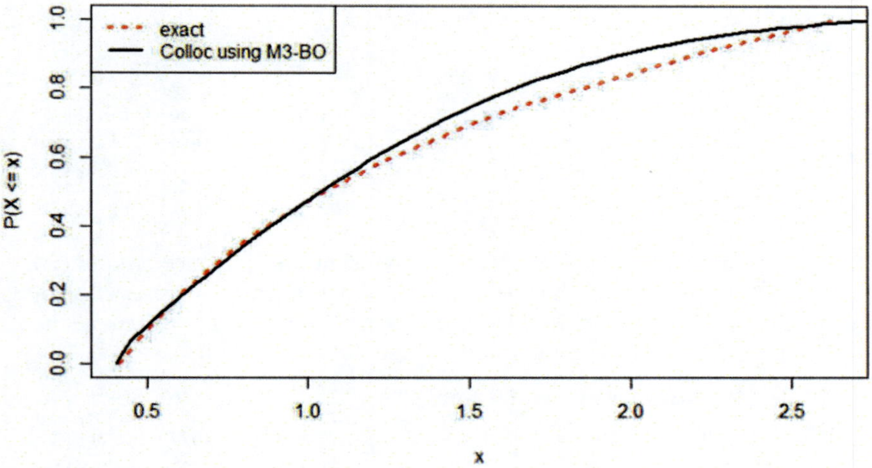

Fig. 3.74 CDF corresponding to $t = 0.5$

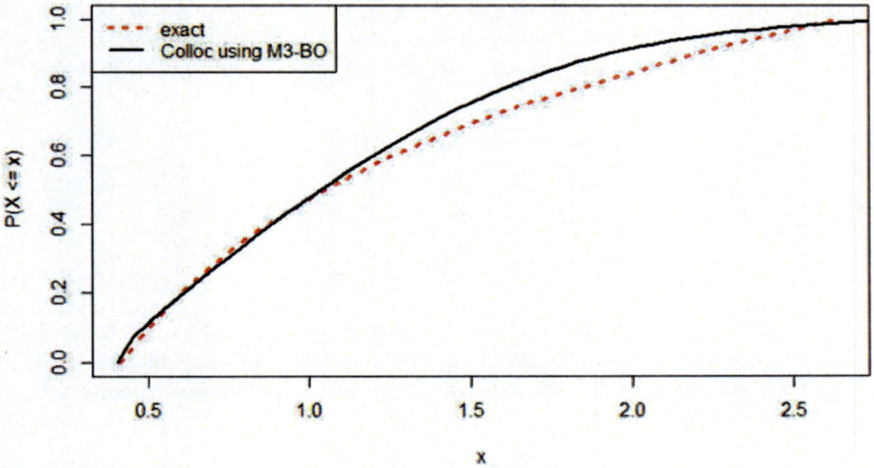

Fig. 3.75 CDF corresponding to $t = 0.25$

Exercises

1. The following data were observed:

U	1.000	0.819	0.670	0.549	0.449	0.368
X	−1.033	−0.578	−0.296	−0.169	−0.097	−0.049
U	0.301	0.247	0.202	0.165	0.135	0.111
X	−0.029	−0.014	−0.009	−0.004	−0.002	−0.001

Assume that the distribution of U is unknown. Introduce an artificial uniformly distributed variable A and use it to determine a representation of U as function of A by M3. Use the representation found to determine the CDF and the PDF of X.

2. The following data were observed:

U	0.0580	0.0609	0.2197	0.3883	0.4677	0.5439
X	0.0000	0.0021	0.0060	0.0078	0.0123	0.0126
U	0.7550	0.8709	0.8761	0.9045	0.9133	0.9778
X	0.0050	0.0022	0.0023	0.0019	0.0020	0.0001

Assume that the distribution of U is unknown. Introduce an artificial uniformly distributed variable A and use it to determine a representation of U as function of A by M3. Use the representation found to determine the CDF and the PDF of X.

3. The following data were observed:

U	0.2138	0.2483	0.2947	0.2951	0.3354	0.3425
X	8.7482	8.7470	7.0831	6.1580	6.8340	5.7317
U	0.3627	0.3654	0.4223	0.5067	0.5422	0.5549
X	5.1731	5.1117	4.3880	4.6680	4.9407	4.6589

Assume that the distribution of U is unknown. Introduce an artificial uniformly distributed variable A and use it to determine a representation of U as function of A by M3. Use the representation found to determine the CDF and the PDF of X.

4. The following data were observed:

U	0.0866	0.4522	0.7285	0.7902	1.1223	1.3086
X	1.1770	2.0239	2.8004	2.9942	4.1943	5.0098
U	1.3672	1.4038	1.5366	1.5799	1.6650	1.6894
X	5.2917	5.4745	6.1855	6.4346	6.9506	7.1057

Assume that the distribution of U is unknown. Introduce an artificial uniformly distributed variable A and use it to determine a representation of U as function of A by M3. Use the representation found to determine the CDF and the PDF of X.

3.5 Multidimensional Expansions

The methods presented extend straightly to the situation where X or U are multidimensional.

3.5.1 Case Where U Is Multidimensional

In the situation where $U = (U_1, \ldots, U_n)$ is a multidimensional vector, the family $\{\varphi_i : 0 \leq i \leq k\}$ is generated by functions $\varphi_{i,j} : \mathbb{R}^n \to \mathbb{R}$. For instance, we may consider a tensor product of basis $\{\varphi_{i,j} : 0 \leq i \leq k_j\}$:

$$\varphi_{\ell_n(i_1, \ldots, i_n)}(U) = \prod_{j=1}^{n} \varphi_{i_j, j}(U_j).$$

The map ℓ_n defines a renumbering of $(i_1, \ldots, i_n)$ which brings the vector to a single integer. For instance,

$$\ell_2(i_1, i_2) = i_1 + i_2 k_1.$$

and

$$\ell_n(i_1, \ldots, i_n) = \ell_{n-1}(i_1, \ldots, i_{n-1}) + i_n \ell_{n-1}(k_1, \ldots, k_{n-1}).$$

For instance,

$$\ell_3(i_1, i_2, i_3) = \ell_2(i_1, i_2) + i_3 \ell_2(k_1, k_2).$$

Then, the approximation reads as

$$X \approx PX = \boldsymbol{\Phi}(U)\boldsymbol{x} = \sum_{\ell=0}^{k} x_\ell \varphi_\ell(U), \; k = \ell_n(k_1, \ldots, k_n); \tag{3.42}$$

id est,

$$X \approx PX = \sum_{i_1=0}^{k_1} \sum_{i_2=0}^{k_2} \cdots \sum_{i_n=0}^{k_n} x_{\ell_n} \varphi_{1, i_1}(U_1) \varphi_{2, i_2}(U_2) \ldots \varphi_{n, i_n}(U_n).$$

An example of creation of a tensor product of two basis is the following:

```
library(R6)
#
tensorbasis2=R6Class("tensorbasis2",
                     public=list(
                       phi1 = NA,
                       phi2 = NA,
                       k1 = NA,
                       k2 = NA,
                       indbasis = NA,
                       initialize=function()
                       {
                       },
                       value= function(k, x){
                         aux = self$invlindex(k)
                         i = aux[[1]]
                         j = aux[[2]]
                         aux = self$phi1$value(i, x[[1]]) * self$phi2$value(j, x[[2]])
                         return(aux)
                       },
                       lindex = function(i,j){
                         val = j + i*(self$k2+1)
                         return(val)
                       },
                       invlindex = function(k){
                         j = k %% self$k2
                         i = (k-j) %/% self$k2
                         val = c(i,j)
                         return(val)
                       }
                     )
)
```

Notice that the class includes methods `lindex` and `invlindex` that correspond to ℓ_2 and ℓ_2^{-1}, respectively.

Example 3.1
Let us consider the situation where the data is the following one:

X	4.45	2.32	2.36	3.28	3.32	2.54	4.45	2.02
U_1	0.95	0.57	0.15	1.00	0.18	0.38	0.95	0.32
U_2	−0.63	0.59	−0.18	0.56	−0.75	−0.07	−0.63	0.44

X	3.28	1.74	1.93	2.79	2.74	4.98	2.19	3.09
U_1	0.65	0.27	0.34	0.30	0.37	0.92	0.24	0.88
U_2	−0.31	0.84	0.64	−0.36	−0.26	−0.90	0.09	0.40

X	1.65	1.86	3.33	3.81	3.73	2.08	2.73	4.03
U_1	0.19	0.24	0.81	0.93	0.29	0.38	0.59	0.78
U_2	0.82	0.52	−0.07	−0.25	−0.87	0.49	0.08	−0.61

Let us consider an approximation by a polynomial family, with $k_1 = k_2 = 3$. Then, taking $a_i = \min(U_i)$, $b_i = \max(U_i)$.

(continued)

Example 3.1 (continued)

$$\varphi_{1,i}(u_1) = \left(\frac{u_1 - a_1}{b_1 - a_1}\right)^i, \quad \varphi_{2,j}(u_2) = \left(\frac{u_2 - a_2}{b_2 - a_2}\right)^j.$$

and

$$\varphi_{j+4i}(u_1, u_2) = \varphi_{1,i}(u_1)\, \varphi_{2,j}(u_2).$$

Here, the numbering is

$$\ell_2(i_1, i_2) = i_2 + i_1(k_2 + 1).$$

The collocation approach furnishes the coefficients

ℓ_2	0	1	2	3	4	5	6	7
i	0	0	0	0	1	1	1	1
j	0	1	2	3	0	1	2	3
X_{ℓ_2}	3.61	−3.96	2.56	−0.61	1.10	−1.30	3.11	−1.96

ℓ_2	8	9	10	11	12	13	14	15
i	2	2	2	2	3	3	3	3
j	0	1	2	3	0	1	2	3
X_{ℓ_2}	0.48	0.56	−1.84	0.98	−0.01	0.25	0.23	−0.17

The RMS error of the approximation is 2.9E-3. We used the code below (the environment contains `x,u1`,u2):

```
source("polynom.R")
source("expansion2D.R")
source("tensorbasis2.R")
tb2 = tensorbasis2$new()
k1 = 3
k2 = 3
a1 = min(u1)
b1 = max(u1)
a2 = min(u2)
b2 = max(u2)
k = (k1+1)*(k2+1)
phi1 = polynom$new(a1,b1,1e-3)
phi2 = polynom$new(a2,b2,1e-3)
tb2$k1 = k1
tb2$k2 = k2
tb2$phi1 = phi1
tb2$phi2 = phi2
he2d2 = expansion2D$new()
he2d2$phi = tb2
coef2 = he2d$colloc(k,0,u,x)
px2 = he2d2$proj(coef2,u)
er = sqrt(sum((x-px)^2)/length(x))
```

3.5.2 Case Where X Is Multidimensional

Let us assume that $\boldsymbol{X} = (X_1, \ldots, X_n)^t$. Then, the approximation reads as

$$\boldsymbol{X} \approx \boldsymbol{PX} = \sum_{j=1}^{k} \boldsymbol{x}_j \varphi_j(U), \boldsymbol{x}_j = (x_{1,j}, \ \ldots, x_{n,j})^t. \tag{3.43}$$

Thus, *each component* is approximated by an expansion analogous to (3.3) or (3.5):

$$X_i \approx PX_i = \sum_{j=0}^{k} x_{i,j} \varphi_j(U), \ i = 1, \ldots, n$$

Consequently, we can apply the methods presented to each component of the variable.

The coefficients $\boldsymbol{x}$ form a matrix $=(x_{i,\ j},\ 1 \leq i \leq n, 0 \leq j \leq k)$ – notice that each line of $\boldsymbol{x}$ contains the coefficients corresponding to the expansion of a component: column j of $\boldsymbol{x}$ contains the coefficients of the expansion of X_i:

$$\boldsymbol{x} = \begin{pmatrix} x_{1,0} & \cdots & x_{1,k} \\ \vdots & \cdots & \vdots \\ x_{n,0} & \cdots & x_{n,k} \end{pmatrix}. \tag{3.44}$$

We may also apply the methods to the whole matrix $\boldsymbol{x}$. For instance, let us consider the approach by collocation: assume that the data are a sample of values

$$\mathcal{S} = \{(\boldsymbol{U}_1, \ \boldsymbol{X}_1), \ \ldots, (\boldsymbol{U}_{ns}, \ \boldsymbol{X}_{ns})\} \tag{3.45}$$

Each $\boldsymbol{X}_i$ is a vector: $\boldsymbol{X}_i = (X_{i,\ 1}, \ldots, X_{i,\ n})^t$. Then, we may range the data in a matrix $=(X_{i,\ j},\ 1 \leq i \leq ns,\ 1 \leq j \leq n\ )$ – notice that each line of $\boldsymbol{\mathcal{X}}$ is an element of the sample: line i of $\boldsymbol{\mathcal{X}}$ contains the values of $\boldsymbol{X}_i$:

$$\boldsymbol{\mathcal{X}} = \begin{pmatrix} X_{1,1} & \cdots & X_{1,n} \\ \vdots & \cdots & \vdots \\ X_{ns,1} & \cdots & X_{ns,n} \end{pmatrix}. \tag{3.46}$$

Let $\boldsymbol{A}$ be the collocation matrix, defined in Eq. (3.10): the coefficients $\boldsymbol{x}$ verify

$$\boldsymbol{Ax}^t = \boldsymbol{\mathcal{X}}. \tag{3.47}$$

The situation is analogous for the variational approach: we consider the matrix $= (\mathcal{B}_{i,j},\ 1 \leq i \leq k,\ 1 \leq j \leq n)$ such that $\mathcal{B}_{ij} = E(X_j \varphi_i(U))$, i.e.:

$$\mathcal{B} = \begin{pmatrix} E(X_1\varphi_1(U)) & \cdots & E(X_n\varphi_1(U)) \\ \vdots & \cdots & \vdots \\ E(X_1\varphi_k(U)) & \cdots & E(X_n\varphi_k(U)) \end{pmatrix}. \tag{3.48}$$

Now, let $\boldsymbol{A}$ be the variational matrix, defined in Eq. (3.24): the coefficients $\boldsymbol{x}$ verify Eq. (3.47) yet.

The extension of Moment Matching to random vectors requests some precaution: indeed, we must approach also the covariances, so that, we must consider the approximation of generalized moments having the form $E\left(\prod_{i=1}^{n} X_i^{m_i}\right)$ – the vector of moments $\boldsymbol{M(X)}$ is replaced by a tensor of moments $\boldsymbol{\mathcal{M}}(\boldsymbol{X}) = \left\{\mathcal{M}_{p_1 \dots p_n} = E\left(\prod_{i=1}^{n} X_i^{p_i}\right), 0 \le p_i \le m_i\right\}$ and we look for the minimum of

$$J(\boldsymbol{x}) = d(\boldsymbol{\mu}(\boldsymbol{x}), \boldsymbol{\mathcal{M}}(\boldsymbol{X})), \tag{3.49}$$

where $\boldsymbol{\mu}(\boldsymbol{x})$ is the tensor of moments of the approximation: $\boldsymbol{\mu}(\boldsymbol{x}) = \left\{\mu_{p_1 \dots p_n} = E\left(\prod_{i=1}^{n} (PX_i)^{p_i}\right), 0 \le p_i \le m_i\right\}$.

Example 3.2

Let us consider the situation where the data is the following one:

U	−0.95	−0.92	−0.92	−0.86	−0.82	−0.81	−0.77	−0.72
X_1	−0.37	−0.37	−0.37	−0.36	−0.36	−0.36	−0.36	−0.35
X_2	−0.16	−0.25	−0.25	−0.42	−0.55	−0.57	−0.65	−0.77

U	−0.64	−1.59	−0.30	−0.26	−0.24	−0.22	0.03	0.05
X_1	−0.34	−0.32	−0.22	−0.20	−0.19	−0.17	0.03	0.05
X_2	−0.90	0.96	−0.82	−0.72	−0.69	−0.63	0.10	0.15

Let us consider an approximation by a polynomial family, with $k = 7$. The collocation approach furnishes the coefficients

$x_{1,j}$	−0.32	−4.56	37.90	−133.58	248.33	−255.26	138.77	−31.22
$x_{2,j}$	0.96	11.72	−103.89	315.23	−547.07	556.80	−293.65	60.05

The RMS error of the approximation is 5E-3.

Exercises

1. The following values of a variable X were observed. Assume that X is a function of U_1 and U_2.

			U_1			
U_2	0,05	0,075	0,1	0,15	0,175	0,2
0,1	8,11	8,79	9,14	9,37	9,53	9,66
0,15	6,99	7,92	8,36	8,60	8,75	8,84
0,2	5,96	7,18	7,74	8,04	8,21	8,31
0,25	4,96	6,48	7,19	7,56	7,77	7,90
0,3	3,97	5,82	6,67	7,12	7,39	7,54

(a) Assume that the distributions of U_1 and U_2 are uniform. Then, determine a representation of X as a function of U and use it to determine the CDF and the PDF of X.

(b) Assume that the distribution of U_1 is triangular $T(0.05,0.1,0.2)$ and the distribution of U_2 is $T(0.1, 0.2, 0.3)$. How this change impacts the representation determine in (a)? Determine the CDF and the PDF of X in this case and compare the results to those found in (a).

2. The following values of a variable X were observed. Assume that X is a function of U_1 and U_2.

			U_1			
U_2	0,05	0,075	0,1	0,15	0,175	0,2
0,1	8,11	8,79	9,14	9,37	9,53	9,66
0,15	6,99	7,92	8,36	8,60	8,75	8,84
0,2	5,96	7,18	7,74	8,04	8,21	8,31
0,25	4,96	6,48	7,19	7,56	7,77	7,90
0,3	3,97	5,82	6,67	7,12	7,39	7,54

(a) Assume that the distributions of U_1 and U_2 are uniform. Then, determine a representation of X as a function of U and use it to determine the CDF and the PDF of X.

(b) Assume that the distribution of U_1 is triangular $T(0.05,0.1,0.2)$ and the distribution of U_2 is $T(0.1, 0.2, 0.3)$. How this change impacts the representation determine in (a)? Determine the CDF and the PDF of X in this case and compare the results to those found in (a).

3. Consider the function $Y = 10\, x_1^\alpha x_2^\beta - 2\, x_1 - 5x_2$

(a) Assume that the distributions of α and β are uniformly distributed on $(0.1, 0.5)$. Determine a representation of Y using a polynomial basis.

(b) Assume $\alpha \sim T(0.1,0.2,0.5)$ and $\beta \sim T(0.1, 0.4, 0.5)$. Determine a representation of Y using a polynomial basis.

3.6 Random Functions

The representation of random functions is an extension of the methods previously presented. To fix the ideas, let us consider a function $X = X(t, U) : t \in (0, T)$, depending upon a random variable U. In such a situation, we look for an expansion having time-dependent coefficients:

$$X(t, U) \approx PX(t, U) = \sum_{j=0}^{k} x_j(t)\varphi_j(U). \tag{3.50}$$

Equation (3.50) for random functions is the equivalent of Eq. (3.3) for random variables. The unknowns to be determined are the functions $x_i(t)$, $t \in (0, T)$.

For instance, assume that a sample from X is available: $\{X_{i,\ j} = X(t_i, U_j)$: $0 \leq i \leq nt, 1 \leq j \leq ns\}$, with $0 = t_0 < t_1 < \ldots < t_{nt} = T$. In such a situation, we may consider $\boldsymbol{X} = (X_0, \ldots, X_{nt})^t$, where $X_i = X(t_i)$. $\boldsymbol{X}$ is a random vector of dimension $nt + 1$, so that we can apply the approach previously introduced for multidimensional random vectors: for each t_i

$$X(t_i, U) \approx PX(t_i, U) = \sum_{j=0}^{k} x_{i,j}\varphi_j(U).$$

The coefficients $x_{i,\ j}$ can be gathered in a matrix (3.44) and be determined by solving (3.47). Equivalently, we can consider a sample

$$S_i = \{(U_1,\ X_{i,1}),\ \ldots,\ (U_{ns},\ X_{i,\ ns})\}$$

and apply to S_i the methods presented in the preceding. We have

$$x_j(t_i) \approx x_{i,j},\ 0 \leq i \leq nt,\ 1 \leq j \leq ns. \tag{3.51}$$

If necessary, the values of $x_j(t)$ for other values may be determined by interpolation.

Example 3.3

Let $X_t = \sin(Ut)$, where U is a random variable. We can determine $\boldsymbol{x}(t) = (x_0(t), \ldots, x_k(t))^t$ by solving the linear system

$$\boldsymbol{Ax}(t) = \boldsymbol{B}(t),$$

where, for $1 \leq i, j \leq k + 1$

(continued)

Example 3.3 (continued)

$$A_{ij} = E\left(\varphi_{i-1}(U)\varphi_{j-1}(U)\right),\ B_i(t) = E(X(t,\ U)\varphi_{i-1}(U)).$$

In practice, the means can be evaluated by numerical integration or by using a sample from U. It is also useful to discretize le time and evaluate the coefficients at discrete times, as previously indicated. For instance, if $t \in (0, T_{max})$, we can conside the discrete times $t_i = i\Delta t$, $i = 0, \ldots, nt$; $\Delta t = T_{max}/nt$.

For instance, assume that the PDF of U is evaluated by a function `fde(u)`, the Hilbert basis is `phi`, the integrals are evaluated on `(UMIN, UMAX)`. Then, A can be evaluated as follows:

```
A = matrix(nrow = (k+1), ncol = (k+1))
for (i in 0:k){
  f1 = function(u){
    aux3 = phi$value(i,u)*phi$value(i,u)*fde(u)
    return(aux3)
  }
  ff1 = Vectorize(f1)
  aux = integrate(ff1,UMIN,UMAX)
  A[[i+1,i+1]] = aux$value
  if (i < k){
    for (j in (i+1):k){
      f1 = function(u){
        aux4 = phi$value(i,u)*phi$value(j,u)*fde(u)
        return(aux4)
      }
      ff1 = Vectorize(f1)
      auxx = integrate(ff1,UMIN,UMAX)
      A[[i+1,j+1]] = auxx$value
      A[[j+1,i+1]] = auxx$value
    }
  }
}
```

Assume that $X(t, U)$ is evaluated by a function `fonc(t,u)` and the coefficients are determined at times given in the vector `tc`. Then, the coefficients can be determined as follows:

```
cc = matrix(nrow = length(tc),ncol = (k+1))
for (i in 1:length(tc)){
  tt = tc[[i]]
  ft = function(u){
    aux =fonc(tt,u)
    return(aux)}
  B = matrix(nrow = (k+1), ncol = 1)
  for (j in 0:k){
    f1 = function(u){
      aux5 = phi$value(j,u)*ft(u)*fde(u)
      return(aux5)
    }
    ff1 = Vectorize(f1)
    auxxx = integrate(ff1,UMIN,UMAX)
    B[[j+1]] = auxxx$value
  }
  cc[i,]=t(solve(A,B))
}
```

(continued)

Example 3.3 (continued)

Let us consider U is uniformly distributed on $(-\pi, \pi)$. In this case, we have

$$E(X_t) = 0, \; V(X_t) = \frac{1}{2} - \frac{\sin(2\pi t)}{4\pi t}$$

For $T_{\max} = 1$, $nt = 25$, $k = 6$, we obtain the results in Fig. 3.76. The corresponding RMS error is 7.6E-6.

Let us consider U is uniformly distributed on $(0, 2\pi)$. In this case, we have

$$E(X_t) = \frac{\sin^2(\pi t)}{\pi t}, \; V(X_t) = \frac{-8\sin^4(\pi t) + \pi t(4\pi t - \sin(4\pi t))}{8\pi^2 t^2}$$

For $T_{\max} = 1$, $nt = 25$, $k = 6$, we obtain the results in Figs. 3.77 and 3.78. The corresponding RMS errors are 6.6E-6 and 7.7E-6, respectively.

Assume that $U \sim N(0, \pi)$ and that the expression of X_t is unknown: we have observations corresponding to $ns = 20$, $nt = 25$, $T_{\max} = 1$. We apply the procedure of collocation to determine a representation using a trigonometrical basis with $k = 24$ – recall that collocation coincides with the variational approach in such a situation, so that it is equivalent to approximate the means by the empirical means on the samples S_i:

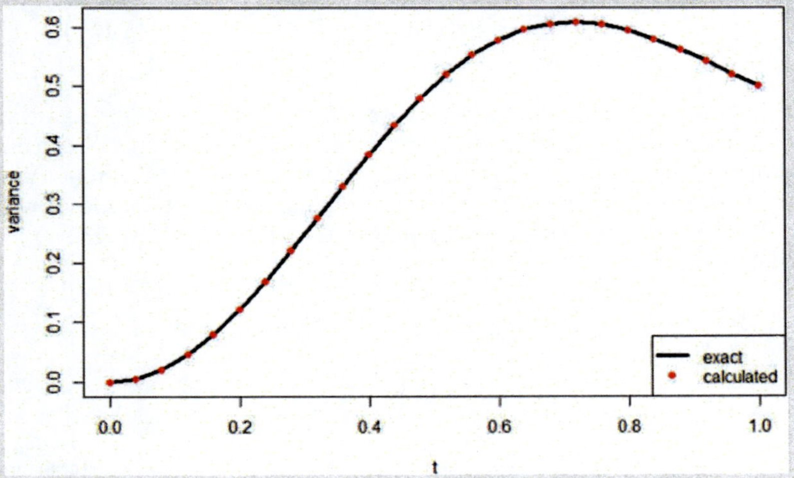

Fig. 3.76 Variance as function of time ($U \sim U(-\pi, \pi)$). The approximation furnishes values close to exact ones

(continued)

Example 3.3 (continued)

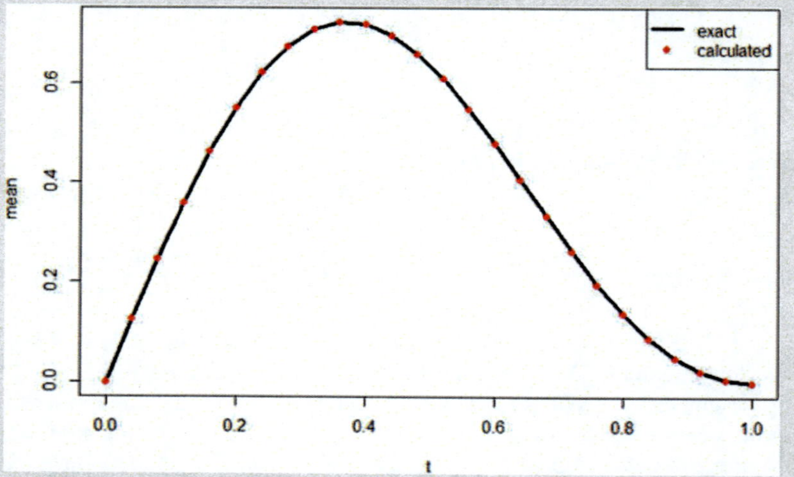

Fig. 3.77 Mean as function of time ($U \sim U(0, 2\pi)$). The approximation furnishes values close to exact ones

$$A_{ij} \approx \frac{1}{ns} \sum_{k=1}^{ns} \varphi_{i-1}(U_k)\varphi_{j-1}(U_k),\ B_i(t) \approx \frac{1}{ns} \sum_{k=1}^{ns} \varphi_{i-1}(U_k) X_t(U_k).$$

With this approach, the RMS error on the approximation of the data was 6E-5. We generated a sample of 10000 variates from $N(0, \pi)$ and evaluated the approximation and the exact value on this new sample. The RMS error on this new data was 7E-5. The RMS error in the approximation of the variance was 1.6E-5. A comparison of the variances is shown in Fig. 3.79. The exact solution is (Fig. 3.79)

$$E(X_t) = 0,\ V(X_t) = \frac{1}{2}\left(1 - e^{-19.7392088021787 2t^2}\right).$$

(continued)

Example 3.3 (continued)

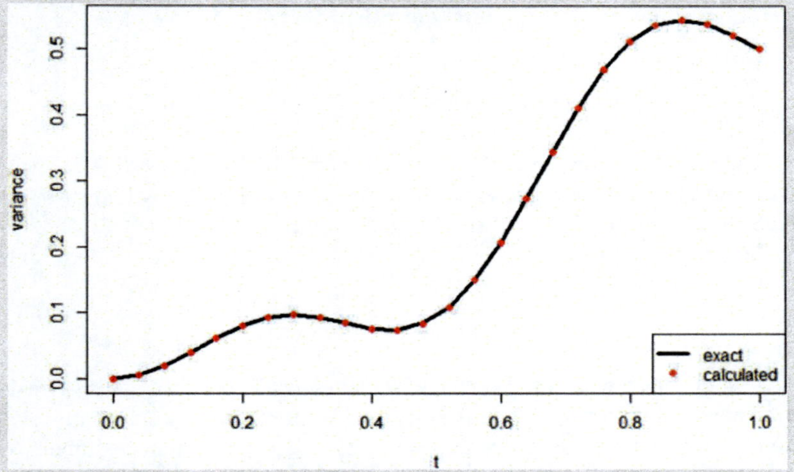

Fig. 3.78 Variance as function of time ($U \sim U(0, 2\pi)$). The approximation furnishes values close to exact ones

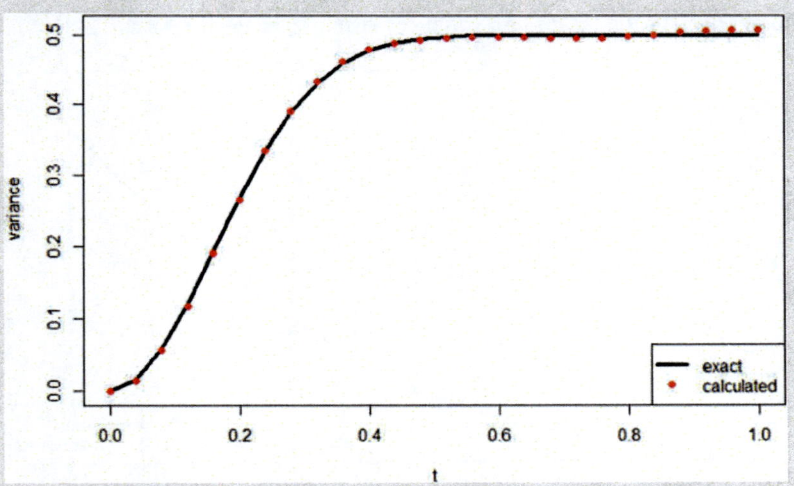

Fig. 3.79 Variance as function of time ($U \sim N(0, \pi)$). The approximation furnishes values close to exact ones

Example 3.4

Let $X_t = e^{Ut}$, where U is uniformly distributed on $(-1, 1)$. We have

$$E(X_t) = \frac{e^t - e^{-t}}{2t}, V(X_t) = \frac{e^{2t} - e^{-2t}}{4t} - (E(X_t))^2$$

The procedure produces the results shown in Figs. 3.80 and 3.81. The mean was calculated at times defined by $nt = 25$, $T_{\max} = 1$. The RMS error in the mean is 1.1E-6 and the RMS error in the variance is 3.5E-6.

Let $U \sim N(0, 1)$. Then

$$E(X_t) = e^{\frac{t^2}{2}}, V(X_t) = e^{2t^2} - e^{t^2}.$$

Assume that we do not have the information that $X_t = e^{tU}$, id est, that we do not know the expression of X_t. In place of this information, we have observations corresponding to $ns = 20$, $nt = 25$, $T_{\max} = 1$. We determine by collocation a representation using a polynomial basis with $k = 6$. The RMS error on the approximation of the data is 1.4E-3. We generated a sample of 10,000 variates from $N(0, 1)$ and evaluated the approximation and the exact value on this new sample. The RMS error on this new data was 8.7E-3. A comparison of the means and variances is shown in Figs. 3.82 and 3.83.

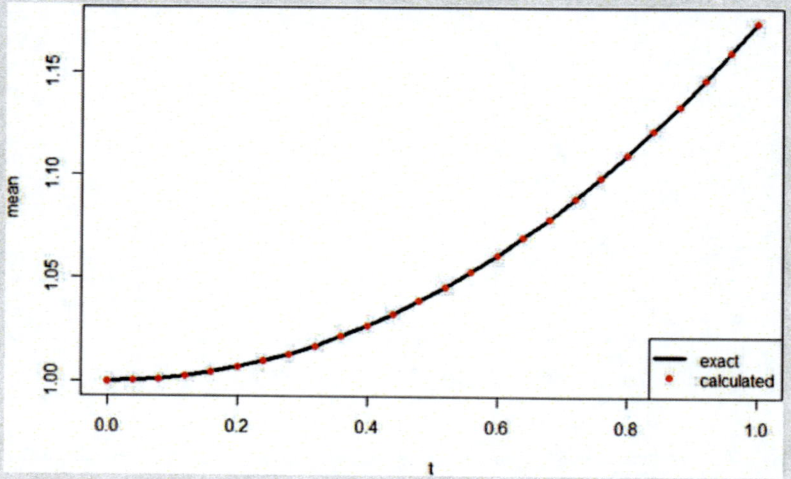

Fig. 3.80 Mean as function of time ($U \sim U(-1, 1)$). The approximation is remarkably close to the exact values

(continued)

Example 3.4 (continued)

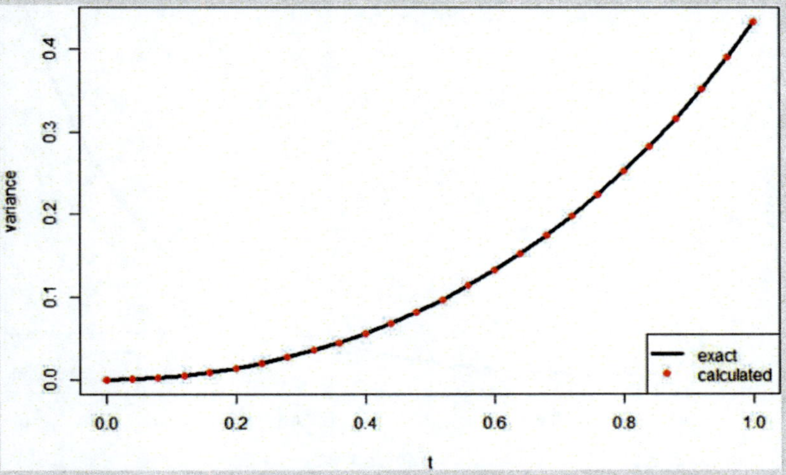

Fig. 3.81 Variance as function of time($U \sim U(-1, 1)$). The approximation is awfully close to the exact values

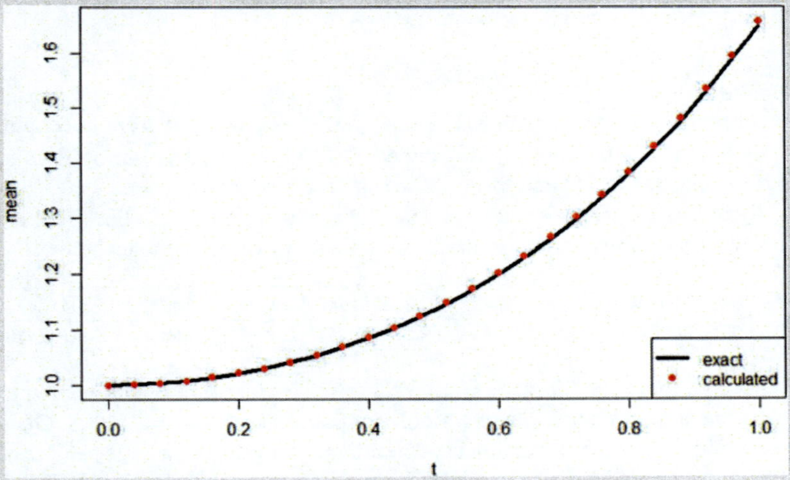

Fig. 3.82 Mean as function of time ($U \sim N(0, 1)$). The approximation is close to the exact values

(continued)

Example 3.4 (continued)

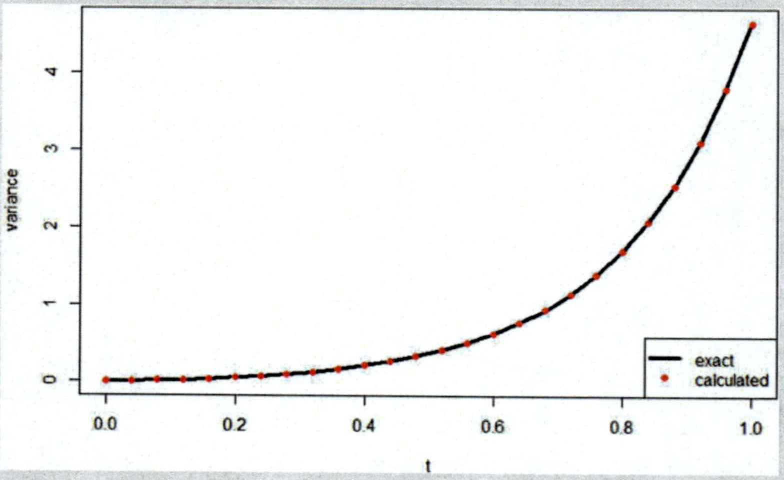

Fig. 3.83 Variance as function of time ($U \sim N(0, 1)$). The approximation is close to the exact values

Example 3.5

Let us consider a capital path $X_t = \left(X_0 - \frac{s}{\delta}\right)e^{-\delta t} + \frac{s}{\delta}$, where s, δ are uncertain. Assume that s is triangularly distributed $T(0.15,0.2,0.3)$, δ is uniformly distributed on (0.05,0.15) and the variables are independent.

Again, assume that we do not know the expression of X_t, but we have observations corresponding to the 25 combinations of the samples below:

s	0.171553736	0.179011042	0.223251145	0.249395849	0.249458937
δ	0.052162242	0.079594785	0.122753751	0.135923868	0.146172518

Assume that $nt = 25$, $T_{\max} = 1$. We apply the procedure of collocation to determine a representation using a polynomial basis with $k_1 = k_2 = 3$. Once the coefficients of the expansion are determined, we generate a large sample of 1E6 variates from U, which is used to generate samples of X_{t_i} (each sample has 1E6 variates). Figures 3.84 and 3.85 show the results for the mean and the variance. The corresponding RMS errors are 1.8E-7 and 1.6E-6, respectively.

(continued)

Example 3.5 (continued)

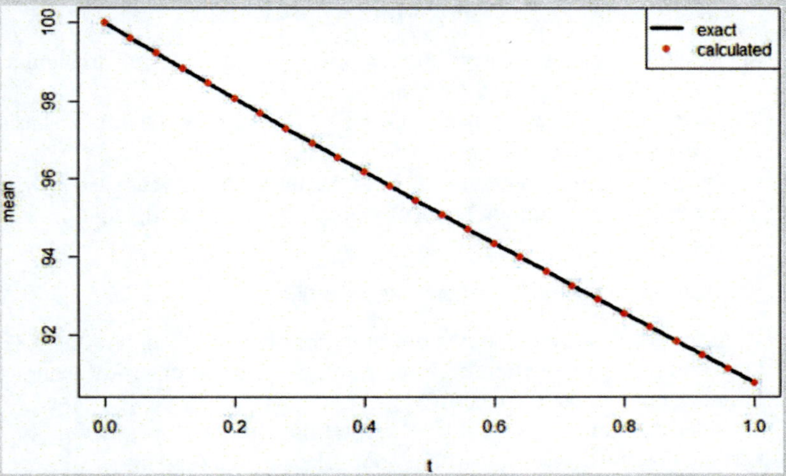

Fig. 3.84 Mean as function of time ($U_1 \sim T(0.15,0.3,0.2), U_2 \sim U(0.05,0.15)$). The approximation is close to the exact values

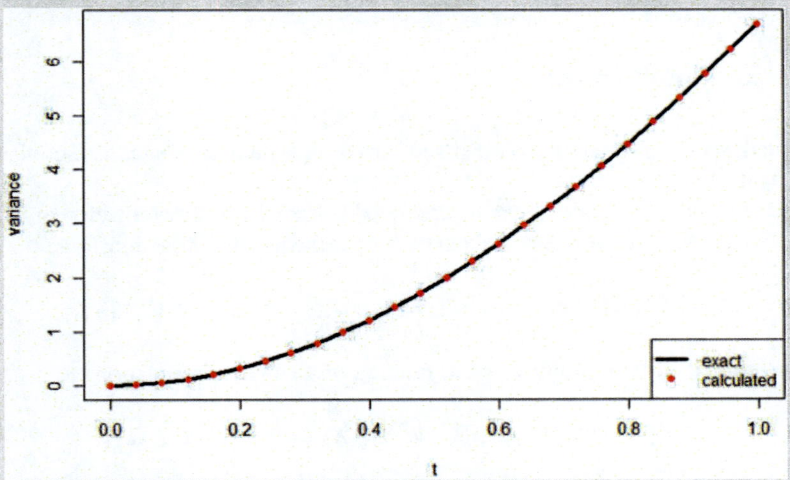

Fig. 3.85 Variance as function of time ($U_1 \sim T(0.15,0.3,0.2), U_2 \sim U(0.05,0.15)$). The approximation is close to the exact values

Exercises

1. Consider $X_t = \left(X_0 - \frac{s}{\delta}\right)e^{-\delta t} + \frac{s}{\delta}$, where s, δ are uncertain.

   (a) Assume that the distributions of s and δ are triangular: $s \sim T(0.1, 0.25, 0.5)$ and $\delta \sim T(0.05, 0.15, 0.25)$.
   (b) Determine a representation of X_t in the polynomial basis $1, u_1, u_2, u_1^2, u_1u_2, u_2^2, \ldots$
   (c) Determine a representation of X_t in the tensor basis generated by the trigonometric functions 1, sin(u), cos (u), sin (u_1), cos (u), sin (2u), cos (2u), ...

2. Consider $X_t = (1-u)^2 \frac{t^2}{4}$, where u is uncertain.

   Assume that u is a random variable, uniformly distributed on (0.2, 0.8). Determine a representation of X_t in the basis of trigonometric functions 1, sin(u), cos (u), sin (2u), cos (2u), . . .
   Consider the family $X_t = u$, if $|t| < u$; $X_t = (|t| - u)^2 + u$, otherwise. Assume that the distribution of u is triangular: $u \sim T(0,0.5,1)$.

   (a) Determine a representation of X_t in the polynomial basis.
   (b) Determine a representation of X_t in the trigonometric basis 1, sin (u), cos (u), sin (2u), cos (2u), . . .

3.7 Random Curves

The analysis of uncertainties on demand curves reveals unsuspected difficulties at first sight.

Let us recall that a curve C can be described either by a parameterization or by an implicit equation. A parameterized curve is described by two functions:

$$C = \left\{X = (X_1,\ X_2) \in \mathbb{R}^2 : X_1 = X_1(t), X_2 = X_2(t), t \in (0,\ T)\right\}.$$

An implicit curve is defined by the solution of an algebraic equation:

$$C = \left\{X = (X_1,\ X_2) \in \mathbb{R}^2 : \psi(X_1,\ X_2) = 0, \text{ with } X \in \Omega\right\}.$$

We must distinguish the equations defining a curve and the set of points forming the curve. Indeed, the set of points is unique, but its representation by a parameterization or an equation is not unique.

As an example, let us consider a curve defined by the equation

$$\psi_1(X, U) = \sqrt{X_1} + \sqrt{X_2} - U = 0, X_1 \geq 0, X_2 \geq 0.$$

We can rewrite this equation under the form

$$\psi_2(X, U) = X_2 - \left(X_1 - 2U\sqrt{X_1} + U^2\right) = 0, X_1 \geq 0.$$

or, equivalently,

$$\psi_3(X, U) = X_1 - \left(X_2 - 2U\sqrt{X_2} + U^2\right) = 0, X_2 \geq 0.$$

We can also generate a parametric representation of the curve:

$$X_1(t) = t^2U^2, x_2(t) = (1-t)^2U^2, \mathrm{t} \in (0, 1)$$

The non-uniqueness of the representation introduces difficulties: according to the description used, the statistical properties may differ (see Fig. 3.86). For instance, let us suppose that U is uniformly distributed on (0, 1). As an example, let us try to evaluate the mean of the family of curves. The mean parameterization $\bar{X}_i(t) = E(X_i(t,\ U))$ furnishes

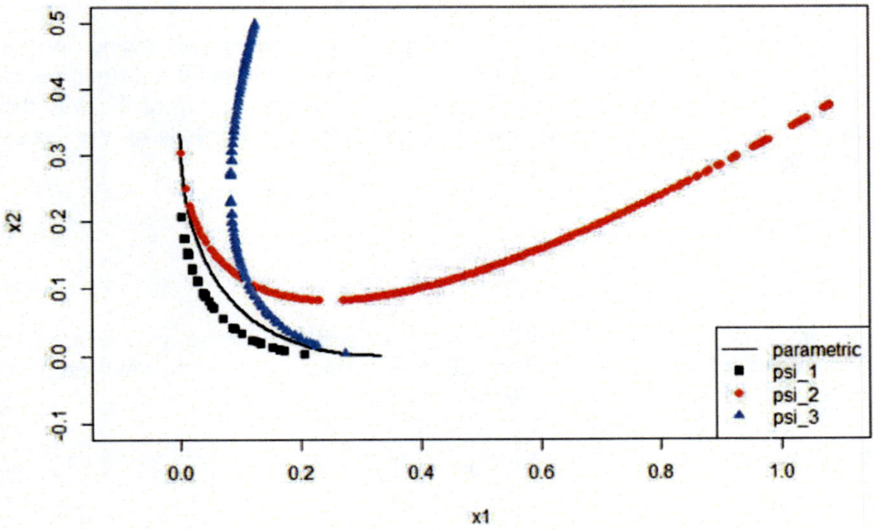

Fig. 3.86 Example of variability of the mean curve: according to the description used, the mean equation leads to a different curve

$$\bar{X_1}(t) = \frac{t^2}{3}, \bar{X_2}(t) = \frac{(1-t)^2}{3}, t \in (0, 1).$$

The mean equations $\bar{\psi_i}(\boldsymbol{X}) = E(\psi_i(\boldsymbol{X}, U)) = 0$ furnish:

$$\bar{\psi_1}(\boldsymbol{X}) = \sqrt{X_1} + \sqrt{X_2} - \tfrac{1}{2} = 0, X_1 \geq 0, X_2 \geq 0.$$

$$\bar{\psi_2}(\boldsymbol{X}) = X_2 - \left(X_1 - \sqrt{X_1} + \tfrac{1}{3}\right) = 0, X_1 \geq 0, X_2 \geq 0.$$

$$\bar{\psi_3}(\boldsymbol{X}) = X_1 - \left(X_2 - \sqrt{X_2} + \tfrac{1}{3}\right) = 0, X_1 \geq 0, X_2 \geq 0.$$

As shown in Fig. 3.86, each description of the curve furnishes a different mean. The same difficulties appear when looking for expansions. For instance, if we consider an expansion in a polynomial basis, using $k = 6$, we get

$$PX = \left(t^2U^2, (1-t)^2U^2\right) \quad \text{(parametric)}$$

$$PX = \left(t^2, t^2 - 2Ut + U^2\right) \quad (\text{implicit } \psi_1)$$

$$PX = \left(t, -U\sqrt{t} + \frac{1}{3}(1+3t)\right) \quad (\text{implicit } \psi_2)$$

$$PX = \left(-U\sqrt{t} + \frac{1}{3}(1+3t), t\right) \quad (\text{implicit } \psi_3)$$

To construct a procedure which is independent of the parameterization or the implicit equation, we may consider the set of points C and build a method based on them: indeed, the set of points C is the same independently of the way the equations defining the curve are written. For instance, we may look for the sets of points generated by the representation

$$\boldsymbol{PX}(t, U) = \sum_{j=0}^{k} \boldsymbol{x}_j(t)\varphi_j(U)$$

and look for the representation that generates the best set of points, in the sense that they minimize the distance to C. Formally, we consider a curve of the family as a set of points, whatever the choice of form to describe them: either

$$C(U) = \left\{\boldsymbol{X} = (X_1, X_2) \in \mathbb{R}^2 : X_1 = X_1(t, U), X_2 = X_2(t, U), t \in (0, T)\right\} \tag{3.52}$$

or

$$C(U) = \left\{\boldsymbol{X} = (X_1, X_2) \in \mathbb{R}^2 : \psi(X_1, X_2, U) = 0\right\}. \tag{3.53}$$

The data will be the sets of points $C(u_1), \ldots, C(u_{ns})$. Then, we consider the curves generated by the representations:

$$\mathcal{D}(\boldsymbol{x}, u) = \left\{ \boldsymbol{PX} = (PX_1, \ PX_2) \in \mathbb{R}^2 : PX_i = \sum_{j=0}^{k} x_{i,j}(s)\varphi_j(U) : s \in (s_{\min}, \ s_{\max}) \right\}.$$

For each element from the sample, we may measure the distance between $\mathcal{D}(\boldsymbol{x}, u_i)$ and $C(u_i)$:

$$d_i(\boldsymbol{x}) = dist(\ \mathcal{D}(\boldsymbol{x}, \ u_i), \ C(u_i)). \tag{3.54}$$

Here, *dist* is an adequate function to measure the distance between sets of points, such as, for instance, the distance from the points of $\mathcal{D}(\boldsymbol{x}, u_i)$ to $C(u)$:

$$dist(A, B) = d_{SH}(A, B) = \sup_{a \in A} \inf_{b \in B} \|a - b\| \tag{3.55}$$

or the Hausdorff distance:

$$dist(A, B) = d_H(A, B) = \max\{d_{SH}(A, \ B), d_{SH}(B, \ A)\} \tag{3.56}$$

The distances are collected in a vector

$$\boldsymbol{d}(\boldsymbol{x}) = (d_1(\boldsymbol{x}), \ \ldots, d_{ns}(x))^t. \tag{3.57}$$

Then, we look for the coefficients $\boldsymbol{x}$ that minimize a measure $g(\boldsymbol{d}(\boldsymbol{x}))$ of the magnitude of $\boldsymbol{d}(\boldsymbol{x})$, such as, for instance, its RMS norm:

$$g(\boldsymbol{d}(\boldsymbol{x})) = \sqrt{\frac{1}{ns} \sum_{i=1}^{ns} d_i^2(\boldsymbol{x})}. \tag{3.58}$$

Alternatively, we may consider other measures g, such as, for instance,

$$g(\boldsymbol{d}(\boldsymbol{x})) = \max\{d_i(\boldsymbol{x}) : 1 \leq i \leq ns\}.$$

Thus, we must consider two situations:

- Either the parameterization or implicit equation appears as clear and undoubtful. Then, we may use it to determine the representation: we consider the components as functions, and we apply the methods previously presented for functions (Sect. 3.3).
- Or it is dubious, and we may use the approach by the Hausdorff distance.

```r
library(R6)
#
curve=R6Class("curve",
              public=list(
                p = NA,
                np = NA,
                initialize=function(p)
                {
                  if (!missing(p)){
                    self$p = p
                    d = dim(p)
                    self$np = d[[2]]
                  }
                },
                distp2p = function(p1, p2){
                  aux = sqrt(sum((p1-p2)^2))
                  return(aux)
                },
                distp2c = function(p1,curv){
                  npc = curv$np
                  d = numeric(length=npc)
                  for (i in 1:npc){
                    d[[i]] = self$distp2p(p1, curv$p[ ,i])
                  }
                  aux = min(d)
                  return(aux)
                },
                distc2c = function(curv1,curv2){
                  np1 = curv1$np
                  d = numeric(length = np1)
                  for (i in 1:np1){
                    p1 = curv1$p[ ,i]
                    d[[i]] = self$distp2c(p1, curv2)
                  }
                  aux = max(d)
                  return(aux)
                },
                disthauss = function(curv1, curv2){
                  d12 = self$distc2c(curv1, curv2)
                  d21 = self$distc2c(curv2, curv1)
                  aux = max(c(d12,d21))
                  return(aux)
                },
                distfam = function(F1,F2){
                  np1 = length(F1)
                  np2 = length(F2)
                  d = numeric(length = np1)
                  for (i in 1:np1){
                    dd = numeric(length = np2)
                    for (j in 1: np2){
                      dd[[j]] = self$disthauss(F1[[i]],F2[[j]])
                    }
                   d[[i]]= min(dd)
                  }
                  return(d)
                },
                distpairedfam = function(F1,F2){
                  np1 = length(F1)
                  d = numeric(length = np1)
                  for (i in 1:np1){
                    d[[i]] = self$disthauss(F1[[i]],F2[[i]])
                  }
                  return(d)
                },
                distx = function(coefs,he,u,F){
                  ncurve = length(F)
                  auxx = c(0,0)
                  dd = dim(coefs)
                  k = dd[[2]]
                  nt = dd[[3]]
                  F1 = list()
                  pac = array(dim = c(2,length(u),nt))
                  for (nnt in 1:nt){
                    c1  = coefs[1, ,nnt]
                    c2  = coefs[2, ,nnt]
                    pac1 = he$proj(c1,u)
                    pac2 = he$proj(c2,u)
                    pac[1, ,nnt]=pac1
                    pac[2, ,nnt]=pac2
                  }
                  for (i in 1:length(u)){
                    curv = list(pac[ ,i, ],length(pac[1,i,]))
                    names(curv) = c("p","np")
                    F1[[i]] = curv
                  }
                  d = self$distpairedfam(F1, F)
                  aux = sqrt(sum(d^2))
                  return(aux)
                }
              )
)
```

Fig. 3.87 Example of code for the evaluation of the distances

The distances can be evaluated by using the class curve defined in Fig. 3.87. In this class, `distp2p` evaluates the Euclidean distance between two points; `distp2c` calculates the distance of a point to a curve (which is a list of points); `distc2c` evaluates d_{SH}; `disthauss` evaluates d_H; `distpairedfam` evaluates $d_i = d_H(F1_i, F2_i)$ for all the elements of two families $F1$, $F2$ of paired curves; `distfam` evaluates the $d_i = \min \{d_H(F1_i, F2_j) : j\}$ for two families of curves $F1$, $F2$; `distX` evaluates the norm of a vector $\boldsymbol{d}(\boldsymbol{x})$ of Haussdorff distances between the curves generated by the coefficients given and a family – `he` is an object of type `expansion1D`, which defines the Hilbert expansion to be used; `u` is vector of random numbers and `F` is the family.

As an example, let us consider two families of circles: `F1` is a family of 5 circles, centered at the origin and having radius 1,2,3,4,5. `F2` is a family of 5 identical circles of radius 5. Each element of each family is discretized with 51 points (angles equally spaced).

We have `dsh(F1[[i]],F2[[i]])` = `dh(F1[[i]],F2[[i]])=5-i`, `dsh(F1[[i]],F1[[j]])` = `dh(F1[[i]],F1[[j]])` = `Abs(i-j)`. The results are the following:

```
> dsh = matrix(nrow = nr, ncol = nr)
> for (i in 1: nr){
+   for (j in 1: nr){
+     dsh[[i,j]] = ccc$distc2c(F1[[i]],F1[[j]])
+   }
+ }
> print(dsh)
     [,1] [,2] [,3] [,4] [,5]
[1,]    0    1    2    3    4
[2,]    1    0    1    2    3
[3,]    2    1    0    1    2
[4,]    3    2    1    0    1
[5,]    4    3    2    1    0
> |
```

```
> dh = matrix(nrow = nr, ncol = nr)
> for (i in 1: nr){
+   for (j in 1: nr){
+     dh[[i,j]] = ccc$disthauss(F1[[i]],F1[[j]])
+   }
+ }
> print(dh)
     [,1] [,2] [,3] [,4] [,5]
[1,]    0    1    2    3    4
[2,]    1    0    1    2    3
[3,]    2    1    0    1    2
[4,]    3    2    1    0    1
[5,]    4    3    2    1    0
> df = ccc$distfam(F1,F2)
> print(df)
[1] 4 3 2 1 0
>
```

To determine a representation of the family `F1`, we may consider the parametric representation

$$x_1(\theta) = U\cos(\theta), x_2(t) = U\sin(\theta), \theta \in (0, 2\pi)$$

We consider $k = 3$ and determine the coefficients $\boldsymbol{x} = (x_{i,\ j}(\theta_\ell) : 1 \le i \le 2, 0 \le j \le k, 0 \le \ell \le 50)$. In the function approach presented in the previous section, the coefficients can be determined by collocation at each angle, using the available data. An example of prediction for $r = 3.5$ is shown in Fig. 3.88. The corresponding RMS error is 2.7E-14.

As an alternative, we can look for the coefficients that minimize the distance (3.58), which corresponds to `distX`. The prediction for $r = 3.5$ is shown in Fig. 3.89. The corresponding RMS error is 2.8E-14.

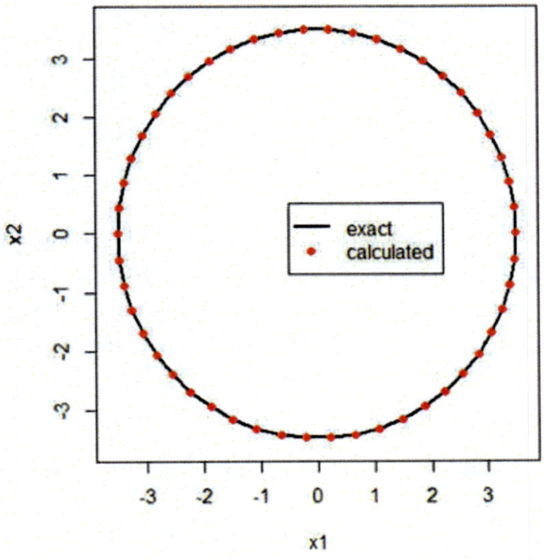

Fig. 3.88 Prediction of the curve for $r = 3.5$, by the function approach, with a polynomial basis, with $k = 3$

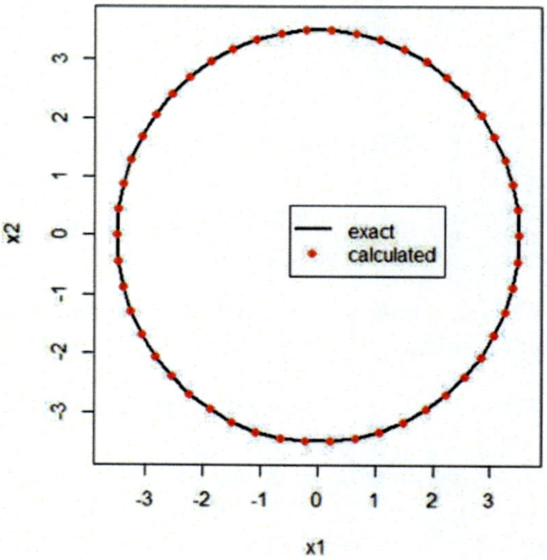

Fig. 3.89 Prediction for $r = 3.5$ by the Haussdorff approach, using a polynomial basis, with $k = 3$

Example 3.6

Let us consider the following multiobjective optimization problem (similar to Binh-Korn presented in Sect. 1.13.3. See also (Bassi et al., 2018)):

$$f_1(\boldsymbol{x}) = 4(x_1 - u_1)^2 + 4(x_2 - u_2)^2$$

$$f_2(\boldsymbol{x}) = \left(x_1 + \frac{u_1}{10} - 5\right)^2 + \left(x_2 - \frac{u_2}{10}\right)^2$$

$$(x_1 - 5)^2 + x_2^2 \leq 25,\ (x_1 - 8)^2 + (x_2 + 3)^2 \geq 7.7,$$

$$0 \leq x_1 \leq 5,\ 0 \leq x_2 \leq 3$$

Assume that $u_1 \sim N(2, 0.1)$ and $u_2 \sim T(0,0.5,1)$ and these variables are independent. In this case, the Pareto front associated to the multiobjective optimization problem is a random curve. To determine a representation, we can consider a sample from the variables. Since they are independent, we may consider independent samples. For instance, let us consider the samples in Table 3.8.

Table 3.8 Two small samples for the representation of a Pareto's front

u_1	1.793437	1.88848	1.999165	2.010893	2.01773
u_2	0.210664	0.407556	0.659623	0.787493	0.823312

(continued)

Example 3.6 (continued)

We have 5 values from each variable: independence allows us to generate a sample formed by 25 couples $\left(u_{1_i}, u_{2_j}\right)$: we generate a sample of 25 Pareto fronts using these values – the bi-objective optimization problem is solved by the method presented in Sect. 1.13.3, with discretized times $t_j = 0.01 * j$, $0 \le j \le 100$. At this stage, we can adopt three different strategies, all based on the fact that the approach presented in Sect. 1.13.3 involves a unique class of parameterization for all the curves: each front is represented by $F_i = \{F_{i,j} = (f_{1,\, i,\, j}, f_{2,\, i,\, j}), 0 \le j \le 100\}$, $f_{k,\, i,\, j} = f_k(x_i(t_j))$, $k = 1, 2$, with

$$x_i(t) = x^*_{1,i} + \left(x^*_{2,i} - x^*_{1,i}\right)t + \sum_{j=1}^{k} c_{i,j}\varphi^0_i(t),\ t \in (0, 1).$$

These parameterizations suggest three approaches:

- *Method 1:* We can collect the points $\{F_{1,\, j}, \ldots, F_{25,\, j}\}$ in a sample and determine a representation for these points. Repeating the procedure for all the 101 points, we obtain a representation of the curves.
- *Method 2:* We can collect the coefficients $\{c_{1,\, j}, \ldots, c_{25,\, j}\}$ in a sample and find a representation for $c_j(U)$. Repeating the procedure for all the indexes, $1 \le i \le k$, we determine a representation of the Pareto fronts.
- *Method 3:* We can collect the points $\{x_1(t_j), \ldots, x_{25}(t_j)\}$ in a sample and find a representation for $x(t_j, U)$. Repeating the procedure for all the 101 values of t_j, we determine a representation of the Pareto fronts.

For each one of these approaches, we determined the coefficients of an expansion with polynomial families and $k_1 = k_2 = 3$. The representations can be used to generate larger samples from the Pareto front. As an example, Figs. 3.90, 3.91, and 3.92 compare the approximation generated by the three methods above with the solution of the multiobjective problem, for the values $u_1 = 1.5$, $u_2 = 0.1$.

Now, assume that $u_1 \sim T(1.5,2,2.5)$ and $u_2 \sim T(0,0.5,1)$:consider the samples in Table 3.9. Figure 3.93 compares the fronts generated by the expansion and by the solution of the multiobjective problem, for the values $u_1 = 1.5$; $u_2 = 0.1$.

(continued)

Example 3.6 (continued)

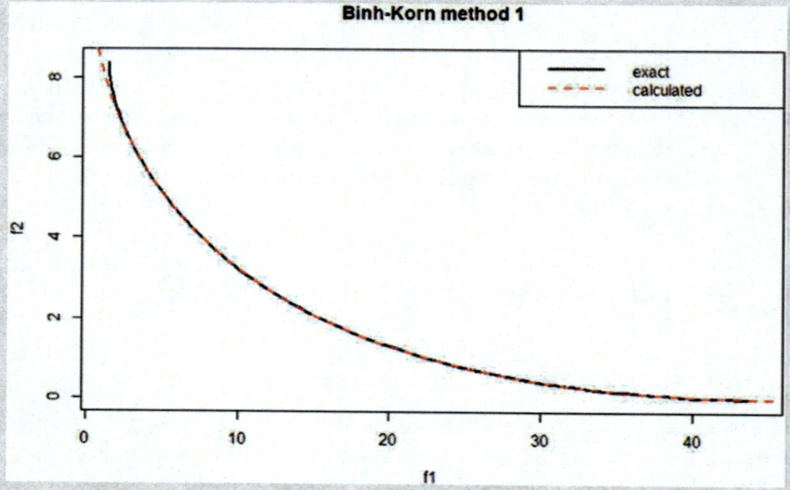

Fig. 3.90 Comparison between the Pareto front generated by the expansion and the solution of the multiobjective optimization problem (method 1)

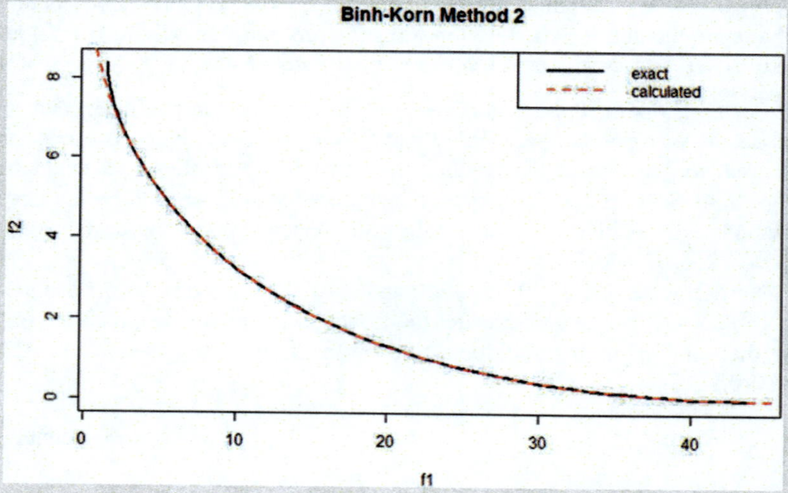

Fig. 3.91 Comparison between the Pareto front generated by the expansion and the solution of the multiobjective optimization problem (method 2)

(continued)

Example 3.6 (continued)

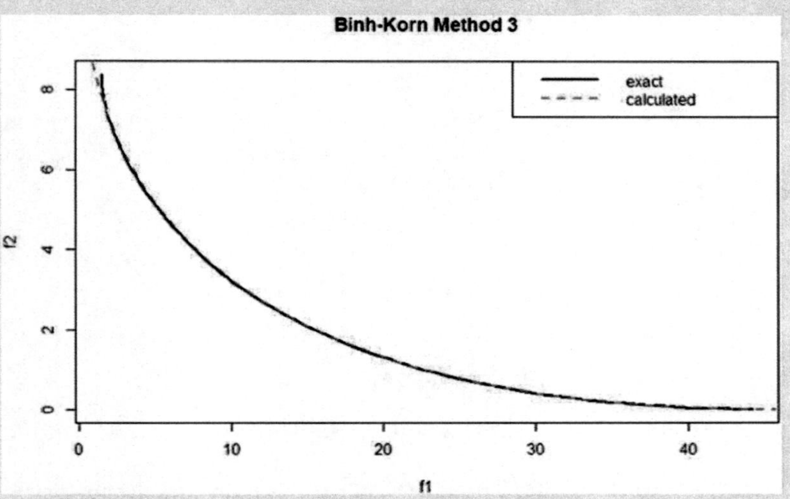

Fig. 3.92 Comparison between the Pareto front generated by the expansion and the solution of the multiobjective optimization problem (method 3)

Table 3.9 A small sample for the representation of Pareto fronts

u_1	1.755633	1.924332	2.021606	2.204238	2.405309
u_2	0.210664	0.407556	0.659623	0.787493	0.823312

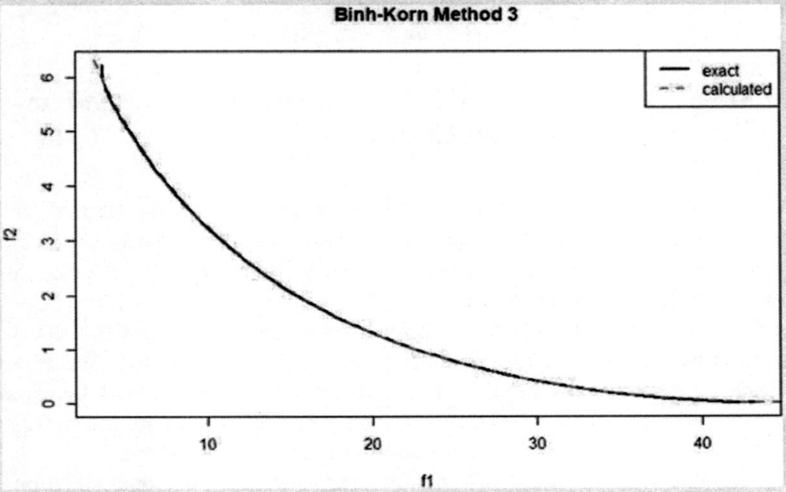

Fig. 3.93 Comparison between the Pareto front generated by the expansion and the solution of the multiobjective optimization problem (method 3)

Exercises

Consider $X = 10\, x_1^{\alpha} x_2^{\beta} - 2\, x_1 - 5x_2$

(a) Assume that the distributions of α and β are uniform. Then, determine an expansion of X as a function of U in a polynomial basis.

(b) Assume that the distribution of α is triangular $T(0.05,0.1,0.2)$ and the distribution of β is $T(0.1, 0.2, 0.3)$. Determine the new expansion. Compare the results to those found in (a).

1. Consider the multiobjective optimization problem where

$$f_1(\boldsymbol{x}) = r_1 x_1 + r_2 x_2 + r_3(1 - x_1 - x_2),$$

$$f_2(\boldsymbol{x}) = \sigma^2 \left(\beta_1^2 r_1^2 x_1^2 + \beta_2^2 r_2^2 x_2^2 + (1 - x_1 - x_2)^2 r_3^2 \beta_3^2\right),$$

$$x_1 \geq 0,\, x_2 \geq 0,\, x_1 + x_2 \leq 1.$$

Let $r_1 = 0.1$, $r_2 = 0.15$, $r_3 = 0.05$, $\sigma = 0.2$. Assume that the coefficients β_1, β_2 are uncertain and independent. Consider the small sample

β_1	0.984164403	0.996937932	1.017274687	1.062638505	1.091857263
β_2	1.12733492	1.187675572	1.230599464	1.236082147	1.272463665

(a) Analogously to Example 3.6, determine a representation of the Pareto front of this problem.

Use the representation to determine the front for $\beta_1 = 1.05$, $\beta_2 = 1.2$.

3.8 Mean, Variance, and Confidence Intervals for Random Functions or Random Curves

In Sect. 2.10, we saw that we may use samples to obtain some statistics of random variables, such as the empirical mean and the variance and to determine confidence intervals that consider the margin of error connected to the fact that the samples do not represent the whole population.

In Sects. 3.2, 3.3, and 3.4 of this chapter, we saw that we may use the samples to generate representations of random variables and then use the representations to generate larger samples and use them to obtain the CDF and the PDF of the variables under consideration. Larger samples may also be used to evaluate the means and improve the evaluation of the variance and of the confidence interval.

In Sect. 3.5, these approaches were extended to multidimensional situations, so that analogous procedures may be used for vectors of random variables.

However, their extension to random functions and curves may face difficulties: as pointed in the preceding section, the determination of the mean of a family of random curves must consider the non-uniqueness of the parametric and the implicit representations. These difficulties may concern also random functions, except for the situations where the parameterization to be used is undoubtful.

Indeed, analogously to Sect. 3.7, we may look for the mean of a family of curves either by considering a parameterization or the Hausdorff approach. Let us recall the definition of the family of curves (Eq. 3.52):

$$\mathcal{C}(u) = \{X = (X_1, X_2) \in \mathbb{R}^2 : X_1 = X_1(t, u), X_2 = X_2(t, u), t \in (0, T)\}$$

The mean of the family is a curve

$$E(\mathcal{C}) = \{\bar{X} = (\bar{X}_1, \bar{X}_2) \in \mathbb{R}^2 : \bar{X}_1 = \bar{X}_1(t), \bar{X}_2 = \bar{X}_2(t), t \in (0, T)\}$$

As shown in the preceding section, the solution consisting in taking the mean representation at each time $\bar{X}_i(t) = E(X_i(t, u))$ may lead to a result that depends on the parameterization used. As an alternative, we can use the variational formulation of the mean: if X is a random variable taking values in $\mathbb{R}$, then

$$E(X) = \arg\min \left\{J(y) = E\left((X - y)^2\right) : y \in \mathbb{R}\right\},$$

id est, $E(X)$ is the real number that minimizes the mean quadratic deviation from X. Analogously, the median is the value that minimizes the mean absolute deviation:

$$med(X) = \arg\min \{I(y) = E(|x - y|) : y \in \mathbb{R}\},$$

The variance is the value of $J(E(X)) = E((X - E(X))^2)$. We may use these properties to determine the mean, the median, and the variance of families of curves. Let

$$\mathcal{D} = \{Y = (Y_1, Y_2) \in \mathbb{R}^2 : Y_1 = Y_1(t), Y_2 = Y_2(t), t \in (0, T)\}$$

Then, we shall look for a curve $E(\mathcal{C})$ such that

$$E(\mathcal{C}) = \arg\min \left\{J(\mathcal{D}) = E\left(dist^2(\mathcal{D}, \mathcal{C}(u))\right)\right\},$$

where $dist$ is a measure of the distance between the curves $\mathcal{D}$ and $\mathcal{C}(u)$. Analogously to Sect. 3.3, we may consider either a distance such as

$$dist^2(\mathcal{D}, C(u)) = \int_0^T \|X(t, u) - \bar{X}(t)\|^2 dt \tag{3.59}$$

or one of the set distances (for instance, the Hausdorf distance (3.56)).

In practice, only samples from X are available. Let $\mathfrak{C} = \{C_i = C(u_i): 1 \leq i \leq ns\}$ be a sample formed by ns variates from $C(u)$: as an approximation, we may look for

$$E(C) \approx \bar{C} = \arg\min \{J(\mathcal{D}) = E(dist^2(\mathcal{D}, C_i)) : C_i \in \mathfrak{C}\}. \tag{3.60}$$

For practical purposes, the distance is often not squared, and we look for

$$E(C) \approx \bar{C} = \arg\min \{I(\mathcal{D}) = E(dist(\mathcal{D}, C_i)) : C_i \in \mathfrak{C}\}. \tag{3.61}$$

Notice that, formally, this definition corresponds to the median of the family, not to the mean. Then, we may proceed into a way analogous to the one used in Sects. 3.1, 3.2, 3.3, 3.4, and 3.5: consider a family $\{\varphi_j : 0 \leq j \leq k\}$ and

$$\mathcal{D}(x) = \left\{ PX = (PX_1, PX_2) \in \mathbb{R}^2 : PX_i = \sum_{j=0}^{k} x_{i,j}\varphi_j(s) : s \in (s_{\min}, s_{\max}) \right\}.$$

Then, we may determine the coefficients x that minimize a measure $g(d(x))$ of the magnitude of $d(x)$, given by

$$d(x) = (d_1(x), \ldots, d_{ns}(x))^t; d_i(x) = dist(C(u_i), \mathcal{D}(x)).$$

For instance, we may take $g(d)$ as the norm of d (as in Eq. (3.58)). If a discretization in s is used, such as $s_{\min} = s_0 < s_1 < \ldots < s_{np} = s_{\max}$, we can look for the values of $PX(s_i)$, $1 \leq i \leq np$, i. e., for 2 np unknown values defining np points in $\mathbb{R}^2$.

However, these formulations present some difficulties yet: indeed, the optimization problems (3.60) and (3.61) may have multiple solutions and it may be necessary to impose supplementary conditions to select the convenient ones.

A simpler solution may be found, which may be employed if the sample under consideration is larger enough and all its elements have the same probability: we can look for the element of the sample which is the best representative of all the sample's population, id est:

$$E(C) \approx \bar{C} = \arg\min \{J(C_i) = E(dist(C_i, C_j)) : C_j \in \mathfrak{C}\}. \tag{3.62}$$

In this case, we determine *one of the curves of the family*, corresponding to the minimal mean Hausdorff distance to the other members of the family – it can be considered as the best representative of the family – the element which is minimally off-centered. One of the interesting features of this approach is that it produces as a result a member of the family, id est, a curve having the same general properties as the other members of the family.

In both the cases, the variance is estimated by $J(\overline{C})$. Confidence intervals with a risk α may be generated by considering quantiles: the confidence interval is formed by the curves of the family that generate a subset $C_{\mathbf{conf}}$ such that $\overline{C} \in C_{\mathbf{conf}}$ and $P(C_{\mathbf{conf}}) = 1 - \alpha$.

Example 3.7
Let us consider the family of circles

$$C(u) = \{\boldsymbol{X}(t,\ u) = r(\sin(2\pi(t+u)),\ \cos(2\pi(t+u))),\ t \in (0,\ 1)\}$$

with u uniformly distributed on $(1, 3)$. All the values of u generate the same circle, centered at the origin and having a radius r. Thus, we expect that the mean circle is the single curve of the family:

$$E(C) = \left\{\overline{\boldsymbol{X}} = (\overline{X}_1,\ \overline{X}_2) \in \mathbb{R}^2 : \overline{X}_1{}^2(t) + \overline{X}_2{}^2(t) = r^2,\ t \in (0,\ 1)\right\}$$

We have $E(X_1(t, u)) = E(X_2(t, u)) = 0$, so that the approach by the mean equation leads to the incorrect result $\overline{X}_1(t) = \overline{X}_2(t) = 0, \forall t \in (0, 1)$. Using Eq. (3.59) leads to the same result. Indeed,

$$E\left(\int_0^1 \|\boldsymbol{X}(t,\ u) - \overline{\boldsymbol{X}}(t)\|^2 dt\right) = \int_0^1 \|\overline{\boldsymbol{X}}(t)\|^2 dt + 1,$$

so that the minimum corresponds to $\overline{\boldsymbol{X}}(t) = 0, \forall t$.

Let us use the Hausdorff distance with a discretization of the solution: we look for the points $\boldsymbol{PX}_i = (PX_{1,\ i}, PX_{2,\ i})$, $1 \leq i \leq np$ that minimize the mean Hausdorff distance to the family. To fix the ideas, let us consider the situation where we consider a single point: $np = 1$, $\boldsymbol{PX}_1 = (a, b)$. We have:

$$d_{SH}(PX_1, C(u)) = |r - \sqrt{a^2 + b^2}|,\ d_{SH}(C(u), PX_1) = r + \sqrt{a^2 + b^2}$$

Thus, the Hausdorff distance between the circle of radius r and the point is $r + \sqrt{a^2 + b^2}$, so that the minimum is attained for $\sqrt{a^2 + b^2} = 0$, id est, when $\boldsymbol{PX}_1 = (0, 0)$. When considering np points, $\boldsymbol{PX}_i = (a_i, b_i)$, $1 \leq i \leq np$, we have

(continued)

Example 3.7 (continued)

$$d_{SH}(PX_i, C(u)) = |r - \sqrt{a_i^2 + b_i^2}|, d_{SH}(C(u), PX_i) = r + \sqrt{a_i^2 + b_i^2},$$

so that the result is similar: the minimum is attained when $a_i^2 + b_i^2 = 0$,id est, when the points coincide at the origin.

Notice that the use of $d_{SH}(PX_i, C(u))$ leads to a more significant result: the distance is $\max\left\{ |r - \sqrt{a_i^2 + b_i^2}| \right\}$, which is minimal when $\sqrt{a_i^2 + b_i^2} = r$, id est, when the points belong to the circle – but they can coincide.

Finally, let us consider the approach (3.62): both d_{SH} and the Hausdorff distance d_H verify

$$dist(C_i, C_j) = 0, \forall i, j.$$

Thus, any member of the family is a solution $E(C) = C_i, \forall i$ – this result corresponds to the expected one (recall that all the C_i generate the same circle). In this case, the variance is equal to zero and the confidence intervals contain the single member of the family (Fig. 3.94).

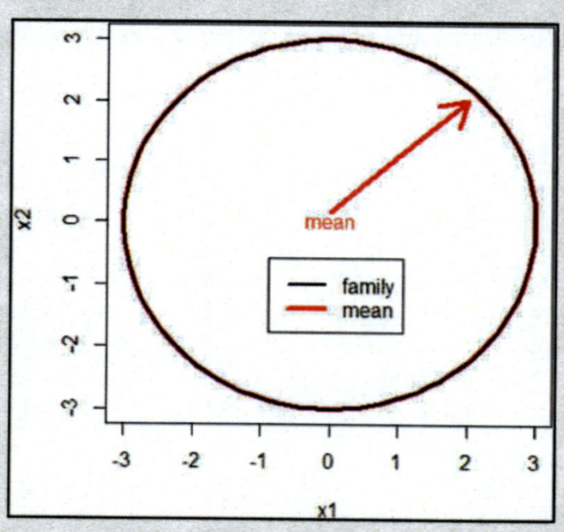

Fig. 3.94 A situation that frustrates several methods: the family is formed by a single circle of radius $r = 3$, with several different parameterizations. We expect the single curve of the family as its mean. However, many approaches furnish (0,0) as the mean

Example 3.8
Let us consider the family of circles

$$C(u) = \{X(t,\ u) = u(\sin(2\pi t),\ \cos(2\pi t)),\ t \in (0,\ 1)\}$$

with u uniformly distributed on $(1, 5)$. The family is formed by concentric circles centered at the origin and having a radius $u \in (1, 5)$. Thus, we expect that the mean circle is the circle of mean radius:

$$E(C) = \left\{\bar{X} = (\bar{X}_1,\ \bar{X}_2) \in \mathbb{R}^2 : \bar{X}_1^{\ 2}(t) + \bar{X}_2^{\ 2}(t) = 3^2,\ t \in (0,\ 1)\right\}$$

We have

$$E(X_1(t,\ u)) = 3\sin(2\pi t),\ E(X_2(t,\ u)) = 3\cos(2\pi t)$$

so that the approach by the mean equation leads to the correct. Using Eq. (3.59) leads to the same result. Indeed,

$$E\left(\int_0^1 \|X(t,\ u) - \bar{X}(t)\|^2 dt\right) = J_1(\bar{X}_1) + J_2(\bar{X}_2),$$

where (c is a convenient constant)

$$J_1(\bar{X}_1) = \int_0^1 \left(\bar{X}_1^{\ 2}(t) - 6\bar{X}_1(t)\sin(2\pi t)\right)dt + c,$$

$$J_2(\bar{X}_2) = \int_0^1 \left(\bar{X}_2^{\ 2}(t) - 6\bar{X}_2(t)\cos(2\pi t)\right)dt + c.$$

Thus, the minimum is attained at $\bar{X}_1(t) = 3\sin(2\pi t)$, $\bar{X}_2(t) = 3\cos(2\pi t)$. Approach (3.62) also furnishes the exact result.

In this situation, the confidence interval with risk α is formed by the circles having radius from $3 - \Delta$ to $3 + \Delta$, $\Delta = 2(1 - \alpha)$. This region contains a proportion $1 - \alpha$ of the family (Fig. 3.95).

(continued)

Example 3.8 (continued)

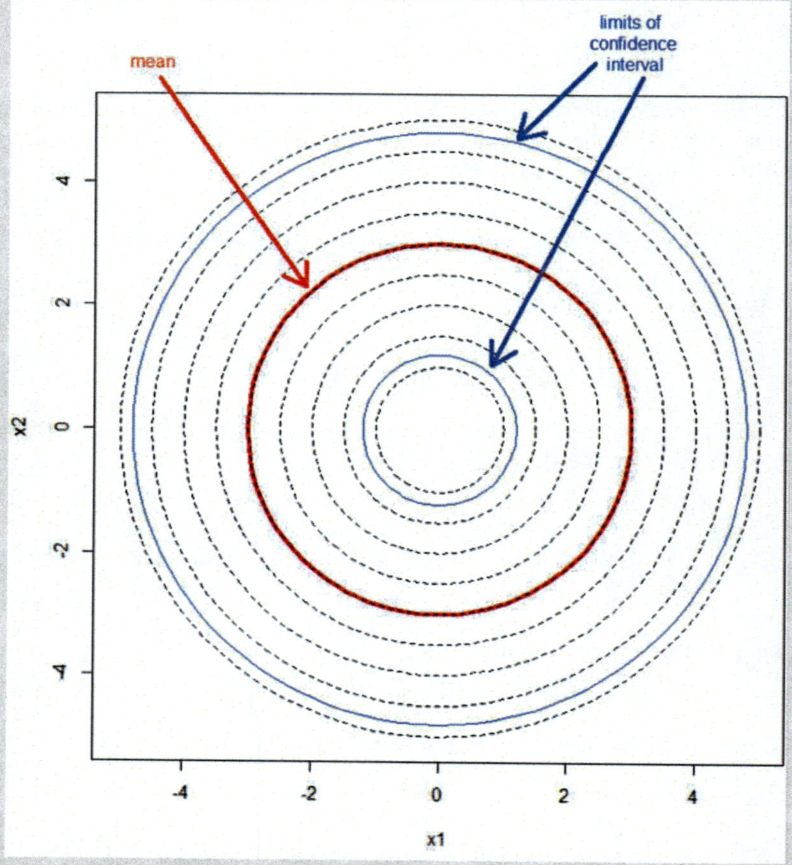

Fig. 3.95 A situation where all the approaches furnish the exact result: the family is formed by concentric circles of radius $u \in (1, 5)$, with the same parameterization. The blue circles delimit the confidence interval for $\alpha = 10\%$

Example 3.9

Let us consider the Pareto Front associated to the Binh-Korn problem involving two independent random variables $u_1 \sim N(2, 0.1)$ and $u_2 \sim T(0,0.5,1)$.

As shown in Example 3.7.1, we may use a small sample and a discretization $t_i = 0.01 * i$, $0 \leq i \leq 100$ to determine a representation of the family of the

(continued)

Example 3.9 (continued)
Pareto fronts. Once the representation is determined, with the same data and parameters as in Example 3.6, we can generate a large sample of 10000 Pareto fronts using samples of 100 variates from each variable u_i, $i = 1, 2$. Then, we determine the mean front as previously mentioned: we determine the most representative front of the family by looking for the front that minimizes the mean distance to the other fronts of the family. After that, we generate a confidence interval by considering the 80% members of the family of fronts which are the closer to the mean front. The results are shown in Fig. 3.96

Figure 3.97 compares the mean front generated by three methods. The first method (`solutions`) consists in solving the multiobjective problem (method of Sect. 1.13.3) for a sample of 2500 variates, generated by two samples of 50 values of each u_i, $i = 1, 2$. Then, we determine the mean of the family of fronts as previously: we evaluate the Hausdorff distance of each member of the family to all the other members of the family and we look for the element that minimizes its own mean distance to the family. The second method (`expansion`) uses the sample of 1E4 fronts previously generated: we calculate its mean as being the most representative element, by the same way (minimization of the mean distance). The third method (`temporal mean`) consists in evaluate the mean point for each t_i for $0 \leq i \leq 100$.

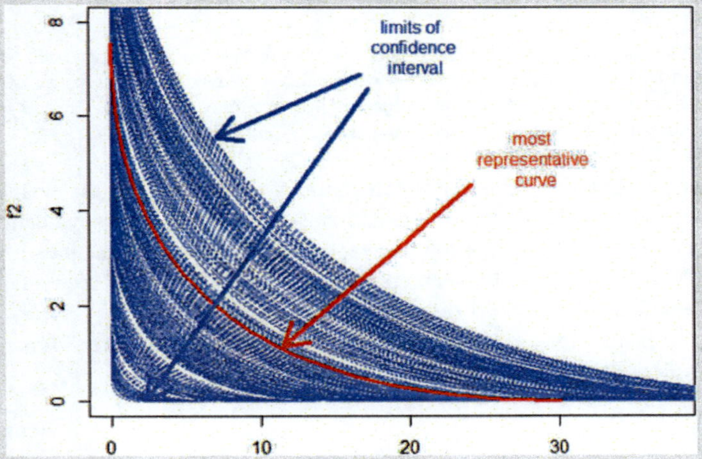

Fig. 3.96 Determination of the Mean Pareto Front and a confidence interval from the expansion, which is used to generate a sample of 10000 Pareto fronts – the cost of the generation of a front is the evaluation of a polynomial at each value of t_i

(continued)

Example 3.9 (continued)

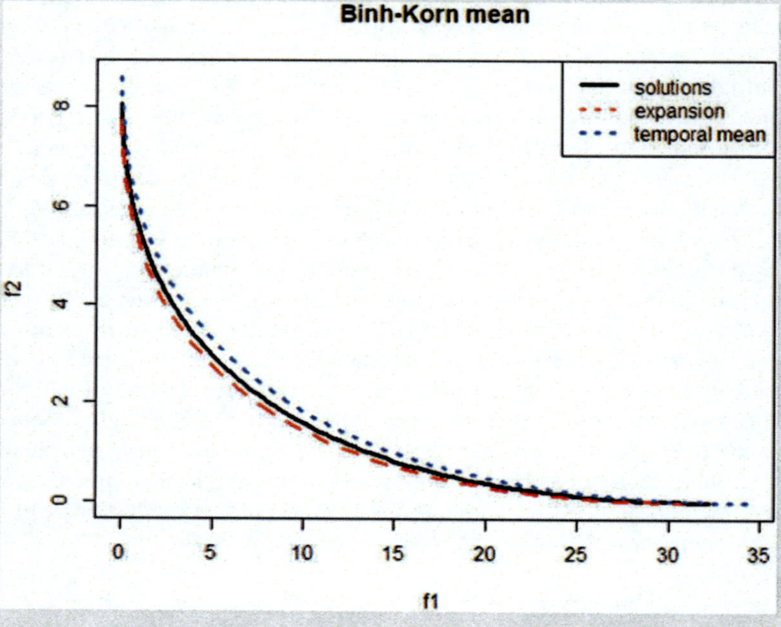

Fig. 3.97 Comparison between two mean Pareto Fronts of a sample of 10000 fronts generated by the expansion and the most representative Pareto front of a sample of 2500 variates (generated by solving the multiobjective problem for each variate). The cost for the expansion is the evaluation of a polynomial at each t_k and $\left(u_{1_i}, u_{2_j}\right)$

When using the expansion, the generation of a front involves the evaluation of a polynomial at each t_i: the computational cost corresponds to evaluation of the polynomial for each t_i. Then, the cost of determination of the mean is the cost of the evaluation of the mean point at each t_i – the computational cost of the evaluation by Hausdorff distance is much higher than it.

The results for $u_1 \sim T(1.5, 2, 2.5)$ and $u_2 \sim T(0, 0.5, 1)$ are shown in Figs. 3.98 and 3.99.

(continued)

Example 3.9 (continued)

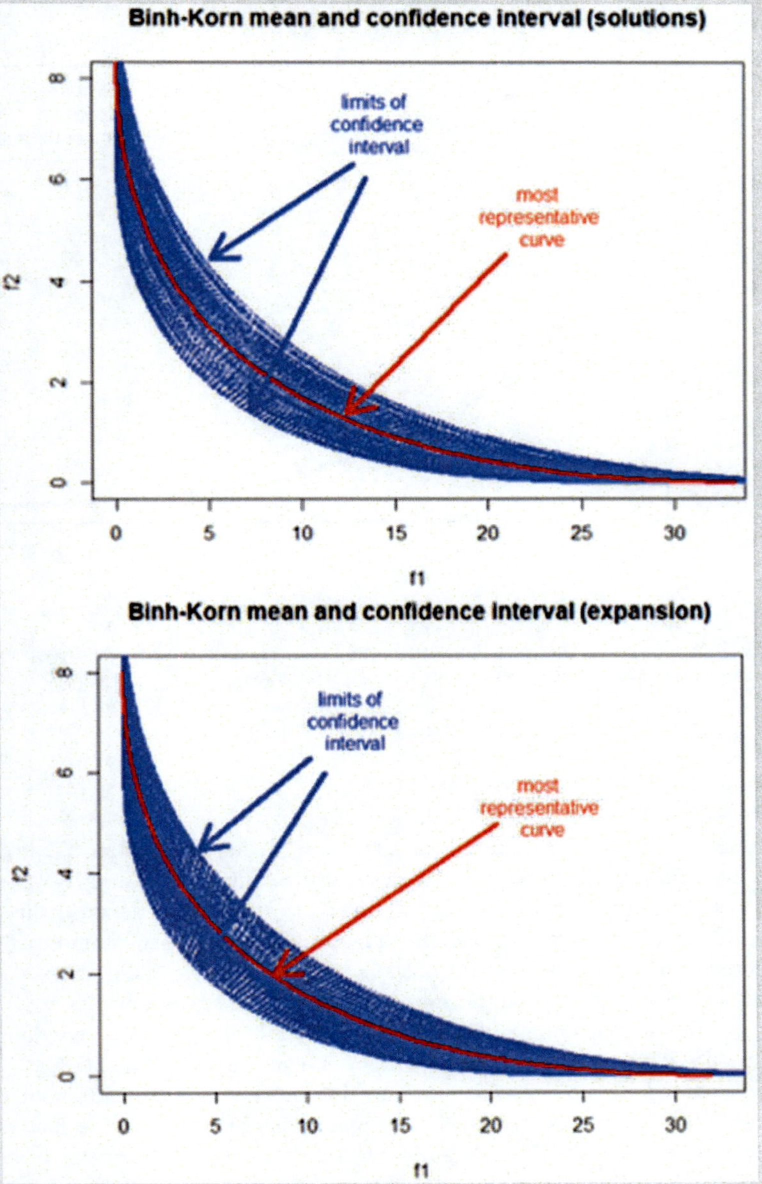

Fig. 3.98 Determination of the Mean Pareto Front and a confidence interval from the expansion, which is used to generate a sample of 10000 Pareto fronts – the cost of the generation of a front is the evaluation of a polynomial at each value of t_i

(continued)

Example 3.9 (continued)

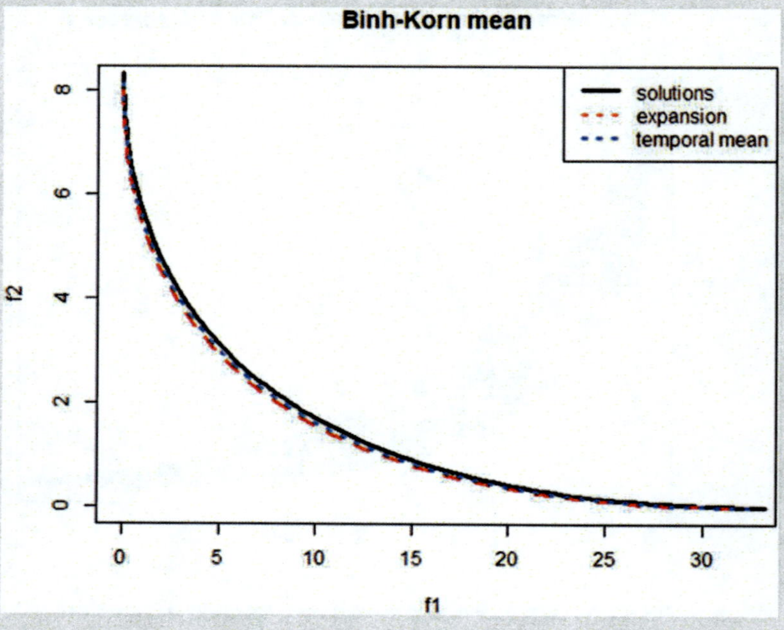

Fig. 3.99 Comparison between the mean Pareto Front generated by the expansion and the most representative Pareto front on a sample of 2500 variates

Example 3.10

Let us consider the indifference curves $x_1^{\alpha} x_2^{\beta} = 1$, when α and β are uncertain: we look for the most representative curve and a confidence interval 80%.

Let us consider a first model where $\beta = 1 - \alpha$ and suppose that at $\alpha \in (0.4, 0.6)$. We start by generating a family of curves: since no information about the distribution of α is given, let us assume a uniform distribution on the intervals. We generate a sample formed by a family of curves corresponding to *na* equally spaced subintervals for α and we apply the procedure previously introduced. The results obtained for $na = 100$ are shown in Fig. 3.100:

A second possibility when not any information is available about the distribution of α is the use of the triangular distribution – for instance, we can consider *T*(0.4,0.5,0.6). Again, let us generate $na = 100$ variates from α and use the associated family of curves to find the most representative element. The result is shown in Fig. 3.101.

(continued)

Example 3.10 (continued)

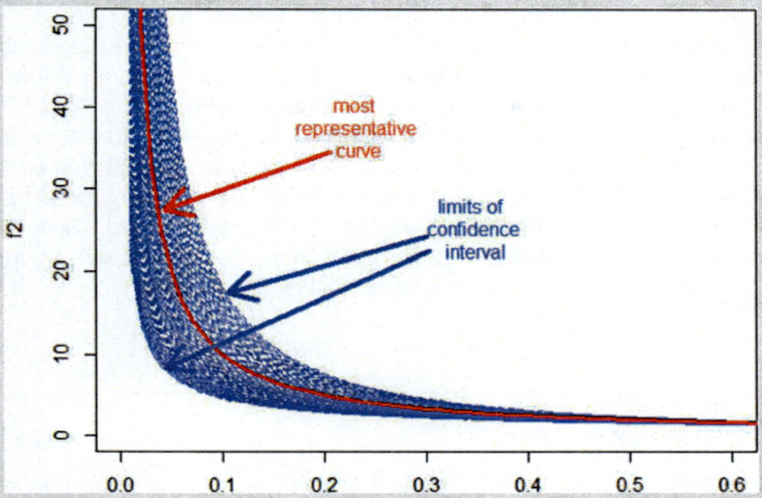

Fig. 3.100 Indifference curves with uncertain exponents $\beta = 1 - \alpha$. α is uniformly distributed on $(0.4, 0.6)$

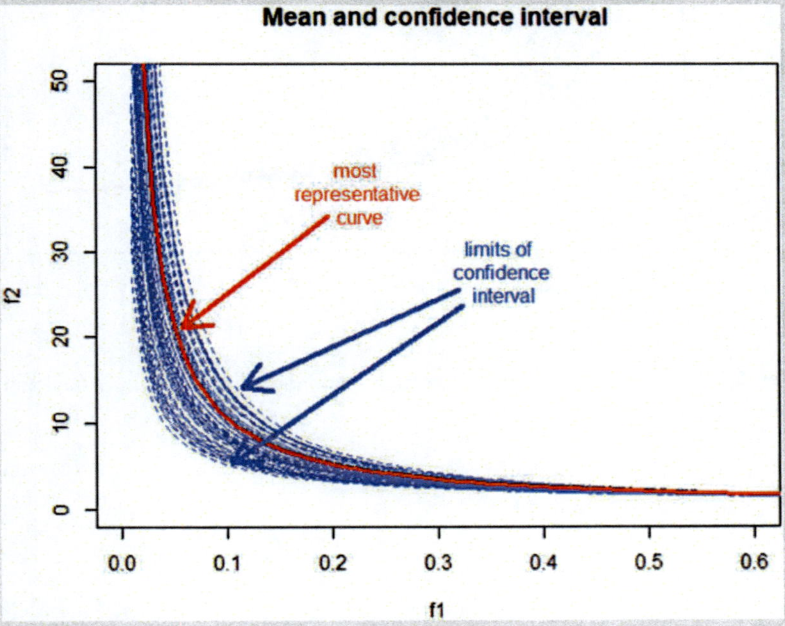

Fig. 3.101 Indifference curves with uncertain exponents $\beta = 1 - \alpha$. α is triangular $T(0.4, 0.5, 0.6)$

(continued)

Example 3.10 (continued)

Consider a second model where α and β are independent and suppose that the experts agree that $\alpha \in (0.4, 0.6)$ and $\beta \in (0.3, 0.5)$. Again, we desire to determine the mean indifference curve and a confidence interval of level 80%. We start by generating a family of curves: since no information about the distribution of α and β is given, let us assume a uniform distribution on the intervals. We generate a sample formed by a family of curves corresponding to na subintervals for α and nb subintervals for β, all equally spaced. The results for $na = nb = 10$ are exhibited in Fig. 3.102. If the distributions are supposed triangular, the results correspond to those in Fig. 3.103.

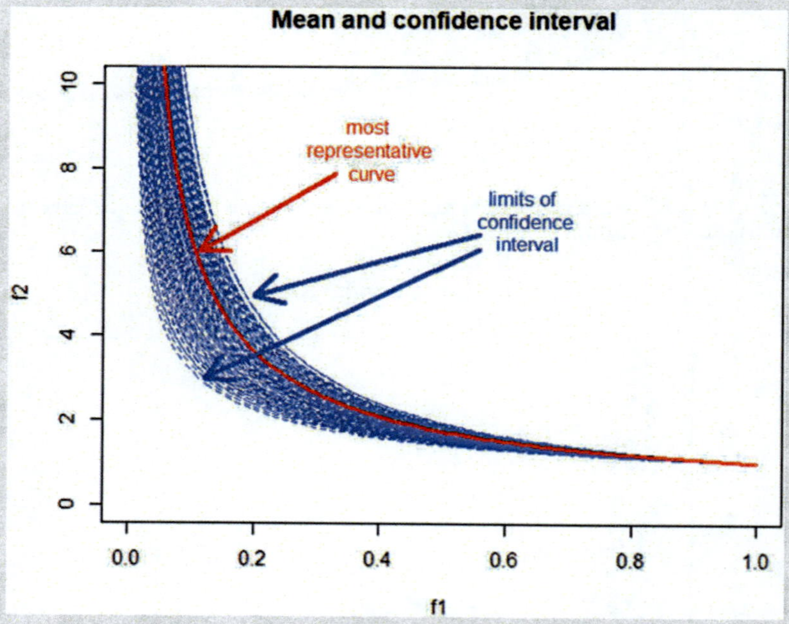

Fig. 3.102 Indifference curves with uncertain independent exponents α, β uniformly distributed

(continued)

Example 3.10 (continued)

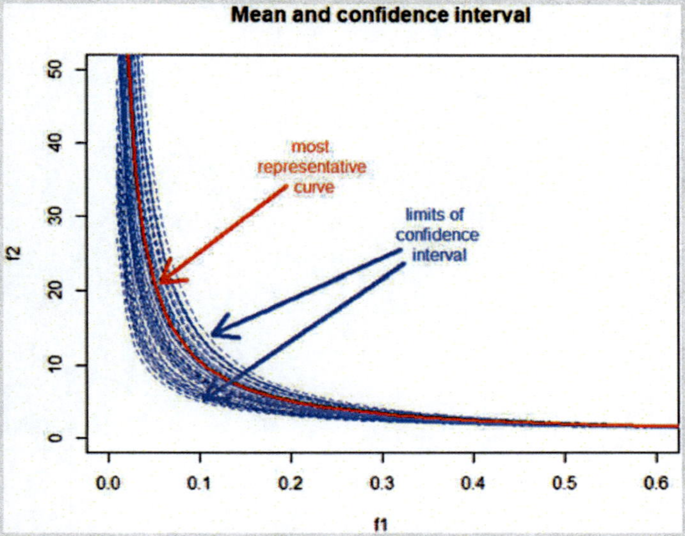

Fig. 3.103 Indifference curves with uncertain independent exponents α, β triangularly distributed $T(0.4, 0.5, 0.6)$ and $T(0.3, 0.4, 0.5)$

Example 3.11
Let us consider the family of random functions

$$X(t, u) = \cos(2\pi ut), t \in (0, 1).$$

with u uniformly distributed on $(-1, 1)$. Here

$$C(u) = \{X(t, u) = (t, \cos(2\pi ut)), t \in (0, 1)\}.$$

We have

$$E(X_1(t, u)) = (t, E(X_2(t, u))) = \frac{\sin(2\pi t)}{2\pi t}$$

Analogously to the first example, this result may be considered as unsatisfactory since the curve is not representative of the family. Let us look for the most representative element of the family. The result is shown in Fig. 3.104.■

(continued)

Example 3.11 (continued)

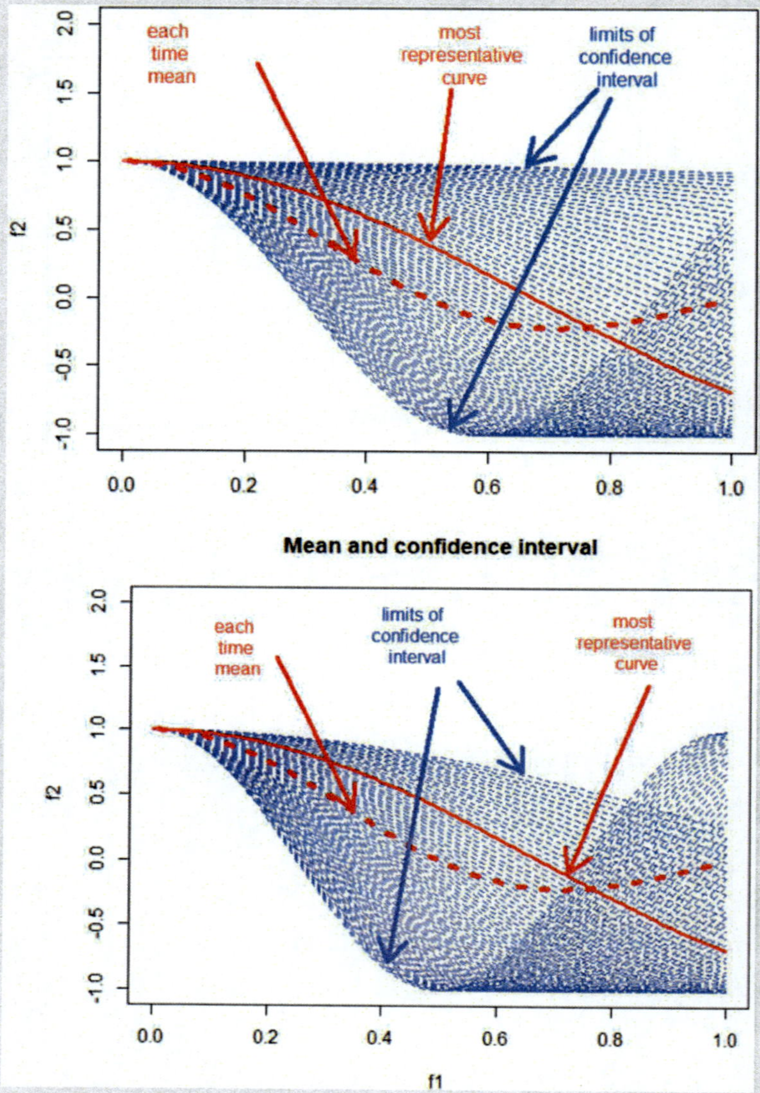

Fig. 3.104 Family of random curves. The mean evaluated by taking the punctual mean at fixed t furnishes a curve that is not representative of the family. We may look for an element that is the most representative in the sense of a distance. The results for two distances are shown in Figure

Exercises

1. Consider

$$X_t = (1-u)^2 \frac{t^2}{4}.$$

   Assume that u is a random variable, uniformly distributed on (0.2, 0.8). Determine the mean path and a confidence interval 80%.

2. Consider $X_t = \left(X_0 - \frac{s}{\delta}\right)e^{-\delta t} + \frac{s}{\delta}$, where s, δ are uncertain.

   Assume that the distributions of s and δ are triangular: $s \sim T(0.1, 0.25, 0.5)$ and $\delta \sim T(0.05, 0.15, 0.25)$.
   Determine the mean path and a confidence interval 80% for $X_0 = 10$.

3. Consider $X_t = \left(X_0 - \frac{u_1}{1-u_2}\right)(u_2)^t + \frac{u_1}{1-u_2}$, where u_1, u_2 are uncertain.

   Assume that the distributions of u_1 and u_2 are triangular: $u_1 \sim T(1, 2, 3)$ and $u_2 \sim T(0.1, 0.25, 0.4)$.
   Determine the mean path and a confidence interval 80% for $X_0 = 10$.

4. Consider the family $X_t = u$, if $|t| < u$; $X_t = (|t| - u)^2 + u$, otherwise.

   Assume that the distribution of u is triangular: $u \sim T(0,0.5,1)$.
   Determine the mean path and a confidence interval 80%.

Chapter 4
Stochastic Processes

Abstract In this chapter, we consider stochastic processes, with a focus on MA, AR, ARMA, diffusion processes, Ito's stochastic integrals, and Ito's stochastic differential equations.

Stochastic processes may be considered as random variables indexed by another variable – usually the time. Thus, a typical stochastic process is a family of random variables depending upon time: $X(t) = X_t$. We can consider also more general indexing sets:

Definition 4.1
Let (Ω, P) be a probability space and $\mathcal{T} \subset \mathbb{R}$ be a non-empty set. A **stochastic process** indexed by $\mathcal{T}$ is family of applications $X = \{X_t : t \in \mathcal{T}\}$, where X_t is a random variable on (Ω, P). We say that $t \in \mathcal{T}$ is the **time**.

Remark 4.1
From the definition above, we see that a stochastic process defines a family of applications $\Omega \ni \omega \to X_t(\omega) \in \mathbb{R}$, indexed by $t \in \mathcal{T}$. Thus, a stochastic process can be seen as a function $X : \Omega \times \mathcal{T} \to \mathbb{R}$, id est, an application $\Omega \times \mathcal{T} \ni (\omega, t) \to X(\omega, t) = X_t(\omega) \in \mathbb{R}$.

It is usual to make a distinction between discrete time (when $\mathcal{T}$ is a discrete set) and continuous time (when $\mathcal{T} = (a, b)$ is an interval).

According to the situation, the process is said to be discrete (when $\mathcal{T}$ is a discrete set) or continuous (when $\mathcal{T} = (a, b)$ is an interval).

In many applications, $\mathcal{T} = \mathbb{N}$ or $\mathcal{T} = \mathbb{Z}$.

Supplementary Information The online version of this chapter (https://doi.org/10.1007/978-3-031-17785-9_4) contains supplementary material, which is available to authorized users.

E. Souza de Cursi, *Uncertainty Quantification using R*, International Series
in Operations Research & Management Science 335,
https://doi.org/10.1007/978-3-031-17785-9_4

Since each X_t is a random variable, we can define statistics of random processes:

Definition 4.2
The **mean** of X is $m_X(t) = E(X_t)$.
The **variance** of X is $V_X(t) = V(X_t) = E(X_t^2) - (E(X_t))^2$.
The **standard deviation** of X is $\sigma_X(t) = \sqrt{V_X(t)}$.
The **moment of order** k of X is $M_{X,k}(t) = E(X_t^k)$.
X is a **second-order process** if and only if $X_t \in L^2(\Omega, P)$, $\forall\, t$.

Second-order processes are often used in practice. We have

Proposition 4.1
Let $X = \{X_t : t \in \mathcal{T}\}$ be a second-order process. Then,

$$\forall t : E(X_t) < \infty, E(|X_t|) < \infty, E(X_t^2) < \infty.$$

Proof
$X_t \in L^2(\Omega, P)$ if and only if $\|X_t\|^2 = E(X_t^2) < \infty$ (see Sect. 2.5). In addition, $E(|X_t|) = (1, |X_t|) \leq |(1, |X_t|)| \leq \|1\| \, \|X_t\| = \|X_t\| < \infty$. Finally, Jensen's inequality (see Proposition 2.6) shows that $|E(X_t)| \leq E(|X_t|) < \infty$.

When dealing with second-order processes, we can consider supplementary definitions:

Definition 4.3
Let $X = \{X_t : t \in \mathcal{T}\}$ be a second-order process.
The **autocorrelation** of X is $R_{XX}(t, s) = E(X_t X_s)$.
The **autocovariance** of X is $C_{XX}(t, s) = E(X_t X_s) - E(X_t)E(X_s)$.
The **cross-correlation** of X and Y is $R_{XY}(t, s) = E(X_t Y_s)$.
The **cross-covariance** of X and Y is $C_{XY}(t, s) = E(X_t Y_s) - E(X_t)E(Y_s)$

Remark 4.2
Notice that $R_{XX}(t, s) = R_{XX}(s, t)$ and $C_{XX}(t, s) = C_{XX}(s, t)$.

4.1 Ergodicity

Definition 4.4

1. X is **weakly stationary** if and only if $V_X(t)$ is finite for any t; m_X is independent from t: $m_X(t) = m_0$, $\forall\, t$ and $R_{XX}(s, t) = R_X(|s - t|)$, $\forall\, s, t$. R_X is extended by symmetry to negative values: $R_X(a) = R_X(-a)$, so that $R_{XX}(s, t) = R_X(s - t) = R_X(t - s)$.
2. If X and Y are weakly stationary, $m_X(t) = m_0$, $m_Y(t) = n_0$, $\forall\, t$. Then, we can evaluate $R_{XY}(s) = E(X_{t+s}Y_t)$, $C_{XY}(s) = E(X_{t+s}Y_t) - m_0 n_0$.
3. X is **stationary** if and only if the distribution of $(X(t_1), \ldots, X(t_n))$ coincides with the distribution of $(X(t_1 + \tau), \ldots, X(t_n + \tau))$ for any n, any $(t_1, \ldots, t_n)$ and any τ. If X is stationary, then X is weakly stationary.
4. Let X be weakly stationary. X is **ergodic in the mean** if and only if $m_0 = \lim\limits_{T\to+\infty} \frac{1}{T}\int\limits_0^T X(s)ds$. It is **ergodic in correlation** if and only if $R_X(\tau) = \lim\limits_{T\to+\infty} \frac{1}{T}\int\limits_0^T X(s)X(s+\tau)ds$.

Notice that, for a weakly stationary process,

$$C_{XX}(t, s) = R_{XX}(t, s) - m_X^2 = R_X(|s - t|) - m_0^2 = C_X(|s - t|)$$

For ergodic processes, we can evaluate the mean and/or the autocorrelation from a single realization, by approximating the limits by the evaluation at a large T. For discrete processes (for instance, $\mathcal{T} = \mathbb{N}$ or $= \mathbb{Z}$), the integrals are replaced by sums:

$$m_0 = \lim_{n\to\infty} \frac{1}{n}\sum_{i=0}^{n} X_i \;\; ; \;\; R_X(s) = \lim_{n\to\infty} \frac{1}{n}\sum_{i=0}^{n} X_i X_{i+s}.$$

In practice, these quantities are estimated by finite sums. Under R, the intrinsic function `mean` evaluates the mean of a signal, `acf` helps you to evaluate R_X and C_X: `acf(X, lag.max=k, type="correlation")` evaluates $R_X^n(s) = R_X(s)/R_X(0), 0 \le s \le k$; `acf(X, lag.max=k, type = "covariance")` evaluates $C_X(s)$, $0 \le s \le k$. Analogously, `ccf` helps you to evaluate R_{XY} and C_{XY}: `ccf (X, y, lag.max = k, type = "correlation")` . evaluates $R_{XY}^n(s) = R_X(s)/\sqrt{R_X(0)R_Y(0)}, 0 \le s \le k$; `ccf(X, y, lag.max = k, type = "covariance")` evaluates $R_{XY}(s)$, $-k \le s \le k$. A parameter `demean` can be set to `TRUE` or `FALSE` to use or not centered variables – the default value is `TRUE`. Both the functions produce a plot by default – it is controlled by a parameter `plot`, which takes the value `TRUE` by default: set it to FALSE if you do not desire a plot. Otherwise, you can use the functions below:

```
autocorr = function(k,x){
        nx = length(x)
        kk = abs(k)
        aux = sum(x[1:(nx-kk)]*x[(kk+1):nx])/(nx - kk)
        return(aux) }
vautocorr = function(k,x){
        aux = numeric(length=k+1)
        for (i in 0:k){aux[[i+1]]= autocorr(i,x)}
        return(aux) }
crosscorr = function(k,y,z){
        ny = length(y)
        nz = length(z)
        nn = min(c(ny,nz))
        if(k < 0){
         kk = abs(k)
         aux=sum(y[1:(nn-kk)]*z[(kk+1):nn])/(nn - kk)
        }else{
            aux =sum(y[(k+1):nn]*z[1:(nn-k)])/(nn - k)
        }
        return(aux) }
vcrosscorr = function(k,y,z){
        aux = numeric(length=2*k+1)
        for (i in seq(-k,k,by=1)){
            aux[[i+k+1]] = crosscorr(i,y,z)
            }
        return(aux) }
autocov = function(k,x){
         aux = autocorr(k,x) - mean(x)^2
        return(aux) }
vautocov = function(k,x){
        aux = numeric(length=k+1)
        for (i in 0:k){aux[[i+1]]= autocov(i,x)}
        return(aux) }
crosscov = function(k,y,z){
        aux = crosscorr(k,y,z) - mean(y)*mean(z)
        return(aux) }
vcrosscov = function(k,y,z){
        aux = numeric(length=k+1)
        for (i in 0:k){aux[[i+1]]= crosscov(i,y,z)}
        return(aux) }
```

These functions are implemented in the class `signal.R`.

Remark 4.3
Notice that the formula used by *R* in the computation of autocorrelations and autocovariances is different from the definition introduced in this text. `acf` evaluates the covariances using the standard statistical formula

$$R_X(k) \approx \frac{1}{n}\sum_{i=1}^{n-k}\left(X_{i+k}-\overline{X}\right)\left(X_i-\overline{X}\right).$$

For instance, consider the signal

(continued)

Remark 4.3 (continued)

$$X_i = \exp\left(\frac{t_i}{4}\right), t_i = \frac{i}{100}, 0 \leq i \leq 1000$$

The code

```
t = seq(0,10,by=0.01)
x = exp(t/4)
acxa = acf(x,lag.max = 100, type = "covariance")
Rxa = acxa$acf
Rxtest = numeric(length = 101)
nx = length(x)
xbar = sum(x)/nx
for (k in 1: 101){
  Rxtest[[k]] = sum((x[(k+1):nx] - xbar)*(x[1:(nx-k)] - xbar))/nx
}
plot.new()
plot(Rxa,type = "n")
lines(Rxa,lty=1,lwd = 2)
lines(Rxtest,lty=2,lwd = 4,col="red")
legend('top',legend = c('acf','direct'),
       col=c('black','red'),lty = c(1,2),lwd = c(2,4))
```

Running this code under RStudio 2021.09.1 Build 372, with R 4.1.3 produces the result in Fig. 4.1. This result is slightly different from the result furnished by the code proposed (function `autocov`), namely when k increases – `autocov` uses a division by $n - k$.

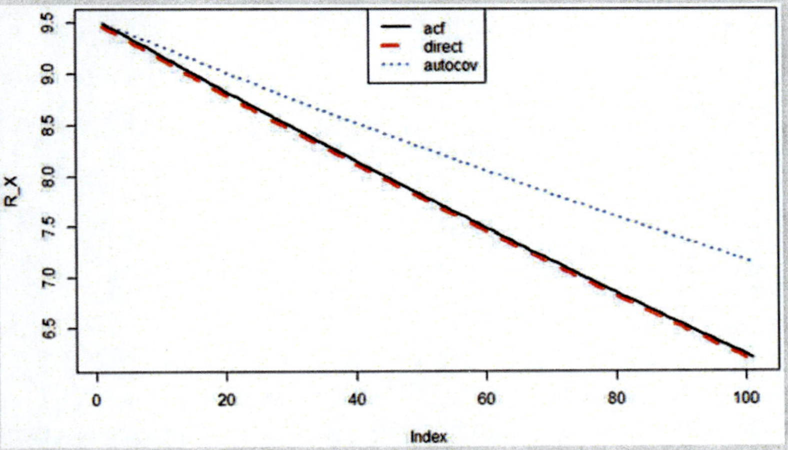

Fig. 4.1 Verification of the formula used by `acf` for covariance and comparison with the result of `autocov`

Analogously, R evaluates correlations using the standard statistical formula

(continued)

Remark 4.3 (continued)

$$R_X(k) = \frac{\frac{1}{n-k}\sum_{i=1}^{n-k}(X_{i+k}-\overline{X})(X_i-\overline{X})}{\sqrt{\frac{1}{n}\sum_{i=1}^{n}(X_i-\overline{X})^2}\sqrt{\frac{1}{n-k}\sum_{i=1}^{n-k}(X_{i+k}-\overline{X})^2}},$$

R evaluates correlations using

$$R_X(k) = \frac{\frac{1}{n-k}\sum_{i=1}^{n-k}(X_{i+k}-\overline{X})(X_i-\overline{X})}{\sqrt{\frac{1}{n}\sum_{i=1}^{n}(X_i-\overline{X})^2}\sqrt{\frac{1}{n-k}\sum_{i=1}^{n-k}(X_i-\overline{X})^2}},$$

while the usual definition in signal analysis is used in this text. To verify these formulae, we can use the code:

```
t = seq(0,10,by=0.01)
x = exp(t/4)
acxa = acf(x,lag.max = 100, type = "correlation")
Rxa = acxa$acf
Rxtest = numeric(length = 101)
nx = length(x)
xbar = sum(x)/nx
sbar = sum((x - xbar)^2)/nx
for (k in 1: 101){
  s1 = sum((x[(k+1):nx] - xbar)*(x[1:(nx-k)] - xbar))/nx
  s2 = sum((x[(k+1):nx] - xbar)^2)/(nx-k)
  Rxtest[[k]] = s1/sqrt(s2*sbar)
}
plot.new()
plot(Rxa,type = "n",ylab = "R_X")
lines(Rxa,lty=1,lwd = 2)
lines(Rxtest,lty=2,lwd = 4,col="red")
legend('top',legend = c('acf','direct'),
       col=c('black','red'),lty = c(1,2),lwd = c(2,4))
```

Running this code under RStudio 2021.09.1 Build 372, with R 4.1.3 produces the result in Fig. 4.2. This result is different from the result furnished by the code proposed (function `autocorr`). We can compare the results by normalizing the results of `autocorr` to meet the statistical definition of correlation.

(continued)

Remark 4.3 (continued)

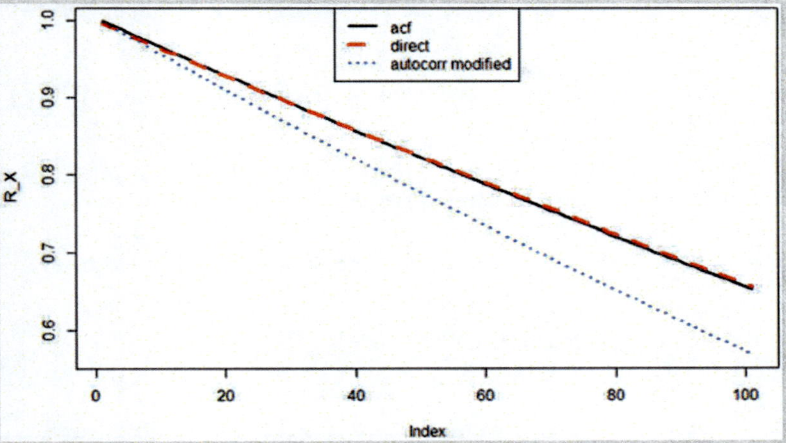

Fig. 4.2 Verification of the formula used by `acf` for covariance and comparison with the result of `autocorr`

Example 4.1

Let us consider a white noise ε (see Sect. 4.3) and the processes defined by

$$X_n = \varepsilon_n + \varepsilon_{n-1} + 2\varepsilon_{n-2} - \varepsilon_{n-3} + 2\varepsilon_{n-4};$$

$$Y_i = \varepsilon_n + \varepsilon_{n-1} + 2\varepsilon_{n-2} + \varepsilon_{n-3} + 2\varepsilon_{n-4} + 3\varepsilon_{n-5} + 4\varepsilon_{n-6}.$$

We generate a trajectory containing 1000 points as follows (see Sect. 4.4):

```
x = arima.sim(model = list(ma = c(1,2,-1,2)),1000)
y = arima.sim(model = list(ma = c(1,2,1,2, 3, 4)),1000)
```

Let us evaluate the autocorrelation $R_X(s)$, for $0 \leq s \leq 100$.

We can use the function `vautocorr`. At right, an example of call using the class `signal`.	`source("signal.R")`
	`s1 = signal$new(x)`
	`Rx = s1$vautocorr(100)`

(continued)

Example 4.1 (continued)

Otherwise, we can use `acf` and correct the result by adding the square of the mean of X:

```
acxa = acf(x,lag.max = 100, type = "covariance")
mx = mean(x)
Rxa = acxa$acf + mx^2
```

Finally, we can use parameter `demean`:

```
acxb = acf(x,lag.max = 100, type = "covariance",demean = FALSE)
Rxb = acxb$acf
```

By default, `acf` generates a plot as in Fig. 4.3.

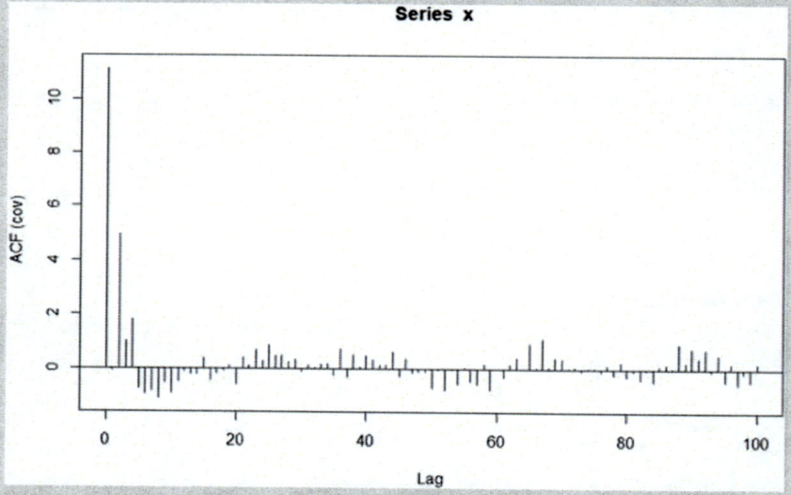

Fig. 4.3 Plot produced by `acf`, by default

If you do not desire the plot, you must use the parameter `plot = FALSE`. For instance,

```
acxa = acf(x,lag.max = 100, type = "covariance",plot = FALSE)
```

(continued)

Example 4.1 (continued)

Here, the parameter `plot = FALSE` indicates that we do not desire to plot the autocovariance. You can plot it by using the standard graphics' instructions from `R`. For instance:

```
plot.new()
plot(tcex,Rx,type = "n", xlab = "s",ylab = "autocorrelation",
     main = "autocorrelation of X")
lines(tcex,Rxexv(tcex),lty = 1, lwd = 6, col = 'black')
lines(tcex,Rx,lty = 2, lwd = 4, col = 'red')
lines(tcex,Rxa,lty = 3, lwd = 3, col = 'blue')
lines(tcex,Rxb,lty = 4, lwd = 2, col = 'blue')
legend('top',legend = c('exact','signal','acf corrected','acf no demean'),
       col=c('black','red','blue','blue'),lty = c(1,2,3,4),lwd = c(6,4,3,2))
```

produces the result shown in Fig. 4.4.

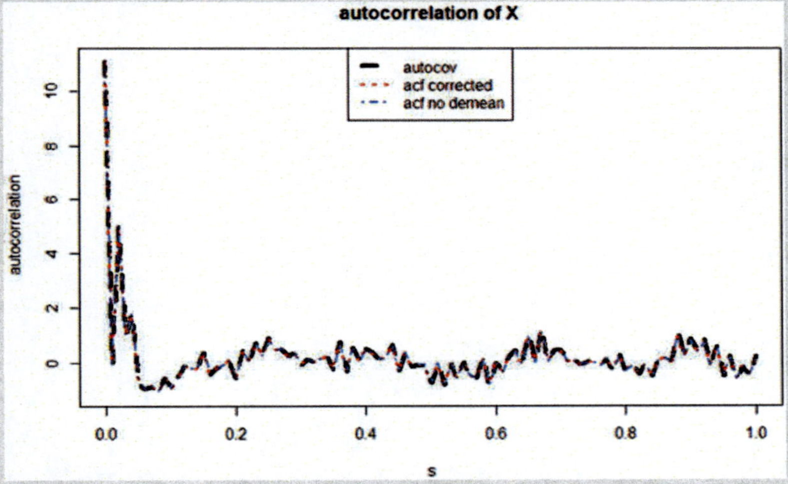

Fig. 4.4 Autocorrelation of X generated by three different methods. The black curve labeled "`autocov`" corresponds to the programs previously presented (implemented in class `signal`)

An example of the plot produced by `ccf by default` is shown in Fig. 4.5. Analogously to the preceding, we can consider

```
plot.new()
plot(tcex,Rx,type = "n", xlab = "s",ylab = "autocorrelation",
     main = "autocorrelation of X")
lines(tcex,Rxexv(tcex),lty = 1, lwd = 6, col = 'black')
lines(tcex,Rx,lty = 2, lwd = 4, col = 'red')
lines(tcex,Rxa,lty = 3, lwd = 3, col = 'blue')
lines(tcex,Rxb,lty = 4, lwd = 2, col = 'blue')
legend('top',legend = c('exact','signal','acf corrected','acf no demean'),
       col=c('black','red','blue','blue'),lty = c(1,2,3,4),lwd = c(6,4,3,2))
```

(continued)

Example 4.1 (continued)

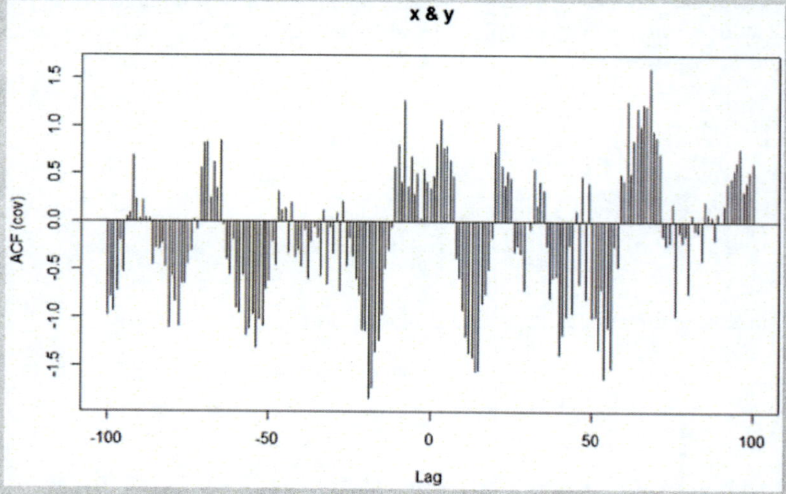

Fig. 4.5 Plot of the results of `ccf`, by default

This code produces the result in Fig. 4.6.

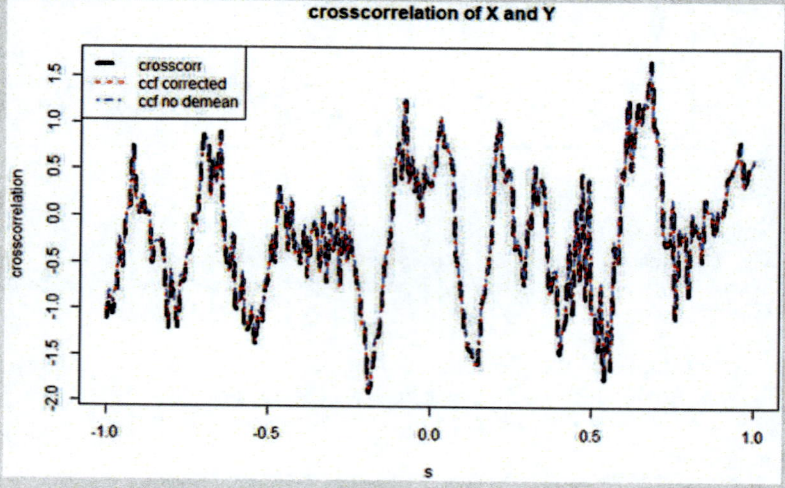

Fig. 4.6 Plot of the results of `ccf`. The black curve labeled "`crosscorr`" corresponds to the programs previously presented (implemented in class `signal`)

Example 4.2
Let us consider the times $t_i = i\Delta t$, $\Delta t = 0.01$, $0 \le i \le 1000$ and two signals defined by

$$X_i = \sin(2\pi t_i),\ \ Y_i = \cos(2\pi t_i).$$

Notice that both are deterministic non-stationary signals. In this case, we have, for $s > 0$,

$$R_X(s) = \frac{1}{10-s}\int_0^{10-s} \sin(2\pi(t+s))\sin(2\pi t)dt,$$

so that

$$R_X(s) = \frac{1}{2}\cos(2\pi s) + \frac{\sin(2\pi s)}{4\pi(10-s)}.$$

For $s < 0$,

$$R_X(s) = \frac{1}{10+s}\int_{-s}^{10} \sin(2\pi(t+s))\sin(2\pi t)dt,$$

and

$$R_X(s) = \frac{1}{10-|s|}\int_{|s|}^{10} \sin\ (2\pi(\mathrm{t}-|s|))\sin(2\pi t)dt,$$

so that ($u = t + |s|$)

$$R_X(s) = \frac{1}{10-|s|}\int_{|s|}^{10} \sin(2\pi \mathrm{u})\sin((2\pi(\mathrm{u}+|s|))du = R_X(|s|).$$

Thus, $R_X(s) = R_x(-s)$. Analogously, for $s > 0$,

$$R_{XY}(s) = \frac{1}{10-s}\int_0^{10-s} \sin(2\pi(t+s))\cos(2\pi t)dt,$$

(continued)

Example 4.2 (continued)

so that

$$R_{XY}(s) = \frac{1}{2}\sin(2\pi s).$$

For $s < 0$,

$$R_{XY}(s) = \frac{1}{10+s}\int_{-s}^{10} \sin(2\pi(t+s))\cos(2\pi t)dt.$$

So $(u = -s)$,

$$R_{XY}(s) = \frac{1}{10-u}\int_{u}^{10} \sin(2\pi(t-u))\cos(2\pi t)dt.$$

Thus,

$$R_{XY}(s) = -\frac{1}{2}\sin(2\pi u) = \frac{1}{2}\sin(2\pi s) \quad (s<0).$$

In addition,

$$E(X) = \frac{1}{10}\int_{0}^{10} \sin(2\pi t)dt = 0$$

$$E(Y) = \frac{1}{10}\int_{0}^{10} \cos(2\pi t)dt = 0$$

We generate the signals by the code at right:

```
t = seq(0,10,by=0.01)
x = sin(2*pi*t)
y = cos(2*pi*t)
```

Let us evaluate the autocorrelation $R_X(s)$, for $0 \le s \le 100$. In this case, we obtain the results in Figs. 4.7 and 4.8.

(continued)

Example 4.2 (continued)

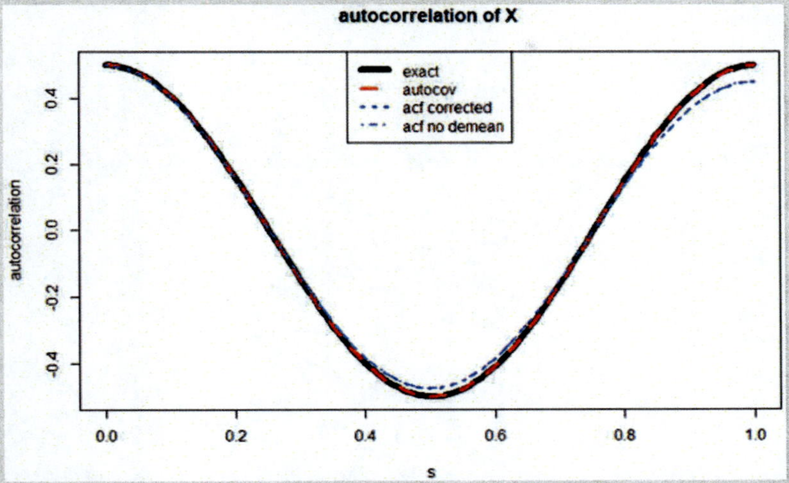

Fig. 4.7 Autocorrelation of X generated by three different methods. The red curve labeled "`autocov`" corresponds to the programs previously presented (implemented in class `signal`)

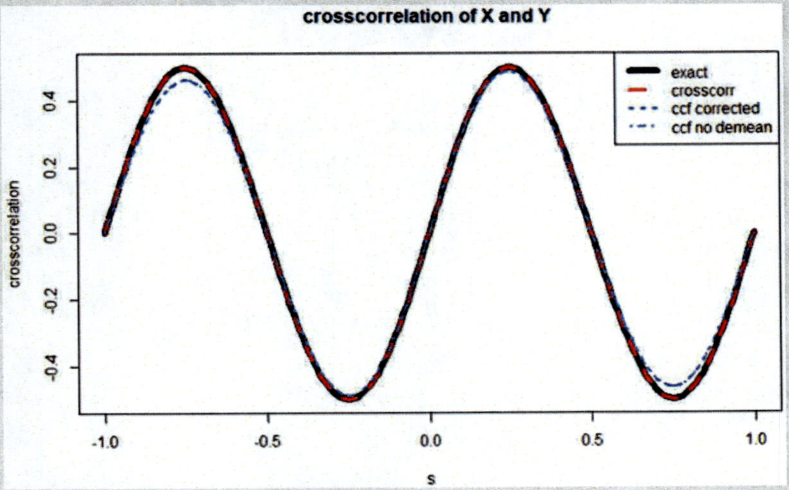

Fig. 4.8 Plot of the results of `ccf`. The red curve labeled "`crosscorr`" corresponds to the programs previously presented (implemented in class `signal`)

Example 4.3
Let us consider the times $t_i = i\Delta t$, $\Delta t = 0.01$, $0 \le i \le 1000$ and two signals defined by

$$X_i = \exp\left(\frac{t_i}{4}\right), \quad Y_i = \exp\left(-\frac{t_i}{10}\right).$$

Notice that both are deterministic non-stationary signals. In this case, we have, for $s > 0$,

$$R_X(s) = \frac{1}{10-s}\int_s^{10} \exp\left(\frac{t+s}{4}\right)\exp\left(\frac{t}{4}\right)dt,$$

so that

$$R_X(s) = \frac{2}{10-s}\left(\exp\left(5-\frac{s}{4}\right) - \exp\left(\frac{s}{4}\right)\right).$$

For $s < 0$, the same transformations used in the preceding example show that $R_X(s) = R_x(-s)$. Let us evaluate the cross-covariance: for $s > 0$:

$$R_{XY}(s) = \frac{1}{10-s}\int_s^{10} \exp\left(\frac{t+s}{4}\right)\exp\left(-\frac{t}{10}\right)dt,$$

so that

$$R_{XY}(s) = \frac{20e^{s/10}\left(e^{3/2} - e^{3s/20}\right)}{3(10-s)}.$$

For $s < 0$:

$$R_{XY}(s) = \frac{1}{10+s}\int_{-s}^{10} \exp\left(\frac{t+s}{4}\right)\exp\left(-\frac{t}{10}\right)dt,$$

id est,

$$R_{XY}(s) = \frac{1}{10-u}\int_u^{10} \exp\left(\frac{t-u}{4}\right)\exp\left(-\frac{t}{10}\right)dt,$$

(continued)

Example 4.3 (continued)

so that,

$$R_{XY}(s) = \frac{20e^{-u/4}\left(e^{3/2} - e^{3u/20}\right)}{3(10-u)} = \frac{20e^{s/4}\left(e^{3/2} - e^{-3s/20}\right)}{3(10+s)}$$

In addition,

$$E(X) = \frac{1}{10}\int_0^{10} \exp\left(\frac{t}{4}\right)dt = \frac{2}{5}\left(-1 + e^{5/2}\right)$$

$$E(Y) = \frac{1}{10}\int_0^{10} \exp\left(-\frac{t}{10}\right)dt = \frac{-1+e}{e}.$$

The results furnished by the programs previously introduced are shown in Figs. 4.9 and 4.10.

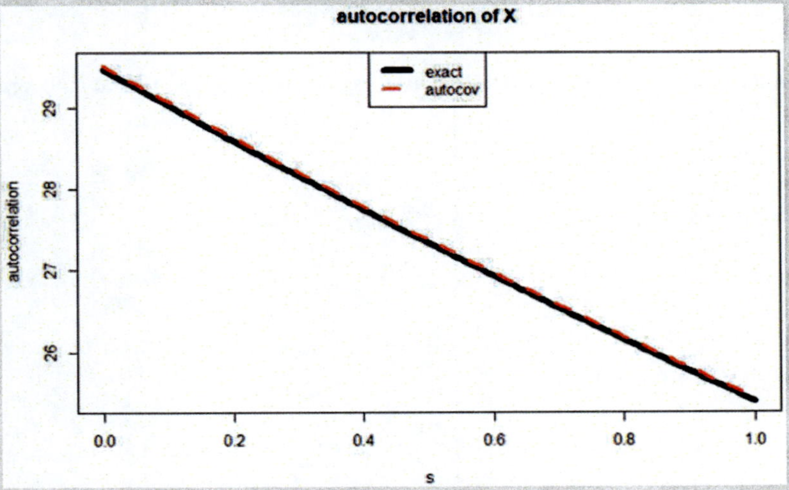

Fig. 4.9 Autocorrelation of *X* generated by the programs previously introduced (available in class `signal`)

(continued)

Example 4.3 (continued)

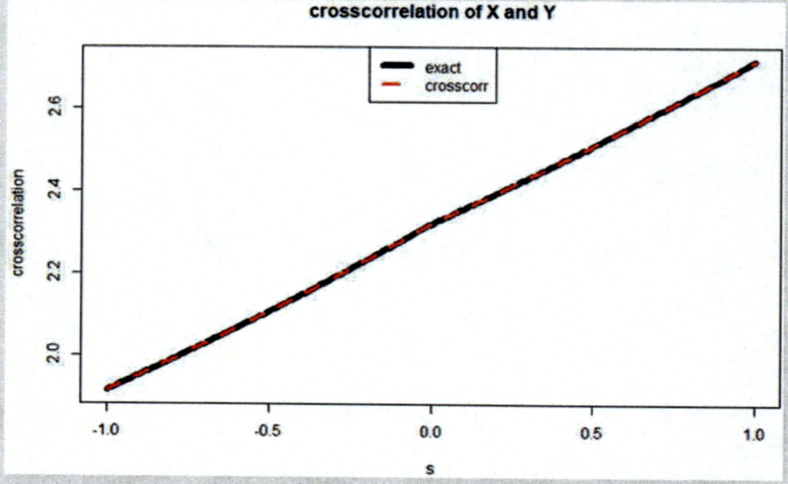

Fig. 4.10 Cross-correlation of X and Y generated by the programs previously introduced (available in class `signal`)

In the analysis of weakly stationary process, we may use the partial autocorrelations, defined below:

Definition 4.5
The **partial autocorrelation** $\rho(s)$ of a discrete process X is

$$\rho(s) = E((X_{t+s} - P_s X_{t+s})(X_t - P_s X_t)),$$

where $P_s X_t$ is the best approximation of X_t by an affine function of $\chi_{s,t} = (X_{t+1}, \ldots, X_{t+s-1})$, *id est*,

$$P_s X_t = \arg\min \left\{ E\left((Y - X_t)^2 \right) : Y \text{ is an affine function } of\ \chi_{s,t} \right\}$$

In addition, $\rho(s)$ coincides with the coefficient of X_t for the best approximation of X_{t+s} by an affine function of $X_t, \ldots, X_{t+s-1}$.

Under `R`, partial autocorrelations are evaluated either by `acf` with parameter `"type = partial"` or by `pacf`. These functions use the approximations introduced in Remark 4.3. If you desire to evaluate partial autocorrelations by your own program, you can consider the general form of the approximation:

$$X_{t+s} \approx PX_{t+s} = c_1 X_{t+s-1} + c_2 X_{t+s-2} + \ldots + c_s X_t + c_{s+1}$$

Let us denote

$$Z_i = X_{t+s-i}, 1 \leq i \leq s; Z_{s+1} = 1.$$

The best approximation PX_{t+s} is the orthogonal projection of X_{t+s} onto the subspace generated by $\{Z_1, Z_2, \ldots, Z_{s+1}\}$. Thus,

$$E((X_{t+s} - PX_{t+s})Z_i) = 0, 1 \leq i \leq s+1.$$

Analogously to the methods of collocation and variational approximation introduced in Chap. 3, the coefficients $C = (c_1, \ldots, c_{s+1})^t$ are the solution of a linear system:

$$AC = B;\ A_{ij} = E(Z_i Z_j),\ B_i = E(X_{t+s} Z_i),\ 1 \leq i, j \leq s+1. \tag{4.1}$$

We have $\rho(s) = c_s$. Thus, an alternative to determine the coefficients is to solve the linear systems (4.1). An example of code for the determination of the partial autocorrelation by this way is the following:

```
vpcorr1 = function(kmax,z,epsi){
 PAC = numeric(length = kmax)
 nz = length(z)
 for (s in 1:kmax){
    ncol = s
    ndata = nz -s
    Y = matrix(z[(s+1):nz], nrow = ndata,ncol = 1)
    X = list()
    for (j in 1:s){
      t_ini = s + 1 - j
      t_end = t_ini + ndata - 1
      X[[j]] = matrix(z[t_ini:t_end], nrow = ndata,ncol = 1)
    }
    X[[s+1]] = matrix(1, nrow = ndata,ncol = 1)
    A = matrix(0, nrow = s+1, ncol = s+1)
    B = matrix(0, nrow = s+1, ncol = 1)
    for (i1 in 1:(s+1)){
       B[[i1,1]] = sum(X[[i1]]*Y)/ndata
       for (j1 in 1:(s+1)){
          A[[i1,j1]] = sum(X[[i1]]*X[[j1]])/ndata
       }
    }
    cof = solve(A +epsi*diag(s+1),B)
    PAC[[s]] = cof[[s]]
  }
  return(PAC)
}
```

You can easily build a variation of this program which eliminates the coefficient c_{s+1}:

```
vpcorr1a = function(kmax,z,epsi){
 PAC = numeric(length = kmax)
 nz = length(z)
 for (s in 1:kmax){
   ncol = s
   ndata = nz -s
   Y = matrix(z[(s+1):nz], nrow = ndata,ncol = 1)
   X = list()
   for (j in 1:s){
     t_ini = s + 1 - j
     t_end = t_ini + ndata - 1
     X[[j]] = matrix(z[t_ini:t_end], nrow = ndata,ncol = 1)
   }
   A = matrix(0, nrow = s, ncol = s)
   B = matrix(0, nrow = s, ncol = 1)
   for (i1 in 1:S){
     B[[i1,1]] = cov(X[[i1]],Y)
     for (j1 in 1:s){
       A[[i1,j1]] = cov(X[[i1]],X[[j1]])
     }
     }
     cof = solve(A +epsi*diag(s),B)
     PAC[[s]] = cof[[s]]
   }
   return(PAC)
}
```

It is also possible to use `lm` to fit a linear model to the data and retrieve the coefficients:

```
vpcorr2 = function(kmax,z){
 PAC = numeric(length = kmax)
 nz = length(z)
 for (s in 1:kmax){
   ncol = s
   ndata = nz -s
   Y = matrix(z[(s+1):nz], nrow = ndata,ncol = 1)
   matrix(0,nrow = ndata,ncol = s+2)
   for (j in 1:s){
     t_ini = s + 1 - j
     t_end = t_ini + ndata - 1
     X[,j+1]=matrix(z[t_ini:t_end], nrow=ndata,ncol=1)
   }
   X[ ,s +2] = c(rep(1,ndata))
   X[ ,1] = c(Y)
   ccc = character(length=s+2)
   ccc[[1]] = "Y"
   for (j in 1: (s+1)){
       ccc[[j+1]] = paste0("X",j)
```

```
    }
    ddd <- data.frame(X)
    colnames(ddd) <- ccc
    mod1 = lm(ddd$Y~.-1, data=ddd)
    PACF[[s]] = mod1$coefficients[[s]]
 }
  return(PAC)
 }
vpcorr2a = function(kmax,z){
 PAC = numeric(length = kmax)
 nz = length(z)
 for (s in 1:kmax){
    ncol = s
    ndata = nz -s
    Y = matrix(z[(s+1):nz], nrow = ndata,ncol = 1)
    matrix(0,nrow = ndata,ncol = s+1)
    for (j in 1:s){
    t_ini = s + 1 - j
    t_end = t_ini + ndata - 1
    X[,j+1]=matrix(z[t_ini:t_end], nrow=ndata,ncol=1)
    X[,j+1] = X[,j+1] - mean(X[,j+1])
  }
  X[ ,1] = c(Y)
  X[,1] = X[,1] - mean(X[,1])
  ccc = character(length=s+1)
  ccc[[1]] = "Y"
  for (j in 1: s){
      ccc[[j+1]] = paste0("X",j)
  }
  ddd <- data.frame(X)
  colnames(ddd) <- ccc
  mod1 = lm(ddd$Y~.-1, data=ddd)
  PACF[[s]] = mod1$coefficients[[s]]
 }
 return(PAC)
}
```

Remark 4.4
If s is small, an alternative is the use of Cramer's formula for the solution of the linear system – since we are interested only in the coefficient c_s, we can determine it directly as:

$$c_s = \frac{\det A_s}{\det A},$$

where A_s is the matrix generated by replacing the column s of A by B, id est,

$$(A_s)_{ij} = A_{ij}, if\ j \neq s \ ; \ (A_s)_{is} = B_i.$$

Example 4.4

Let us consider a white noise ε (see Sect. 4.3) and the processes defined by

$$X_n = \varepsilon_n + \varepsilon_{n-1} + 2\varepsilon_{n-2} - \varepsilon_{n-3} + 2\varepsilon_{n-4};$$

$$Y_i = \varepsilon_n + \varepsilon_{n-1} + 2\varepsilon_{n-2} + \varepsilon_{n-3} + 2\varepsilon_{n-4} + 3\varepsilon_{n-5} + 4\varepsilon_{n-6}.$$

Let us evaluate the partial autocorrelations $\rho_X(s)$, for $1 \leq s \leq 20$: we can use either

```
pacx = acf(x,lag.max = 20, type = "partial")
```

or

```
pacx = pacf(x,lag.max = 20)
```

These instructions create a list `pacx` having a field named `"acf"`, which contains the values of the partial autocorrelation and a field named "`lags`", containing the lags s. By default, the instruction generates a plot as in Fig. 4.11.

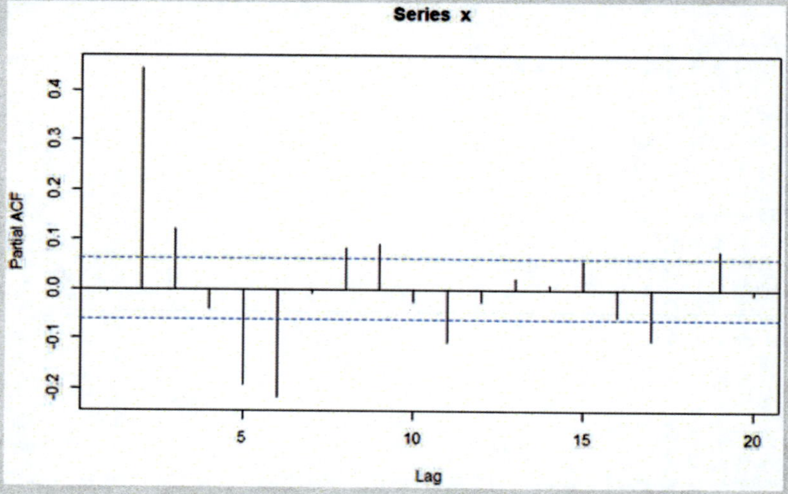

Fig. 4.11 Plot produced by `pacf` (or `acf` with `type = "partial")`, by default

Analogously to the preceding example, if you do not desire the plot, you must use the parameter `plot = FALSE`. For instance,

```
pacx = pacf(x,lag.max = 20,plot = FALSE)
```

generates the list `pacx`, but does not plot the results. Again, you can plot it by later. In Figs. 4.12 and 4.13, we compare the results of the different methods of

(continued)

Example 4.4 (continued)

evaluation presented (`acf partial`, `pacf`, `vpcorr1`, `vpcorr1a`, `vpcorr2`, `vpcorr2a`).

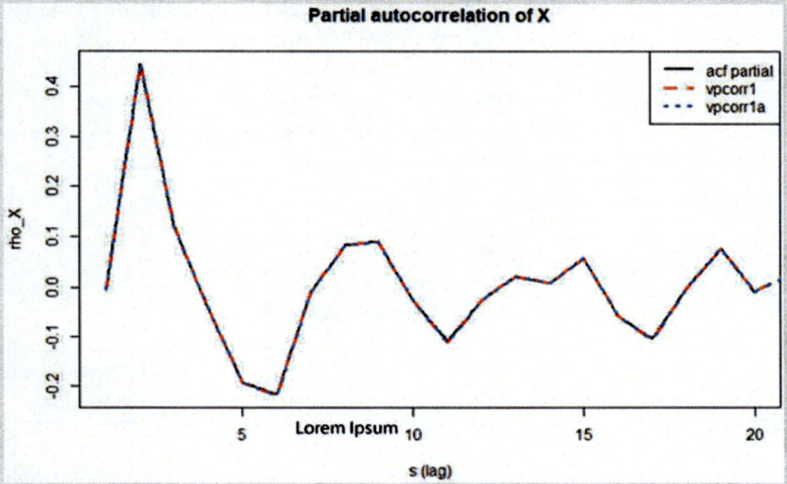

Fig. 4.12 Plot of the results of `acf partial`, `vpcorr1`, `vpcorr1a`

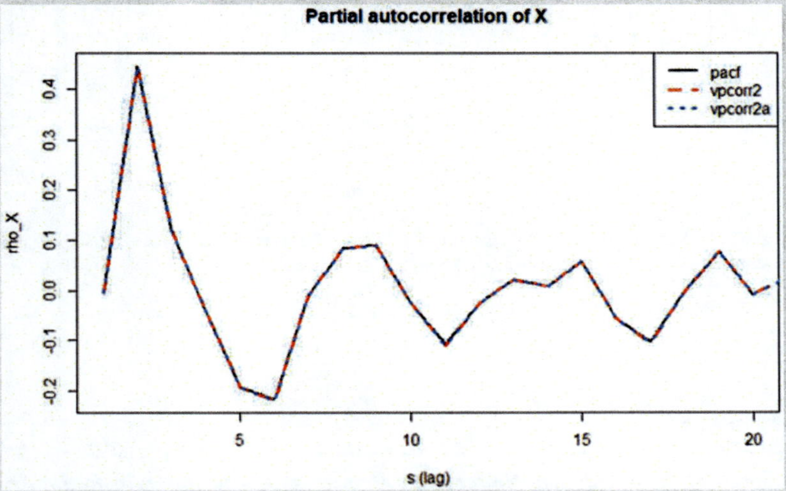

Fig. 4.13 Plot of the results of `pacf`, `vpcorr2`, `vpcorr2a`

Example 4.5
Let us consider the times $t_i = i\Delta t$, $\Delta t = 0.01$, $0 \le i \le 1000$ and the signal defined by $X_i = \sin(2\pi t_i)$. ***Notice that it is a deterministic non-stationary signal***. Let us evaluate the partial autocorrelations $\rho_X(s)$, for $1 \le s \le 20$. In this case, the methods produce results that are significantly different, due to the ill conditioning of the matrices A in `vpcorr1`, `vpcorr1a`. We use a parameter of regularization:

```
pacx_acf  = acf(x,lag.max = 20, type = "partial")
pacx_pacf = pacf(x,lag.max = 20)
pacx1  = ss1$vpcorr1(20,x,1e-9)
pacx1a = ss1$vpcorr1a(20,x,1e-9)
```

The results appear in Fig. 4.14.

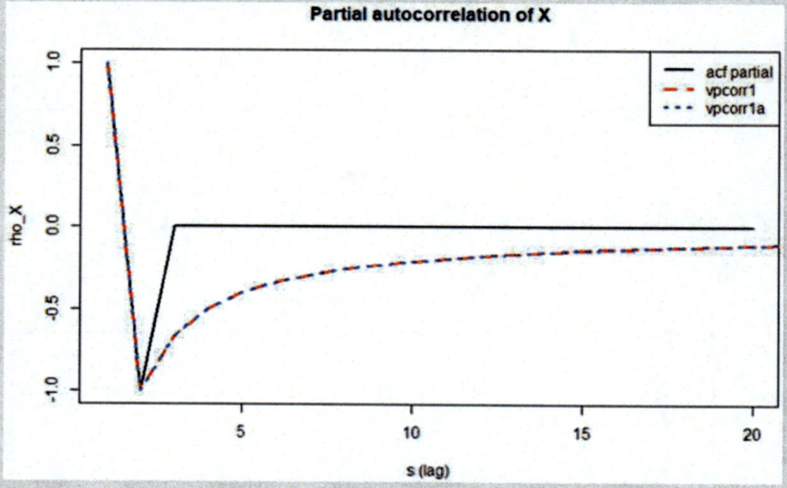

Fig. 4.14 Results obtained in Example 4.5: the matrices of the linear systems are ill-conditioned in `vpcorr1` and `vpcorr1a`

Example 4.6
Let us consider the times $t_i = i\Delta t$, $\Delta t = 0.01$, $0 \le i \le 1000$ and the signal defined by $X_i = \exp\left(\frac{t_i}{4}\right)$. ***Notice that it is a deterministic non-stationary***

(continued)

Example 4.6 (continued)

signal. Let us evaluate the partial autocorrelations $\rho_X(s)$, for $1 \leq s \leq 20$. In this case, the methods produce results that are significantly different, due to the ill conditioning of the matrices A in `vpcorr1`, `vpcorr1a`. We use a parameter of regularization:

```
pacx_acf = acf(x,lag.max = 20, type = "partial")
pacx_pacf = pacf(x,lag.max = 20)
pacx1 = ss1$vpcorr1(20,x,1e-9)
pacx1a = ss1$vpcorr1a(20,x,1e-9)
```

The results appear in Fig. 4.15.

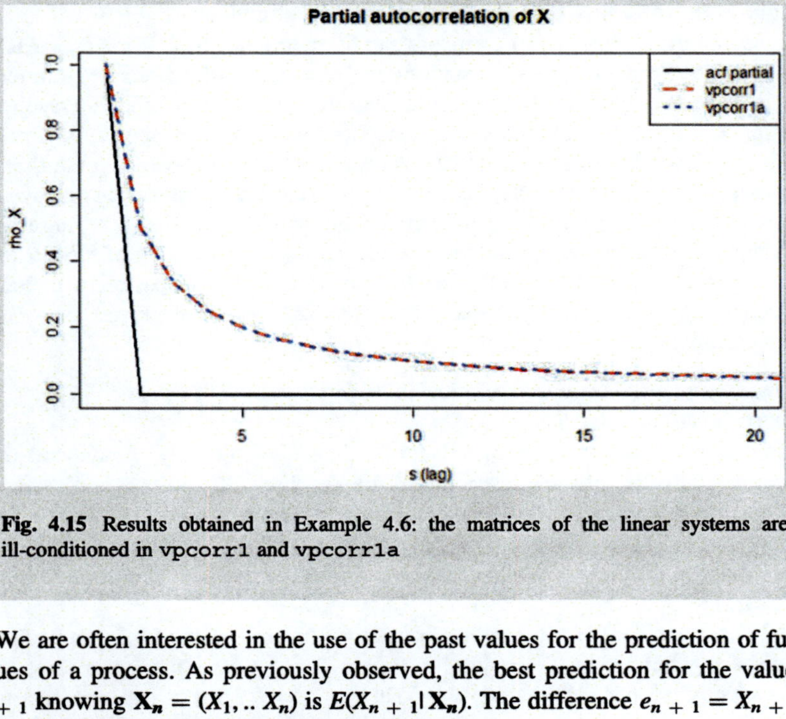

Fig. 4.15 Results obtained in Example 4.6: the matrices of the linear systems are ill-conditioned in `vpcorr1` and `vpcorr1a`

We are often interested in the use of the past values for the prediction of future values of a process. As previously observed, the best prediction for the value of X_{n+1} knowing $\mathbf{X}_n = (X_1, .. X_n)$ is $E(X_{n+1} | \mathbf{X}_n)$. The difference $e_{n+1} = X_{n+1} - E(X_{n+1} | \mathbf{X}_n)$ is the **innovation process**. We have

$$E(e_n) = 0; \quad E(e_m e_n) = 0 \text{ if } m \neq n.$$

We say that the process is **deterministic** if and only if $e_n = 0, \forall n$. Notice that a deterministic process remains random: it *is not* deterministic in the sense where X_n is not a random variable: in the context of stochastic processes, the expression deterministic means that the innovation is null.

It is usual to separate the processes into two parts: the trend and the fluctuations. In general, the trend is first determined and then the fluctuations are studied. Trend may be deterministic or stochastic.

To achieve such a decomposition, we write $X_t = f(t) + \delta X_t$, where f is a deterministic function (for instance, $f(t) = m_X(t)$). If there exists f such that δX_t is stationary, then we say that X_t is **stationary in trend.**

We say that the trend is **seasonal** when the trend is a periodic function.

As an alternative, we may write $X_{t+1} = X_t + \delta X_t$. If $\delta X_t = X_{t+1} - X_t$ is stationary, we say that the process is **stationary in difference** or **integrated.**

More generally, we may consider $\delta^2 X_t = \delta X_{t+1} - \delta X_t, \ldots, \delta^n X_t$ processes that are **stationary in difference of order k** or **integrated of order k.**

Checking the stationarity of a process can be an arduous task. For AR or ARMA processes (see Sects. 4.5 and 4.6), stationarity tests exist in the literature, referred as "unit root tests", such as the Dickey–Fuller test (1979), the Philips–Perron test (Phillips & Perron, 1988), the LMC test (Leybourne & McCabe, 1994), or the KPSS test (Kwiatkowski et al., 1992). As mentioned, these tests assume that the process is AR (see Sect. 4.5). Sometimes, they also use a trend model: in practice, the trend model is either assumed or estimated from data – in the first case, arguments are furnished to justify the choice; in the second one, a procedure of estimation is used (for instance, least squares). The R repositories propose packages including functions to perform some of these tests. For instance, the packages `tseries`, `urca`, or `aTSA`.

Example 4.7

Let $X_{n+1} = X_n + \varepsilon_n$, where $X_0 = 0$ and $\{\varepsilon_n : n \in N\}$ is a family of independent gaussian variables of same distribution $N(0, 1)$. Then, $\delta X_n = X_{n+1} - X_n = \varepsilon_n$. Thus, the distribution of δX_n is independent from t: we have $m_{\delta X}(t) = 0$, $V_{\delta X}(t) = 1$, $R_{\delta X \delta X}(s, t) = 0$, *if* $s \neq t$, $R_{\delta X \delta X}(t, t) = 1$. Thus, the process is stationary in difference. We have $m_X(t) = 0$, $V_X(t) = t$, $R_{XX}(t, s) = \min\{t, s\}$, $C_{XX}(t, s) = \min\{t, s\}$.

Let us consider the partial autocorrelations: the projection of X_{n+1} onto the subspace generated by $\mathcal{X}_{1,n} = \{X_n\}$ is $P_n X_{n+1} = X_n$ (since their coefficient of linear correlation is $\rho(X_{n+1}, X_n) = 1$). Thus, $\rho(1) = 1$. For $k > 1$,the projection of X_{n+k} onto the subspace generated by $\mathcal{X}_{k,n} = \{X_n, \ldots, X_{n+k-1}\}$ is $P_n X_{n+k} = X_{n+k-1}$, so that $\rho(k) = 0$.

(continued)

Example 4.7 (continued)

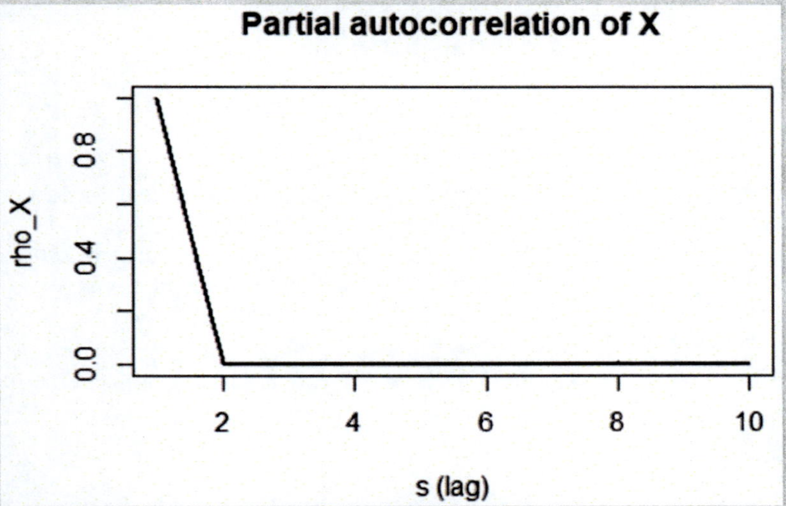

Fig. 4.16 Results furnished by `pacf`

Indeed, the code below

```
X = cumsum(rnorm(1e5))
pcx = pacf(X,10,plot=FALSE)
plot.new()
plot(pcx$lag,pcx$acf, type = "n", xlab = "s (lag)", ylab = "rho_X",
     main = "Partial autocorrelation of X")
lines(pcx$lag,pcx$acf,lwd=2, ylim=c(-1,1))
```

produces the results in Fig. 4.16.

Let us use the *R* libraries to test the stationarity. For instance, the code

```
library(tseries)
resu = adf.test(X)
```

performs the Dickey–Fuller test for the hypothesis of non-stationarity. It creates an object `resu` containing fields `statistic` (the value of the test statistic), `parameter` (the lag order of the autoregressive model), `p.value` (the odds of the hypothesis being tested) – the hypothesis is rejected if this value is inferior to the significance level (often 0.05). Here, the result is

(continued)

Example 4.7 (continued)

```
> resu

        Augmented Dickey-Fuller Test

data:  X
Dickey-Fuller = -2.3745, Lag order = 46, p-value = 0.4189
alternative hypothesis: stationary
```

As we see, the test parameter `p.value` is equal to 0.4189: the test does not reject the null hypothesis, so that that X is not stationary.

We can also use the KPSS test: now, the null hypothesis is that the series is stationary. To use this test, we must choose if there is a trend or not.

For instance, the instructions

```
resu2 = kpss.test(X, null = "Trend")

resu3 = kpss.test(X, null = "Level")
```

```
> resu2$p.value [1] 0.01
> resu3$p.value [1] 0.01
```

produce the results at right.

The test rejects the stationarity in both the cases.

Example 4.8

Let $X_{n+1} = h(n) + \frac{1}{2}X_n + \varepsilon_n$, where $X_0 = 0$ and $\{\varepsilon_n : n \in \mathcal{N}\}$ is a family of independent gaussian variables of same distribution $N(0, 1)$. Then $X_{n+1} = \sum_{k=0}^{n} \frac{h(n-k)}{2^k} + \sum_{k=0}^{n} \frac{\varepsilon_{n-k}}{2^k}$. Thus, the trend is $f(n) = \sum_{k=0}^{n} \frac{h(n-k)}{2^k}$ and $\delta X_n = \sum_{k=0}^{n} \frac{\varepsilon_{n-k}}{2^k}$. We have $m_{\delta X}(n) = 0$, $V_{\delta X}(n) = \frac{4}{3}\left(1 - \frac{1}{4^{n+1}}\right)$. Thus, the process is not stationary in trend.

Example 4.9

Let $X_{n+1} = h(n) + \frac{1}{2}X_n + \varepsilon_n$, where $\{\varepsilon_n : n \in \mathbb{Z}\}$ is a family of independent gaussian variables of same distribution $N(0, 1)$. There is a trend-stationary version of the process, by taking $X_n = \sum_{k=0}^{+\infty} \frac{h(n-k)}{2^k} + \sum_{k=0}^{+\infty} \frac{\varepsilon_{n-k}}{2^k}$. Then,

(continued)

Example 4.9 (continued)

the trend is $f(n)=\sum_{k=0}^{+\infty}\frac{h(n-k)}{2^k}$ and $\delta X_n=\sum_{k=0}^{+\infty}\frac{\varepsilon_{n-k}}{2^k}$. We have $m_{\delta X}(n)=0$, $V_{\delta X}(t)=\frac{4}{3}$. The distribution of δX_n is $N\left(0,\frac{2}{\sqrt{3}}\right)$,$\forall n$.

Exercises

1. Let X_t be a discrete variable Poisson distributed with parameter λt, id est,

$$P(X_t=n)=\frac{(\lambda t)^n}{n!}e^{-\lambda t}.$$

 (a) Show that $m_X(t)=\lambda t$
 (b) Show that $V_X(t)=\lambda t$.
 (c) Show that $R_{XX}(s,t)=\lambda\min(s,t)+\lambda^2 st$.
 (d) Determine $C_{XX}(s,t)$
 (e) Is this process stationary?

2. Let $X_t=A_1\cos(\omega t)+A_2\sin(\omega t)$, where $\omega\in\mathbb{R}$, while A_1, A_2 are independent random variables such that $E(A_i)=m_i, V(A_i)=\sigma_i^2$.

 (a) Show that $m_X(t)=m_1\cos(\omega t)+m_2\sin(\omega t)$;
 (b) Show that $V_X(t)=\sigma_1^2\cos^2(\omega t)+\sigma_2^2\sin^2(\omega t)$;
 (c) Show that $C_{XX}(s,t)=\sigma_1^2\cos(\omega t)\cos(\omega s)+\sigma_2^2\sin(\omega t)\sin(\omega s)$.
 (d) Determine $R_{XX}(s,t)$
 (e) Is this process stationary?

3. Let $X_{n+1}=a+X_n+\varepsilon_n$, where $X_0=0$ and $\{\varepsilon_n : n\in N\}$ is a family of independent gaussian variables of same distribution $N(0,1)$.

 (a) Show that $X_{n+1}=(n+1)a+\sum_{k=0}^{n}\varepsilon_k$
 (b) Show that $m_X(n)=na$.
 (c) Show that $V_X(n)=n$.
 (d) Show that $R_{XX}(m,n)=mna^2+\min\{m,n\}$
 (e) Analyze the process $Y_n=X_{n+1}-X_n$.
 (f) Is this process stationary of some kind?
 (g) Generate a realization from X and use it to evaluate the partial autocorrelation $\rho_X(s)$.

(continued)

4. Let $\{\varepsilon_n : n \in \mathbb{N}\}$ be a family of independent gaussian variables of same distribution $N(0, 1)$ and $X_{2k} = 0$, $X_{2k+1} = 2\varepsilon_{2k-1}$
   (a) Show that $m_X(n) = 0$.
   (b) Show that $V_X(n) = 0$, *if n pair*; $V_X(n) = 4$, *ifn odd.*
   (c) Show that $R_{XX}(m, n) = 0$ *if* $m \neq n$;
   (d) Determine $R_{XX}(m, m)$
   (e) Analyze the process $Y_n = X_n - X_{n-1}$.
   (f) Is this process stationary of some kind?
   (g) Generate a realization from X and use it to evaluate the partial autocorrelation $\rho_X(s)$.
5. Let $\{\varepsilon_n : n \in \mathbb{N}\}$ be a family of independent gaussian variables of same distribution $N(0, 1)$ and $X_n = \varepsilon_n \varepsilon_{n+1} \varepsilon_{n+2}$
   (a) Show that $m_X(n) = 0$.
   (b) Show that $V_X(n) = 1$.
   (c) Show that $R_{XX}(m, n) = 0$ *if* $m \neq n$
   (d) Determine $R_{XX}(m, m)$
   (e) Is this process stationary of some kind?
   (f) Generate a realization from X and use it to evaluate the partial autocorrelation $\rho_X(s)$.
6. Let $\{\varepsilon_n : n \in \mathbb{N}\}$ be a family of independent gaussian variables of same distribution $N(0, 1)$ and $X_n = \varepsilon_n + \varepsilon_{n+1} + \varepsilon_{n+2}$
   (a) Show that $m_X(n) = 0$.
   (b) Show that $V_X(n) = 3$.
   (c) Show that $R_{XX}(m, n) = 0$ *if* $|m - n| \geq 3$
   (d) Determine $R_{XX}(m, m)$, $R_{XX}(m, m + 1)$, $R_{XX}(m, m + 2)$
   (e) Is this process stationary of some kind?
   (f) Generate a realization from X and use it to evaluate the partial autocorrelation $\rho_X(s)$.
7. Let $\{\varepsilon_n : n \in \mathbb{N}\}$ be a family of independent gaussian variables of same distribution $N(0, 1)$ and $X_n = \varepsilon_n + a\varepsilon_{n+1}$.
   (a) Show that $m_X(n) = 0$.
   (b) Show that $V_X(n) = 1 + a^2$.
   (c) Show that $R_{XX}(m, n) = 0$ *if* $|m - n| \geq 1$
   (d) Determine $R_{XX}(m, m)$, $R_{XX}(m, m + 1)$
   (e) Is this process stationary of some kind?
   (f) Generate a realization from X and use it to evaluate the partial autocorrelation $\rho_X(s)$.

(continued)

8. Let $\{\varepsilon_n : n \in \mathbb{N}\}$ be a family of independent gaussian variables of same distribution $N(0, 1)$ and $X_n = an + b + a_0\varepsilon_n$.
   (a) Determine $m_X(n)$.
   (b) Determine $V_X(n)$.
   (c) Is this process stationary?
   (d) Generate a realization from X and use it to evaluate the partial autocorrelation $\rho_X(s)$.
9. Let $\{\varepsilon_n : n \in \mathbb{N}\}$ be a family of independent gaussian variables of same distribution $N(0, 1)$ and $X_n = an + b + a_0\varepsilon_n + a_1\varepsilon_{n-1}$.
   (a) Determine $m_X(n)$.
   (b) Determine $V_X(n)$.
   (c) Is this process stationary?
   (d) Generate a realization from X for and use it to evaluate the partial autocorrelation $\rho_X(s)$.
10. Let $\{\varepsilon_n : n \in \mathbb{N}\}$ be a family of independent gaussian variables of same distribution $N(0, 1)$ and $X_n = h(n) + a_0\varepsilon_n + a_1\varepsilon_{n-1}$ and $Y_n = X_n - h(n)$
    (a) Determine $m_Y(n)$. Conclude $m_X(n)$
    (b) Determine $V_Y(n)$. Infer the value of $V_X(n)$.
    (c) Let $c_0, c_1, \ldots, c_k$ be such that $E\left(\sum_t (X_n - p(n))^2\right)$ is minimal, with $p(n) = c_0 + c_1 n + \ldots + c_k n^k$. Show that $\sum_t (h(n) - p(n))^2$ is minimal.

4.2 Determination of the Distribution of a Stationary Process

If X is stationary, then the probability distribution of X_n is the same for all the values of n. In such a situation, we may use a trajectory to determine the common distribution, by considering the points in the trajectory as a sample from this common distribution. For instance, we may determine a representation of X by one of the methods presented in Chap. 3. Once the representation is determined, if furnishes the CDF and the PDF of X, as shown in Chap. 3.

Example 4.10
Let us consider the situation where the available data consist in the values of $X_n = 1 + \varepsilon_n - 2\varepsilon_{n-1} + 0.5\varepsilon_{n-2}$, for $n \leq 250$, where $\{\varepsilon_n : n \in \mathbb{Z}\}$ is a

(continued)

Example 4.10 (continued)

family of independent gaussian variables of same distribution $N(0, 1)$ – it is a gaussian white noise (see Sect. 4.3). The data used are represented in Fig. 4.17.

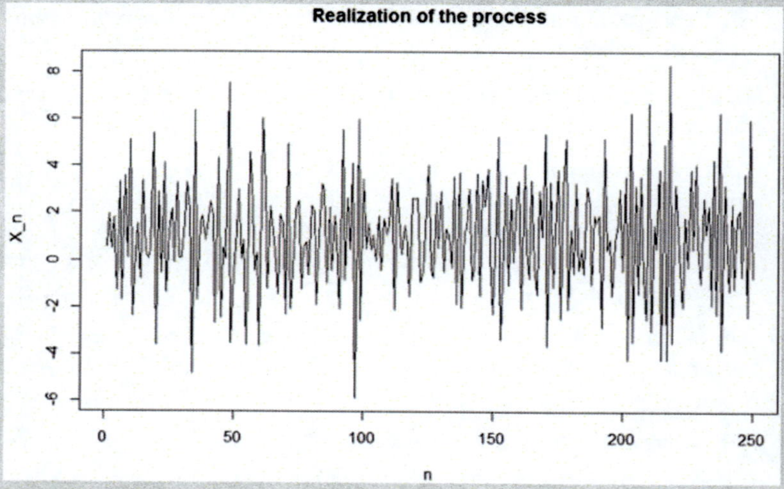

Fig. 4.17 Data used in this example

We use the data to generate a representation $X_n = \boldsymbol{\Phi}(A)\boldsymbol{x}$, involving an artificial variable A. In a first attempt, we use $A \sim U(0, 1)$ and a polynomial representation with $k = 6$. The corresponding CDF is shown in Fig. 4.18 and the PDF is exhibited in Fig. 4.19. The results can be improved by generating nr CDFs (by changing the sample of A used in the approximation) and taking the punctual mean of the values obtained. For instance, the CDF generated by $nr = 100$ appears in Fig. 4.20 and the corresponding PDF appears in Fig. 4.21. The figures suggest that a Gaussian artificial variable $A \sim N(0, 1)$ is more adapted: we present the corresponding results in Figs. 4.22 and 4.23. The mean of $nr = 100$ CDFs generated with $A \sim N(0, 1)$ appears in Fig. 4.24 and the corresponding PDF appears in Fig. 4.25.

(continued)

Example 4.10 (continued)

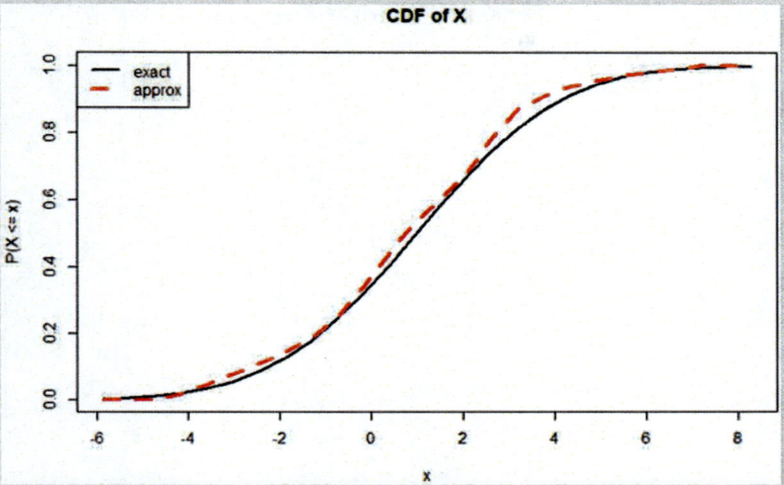

Fig. 4.18 CDF obtained using $A \sim U(0, 1)$

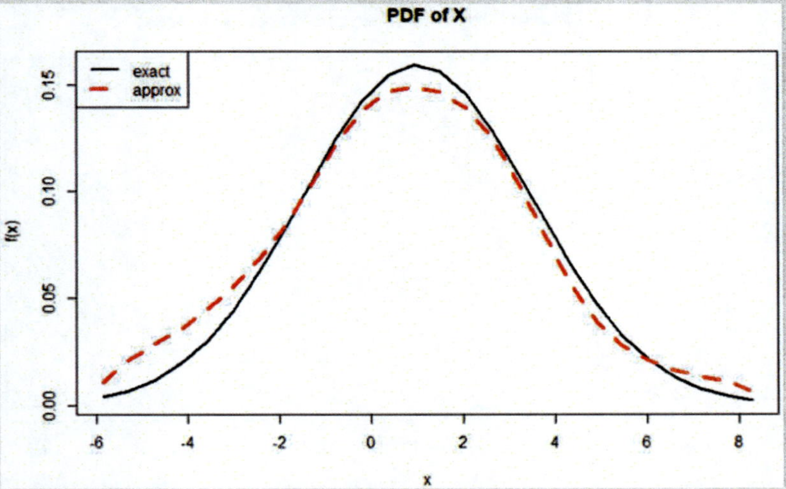

Fig. 4.19 PDF generated by the SPH derivative of the CDF in Fig. 4.20 ($A \sim U(0, 1)$)

(continued)

Example 4.10 (continued)

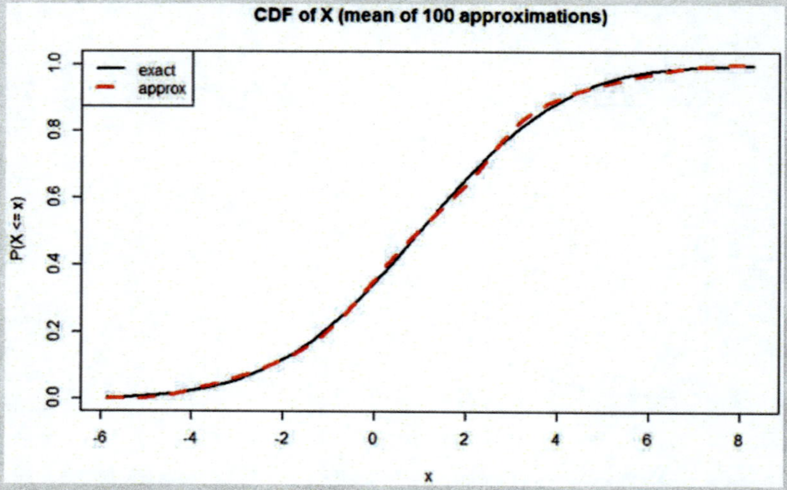

Fig. 4.20 Mean of 100 CDFs generated using $A \sim U(0, 1)$

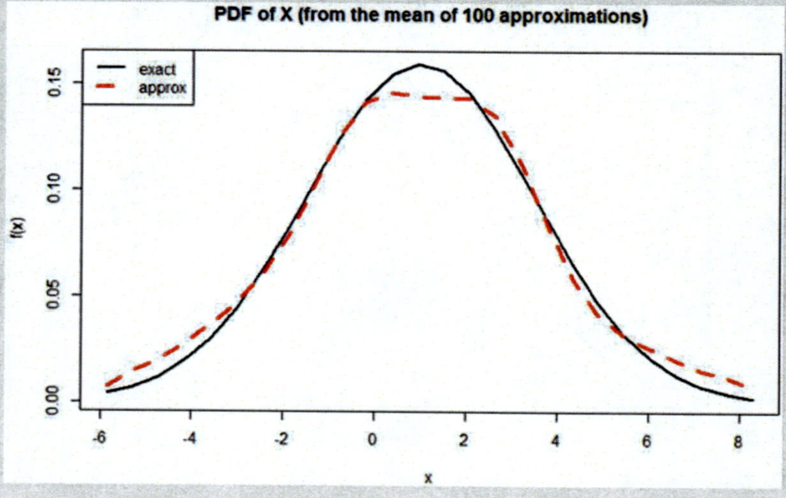

Fig. 4.21 PDF generated by the SPH derivative of the CDF in Fig. 4.20 ($A \sim U(0, 1)$)

(continued)

Example 4.10 (continued)

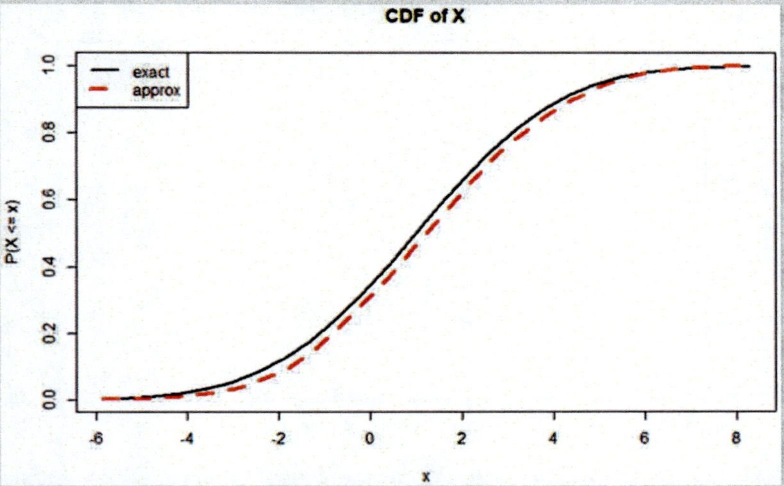

Fig. 4.22 CDF obtained using $A \sim N(0, 1)$

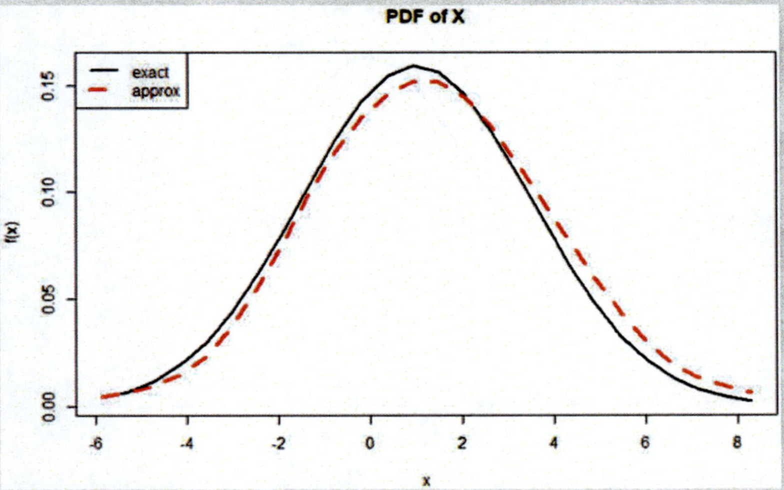

Fig. 4.23 PDF generated by the SPH derivative of the CDF in Fig. 4.22 ($A \sim N(0, 1)$)

(continued)

Example 4.10 (continued)

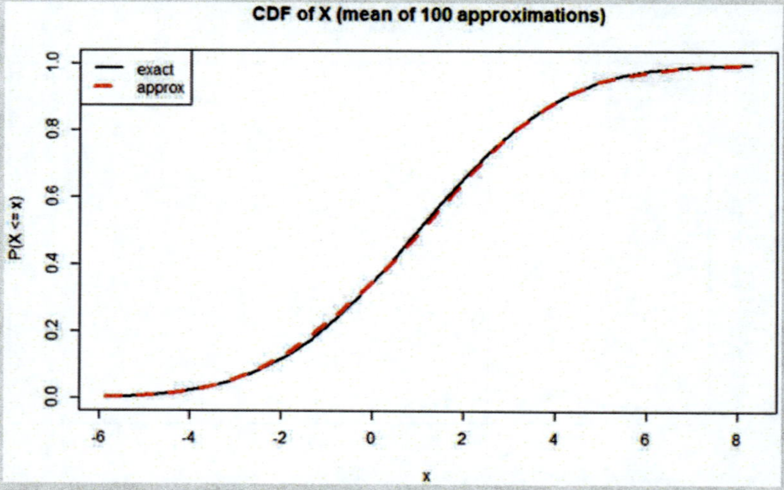

Fig. 4.24 Mean of 100 CDFs generated using $A \sim N(0, 1)$

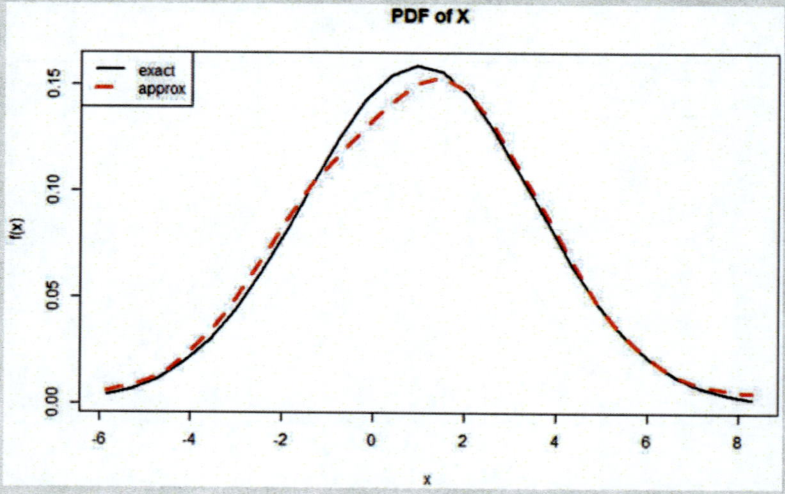

Fig. 4.25 PDF generated by the SPH derivative of the CDF in Fig. 4.20 ($A \sim U(0, 1)$)

4.3 White Noise

Definition 4.6
X is a **white noise process** if and only if X is ergodic and

$$m_X(t) = 0 \text{ and } R_X(\tau) = \sigma^2 \delta(\tau),$$

where δ is the Dirac's mass and $\sigma > 0$.

Notice that the definition does not mention the distribution of X_t: indeed, a white noise may be generated by different distributions. For instance, a white noise generated by an uniform distribution is an *uniform white noise*. Analogously, a white noise generated by a gaussian distribution is a *gaussian white noise*. To generate a white noise, we must generate independent variables for each t. For instance, we may consider X_t as uniformly distributed on $\left(-\sigma\sqrt{3}, \sigma\sqrt{3}\right)$ or normally distributed $N(0, \sigma)$. Otherwise, R offers the possibility of simulating a white noise using the function `arima.sim`.

Example 4.11
As an example, we may generate $1E4$ variates from the uniform distribution on $\left(-\sqrt{3}, \sqrt{3}\right)$. The random numbers generated simulate a white noise on the time interval $(0, 10)$, with a time step $\Delta t = 0.001$: the element number i of the sample corresponds to $X(t_i)$, $t_i = (i - 1)\Delta t$. We use this data to evaluate the autocorrelation for $t \in (-1, 1)$ – the results on $(0, 1)$ are extended to $(-1, 0)$ by symmetry. The results are shown in Fig. 4.26.

```
x = runif(1e4,-sqrt(3),sqrt(3))
aux = acf(x,lag.max=1000,type="correlation",plot = FALSE)
tc = 0.001*c(-rev(aux$lag[2:1001]), aux$lag)
Rx = c(rev(aux$acf[2:1001]), aux$acf)
```

(continued)

Example 4.11 (continued)

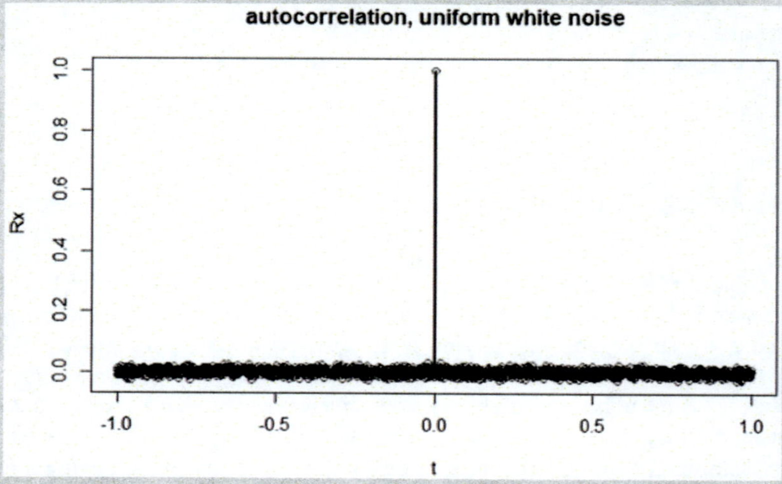

Fig. 4.26 Autocorrelation of an uniform white noise generated by `runif`

We can use `arima.sim` as follows:

```
wn1 = arima.sim(model = list(order = c(0, 0, 0)), n = 1e4, rand.gen = runif)
aux = acf(wn1,lag.max=1000,type="correlation",plot = FALSE)
tc = 0.001*c(-rev(aux$lag[2:1001]), aux$lag)
Rx = c(rev(aux$acf[2:1001]), aux$acf)
```

The result is shown in Fig. 4.27 – it is similar to the preceding one

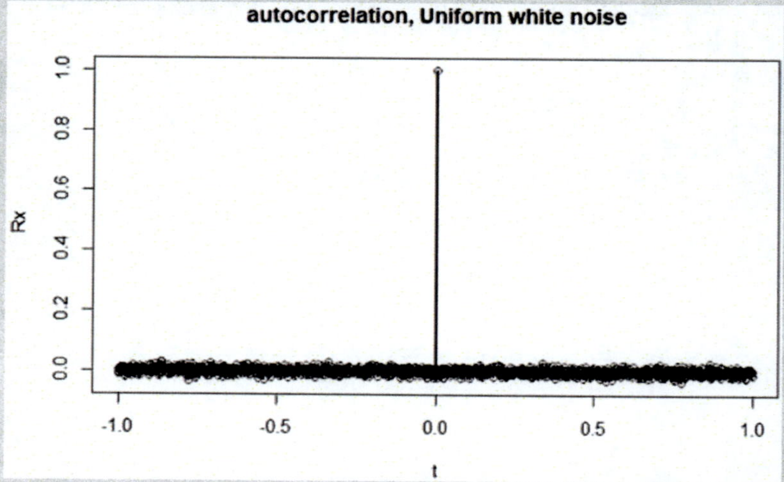

Fig. 4.27 Autocorrelation of an uniform white noise generated by `arima.sim`

(continued)

Example 4.11 (continued)

To generate a gaussian white noise, we can consider $1E4$ variates from $N(0, 1)$. The results are shown in Fig. 4.28.

```
e = rnorm(1e4)
aux = acf(x,lag.max=1000,type="correlation",plot = FALSE)
tc = 0.001*c(-rev(aux$lag[2:1001]), aux$lag)
Rx = c(rev(aux$acf[2:1001]), aux$acf)
```

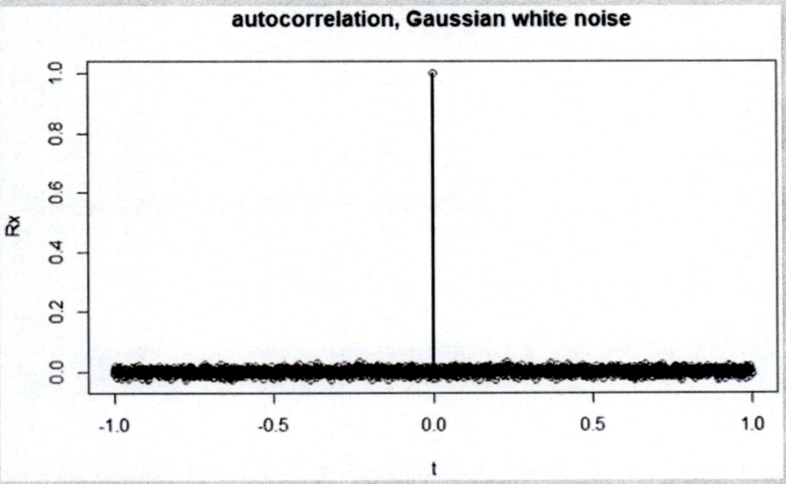

Fig. 4.28 Autocorrelation of a gaussian white noise

Analogously, we can use `arima.sim`:

```
wn2 = arima.sim(model = list(order = c(0, 0, 0)), n = 1e4, rand.gen = rnorm)
aux = acf(wn2,lag.max=1000,type="correlation",plot = FALSE)
tc = 0.001*c(-rev(aux$lag[2:1001]), aux$lag)
Rx = c(rev(aux$acf[2:1001]), aux$acf)
```

(continued)

Example 4.11 (continued)

The result appears in Fig. 4.29 – it is similar to the preceding one.

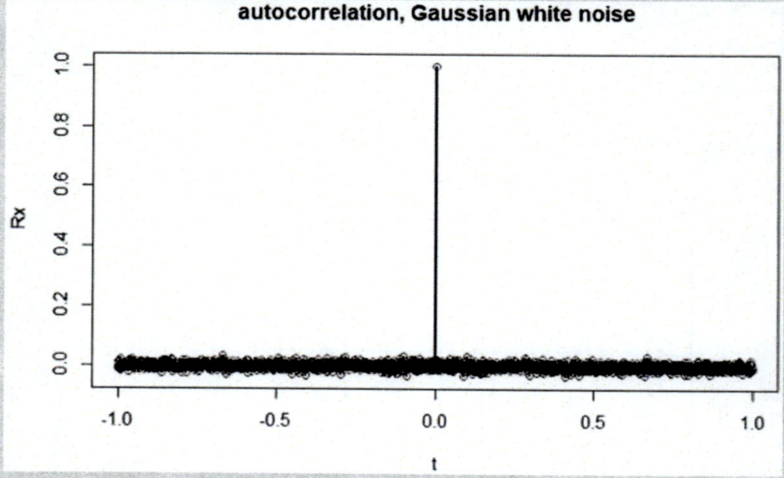

Fig. 4.29 Autocorrelation of a gaussian white noise generated by `arima.sim`

Proposition 4.2
Let $\{e_n : n \in \mathbb{N}\}$ be an innovation process and $E\left(e_n^2\right) = \sigma^2, \forall n$; then, e_n is a white noise.

Proof
We have, for any innovation process,

$$E(e_n) = 0 = \mathrm{m}_0; \quad E(e_m e_n) = 0 \textit{ if } m \neq n.$$

In addition, $E\left(e_n^2\right) = \sigma^2$. Thus, the process is weakly stationary. We have $R_e(k) = 0$, if $k \neq 0$, $R_e(k) = \sigma^2$. Since

(continued)

$$k \neq 0 \Rightarrow E\left(\frac{1}{n+1}\sum_{k=0}^{n} e_n e_{n+k}\right) = \frac{1}{n+1}\sum_{k=0}^{n} E(e_n e_{n+k}) = 0 = R_e(k)$$

$$k = 0 \Rightarrow E\left(\frac{1}{n+1}\sum_{k=0}^{n} e_n e_{n+k}\right) = \frac{1}{n+1}\sum_{k=0}^{n} E(e_n^2) = \sigma^2 = R_e(0)$$

$$E\left(\frac{1}{n+1}\sum_{k=0}^{n} e_n\right) = \frac{1}{n+1}\sum_{k=0}^{n} E(e_n) = 0 = m_0$$

Thus, the process is ergodic in mean, variance, covariance and correlation.

4.4 Moving Average Processes

Definition 4.7
X is a **moving average of order k**, denoted $MA(k)$, if and only if there exists a discrete white noise $\varepsilon = \{\varepsilon_n : n \in \mathbb{Z}\}$, a number $m \in \mathbb{R}$, and a vector of coefficients $\boldsymbol{a} = (a_1, \ldots, a_k)$ such that

$$\forall n \in \mathbb{Z} : X_n = m + \varepsilon_n + \sum_{i=1}^{k} a_i \varepsilon_{n-i} = c + \sum_{i=0}^{k} a_i \varepsilon_{n-i} \ (a_0 = 1).$$

A moving average process is weakly stationary. Then,

$$E(X_n) = m, \ V(X_n) = \sigma^2 \sum_{i=0}^{k} a_i^2,$$

where σ^2 is the variance of the white noise ε. The autocorrelation of $MA(k)$ is

$$R_X(p) = E(X_n X_{n+p}) = \begin{cases} m^2 + \sigma^2 \sum_{j=0}^{k-|p|} a_j a_{j+|p|}, & \text{if } |p| < k \\ m^2, \text{ otherwise.} \end{cases}$$

This equation is often used to determine the order k of a process $MA(k)$: indeed, $R_X(p)$ is constant for $p \geq k$. We can use the normalized procedure of `acf(..., type = correlation)` instead of R_X: then the autocorrelation is near zero for $p \geq k$.

The simulation of a MA process for given coefficients m, a_1, ..., a_k can be performed by generating the white noise (see Sect. 4) and using the expression of X_n. Otherwise, you can use `arima.sim`.

Example 4.12

Let us consider the process $X_n = 1 + \varepsilon_n - 2\varepsilon_{n-1} + 0.5\varepsilon_{n-2}$. We can simulate the process as follows:

```
sim.ma = function(m,a,ns) {
  k = length(a)
  e = rnorm(ns+k+1)
  xs = numeric(length = ns)
  for (i in (1:ns)){
    s = m + e[[i+k+1]]
    for (j in 1:k){
      s = s + a[[j]]*e[[i + k-j+1]]
    }
    xs[[i]] = s
  }
  return(xs)
}
```

```
a1 = -2
a2 = 0.5
m = 1
a = c(a1, a2)

ns = 1000
x1 = sim.ma(m,a,ns)
```

An example of result is shown in Fig. 4.30. Notice that R proposes an intrinsic function `ts.plot` (or `plot.ts`) to plot time series – the graphics 4.30 ***is not*** generated by these functions.

To examine the autocorrelation, we generate a larger sample of $ns = 1E5$ variates and we use `acf` to find the autocorrelation (notice that we use `type = "correlation"` to get normalized values).

```
ns = 1e5
x = sim.ma(m,a,ns)
aux = acf(x,lag.max=10,type="correlation")
Rx = aux$acf
```

The result is exhibited in Fig. 4.31 – we observe that the autocorrelation is near zero for $k \geq 3$, so that the order of the process is 2. Notice that `acf` eliminates the mean m. To keep the mean, you must use the parameter `demean = FALSE`.

(continued)

Example 4.12 (continued)

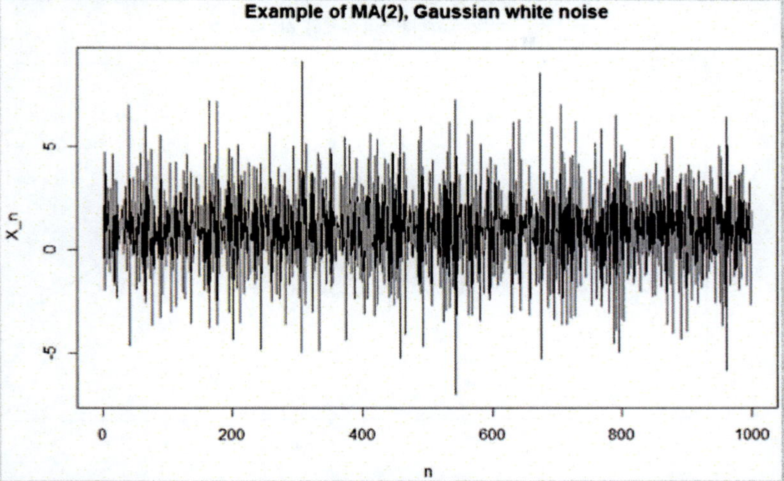

Fig. 4.30 Example of realization of the process $X_n = 1 + \varepsilon_n - 2\varepsilon_{n-1} + 0.5\varepsilon_{n-2}$

We can use `arima.sim`:

```
ma1 = arima.sim(model = list(ma = c(-2,0.5)), n = ns)
x1 = 1 + ma1
aux = acf(x1,lag.max=10,type="correlation",plot = FALSE)
Rx1 = aux$acf
```

The autocorrelation is shown in Fig. 4.32 – it is like the preceding one (the mean c is eliminated, since we did not set `demean = FALSE`)

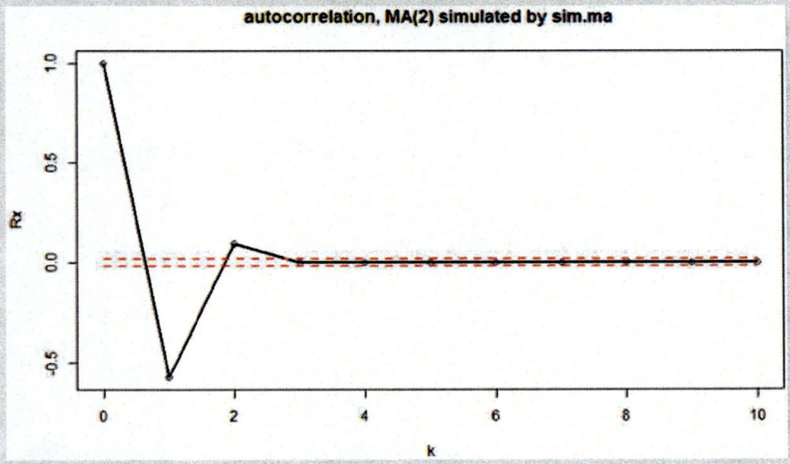

Fig. 4.31 Autocorrelation of the process $X_n = 1 + \varepsilon_n - 2\varepsilon_{n-1} + 0.5\varepsilon_{n-2}$. The values are near zero for $k \geq 3$, what indicates a process of order 2

(continued)

Example 4.12 (continued)

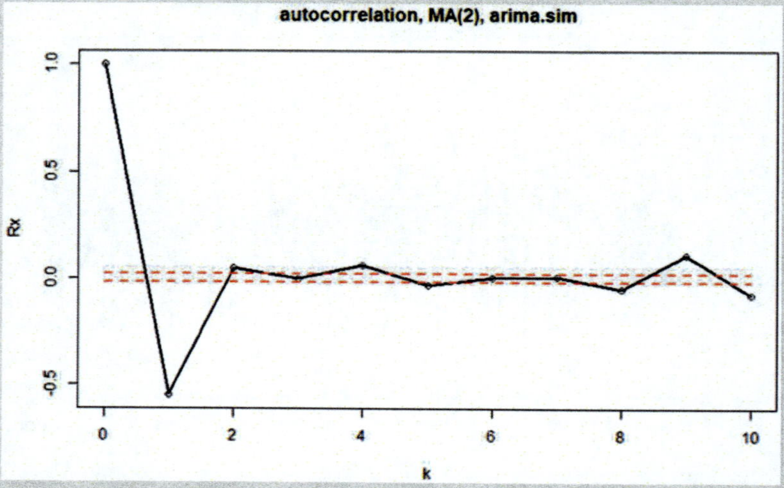

Fig. 4.32 Autocorrelation of the process $X_n = 1 + \varepsilon_n - 2\varepsilon_{n-1} + 0.5\varepsilon_{n-2}$, generated by `arima.sim`

In practice, the situation is often the reverse one: we have the data concerning observations of the process, but we do not know the coefficients and the order – it is necessary to identify the process using the available data. To do this, we start by evaluating the autocorrelations, which furnish the order k of the process. Then, the coefficients of the approximation may be determined from a trajectory of X_n: indeed, we can determine m_X and $R_X$from the trajectory. Since

$$E(X_n) = \mathrm{E}\left(m + \sum_{i=0}^{k} a_i \varepsilon_{n-i}\right) = m, \tag{4.1}$$

so that the mean of the trajectory estimates the constant c. In addition,

$$E(X_n X_{n+p}) = E(m^2) + E\left(\sum_{i,\,j=0}^{k} a_i a_j \varepsilon_{n-i} \varepsilon_{n+p-j}\right),$$

so that, for $0 \le p \le k$:

$$R_X(p) = m^2 + \sigma_\varepsilon^2 \sum_{i=0}^{k-p} a_i a_{i+p},$$

Taking a white noise of variance $\sigma_\varepsilon^2 = 1$, we have:

$$\sum_{i=0}^{k-p} a_i a_{i+p} + \left(m^2 - R_X(p)\right) = 0. \tag{4.2}$$

These equations connect the constant c, the coefficients, and the autocorrelation: they can be used to determine the coefficients: the values of R_X are determined from the data and the coefficients $m, a_1, \ldots, a_k$ are the solution of these equations, to be determined by an adequate method. For instance, we can solve the nonlinear system or minimize the norm of the equations. Under R, the instruction `arima` estimates the coefficients.

Example 4.13
Let us consider again the process $X_n = 1 + \varepsilon_n - 2\varepsilon_{n-1} + 0.5\varepsilon_{n-2}$.
If the coefficients are known, we can use Eq. (4.3) to determine R_X

For instance, we can use the code at right. This code creates a function that receives as arguments a numeric vector $a = (a_1, \ldots, a_k)$ and the mean m. It returns a numeric vector containing $R_X(0), \ldots, R_X(kmax)$.

```
calcr = function(a,mx, kmax){
  Rxth = numeric(length=kmax+1)
  k = length(a)
  aaux = c(1,a)
  for (p in 0: kmax){
    if (k <= (p-1)){
      aux = mx*mx
    }else{
      a1 = aaux[1:(k-p+1)]
      a2 = aaux[(p+1):(k+1)]
      aux = sum(a1*a2) + mx*mx
    }
    Rxth[[p+1]]=aux
  }
  return(Rxth)
}
```

With $a = (-2, 0.5)$, $m = 1$, we obtain $R_X(0) = 6.25$, $R_X(1) = R_X(-1) = -2$, $R_X(2) = R_X(-2) = 1.5$, $R_X(p) = 1$, if $|p| > 2$:

```
> Rx = calcr(c(-2,0.5),1,8)
> print(Rx)
[1]  6.25 -2.00  1.50  1.00  1.00  1.00  1.00  1.00  1.00
```

Assume that the coefficients are unknown, but we have a trajectory of 1000 values:

```
ns = 1000
e = rnorm(ns+2)
ma0 =  e[3:(ns+2)] - 2*e[2:(ns+1)] + 0.5*e[1:ns]
x = ma0 + 1
```

(continued)

Example 4.13 (continued)

To determine the coefficients, we start by evaluating the mean and the autocorrelations:

```
mx = mean(x)
LMAX = 20
source("signal.R")
ss1 = signal$new(x)
Rx = ss1$vautocorr(LMAX)
plot(Rx,type="n",ylab="R_X",xlab = "lag")
lines(Rx)
lines(rep(mx*mx,length(Rx)),col='red',lwd = 3)
```

This code generates the graphics shown in Fig. 4.33

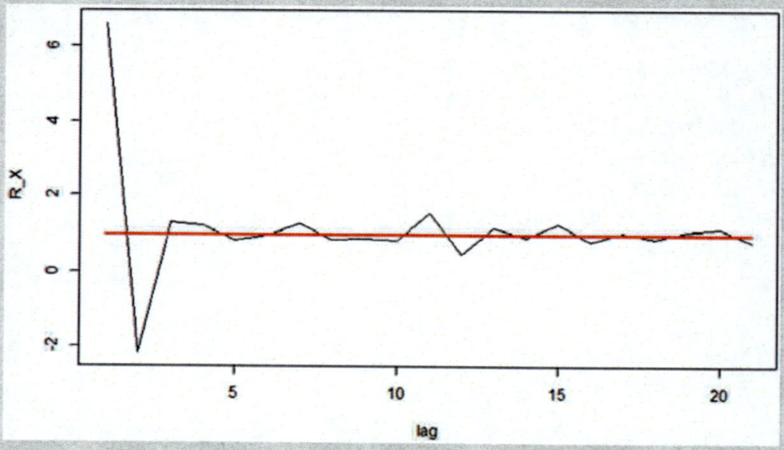

Fig. 4.33 Autocorrelation of X. The values are approximately constant and equal to m_X^2 for $k \geq 3$, what indicates a process of order 2

Alternatively, you can use the code below, which produces a graphics with normalized values. The results appear in Fig. 4.34. In both the cases, the process appears as having order $k = 2$

```
> acf(x,lag.max = 20,type="correlation")
```

(continued)

Example 4.13 (continued)

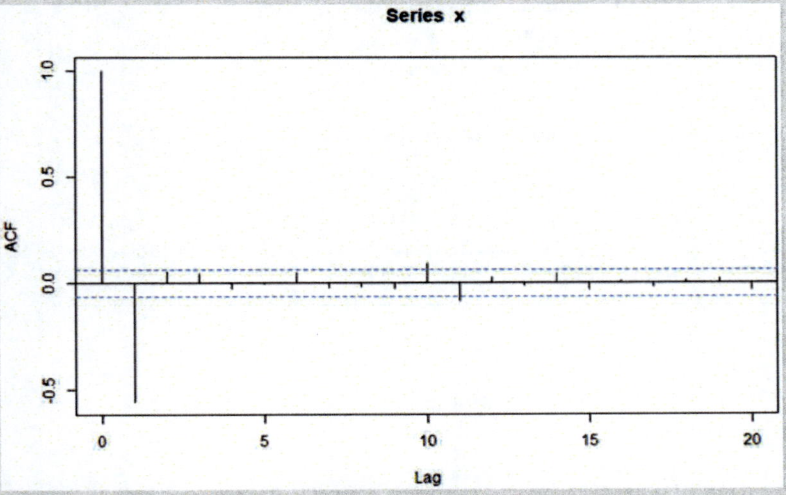

Fig. 4.34 `acf` of X. The values are near zero for $k \geq 3$, what indicates a process of order 2

We must solve the equations

$$\sum_{i=0}^{k-p} a_i a_{i+p} + \left(m^2 - R_X(p)\right) = 0, a_0 = 1.$$

Since the order is $k = 2$ and $a_0 = 1$, the equations reduce to

$$\begin{aligned} 1 + a_1^2 + a_2^2 + m^2 - R_X(0) &= 0 \\ a_1 + a_1 a_2 + m^2 - R_X(1) &= 0, \end{aligned}$$

m is estimated as the empirical mean of data: $m \approx$mx and the values of R_X are estimated from the data:

```
> mx
[1] 0.9913316
```

```
> Rx[1:2]
[1]  6.611236 -2.156701
```

(continued)

Example 4.13 (continued)

We make a first estimation using `arima`:

```
> sol = arima(x,order = c(0,0,2))
> a_ini = sol$coef
> a_ini
       ma1        ma2  intercept
-0.9036008  0.1735254  0.9910963
```

When calling `arima`, you must give the orders `order = c(p,d,q)` for a model *AR*(p), *MA*(q), with differentiation order d. Notice that `arima` also furnish other information, such as the AIC:

```
> sol

Call:
arima(x = x, order = c(0, 0, 2))

Coefficients:
          ma1     ma2  intercept
      -0.9036  0.1735     0.9911
s.e.   0.0320  0.0322     0.0150

sigma^2 estimated as 3.084:  log likelihood = -1982.46,  aic = 3972.93
>
```

Notice also that you can generate confidence intervals for the coefficients:

```
> confint(sol)
               2.5 %     97.5 %
ma1       -0.9663356 -0.8408661
ma2        0.1103820  0.2366689
intercept  0.9616576  1.0205349
```

We use the solution found as initial point to solve the equations: initially, we create the equations, and we evaluate their norm:

```
eqs = function (a){
  Rxa = calcr(a,mx,1)
  aux = sqrt(sum((Rx[1:2] - Rxa[1:2])^2)/2)
  return(aux)
}
```

This function evaluates the RMS residual of the equations. We minimize it, using `a_ini` as starting point:

```
> library(pracma)
> sol = fminunc(eqs,x0=a_ini[1:2])
> sol$par
       ma1        ma2
-2.0078659  0.4855266
```

(continued)

Example 4.13 (continued)

The method estimates $a_1 \approx -2.0$, $a_2 \approx 0.49$. Of course, you can also use `fsolve`:

```
eqs = function (a){
  Rxa = calcr(a,mx,1)
  aux = Rx[1:2] - Rxa[1:2]
  return(aux)
}
```

```
> library(pracma)
> sol = fsolve(eqs,x0=a_ini[1:2])
> sol$x
[1] -2.0078659  0.4855266
```

Alternatively, we can consider m as a variable to be determined: we consider `a[[3]]` $=m$, to be found by `fminunc` or `fsolve` – in this case, we need a supplementary equation for $R_X(2)$, to get 3 equations for 3 unknowns:

```
eqs = function (a){
  Rxa = calcr(a[1:2],a[[3]],2)
  aux = Rx[1:3] - Rxa[1:3]
  return(aux)
}
```

```
> library(pracma)
> sol = fsolve(eqs,x0=a_ini)
> sol$x
[1] -2.0  0.5  1.0
```

An important result concerning *MA* processes is the following one:

Theorem 4.1

Let $\{X_n : n \in \mathbb{Z}\}$ be a second-order stationary process. Then, X_n corresponds to a process $MA(\infty)$: there exists a deterministic process $\{D_n : n \in \mathbb{N}\}$ and a sequence of real coefficients $\{a_n : n \in \mathbb{N}\}$ such that

$$\forall n \in \mathbb{Z} : X_n = D_n + \sum_{i=0}^{+\infty} a_i \varepsilon_{n-i}$$

with $a_0 = 1$, $\sum_{i=0}^{+\infty} a_i^2 < \infty$, $cov(D, \varepsilon_i) = 0$, $\forall\, i$.

Wold's theorem suggests that second-order stationary processes may be approximated by considering truncations $MA(k)$ of the infinite series $MA(\infty)$:

$$X_n \approx D_n + \sum_{i=0}^{k} a_i \varepsilon_{n-i}. \tag{4.3}$$

In practice, we can use $D_n = E(X_n)$. The coefficients can be determined by solving Eq. (4.3). Under *R*, the instruction `arima` furnishes an estimation of the coefficients. The package `forecast` proposes an instruction `auto.arima` which determines automatically a model for the time series.

Example 4.14

Let us consider again the process $X_n = 1 + \varepsilon_n - 2\varepsilon_{n-1} + 0.5\varepsilon_{n-2}$. Assume that the coefficients are unknown and that we do not know that it is a *MA*(2)-process, but we have a trajectory of 1000 values and we want to determine a *MA*(*k*) approximation. We can adopt an approach analogous to the one presented in Example 4.13, but with a varying *k*: solve the equations

$$\sum_{i=0}^{k-p} a_i a_{i+p} + \left(m^2 - R_X(p)\right) = 0, a_0 = 1$$

For different values of *k* and choose the best result. Initially, we create a function that evaluates the RMS of the residuals (function `calcr` is defined in Example 4.13):

```
eqs = function (a, k){
  Rxa = calcr(a,mx,k)
  aux = sqrt(sum((Rx[1:k] - Rxa[1:k])^2/k))
  return(aux)
}
```

Then, we find the solution for several values of *k*:

```
kmax = 10
resa = numeric(length = kmax)
a_ini = list()
res = numeric(length = kmax)
a_sol = list()
for (k in 1:kmax){
  mod1 = arima(x, order = c(0,0,k))
  if (k > 1){
  a_ini[[k]] = mod1$coef[1:k]
  }else{a_ini[[k]]= mod1$coef}
  fobj = function(a){eqs(a,k)}
  sol = fminunc(fobj, x0 = a_ini[[k]])
  res[[k]] = sol$value
  a_sol[[k]] = sol$par
}
```

We obtain the residuals shown in Fig. 4.35: the residuals are very close for all the values of *k*: the vertical axis has a maximum value of 2.5E-11. Indeed, the coefficients a_1, a_2 are approximately the same for all the values of *k*, while the other coefficients are near zero. In Fig. 4.36, we show the values of a_1, a_2, and the RMS norm of the other coefficients.

(continued)

Example 4.14 (continued)

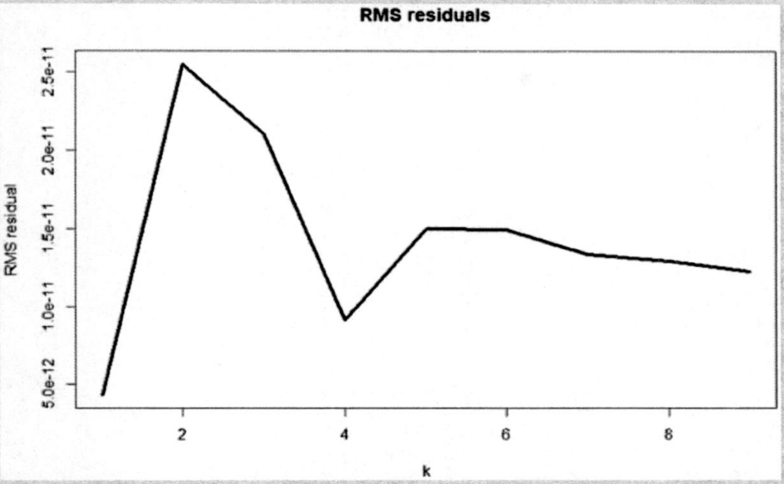

Fig. 4.35 RMS residuals. Notice that the vertical axis is from 5E-12 to 2.5E-11. Although the curve seems to show differences, all the values are very close

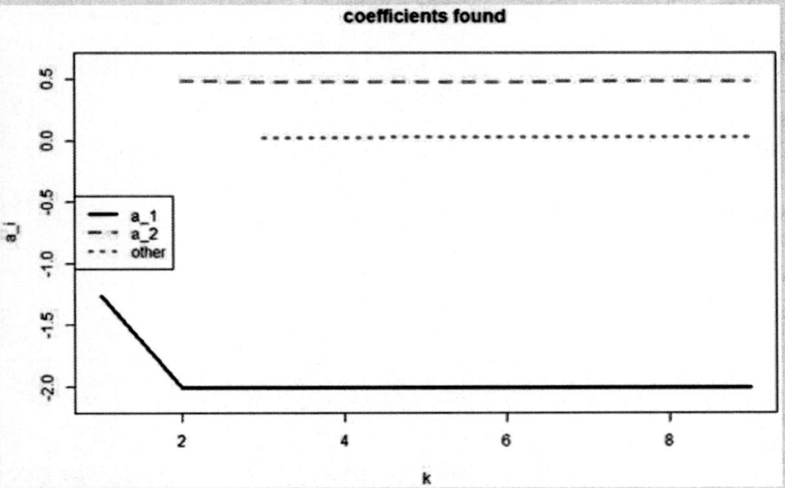

Fig. 4.36 Coefficients found: $a_1 \approx -2$, $a_2 \approx 0.5$ and the RMS norm of the other coefficients is near zero

(continued)

Example 4.14 (continued)

Analogously to Example 4.13, we can consider m as an unknown to be determined. Then, we use the code below, which furnishes the results shown in Figs. 4.37 and 4.38.

```
eqs = function (a, k){
  Rxa = calcr(a[1:k],a[[k+1]],k)
  aux = sqrt(sum((Rx[1:(k+1)] - Rxa[1:(k+1)])^2/(k+1)))
  return(aux)
}

kmax = 10
resa = numeric(length = kmax)
a_ini = list()
res = numeric(length = kmax)
a_sol = list()
for (k in 1:kmax){
  mod1 = arima(x, order = c(0,0,k))
  a_ini[[k]] = mod1$coef
  fobj = function(a){eqs(a,k)}
  sol = fminunc(fobj, x0 = a_ini[[k]])
  res[[k]] = sol$value
  a_sol[[k]] = sol$par
}
```

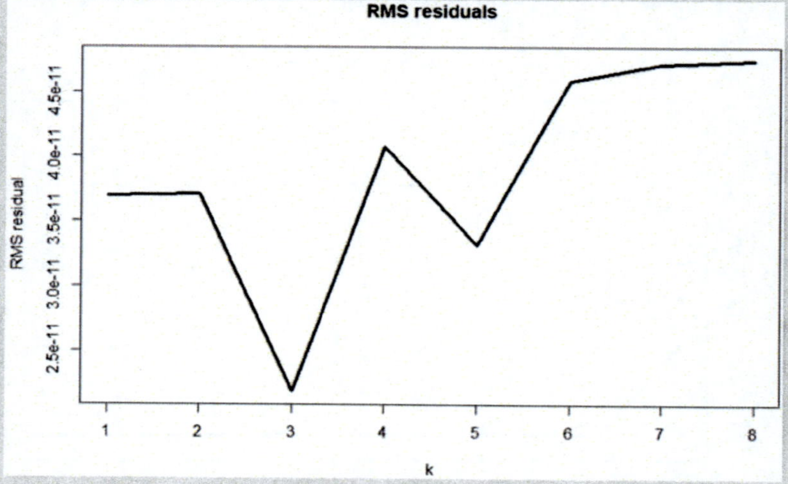

Fig. 4.37 RMS residuals. Notice that the residuals go from 2E-11 to 4E-11. Although the curve seems to show differences, all the values are very close

(continued)

Example 4.14 (continued)

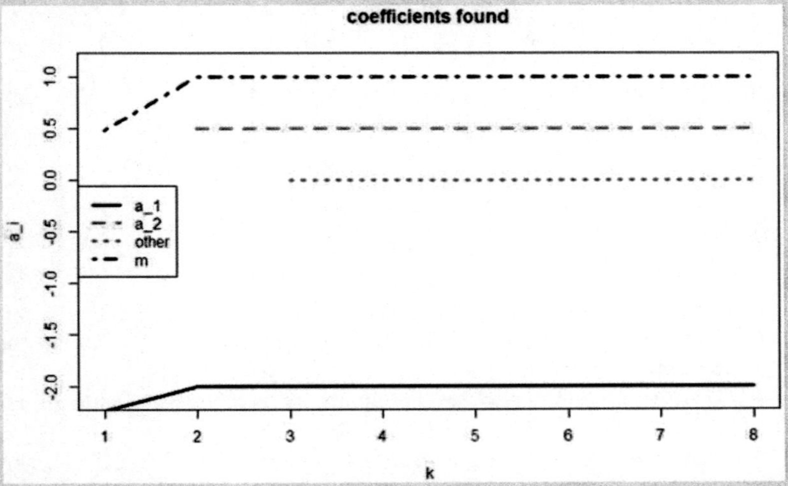

Fig. 4.38 Coefficients found: $a_1 \approx -2, a_2 \approx 0.5, m \approx 1$ and the RMS norm of the other coefficients is near zero

Remark 4.5

The analysis of MA processes is often made using ***delay (or lag) operators***. The basic delay operator is $L(X_t) = X_{t-1}$. Thus, $X_{n-k} = L^k X_n$ and we have

$$X_n = c + \varepsilon_n + \sum_{i=1}^{k} a_i \varepsilon_{n-i} \Leftrightarrow X_n - c = B(L)(\varepsilon_n),\ B(x) = 1 - \sum_{i=1}^{k} a_i x^i.$$

The polynomial B and its roots play an essential rule in the theory of MA and AR processes (see Sect. 4.5 for AR). Indeed, let x_i, $i = 1, \ldots, n$ be the roots of p. Then, $B(L) = a_k(x_1 - L)\ldots(x_k - L)$. Recalling that product of the roots is equal to $\frac{1}{a_k}$, we have $B(L) = \left(Id - \frac{L}{x_1}\right) \ldots \left(Id - \frac{L}{x_k}\right)$, where Id is the identity operator. Thus, $B(L)^{-1}$ exists if and only if not any root of $B(L)$ has a modulus equal to 1: $|x_i| \neq 1, i = 1, \ldots, n$. If $|x_i| > 1$, we have (see, for instance Dautray & Lions, 2012).

(continued)

Remark 4.5 (continued)

$$\left(Id - \frac{L}{x_i}\right)^{-1} = \sum_{n=0}^{+\infty} \frac{1}{x_i^{n+1}} L^n$$

Thus, if $|x_i| > 1$, $i = 1, \ldots, n$, then $\varepsilon_n = B(L)^{-1}(X_n - c)$.

Exercises

1. Let $X_n = 1 + \varepsilon_n$.
   (a) Determine m_X, V_X, R_X, C_X.
   (b) Generate a sample of 1000 values of $N(0, 1)$ and use it to simulate the process.
   (c) Use the data furnished by the simulation to determine approximate values of m_X, $R_X(p)$, $p = 0, 1, 2, 3, 4$.
   (d) Use these approximated values to fit processes MA(0), MA(1), MA (2) to the data.
2. Let $X_n = 1 + \varepsilon_n + \frac{1}{2}\varepsilon_{n-1}$.
   (a) Determine m_X, V_X, R_X, C_X.
   (b) Generate a sample of 1000 values of $N(0, 1)$ and use it to simulate the process.
   (c) Use the data furnished by the simulation to determine approximate values of m_X, $R_X(p)$, $p = 0, 1, 2, 3, 4$.
   (d) Use these approximated values to fit processes MA(0), MA(1), and MA (2) to the data.
3. Let $X_n = \mu + \varepsilon_n + \alpha\varepsilon_{n-1}$.
   (a) Determine m_X, V_X, R_X, C_X.
   (b) Generate a sample of 1000 values of $N(0, 1)$ and use it to simulate the process.
   (c) Use the data furnished by the simulation to determine approximate values of m_X, $R_X(p)$, $p = 0, 1, 2, 3, 4$.
   (d) Use these approximated values to fit processes MA(0), MA(1), and MA (2) to the data.
4. Let $X_n = \varepsilon_n - \varepsilon_n - \alpha\varepsilon_{n-2}$.
   (a) Determine m_X, V_X, R_X, C_X.
   (b) Generate a sample of 1000 values of $N(0, 1)$ and use it to simulate the process.

(continued)

(c) Use the data furnished by the simulation to determine approximate values of m_X, $R_X(p)$, $p = 0, 1, 2, 3, 4$.
(d) Use these approximated values to fit processes MA(0), MA(1), and MA (2) to the data.

5. Let $X_n = \varepsilon_n - \alpha\varepsilon_{n-2}$.

   (a) Determine m_X, V_X, R_X, C_X.
   (b) Generate a sample of 1000 values of $N(0, 1)$ and use it to simulate the process.
   (c) Use the data furnished by the simulation to determine approximate values of m_X, $R_X(p)$, $p = 0, 1, 2, 3, 4$.
   (d) Use these approximated values to fit processes MA(0), MA(1), and MA (2) to the data.

4.5 Autoregressive Processes

Definition 4.8
X is **autoregressive of order k**, denoted $AR(k)$, if and only if X is weakly stationary and there exists a discrete white noise $\varepsilon = \{\varepsilon_n : n \in \mathbb{Z}\}$, a number $m \in \mathbb{R}$, and a vector of coefficients $\boldsymbol{a} = (a_1, \ldots, a_k)$ such that

$$\forall n \in \mathbb{Z} : X_n = m + \varepsilon_n + \sum_{i=1}^{k} a_i X_{n-i}.$$

A condition to verify if the process is stationary is furnished by the result below

Theorem 4.2
The process $AR(k)$ given by $X_n = m + \varepsilon_n + \sum_{i=1}^{k} a_i X_{n-i}$ is stationary if and only if the polynomial $A(x) = 1 - \sum_{i=1}^{k} a_i x^i$ has not any root of modulus inferior or equal to 1: $A(x) = 0 \Rightarrow |x| > 1$.

Remark 4.6
The partial autocorrelation ρ_X of X verifies $\rho_X(s) = 0$, for $s > k$. Indeed, $\rho_X(s)$ is the coefficient of X_{n-s} for the best approximation PX_n of X_n by an affine function of $X_{n-s}, \ldots, X_{n-1}$ and $PX_n = \sum_{i=1}^{k} a_i X_{n-i}$, so that the coefficient is null for $s > k$.

Let us denote $E(X_n) = \mu$. We have

$$\mu = m + \sum_{i=1}^{k} a_i \mu,$$

So that, for $1 - \sum_{i=1}^{k} a_i \neq 0$,

$$\mu = \frac{m}{1 - \sum_{i=1}^{k} a_i}. \tag{4.4}$$

Assume that the white noise has a variance $\sigma^2 = 1$. Then,

$$E(X_n X_n) = m\mu + 1 + \sum_{i=1}^{k} a_i E(X_n X_{n-i})$$

and, for $p > 0$,

$$E(X_n X_{n+p}) = m\mu + \sum_{i=1}^{k} a_i E(X_n X_{n+p-i}).$$

Thus,

$$R_X(p) = m\mu + \delta_{p0} + \sum_{i=1}^{k} a_i R_X(i-p). \tag{4.5}$$

Equation (4.6) are usually referred as the Yule-Walker equations. Taking the values $p = 0, 1, \ldots, k$, we obtain $k + 1$ linear equations for the determination of the quantities $m\mu, a_1, \ldots, a_k$:

$$\begin{pmatrix} R_X(1) & \cdots & R_X(k) & 1 \\ R_X(0) & \cdots & R_X(k-1) & 1 \\ \vdots & \vdots & \vdots & \vdots \\ R_X(2-k) & \cdots & R_X(1) & 1 \\ R_X(1-k) & \cdots & R_X(0) & 1 \end{pmatrix} \begin{pmatrix} a_1 \\ \vdots \\ a_k \\ m\mu \end{pmatrix} = \begin{pmatrix} R_X(0)-1 \\ R_X(1) \\ \vdots \\ R_X(k-1) \\ R_X(k) \end{pmatrix}$$

Once $c\mu$ is known, c is determined by Eq. (4.5). We have

$$m = \sqrt{m\mu \times \left(1 - \sum_{i=1}^{k} a_i\right)}$$

The values of the coefficients can also be determined by a least-squares procedure. Indeed, we have

$$\begin{pmatrix} 1 & X_k & \cdots & X_1 \\ 1 & X_{k+1} & \cdots & X_2 \\ \vdots & \vdots & \vdots & \vdots \\ 1 & X_{k+n} & \cdots & X_{n+1} \end{pmatrix} \begin{pmatrix} m \\ a_1 \\ \vdots \\ a_k \end{pmatrix} = \begin{pmatrix} X_{k+1} \\ X_{k+2} \\ \vdots \\ X_{k+n+1} \end{pmatrix} - \begin{pmatrix} \varepsilon_{k+1} \\ \varepsilon_{k+2} \\ \vdots \\ \varepsilon_{k+n+1} \end{pmatrix}.$$

Thus, the vector of coefficients $(m, a_1, \ldots, a_k)^t$ is the solution of a noisy linear system. Since the noise is gaussian, the maximum-likelihood estimator of the coefficients coincides with the least-squares solution of the linear system where the vector of noise is neglected, id est, the solution of

$$\begin{pmatrix} 1 & X_k & \cdots & X_1 \\ 1 & X_{k+1} & \cdots & X_2 \\ \vdots & \vdots & \vdots & \vdots \\ 1 & X_{k+n} & \cdots & X_{n+1} \end{pmatrix} \begin{pmatrix} m \\ a_1 \\ \vdots \\ a_k \end{pmatrix} = \begin{pmatrix} X_{k+1} \\ X_{k+2} \\ \vdots \\ X_{k+n+1} \end{pmatrix}. \quad (4.6)$$

Notice that the linear system is overdetermined for $n > k + 1$.

Since the process is weakly stationary, $R_X(p - i) = R_X(i - p)$ and we may use the Yule–Walker equations to determine the values of μ, $R_X(p)$, when the values of m, a_1, ..., a_k are known. Indeed, we start by the determination of μ by using Eq. (4.5). The second step is the use of Eq. (4.6): for a stationary process, so that Eq. (4.6) form a linear system for $R_X(p)$, $p = 0, \ldots k$. For instance, for $k = 1$:

$$\begin{pmatrix} 1 & -a_1 \\ -a_1 & 1 \end{pmatrix} \begin{pmatrix} R_X(0) \\ R_X(1) \end{pmatrix} = \begin{pmatrix} m\mu + 1 \\ c\mu \end{pmatrix}$$

For $k = 2$:

$$\begin{pmatrix} 1 & -a_1 & -a_2 \\ -a_1 & 1 - a_2 & 0 \\ -a_2 & -a_1 & 1 \end{pmatrix} \begin{pmatrix} R_X(0) \\ R_X(1) \\ R_X(2) \end{pmatrix} = \begin{pmatrix} m\mu + 1 \\ m\mu \\ m\mu \end{pmatrix}$$

For $k = 3$:

$$\begin{pmatrix} 1 & -a_1 & -a_2 & -a_3 \\ -a_1 & 1-a_2 & -a_3 & 0 \\ -a_2 & -a_1-a_3 & 1 & 0 \\ -a_3 & -a_2 & -a_1 & 1 \end{pmatrix} \begin{pmatrix} R_X(0) \\ R_X(1) \\ R_X(2) \\ R_X(3) \end{pmatrix} = \begin{pmatrix} m\mu+1 \\ m\mu \\ cm \\ m\mu \end{pmatrix}$$

For $k = 4$:

$$\begin{pmatrix} 1 & -a_1 & -a_2 & -a_3 & -a_4 \\ -a_1 & 1-a_2 & -a_3 & -a_4 & 0 \\ -a_2 & -a_1-a_3 & 1-a_4 & 0 & 0 \\ -a_3 & -a_2-a_4 & -a_1 & 1 & 0 \\ -a_4 & -a_3 & -a_2 & -a_1 & 1 \end{pmatrix} \begin{pmatrix} R_X(0) \\ R_X(1) \\ R_X(2) \\ R_X(3) \\ R_X(4) \end{pmatrix} = \begin{pmatrix} m\mu+1 \\ m\mu \\ m\mu \\ m\mu \\ m\mu \end{pmatrix}$$

Finally, we may determine $R_X(p)$, for $p > k$, by using Eq. (4.6) again:

$$R_X(k+n) = m\mu + \sum_{i=n}^{k+n-1} a_{k+n-i} R_X(i). \tag{4.7}$$

For a process $AR(k)$, $\rho(s)$ is evaluated by using the orthogonal projection property: indeed,

$$X_{t+s} = m + \varepsilon_{t+s} + \sum_{i=1}^{k} a_i X_{t+s-i}.$$

Thus, only $X_{t+s-k}, \ldots, X_{t+s-1}$ appear in the expression X_{t+s} as an affine function of $X_{s,t} = (X_t, \ldots, X_{t+s-1})$. Therefore, on the one hand, $\rho(s) = 0$, for $s > k$ (in this case, X_t does not appear in the affine expression of X_{t+s}); on the other hand, $\rho(k) = a_k$ (since $t + k - i = t$ for $i = k$).

Simulations of a process $AR(k)$ may be generated by two ways:

1. If k values $X_0, \ldots, X_{k-1}$ are given, we may generate a sample from the white noise ε and use the recurrence in the definition of the $AR(k)$ to determine X_k, $X_{k+1}, \ldots$;
2. Otherwise, it is necessary to determine the inverse $A(L)^{-1}$ of $A(L) = Id - \sum_{i=1}^{k} a_i L^i$. We have:

$$A(L)^{-1} = \sum_{i=0}^{+\infty} p_i L^i. \tag{4.8}$$

3. The unknown coefficients are determined by using that $A(L)^{-1}A(L) = Id$: we have

$$p_0 = 1; \; p_n = \sum_{j=1}^{\min\{k,\, n\}} p_{n-j} a_j, \; n > 0 \; ; \tag{4.9}$$

4. What determines the coefficients p_i. Then,

$$X_n = A(L)^{-1}(m + \varepsilon_n) \; . \tag{4.10}$$

Of course, R proposes intrinsic functions for the simulation and the identification of AR processes. For instance (see Example 4.15):

- `ar` identifies the coefficients of the AR process. Options allow to choose the method (Yule–Walker is default);
- `arima.sim` simulates the process;
- `arima` identifies the coefficients.

Remark 4.7
The elimination of the mean m is possible in the simulation of an AR process. Indeed, let $Y_n = X_n - \mu$. Then,

$$Y_n = \varepsilon_n + \sum_{i=1}^{k} a_i Y_{n-i}$$

Thus, Y_n verifies the Yule-walker equations with $m = 0$:

$$R_Y(p) = \delta_{p0} + \sum_{i=1}^{k} a_i R_Y(i-p).$$

For instance,

$$\begin{pmatrix} R_Y(1) & \cdots & R_Y(k) \\ R_Y(0) & \ldots & R_Y(k-1) \\ \vdots & \vdots & \vdots \\ R_Y(2-k) & \ldots & R_Y(1) \end{pmatrix} \begin{pmatrix} a_1 \\ \vdots \\ a_k \end{pmatrix} = \begin{pmatrix} R_Y(0)-1 \\ R_Y(1) \\ \vdots \\ R_Y(k-1) \end{pmatrix}$$

Analogously, the least-squares approach leads to the linear system:

$$\begin{pmatrix} Y_k & \cdots & Y_1 \\ \vdots & \cdots & \vdots \\ Y_{k+n} & \cdots & Y_{n+1} \end{pmatrix} \begin{pmatrix} a_1 \\ \vdots \\ a_k \end{pmatrix} = \begin{pmatrix} Y_{k+1} \\ \vdots \\ Y_{k+n+1} \end{pmatrix}.$$

The characteristic polynomial A is the same for X and Y, so that we have $Y_n = A(L)^{-1}(\varepsilon_n)$.

Example 4.15
Let us consider the process $X_n = \varepsilon_n + \frac{5}{6}X_{n-1} - \frac{1}{6}X_{n-2}$. We have $m = 0, k = 2, a_1 = \frac{5}{6}, a_2 = -\frac{1}{6}$. Thus, $\mu = 0$. Since the roots of the polynomial $A(x) = 6 - 5x + x^2$ are $x_1 = 2, x_2 = 3$, the process is stationary. We have:

(continued)

Example 4.15 (continued)

$$\begin{pmatrix} 1 & -\frac{5}{6} & \frac{1}{6} \\ -\frac{5}{6} & \frac{7}{6} & 0 \\ \frac{1}{6} & -\frac{5}{6} & 1 \end{pmatrix} \begin{pmatrix} R_X(0) \\ R_X(1) \\ R_X(2) \end{pmatrix} = \begin{pmatrix} 1 \\ 0 \\ 0 \end{pmatrix}$$

Thus, $R_X(0) = \frac{21}{10}$, $R_X(1) = \frac{3}{2}$, $R_X\ (2) = \frac{9}{10}$. Now, we use Eq. (4.8) to determine the other values of R_X:

$$R_X(3) = R_X(2+1) = a_2 R_X(1) + a_1 R_X(2) = \frac{1}{2};$$

$$R_X(4) = R_X(2+2) = a_2 R_X(2) + a_1 R_X(3) = \frac{4}{15};$$

$$R_X(5) = R_X(2+3) = a_2 R_X(3) + a_1 R_X(4) = \frac{5}{36} \dots$$

The instruction

```
Rxex = ARMAacf(ar = c(5/6 , -1/6), lag.max = 10)
```

generates the exact values of $R_X(i)/R_X(0)$ for $0 \leq i \leq 10$. We present in Fig. 4.39 a comparison between the Yule–Walker formulae (4.6) (as above).

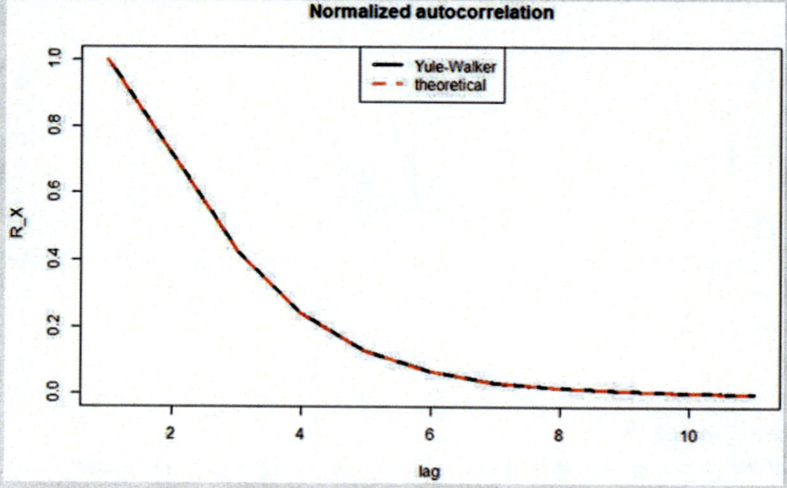

Fig. 4.39 Values of $R_X(i)/R_X(0)$. Yule–Walker's formula and R furnish the same results

(continued)

Example 4.15 (continued)

Now, assume that our data are the values of $R_X(4), \ldots, R_X(8)$. We use the Yule–Walker Eq. (4.6) to determine $m\mu$, a_1, a_2:

$$\begin{pmatrix} R_X(5) & R_X(4) & 1 \\ R_X(6) & R_X(5) & 1 \\ R_X(7) & R_X(6) & 1 \end{pmatrix} \begin{pmatrix} a_1 \\ a_2 \\ m\mu \end{pmatrix} = \begin{pmatrix} R_X(6) \\ R_X(7) \\ R_X(8) \end{pmatrix}$$

Of course, we may consider also

$$\begin{pmatrix} R_X(-1) & R_X(-2) & 1 \\ R_X(0) & R_X(-1) & 1 \\ R_X(1) & R_X(0) & 1 \end{pmatrix} \begin{pmatrix} a_1 \\ a_2 \\ c\mu \end{pmatrix} = \begin{pmatrix} R_X(0)-1 \\ R_X(1) \\ R_X(2) \end{pmatrix}$$

The results are almost exact:

```
          [,1]       [,2]         [,3]
[1,] 0.8333333 -0.1666667 1.138412e-18
```

A simulation of X_n can be made by calling `arima.sim`:

```
ns = 1000
x = arima.sim(model = list(ar = c(5/6 , -1/6)),ns)
```

An example of simulation appears in Fig. 4.40.

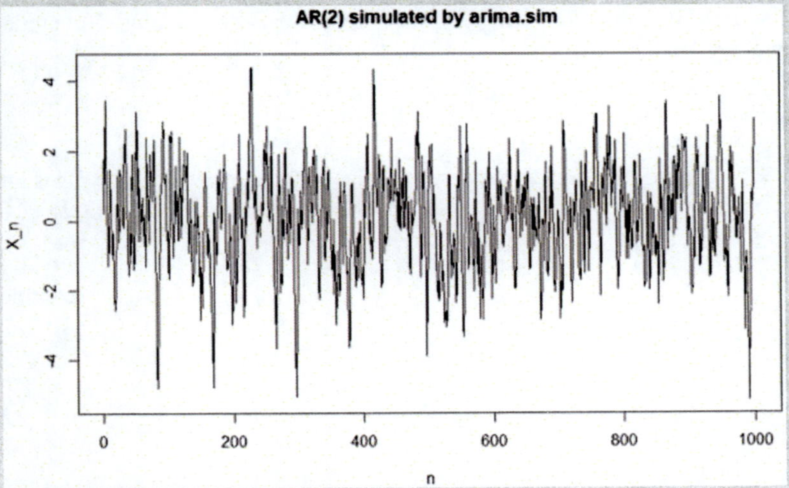

Fig. 4.40 Simulation of the process by `arima.sim`

(continued)

Example 4.15 (continued)

We can determine the empirical values of $R_X(i)/R_X(0)$ by the code at right. Here, `Rx_normalized` contains the normalized empirical values. A comparison is made in Fig. 4.41.

```
source("signal.R")
ss1 = signal$new(x)
Rx = ss1$vautocorr(9)
Rx_normalized = Rx/Rx[[1]]
```

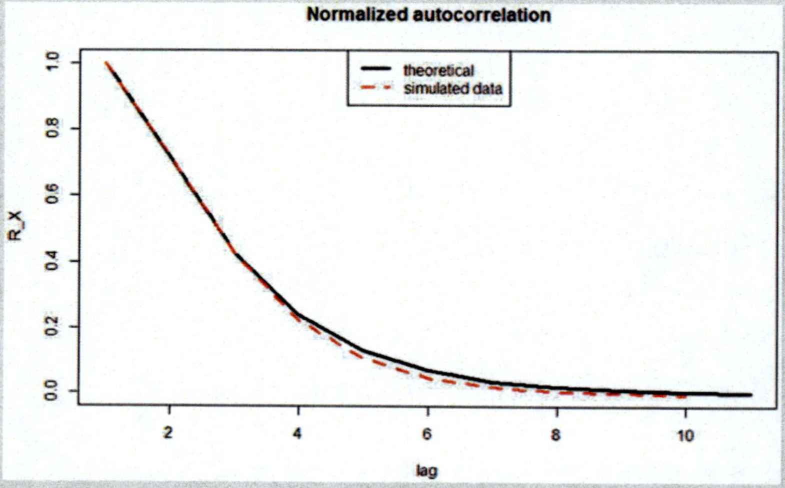

Fig. 4.41 Values of $R_X(i)/R_X(0)$

We can verify the order of the process by determining its partial autocorrelations:

```
> rhox=pacf(x,lag.max=20)
```

The command produces the result in Fig. 4.42: as expected, the values are near zero for $s > 2$. Analogous results are generated by `vpcorr1a` (Fig. 4.43)

(continued)

Example 4.15 (continued)

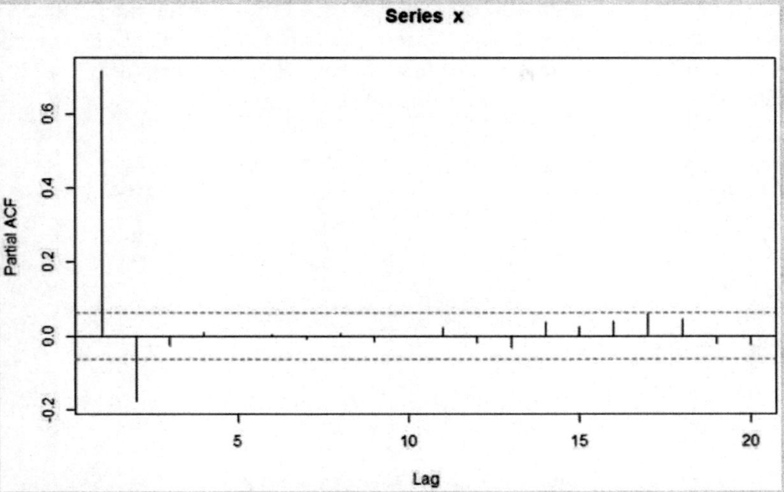

Fig. 4.42 Values of ρ_X generated by `pacf`. As expected, the values are near zero for $s > 2$

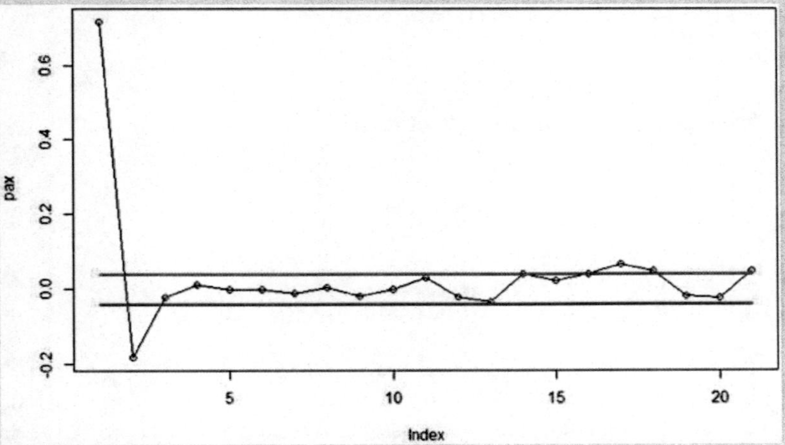

Fig. 4.43 Values of ρ_X generated by `vpcorr1a`

(continued)

Example 4.15 (continued)

Now, assume that we desire to identify the coefficients from the available data (the trajectory with 1000 values, generated by `arima.sim`). Since we identified that the order is $k = 2$ from the partial autocorrelations, we call `arima` to furnish a first estimation of the coefficients and use it as first estimate for a least-squares estimation of the coefficients:

```
sol1 = arima(x,order = c(2,0,0))
aa_ini = sol1$coef
```

```
library("pracma")
sol = fminunc(aa_ini,eqs)
```

The function `eqs` evaluates the residual $\|X_n - a_1X_{n-1} - a_2X_{n-2} - a_3\|$ a_3 estimates m.

```
eqs = function (aa){
  a = aa[1:(length(aa)-1)]
  mm = aa[length(aa)]
  nn = length(a)
  xx = list()
  for (ii in 1: (nn+1)){
    xx[[ii]] = x[(nn - ii + 2):(length(x)-ii + 1)]
  }
  aux = mm + xx[[1]]
  for (i in 1: nn){
    aux = aux - a[[i]]*xx[[i+1]]
  }
  resi = sqrt(sum(aux^2)/length(aux))
  return(resi)
}
```

Running this code, an example of result is

```
> sol1$coef
        ar1         ar2   intercept
 0.84879622 -0.18172777  0.07985101
> sol$par
        ar1         ar2   intercept
 0.84920758 -0.18163349 -0.02500087
```

Thus, the coefficients are estimated as $a_1 \approx 0.85$, $a_2 \approx -0.18$, $m \approx -0.03$. The exact values are $a_1 = 5/6 \approx 0.83$, $a_2 = 1/6 \approx -0.17$, $m = 0 \approx 0.00$. Notice that `arima` furnishes good results, which are slightly improved by `fminunc`. In addition, `arima` furnishes the AIC value and you can generate confidence intervals for the coefficients:

(continued)

Example 4.15 (continued)

```
> sol1

Call:
arima(x = x, order = c(2, 0, 0))

Coefficients:
         ar1      ar2  intercept
      0.8488  -0.1817     0.0799
s.e.  0.0312   0.0312     0.0928

sigma^2 estimated as 0.9566:  log likelihood = -1397.13,  aic = 2802.26
> confint(sol1)
                2.5 %     97.5 %
ar1        0.7877055  0.9098869
ar2       -0.2428455 -0.1206100
intercept -0.1019690  0.2616710
>
```

As indicated previously, we can simulate the process by using (4.11) directly. In this case, we must start by finding the values of p_n: Eq. (4.9) furnishes the values, which can be determined by the function calc_p at right. The instructions

```
a = c(5/6,-1/6)
p = calc_p(a,nmax)
```

determine $p_0, \ldots, p_{nmax}$

```
calc_p = function(a,nmax){
  k = length(a)
  resu = numeric(length = (nmax+1))
  resu[[1]] = 1
  resu[[2]] = resu[[1]]*a[[1]]
  if (nmax > 1){
    for (i in (2:nmax)){
      s = 0
      for (j in (1: min(c(k,i)))){
        s = s + resu[[i-j+1]]*a[[j]]
      }
      resu[[i+1]] = s
    }
  }
  return(resu)
}
```

Taking $nmax = 20$, we obtain (p[[i+1]] $= p_i$)

```
> p
 [1] 1.000000e+00 8.333333e-01 5.277778e-01 3.009259e-01 1.628086e-01
 [6] 8.551955e-02 4.413152e-02 2.252301e-02 1.141392e-02 5.757764e-03
[11] 2.895817e-03 1.453554e-03 7.286585e-04 3.649565e-04 1.826873e-04
[16] 9.141335e-05 4.572991e-05 2.287270e-05 1.143893e-05 5.720325e-06
[21] 2.860449e-06
```

(continued)

Example 4.15 (continued)

Then, the simulation is generated by the function `sim.ar` below:

```
sim.ar = function(m,a,nmax,ns) {
  k = length(a)
  p = calc_p(a,nmax)
  e = rnorm(ns+nmax+1)
  xs = numeric(length = ns)
  for (i in (1:ns)){
    s = 0
    for (j in 0:nmax){
      s = s + p[[j+1]]*(m + e[[i + nmax-j]])
    }
    xs[[i]] = s
  }
  return(xs)
}
```

For instance, we can call this function as shown at right. This call generates a trajectory `x1` formed by $ns = 1000$ points.

```
m = 0
a = c(5/6, -1/6)
ns = 1000
x1 = sim.ar(m,a,nmax,ns)
```

The trajectory is plotted in Fig. 4.44. The partial autocorrelation is plotted in Fig. 4.45. The normalized autocorrelations appear in Fig. 4.46. If m is estimated from the trajectory, we obtain $m \approx 5E - 3$.

```
> mx1=(1 - sum(a))*mean(x1)
> mx1
[1] -0.004960801
```

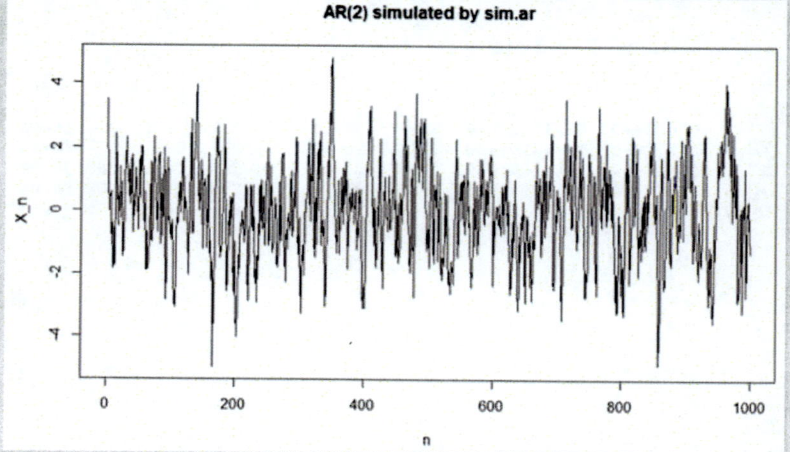

Fig. 4.44 A realization of trajectory generated by `sim.ar`

(continued)

Example 4.15 (continued)

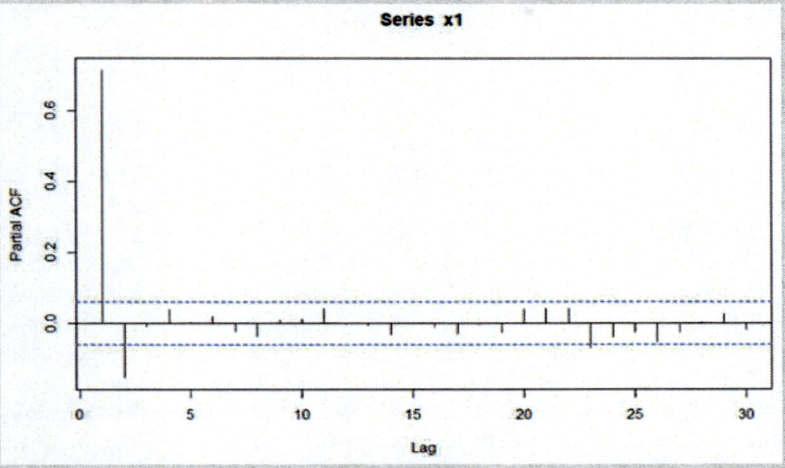

Fig. 4.45 Values of ρ_X of the trajectory generated by `sim.ar`. As expected, they are near zero for a lag > 2

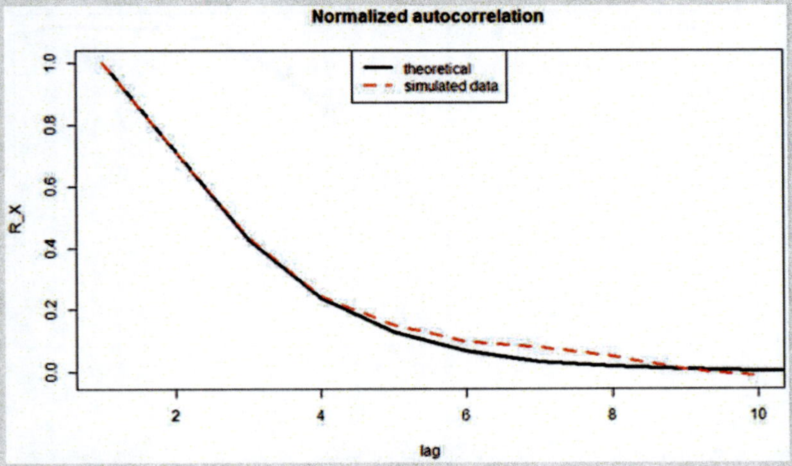

Fig. 4.46 Values of $R_X(i)/R_X(0)$for the trajectory generated by sim.ar

(continued)

Example 4.15 (continued)

For this simulated data, the identification procedure produces

```
> sol1$coef
        ar1         ar2   intercept
 0.83038322 -0.15604967 -0.01123874
> sol$par
         ar1          ar2    intercept
 0.826618063 -0.155432046  0.009070908
```

Thus, the estimates are $a_1 \approx 0.83$, $a_2 \approx -0.16$, $m \approx 0.00$.

The simulations can be used to determine the distribution of X_n – since the process is stationary, the distribution is independent from n. The determination be done either from the data or from Eq. (4.11). To use the data, we can apply the methods from Chap. 3 and introduce an artificial gaussian variable to be used to determine the distribution. Otherwise, Eq. (4.11) furnishes a representation of X_n as a sum of terms of a white noise – analogous to an MA process – which can be used to determine the exact distribution of X_n: the distribution is gaussian, with mean $m\sum p_i$ and variance $\sum p_i^2 - m$ is estimated. The results are compared in Figs. 4.47 and 4.48. Observe that (4.11) furnishes excellent results.

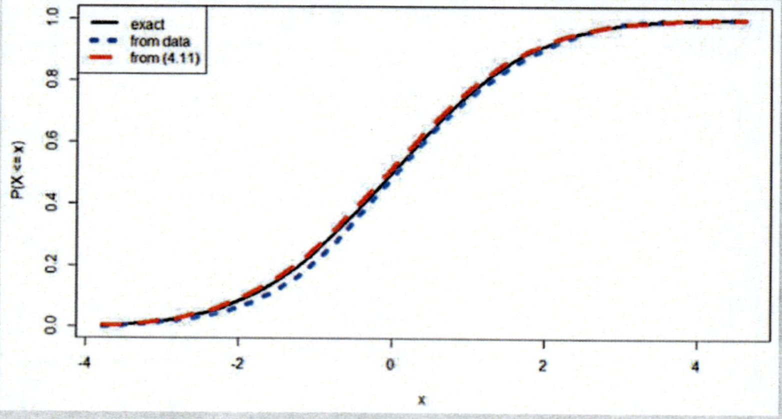

Fig. 4.47 Cumulative function and probability density of process $X_n = \varepsilon_n + \frac{5}{6}X_{n-1} - \frac{1}{6}X_{n-2}$, corresponding to a simulation of 1000 variates. Continuous line is the exact value

(continued)

Example 4.15 (continued)

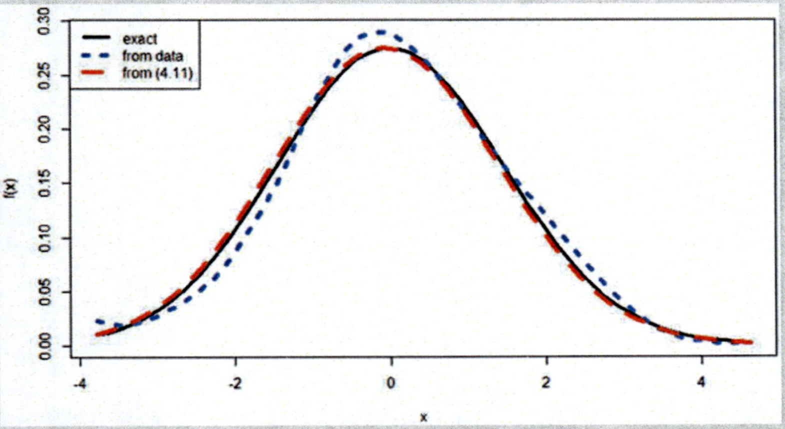

Fig. 4.48 PDF of process $X_n = \varepsilon_n + \frac{5}{6}X_{n-1} - \frac{1}{6}X_{n-2}$, corresponding to a simulation of 1000 variates. Continuous line is the exact. When using data, the PDF is obtained by SPH derivative of the CDF using a gaussian kernel

Exercises

1. Let $X_n = \alpha X_{n-1} + \varepsilon_n$.
   (a) Show that the process is stationary if $|\alpha| < 1$.
   (b) Assuming that $|\alpha| < 1$, find the coefficients p_n defining $A(L)^{-1}$.
   (c) Determine the values of $R_X(p)$, $p = 0, 1, 2, 3, 4$.
   (d) Determine $\rho_X(p)$, $p = 0, 1, 2, 3, 4$.
   (e) Generate a sample of 1000 values of $N(0, 1)$ and use it to simulate the process
   (f) Use these data to identify the process.
2. Let $X_n = \varepsilon_n + \frac{5}{6}X_{n-1} - \frac{1}{6}X_{n-2}$.
   (a) Show that the process is stationary.
   (b) Find the coefficients p_n defining $A(L)^{-1}$.
   (c) Determine the values of $R_X(p)$, $p = 0, 1, 2, 3, 4$.
   (d) Determine $\rho_X(p)$, $p = 0, 1, 2, 3, 4$.

(continued)

(e) Generate a sample of 1000 values of $N(0, 1)$ and use it to simulate the process.
(f) Use these data to identify the process.

3. Let $X_n = \varepsilon_n + \frac{7X_{n-1}}{6} - \frac{X_{n-2}}{3}$.
   (a) Show that the process is stationary.
   (b) Find the coefficients p_n defining $A(L)^{-1}$.
   (c) Determine the values of $R_X(p)$, $p = 0, 1, 2, 3, 4$.
   (d) Determine $\rho_X(p)$, $p = 0, 1, 2, 3, 4$.
   (e) Generate a sample of 1000 values of $N(0, 1)$ and use it to simulate the process.
   (f) Use these data to identify the process.
4. Let $X_n = \varepsilon_n + \frac{13X_{n-1}}{12} - \frac{3X_{n-2}}{8} + \frac{X_{n-3}}{24}$.
   (a) Show that the process is stationary.
   (b) Find the coefficients p_n defining $A(L)^{-1}$.
   (c) Determine the values of $R_X(p)$, $p = 0, 1, 2, 3, 4$.
   (d) Determine $\rho_X(p)$, $p = 0, 1, 2, 3, 4$.
   (e) Generate a sample of 1000 values of $N(0, 1)$ and use it to simulate the process.
   (f) Use these data to identify the process.

4.6 ARMA Processes

A process ARMA is the combination between an autoregressive and a moving average processes:

Definition 4.9
X is ***ARMA***(r, s) if and only if X is weakly stationary and there exists a discrete white noise $\varepsilon = \{\varepsilon_n : n \in \mathbb{Z}\}$, a number $m \in \mathbb{R}$ and two vectors of coefficients $\boldsymbol{a} = (a_1, \ldots, a_r)$ and $\boldsymbol{b} = (b_1, \ldots, b_s)$ such that

$$\forall n \in \mathbb{Z} : X_n = m + \varepsilon_n + \sum_{i=1}^{r} a_i X_{n-i} + \sum_{i=1}^{s} b_i \varepsilon_{n-i}.$$

The instruction `arima.sim` can simulates ARMA processes (see Example 4.16). If you desire to make your own program of simulation, it is necessary to use an approach analogous to the one introduced for AR processes: let L be the delay operator (see Remark 4.5) and let us denote

$$A(L) = Id - \sum_{i=1}^{r} a_i L^i, B(L) = Id + \sum_{i=1}^{s} b_i L^i. \tag{4.11}$$

Then,

$$A(L)X_n = m + B(L)\varepsilon_n$$

and

$$X_n = A(L)^{-1}m + A(L)^{-1}B(L)\varepsilon_n \tag{4.12}$$

The determination of $A(L)^{-1}$ is made by Eqs. (4.9) and (4.10). In addition, let us set

$$A(L)^{-1}B(L) = \sum_{n=0}^{+\infty} q_n L^n.$$

Let us take $b_0 = 1$. Then

$$A(x)^{-1}B(x) = \sum_{i=0}^{+\infty} p_i x^i \times \sum_{j=0}^{s} b_j x^j .$$

Thus,

$$A(x)^{-1}B(x) = \sum_{i=0}^{+\infty} \sum_{j=0}^{s} p_i b_j x^{i+j},$$

id est,

$$A(x)^{-1}B(x) = \sum_{n=1}^{+\infty} x^n \left(\sum_{j=0}^{\min(s,\ n)} p_{n-j} b_j \right).$$

Hence,

$$q_n = \sum_{j=0}^{\min(s,\ n)} p_{n-j} b_j . \tag{4.13}$$

This equation is the equivalent for $A(L)^{-1}B(L)$ of Eq. (4.10) for $A(L)^{-1}$. Thus, the simulation of a process ARMA analogous to the simulation of a process AR:

1. If r values $X_0, \ldots, X_{r-1}$ are given, you can generate a sample from the white noise ε and use the recurrence in the definition of the $ARMA(r, s)$ to determine X_r, $X_{r+1}, \ldots$;
2. Otherwise, it is necessary to determine $A(L)^{-1}$ and $A(L)^{-1}B(L)$ by using Eqs. (4.10) and (4.14). Then, you can generate a sample from the white noise and use Eq. (4.13) to generate X.

Analogously to AR processes, ARMA ones satisfy a family of Yule-Walker equations :

$$\mu\left(1-\sum_{i=1}^{r} a_i\right)=m; \tag{4.14}$$

$$R_X(p)=m\mu+\sum_{i=1}^{r} a_i R_X(i-p), \text{if } p>s. \tag{4.15}$$

By the same way, these equations may be used to determine the quantities $m\mu$, a_1, $\ldots, a_r$. Taking the values $=s+1, \ldots, s+r$, we obtain $k+1$ linear equations for the determination of the quantities $m\mu$, $a_1, \ldots, a_k$:

$$\begin{pmatrix} R_X(s) & \cdots & R_X(s-r+1) & 1 \\ R_X(s+1) & \ldots & R_X(s-r+2) & 1 \\ \vdots & \vdots & \vdots & \vdots \\ R_X(s+r-1) & \ldots & R_X(s) & 1 \\ R_X(s+r) & \cdots & R_X(s+1) & 1 \end{pmatrix} \begin{pmatrix} a_1 \\ \vdots \\ a_r \\ m\mu \end{pmatrix} = \begin{pmatrix} R_X(s+1) \\ R_X(s+2) \\ \vdots \\ R_X(s+r) \\ R_X(s+r+1) \end{pmatrix}$$

Once the system is solved, we may determine m, analogously to an AR process. As an alternative, the identification of the AR part of the process may also be made by the least-squares approach, by solving the linear system (4.7).

Finally, we observe that

$$Y_n=X_n-m-\sum_{i=1}^{r} a_i X_{n-i}=\varepsilon_n+\sum_{i=1}^{s} b_i \varepsilon_{n-i}. \tag{4.16}$$

Thus, Y_n is a process $MA(s)$ having a mean equal to zero. Thus, the coefficients b_i, $i = 1, \ldots, s$ verify Eqs. analogous to Eq. (4.9):

$$\sum_{i=0}^{s-p} b_i b_{i+p}=R_Y(p). \tag{4.17}$$

Taking $p = 0, \ldots, s-1$, we obtain a system of nonlinear equations which determines the coefficients $b_1, \ldots, b_s$.

The main difficult in the identification of ARMA processes is the determination of the orders (r, s). If these orders are known, the methods for the identification of AR and MA may be applied. But the selection a convenient couple (r, s) may involve difficulties. One of the methods to choose the values of (r, s) is the use of model selection tools, such as AIC (Akaike, 1974), BIC (Schwarz, 1978), or HQ (Hannan & Quinn, 1979). The reader will find in the literature other methods pf identification of ARMA processes (see, for instance, Choi, 1992). When using R, `arima` returs the values of AIC. In addition, the package `forecast` proposes a function `auto.arima` which automatically chooses the best model according to AIC or BIC values. It proposes also a function `Arima` which returns AIC end BIC values, and a function `forecast`, which can be used for simulation.

Example 4.16

Let us consider the process $X_n = \frac{16X_{n-1}}{15} - \frac{4X_{n-2}}{15} + \varepsilon_n + \frac{5\varepsilon_{n-1}}{6} + \frac{\varepsilon_{n-2}}{6}$. We have m=0, $r = s = 2$, $a_1 = \frac{16}{15}$,$a_2 = -\frac{4}{15}$, $b_1 = \frac{5}{6}$,$b_2 = \frac{1}{6}$. Thus, $\mu = 0$. We have also

$$A(x) = 1 - \frac{16x}{15} + \frac{4x^2}{15}, B(x) = 1 + \frac{5x}{6} + \frac{x^2}{6}$$

The roots of the polynomial A are $x_1 = \frac{3}{2}$,$x_2 = \frac{5}{2}$, so that the process is stationary. We can simulate the process using `arima.sim`:

```
ar1 = c(16/15, -4/15)
ma1 = c(5/6, 1/6)

x = arima.sim(model= list(ar=ar1,ma=ma1), 1000)
```

An example of result appears in Fig. 4.49. Assuming that we know the order of the process, we can use `arima` to identify the coefficients:

```
sol = arima(x, order = c(2,0,2))
```

We obtain

```
> sol = arima(x, order = c(2,0,2))
> sol

Call:
arima(x = x, order = c(2, 0, 2))

Coefficients:
         ar1      ar2     ma1     ma2  intercept
      1.1263  -0.3409  0.7310  0.0882     0.1748
s.e.  0.0914   0.0789  0.0957  0.0810     0.2544

sigma^2 estimated as 0.9056:  log likelihood = -1371.15,  aic = 2754.31
```

(continued)

Example 4.16 (continued)

Thus, the estimates are $a_1 \approx 1.13$, $a_2 \approx -0.34$, $b_1 \approx 0.73$, $b_2 \approx 0.09$, $m \approx 0.17$. The exact values are $a_1 \approx 1.07$, $a_2 \approx -0.27$, $b_1 \approx 0.83$, $b_2 \approx 0.17$, $m \approx 0.00$. Confidence intervals can be generated by `confint`:

```
> confint(sol)
                 2.5 %     97.5 %
ar1         0.94720953  1.3053328
ar2        -0.49548386 -0.1863633
ma1         0.54334050  0.9186451
ma2        -0.07056965  0.2469789
intercept  -0.32384077  0.6733933
```

Notice that the standard errors can be used to evaluate t-values and p-values:

```
> tval = abs(sol$coef)/sqrt(diag(sol$var.coef))
> tval
       ar1        ar2        ma1        ma2  intercept
12.3278841  4.3232198  7.6349699  1.0888280  0.6870106
> pval = 2*(1-pnorm(abs(sol$coef)/sqrt(diag(sol$var.coef))))
> pval
         ar1          ar2          ma1          ma2    intercept
0.000000e+00 1.537684e-05 2.264855e-14 2.762297e-01 4.920760e-01
```

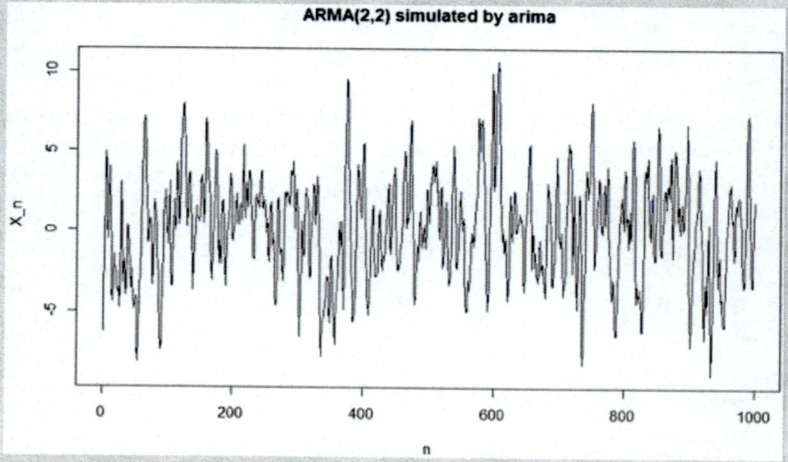

Fig. 4.49 Example of trajectory generated by `arima.sim`

(continued)

Example 4.16 (continued)

Let us use `auto.arima`:

```
> library(forecast)
> sol2 = auto.arima(x)
> sol2
Series: x
ARIMA(3,0,1) with zero mean

Coefficients:
         ar1      ar2     ar3     ma1
      1.3116  -0.5889  0.1119  0.5746
s.e.  0.0550   0.0881  0.0479  0.0481

sigma^2 = 1.031:  log likelihood = -1434.27
AIC=2878.53   AICc=2878.59   BIC=2903.07
```

Thus, the process is identified as ARMA(3,1).

If you desire to use your own program of simulation, you need to determine the coefficients p_n and q_n. Equation (4.10) furnish p_n:

```
> p
 [1] 1.0000000000 1.0666666667 0.8711111111 0.6447407407
 [5] 0.4554271605 0.3138581070 0.2133347380 0.1438615587
 [9] 0.0965630658 0.0646375212 0.0431965384 0.0288396353
[13] 0.0192432007 0.0128355114 0.0085596919 0.0057075350
[17] 0.0038054529 0.0025371404 0.0016914956 0.0011276912
[21] 0.0007518052
```

Then, Eq. (4.16) furnishes q_n:

```
> q
 [1] 1.000000000 1.900000000 1.760000000 1.370666667
 [5] 0.992711111 0.693380741 0.474883160 0.321640507
 [9] 0.216447698 0.145106743 0.097061139 0.064836751
[13] 0.043276230 0.028871512 0.019255951 0.012840612
[17] 0.008561732 0.005708351 0.003805779 0.002537271
[21] 0.001691548
```

These quantities can be evaluated as follows:

(continued)

Example 4.16 (continued)

```
calc_p = function(ar,nmax){
  k = length(ar)
  resu = numeric(length = (nmax+1))
  resu[[1]] = 1
  resu[[2]] = resu[[1]]*ar[[1]]
  if (nmax > 1){
    for (i in (2:nmax)){
      s = 0
      for (j in (1: min(c(k,i)))){
        s = s + resu[[i-j+1]]*ar[[j]]
      }
      resu[[i+1]] = s
    }
  }
  return(resu)
}


calc_q = function(ma,nmax,p){
  s = length(ma) + 1
  mma = c(1, c(ma))
  resu = numeric(length = (nmax+1))
  resu[[1]] = p[[1]]*mma[[1]]
  resu[[2]] = mma[[1]]*p[[2]] + p[[1]]*mma[[2]]
  if (nmax > 1){
    for (i in (2:nmax)){
      s = 0
      for (j in (1: min(c(s,i+1)))){
        s = s + mma[[j+1]]*p[[i-j+1]]
      }
      resu[[i+1]] = s
    }
  }
  return(resu)
}
```

and the process is simulated as follows:

```
sim.arma = function(m,a,b,nmax,ns) {
  k = length(a)
  p = calc_p(a,nmax)
  s = length(b)
  q = calc_q(b,nmax,p)
  e = rnorm(ns+nmax+1)
  xs = numeric(length = ns)
  for (i in (1:ns)){
    s = 0
    for (j in 0:nmax){
      s = s + q[[j+1]]*e[[i + nmax-j]]
    }
    xs[[i]] = s + sum(p)*m
  }
  return(xs)
}
```

(continued)

Example 4.16 (continued)

An example of simulation is given in Fig. 4.50 It uses the code

```
a1 = 16/15
a2 = -4/15
b1 = 5/6
b2 = 1/6
m = 0
```

```
nmax = 30
a = c(a1,a2)
b = c(b1,b2)
ns = 1000
x1 = sim.arma(m,a,b,nmax,ns)
```

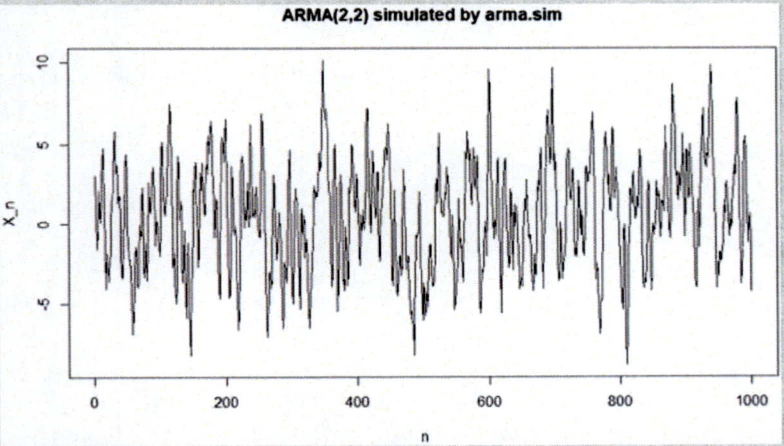

Fig. 4.50 Example of trajectory generated by `arma.sim`

Let us use the data to identify the coefficients. Using `arima`:

```
> sol1 = arima(x1, order = c(2,0,2))
> sol1

Call:
arima(x = x1, order = c(2, 0, 2))

Coefficients:
         ar1      ar2     ma1      ma2  intercept
      1.1303  -0.3281  0.7880  -0.0703     0.5224
s.e.  0.1036   0.0867  0.1072   0.1006     0.2721

sigma^2 estimated as 0.9875:  log likelihood = -1414.74,  aic = 2841.49
```

Thus, the estimates are $a_1 \approx 1.13$, $a_2 \approx -0.33$, $b_1 \approx 0.79$, $b_2 \approx 60.07$, $m \approx 0.5$. Let us use `auto.arima`:

(continued)

Example 4.16 (continued)

```
> library(forecast)
> sol2 = auto.arima(x1)
> sol2
Series: x1
ARIMA(3,0,1) with non-zero mean

Coefficients:
         ar1      ar2      ar3     ma1    mean
      1.0432  -0.2280  -0.0323  0.8730  0.5227
s.e.  0.0392   0.0602   0.0382  0.0227  0.2704

sigma^2 = 0.9922:  log likelihood = -1414.63
AIC=2841.26   AICc=2841.35   BIC=2870.71
```

Thus, `auto.arima` identifies a process ARMA(3,1). The results improve by increasing nmax. For nmax=100, `arima` furnishes

```
> sol1 = arima(x1, order = c(2,0,2))
> sol1

Call:
arima(x = x1, order = c(2, 0, 2))

Coefficients:
         ar1      ar2     ma1     ma2  intercept
      1.0297  -0.2642  0.8711  0.0429     0.0114
s.e.  0.1063   0.0887  0.1089  0.0998     0.2530

sigma^2 estimated as 0.9662:  log likelihood = -1403.7,  aic = 2819.4
```

and `auto.arima` furnishes

```
> library(forecast)
> sol2 = auto.arima(x1)
> sol2
Series: x1
ARIMA(2,0,1) with zero mean

Coefficients:
         ar1      ar2     ma1
      1.0725  -0.2988  0.8252
s.e.  0.0337   0.0336  0.0207

sigma^2 = 0.9693:  log likelihood = -1403.79
AIC=2815.59   AICc=2815.63   BIC=2835.22
```

These results are close to the exact values. `auto.arima` identifies a process AR(2,1), which can be considered as AR(2,2) with $b_2 = 0$ instead of the exact value 0.04. We can use these values to determine the distribution of X_n.

(continued)

Example 4.16 (continued)

Indeed, we can determine the coefficients: associated to the process identified:

```
ae = sol2$model$phi
be = sol2$model$theta
pe = calc_p(ae,100)
qe = calc_q(be,100,pe)
```

The mean is identified by `auto.arima` as zero (otherwise it is the value of the last coefficient) and the variance is $\sigma^2 = \sum q_i^2$. Thus, the distribution is $N\left(0, \sqrt{\sum pe_i^2}\right)$. A comparison with the exact one is made in Figs. 4.51 and 4.52.

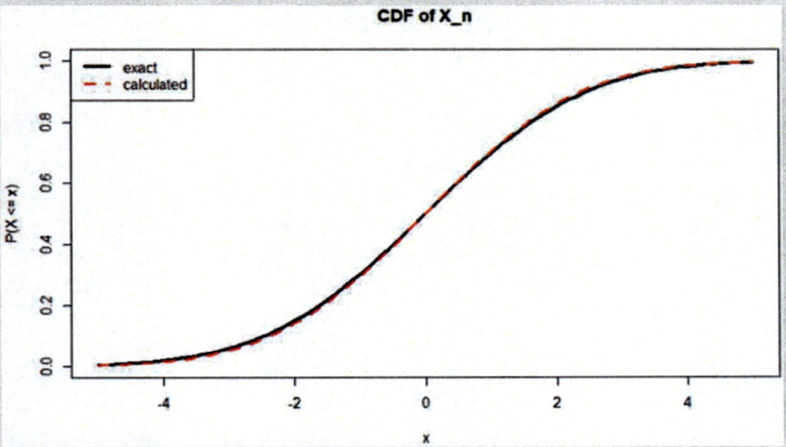

Fig. 4.51 CDF identified by `auto.arima` from data generated by `sim.arma`

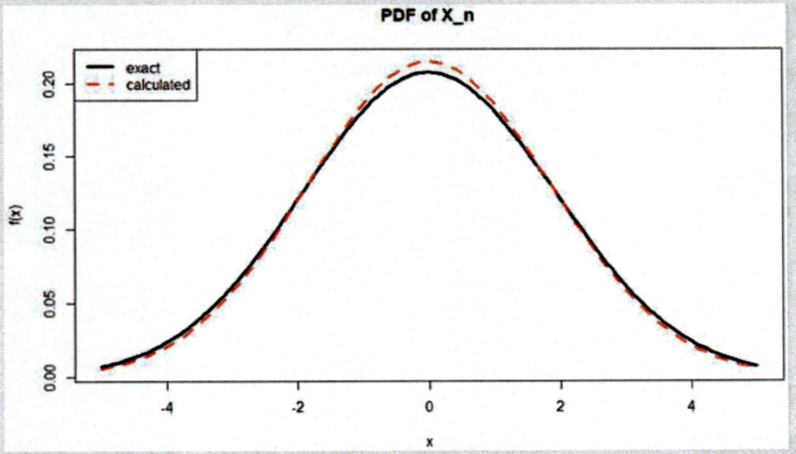

Fig. 4.52 CDF identified by `auto.arima` from data generated by `sim.arma`

Exercises

1. Let $X_n = \frac{X_{n-1}}{2} + \varepsilon_n + \frac{\varepsilon_{n-1}}{3}$.
   (a) Determine $A(x)$, $B(x)$.
   (b) Verify that the process is stationary.
   (c) Determine the coefficients p_n of the expansion of $A(x)^{-1}$
   (d) Determine the coefficients q_n of the expansion of $A(x)^{-1}B(x)$
   (e) Generate a sample of 1000 values of $N(0, 1)$ and use it to simulate X.
   (f) Use the data furnished by the simulation to determine approximate values of m_X, $R_X(p)$, $p = 0, 1, 2, 3, 4$.
   (g) Fit processes $ARMA(1, 1)$, $ARMA(2, 1)$, $ARMA(1, 2)$ to the data.
   (h) Assume that the orders are unknown. Use `auto.arima` to fit an ARMA process to the data.

   (a) Let $X_n = \dfrac{X_{n-1}}{3} + \varepsilon_n + \dfrac{\varepsilon_{n-1}}{2}$.
   (b) Determine $A(x)$, $B(x)$.
   (c) Verify that the process is stationary.
   (d) Determine the coefficients p_n of the expansion of $A(x)^{-1}$
   (e) Determine the coefficients q_n of the expansion of $A(x)^{-1}B(x)$
   (f) Generate a sample of 1000 values of N(0,1) and use it to simulate X.
   (g) Use the data furnished by the simulation to determine approximate values of m_X, $R_X(p)$, $p = 0, 1, 2, 3, 4$.
   (h) Fit processes $ARMA(1, 1)$, $ARMA(2, 1)$, $ARMA(1, 2)$ to the data.
   (i) Assume that the orders are unknown. Use `auto.arima` to fit an ARMA process to the data.

   (a) Let $Y_n = 2 + \dfrac{3Y_{n-1}}{4} - \dfrac{Y_{n-2}}{8} + \varepsilon_n - \dfrac{\varepsilon_{n-1}}{16}$.
   (b) Show that $m_X = \dfrac{16}{3}$.
   (c) Let $X_n = Y_n - m_X$. Show that $X_n = \frac{3}{4}X_{n-1} - \frac{1}{8}X_{n-2} + \varepsilon_n - \frac{1}{16}\varepsilon_{n-1}$
   (d) Determine $A(x)$, $B(x)$ for the process X_n
   (e) Verify that the process X_n is stationary
   (f) Determine the coefficients p_n of the expansion of $A(x)^{-1}$
   (g) Determine the coefficients q_n of the expansion of $A(x)^{-1}B(x)$
   (h) Generate a sample of 1000 values of N(0,1) and use it to simulate X.
   (i) Use the data furnished by the simulation to determine approximate values of m_X, $R_X(p)$, $p = 0, 1, 2, 3, 4$.
   (j) Fit processes $ARMA(1, 1)$, $ARMA(2, 1)$, $ARMA(1, 2)$ to the data.
   (k) Assume that the orders are unknown. Use `auto.arima` to fit an ARMA process to the data.

4.7 Markov Processes

Markov processes – also called Markov chains – are processes without memory: the future value X_{n+1} depends only on the actual state X_n. In other words, the conditional probability of X_{n+1} knowing the past values $X_0, \ldots, X_n$ depends only upon X_n:

$$P(X_{n+1} < x_{n+1} | X_i < x_i, 0 \le i \le n) = P(X_{n+1} < x_{n+1} | X_n < n).$$

Therefore, a Markov process is fully characterized by its probability of transition:

$$P(t, x, s, y) = P(X_t < x | X_s < y).$$

The associated probability density is $p(t, x, s, y) = \frac{\partial^2 P(t, x, s, y)}{\partial x \partial y}$. The process is stationary if and only if $P(t, x, s, y) = P(t - s, x, y)$ and $p(t, x, s, y) = p(t - s, x, y)$. As previously explained, it is usual to distinguish between continuous time end discrete time, according to the possible values of t and s.

R propose many packages to deal with Markov processes. For instance, `markovchain` (Discrete time Markov Chains) and `msm` (Markov chains in continuous time). You will find in the R repositories many other packages for the simulation, analysis, and modeling of Markov processes – for instance, `hmm` for Hidden Markov Chains – id est, chains where our knowledge is incomplete: values of probabilities or states are unknown.

Let us consider a stationary Markov chain where X can take a finite number of values $\{x_1, \ldots, x_k\}$. In the language of Markov chains, x_i is referred as a possible state of the chain. Then,

$$p(n, x_i, x_j) = \mathrm{P}(X_{n+1} = x_i | X_n = x_j) = \pi_{ij}(n).$$

We have

$$\pi_{ij}(n) \ge 0, \quad \sum_{i=1}^{k} \pi_{ij}(n) = 1$$

These probabilities are collected into a matrix $\Pi(n) = (\pi_{ij}(n) : 1 \le i, j \le k)$. In the framework of Markov processes, the matrix $\mathbf{\Pi(1)}$ is simply denoted $\mathbf{\Pi}$ and is called *transition matrix*. The chain may be represented by a graph where the arcs represent the transitions. As an example, we represent the case where $k = 4$ in Fig. 4.53.

We have

$$P(X_{n+1} = x_i) = \sum P(X_{n+1} = x_i | X_n = x_j) P(X_n = x_j)$$

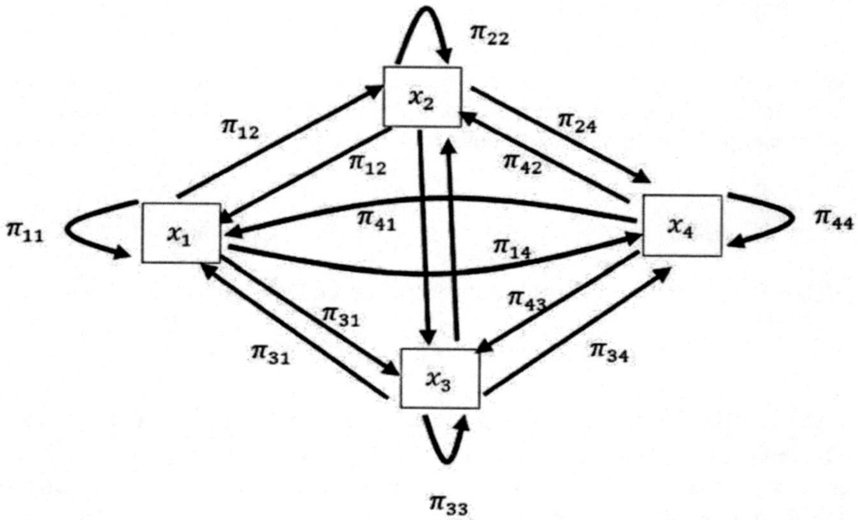

Fig. 4.53 Graphical representation of a Markov Chain with 4 states

Let us introduce

$$p^{(s)} = \left(p_1^{(s)}, \ldots, p_k^{(s)}\right)^t, \; P(X_s = x_i) = p_i^{(s)}.$$

Then,

$$\boldsymbol{p}^{(1)} = \boldsymbol{\Pi p}^{(0)}.$$

Thus, by recurrence

$$\boldsymbol{p}^{(n)} = \boldsymbol{\Pi}^n \boldsymbol{p}^{(0)} \Leftrightarrow \boldsymbol{\Pi}(\mathrm{n}) = \boldsymbol{\Pi}^n.$$

The states of a Markov chain are usually classified into recurrent, transient, and absorbing.

A state is recurrent when the probability of visiting at some time is equal to 1: $P(\exists n \in N : X_n = x_i) = 1$. Markov chains with a finite number of states have at least one recurrent state.

A state is transient when such a probability is strictly inferior to 1: $P(\exists n \in N : X_n = x_i) < 1$.

A state is absorbing when it cannot be leaved, id est, $\exists n \in N : P(X_{n+s} = x_i | X_n = x_i) = 1, \forall s \geq 0$. Not all the Markov chains have absorbing states. An absorbing state is characterized by the fact that $\pi_{ii}(n) = 1$, so that the transition matrix $\Pi(n)$ contains columns of an identity matrix. A Markov chain that has an absorbing state is said an absorbing chain. For such a chain, it is usual to renumber the states to write the transition matrix under the canonical form

$$\mathbf{\Pi} = \begin{pmatrix} \mathbf{Id} & \mathbf{0} \\ \mathbf{A} & \mathbf{B} \end{pmatrix}$$

Under the canonical form, $\mathbf{A}$ gives the probability of transition from a non-absorbing state to an absorbing state, while $\mathbf{B}$ gives the probability of transitions between non-absorbing states. Matrix $\mathbf{M} = (\mathbf{Id} - \mathbf{B})^{-1}$ plays an essential rule in the theory of Markov processes. For instance, M_{ij} gives the expected number of visits to state i when starting at state j. Analogously, $M_{\blacksquare j} = \sum_{i=1}^{k} M_{ij}$ gives the expected number of steps before absorption when starting at state j. The matrix $\mathbf{N} = \mathbf{MA}$ gives the probability N_{ij} of an absorption by the $ith.$ absorbing state when the initial state is the jth non-absorbing state.

We say that $\mathbf{p}$ is an stationary probability if and only if

$$\mathbf{p} = \mathbf{\Pi p}.$$

A stationary probability is a fixed point of the chain: if the chain reaches such a probability, then the probability distribution remains constant for all the posterior times. If $\mathbf{p}^{(n)} \to \mathbf{p}$ for $n \to +\infty$ then $\mathbf{p} = \mathbf{\Pi p}$, so that $\mathbf{p}$ is a steady state probability. Notice that $\mathbf{p}$ corresponds to an eigenvector of $\mathbf{\Pi}$, associated to the eigenvalue 1, so that it is a solution of the overdetermined linear system

$$\mathbf{Cp} = \mathbf{D}, \mathbf{C} = \begin{pmatrix} \mathbf{Id} - \mathbf{\Pi} \\ \mathbf{1} \end{pmatrix}, \ \mathbf{D} = \begin{pmatrix} \mathbf{0} \\ \mathbf{1} \end{pmatrix}.$$

This linear system may be solved by least squares:

$$\mathbf{C}^t\mathbf{Cp} = \mathbf{C}^t\mathbf{D}.$$

To verify if the least-squares solution is a stationary probability, we must evaluate the error $e = \|\mathbf{Cp} - \mathbf{D}\|$. If $e = 0$, the solution found is a stationary probability. Otherwise, not.

From Perron–Frobenius theorem, it is expected that $\mathbf{\Pi}^n$ converges to a matrix having all the columns equal to $\mathbf{p}$, i. e.,

$$\mathbf{\Pi}^n \to (\mathbf{p}, \ldots, \mathbf{p}).$$

Thus, a second way to find a stationary probability consists in calculating powers of $\mathbf{\Pi}$ until convergence to a matrix having the form above.

To simulate the chain, we must generate a sequence $X_1, \ldots, X_n$ from an initial value $X_0 = x_{i_0}$. To do this, we choose an initial vector $\overline{\mathbf{p}}^{(0)}$ having a single component equal to 1 at position i_0 and the other components null. Then, we evaluate

$\boldsymbol{p}^{(1)} = \boldsymbol{\Pi}\overline{\boldsymbol{p}}^{(0)}$ and we choose the value x_{i_1} at random, among the values $x_1, \ldots, x_k$, according to the probabilities $p_1^{(1)}, \ldots, p_1^{(k)}$. For instance, we may generate a variate u_1 from the uniform distribution on (0,1) and find its position in the cumulative distribution associated to $\boldsymbol{p}^{(1)}$: we take the value x_{i_1} such that $p_1^{(1)} + \ldots + p_1^{(i_1-1)} < u_1 < p_1^{(1)} + \ldots + p_1^{(i_1)}$ Then, we set $\overline{\boldsymbol{p}}^{(1)}$ as the vector having a single component equal to 1 at position i_1 and the other components null and we make the same operations again.

Remark 4.8
Under *R*, package `markovchain` proposes a complete set of instructions to deal with Discrete Markov Chains. The formulation used in the package is slightly different, since the probabilities of transition are transposed, id est, $p(n, x_i, x_j) = \pi_{ji}(n)$, except if you set parameter `byrow` = TRUE

Example 4.17
Let us consider a stationary Markov chain such that

$$\boldsymbol{\Pi} = \begin{pmatrix} \frac{1}{2} & 0 & \frac{1}{2} \\ 0 & \frac{1}{2} & \frac{1}{2} \\ \frac{1}{2} & \frac{1}{2} & 0 \end{pmatrix}$$

The matrix $\boldsymbol{\Pi}$ is symmetric, so that we can work by rows (probabilities of transition are given by a row: parameter `byrow` = TRUE) or not (probabilities of transition are given by a column: parameter `byrow` = FALSE). We can create a Markov chain *with probabilities of transition in rows* as shown at right:

```
matPi = matrix(
  c(0.5,0,0.5,
    0, 0.5, 0.5,
    0.5, 0.5, 0
    ), byrow=TRUE,
  nrow=3,ncol=3)

lst = c("st 1","st 2","st 3" )

library("markovchain")
mc1 = new("markovchain",states=lst,
          transitionMatrix=matPi,
          name = "Test Chain",
          byrow=TRUE)
```

To verify the transition matrix, use `print` (a more complete view is offered by `show`).

```
> print(mc1)
     st 1 st 2 st 3
st 1  0.5  0.0  0.5
st 2  0.0  0.5  0.5
st 3  0.5  0.5  0.0
```

To find the stationary probabilities, use `steadyStates`:

```
> steadyStates(mc1)
          st 1      st 2      st 3
[1,] 0.3333333 0.3333333 0.3333333
```

To find the absorbing states, use `absorbingStates` – this chain has not any absorbing state.

```
> absorbingStates(mc1)
character(0)
```

For the recurrent and transient states, there are analogous instructions. For this chain, all the states are recurrent and none is transient.

```
> recurrentStates(mc1)
[1] "st 1" "st 2" "st 3"
>
> transientStates(mc1)
character(0)
```

Fig. 4.54 Plot of the chain

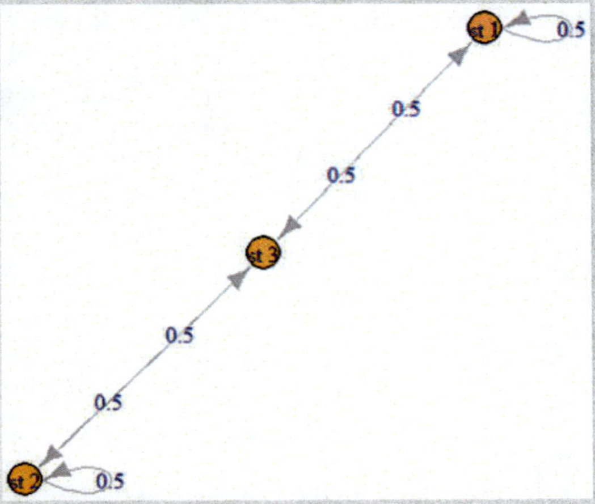

The package offers many functions, allowing you to determine conditional probabilities, hitting probabilities, analyzing first passages, etc. For instance,

(continued)

Example 4.17 (continued)
you can plot the chain: the instruction `plot(mc1)` produces the result in Fig. 4.54.

To simulate the chain, you can use `rmarkovchain`. For instance, let us simulate 20 states, starting from the initial state "`st 2`":

```
> sim1 <- rmarkovchain(n = 20, object = mc1, t0 = "st 2")
> sim1
 [1] "st 3" "st 2" "st 3" "st 1" "st 1" "st 1" "st 3"
 [8] "st 2" "st 3" "st 1" "st 1" "st 3" "st 1" "st 1"
[15] "st 1" "st 3" "st 1" "st 3" "st 2" "st 3"
```

The package offers also functions to fit a Markov chain. For instance, let us create a sample of 100 variates from the uniform distribution on $\{1, 2, 3\}$:

```
> s1 = sample(1:3,size=100,replace=TRUE)
> data1 = as.character(s1)
```

Let us consider `data1` as a sequence of observed data. We start by testing the Markov property of the data:

```
> verifyMarkovProperty(data1)
Testing markovianity property on given data sequence
Chi - square statistic is: 11.51377
Degrees of freedom are: 19
And corresponding p-value is: 0.905345
```

Thus, the odds that the data do not correspond to a Markov chain are about 10 %. The instruction `createSequenceMatrix` counts the transitions:

```
> mat1 = createSequenceMatrix(data1,toRowProbs = TRUE)
> mat1
          1         2         3
1 0.2424242 0.3333333 0.4242424
2 0.3214286 0.3214286 0.3571429
3 0.3947368 0.2368421 0.3684211
```

You can also use `markovchainFit`:

```
> mc2 = markovchainFit(data1,name = "MC Fit")
> mc2$estimate
MC Fit
 A  3 - dimensional discrete Markov Chain defined by the following states:
 1, 2, 3
 The transition matrix  (by rows)  is defined as follows:
          1         2         3
1 0.2424242 0.3333333 0.4242424
2 0.3214286 0.3214286 0.3571429
3 0.3947368 0.2368421 0.3684211
```

(continued)

Example 4.17 (continued)

Here, mc2 contains other fields, such as the standard errors and confidence intervals.

You can generate predictions from the empirical data. For instance, let us predict the 4 next states knowing that the three first ones are 1, 1, 2:

```
> predict(mc2$estimate,newdata=c("1","1","2"),n.ahead = 4)
[1] "3" "1" "3" "1"
```

Example 4.18

Let us consider a stationary Markov chain such that

$$\mathbf{\Pi} = \begin{pmatrix} 0 & 1 & 0 \\ 0 & 0 & 1 \\ 1 & 0 & 0 \end{pmatrix}$$

We create a Markov Chain *using the columns* as probabilities of transition: here, $P(X_{n+1} = x_i | X_n = x_j) = \pi_{ji}(n)$

```
matPi = matrix(
  c(0,1,0,
    0, 0, 1,
    1, 0, 0
    ), byrow=TRUE,
  nrow=3,ncol=3)
lst = c("st 1","st 2","st 3" )
library("markovchain")
mc1 = new("markovchain",states=lst,
          transitionMatrix=matPi,
          name = "Test Chain")
```

Analyze the states :

```
> steadyStates(mc1)
          st 1      st 2      st 3
[1,] 0.3333333 0.3333333 0.3333333
> absorbingStates(mc1)
character(0)
> recurrentStates(mc1)
[1] "st 1" "st 2" "st 3"
> transientStates(mc1)
character(0)
```

(continued)

Example 4.18 (continued)
`plot(mc1)` plots the chain: the result appears in Fig. 4.55.

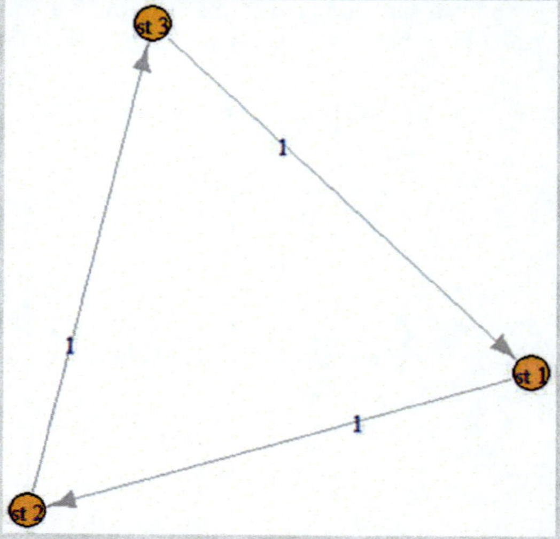

Fig. 4.55 Plot of the chain

Let us simulate $1e4$ states, starting from the initial state "`st 1`", check the Markov property on the simulated data and fit a transition matrix to it

```
> sim1 <- rmarkovchain(n = 10000, object = mc1, t0 = "st 1")
> verifyMarkovProperty(sim1)
Testing markovianity property on given data sequence
Chi - square statistic is: 0
Degrees of freedom are: 2
And corresponding p-value is: 1
> mat1 = createSequenceMatrix(sim1,toRowProbs = TRUE)
> mat1
     st 1 st 2 st 3
st 1    0    1    0
st 2    0    0    1
st 3    1    0    0
```

Example 4.19
Let us consider a stationary Markov chain such that

$$\mathbf{\Pi} = \begin{pmatrix} 0.2 & 0 & 0.8 & 0 \\ 0 & 1 & 0 & 0 \\ 0.4 & 0.2 & 0.1 & 0.3 \\ 0 & 0 & 0 & 1 \end{pmatrix}$$

This Markov chain has 2 absorbing states: 2 and 4. By renumbering the states: $1 \to 3$, $2 \to 1$, $3 \to 4$, $4 \to 2$, we obtain the canonical form

$$\mathbf{\Pi}_{\mathbf{can}} = \begin{pmatrix} 1 & 0 & 0 & 0 \\ 0 & 1 & 0 & 0 \\ 0 & 0 & 0.2 & 0.8 \\ 0.2 & 0.3 & 0.4 & 0.1 \end{pmatrix}.$$

Thus,

$$\boldsymbol{A} = \begin{pmatrix} 0 & 0 \\ 0.2 & 0.3 \end{pmatrix}, \boldsymbol{B} = \begin{pmatrix} 0.2 & 0.8 \\ 0.4 & 0.1 \end{pmatrix}.$$

We create the Markov Chain, *using the rows as probabilities of transition*:

```
matPi = matrix(
  c(0.2,0,0.8,0,
    0,1,0,0,
    0.4,0.2,0.1,0.3,
    0,0,0,1
    ),
  nrow=4,ncol=4,
  byrow = TRUE)
lst = c("st 1","st 2","st 3","st 4" )
library("markovchain")
mc1 = new("markovchain",states=lst,
          transitionMatrix=matPi,
          name = "Test Chain",
          byrow = TRUE)
```

Analyze the states:

```
> steadyStates(mc1)
     st 1 st 2 st 3 st 4
[1,]    0    0    0    1
[2,]    0    1    0    0
> absorbingStates(mc1)
[1] "st 2" "st 4"
> recurrentStates(mc1)
[1] "st 2" "st 4"
> transientStates(mc1)
[1] "st 1" "st 3"
```

(continued)

Example 4.19 (continued)

The chain is plotted in Fig. 4.56.

Fig. 4.56 Plot of the chain

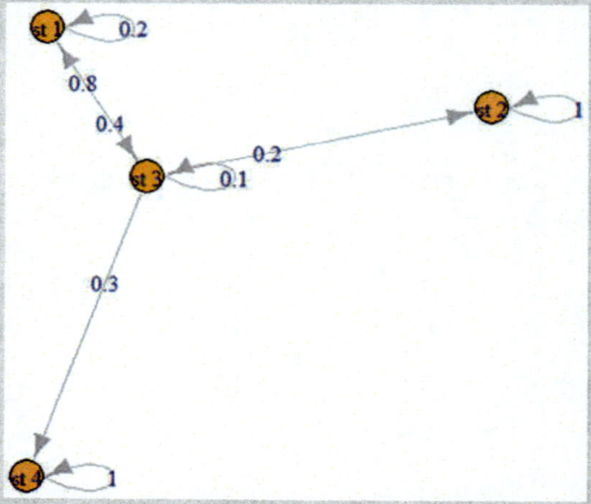

Let us determine the canonical form using *R*:

```
> mc1can = canonicForm(mc1)
> matPican = mc1can@transitionMatrix
> matPican
     st 2 st 4 st 1 st 3
st 2  1.0  0.0  0.0  0.0
st 4  0.0  1.0  0.0  0.0
st 1  0.0  0.0  0.2  0.8
st 3  0.2  0.3  0.4  0.1
```

Determine M. The expected number of steps before absorption are $M_{\blacksquare 1}\cdot = 4.25$, $M_{\blacksquare 3} = 3$

```
> B = matPican[3:4,3:4]
> M = solve(diag(2)-B)
> M
     st 1 st 3
st 1 2.25    2
st 3 1.00    2
```

You can get the expected number of steps before absorption directly:

```
> meanAbsorptionTime(mc1)
st 1 st 3
4.25 3.00
```

Evaluate N:

```
> A = matPican[3:4,1:2]
> N = M %*% A
> N
     st 2 st 4
st 1  0.4  0.6
st 3  0.4  0.6
```

You can find the probabilities of absorption directly:

```
> absorptionProbabilities(mc1)
     st 2 st 4
st 1  0.4  0.6
st 3  0.4  0.6
```

Exercises

1. Let us consider a Markov chain having states $\{1, 2\}$ and transition matrix ($0 \le a, b \le 1$)

$$\mathbf{\Pi} = \begin{pmatrix} a & b \\ 1-a & 1-b \end{pmatrix}$$

   (a) Draw the graphical representation of the chain.
   (b) Determine the values of $P(X_{n+1} = 1 | X_n = 2)$ and $P(X_{n+1} = 2 | X_n = 1)$
   (c) Determine the values of a, b such that the chain is absorbing.
   (d) Determine the values of a, b such that the chain is periodical.
   (e) Find a general expression of the eigenvector $\boldsymbol{v}$ of $\mathbf{\Pi}$ associated to the eigenvalue 1.
   (f) Determine the conditions on $\boldsymbol{v}$ such that $\boldsymbol{v} = \boldsymbol{p}$.
   (g) Simulate the chain for convenient values of a, b.

2. Let us consider a Markov chain having states $\{1, 2, 3, 4\}$ and transition matrix

$$\mathbf{\Pi} = \begin{pmatrix} 0,02 & 0,12 & 0 & 0,7 \\ 0,91 & 0,4 & 0,6 & 0 \\ 0 & 0,48 & 0,4 & 0 \\ 0,07 & 0 & 0 & 0,3 \end{pmatrix}$$

   (a) Draw a graphical representation of the chain
   (b) Determine the values of $P(X_{n+1} = 2 | X_n = 3)$ and $P(X_{n+1} = 3 | X_n = 2)$
   (c) Determine the stationary probability.
   (d) Simulate the chain.

(continued)

3. This exercise comes from (Sinclair, 2005), under Creative Commons Attribution License 1.0 license. Consider a machine that can be in the states idle, busy, waiting, broken , in repair – numbered respectively 1–5. The transition matrix is

$$\Pi = \begin{pmatrix} 0,05 & 0,1 & 0 & 0 & 0,75 \\ 0,93 & 0,43 & 0,6 & 0 & 0 \\ 0 & 0,43 & 0,35 & 0 & 0 \\ 0,02 & 0,04 & 0,05 & 0,8 & 0 \\ 0 & 0 & 0 & 0,2 & 0,25 \end{pmatrix}$$

(a) Draw a graphical representation of the chain
(b) Determine the values of $P(X_{n+1} = 2 | X_n = 4)$ and $P(X_{n+1} = 4 | X_n = 2)$
(c) Determine the stationary probability.
(d) Simulate the chain.

4. This exercise comes from (Chen, 2008), under Creative Commons Attribution License 1.0 license. Consider a Markov chain to generate music: the states are the notes A, A#, B, C, D, E, F, G, G# – numbered respectively 1–9. The transition matrix is

$$\Pi = \begin{pmatrix} \frac{4}{19} & 1 & \frac{7}{15} & 0 & 0 & \frac{4}{19} & 0 & \frac{1}{5} & 0 \\ 0 & 0 & 0 & 0 & 0 & \frac{1}{19} & 0 & 0 & 0 \\ \frac{3}{19} & 0 & \frac{1}{15} & \frac{6}{15} & 0 & 0 & \frac{1}{5} & \frac{1}{5} & \frac{3}{4} \\ 0 & 0 & \frac{4}{15} & \frac{3}{15} & \frac{3}{11} & \frac{3}{19} & 0 & \frac{2}{5} & 0 \\ \frac{2}{19} & 0 & 0 & \frac{6}{15} & \frac{3}{11} & 0 & 0 & 0 & 0 \\ \frac{1}{19} & 0 & \frac{3}{15} & 0 & \frac{5}{11} & \frac{5}{19} & \frac{1}{5} & 0 & \frac{1}{4} \\ 0 & 0 & 0 & 0 & 0 & \frac{4}{19} & 0 & 0 & 0 \\ \frac{6}{19} & 0 & 0 & 0 & 0 & \frac{1}{19} & \frac{3}{5} & \frac{1}{5} & 0 \\ \frac{3}{19} & 0 & 0 & 0 & 0 & \frac{1}{19} & 0 & 0 & 0 \end{pmatrix}$$

(a) Determine the values of $P(X_{n+1} = 2 | X_n = 6)$ and $P(X_{n+1} = 6 | X_n = 2)$
(b) Determine the stationary probability.
(c) Simulate the chain.

4.8 Diffusion Processes

Stochastic differential equations (SDE) are a common tool in Mathematical Finance, namely the famous model of Fischer Black and Myron Scholes (1973) for the price of options, initially introduced as a partial differential equation and reinterpreted by Robert Cox Merton as a SDE (Merton, 1973). Previous works tending to introduce probability in the analysis of finance were made by (Regnault, 1863; Cowles, 1933) and namely by (Bachelier, 1900), which firstly introduced Brownian motion in finance.

Since these works, SDE has become a central tool in finance, but also in other fields such as, for instance, stochastic optimization and reliability. The mathematical foundations are connected to Itô's Calculus (Itô, 1944, Itô, 1946, 1950, 1951).

The solutions of SDE are usually referred as diffusion processes. You can find in the R repositories packages for the simulation of SDE, such as `Sim.DiffProc` and `sde`.

4.8.1 *Time Integral and Derivative of a Process*

Standard integrals are defined as limits of finite sums. For instance, to define $\int_a^b f(t)dt$, we may consider a partition

$$\mathcal{T}=\{a=t_0<t_1<\ldots<t_{n-1}<t_n=b\} \tag{4.18}$$

of the interval (a, b) and the finite sum

$$S(\mathcal{T},f)=\sum_{i=1}^{n}(t_i-t_{i-1})f(t_{i-1}). \tag{4.19}$$

For a regular function f,

$$S(\mathcal{T},f)\to\int_a^b f(t)dt$$

when $n\to+\infty$ and $\max\{t_i-t_{i-1}\}\to 0$.

We may define the integral of a stochastic process with respect to time into an analogous way. For instance, let us consider a stochastic process $\{X(t), t\in(a,b)\}$. We may consider the finite sum

$$I(\mathcal{T},\mathrm{X})=\sum_{i=1}^{n}(t_i-t_{i-1})X(t_{i-1}). \tag{4.20}$$

If the process has integrable variances, id est, if

$$\int_a^b E\left(X^2(t)\right)dt < \infty ,$$

then the finite sum converges for $n \to +\infty$ and $\max\{t_i - t_{i-1}\} \to 0$. In this case, we have, analogously the standard functions (see, for instance Föllmer, 1981):

$$I(\mathcal{T}, X) \to I(X) = \int_a^b X(t)dt. \tag{4.21}$$

$I(X)$ is called the integral of the process X with respect to the time. It is linear:

$$I(\alpha X + \beta Y) = \alpha I(X) + \beta I(Y), \ \forall \alpha, \beta \in \mathbb{R}.$$

and

$$\int_a^c X_t dt + \int_c^b X_t dt = \int_a^b X_t dt. \tag{4.22}$$

From Eq. (4.21):

$$E(I(X)) = \int_a^b m_X(t)dt, E\left(I(X)^2\right) = \iint_a^b R_{XX}(s,t)dsdt. \tag{4.23}$$

Let us define

$$dY_t = X_t dt \Leftrightarrow Y_t = Y_0 + \int_0^t X_s ds, \forall t > 0. \tag{4.24}$$

Then, X_t is the derivative of Y_t with respect to time. We have

$$m_Y(t) = m_Y(0) + \int_0^t m_X(s)ds, \forall t > 0 \Rightarrow m_Y = \frac{d}{dt}m_X. \tag{4.25}$$

Assuming that Y_0 and X are independent, we have

$$R_{YY}(s,t)=E\left(Y_0^2\right)+m_Y(0)\left(\int_0^s m_X(\tau_1)d\tau_1+\int_0^t m_X(\tau_2)d\tau_2\right)$$
$$+\int_{\tau_1=0}^{\tau_1=s}\int_{\tau_2=0}^{\tau_2=t} R_{XX}(\tau_1,\tau_2)d\tau_1 d\tau_2, \tag{4.26}$$

so that

$$R_{XX}(s,t)=\frac{\partial^2}{\partial s\partial t}R_{YY}(s,t). \tag{4.27}$$

Into an analogous way, we have

$$R_{YX}(s,t)=\frac{\partial}{\partial t}R_{YY}(s,t). \tag{4.28}$$

Example 4.20

Let us consider the process $Y_t=A_1\cos(\omega t)+A_2\sin(\omega t)$, where $\omega\in\mathbb{R}$, while A_1, A_2 are independent random variables such that $E(A_i)=0$, $V(A_i)=\sigma^2, i=1,2$. We have

$$m_Y(t)=0, V_Y(t)=\sigma^2, R_{YY}(s,t)=\sigma^2\cos(\omega(s-t)).$$

Let $dY_t=X_t dt$, $Y_0=A_1$. We have

$$R_{XX}(s,t)=\frac{\partial^2}{\partial s\partial t}R_{YY}(s,t)=\omega^2\sigma^2\cos(\omega(s-t)).$$
$$R_{YX}(s,t)=\frac{\partial}{\partial t}R_{YY}(s,t)=\omega\sigma^2\sin(\omega(s-t)).$$

Indeed,

$$Y_t=A_1-A_1\omega\int_0^t\sin(\omega s)ds+A_2\omega\int_0^t\cos(\omega s)ds,$$

(continued)

Example 4.20 (continued)

So that, we have

$$X_t = -A_1\omega \sin(\omega t) + A_2\omega \cos(\omega t).$$

Example 4.21

Let us consider the process $X_t = A_1 \cos(\omega t) + A_2 \sin(\omega t)$, where $\omega \in \mathbb{R}$, while A_1, A_2 are independent random variables such that $E(A_i) = 0$, $V(A_i) = \sigma^2$, $i = 1, 2$. We have

$$m_X(t) = 0, V_X(t) = \sigma^2, R_{XX}(s, t) = \sigma^2 \cos(\omega(s - t)).$$

Let $dY_t = X_t dt$, $Y_0 = 0$. We have

$$m_Y(t) = 0$$

and

$$R_{YY}(s, t) = \int_{\tau_1=0}^{\tau_1=s} \int_{\tau_2=0}^{\tau_2=t} R_{XX}(\tau_1, \tau_2) d\tau_1 d\tau_2,$$

so that

$$R_{YY}(s, t) = \sigma^2 \frac{\cos(\omega(s - t)) - \cos(\omega s) - \cos(\omega t) + 1}{\omega^2},$$

Indeed, let

$$Y_t = A_1 \int_0^t \sin(\omega s) ds + A_2 \int_0^t \cos(\omega s) ds.$$

(continued)

Example 4.21 (continued)

Then,

$$Y_t = \frac{A_1}{\omega}(1 - \cos(\omega t)) + \frac{A_2}{\omega}\sin(\omega t)$$

and we have

$$E(Y_t Y_s) = \sigma^2 \frac{\cos(\omega(s-t)) - \cos(\omega s) - \cos(\omega t) + 1}{\omega^2}.$$

Example 4.22

Let us consider the process X_t, such that each X_t is a random variable having the uniform distribution on (a, b) and X_t is independent from X_s for $t \neq s$. We have

$$m_X(t) = \frac{a+b}{2}, V_X(t) = \frac{(b-a)^2}{12}, R_{XX}(s, t) = \frac{a^2 + ab + b^2}{3}\delta(t-s).$$

Let us evaluate $Y_t = \int_0^t X_s ds$. We have $E(Y_t) = tm_X$, $V(Y_t) = 0$. We generate a sample of 1E4 variates from Y_t and we evaluate the empirical mean and the empirical variance. The results appear in Figs. 4.57 and 4.58.

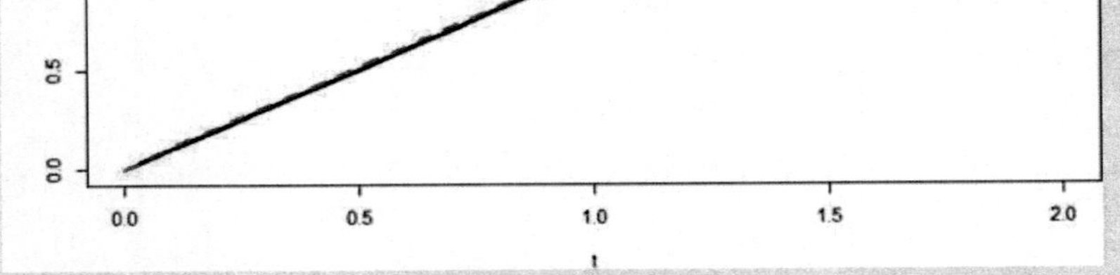

Fig. 4.57 Results from a sample of 1E4 trajectories of the process Y_t

(continued)

Example 4.22 (continued)

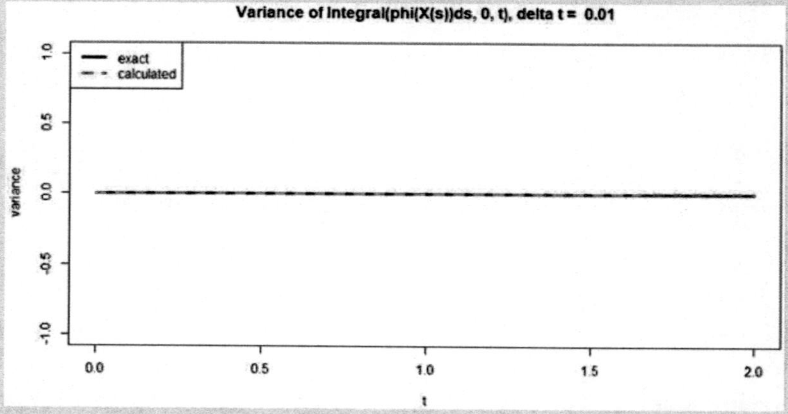

Fig. 4.58 Results from a sample of 1E4 trajectories of the process Y_t

Exercises

1. Consider the process $X_t = A_1 + A_2 t$, where A_1, A_2 are independent random variables such that $E(A_i) = 0$, $V(A_i) = 1$, $i = 1, 2$.

   (a) Determine $R_{XX}(s, t)$.
   (b) Let $Y_t = X_t dt$, $Y_0 = 0$. Determine $R_{YY}(s, t)$.
   (c) Let $dX_t = Y_t dt$. Determine $R_{YY}(s, t)$.

2. Consider the process $X_t = A_1 + A_2 \exp(-t)$, where A_1, A_2 are independent random variables such that $E(A_i) = 0$, $V(A_i) = 1$, $i = 1, 2$.

   (a) Determine $R_{XX}(s, t)$.
   (b) Let $Y_t = X_t dt$, $Y_0 = 0$. Determine $R_{YY}(s, t)$.
   (c) Let $dX_t = Y_t dt$. Determine $R_{YY}(s, t)$.

3. Consider a process X_t such that $R_{XX}(s, t) = \exp(-|s - t|)$.

   (a) Let $Y_t = X_t dt$, $Y_0 = 0$. Determine $R_{YY}(s, t)$.
   (b) Let $dX_t = Y_t dt$. Determine $R_{YY}(s, t)$.

4. Consider a process X_t such that $R_{XX}(s, t) = 1 - |s - t|$, *if* $|s - t| < 1$; $R_{XX}(s, t) = 0$, *otherwise*.

   (a) Let $Y_t = X_t dt$, $Y_0 = 0$. Determine $R_{YY}(s, t)$.
   (b) Let $dX_t = Y_t dt$. Determine $R_{YY}(s, t)$.

4.8.2 Simulation of the Time Integral of a White Noise

The simulation of the process defined by the time integral (4.22) or (4.25) can request, on the one hand, special simulation features and, on the other hand, special quadrature formulae.

Indeed, a simulation is made by choosing a final time of simulation $T > 0$ and a partition of the interval $(0, T)$ in a maximal number of steps $ns > 0$: $0 = t_0 < t_1 < \ldots < t_{ns-1} < t_{ns} = T$. For instance, we may take $t_i = i\Delta t$, $\Delta t = \frac{T}{ns}$, $0 \le i \le ns$. Then, we generate a sequence of values $X_1, \ldots, X_{ns}$ such that $X_i = X(t_i)$ and use them to evaluate

$$Y_{n+1} = Y_n + \int_{t_n}^{t_{n+1}} X_s ds, 0 \le n \le ns - 1. \tag{4.29}$$

By this way, we generate $Y_1, \ldots, Y_{ns}$ where $Y(t_j) = Y_j$. Such a procedure requests some precautions. On the one hand, the generation of $X_1, \ldots, X_n, \ldots$ must be consistent with the statistical properties of X_t – namely $E(X_n) = m_X(t_n)$ and $E(X_iX_j) = R_{XX}(t_i, t_j)$. On the other hand, the evaluation of the integral at the right-hand side must be consistent with Eqs. (4.24) and (4.27). For instance, let us consider the approximation

$$\int_{t_n}^{t_n+\Delta t} X_s ds \approx \Delta t(\alpha X_n + \beta). \tag{4.30}$$

The conditions (4.24) yield that

$$\alpha m_X(t_n) + \beta = \frac{1}{\Delta t} \int_{t_n}^{t_n+\Delta t} m_X(s) ds$$

$$\alpha^2 R_{XX}(t_n, t_n) + 2\alpha\beta m_X(t_n) + \beta^2 = \frac{1}{\Delta t^2} \int_{t_n}^{t_n+\Delta t} \int_{t_n}^{t_n+\Delta t} R_{XX}(s, t) ds dt$$

If the process is weakly stationary, these equations read as:

$$\alpha m_X + \beta = m_X$$

$$\alpha^2 R_X(0) + 2\alpha\beta m_X + \beta^2 = \frac{1}{\Delta t^2} \int_{t_n}^{t_n+\Delta t} \int_{t_n}^{t_n+\Delta t} R_X(|s - t|) ds dt$$

id est,

$$\beta = (1 - \alpha) m_X$$

$$\alpha^2 C_X(0) + m_X^2 = \frac{1}{\Delta t^2} \int_{t_n}^{t_n+\Delta t} \int_{t_n}^{t_n+\Delta t} R_X(s-t) ds dt$$

If R_X is a regular function, we can take as approximation

$$\int_{t_n}^{t_n+\Delta t} \int_{t_n}^{t_n+\Delta t} R_X(s-t) ds dt \approx \Delta t^2 R_X(0),$$

so that $\alpha \approx 1, \beta \approx 0$ are convenient values. But these values may be not convenient if R_X is not a regular function. For instance, if $R_X(s) = \sigma^2 \delta(s)$, we obtain

$$\alpha = \sqrt{\frac{1}{\Delta t} - \frac{m_X^2}{\sigma^2}}\ ,\beta = (1-\alpha) m_X.$$

If $X_t = \varepsilon_t$ is a white noise, we have $m_X = 0$, so that

$$\alpha = \frac{1}{\sqrt{\Delta t}}\ ,\ \beta = 0\ . \tag{4.31}$$

In the sequel, we shall consider the time integral of a white noise:

$$E_t = E_0 + \int_0^t \varepsilon_s ds. \tag{4.32}$$

According to Eq. (4.32), the simulations use the approximation

$$\int_{t_n}^{t_n+\Delta t} \varepsilon_s ds \approx \varepsilon_n \sqrt{\Delta t}. \tag{4.33}$$

We have $m_E(t) = 0$ and

$$R_{EE}(s, t) = \sigma^2 \int_{\tau_1=0}^{\tau_1=s} \int_{\tau_2=0}^{\tau_2=t} \delta(\tau_1 - \tau_2) d\tau_1 d\tau_2 = \sigma^2 \min\{s, t\}. \tag{4.34}$$

Taking into account Eqs. (4.20) and (4.21) and the Theorem 2.5, E_t is gaussian. In addition,

$$0 \le r \le s \le t \Rightarrow E((E_t - E_s)E_r) = \sigma^2 r - \sigma^2 r = 0. \tag{4.35}$$

Thus, on the one hand, $E_t \sim N\left(0, \sigma\sqrt{t}\right)$ and, on the other hand, $(E_t - E_s)$ is independent from E_r if $0 \le r \le s \le t$. We have, for $0 \le s \le t$:

$$E\left((E_t - E_s)^2\right) = \sigma^2(t-s).$$

Thus, $E_t - E_s \sim N\left(0, \sigma\sqrt{t-s}\right)$ for $0 \le s \le t$.

Let us consider the process $X_t = \varepsilon_{t\,+\,\delta t}$ and

$$Y_t = \int_0^t \varepsilon_{s+\delta t} ds = \int_{\delta t}^{t+\delta t} \varepsilon_s ds = \int_0^{t+\delta t} \varepsilon_s ds - \int_0^{\delta t} \varepsilon_s ds = E_{t+\delta t} - E_{\delta t} = \delta E_t.$$

Let us introduce $\delta_k E_t = E_{t+\delta t_k} - E_{\delta t_k}$. Then, $R_{ij}(s, t) = E(\delta_i E_t \times \delta_j E_s)$ verifies

$$R_{ij}(s, t) = \sigma^2\big(\min\left\{t + \delta t_i, s + \delta t_j\right\} - \min\left\{\delta t_i, s + \delta t_j\right\} \\ - \min\left\{t + \delta t_i, \delta t_j\right\} + \min\left\{\delta t_i, \delta t_j\right\}\big). \tag{4.36}$$

Thus, for a process

$$X_t = \sum_{i=0}^{k} a_i \varepsilon_{t+\delta ti} \;,$$

$Y_t = \int_0^t X_t ds$ verifies

$$Y_t = \sum_{i=0}^{k} a_i \delta_i E_t, \delta_i E_t = E_{t+\delta t_i} - E_{\delta t_i}$$

and

$$R_{YY}(s, t) = E(Y_t Y_s) = \sum_{i=0}^{k}\sum_{j=0}^{k} a_i a_j R_{ij}(s, t). \tag{4.37}$$

A process

$$X_t = \sum_{i=0}^{k} a_i \varepsilon_{t\,-\,\delta ti}$$

may be analyzed into an analogous way: notice that

$$\int_{\delta t}^{t+\delta t} \varepsilon_s ds \sim \int_{-\delta t}^{t-\delta t} \varepsilon_s, \tag{4.38}$$

so that Eqs. (4.36) and (4.37) apply yet.

Example 4.23
Let us consider a white noise $\{\varepsilon_t : t \geq 0\}$ of variance 1 and the process $X_t = \varepsilon_t + \varepsilon_{t+1}$. Let $Y_t = \int_0^t X_s ds$. We have $k = 1$, $a_0 = a_1 = 1$, $\delta_0 t = 0$, $\delta_1 t = 1$. Thus,

$$Y_t = E_t + E_{t+1} - E_1.$$

$$R_{00}(s, t) = \min\{t, s\} - \min\{0, s\} - \min\{t, 0\} + \min\{0, 0\} = \min\{t, s\};$$

$$R_{01}(s, t) = \min\{t, s+1\} - \min\{0, s+1\} - \min\{t, 1\} + \min\{0, 1\} = \min\{t, s+1\} - \min\{t, 1\};$$

$$R_{11}(s, t) = \min\{t+1, s+1\} - \min\{1, s+1\} - \min\{t+1, 1\} + \min\{1, 1\} = \min\{t+1, s+1\} - 1 = \min\{t, s\};$$

For $t = s$,

$$R_{00}(t, t) = t, R_{01}(t, t) = t - \min\{t, 1\}, R_{11}(t) = t$$

Thus,

$$V_Y(t) = 2t, \text{if } t < 1; \quad V_Y(t) = 4t - 2, \text{if } t > 1.$$

Figure 4.59 shows the results of a numerical evaluation of the variances for $t \in (0, 2)$ by using 10000 simulations of Y_t with a step $\Delta t = 0.01$.

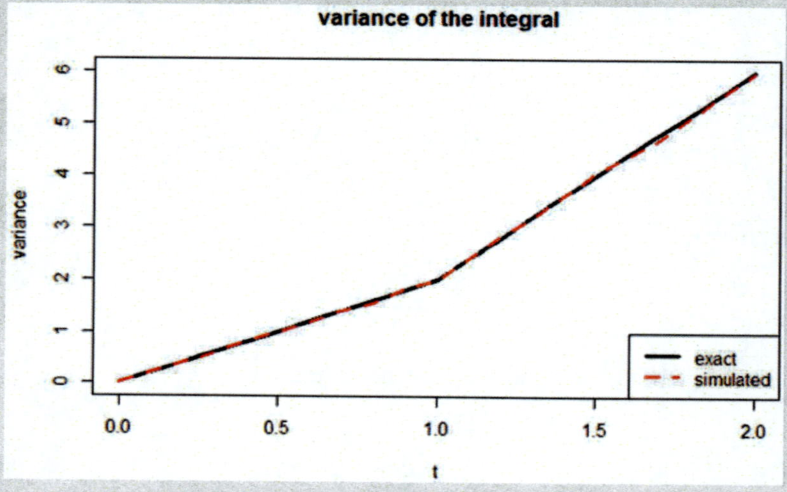

Fig. 4.59 Results obtained by simulation of the process Y_t

Example 4.24
Let us consider a white noise $\{\varepsilon_t : t \geq 0\}$ of variance 1 and the process $X_t = \varepsilon_t - \varepsilon_{t-1}$. Let $Y_t = \int_0^t X_s ds$. We have $k = 1, a_0 = 1, a_1 = -1, \delta_0 t = 0, \delta_1 t = -1$.
We apply the analogy given in Eq. (4.38)

$$E_{t-1} - E_{-1} \sim E_{t+1} - E_1 \Rightarrow Y_t \sim E_t - (E_{t+1} - E_1).$$

Thus,

$$V_Y(t) = 2t, \text{if } t < 1; \quad V_Y(t) = 2, \text{if } t > 1.$$

Figure 4.60 shows the results of a numerical evaluation of the variances for $t \in (0, 2.5)$ by using 10000 simulations of Y_t with a step $\Delta t = 0.01$.

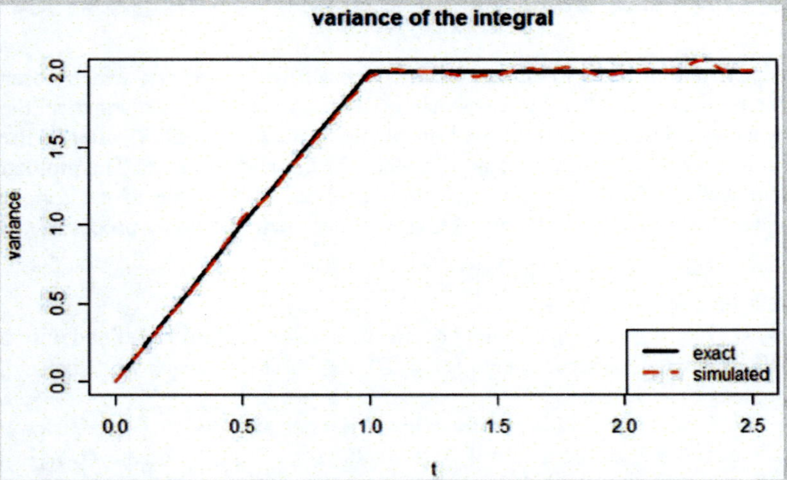

Fig. 4.60 Results obtained by simulation of the process Y_t

Exercises

1. Consider a white noise $\{\varepsilon_t : t \geq 0\}$ of variance 1 and the process $X_t = \varepsilon_t + \varepsilon_{t-2}$.

   (a) Identify the coefficients a_0, a_1, a_2.
   (b) Determine $R_{ij}(t, t)$, $0 \leq i, j \leq 2$.
   (c) Evaluate the variance of Y_2 by using simulations of the process Y_t.

2. Consider a white noise $\{\varepsilon_t : t \geq 0\}$ of variance 1 and the process $X_t = \varepsilon_t - \varepsilon_{t+2}$.

   (a) Identify the coefficients a_0, a_1, a_2.
   (b) Determine $R_{ij}(t, t)$, $0 \leq i, j \leq 2$.

   Evaluate the variance of Y_2 by using simulations of the process Y_t.

4.8.3 Brownian Motion

Brownian motion takes his name from Robert Brown, a botanist which observed chaotic motions of particles in suspension (Brown, 1828). The mathematical theory of Brownian motion was established much later (Paley et al., 1933), after its use by (Bachelier, 1900), the questions by (Einstein, 1905), and (Pearson, The problem of random walk, 1905) and the model by (Langevin, 1908). Due to the formalism introduced by Wiener, Brownian motion is also known as Wiener process.

Definition 4.10
$\{W_t = W(t) : t \geq 0\}$ is a **Wiener process** (or **Brownian motion**) if and only if $W(0) = 0$ and $dW_t = \varepsilon_t dt$, where $\{\varepsilon_t = \varepsilon(t) : t \in \mathbb{R}\}$ is a gaussian white noise of variance $\sigma^2 = 1$.

$\{\mathbf{W}_t = \mathbf{W}(t) : t > 0\}$ is a ***n*-dimensional Wiener process** (or ***n*-dimensional** Brownian motion) if and only if $\mathbf{W}(t) = (W_1(t), \ldots, W_n(t))$, where each W_i is a Wiener process and W_i is independent from W_j for $i \neq j$.

Thus, a Brownian motion corresponds to an integral of a white noise. As shown in Sect. 4.8, we have

$$m_W(t) = 0, R_{WW}(s, t) = \min\{s, t\}, E\left(W_t^2\right) = t. \tag{4.39}$$

$$0 \leq r \leq s \leq t \Rightarrow E((W_t - W_s)W_r) = r - r = 0. \tag{4.40}$$

$$W_t - W_s \sim N\left(0, \sqrt{t-s}\right), E\left((W_t - W_s)^2\right) = t - s, for\ 0 \leq s \leq t. \tag{4.41}$$

Let $\varphi : \mathbb{R} \to \mathbb{R}$ be a regular function and

$$I_t(\varphi) = \int_0^t \varphi(W_s)ds. \tag{4.42}$$

For a regular φ:

$$E(I_t(\varphi)) = \int_0^t E(\varphi(W_s))ds. \tag{4.43}$$

The simulation of a Brownian motion can be made by the method previously presented (Sect. 4.8):

$$W_{n+1} = W_n + \sqrt{\Delta t}\, Z_n, 0 \le n \le ns - 1, Z_n \sim N(0, 1). \tag{4.44}$$

Example 4.25
Let us consider $T = 4$. We desire to simulate a bidimensional Brownian motion with $ns = 4000$, so that $\Delta t = 1E - 3$. For instance, we can use `Sim.DiffProc`:

```
library(Sim.DiffProc)
tmax = 4
ns = 4000
x = as.vector(BM(N=ns, T=tmax))
y = as.vector(BM(N=ns, T=tmax))
```

An example of result is shown in Fig. 4.61.

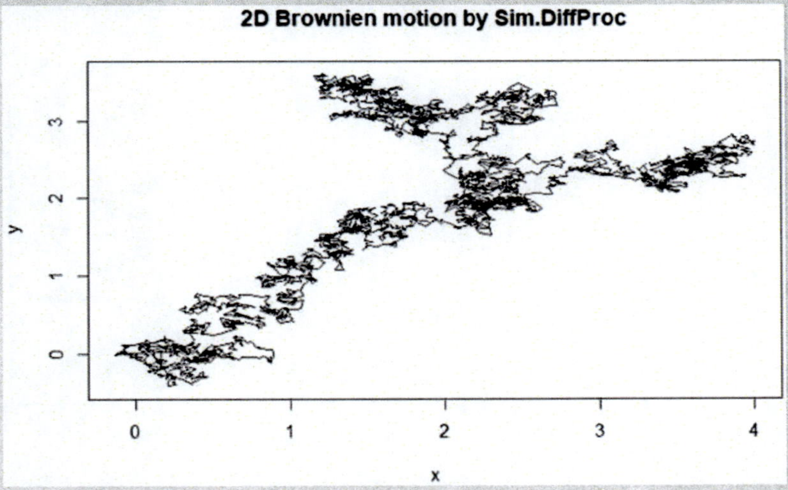

Fig. 4.61 Simulation of a Brownian Motion until $T = 4$, with $ns = 4000$ steps

(continued)

Example 4.25 (continued)

We evaluate $E(\|X(T)\|^2)$ for different values of *ns* and different numbers of trajectories. The theoretical value is $E(\|X(T)\|^2) = 8$. The results are shown in Tables 4.1 and 4.2.

Table 4.1 Results furnished by the simulation of a Brownian motion until $T = 4$, sample of 10000 trajectories generated by `Sim.DiffProc`

ns	400	4000	40000
$E(\|X(T)\|^2)$	8.1	7.9	8.1
$\sigma(\|X(T)\|^2)$	8.1	8.0	8.1

Table 4.2 Results furnished by the simulation of a Brownian motion until $T = 4$, with $ns = 4000$. Trajectories generated by `Sim.DiffProc`

Simulations	100	1000	10000
$E(\|X(T)\|^2)$	8.8	8.0	8.0
$\sigma(\|X(T)\|^2)$	10.2	8.4	8.0

We can also use the package `sde`:

```
library(sde)
tmax = 4
ns = 4000
x = as.vector(BM(N=ns, T=tmax))
y = as.vector(BM(N=ns, T=tmax))
```

The results appear in Fig. 4.64.

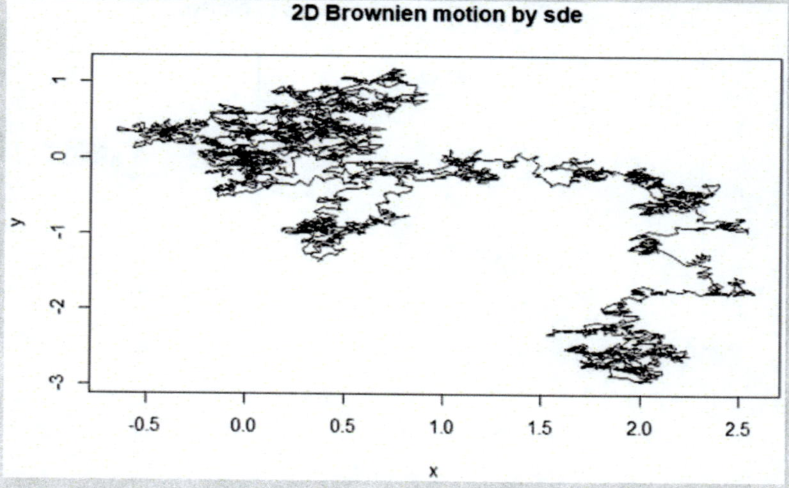

Fig. 4.64 Simulation of a Brownian Motion until $T = 4$, with $ns = 4000$ steps

(continued)

Example 4.25 (continued)

Tables 4.3 and 4.4 show the estimations of $E(\|X(T)\|^2)$ for different values of *ns* and different numbers of trajectories.

Table 4.3 Results furnished by the simulation of a Brownian motion until $T = 4$, sample of 10000 trajectories generated by `sde`

ns	400	4000	40000
$E(\|X(T)\|^2)$	8.0	7.9	7.9
$\sigma(\|X(T)\|^2)$	8.0	7.9	8.0

Table 4.4 Results furnished by the simulation of a Brownian motion until $T = 4$, with $ns = 4000$. Trajectories generated by `sde`

Simulations	100	1000	10000
$E(\|X(T)\|^2)$	6.9	8.0	7.8
$\sigma(\|X(T)\|^2)$	6.8	7.8	7.8

If you do not want to use the packages, you can easily build your own program. For example:

```
ns = 4000
dt = tmax / ns
h = sqrt(dt)
x = numeric(length=ns+1)
y = numeric(length=ns+1)
x[[1]] = 0
y[[1]] = 0
for (i in 1:ns){
  dx = rnorm(2)
  x[[i+1]] = x[[i]] + h * dx[[1]]
  y[[i+1]] = y[[i]] + h * dx[[2]]
}
```

An example of trajectory is plotted in Fig. 4.65.

(continued)

Example 4.25 (continued)

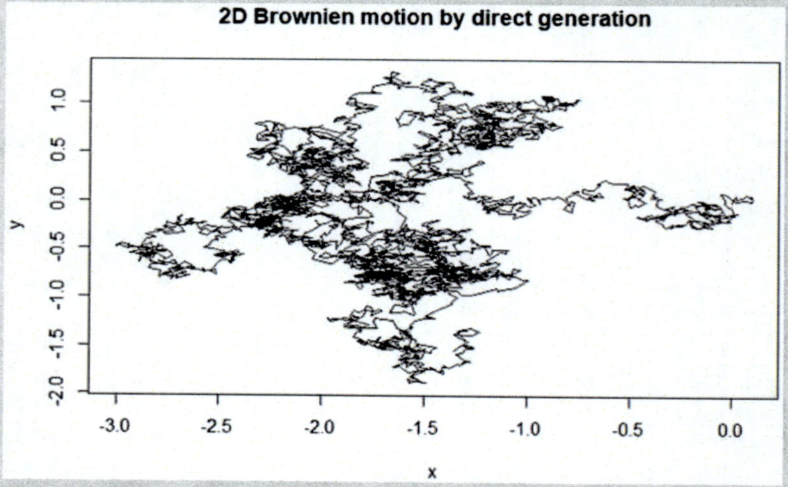

Fig. 4.65 Simulation of a Brownian Motion until $T = 4$, with $ns = 4000$ steps

Example 4.26

Let us consider $\varphi(W_s) = W_s^2$.Then, $E(I_t(\varphi)) = t^2/2$.

To generate one variate of $\int_{t_{min}}^{t_{max}} \varphi(t)dt$, we use the code at right:

```
intbrownsim = function(tmin,tmax, dt){
  library(Sim.DiffProc)
  ddt = tmax-tmin
  nmax = ceiling(tmax/dt)
  W = BM(N=nmax,  T=tmax)
  nmin = floor(tmin/dt)
  s = 0
  tt = tmin
  i = nmin
  while (tt < tmax){
    i = i + 1
    s = s + phi(W[[i]])
    tt = tt + dt
  }
  s = s * dt
  return(s)
}
```

(continued)

Example 4.26 (continued)

We consider $T = 2$, $t_i = (i - 1)/20$ evaluate successively the integrals $\int_{t_i}^{t_{i+1}} \varphi(W_t)dt$ over (t_i, t_{i+1}), $i = 1, \ldots, 20$ – each integral is evaluated using $\Delta t = 0.01$. Then $I_{t_i}(\varphi) \approx \sum_{j=1}^{i-1} \int_{t_j}^{t_{j+1}} \varphi(W_t)dt$.

We generate a sample of 1000 variates from $I_{t_i}(\varphi), 1 \leq i \leq 21$ and we use the sample to estimate the mean $E(I_{ti}(\varphi))$. The results are shown in Fig. 4.66. The results for 1E4 simulations appear in Fig. 4.67.

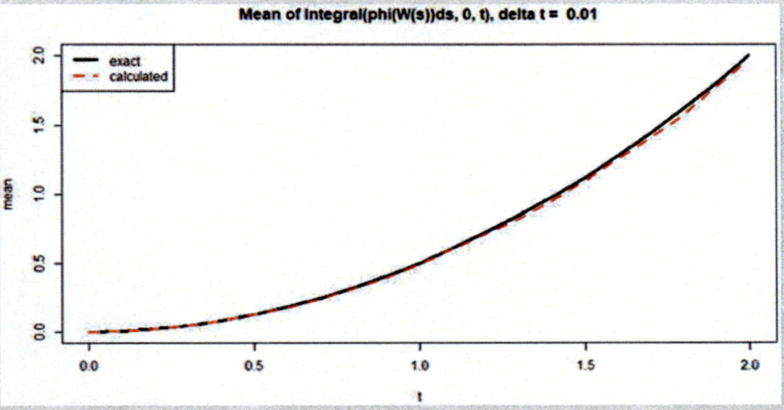

Fig. 4.66 Empirical mean of 1E3 simulations of the time integral of $\varphi(W_t) = W_t^2$, with $\Delta t = 1E - 2$. Brownian motion generated by `Sim.DiffProc`

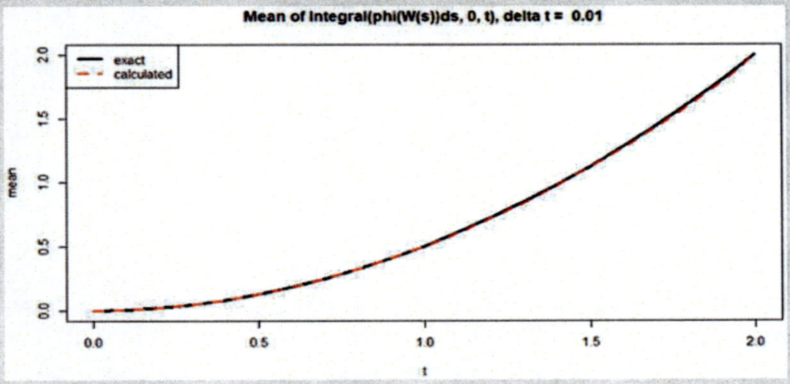

Fig. 4.67 Empirical mean of 1E4 simulations of the time integral of $\varphi(W_t) = W_t^2$, with $\Delta t = 1E - 2$. Brownian motion generated by `Sim.DiffProc`

(continued)

Example 4.26 (continued)

Of course, you can use `sde` instead of `Sim.DiffProc` : the results are shown in Figs. 4.68 and 4.69.

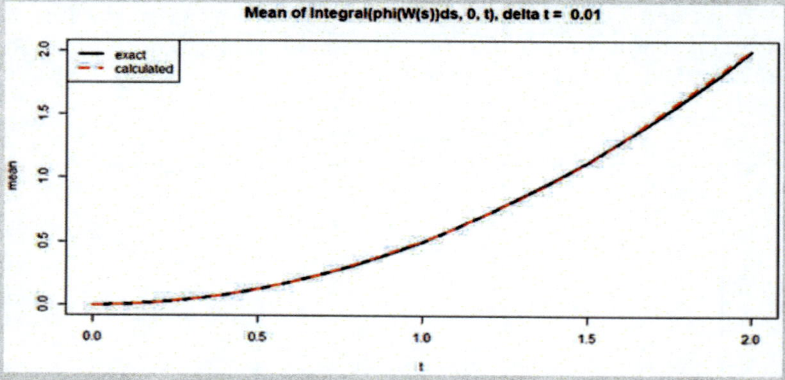

Fig. 4.68 Empirical mean of 1E3 simulations of the time integral of $\varphi(W_t) = W_t^2$, with $\Delta t = 1E - 2$. Brownian motion generated by `sde`

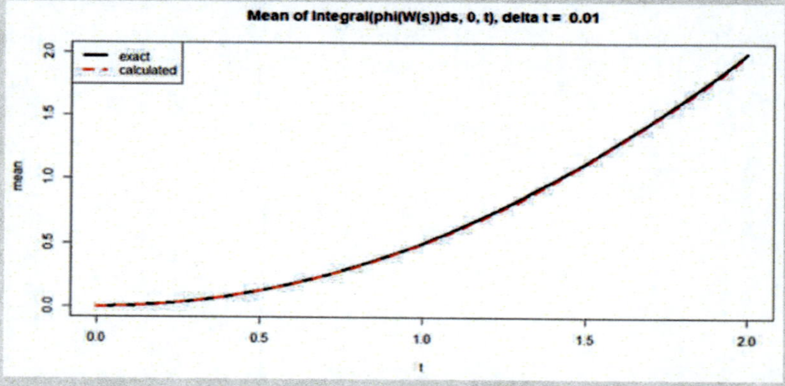

Fig. 4.69 Empirical mean of 1E4 simulations of the time integral of $\varphi(W_t) = W_t^2$, with $\Delta t = 1E - 2$. Brownian motion generated by `sde`

Exercises

1. For each function below, generate a sample of 1000 variates from $I_t(\varphi)$ for $t = 1$ and use the sample to estimate the mean value of $I_t(\varphi)$.

   (a) $\varphi(W_s) = \exp(W_s)$.
   (b) $\varphi(W_s) = \sin(2\pi W_s)$.
   (c) $\varphi(W_s) = W_s^4$.

4.8.4 Random Walks

Random walks generalize Brownian motion: a random walk is a Markov process which generates trajectories in the space $\mathbb{R}^n$. A discrete random walk generates points $X_1, \ldots, X_n, \ldots$ in $\mathbb{R}^n$ and the trajectory connects the points generated. A continuous random walk generates a trajectory $\{X_t = X(t) : t \geq 0; X(0) = X_0\}$.

In general, a discrete random walk is defined by the distribution of the displacements $\Delta X_n = X_{n+1} - X_n$ and the starting point X_0. The displacements ΔX_n are often multiples of a random vector having a known distribution: $\Delta X_n = hZ_n$, $h > 0$, Z_n known. In practice, the distribution of Z_n is often taken as independent of n, what is equivalent to take $Z_n \sim Z$. For instance, a discrete Brownian motion is a random walk where Z is a gaussian variable $N(0, 1)$ – parameter h is connected to the time discretization: $h = \sqrt{\Delta t}$.

Let us assume that $E(Z) = 0$, $V(Z) = \sigma^2$ and that Z_i, Z_j are independent for $i \neq j$. Noticing that

$$X_n = X_0 + \sum_{i=1}^{n} \Delta X_i, \tag{4.45}$$

We have

$$E(X_n) = X_0, V(X_n) = nh^2\sigma^2. \tag{4.46}$$

To simulate a discrete random walk, we need to know the values of X_0, h and the distribution of Z. Then, we choose a finite number of steps $ns > 0$ and we calculate $X_1, \ldots, X_{ns}$ by the recurrence (Z_n is a variate from Z).

$$X_{n+1} = X_n + hZ_n,\ 0 \leq n \leq ns - 1. \tag{4.47}$$

Example 4.27

Let us consider a bidimensional random walk starting at $X_0 = (0, 0)$ defined by $\Delta X_n = hZ_n$, $Z_n \sim Z$, $Z = (\cos\theta, \sin\theta)$, $P(\theta = 0) = P(\theta = \frac{\pi}{2}) = P(\theta = \pi) = P(\theta = \frac{3\pi}{2}) = \frac{1}{4}$. We generate a sample of $ns = 10000$ variates of Z and the corresponding random walk as follows:

```
dt=0.1
h = sqrt(dt)
angles = c(0,pi/2,pi, 3*pi/2)
sang = sample(angles,10000,replace=TRUE)
dx = c(0,h*cos(sang))
dy = c(0, h*sin(sang))
x = cumsum(dx)
y = cumsum(dy)
```

The resulting path is shown in Fig. 4.70.

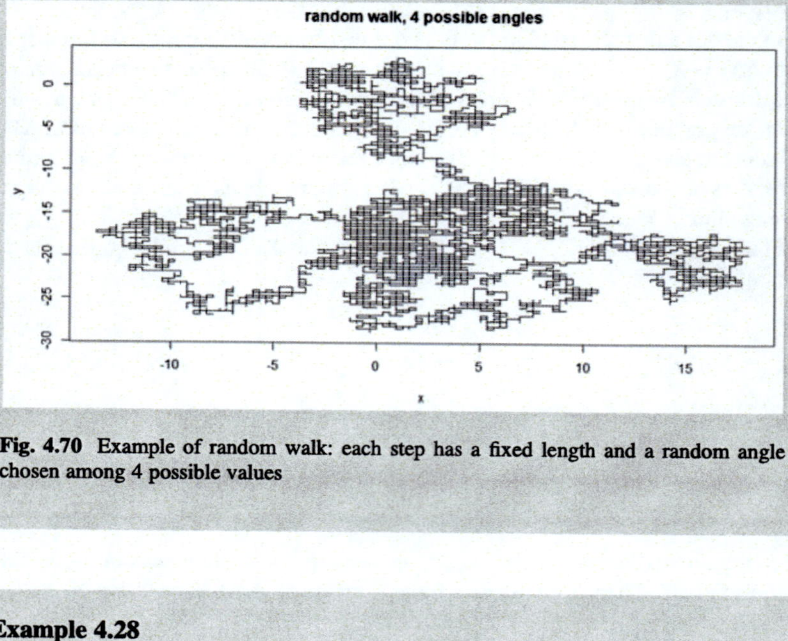

Fig. 4.70 Example of random walk: each step has a fixed length and a random angle chosen among 4 possible values

Example 4.28

Let us consider a bidimensional random walk starting at $X_0 = (0, 0)$ defined by $dX_t = Z_t$, $Z_t \sim Z$, $Z = (\cos\theta, \sin\theta)$, θ uniformly distributed on $(0, 2\pi)$. We generate a sample of $ns = 10000$ variates of Z and the corresponding random walk as follows:

(continued)

Example 4.28 (continued)

```
dt=0.1
h = sqrt(dt)
angles = c(0,pi/2,pi, 3*pi/2)
sang = sample(angles,10000,replace=TRUE)
dx = c(0,h*cos(sang))
dy = c(0, h*sin(sang))
x = cumsum(dx)
y = cumsum(dy)
```

The resulting path is shown in Fig. 4.71.

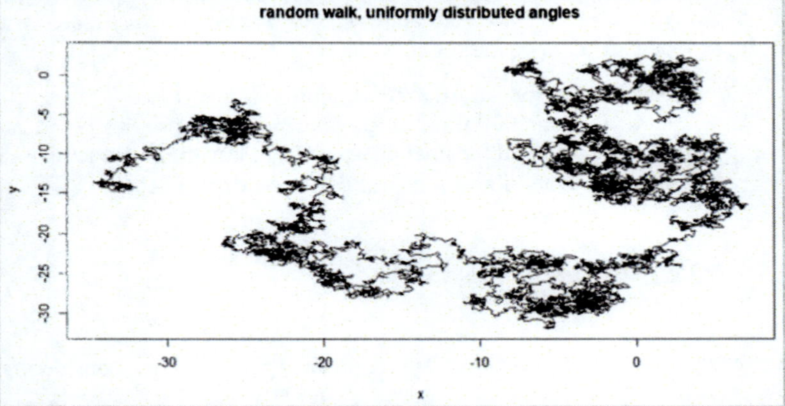

Fig. 4.71 Example of random walk: each step has a fixed length and a random angle uniformly distributed on $(0, 2\pi)$

Remark 4.9
Continuous random walks are generally defined by their initial point X_0 and their derivatives:

$$dX_t = Z_t dt, \ X(0) = X_0. \tag{4.48}$$

Thus (see Sect. 4.8.1):

$$X_t - X_0 = \int_0^t Z_s ds, \forall t > 0. \tag{4.49}$$

Analogously to the discrete case, the distribution of Z_t is often chosen as independent of t: $Z_n \sim Z$. For instance, a continuous Brownian motion corresponds to $Z \sim N(0, 1)$.

Exercises

1. Consider a bidimensional random walk starting at $X_0 = (0, 0)$ defined by $\Delta X_n = hZ_n$, $Z_n \sim Z$, $Z = (\cos\theta, \sin\theta)$, $P\left(\theta = (i-1)\frac{2\pi}{n}\right) = \frac{1}{n}, 1 \le i \le n$, with $n = 10$.
   (a) Simulate a trajectory of $ns = 1000$ steps with $h = 0.1$.
   (b) Simulate $nsim = 1000$ trajectories of $ns = 100$ steps with $h = 1$ and collect the data about the final distance d_{ns} to the origin. Determine the empirical CDF and the empirical PDF of the data collected.
2. Consider a bidimensional random walk starting at $X_0 = (0, 0)$ defined by $\Delta X_n = hZ_n$, $Z_n \sim Z$, $Z = (\cos\theta, \sin\theta)$, where θ is normally distributed with mean 0 and standard deviation π.
   (a) Simulate a trajectory of $ns = 1000$ steps with $h = 0.1$.
   (b) Simulate $nsim = 1000$ trajectories of $ns = 100$ steps with $h = 1$ and collect the data about the final distance d_{ns} to the origin. Determine the empirical CDF and the empirical PDF of the data collected.

4.8.5 Itô's Integrals

Analogously to $\int_a^b f(t)dt$, we may approximate $\int_a^b f(t)dg(t)$ by considering the partition given in Eq. (4.19) and the finite sum

$$S(\mathcal{T}, f, g) = \sum_{i=1}^{n} (g(t_i) - g(t_{i-1}))f(t_{i-1}). \tag{4.50}$$

For regular functions f and g,

$$SS(\mathcal{T}, f, g) \to \int_a^b f(t)dg(t)$$

when $n \to +\infty$ and $\max\{t_i - t_{i-1}\} \to 0$.

Let us consider two stochastic processes $\{X(t), t \in (a, b)\}$, $\{Y(t), t \in (a, b)\}$ and the finite sum

$$I(\mathcal{T}, X, Y) = \sum_{i=1}^{n} (Y(t_i) - Y(t_{i-1}))X(t_{i-1}). \tag{4.51}$$

If both processes have integrable variances, id est, if

$$\int_a^b E\big(X^2(t)\big)dt < \infty \ \ and \ \int_a^b E\big(Y^2(t)\big)dt < \infty,$$

then this finite sum converges for $n \to +\infty$ and $\max\{t_i - t_{i-1}\} \to 0$. We have analogously the standard functions (see, for instance, Föllmer, 1981):

$$I(\mathcal{T}, \mathrm{X}, \mathrm{Y}) \to I(X, Y) = \int_a^b X(t)dY(t). \tag{4.52}$$

$I(X, Y)$ is referred as Itô's integral of the process X with respect to the process Y. Itô's integral is linear:

$$I(\alpha X + \beta Y, Z) = \alpha I(X, Z) + \beta I(Y, Z),$$

$$I(X, \alpha Y + \beta Z) = \alpha I(X, Y) + \beta I(X, Z).$$

When $Y = W$ and $X = \varphi(W)$, supplementary properties may be established. Let us introduce

$$I_W(\varphi) = I(\varphi(W), W) = \int_a^b \varphi(W(t))dW(t). \tag{4.53}$$

Then,

$$E(I_W(\varphi)) = 0, \tag{4.54}$$

$$E(I_W(\varphi)I_W(\psi)) = \int_a^b E(\varphi(W(t))\psi(W(t)))dt, \tag{4.55}$$

$$E\Big(I_W(\varphi)^2\Big) = \int_a^b E\Big(\varphi(W(t))^2\Big)dt. \tag{4.56}$$

The reader will find in the literature methods for the numerical evaluation of $I_W(\varphi)$ – for instance, Chorin (1973) and Blankenship and Baras (1981).

A supplementary property of $I_W(\varphi)$ is Itô's formula: for a regular function $u : \mathbb{R} \to \mathbb{R}$ (see, for instance, Föllmer, 1981)

$$I_W(u') = (W(b)) - u(W(a)) - \frac{1}{2} I_t(u''),$$

id est,

$$\int_a^b u'(W(t))dW(t) = u(W(b)) - u(W(a)) - \frac{1}{2}\int_a^b u''(W(t))dt.$$

Itô's formula extends to multidimensional situations where $u : \mathbb{R}^p \to \mathbb{R}$ as

$$\int_a^b \nabla u(\mathbf{W}(t))d\mathbf{W}(t) = u(\mathbf{W}(b)) - u(\mathbf{W}(a)) - \frac{1}{2}\int_a^b \Delta u(\mathbf{W}(t))dt.$$

Ito's formula extends also to functions that derive from Itô's integrals: if

$$\boldsymbol{X}(t) - \boldsymbol{X}(0) = \int_0^t \boldsymbol{a}(s, \boldsymbol{X}(s))d\boldsymbol{W}(s) + \int_0^t \boldsymbol{b}(s, \boldsymbol{X}(s))dt,$$

then

$$\int_a^b \nabla u(\boldsymbol{X}(t)).d\boldsymbol{X}(t) = u(\boldsymbol{X}(b)) - u(\boldsymbol{X}(a)) - \frac{1}{2}\int_a^b \boldsymbol{a}^t\mathbf{D}^2\boldsymbol{u}.\boldsymbol{a}dt.$$

Here, $\mathbf{D}^2\boldsymbol{u}$ is the Hessian matrix:

$$\mathbf{D}^2\boldsymbol{u} = \left(\frac{\partial^2 u}{\partial x_i \partial x_j}\right),$$

so that

$$\boldsymbol{a}^t\mathbf{D}^2\boldsymbol{u}.\boldsymbol{a} = \sum_{i,j} \boldsymbol{a}_i(t, \boldsymbol{X}_t) D^2_{ij}(\boldsymbol{X}(t))\boldsymbol{a}_j(t, \boldsymbol{X}_t).$$

In the next section, these formulae are considered as stochastic differential equations.

Remark 4.10
For standard integrals, we may use different forms of discretization to evaluate $\int_a^b f(t)dg(t)$. For instance, we may consider

$$S_3(\mathcal{T}, f, g) = \sum_{i=1}^{n} (g(t_i) - g(t_{i-1}))\frac{f(t_{i+1}) + f(t_i)}{2}$$

For regular functions, such a finite sum converges to the same value $\int_a^b f(t)dt$. However, such a property is not verified for stochastic integrals. Indeed,

$$I_3(\mathcal{T}, X, Y) = \sum_{i=1}^{n} (Y(t_i) - Y(t_{i-1}))\frac{X(t_{i+1}) + X(t_i)}{2}$$

leads to the Stratonovich's integral (Stratonovich, 1966; Fisk, 1963), which is a different stochastic integral, with distinct properties.

Example 4.29
The evaluation of integrals $\int_a^b \varphi(W(t))dW(t)$ can be made using the package `sim.DiffProc`. For instance, let us consider the evaluation of $I(t) = \int_0^t W_s^2 dW_s$. In this case, $E(I(t)) = 0$ and $V(I(t)) = t^3$. You can simulate 10000 trajectories of the integral (for $0 \leq t \leq 1$) as follows:

```
phi = function(t,w){w^2}
f = expression(phi(t,w))
library(Sim.DiffProc)
resu = st.int(expr = f,M=10000,
              lower=0,upper=1,type="ito")
```

To use Stratonovich's integration, you must use the option `type="str"`. The result `resu` is a list containing a time series with the values of the integral and a vector of times. The results are shown in Figs. 4.72 and 4.73.

(continued)

Example 4.29 (continued)

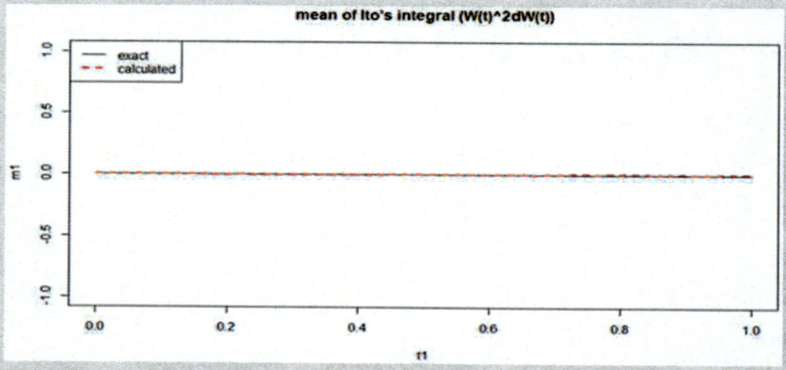

Fig. 4.72 Mean of 1E4 variates of $\int_0^t W_s^2 dW_s$, generated by sim.DiffProc

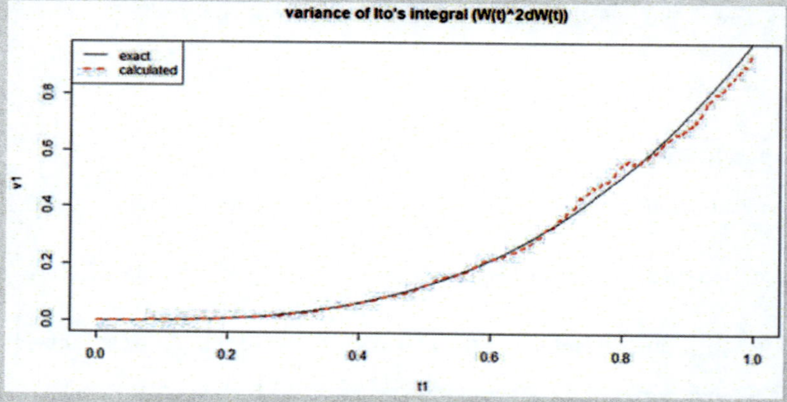

Fig. 4.73 Variance of 1E4 variates of $\int_0^t W_s^2 dW_s$, generated by sim.DiffProc

Example 4.30

The evaluation of integrals $\int_a^b \varphi(W(t))dW(t)$ can be made using the package `sim.DiffProc`. For instance, let us consider the evaluation of $I(t) =$

(continued)

Example 4.30 (continued)

$\int_0^t W_s^2 dW_s$. In this case, $E(I(t)) = 0$ and $V(I(t)) = t^3$. You can simulate 10000 trajectories of the integral (for $0 \leq t \leq 1$) as follows:

```
phi = function(t,w){w^2}
f = expression(phi(t,w))
library(Sim.DiffProc)
resu = st.int(expr = f,M=10000,
              lower=0,upper=1,type="ito")
```

To use Stratonovich's integration, you must use the option `type="str"`. The result `resu` is a list containing a time series with the values of the integral and a vector of times. The results are shown in Figs. 4.74 and 4.75.

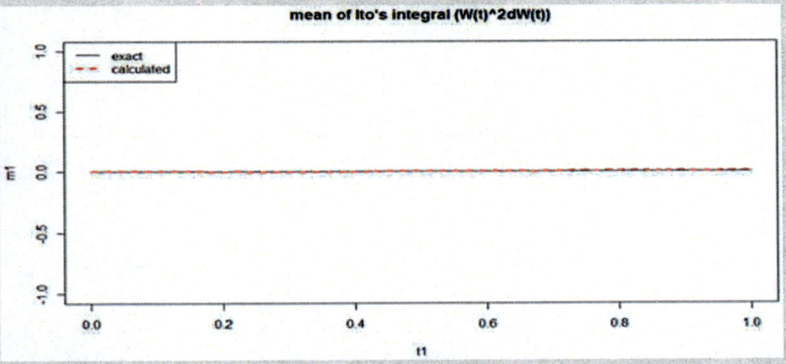

Fig. 4.74 Mean of 1E4 variates of $\int_0^t W_s^2 dW_s$, generated by sim.DiffProc

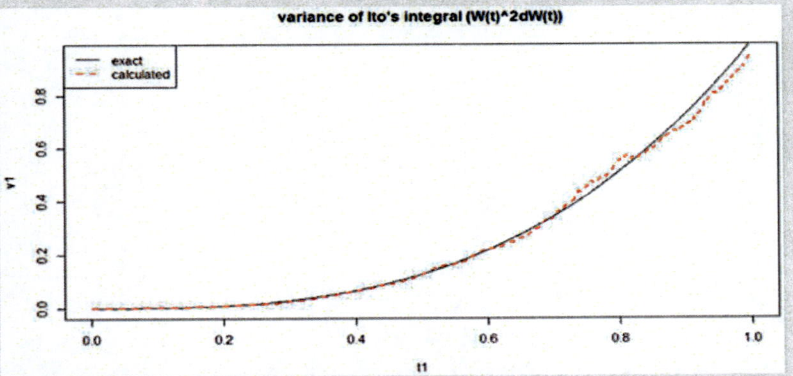

Fig. 4.75 Variance of 1E4 variates of $\int_0^t W_s^2 dW_s$, generated by sim.DiffProc

Example 4.31
Let us consider the process X_t, such that each X_t is a random variable having the uniform distribution on (a, b) and X_t is independent from X_s for $t \neq s$. We have

$$m_X(t) = \frac{a+b}{2},\ V_X(t) = \frac{(b-a)^2}{12},\ R_{XX}(s, t) = \frac{a^2+ab+b^2}{3}\delta(t-s).$$

Let us evaluate $Y_t = \int_0^t X_s dW_s$. We have $E(Y_t) = 0, V(Y_t) = t\frac{a^2+ab+b^2}{3}$. We generate a sample of 1E4 variates from Y_t and we evaluate the empirical mean and the empirical variance. The results appear in Figs. 4.76 and 4.77.

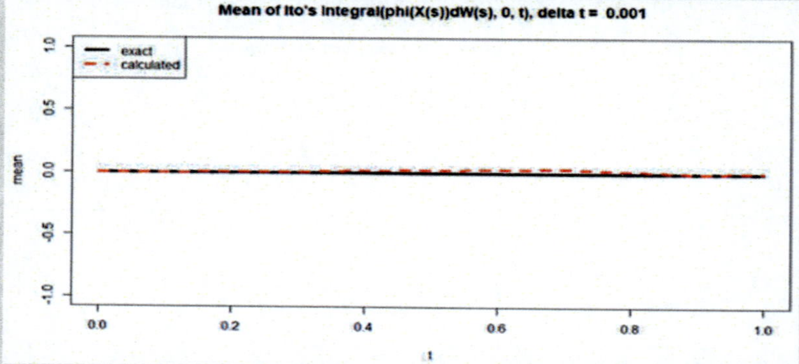

Fig. 4.76 Results from a sample of 1E4 trajectories of the process Y_t

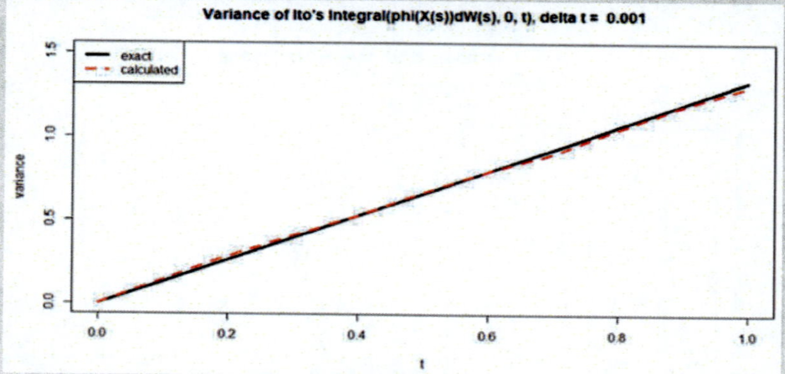

Fig. 4.77 Results from a sample of 1E4 trajectories of the process Y_t

(continued)

Example 4.31 (continued)

The variates were generated by `sim.ito.int` given in the code below. We used `dt_in` = `0.001`, `dt_out` = `0.1`, `tmax` = `1`,`pphi` = `phi`.

```
phi = function(t,x){x}

itointsim = function(tmin,tmax, dt, pphi){
  if (tmax < dt){return(0)}else{
    library(sde)
    ddt = tmax-tmin
    nmax = ceiling(tmax/dt)
    W = BM(N=nmax+1,  T=tmax)
    X = runif(nmax+1,a,b)
    nmin = floor(tmin/dt)
    s = 0
    tt = tmin
    i = nmin+1
    while (tt < tmax){
      i = i + 1
      s = s + pphi(tt,X[[i]])*(W[[i]]-W[[i-1]])
      tt = tt + dt
    }
    return(s)
  }
}

sim.ito.int = function(tmax, dt_in,dt_out,pphi){
  ns = ceiling(tmax / dt_out)
  tc = (0:ns)*dt_out
  intphi= numeric(length = ns+1)
  intphi[[1]] = 0
  for (i in 1 : ns){
    aux = itointsim(tc[[i]],tc[[i+1]], dt_in,pphi)
    intphi[[i+1]] = intphi[[i]] + aux
  }
  return(intphi)
}
```

4.8.6 Itô's Calculus

Itô's calculus manipulates stochastic differential equations (SDE) having the general form

$$dX_t = a(t, X_t)dW_t + b(t, X_t)dt. \tag{4.57}$$

Equation (4.58) defines a **Itô's diffusion.** a is the diffusion coefficient; b is the drift. Notice that the Brownian motion is not differentiable, so that (4.58) must be interpreted as an integral equation:

$$X(t) - X(0) = \int_0^t a(s, X(s))dW(s) + \int_0^t b(s, X(s))dt, \forall s > 0.$$

For instance, Itô's formula may be rewritten as a SDE:

$$du(W_t) = u'(W_t)dW_t + \frac{1}{2}u''(W_t)dt. \tag{4.58}$$

For a process given by (4.58),

$$du(X_t) = u'(X_t)dW_t + \frac{1}{2}u''(W_t)a^2(t, X_t)dt. \tag{4.59}$$

In the multidimensional situation, $\boldsymbol{Y}_t = \boldsymbol{F}(t, \boldsymbol{X}_t)$, $\boldsymbol{F} : \mathbb{R} \times \mathbb{R}^n \to \mathbb{R}^m$. Then, we have

$$d(Y_i)_t = \frac{\partial F_i}{\partial t}(t, \boldsymbol{X}_t) + (\boldsymbol{A}.\boldsymbol{dX}_t)_i + \frac{1}{2}\boldsymbol{dX}_t^t\boldsymbol{B}_i\boldsymbol{dX}_t, \tag{4.60}$$

with

$$A_{ij} = \frac{\partial F_i}{\partial x_j}, B_{ijk} = \frac{\partial^2 F_i}{\partial x_j \partial x_k}.$$

If $\boldsymbol{dX}_t = \boldsymbol{a}(t, \boldsymbol{X}_t)\boldsymbol{dW}_t + \boldsymbol{b}(t, \boldsymbol{X}_t)dt$, then

$$d(Y_i)_t = \frac{\partial F_i}{\partial t}(t, \boldsymbol{X}_t) + (\boldsymbol{A}.\boldsymbol{dX}_t)_i + \frac{1}{2}\boldsymbol{a}^t\boldsymbol{B}_i\boldsymbol{a}, \tag{4.61}$$

For $n = m = 1$, $Y_t = F(t, X_t)$ and $dX_t = a(t, X_t)dW_t + b(t, X_t)dt$. Then

$$dZ_t = \frac{\partial F}{\partial t}(t, X_t)dt + \frac{\partial F}{\partial x}(t, X_t)dX_t + \frac{1}{2}\frac{\partial^2 F}{\partial x^2}(t, X_t)(a(t, W_t))^2 dt. \tag{4.62}$$

SDE follow some standard rules. For instance, let $dX_t = a_X(t, X_t)dW_t + b_X(t, X_t)dt$, $dY_t = a_Y(t, Y_t)dW_t + b_Y(t, Y_t)dt$ and $Z_t = \alpha X_t + \beta Y_t$. Then, for $\alpha, \beta \in \mathbb{R}$:

$$dZ_t = (\alpha a_X(t, X_t) + \beta a_Y(t, Y_t))dW_t + (\alpha b_X(t, X_t) + \beta b_Y(t, Y_t))dt.$$

Indeed, consider $S_t = (X_t, Y_t)$ and $F(S) = \alpha S_1 + \beta S_2$. Then,

$$\frac{\partial F}{\partial t} = 0, \quad \frac{\partial F}{\partial S_1} = \alpha, \frac{\partial F}{\partial S_2} = \beta, \quad \frac{\partial^2 F}{\partial S_j \partial S_k} = 0,$$

so that (4.61) yields the result.

If $Z_t = X_t Y_t$, then

$$dZ_t = X_t dY_t + Y_t dX_t + (a_X(t,\ X_t) a_Y(t,\ Y_t))dt. \tag{4.63}$$

Indeed, taking $S_t = (X_t, Y_t)$ and $F(S) = S_1 S_2$, we have

$$\frac{\partial F}{\partial t} = 0, \quad \frac{\partial F}{\partial S_1} = S_2, \frac{\partial F}{\partial S_2} = S_1, \quad \frac{\partial^2 F}{\partial S_j \partial S_k} = \begin{pmatrix} 0 & 1 \\ 1 & 0 \end{pmatrix},$$

and (4.61) establishes (4.64).

Let $X_t = x(W_t)$, $Y_t = y(W_t)$ and $Z_t = X_t Y_t$. Then, (4.59) and (4.64) yield that

$$dZ_t = X_t dY_t + Y_t dX_t + x'(W_t) y'(W_t) dt. \tag{4.64}$$

If $Z_t = y(X_t)$, then

$$dZ_t = y'(X_t) dX_t + \frac{1}{2} y''(X_t)(x'(W_t))^2 dt. \tag{4.65}$$

Indeed, $Z_t = u(X_t)$, $u(s) = y(x(s))$. Thus, $u'(s) = y'(x(s))x'(s)$, $u''(s) = y'(x(s))x''(s) + y''(x(s))(x'(s))^2$, so that

$$dZ_t = y'(x(W_t))\left(x'(W_t)dW_t + \frac{1}{2} x''(W_t)dt\right) + \frac{1}{2} y''(x(W_t))(x'(W_t))^2 dt$$

and we obtain (4.66).

These rules are often used to manipulate SDE to get explicit solutions or to generate representations of the solutions of differential equations, such as Feynman–Kac formulae. The reader will find in the literature many representations of the solutions of differential equations and partial differential equations: see, for instance Elepov and Mikhailov (1969), Booth (1982), Souza de Cursi (1994), Morillon (1997), Hwang and Mascagni (2001), Hwang et al. (2003), Milstein and Tretyakov (2004, 2012, 2013, 2020), Kharroubi and Pham (2015), Zhou and Cai, Numerical Solution of the Robin Problem of Laplace Equations with a Feynman-Kac Formula and Reflecting Brownian Motions (2016).

Example 4.32
The classical Black-Scholes model for the evolution of the price of an option reads as

$$dS_t = \sigma S_t dW_t + \mu S_t dt.$$

Let

$$F(t,x) = F_0 \exp\left(\mu t - \frac{\sigma^2}{2}t + \sigma x\right).$$

We have

$$\frac{\partial F}{\partial t} = \left(\mu - \frac{\sigma^2}{2}\right)F(t,x)\ , \frac{\partial F}{\partial x} = \sigma F(t,x), \frac{\partial^2 F}{\partial x^2} = \sigma^2 F(t,x).$$

Thus, if $Y_t = F(t, X_t)$ and $dX_t = a(t, W_t)dW_t + b(t, W_t)dt$, we have

$$dY_t = \left(\mu - \frac{\sigma^2}{2}\right)Y_t dt + \sigma Y_t dX_t + \frac{1}{2}\sigma^2 Y_t (a(t,\ W_t))^2 dt$$

Let us choose $X_t = W_t$. Then, $dX_t = dW_t$ and $a(t, W_t) = 1$, $b(t, W_t) = 0$. Consequently,

$$dY_t = \mu Y_t dt + \sigma Y_t dW_t$$

Thus, the solution is

$$S_t = S_0 \exp\left(\mu t - \frac{\sigma^2}{2}t + \sigma W_t\right).$$

Notice that $S_t = A_t U_t$, $A_t = S_0 \exp\left(\mu t - \frac{\sigma^2}{2}t\right)$, $U_t = \exp(\sigma W_t)$. A_t is deterministic and U_t is a lognormal variable (recall that $W_t \sim N\left(0, \sqrt{t}\right)$). We have

$$E(S_t) = A_t E(U_t), V(S_t) = A_t^2 V(U_t),$$

with

$$E(U_t) = \exp\left(\sigma^2 t\right), V(U_t) = \exp\left(2\sigma^2 t\right) - \exp\left(\sigma^2 t\right).$$

Thus,

$$E(S_t) = S_0 \exp(\mu t), V(S_t) = S_0^2 \exp(2\mu t)\left(\exp\left(\sigma^2 t\right) - 1\right).$$

Example 4.33
The Vasicek's model for the evolution of the interest rates reads as

$$dX_t = \sigma dW_t + \theta(\mu - X_t)dt.$$

Let

$$F(t, x) = xe^{\theta t}.$$

We have

$$\frac{\partial F}{\partial t} = \theta F \ , \frac{\partial F}{\partial x} = e^{\theta t}, \frac{\partial^2 F}{\partial x^2} = 0.$$

Thus, if $Y_t = F(t, X_t)$ and $dX_t = a(t, W_t)dW_t + b(t, W_t)dt$, we have

$$dY_t = \theta X_t e^{\theta t} dt + e^{\theta t} dX_t = \sigma e^{\theta t} dW_t + \theta \mu e^{\theta t} dt$$

Thus,

$$Y_t = Y_0 + \sigma \int_0^t e^{\theta s} dW_s + \theta \mu \int_0^t e^{\theta s} ds$$

Thus, the solution is

$$X_t = X_0 e^{-\theta t} + \mu\left(1 - e^{-\theta t}\right) + \sigma e^{-\theta t} \int_0^t e^{\theta s} dW_s.$$

We have

$$E(S_t) = X_0 e^{-\theta t} + \mu\left(1 - e^{-\theta t}\right), V(S_t) = \frac{\sigma^2}{2\theta}\left(1 - e^{-2\theta t}\right).$$

Vasicek's model is an example of Ornstein–Uhlenbeck process.

Example 4.34
Let us consider the partial differential equation

$$\frac{\partial u}{\partial t} - \frac{a^2}{2} \Delta u = f \ \ on \ Q = \Omega \times (0, T),$$

(continued)

Example 4.34 (continued)

$$u(x,0)=u_0(x),x\in\Omega \quad ,u(x,t)=u_0(x,t),x\in\partial\Omega,t\in(0,T)$$

Let us consider

$$dX_t=a dW_t,dT_t=-dt,X_0=\boldsymbol{x},T_0=t,Y_t=u(X_t,T_t).$$

Let us introduce $\mathbf{Z}_t=(X_t,T_t,)$. Then,

$$d\mathbf{Z}_t=\begin{pmatrix} a & 0 & \cdots & 0\\ 0 & \ddots & \cdots & \vdots\\ \vdots & \vdots & a & 0\\ 0 & \cdots & \cdots & 0\end{pmatrix} d\boldsymbol{W}_t+\begin{pmatrix}0\\0\\\vdots\\-1\end{pmatrix}dt.$$

Let $F(\mathbf{Z}_t)=u(X_t,T_t)$. We have

$$A_{1i}=\frac{\partial F}{\partial Z_i}=\frac{\partial u}{\partial x_i},A_{1,n+1}=\frac{\partial F}{\partial Z_{i+1}}=\frac{\partial u}{\partial t},$$

$$B_{1ij}=\frac{\partial^2 F}{\partial Z_i\partial Z_j}=\frac{\partial^2 u}{\partial x_i\partial x_j},$$

$$B_{1i,n+1}=\frac{\partial^2 F}{\partial Z_i\partial Z_{n+1}}=\frac{\partial^2 u}{\partial x_i\partial t},$$

$$B_{1,n+1,n+1}=\frac{\partial^2 F}{\partial Z_{n+1}^2}=\frac{\partial^2 u}{\partial t^2}.$$

$$dY_t=\frac{\partial u}{\partial t}dT_t+\sum\frac{\partial u}{\partial x_i}d(X_i)_t+\frac{1}{2}a^2\times\Delta u\,dt,$$

id est,

$$dY_t=\left(\frac{a^2}{2}\Delta u-\frac{\partial u}{\partial t}\right)dt+\sum\frac{\partial u}{\partial x_i}a\,dW_i.$$

Thus

$$dY_t=-f(X_t,T_t)dT_t+a\nabla u.dW$$

and

(continued)

Example 4.34 (continued)

$$Y_t - Y_0 = -\int_0^t f(X_s, T_s)ds + a\int_0^t \nabla u(X_s, T_s).dW$$

Let

$$\tau(x) = \inf\{t : Z_t \notin Q\}.$$

Then, $Y_{\tau(x)} = u_0(X_{\tau(x)}, T_{\tau(x)})$, $Y_0 = u(t, x)$ and

$$u(x, t) = u_0\left(X_{\tau(x)}, T_{\tau(x)}\right) + \int_0^{\tau(x)} f(X_s, T_s)ds - a\int_0^t \nabla u(X_s, T_s).dW.$$

Since $E\left(\int_0^t \nabla u(X_s, T_s).dW\right) = 0$, we have

$$u(t, x) = E\left(u_0\left(X_{\tau(x)}, T_{\tau(x)}\right) + \int_0^{\tau(x)} f(X_s, T_s)ds\right).$$

This equality is an example of Feynman–Kac formula.

Example 4.35

Let us consider the partial differential equation

$$-\frac{a^2}{2}\Delta u = f \quad on\ \Omega,$$
$$u(x) = u_0(x), x \in \partial\Omega$$

Let us consider

$$dX_t = a dW_t, X_0 = x, Y_t = u(X_t).$$

Let $F(Z_t) = u(X_t)$. We have

$$dY_t = \sum \frac{\partial u}{\partial x_i} d(X_i)_t + \frac{1}{2} a^2 \times \Delta u \, dt,$$

(continued)

Example 4.35 (continued)

id est,

$$dY_t = dt\frac{a^2}{2}\Delta u + \sum \frac{\partial u}{\partial x_i}\, a\, dW_i.$$

Thus

$$dY_t = -f(X_t)dX_t + a\nabla u.dW$$

and

$$Y_t - Y_0 = -\int_0^t f(X_s)ds + a\int_0^t \nabla u(X_s).dW$$

Let

$$\tau(x) = \inf\{t : Z_t \notin \Omega\}.$$

Then, $Y_{\tau(x)} = u_0(X_{\tau(x)})$, $Y_0 = u(x)$ and

$$u(x) = u_0(X_{\tau(x)}) + \int_0^{\tau(x)} f(X_s)ds - a\int_0^t \nabla u(X_s).dW.$$

Since $E\left(\int_0^t \nabla u(X_s).dW\right) = 0$, we have

$$u(x) = E\left(u_0(X_{\tau(x)}) + \int_0^{\tau(x)} f(X_s)ds\right).$$

This equality is another example of Feynman–Kac formula.

4.8.7 *Numerical Simulation of Stochastic Differential equations*

Let us consider the following SDE;

$$dX_t = a(t, X_t)dW_t + b(t, X_t)dt, X_0 = x_0. \tag{4.66}$$

a is usually referred as diffusion coefficient and *b* as drift coefficient. The reader will find in the literature many works concerning methods for the simulation of SDE analogous to (4.66). For instance, Milstein (1973), Pardoux and Talay (1985), Milstein and Tretyakov (2004), Graham and Talay (2013), Kloeden and Platen (1989), Talay (2015), Boukhetala and Guidoum (2011) and Iacus (2008). One of the most popular ones is Euler's method: let us introduce a maximum time of simulation $T > 0$, a maximum number of time steps $ns > 0$, $\Delta t = T/ns$, $t_i = i\Delta t$, $i \geq 0$, $W_i = W(t_i)$, $X_i = X(t_i)$. Then, Eq. (4.67) is discretized as

$$\begin{aligned} X_{n+1} - X_n &= a_n \Delta W_n + b_n \Delta t, \quad . \\ a_n = a(t_n, X_n), b_n &= b(t_n, X_n), \Delta W_n = (W_{n+1} - W_n) \end{aligned} \tag{4.67}$$

A simple way to generate ΔW_n consists in using that

$$\Delta W_n \sim N\left(0, \sqrt{\Delta t}\right). \tag{4.68}$$

Thus,

$$\begin{aligned} X_{n+1} &= X_n + a_n \sqrt{\Delta t} Z_n + b_n \Delta t \\ W_{n+1} &= W_n + \sqrt{\Delta t} Z_n, Z_n \sim N(0,1) \end{aligned} \tag{4.69}$$

A second popular method of discretization is Milstein's method:

$$\begin{aligned} X_{n+1} - X_n &= a_n \Delta W_n + \left(b_n \Delta t + c_n \left(\Delta W_n^2 - \Delta t\right)\right), \\ c_n &= \frac{1}{2} a_n \frac{\partial a_n}{\partial x}, \frac{\partial a_n}{\partial x} = \frac{\partial a}{\partial x}(t_n, X_n). \end{aligned} \tag{4.70}$$

Analogously to Euler's method, we may generate $W_{n+1} - W_n$ by (4.69), so that

$$\begin{aligned} X_{n+1} - X_n &= a_n \sqrt{\Delta t} Z_n + a_n \Delta W_n + \Delta t\left(b_n + c_n\left(Z_n^2 - 1\right)\right) \# \\ W_{n+1} &= W_n + \sqrt{\Delta t} Z_n, Z_n \sim N(0,1) \end{aligned} \tag{4.71}$$

Some R packages include instruction for the simulation of SDE. For instance, `sde` proposes `sde.sim` and `Sim.DiffProc` proposes `snssde1d`, `snssde2d`, `snssde3d`.

Example 4.36
The classical Black-Scholes model for the evolution of the price of an option reads as

$$dS_t = \sigma S_t dW_t + \mu S_t dt.$$

Let us simulate this SDE with $\mu = 0.1$, $\sigma = 0.3$, $S_0 = 1$. We use package `Sim.DiffProc` as follows:

(continued)

Example 4.36 (continued)

```
mu = 0.1
sig = 0.3
S0 = 1
ns = 1E4
npt = 1000
tmax = 1;
dt = tmax/npt

library(Sim.DiffProc)
a <- expression(sig*x )
b <- expression(mu*x)
mod1 <- snssde1d(drift=b,diffusion=a,M=ns,x0=S0,
            Dt=dt,T=tmax,type="ito",method="euler")
mod2 <- snssde1d(drift=b,diffusion=a,M=ns,x0=S0,
            Dt=dt,T=tmax,type="ito",method="milstein")
mod3 <- snssde1d(drift=b,diffusion=a,M=ns,x0=S0,
            Dt=dt,T=tmax,type="ito",method="rk3")
```

This code generates 1E4 trajectories for $0 \le t \le 1$, using 3 different methods of discretization (Euler, Milstein, Runge-Kutte RK3), with $\Delta t = 0.001$. The result is a list containing the results in a field `$X` (for instance, `mod1$X`), where each column is a simulation. We evaluate the mean for each time $t_i = (i-1)\Delta t$. The results are exhibited in Fig. 4.78 and Table 4.5. The variances for each time are compared in Fig. 4.79 and Table 4.6.

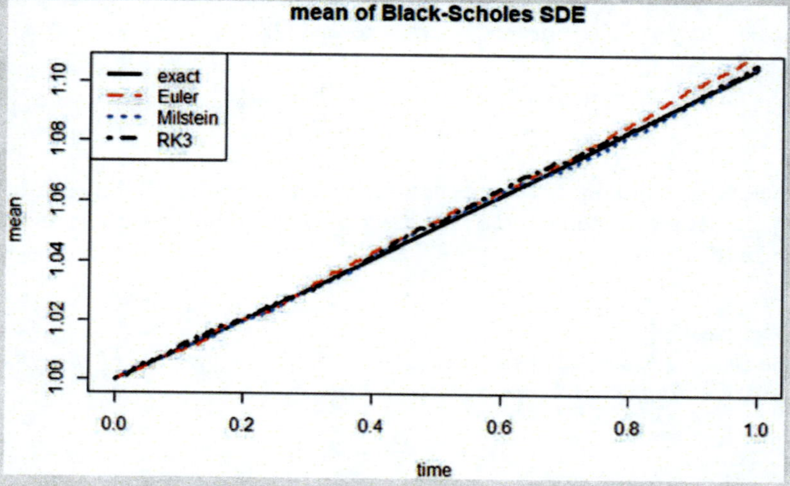

Fig. 4.78 Simulation of Black-Scholes SDE by `Sim.DiffProc`, with $\Delta t = 0.001$. Mean of a sample of 10000 realizations

(continued)

Example 4.36 (continued)

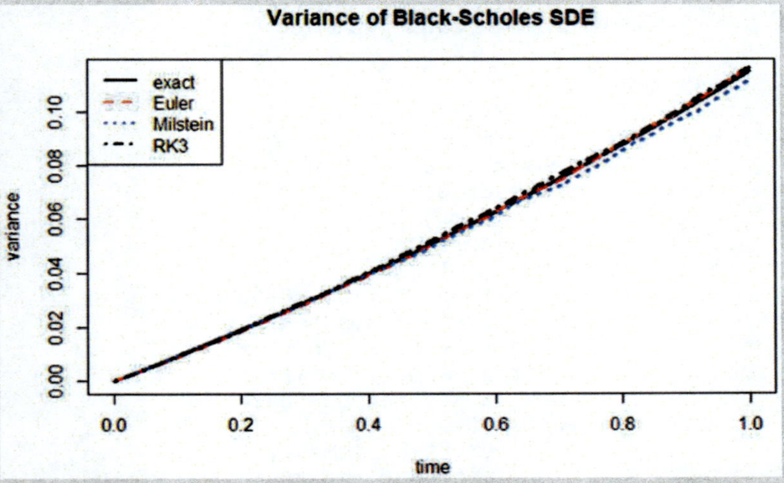

Fig. 4.79 Simulation of Black-Scholes SDE by `Sim.DiffProc`, with $\Delta t = 0.001$. Variance of a sample of 10000 variates

Table 4.5 Estimation of $E(S_T)$ by `Sim.DiffProc`

	$\Delta t = 0.01$			$\Delta t = 0.001$		
ns	1000	5000	10000	1000	5000	10000
Euler	1.1150	1.1021	1.1056	1.1091	1.1057	1.1117
Milstein	1.1119	1.1116	1.1105	1.1063	1.1072	1.1054
RK3	1.1022	1.1074	1.0994	1.0956	1.1023	1.1065
Exact	1.1052	1.1052	1.1052	1.1052	1.1052	1.1052

Table 4.6 Estimation of $V(S_T)$ by `Sim.DiffProc`

	$\Delta t = 0.01$			$\Delta t = 0.001$		
ns	1000	5000	10000	1000	5000	10000
Euler	0.1170	0.1112	0.1182	0.1115	0.1147	0.1168
Milstein	0.1191	0.1146	0.1190	0.1127	0.1131	0.1116
RK3	0.1192	0.1140	0.1110	0.1164	0.1205	0.1164
Exact	0.1150	0.1150	0.1150	0.1150	0.1150	0.1150

(continued)

Example 4.36 (continued)

We use package sde as follows (sde proposes an option to simulate Black-Scholes directly, but we do not use it).

```
library(sde)
a <- expression(sig*x )
b <- expression(mu*x)
a.x <- expression(sig)
a.xx <- expression(0)
b.x <- expression(mu)
b.xx <- expression(0)
b.t <- expression(0)
mod1 <- sde.sim(t0=0, T=tmax,X0=S0,delta = dt,
          drift=b,sigma=a,sigma.x = a.x,
          sigma.xx = a.xx, drift.x = b.x,
          drift.xx = b.xx, drift.t = b.t,
          M=ns, N=npt,method="euler")
mod2 <- sde.sim(t0=0, T=tmax,X0=S0,delta = dt,
                drift=b,sigma=a,sigma.x = a.x,
                sigma.xx = a.xx, drift.x = b.x,
                drift.xx = b.xx, drift.t = b.t,
                M=ns, N=npt, method="milstein")
mod3 <- sde.sim(t0=0, T=tmax,X0=S0,delta = dt,
                drift=b,sigma=a,sigma.x = a.x,
                sigma.xx = a.xx, drift.x = b.x,
                drift.xx = b.xx, drift.t = b.t,
                M=ns, N=npt, method="KPS")
```

This code generates ns trajectories for $0 \leq t \leq t\max$, using 3 different methods of discretization (Euler, Milstein, KPS), with $\Delta t = dt$. The result is a matrix where each column is a simulation. We evaluate the mean for each time $t_i = (i - 1)\Delta t$. The results are exhibited in Fig. 4.80 and Table 4.7. The variances for each time are compared in Fig. 4.81 and Table 4.8.

(continued)

Example 4.36 (continued)

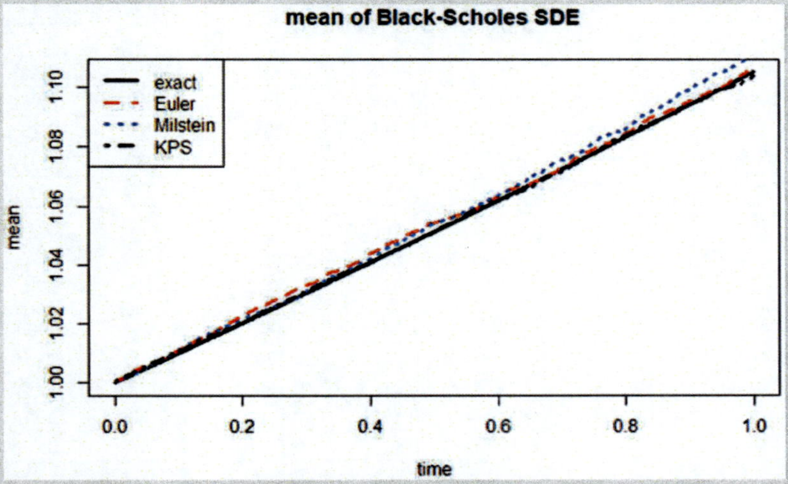

Fig. 4.80 Simulation of Black-Scholes SDE by `sde`, with $\Delta t = 0.001$. Mean of a sample of 10000 realizations

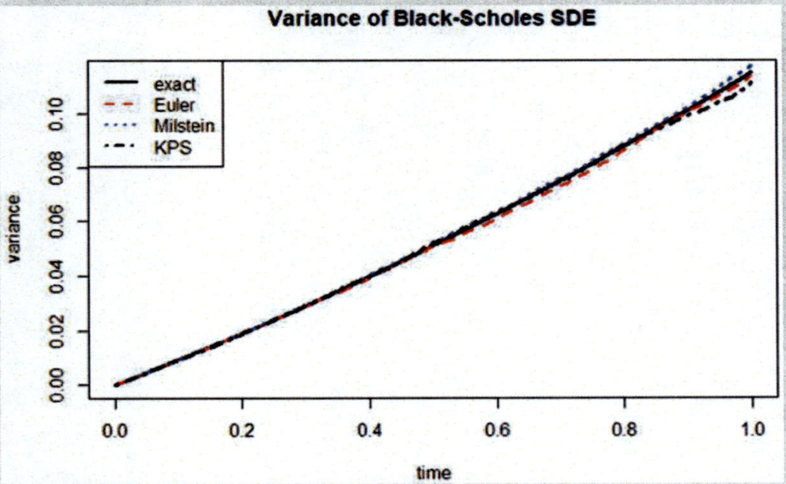

Fig. 4.81 Simulation of Black-Scholes SDE by `sde`, with $\Delta t = 0.001$. Variance of a sample of 10000 variates

(continued)

Example 4.36 (continued)

Table 4.7 Estimation of $E(S_T)$ by sde

	$\Delta t = 0.01$			$\Delta t = 0.001$		
ns	1000	5000	10000	1000	5000	10000
Euler	1.1101	1.1144	1.1062	1.1040	1.1002	1.1064
Milstein	1.1098	1.0993	1.1088	1.1087	1.0999	1.1105
KPS	1.1001	1.1072	1.1015	1.1048	1.1097	1.1037
Exact	1.1052	1.1052	1.1052	1.1052	1.1052	1.1052

Table 4.8 Estimation of $V(S_T)$ by sde

	$\Delta t = 0.01$			$\Delta t = 0.001$		
ns	1000	5000	10000	1000	5000	10000
Euler	0.1114	0.1198	0.1155	0.1205	0.1117	0.1150
Milstein	0.1268	0.1137	0.1173	0.1305	0.1123	0.1140
KPS	0.1063	0.1154	0.1151	0.1304	0.1182	0.1175
Exact	0.1150	0.1150	0.1150	0.1150	0.1150	0.1150

We observe that all the results are close to the exact ones.

Example 4.37
The Vasicek's model (Vasicek, 1977) for the evolution of the interest rates reads as

$$dX_t = \sigma dW_t + \theta(\mu - X_t)dt.$$

We can use `Sim.DiffProc` to simulate this SDE with $\mu = 0.05$, $\sigma = 0.3$, $\theta = 2$, $X_0 = 0.1$:

```
mu = 0.05
sig = 0.3
teta = 2
XX0 = 0.1
ns = 1E4
npt = 2000
tmax = 2;
dt = tmax/npt

library(Sim.DiffProc)
a <- expression(sig)
b <- expression(teta*(mu-x))
mod1 <- snssde1d(drift=b,diffusion=a,M=ns,x0=XX0,
          Dt=dt,N=npt,type="ito",method="euler")
mod2 <- snssde1d(drift=b,diffusion=a,M=ns,x0=XX0,
          Dt=dt,N=npt,type="ito",method="smilstein")
mod3 <- snssde1d(drift=b,diffusion=a,M=ns,x0=XX0,
          Dt=dt,N=npt,type="ito",method="heun")
```

(continued)

Example 4.37 (continued)

Analogously to the preceding example, we generate a sample of *ns* variates from X_T to evaluate the mean $E(X_T)$ and $V(X_T)$ for several times T. Figures 4.82 and 4.83, Tables 4.9 and 4.10 show examples of results.

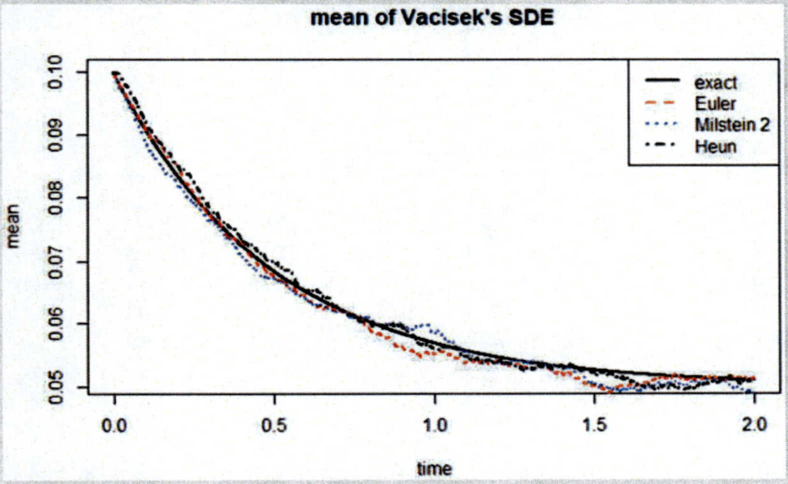

Fig. 4.82 Simulation of Vasicek SDE by `Sim.DiffProc`, with $\Delta t = 0.001$. Mean of a sample of 10000 realizations

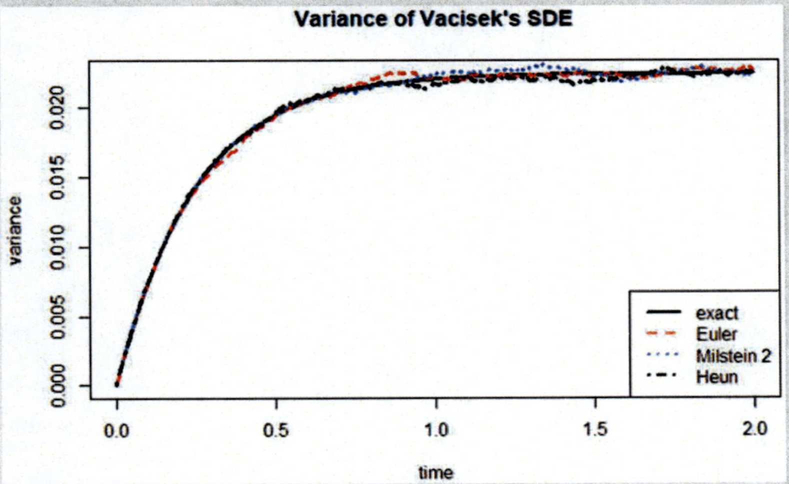

Fig. 4.83 Simulation of Vasicek SDE by `Sim.DiffProc`, with $\Delta t = 0.001$. Variance of a sample of 10000 variates

(continued)

Example 4.37 (continued)

Table 4.9 Estimation of $E(X_T)$ by `Sim.DiffProc`

	$\Delta t = 0.01$			$\Delta t = 0.001$		
ns	1000	5000	10000	1000	5000	10000
Euler	0.0531	0.0533	0.0530	0.0473	0.0495	0.0513
Milstein 2	0.0509	0.0503	0.0551	0.0442	0.0536	0.0486
Heun	0.0508	0.0496	0.0501	0.0530	0.0475	0.0505
Exact	0.0592	0.0592	0.0592	0.0592	0.0592	0.0592

Table 4.10 Estimation of $V(X_T)$ by `Sim.DiffProc`

	$\Delta t = 0.01$			$\Delta t = 0.001$		
ns	1000	5000	10000	1000	5000	10000
Euler	0.0235	0.0225	0.0225	0.0214	0.0226	0.0227
Milstein 2	0.0238	0.0230	0.0226	0.0231	0.0219	0.0225
Heun	0.0243	0.0238	0.0232	0.0217	0.0223	0.0224
Exact	0.0249	0.0249	0.0249	0.0249	0.0249	0.0249

Example 4.38
The Cox-Ingersoll-Ross model (Cox et al., 1985) for the instantaneous interest rates is the solution of the SDE:

$$dX_t = \sigma\sqrt{X_t}dW_t + \theta(\mu - X_t)dt.$$

Then (Jafari & Abbasian, 2017)

$$E(X_t) = X_0 e^{-\theta t} + \mu\left(1 - e^{-\theta t}\right),$$

$$V(X_t) = \frac{\sigma^2}{\theta} X_0\left(e^{-\theta t} - e^{-2\theta t}\right) + \frac{\mu\sigma^2}{2\theta}\left(1 - e^{-\theta t}\right)^2.$$

We use Sim.DiffProc to simulate this SDE, with $\mu = 0.1$, $\sigma = 0.3$, $\theta = 2$, $X_0 = 1$:

(continued)

Example 4.38 (continued)

```
mu = 0.1
sig = 0.3
teta = 2
XX0 = 1
ns = 1E4
npt = 2000
tmax = 2;
dt = tmax/npt

library(Sim.DiffProc)
a <- expression(sig*sqrt(x))
b <- expression(teta*(mu-x))
mod1 <- snssde1d(drift=b,diffusion=a,M=ns,x0=XX0,
           Dt=dt,N=npt,type="ito",method="euler")
mod2 <- snssde1d(drift=b,diffusion=a,M=ns,x0=XX0,
           Dt=dt,N=npt,type="ito",method="taylor")
mod3 <- snssde1d(drift=b,diffusion=a,M=ns,x0=XX0,
           Dt=dt,N=npt,type="ito",method="rk2")
```

Analogously to the preceding examples, we generate a sample of *ns* variates from X_T to evaluate the mean $E(X_T)$ and $V(X_T)$ for several times T. Figures 4.84 and 4.85, Tables 4.11 and 4.12 show examples of results.

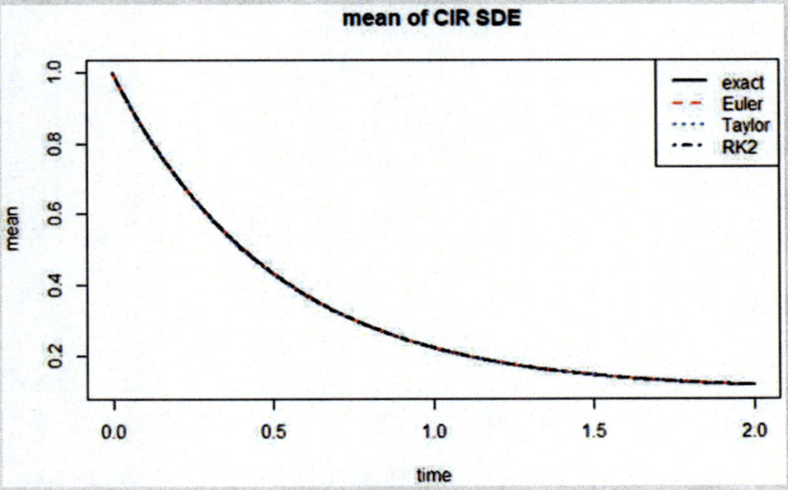

Fig. 4.84 Simulation of Cox-Ingersoll-Ross SDE by `Sim.DiffProc`, with $\Delta t = 0.001$. Mean of a sample of 10000 realizations

(continued)

Example 4.38 (continued)

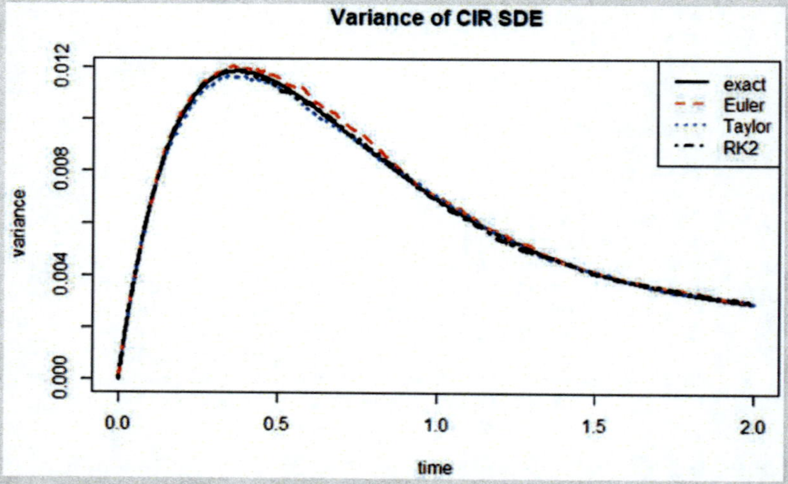

Fig. 4.85 Simulation of Cox-Ingersoll-Ross SDE by `Sim.DiffProc`, with $\Delta t = 0.001$. Variance of a sample of 10000 variates

Table 4.11 Estimation of $E(X_T)$ by `Sim.DiffProc`

	$\Delta t = 0.01$			$\Delta t = 0.001$		
ns	1000	5000	10000	1000	5000	10000
Euler	0.1159	0.1161	0.1155	0.1177	0.1170	0.1173
Taylor	0.1178	0.1162	0.1177	0.1173	0.1169	0.1162
RK2	0.1152	0.1163	0.1153	0.1166	0.1155	0.1173
Exact	0.1164	0.1164	0.1164	0.1164	0.1164	0.1164

Table 4.12 Estimation of $V(X_T)$ by `Sim.DiffProc`

	$\Delta t = 0.01$			$\Delta t = 0.001$		
ns	1000	5000	10000	1000	5000	10000
Euler	3.0E-3	3.0E-3	3.0E-3	3.0E-3	3.0E-3	3.0E-3
Taylor	2.9E-3	2.8E-3	3.1E-3	3.1E-3	3.1E-3	3.0E-3
RK2	3.0E-3	2.9E-3	3.0E-3	2.7E-3	2.8E-3	3.0E-3
Exact	3.0E-3	3.0E-3	3.0E-3	3.0E-3	3.0E-3	3.0E-3

Example 4.39
Consider the Partial Differential Equation

$$-\Delta u = f \ \ on\ \Omega, u(x) = u_0(x) \text{ on } \partial\Omega,$$

with

$$\Omega = \{(x_1,\ x_2,\ x_3) : x_1 + x_2 + x_3 \leq 1, x_1 \geq 0, x_2 0, x_3 \geq 0\},$$

$$f(x) = -6, u_0(x) = x_1^2 + x_2^2 + x_3^2.$$

Let us consider

$$dX_t = \sqrt{2} dW_t, X_0 = x = \left(\frac{1}{3}, 0, \frac{1}{2}\right).$$

We evaluate

$$u(x) \approx \frac{1}{nsim} \sum_{i=1}^{nsim} Y_i, Y_i = u_0\left(X_{\tau_i(x)}\right) + \int_0^{\tau_i(x)} f(X_s) ds$$

The exact value is $u(X_0) = 13/36 \approx 0, 361111$. We can generate a sample from Y as follows:

Firstly, define f and u_0:

```
f = function(x){-6}
u0 = function(x){sum(x^2)}
```

Then define Ω: the function `region` takes the value `TRUE` when the point is in Ω and `FALSE` otherwise.

```
region = function(x){
  aux = TRUE
  if (sum(x) >= 1){aux = FALSE}
  for (i in 1: length(x)){
    if (x[[i]]<= 0){aux = FALSE}
  }
  return(aux)
}
```

Create a function that generates one variate from *Y*:

```
onewalk = function(xc,dt,f,u0){
  sig = sqrt(2*dt)
  interior = TRUE
  s = 0
  x = xc
  while (interior){
    s = s + f(x)
    dx = rnorm(length(x))
    dim(dx) <- dim(x)
    x = x + sig*dx
    interior = region(x)
  }
  s = s*dt + u0(x)
  return(s)
}
```

Use it to create a function that generates the mean value of *nw* variates from *Y*:

```
manywalks = function(xc,dt,f,u0,nw){
  s = 0
  for (i in 1:nw){
    aux = onewalk(xc,dt,f,u0)
    s = s + aux
  }
  s = s/nw
  return(s)
}
```

Give a point `xc` for the evaluation, the number `nw` of variates of *Y*, the time step `dt` and generate the estimation of *u* at the point `xc`:

```
nw = 1E3
dt = 1e-3
xc = c(1/3,0,1/2)
uest = manywalks(xc,dt,f,u0,nw)
```

We obtain

```
> print(paste("uest = ",uest, "error = ",u0(xc)-uest))
[1] "uest =  0.359327473715289 error =  0.00178363739582205"
```

In Tables 4.13, 4.14, 4.15, 4.16, 4.17, and 4.18, we present the results for different values of `nw` and `dt`. $\overline{u}$ is the mean value of `uest` and $\overline{e}$ is the mean absolute error $e = |uest - u(xc)|$. $s(u)$, $s(e)$ are the standard deviations of the estimation `uest` and of the absolute error, respectively

Table 4.13 Statistics of 100 estimations of *u* by simulation of the SDE ($nw = 1000$)

Δt	1E-2	1E-3	1E-4	1E-5	1E-6	1E-7	1E-8
$\overline{u}$	0.362	0.361	0.361	0.361	0.361	0.361	0.361
$s(u)$	7E-3	2E-3	1E-3	7E-4	5E-4	3E-4	1E-4

(continued)

Example 4.39 (continued)

Table 4.14 Statistics of the absolute errors in 100 estimations of u by simulation of the SDE ($nw = 1000$)

Δt	1E-2	1E-3	1E-4	1E-5	1E-6	1E-7	1E-8
$\bar{e}$	5E-3	2E-3	1E-3	6E-4	4E-4	2E-4	9E-5
$s(e)$	4E-3	2E-3	1E-3	4E-4	3E-4	2E-4	9E-5

Table 4.15 Statistics of 100 estimations of u by simulation of the SDE ($\Delta t = 1E - 3$)

nw	1000	2500	5000	10000	25000	50000	100000
$\bar{u}$	0.361	0.361	0.361	0.361	0.361	0.361	0.361
$s(u)$	3E-3	2E-3	1E-3	1E-3	6E-4	4E-4	3E-4

Table 4.16 Statistics of the absolute errors in 100 estimations of u by simulation of the SDE ($\Delta t = 1E - 3$)

nw	1000	2500	5000	10000	25000	50000	100000
$\bar{e}$	2E-3	1E-3	9E-4	8E-4	5E-4	3E-4	2E-4
$s(e)$	2E-3	8E-4	9E-4	6E-4	4E-4	2E-4	2E-4

Table 4.17 Statistics of 100 estimations of u by simulation of the SDE ($\Delta t = 1E - 4$)

nw	1000	2500	5000	10000	25000	50000	100000
$\bar{u}$	0.361	0.361	0.361	0.361	0.361	0.361	0.361
$s(u)$	1E-3	9E-4	7E-4	5E-4	3E-4	2E-4	1E-4

Table 4.18 Statistics of the absolute errors in 100 estimations of u by simulation of the SDE ($\Delta t = 1E - 4$)

nw	1000	2500	5000	10000	25000	50000	100000
$\bar{e}$	1E-3	7E-4	6E-4	4E-4	2E-4	2E-4	1E-4
$s(e)$	8E-4	5E-4	5E-4	3E-4	2E-4	1E-4	8E-5

Example 4.40
Consider the Partial Differential Equation

$$\frac{\partial u}{\partial t} - \Delta u = f \quad on\ \Omega \times (0, T),$$

(continued)

Example 4.40 (continued)

$$u(x,t)=u_0(x,t)\ on\ \partial\Omega\times(0,\mathrm{T}),u(x,0)=u_0(x,0)\ on\ \Omega$$

With

$$\Omega=\{(x_1,\ x_2,\ x_3):x_1+x_2+x_3\leq 1,x_1\geq 0,x_20,x_3\geq 0\},$$
$$f(x)=(x_1^2+x_2^2+x_3^2-6)e^t,\ u_0(x)=x_1^2+x_2^2+x_3^2.$$

Let us consider

$$dX_t=\sqrt{2}dW_t,dT_t=-dt,X_0=x=\left(\frac{1}{3},0,\frac{1}{2}\right),T_0=1.$$

We evaluate

$$u(x,t)\approx\frac{1}{nsim}\sum_{i=1}^{nsim}Y_i,Y_i=u_0\left(X_{\tau_i(x)},T_{\tau_i(x)}\right)+\int_0^{\tau_i(x)}f(X_s,T_s)ds$$

The exact value is $u(\mathbf{X_0},T_0)=13e/36\approx 0.9816018$. We can generate a sample from Y as follows:

Firstly, define f and u_0:

```
f = function(x,t){(sum(x^2)-6)*exp(t)}
u0 = function(x,t){sum(x^2)*exp(t)}
```

Then adapt the function `region` to take the value `TRUE` when the point is in Ω and $t>0$. It will take the value `FALSE` otherwise.

```
region = function(x,t){
  aux = TRUE
  if (t <=0){aux = FALSE}
  if (sum(x) >= 1){aux = FALSE}
  for (i in 1: length(x)){
    if (x[[i]]<= 0){aux = FALSE}
  }
  return(aux)
}
```

Create a function that generates one variate from *Y*:

```
onewalk = function(xc,tc,dt,f,u0){
  sig = sqrt(2*dt)
  interior = TRUE
  s = 0
  x = xc
  t = tc
  while (interior){
    s = s + f(x,t)
    dx = rnorm(length(x))
    dim(dx) <- dim(x)
    x = x + sig*dx
    t = t - dt
    interior = region(x,t)
  }
  s = s*dt + u0(x,t)
  return(s)
}
```

Use it to create a function that generates the mean value of *nw* variates from *Y*:

```
manywalks = function(xc,tc,dt,f,u0,nw){
  s = 0
  for (i in 1:nw){
    aux = onewalk(xc,tc,dt,f,u0)
    s = s + aux
  }
  s = s/nw
  return(s)
}
```

Give a point (`xc,tc`) for the evaluation, the number `nw` of variates of *Y*, the time step `dt` and generate the estimation of *u* at the point (`xc,tc`):

```
dt = 1e-3
xc = c(1/3,0,1/2)
tc = 1
nw = 1e3
uest = manywalks(xc,tc,dt,f,u0,nw)
```

We obtain

```
> print(paste("uest = ", uest,", error = ", u0(xc,tc) - uest))
[1] "uest =  0.973799887743801 , error =  0.00780188364418777"
```

In Tables 4.19, 4.20, 4.21, 4.22, 4.23, and 4.24, we present the results for different values of `nw` and `dt`. $\overline{u}$ is the mean value of `uest` and $\overline{e}$ is the mean absolute error $e = |uest - u(xc)|$. $s(u)$, $s(e)$ are the standard deviations of the estimation `uest` and of the absolute error, respectively

(continued)

Example 4.40 (continued)

Table 4.19 Statistics of 100 estimations of u by simulation of the SDE ($nw = 1000$)

Δt	1E-2	1E-3	1E-4	1E-5	1E-6	1E-7	1E-8
$\bar{u}$	0.974	0.982	0.982	0.981	0.981	0.982	0.982
$s(u)$	2E-2	8E-3	4E-3	2E-3	1E-3	7E-4	4E-4

Table 4.20 Statistics of the absolute errors in 100 estimations of u by simulation of the SDE ($nw = 1000$)

Δt	1E-2	1E-3	1E-4	1E-5	1E-6	1E-7	1E-8
$\bar{e}$	2E-2	7E-3	3E-3	2E-3	1E-3	6E-4	3E-4
$s(e)$	1E-2	4E-3	2E-3	1E-3	9E-4	4E-4	3E-4

Table 4.21 Statistics of 100 estimations of u by simulation of the SDE ($\Delta t = 1E - 3$)

nw	1000	2500	5000	10000	25000	50000	100000
$\bar{u}$	0.982	0.981	0.982	0.981	0.982	0.982	0.982
$s(u)$	9E-3	5E-3	3E-3	2E-3	2E-3	1E-3	9E-4

Table 4.22 Statistics of the absolute errors in 100 estimations of u by simulation of the SDE ($\Delta t = 1E - 3$)

nw	1000	2500	5000	10000	25000	50000	100000
$\bar{e}$	7E-3	4E-3	3E-3	2E-3	1E-3	9E-4	7E-4
$s(e)$	5E-3	3E-3	2E-3	2E-3	9E-4	7E-4	5E-4

Table 4.23 Statistics of 100 estimations of u by simulation of the SDE ($\Delta t = 1E - 4$)

nw	1000	2500	5000	10000	25000	50000	100000
$\bar{u}$	0.982	0.982	0.981	0.982	0.982	0.982	0.982
$s(u)$	4E-3	2E-3	2E-3	1E-3	9E-4	6E-4	4E-4

Table 4.24 Statistics of the absolute errors in 100 estimations of u by simulation of the SDE ($\Delta t = 1E - 4$)

nw	1000	2500	5000	10000	25000	50000	100000
$\bar{e}$	3E-3	2E-3	1E-3	1E-3	7E-4	4E-4	4E-4
$s(e)$	3E-3	2E-3	1E-3	8E-4	5E-4	3E-4	2E-4

Remark 4.11
The reader will find in the literature other methods for the solution of partial differential equations by means of SDE. For instance, the method usually referred as "Walk on Spheres": see, for instance, Elepov and Mikhailov (1969), Booth (1982), Hwang and Mascagni, Efficient modified "walk on spheres" algorithm for the linearized Poisson–Bolzmann (2001), Hwang et al. (2003), and Zhou and Cai (2016).

Chapter 5
Uncertain Algebraic Equations

Abstract In this chapter, we consider the situation where an unknown n-dimensional vector X has to be determined by solving a system of equations having the form F(X, U) = 0, where F is a mapping from the n-dimensional Euclidean space on itself and U is a random k-dimensional vector. We focus on the numerical determination of the distribution of solution X, which is also a random variable. We consider linear and nonlinear situations. We consider also the determination of the distribution for eigenvalues of matrices.

Solving equations is a basic activity in most fields of knowledge. As an example, let us consider the modeling of travel distributions among different locations. The travels are described by a matrix $\boldsymbol{T} = (T_{ij}, 1 \leq i,j \leq n)$ connecting inputs and outputs of the different locations: T_{ij} is the number of travels from location i to location j–i is the origin and j is the destination. The matrix $\boldsymbol{T}$ is usually referred as *Origin/ Destination matrix*, or simply *O/D matrix*. In general, it takes the form presented in Table 5.1.

Such a matrix synthetizes the empirical data about the travels between the locations: the sum of the row i gives the number O_i of travels leaving the location i, while the sum of the column j gives the number of travels arriving at the location j. We have:

$$O_i = T_{i\blacksquare} = \sum_{j=1}^{n} T_{ij}, \quad D_j = T_{\blacksquare j} = \sum_{i=1}^{n} T_{ij}. \tag{5.1}$$

One of the main uses of these data is the prediction of the future numbers of travels if some modification intervenes in the transportation network (for instance, infrastructure evolution, such as the opening of a new road), in the intervening opportunities (for instance, the installation of a new facility – school, hospital, and

Supplementary Information The online version of this chapter (https://doi.org/10.1007/978-3-031-17785-9_5) contains supplementary material, which is available to authorized users.

E. Souza de Cursi, *Uncertainty Quantification using R*, International Series in Operations Research & Management Science 335,
https://doi.org/10.1007/978-3-031-17785-9_5

Table 5.1 Example of O/D matrix

	Location 1	...	Location n
Location 1	From 1 to 1	...	From 1 to n
...	...	...	...
Location n	From n to 1	...	From n to n

Table 5.2 An example of flux matrix $\boldsymbol{T}$

Destination							
		1	2	3	4	5	6
O	1	0	211	95	382	236	5
R	2	244	0	42	51	58	2
I	3	189	72	0	119	36	1
G	4	399	46	63	0	94	2
I	5	193	41	15	74	0	4
N	6	53	20	6	18	49	0

Table 5.3 An example of distance matrix $\boldsymbol{D}$

Destination							
		1	2	3	4	5	6
O	1	0	1	0.9	0.8	1.1	1.9
R	2	1	0	1.2	1.8	1.6	1.7
I	3	0.9	1.2	0	1.2	2.0	2.6
G	4	0.8	1.8	1.2	0	1.4	2.5
I	5	1.1	1.6	2.0	1.4	0	1.4
N	6	1.9	1.7	2.6	2.5	1.4	0

commerce) or in the distribution of the population (relocations and moves of people). To make such a prediction, we need to use the data to calibrate a model describing it, id est, a *spatial interaction model* describing the flux of population. The literature proposes many models of this type, for instance, (Ravenstein, 1885; Reilly, 1929, 1931; Hoyt, 1939; Stouffer, 1940; Zipf, 1946; Stewart, 1947; Stewart, 1948; Converse, 1949; Zipf, 1949; Clark, 1951; Isard, 1956; Hansen, 1959; Huff, 1964; Wilson, 1967, 1970; Tobler, 1976; Wilson, 2010). In general, all the models include parameters to be determined from the available data.

As an example, let us consider the fluxes given in Table 5.2. We can consider the model

$$T_{ij} = A_i O_i B_j D_j \exp\left(-\alpha d_{ij} - \beta c_{ij}\right), \quad i \neq j. \tag{5.2}$$

where $\boldsymbol{D}$ is a distance matrix and $\boldsymbol{C}$ is an opportunity matrix – examples are given in Tables 5.3 and 5.4. The unknowns to be determined are α, β, A_i, B_i, $i = 1, \ldots 6$.

The reader can find in the literature many works considering the determination of these unknowns from observed data – id est, about the calibration of the model. A popular method is the method of Furness (1965), which was adapted by many authors – including the author of this book (Gonçalves & Souza de Cursi, 2001). However, our purpose here is not to study calibration methods, but those of

Table 5.4 An example of opportunity matrix C

	Destination						
		1	2	3	4	5	6
O	1	0	1.5	2.5	1	2	3
R	2	1.5	0	2.5	4.5	4	4
I	3	2.5	2.5	0	2	5	5
G	4	1	4.5	2	0	3.5	4.5
I	5	2	4	5	3.5	0	2
N	6	3	4	5	4.5	2	0

uncertainty quantification, namely for algebraic equations. Thus, we shall adopt a simplified method of calibration, more useful for our purposes: we shall look for the coefficients α, β, a_i, b_i, $i = 1, \ldots 6$ such that

$$T_{ij} = a_i b_j \exp\left(-\alpha d_{ij} - \beta c_{ij}\right), \quad i \neq j \tag{5.3}$$

Of course, Eqs. (5.2) and (5.3) coincide if $A_i = a_i/O_i$, $B_j = b_j/D_j$, but (5.3) leads to a simpler procedure. Notice that

$$-d_{ij}\alpha - c_{ij}\beta + \log(a_i) + \log(b_j) = \log(T_{ij}), \quad i \neq j\,. \tag{5.4}$$

so that we can take as unknowns α, β, $\log(A_i)$, $\log(B_i)$, $i = 1, \ldots 6$. In addition, we can set

$$\boldsymbol{X} = (\alpha, \beta, a_1, \ldots, a_6, b_1, \ldots, b_6)^t$$

and we rewrite Eq. (5.4) as a linear system

$$\boldsymbol{MX} = \boldsymbol{N}. \tag{5.5}$$

where for a convenient $k \in \{1, \ldots, 30\}$:

$$M_{k1} = -d_{ij}, \quad M_{k2} = -c_{ij}, \quad M_{k,i+2} = 1, \quad M_{k,i+8} = 1, \quad N_k = \log(T_{ij}).$$

The other elements from $\boldsymbol{M}$ are null. Notice that $\boldsymbol{M}$ is a 30×14 – matrix, so that the linear system (5.5) is overdetermined: we have 14 unknowns and 30 values of T_{ij}, $i \neq j$, so that there are more equations than unknowns. Consequently, it must be solved by an adapted approach. For instance, by linear least squares with a regularizing parameter $\varepsilon > 0$:

$$(\boldsymbol{M}^t\boldsymbol{M} + \varepsilon \boldsymbol{Id})\boldsymbol{X} = \boldsymbol{M}^t\boldsymbol{N}. \tag{5.6}$$

In general, the values determined by (5.6) do not verify the conditions (5.1) and it is necessary to make a second step involving constrained nonlinear optimization to get results that satisfy (5.1) and $\alpha, \beta \geq 0$.

Notice that this second step involves a hard optimization problem: the results can vary significantly according to the method and parameters of optimization chosen by the user. In the sequel, we call `fmincon` with the restrictions $0 \le \alpha, \beta \le 1$. On error, two other calls to `fmincon` are made, with a different starting point (a random perturbation of the bsolution of the solution of (5.6)). If all these calls produce an error, we call `fminsearch` (package `pracma`) followed by a call to `GA`, with a penalized objective function (quadratic penalty) – the penalty coefficient is chosen increasingly as 1E4, 1E5, . . ., 1E10, using as initial point the preceding solution. If this procedure is applied to the data in Tables 5.2, 5.3, and 5.4, we obtain in the first step (linear least squares solution with regularizing parameter $\varepsilon = 1E - 8$):

```
         [,1]       [,2]     [,3]     [,4]
[1,] 3.609157 -0.4397933 3.909869 3.701051
         [,5]     [,6]     [,7]     [,8]     [,9]
[1,] 3.891691 4.017089 3.718676 4.385628 3.909869
        [,10]    [,11]    [,12]    [,13]    [,14]
[1,] 3.701051 3.891691 4.017089 3.718676 4.385629
```

which corresponds to

$$\alpha = 3.609157,\ \beta = -0.4397933,$$

$$\boldsymbol{a} = (49.89239, 40.48984, 48.99367, 55.53921, 41.20979, 80.28867)$$

$$\boldsymbol{b} = (49.89241, 40.48984, 48.99366, 55.53921, 41.20980, 80.28867)$$

The second step furnishes

```
     [,1] [,2]     [,3]     [,4]    [,5]     [,6]
[1,] 0.45 0.25 3.472284 2.787081 2.73595 3.063591
         [,7]     [,8]     [,9]    [,10]   [,11]
[1,] 2.720277 2.032526 3.535871 2.706611 2.11461
        [,12]    [,13]      [,14]
[1,] 3.082427 2.985926 -0.2304002
```

which corresponds to

$$\alpha = 0.45, \quad \beta = 0.25,$$

$$\boldsymbol{a} = (32.210238, 16.233560, 15.424393, 21.404289, 15.184532, 7.633345)$$

$$\boldsymbol{b} = (34.324890, 14.978429, 8.286350, 21.811274, 19.804841, 0.794216)$$

With these values, we have ($\boldsymbol{T}(\boldsymbol{x})$ is the matrix of fluxes generated by $\boldsymbol{x}$):

$$\max\left\{\left|T_{ij} - T_{ij}(\boldsymbol{x})\right|,\ 1 \le i, j \le 6\right\} \le 0.5,$$

so that the relative error between $\boldsymbol{T}(\boldsymbol{x})$ and the data $\boldsymbol{T}$ of Table 5.2 is about 0.1%. As previously indicated, other approaches exist in the literature. Whatever the method adopted, matrices $\boldsymbol{T}, \boldsymbol{C}$, and $\boldsymbol{D}$ are generally determined from empirical data and observations – thus, they contain errors, and their values are uncertain, due to the error margins.

As an example, assume that the fluxes in $\boldsymbol{T}$ result from measurements and are affected by some variability and errors: for instance, the number of travels can vary

Table 5.5 An example of variability of T

T_{12}	214	219	220	209	229	210	194	205	220	231
T_{21}	267	258	261	267	266	233	255	226	236	225
T_{12}	194	195	193	224	203	202	224	232	216	213
T_{21}	245	235	245	266	243	250	245	256	238	264

from a day to another, from a week to another and so on – a new measurement probably will lead to different values. Thus, each observed value T_{ij} can be considered as a fluctuation around a mean value $\overline{T_{ij}}$, id est, a realization of a random variable: $T_{ij} = \overline{T_{ij}} + u_{ij}$ (additive model of error/variability) or $T_{ij} = (1 + u_{ij})\overline{T_{ij}}$ (multiplicative model of error/variability), where u_{ij} is a random variable. Analogously, the values of C can be erroneous or badly evaluated.

To consider the variability of the data and analyze its impact on the determination of the coefficients, we can use the approaches previously presented in Chap. 3. Let us exemplify such an analysis by considering the (very!) simple situation where only the values of T_{12} and T_{21} are affected by uncertainty: $T_{12} = (1 + u_1)\overline{T_{12}}$ and $T_{21} = (1 + u_2)\overline{T_{21}}$. Assume that we get the following data (Table 5.5):

We start by estimating $\overline{T_{12}}$, $\overline{T_{21}}$ by the empirical mean of the data:

```
> T12 = c(214,219,220,209,229,210,194,205,220,231,
+         194,195,193,224,203,202,224,232,216,213)
> T21 = c(267,258,261,267,266,233,255,226,236,225,
+         245,235,245,266,243,250,245,256,238,264)
> T12bar = mean(T12)
> T21bar = mean(T21)
> T12bar
[1] 212.35
> T21bar
[1] 249.05
```

Then, we determine the values of u_1, u_2:

```
> u1 = T12/T12bar  - 1
> u2 = T21/T21bar  - 1
> u1
 [1]  0.007770191  0.031316223  0.036025430 -0.015775842
 [5]  0.078408288 -0.011066635 -0.086413939 -0.034612668
 [9]  0.036025430  0.087826701 -0.086413939 -0.081704733
[13] -0.091123146  0.054862256 -0.044031081 -0.048740287
[17]  0.054862256  0.092535908  0.017188604  0.003060984
> u2
 [1]  0.072073881  0.035936559  0.047982333  0.072073881
 [5]  0.068058623 -0.064444891  0.023890785 -0.092551696
 [9] -0.052399117 -0.096566954 -0.016261795 -0.056414375
[13] -0.016261795  0.068058623 -0.024292311  0.003814495
[17] -0.016261795  0.027906043 -0.044368601  0.060028107
```

Then, we determine the values of α and β by the simplified procedure described:

```
> alfa
 [1] 0.5145731 0.5370590 0.5128242 0.5114287 0.5078652 0.5823512
 [7] 0.4500000 0.5288458 0.4986681 0.5342626 0.4508116 0.4500000
[13] 0.5176035 0.4500000 0.4500000 0.5579844 0.4987811 0.5260519
[19] 0.3906956 0.5158433
> beta
 [1] 0.2459009 0.2404114 0.2293481 0.2442739 0.2378900 0.2142586
 [7] 0.2500000 0.2190718 0.2352101 0.2368665 0.2881994 0.2500000
[13] 0.2303162 0.2500000 0.2500000 0.2221037 0.2377189 0.2238295
[19] 0.2612784 0.2566798
```

Now, we have a sample from (α, β) and the corresponding values of $u = (u_1, u_2)$: we can apply the methods presented in Chap. 3, namely in Sect. 3.5.1. For instance, we can look for 2D-expansions with $k_1 = k_2 = 3$: we obtain the coefficients `c1` for the expansion of α and the coefficients `c2` for the expansion of β:

```
> t(c1)
          [,1]      [,2]     [,3]      [,4]
[1,] 0.7243757 -2.814065 7.711517 -6.046063
          [,5]     [,6]      [,7]     [,8]
[1,] -4.458522 38.69808 -84.49102 55.6184
         [,9]     [,10]    [,11]     [,12]
[1,] 16.53952 -133.3726 267.0318 -159.9265
         [,13]    [,14]     [,15]    [,16]
[1,] -12.4853 103.5079 -203.9016 118.3201
```

```
> t(c2)
          [,1]      [,2]     [,3]      [,4]
[1,] 0.7243757 -2.814065 7.711517 -6.046063
          [,5]     [,6]      [,7]     [,8]
[1,] -4.458522 38.69808 -84.49102 55.6184
         [,9]     [,10]    [,11]     [,12]
[1,] 16.53952 -133.3726 267.0318 -159.9265
         [,13]    [,14]     [,15]    [,16]
[1,] -12.4853 103.5079 -203.9016 118.3201
```

If the distribution of u is known, the knowledge of the coefficients allows the determination of the distribution of (α, β). For instance, assume that u_1 and u_2 are independent, uniformly distributed. Then, we can generate a large sample from u and use it to estimate the CDF and the PDF of the variables. For instance, we can generate 400 variates from each u_i on the interval $(\min u_i, \max u_i)$, forming a sample of 1.6E5 variates $\left(u_{1_i}, u_{2_j}\right)$ from (u_1, u_2). Calculating the 2D expansion on this sample furnishes a sample of 1.6E5 variates from (α, β). Notice that the expansion can generate non-compliant values, such as negative values for the coefficients. In Fig. 5.1a–d, we present the crude results, where the non-compliant values are kept. In Fig. 5.2a–d, the non-compliant values were eliminated – observe that the results are similar to the results generated by the whole set of results, including non-compliant values.

If a supplementary information on the support of the variables is available, we can generate variates on the whole support. For instance, assume that they variables have as support $(-0.1, 0.1)$. In this case, we generate a large sample of (u_1, u_2) on this interval. Since the data are limited to $(-0.0911, 0.0925)$ for u_1 and $(-0.0966, 0.0721)$ for u_2, this procedure implies extrapolation, what introduces errors, such as negative values for the coefficients. The results presented in Fig. 5.3 are crude results, where we kept the non-compliant values. If the non-compliant values are eliminated, we obtain the results in Fig. 5.4.

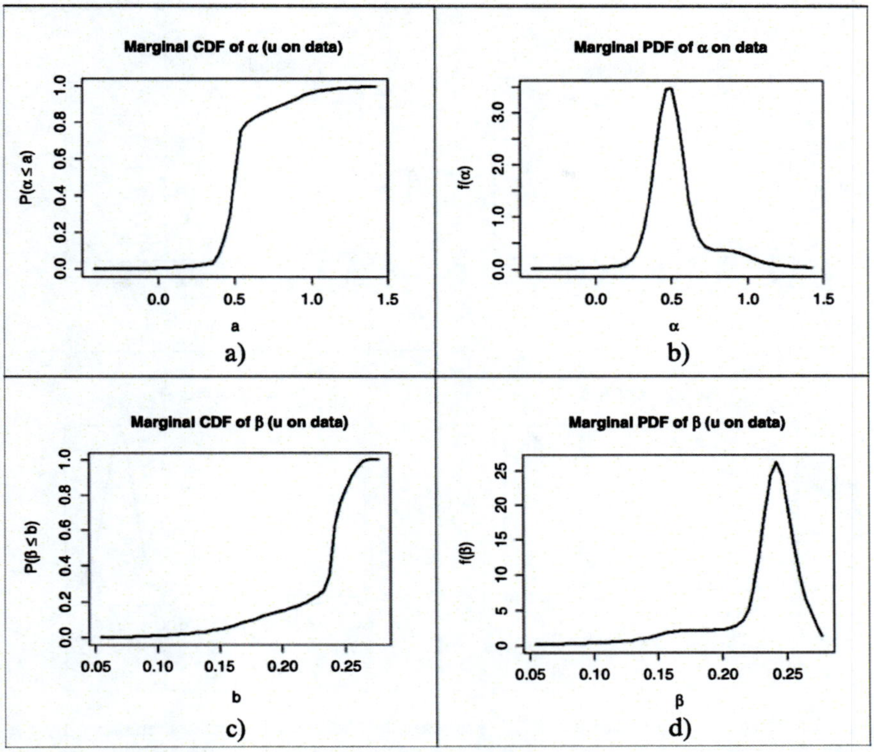

Fig. 5.1 Crude results for the CDF and PDF, generated by a polynomial expansion with $k_1 = k_2 = 3$, with the sample in Table 5.5. The CDF is generated by a sample 401 variates from each u_i, forming a sample of 1.6E5 variates from (u_1, u_2). The PDF is the SPH derivative of the empirical CDF. (**a**) CDF of α. (**b**) PDF of α. (**c**) CDF of β. (**d**) PDF of β

As a second example, let us consider that matrix $\boldsymbol{C}$ is affected by uncertainty. Let us consider the simple situation where only C_{12} and C_{13} are affected by uncertainty. For instance, assume that different evaluations produced the values in Table 5.6.

We adopt an analogous procedure:

```
C12 = c(1.8,1.5,1.8,1.3,2.0,1.5,1.9,1.0,1.3,1.6,
        1.4,1.0,1.3,1.6,1.9,1.2,1.8,1.1,1.3,1.3)
C13 = c(2.2,2.9,2.8,2.0,2.1,2.1,2.6,2.1,2.7,2.7,
        3.0,2.3,2.8,2.2,2.3,2.5,2.8,2.7,2.4,2.3)
C12bar = mean(C12)
C13bar = mean(C13)
u1 = C12 - C12bar
u2 = C13 - C13bar
```

Then, we generate the sample from (α, β):

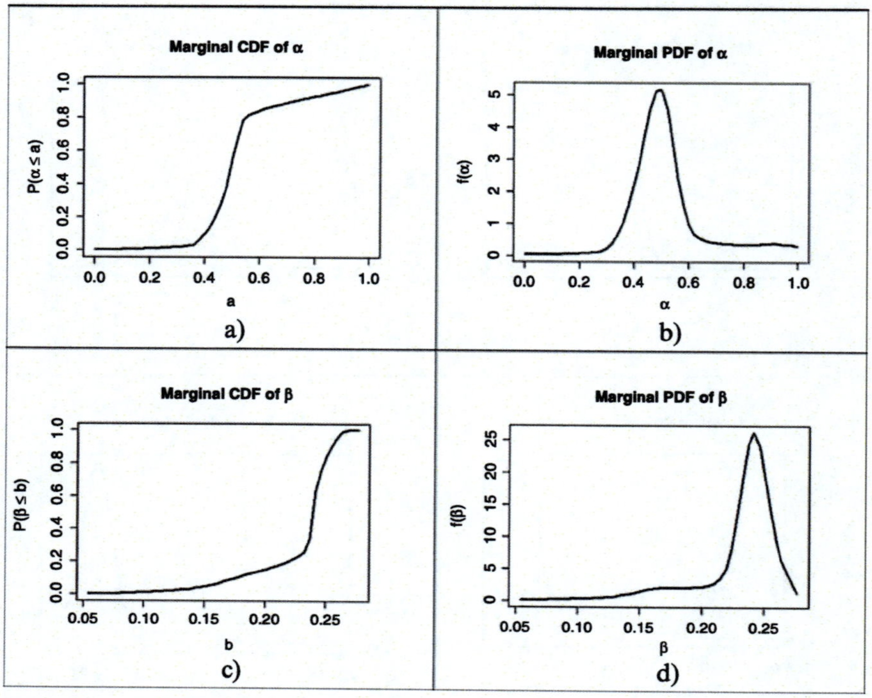

Fig. 5.2 Results for the CDF and PDF, generated by the compliant values furnished by the polynomial expansion with $k_1 = k_2 = 3$ used to generate Fig. 5.1. The PDF is the SPH derivative of the empirical CDF. **(a)** CDF of α. **(b)** PDF of α. **(c)** CDF of β. **(d)** PDF of β

```
> alfa
 [1] 0.5439172 0.4200267 0.5257403 0.6020248 0.5409932 0.4954788 0.6197539 0.4773853
 [9] 0.4525902 0.5144931 0.7247400 0.5153820 0.5061634 0.5367320 0.5412817 0.7402195
[17] 0.5904507 0.5299388 0.4500000 0.5106252
> beta
 [1] 0.2231474 0.2680643 0.2076079 0.2154650 0.2265895 0.2513701 0.2001073 0.2458874
 [9] 0.2324983 0.2489297 0.1050983 0.2371204 0.2793724 0.2289177 0.2177009 0.1972839
[17] 0.2109956 0.2427364 0.2500000 0.2794554
```

```
> t(c1)
         [,1]      [,2]      [,3]      [,4]      [,5]
[1,] 2.045901 -23.27752 83.51725 -76.23842 -8.588038
         [,6]      [,7]      [,8]      [,9]     [,10]
[1,] 134.3865 -499.4539 464.9864 15.12454 -243.8657
        [,11]     [,12]     [,13]     [,14]     [,15]
[1,] 936.3975 -888.4318 -8.487049 139.6676 -547.6423
        [,16]
[1,] 527.1326
```

```
> t(c2)
          [,1]     [,2]      [,3]     [,4]     [,5]
[1,] -0.3813355 9.279464 -33.10258 30.20486 3.396714
         [,6]     [,7]      [,8]      [,9]   [,10]
[1,] -51.75106 194.6427 -182.8234 -5.582225 89.2949
        [,11]    [,12]    [,13]     [,14]    [,15]
[1,] -355.3761 343.9118 2.964895 -49.51822 204.8005
        [,16]
[1,] -202.5715
```

The distributions obtained are shown in Fig. 5.5a–d.

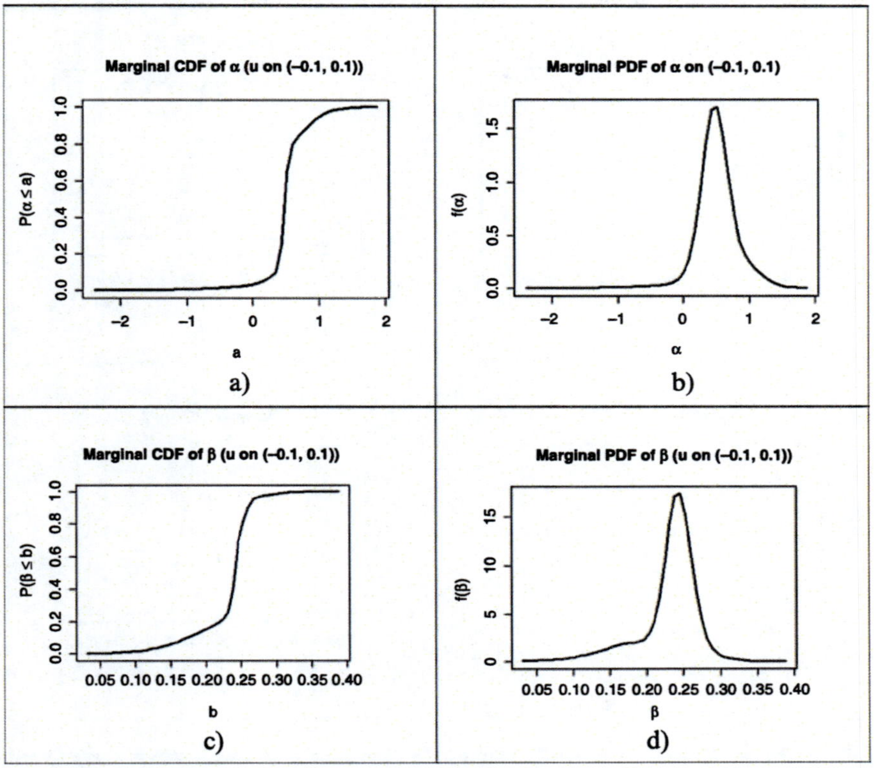

Fig. 5.3 Crude results for the CDF and PDF, generated by a polynomial expansion with $k_1 = k_2 = 3$ and $u_i \in (-0.1, 0.1)$. (**a**) CDF of α. (**b**) PDF of α. (**c**) CDF of β. (**d**) PDF of β

5.1 Uncertain Linear Systems

The previous examples show that we can use the approaches introduced in Chap. 3 to determine the distributions of the solutions of algebraic equations, but there are also some alternative approaches. Let us examine a second method: we consider a linear system

$$\boldsymbol{A}(\boldsymbol{U})\boldsymbol{X} = \boldsymbol{B}(\boldsymbol{U}), \tag{5.7}$$

where $\boldsymbol{U}$ is a random vector, $\boldsymbol{A}$ is a $n \times n$ matrix, $\boldsymbol{B}$ is a $n \times 1$ vector. Let us consider an expansion analogous to Eq. (3.3) of Chap. 3:

$$\boldsymbol{X} \approx \boldsymbol{P}\boldsymbol{X} = \sum_{j=1}^{k} \boldsymbol{x}_j \varphi_j(\boldsymbol{U}), \boldsymbol{x}_j = \left(x_{1,j}, \ \ldots, x_{n,j}\right)^{\mathrm{t}}. \tag{5.8}$$

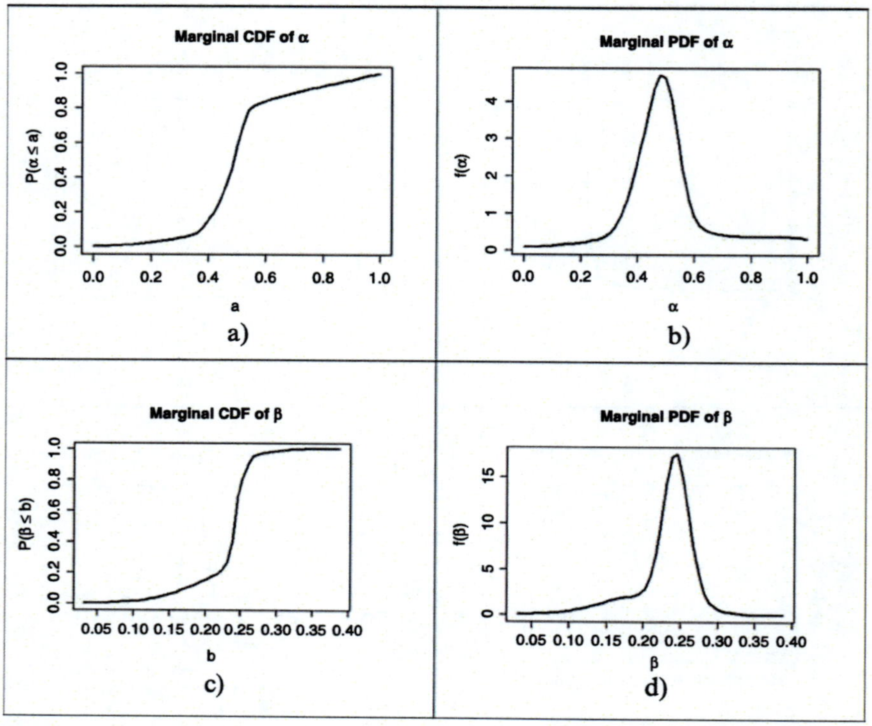

Fig. 5.4 Results for the CDF and PDF, generated by a polynomial expansion with $k_1 = k_2 = 3$ and $u_i \in (-0.1, 0.1)$, after elimination of non-compliant values. (**a**) CDF of α. (**b**) PDF of α. (**c**) CDF of β. (**d**) PDF of β

Table 5.6 An example of uncertainty on $\boldsymbol{C}$

C_{12}	1.8	1.5	1.8	1.3	2.0	1.5	1.9	1.0	1.3	1.6
C_{13}	2.2	2.9	2.8	2.0	2.1	2.1	2.6	2.1	2.7	2.7
C_{12}	1.4	1.0	1.3	1.6	1.9	1.2	1.8	1.1	1.3	1.3
C_{13}	3.0	2.3	2.8	2.2	2.3	2.5	2.8	2.7	2.4	2.3

Then, using Eq. (5.7):

$$\boldsymbol{A}(\boldsymbol{U})\boldsymbol{P}\boldsymbol{X} \approx \boldsymbol{B}(\boldsymbol{U}),$$

id est,

$$\sum_{j=1}^{k} \boldsymbol{A}(\boldsymbol{U})\boldsymbol{x}_j\varphi_j(\boldsymbol{U}) \approx \boldsymbol{B}(\boldsymbol{U}).$$

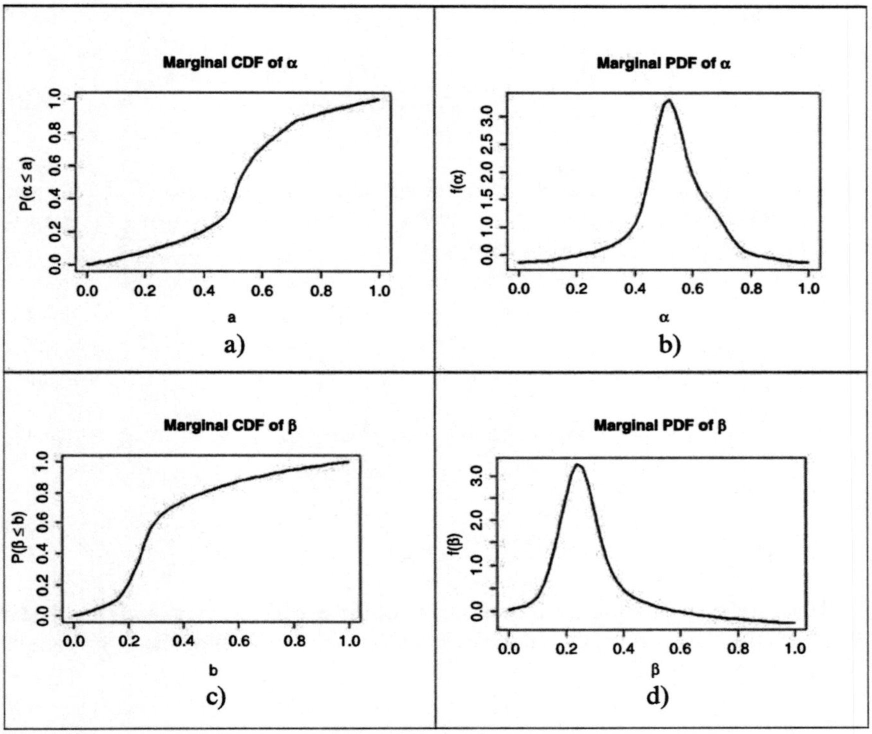

Fig. 5.5 Empirical CDF and PDF of the compliant values generated by a polynomial expansion with $k_1 = k_2 = 3$ and $u_i \in (-1, 1)$ (variability of C_{12}, C_{13}). The PDF is the SPH derivative of the CDF. (**a**) CDF of α. (**b**) PDF of α. (**c**) CDF of β. (**d**) PDF of β

Thus, for $1 \leq r \leq n$

$$\sum_{j=1}^{k} \sum_{s=1}^{n} A_{rs}(\boldsymbol{U}) x_{s,j} \varphi_j(\boldsymbol{U}) \approx B_r(\boldsymbol{U}),$$

so that, for $1 \leq i \leq k$

$$\sum_{j=1}^{k} \sum_{s=1}^{n} \varphi_i(\boldsymbol{U}) A_{rs}(\boldsymbol{U}) x_{s,j} \varphi_j(\boldsymbol{U}) \approx \varphi_i(\boldsymbol{U}) B_r(\boldsymbol{U}),$$

Let us denote

$$\mathcal{A}_{risj} = E\big(\varphi_i(\boldsymbol{U}) A_{rs}(\boldsymbol{U}) \varphi_j(\boldsymbol{U})\big), \quad \mathcal{B}_{ri} = E(\varphi_i(\boldsymbol{U}) B_r(\boldsymbol{U})) \tag{5.9}$$

Then, for $1 \le r \le n$, $1 \le i \le k$:

$$\sum_{j=1}^{k} \sum_{s=1}^{n} \mathcal{A}_{risj} x_{s,j} \approx \mathcal{B}_{ri}. \tag{5.10}$$

Equation (5.10) form a linear system for the unknown coefficients $(x_{s,j} : 1 \le s \le n, 1 \le j \le k)$. We may transform it in a classical linear system by using the map ℓ_2 introduced in Sect. 3.5.1. For instance, set

$$p = i + (r-1)k, q = j + (s-1)k. \tag{5.11}$$

Then, taking

$$\mathbb{A}_{pq} = \mathcal{A}_{risj}, \mathbb{B}_p = \mathcal{B}_{ri}, \mathbb{X}_q = x_{s,j}, \tag{5.12}$$

We have

$$\mathbb{A}\mathbb{X} = \mathbb{B}. \tag{5.13}$$

Thus, we may determine the coefficients of the expansion by solving the linear system (5.13). In practice, the information on $\boldsymbol{U}$ is frequently furnished by a sample $(\boldsymbol{U}_m : 1 \le m \le ns)$. Then,

$$\mathcal{A}_{risj} \approx \frac{1}{ns} \sum_{m=1}^{ns} \varphi_i(\boldsymbol{U}_m) A_{rs}(\boldsymbol{U}_m) \varphi_j(\boldsymbol{U}_m), \;\; \mathcal{B}_{ri} \approx \frac{1}{ns} \sum_{m=1}^{ns} \varphi_i(\boldsymbol{U}_m) B_r(\boldsymbol{U}_m) \tag{5.14}$$

An example of code for the generation of matrices $\mathbb{A}$ and $\mathbb{B}$ is given in class `uls`:

```
varlinsys = function(us,k){
  n = self$ndim
  nn = k * n
  aa = matrix(0,nrow=nn,ncol=nn)
  bb = matrix(0,nrow = nn, ncol = 1)
  aux = dim(us)
  dimU = aux[[1]]
  ns = aux[[2]]
  uu = matrix(0, nrow = dimU,ncol = 1)
  for (m in 1 : ns){
    for (ii in 1 : dimU){
      uu[[ii]] = u[[ii,m]]
    }
    Auu = self$A(uu)
    Buu = self$B(uu)
    for (r in 1 : n){
      for (i in 1 : k){
        p = i + (r - 1) * k
        aux1 = self$basis$value(i-1, uu)
        bb[[p]] = bb[[p]] + Buu[[r]] * aux1
        for (s in 1 : n){
          for (j in 1 : k){
            q = j + (s - 1) * k
            aux2 = self$basis$value(j-1, uu)
            aa[[p, q]] = aa[[p, q]] + aux1 * Auu[[r, s]] * aux2
          }
        }
      }
    }
  }
  aa = aa/ns
  bb = bb/ns
  resu = list(aa,bb)
  names(resu) = c("A","B")
  return(resu)
},
```

Class `uls` has as properties functions calculating $A(u)$, $B(u)$, the dimension n and the basis φ.

```
library(R6)
#
uls=R6Class("uls",
public=list(
  A = NA, # a function A(u)
  B = NA, # a function B(u)
  ndim = NA, # dimension n
  basis = NA, # phi
  initialize=function(A,B,ndim,basis)
```

Example 5.1

Let us consider the input/output analysis of economy, (Leontieff, 1936, 1937, 1986; Arrous, 2000), which is described by the linear system

$$\boldsymbol{AX}=\boldsymbol{B},\ \boldsymbol{A}=\boldsymbol{Id}-\boldsymbol{T},\ \boldsymbol{W}=\boldsymbol{diag}(\boldsymbol{L})\boldsymbol{X} \tag{5.15}$$

$\boldsymbol{T}$ is the structural matrix : T_{ij} gives the fraction of the production of the economic sector i consumed by the economic sector j by unit of production of sector j; $\boldsymbol{L}$ is the Labor vector: L_i is the quantity of Labor requested in field i for the production of an unity; $\boldsymbol{B}$ is the demand vector: B_i is the quantity of goods available to consumers and exportation; $\boldsymbol{W}$ is the quantity of Labor necessary to the production. $\boldsymbol{diag}(\boldsymbol{L})$ denotes a diagonal matrix having $\boldsymbol{L}$ as diagonal.

In practice, $\boldsymbol{T}$ and $\boldsymbol{L}$ result from empirical data and observations, while $\boldsymbol{B}$ is a prevision. Consequently, the model involves uncertainty. Let us consider a model situation in dimension $n = 2$, with $B_1 = 125 + u_1$, u_1 uniformly distributed on $(-10, 10)$, $B_2 = 110 + u_2$, u_2 uniformly distributed on $(-5, 15)$;

$$\boldsymbol{T}=\begin{pmatrix} 0.3 & 0.3125 \\ 0.128 & 0.075 \end{pmatrix}$$

We use the data furnished by a sample of 36 values from $\boldsymbol{U} = (u_1, u_2)$, formed by the pairs $(u_{1,\,i}, u_{2,\,j})$, $1 \le i, j \le 6$, given in Table 5.7.

Table 5.7 Data used for the generation of the sample

u_1	−9.36334	−9.07657	−8.05736	−6.57627	−4.46154	4.120922
u_2	−4.31108	1.341990	3.774887	8.89657	11.46916	14.00444

In this case, $\boldsymbol{X}$ verifies Eq. (5.15), which is analogous to Eq. (5.7). Since $\boldsymbol{U} = (u_1, u_2)$, we use the expansion (5.8), with two polynomial families of degree $k_1 = k_2 = 2$ (see Chap. 3, Sect. 3.5.1).

To determine the coefficients, we start by the creation of the basis:

```
a1 = -10
b1 = 10
a2 = -5
b2 = 15
source("polynom.R")
phi1 = polynom$new(a1,b1,1e-3)
phi2 = polynom$new(a2,b2,1e-3)
k1 = 2
k2 = 2
k = (k1+1)*(k2+1)
```

```
source("tensorbasis2.R")
tb2 = tensorbasis2$new()
tb2$phi1 = phi1
tb2$phi2 = phi2
tb2$k1 = k1
tb2$k2 = k2
he2d2 = expansion2D$new()
he2d2$phi = tb2
uls2 = uls$new(matA,matB,2,he2d2$phi)
```

Then, we determine the coefficients of the expansion:

```
c2 = uls2$solution(u,k)
```

We obtained the coefficients shown in Table 5.8. Figures 5.6 and 5.7 show the marginal distribution of each component of the solution: the accordance is particularly good.

Table 5.8 Coefficients obtained

$x_{1,j}$	229.12	10.29	0.00	15.23	0.00	0.00	0.00	0.00	0.00
$x_{2,j}$	145.22	23.05	0.00	2.11	0.00	0.00	0.00	0.00	0.00

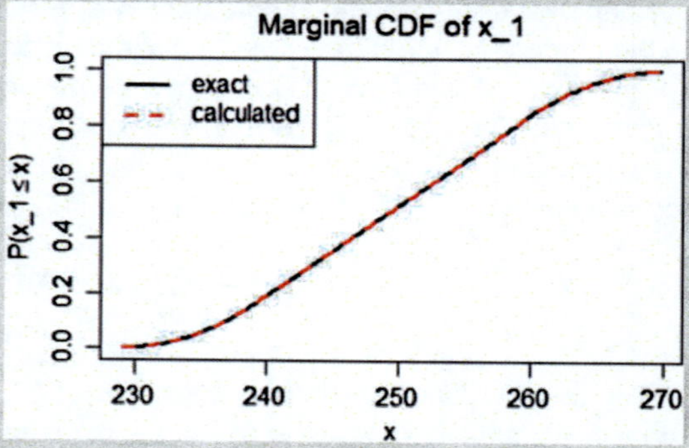

Fig. 5.6 CDF of the first component of the solution (marginal CDF of x_1)

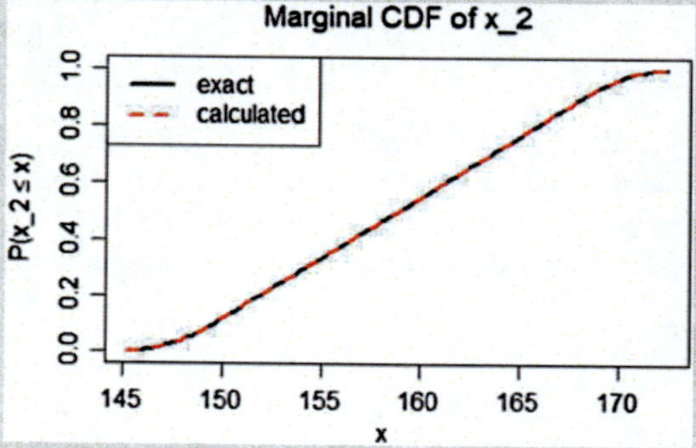

Fig. 5.7 CDF of the second component of the solution (marginal CDF of x_2)

Example 5.2
Let us consider a second situation corresponding to Eq. (5.15). where $B_1 = 125, B_2 = 110$;

$$T = \begin{pmatrix} 0.3 - u_1 & 0.3125 \\ 0.128 + u_2 & 0.075 \end{pmatrix}$$

where u_1 is uniformly distributed on $(0, 0.1)$ and u_2 is uniformly distributed on $(0, 0.05)$.

We use the data furnished by a sample of 36 values from $U = (u_1, u_2)$, formed by the pairs $(u_{1,i}, u_{2,j})$, $1 \le i, j \le 6$, given in Table 5.9.

Table 5.9 Data used for the generation of the sample

u_1	0.00086	0.040390	0.049284	0.072618	0.074695	0.096776
u_2	0.000932	0.030252	0.033205	0.038672	0.043191	0.046179

Again, X verifies Eq. (5.15), which is analogous to Eq. (5.7). As in the preceding example, we use the expansion (5.8), with two polynomial families of degree $k_1 = k_2 = 2$.

We obtained the coefficients shown in Table 5.10. Figures 5.8 and 5.9 show the marginal distribution of each component of the solution: again, the accordance is extremely good.

Table 5.10 Coefficients obtained

$x_{1,j}$	246.90	6.35	0.17	−37.22	−1.90	−0.08	4.62	0.33	0.02
$x_{2,j}$	153.09	14.22	0.38	−5.15	−2.27	−0.12	0.64	0.30	0.02

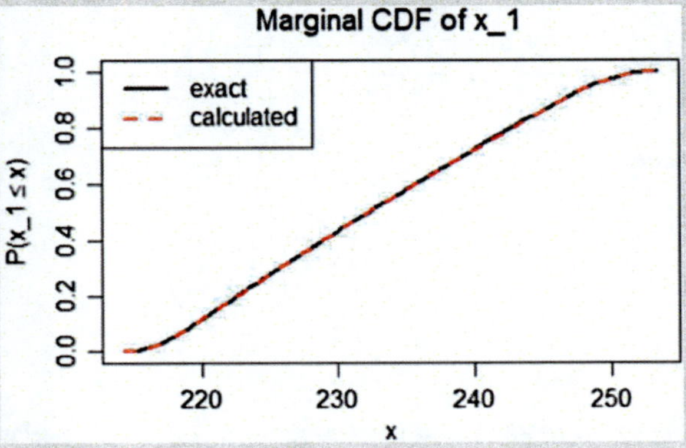

Fig. 5.8 CDF of the second component of the solution (marginal CDF of x_2)

(continued)

Example 5.2 (continued)

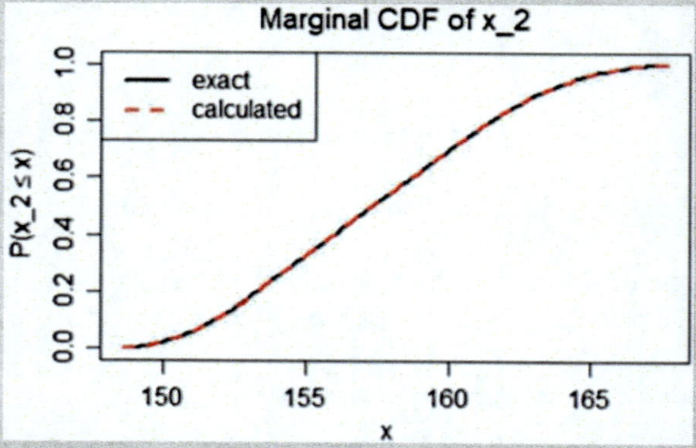

Fig. 5.9 CDF of the second component of the solution (marginal CDF of x_2)

5.1.1 *Very Small Linear Systems*

If your linear system has at just a few variables, you may adopt a simplified approach. Indeed, Cramer's rule may be used if you have just 2 or 3 variables. In his book (Cramer, 1750), Cramer gives the explicit solution of a linear system for 2 or 3 variables. Using Cramer's notation, a linear system

$$\begin{pmatrix} X^1 & Y^1 \\ X^2 & Y^2 \end{pmatrix}\begin{pmatrix} x \\ y \end{pmatrix} = \begin{pmatrix} B^1 \\ B^2 \end{pmatrix}$$

has as solution

$$x = \frac{-B^2Y^1 + B^1Y^2}{-X^2Y^1 + X^1Y^2}, \quad y = \frac{B^2X^1 - B^1X^2}{-X^2Y^1 + X^1Y^2}.$$

A linear system

$$\begin{pmatrix} X^1 & Y^1 & Z^1 \\ X^2 & Y^2 & Z^2 \\ X^3 & Y^3 & Z^3 \end{pmatrix}\begin{pmatrix} x \\ y \\ z \end{pmatrix} = \begin{pmatrix} B^1 \\ B^2 \\ B^3 \end{pmatrix}$$

has as solution

$$x = \frac{B^3Y^2Z^1 - B^2Y^3Z^1 - B^3Y^1Z^2 + B^1Y^3Z^2 + B^2Y^1Z^3 - B^1Y^2Z^3}{X^3Y^2Z^1 - X^2Y^3Z^1 - X^3Y^1Z^2 + X^1Y^3Z^2 + X^2Y^1Z^3 - X^1Y^2Z^3},$$

$$y = \frac{B^3X^2Z^1 - B^2X^3Z^1 - B^3X^1Z^2 + B^1X^3Z^2 + B^2X^1Z^3 - B^1X^2Z^3}{-X^3Y^2Z^1 + X^2Y^3Z^1 + X^3Y^1Z^2 - X^1Y^3Z^2 - X^2Y^1Z^3 + X^1Y^2Z^3},$$

$$z = \frac{B^3X^2Y^1 - B^2X^3Y^1 - B^3X^1Y^2 + B^1X^3Y^2 + B^2X^1Y^3 - B^1X^2Y^3}{X^3Y^2Z^1 - X^2Y^3Z^1 - X^3Y^1Z^2 + X^1Y^3Z^2 + X^2Y^1Z^3 - X^1Y^2Z^3}.$$

Thus, for very small linear systems involving 2 or 3 unknowns, the explicit solution above may be used to generate samples. If the distributions of the terms X^i, Y^j, Z^k, B^ℓ are known, then large samples of the solutions may be generated. If small samples of these variables are given, then small samples of the solutions can be generated and a method of representation (Chap. 3) must be employed.

Example 5.3

Consider again the situation corresponding to Example 5.1, where $B_1 = 125 + u_1$, $u_1 \sim \mathcal{U}(-10, 10)$, $B_2 = 110 + u_2$, $u_2 \sim \mathcal{U}(-5, 15)$;

$$\boldsymbol{T} = \begin{pmatrix} 0.3 & 0.3125 \\ 0.128 & 0.075 \end{pmatrix}$$

In this case, $= \boldsymbol{Id} - \boldsymbol{T}$, so that

$$X^1 = 0.7, \quad X^2 = -0.128, \quad Y^1 = -0.3125, \quad Y^2 = 0.925.$$

Thus,

$$-X^2Y^1 + X^1Y^2 = 0.6075.$$

Since $B^1 = 125 + u_1$, $B^2 = 110 + u_2$, we have

$$-B^2Y^1 + B^1Y^2 = 150 + 0.925u_1 + 0.3125u_2,$$

$$B^2X^1 - B^1X^2 = 93 + 0.128u_1 + 0.7u_2.$$

So, we have

$$x_1 = \frac{150 + 0.925u_1 + 0.3125u_2}{0.6075},$$

$$x_2 = \frac{93 + 0.128u_1 + 0.7u_2}{0.6075}.$$

Finally

$$x_1 \approx 246.9136 + 1.522634\,u_1 + 0.514403\,u_2,$$

$$x_2 \approx 153{,}0864 + 0.2107\,u_1 + 1.152263\,u_2.$$

(continued)

Example 5.3 (continued)

Thus, Cramer's rule furnishes the same result as Table 5.8. If data in Table 5.7 are given, we can generate a sample from x_1:

u_2 \ u_1	−9.36334	−9.07657	−8.05736	−6.57627	−4.46154	4.120922
−4.31108	230.4390	230.8757	232.4275	234.6827	237.9027	250.9706
1.34199	233.3470	233.7836	235.3355	237.5907	240.8106	253.8786
3.774887	234.5985	235.0351	236.5870	238.8421	242.0621	255.1300
8.89657	237.2331	237.6697	239.2216	241.4768	244.6967	257.7647
11.46916	238.5564	238.9931	240.5449	242.8001	246.0201	259.0880

This sample can be used to determine an expansion representing x_1 as a function of $\boldsymbol{U} = (u_1, u_2)$. An analogous table can be generated for x_2. Since the results are the same as in Example 5.1, we shall obtain the same results as in this Example.

Example 5.4

Consider again the situation corresponding to Eq. (**Erreur ! Source du renvoi introuvable.**), where $B_1 = 125$, $B_2 = 110$, $u_1 \sim \mathcal{U}(0, 0.1)$, $u_2 \sim \mathcal{U}(0, 0.05)$;

$$\boldsymbol{T} = \begin{pmatrix} 0.3 - u_1 & 0.3125 \\ 0.128 + u_2 & 0.075 \end{pmatrix}$$

In this case, $\boldsymbol{A} = \boldsymbol{Id} - \boldsymbol{T}$ and

$$X^1 = 0.7 + u_1, X^2 = -0.128 - u_2, Y^1 = -0.3125, Y^2 = 0.925.$$

Thus,

$$-X^2Y^1 + X^1Y^2 = 0.6075 + 0.9025\, u_1 - 0.3125\, u_2.$$

Since $B^1 = 125$, $B^2 = 110$, we have

$$-B^2Y^1 + B^1Y^2 = 150,$$

$$B^2X^1 - B^1X^2 = 93 + 110\, u_1 + 125\, u_2.$$

So, we have

(continued)

Example 5.4 (continued)

$$x_1 = \frac{150}{0.6075 + 0.9025\,u_1 - 0.3125\,u_2},$$

$$x_2 = \frac{93 + 110\,u_1 + 125\,u_2}{0.6075 + 0.9025\,u_1 - 0.3125\,u_2}.$$

If data in Table 5.9 are given, we can generate a sample from x_2:

u_2 \ u_1	0.00086	0.04039	0.049284	0.072618	0.074695	0.096776
0.000932	153.289	163.767	166.185	172.637	173.219	179.488
0.030252	152.673	162.693	165.002	171.161	171.717	177.694
0.033205	152.614	162.590	164.889	171.020	171.573	177.522
0.038672	152.505	162.401	164.681	170.761	171.309	177.208
0.043191	152.417	162.247	164.512	170.550	171.094	176.952

As indicated in the preceding example, this sample can be used to determine an expansion representing x_2 as a function of $U = (u_1, u_2)$. An analogous table can be generated for x_1. Again, the results obtained by this way will be analogous to those obtained in Example 5.2.

Exercises

1. Consider an input-output model where the structural matrix T and the Labor matrix are

$$T = \begin{pmatrix} 0.3+u_1 & 0.5-u_1 & 0.3 \\ 0.2 & 0.2 & 0.3 \\ 0.4 & 0.2 & 0.3 \end{pmatrix}, \quad L = \begin{pmatrix} 300 \\ 150+u_2 \\ 400 \end{pmatrix}$$

Assume that all the variables u_i are independent and triangularly distributed $T(-0.1, 0, 0.1)$.

(a) Generate a sample of 6 variates from each random variable u_i.
(b) Use the values generate to generate a sample of 36 variates from x_1.
(c) Use the sample from x_1 to determine a representation of x_1 as function of $U = (u_1, u_2)$.
(d) Use `varlinsys` to determine a representation of x_1 as function of $U = (u_1, u_2)$.
(e) Compare the results obtained in (c) and (d).
(f) Determine the Marginal distribution of x_1.

(continued)

2. Consider a model for the equilibrium between two Markets characterized by two variables (x_1, x_2) such that

$$\begin{pmatrix} 1 & 4 \\ 1 & -10 \end{pmatrix} \begin{pmatrix} x_1 \\ x_2 \end{pmatrix} = \begin{pmatrix} 200 - u_1 + u_2 \\ 300 + u_1 \end{pmatrix}$$

Assume that the distributions of u_1 and u_2 are triangular: $u_1 \sim T(0,25,50)$ and $u_2 \sim T(-50, 0, 50)$.

   (a) Generate a sample of 6 variates from each random variable u_i.
   (b) Use the values generate to generate a sample of 36 variates from x_1.
   (c) Use the sample from x_1 to determine a representation of x_1 as function of $\boldsymbol{U} = (u_1, u_2)$.
   (d) Use `varlinsys` to determine a representation of x_1 as function of $\boldsymbol{U} = (u_1, u_2)$.
   (e) Compare the results obtained in (c) and (d).
   (f) Determine the Marginal distribution of x_1.

3. Consider a model for the equilibrium between two Markets characterized by two variables (x_1, x_2) such that

$$\begin{pmatrix} 0.2 & 1 \\ 1 & -10 \end{pmatrix} \begin{pmatrix} x_1 \\ x_2 \end{pmatrix} = \begin{pmatrix} 50 + u_1 \\ 300 + u_2 \end{pmatrix}$$

Assume that the distributions of u_1 and u_2 are triangular: $u_1 \sim T(0, 20, 40)$ and $u_2 \sim T(-50, 0, 50)$.

   (a) Generate a sample of 6 variates from each random variable u_i.
   (b) Use the values generate to generate a sample of 36 variates from x_1.
   (c) Use the sample from x_1 to determine a representation of x_1 as function of $\boldsymbol{U} = (u_1, u_2)$.
   (d) Use `varlinsys` to determine a representation of x_1 as function of $\boldsymbol{U} = (u_1, u_2)$.
   (e) Compare the results obtained in (c) and (d).
   (f) Determine the Marginal distribution of x_1.

4. Consider the trip distribution model (5.3) with

$$T = \begin{pmatrix} 0 & 169 & 173 \\ 195 & 0 & 31 \\ 146 & 53 & 0 \end{pmatrix}, \quad D = \begin{pmatrix} 0 & 1 & 0.9 \\ 1 & 0 & 1.2 \\ 0.9 & 1.2 & 0 \end{pmatrix}, \quad C = \begin{pmatrix} 0 & 1.5 & 2.5 \\ 1.5 & 0 & 2.5 \\ 2.5 & 2.5 & 0 \end{pmatrix}$$

(continued)

(a) Solve Eq. (5.4) to determine the coefficients.
(b) Assume that T_{12} is triangularly distributed $T(160,170,180)$. Determine the distribution of the coefficients α, β
(c) Assume that T_{21} is triangularly distributed $T(170,180,210)$. Determine the distribution of the coefficients α, β

5. Consider the trip distribution model (5.3) with

$$T=\begin{pmatrix} 0 & 278 & 89 \\ 321 & 0 & 41 \\ 176 & 70 & 0 \end{pmatrix},\quad D=\begin{pmatrix} 0 & 0.5 & 1 \\ 0.5 & 0 & 1.2 \\ 1 & 1.2 & 0 \end{pmatrix},\quad C=\begin{pmatrix} 0 & 1 & 2 \\ 1 & 0 & 2 \\ 2 & 2 & 0 \end{pmatrix}$$

(a) Solve Eq. (5.4) to determine the coefficients.
(b) Assume that T_{12} is triangularly distributed $T(250,270,290)$. Determine the distribution of the coefficients α, β
(c) Assume that T_{21} is triangularly distributed $T(300,320,340)$. Determine the distribution of the coefficients α, β

6. Consider the trip distribution model (5.3) with

$$T=\begin{pmatrix} 0 & 278 & 89 \\ 321 & 0 & 42 \\ 176 & 72 & 0 \end{pmatrix},\quad D=\begin{pmatrix} 0 & 0.5 & 1 \\ 0.5 & 0 & 1.2 \\ 1 & 1.2 & 0 \end{pmatrix},\quad C=\begin{pmatrix} 0 & 1 & 2 \\ 1 & 0 & 2 \\ 2 & 2 & 0 \end{pmatrix}$$

(a) Solve Eq. (5.4) to determine the coefficients.
(b) Assume that T_{23} is triangularly distributed $T(30,40,50)$. Determine the distribution of the coefficients α, β
(c) Assume that T_{32} is triangularly distributed $T(60,70,80)$. Determine the distribution of the coefficients α, β

5.2 Nonlinear Equations and Adaptation of an Iterative Code

The approach presented in the preceding section extends to nonlinear systems of equations:

$$\boldsymbol{f}(\boldsymbol{x})=\boldsymbol{0} \iff \begin{cases} f_1(\boldsymbol{x})=0 \\ \quad\vdots \\ f_n(\boldsymbol{x})=0 \end{cases} \tag{5.16}$$

Here, $\boldsymbol{f} : \mathbb{R}^n \longrightarrow \mathbb{R}$ defines the equations to be solved for the unknowns $\boldsymbol{x} = (x_1, \ldots, x_n)^t$. Equation (5.16) are generally solved by iterative methods, such as Newton–Raphson's method (see Sect. 1.14.2).

Codes for the iterative solution of algebraical equations are often available and one of the interesting possibilities offered by UQ is that of being able to study uncertainties using existing codes, without modifying them and without rewriting any of their parts.

As in the preceding situation, such an analysis may be made by generating a sample and then using one of the preceding approaches to determine the probability distribution of the results furnished by the code.

Nevertheless, there is an alternative approach – called the *method of adaptation*, which allows to directly obtain the representation of the variability of the code output as a function of the parameters affected by the variability.

As an example, let us consider an iterative method which reads as

$$\boldsymbol{X}^{(r+1)} = \boldsymbol{\Psi}\left(\boldsymbol{X}^{(r)}, U\right) \tag{5.17}$$

$\boldsymbol{\Psi}$ is referred as the *iteration function*. Notice that a direct method may be considered as an iterative method with a single iteration.

If $\boldsymbol{X} = (X_1, \ldots, X_n)$, then $\boldsymbol{\Psi} = (\psi_1, \ldots, \psi_n)$ and we can describe the iterations component by component:

$$\boldsymbol{X}_i^{(r+1)} = \psi_i\left(\boldsymbol{X}^{(r)}, U\right), \quad 1 \leq i \leq n. \tag{5.18}$$

To adapt this iterative method, we can consider approximations analogous to the preceding ones:

$$\boldsymbol{X}^{(p)} \approx \boldsymbol{P}\boldsymbol{X}^{(p)} = \sum_{j=1}^{k} \boldsymbol{x}_j^{(p)} \varphi_j(U), \tag{5.19}$$

i.e.,

$$X_\ell^{(p)} \approx PX_\ell^{(p)} = \sum_{j=1}^{k} x_{j,\ell}^{(p)} \varphi_j(U), \quad \ell = 1, \ldots, n \tag{5.20}$$

Then, we can write

$$\boldsymbol{P}\boldsymbol{X}^{(r+1)} \approx \boldsymbol{\Psi}\left(\boldsymbol{P}\boldsymbol{X}^{(r)}, U\right). \tag{5.21}$$

Thus,

$$\varphi_i(U)\boldsymbol{P}\boldsymbol{X}^{(r+1)} \approx \varphi_i(U)\boldsymbol{\Psi}\left(\boldsymbol{P}\boldsymbol{X}^{(r)}, U\right), \quad 1 \leq i \leq k$$

and

$$E\left(\varphi_i(U)\boldsymbol{P}\boldsymbol{X}^{(r+1)}\right) \approx E\left(\varphi_i(U)\boldsymbol{\Psi}\left(\boldsymbol{P}\boldsymbol{X}^{(r)}, U\right)\right), \quad 1 \leq i \leq k$$

We have

$$E\left(\varphi_j(U)\boldsymbol{P}\boldsymbol{X}^{(r+1)}\right) = \sum_{j=1}^{k} A_{i,j}\boldsymbol{x}_j^{(r+1)}, A_{i,j} = E(\varphi_i(U)\varphi_j(U)), \tag{5.22}$$

id est,

$$E\left(\varphi_j(U)PX_\ell^{(r+1)}\right) = \sum_{j=1}^{k} A_{i,j}x_{j,\ell}^{(r+1)},\ A_{i,j} = E(\varphi_i(U)\varphi_j(U)), \tag{5.23}$$

Let

$$B_{i,\ell}^{(k)} = E\left(\varphi_i(U)\psi_\ell\left(PX^{(k)},\ U\right)\right). \tag{5.24}$$

Then:

$$\sum_{j=1}^{k} A_{i,j}x_{j,\ell}^{(r+1)} = B_{i,\ell}^{(r)},\ 1 \le i \le k \text{ and } 1 \le \ell \le n. \tag{5.25}$$

This is a linear system for the variables $x_{j,\ell}^{(r+1)}$. It can be seen as a sequence of n linear systems of the same matrix A and second member $B_\ell^{(r)} = \left(B_{1,\ell}^{(r)},\ \ldots,\ B_{k,\ell}^{(r)}\right)^t$: the solution of $AY_\ell^{(r)} = B_\ell^{(r)}$ furnishes $Y_\ell^{(r)} = \left(x_{1,\ell}^{(r+1)},\ \ldots,\ x_{k,\ell}^{(r+1)}\right)^t$.

This approach does not require any modification to the code evaluating Ψ: indeed, the calculation of $B_{i,\ell}^{(e)}$ requires only calls to the code that evaluates Ψ – it can be used "as it is", without any modification.

Such a method is non-intrusive, since it does not request the user to plunge into the meanders of the code to modify it – it requests only calls to the code. An example of algorithm for the adaptation is given below:

Adaptation of an Iterative Method

Input: a sample from $U : (U_1, \ldots, U_{ns})$ is given, as a program which determines the value of $\Psi(X, U)$. The basis φ_i and the degree of approximation k are given. Give the stopping conditions (for instance: maximum iteration number and minimal step at each iteration)

Output: the coefficients of the representation PX of X.

Initialization: Determine the matrix A and give an initial guess $\left(x_{j,\ell}^{(0)} : 1 \le j \le k,\ 1 \le \ell \le n\right)$. Set the iteration number to zero: $r \leftarrow 0$.

1. Determine $Z_m = PX^{(r)}(U_m)$, for $1 \le m \le ns$.
2. Evaluate $T_m = \Psi(Z_m, U_m)$, for $1 \le m \le ns$. Let $T_m = (T_{1,\ m}, \ldots, T_{n,\ m})$
3. Determine $V_{m,\ i} = \varphi_i(U_m)$, $1 \le m \le ns$, $1 \le i \le k$.
4. For $1 \le \ell \le n$:
   (a) Evaluate $B_\ell^{(r)} : B_{i,\ell}^{(r)} \approx \frac{1}{ns} \sum_{m=1}^{ns} V_{m,i}T_{\ell,m}$,
   (b) Solve the linear system $AY_r^{(r)} = B_r^{(r)}$,
   (c) Set $x_{i,\ell}^{(r+1)} = Y_{i,\ell}^{(r)}$, $1 \le i \le k$.

(continued)

5. Increment the iteration number $r \longleftarrow r + 1$.
6. Test for the stopping conditions. If the iterations continue, then go to 1, else estimate $\boldsymbol{x} \approx \boldsymbol{x}^{(r)}$.

This approach is quite general and can be used for any situation where the code performing the iterations $\boldsymbol{X}^{(k+1)} = \boldsymbol{\Psi}(\boldsymbol{X}^{(k)}, U)$ can be executed.

This algorithm is implemented in class `uae`.

```
library(R6)
#
uae=R6Class("uae",
            public=list(
              initialize=function(){
              },
```

The method `iteration_adaptation` performs one iteration:

```
iteration_adaptation = function(psi,Xold,A,V,phi,u,ndim,nb){
  Z = self$proj(Xold,u,ndim,nb, phi)
  aux = dim(u)
  dimU = aux[[1]]
  ns = aux[[2]]
  T = matrix(0,nrow=ndim,ncol=ns)
  for (m in 1: ns){
    aaux = matrix(Z[,m], nrow = ndim, ncol = 1)
    bbux = matrix(u[,m], nrow = dimU, ncol = 1)
    T[ ,m] = psi(aaux,bbux)
  }
  aux = dim(Xold)
  Xnew = matrix(0,nrow=aux[[1]],ncol=aux[[2]])
  for (r in 1: ndim){
    Br = matrix(0,nrow=nb,ncol=1)
    for (i in 1: nb){
      s = 0
      for (m in 1: ns){
        s = s + V[[m,i]]*T[[r,m]];
      }
      Br[[i]]=s/ns
    }
    Yr = solve(A,Br)
    for (i in 1: nb){
      Xnew[[r,i]]=Yr[[i]]
    }
  }
  return(Xnew)
},
```

In this code, `self$proj` is a method that evaluates the expansion having coefficients `Xold` and the basis `phi`, with degree `nb-1` (nb is the number of coefficients).

The expansion is evaluated in the values of `u` (each column of `u` is a variate) and `ndim` is the dimension of $\boldsymbol{x}$. For instance, we can use the code at right.

```
proj = function(X,u,ndim,nb, phi){
  aux = dim(u)
  ns = aux[[2]]
  px = matrix(0,nrow=ndim,ncol=ns)
  for (m in 1: ns){
    uu = u[ ,m]
    s = matrix(0,nrow=ndim,ncol=1)
    for (j in 1: nb){
      s = s + X[ ,j]*phi$value(j-1,uu)
    }
    px[ ,m] = s
  }
  return(px)
}
```

Matrices `A` and `V` are calculated by other methods of the class

```
basis_sp = function(us,phi,nb){
  A = matrix(0, nrow=nb,ncol=nb)
  for  (m in 1: nb){
    phim = function(u) {phi$value(m-1,u)}
    ff = function(x) {phim(x)^2}
    A[[m,m]] = self$moyenne_echantillon(ff,us)
    if ( m < nb ){
      for (n in (m+1): nb){
        phin = function(u) {phi$value(n-1,u)}
        ff = function(x) {phim(x)*phin(x)}
        aux = self$moyenne_echantillon(ff,us)
        A[[m,n]] = aux
        A[[n,m]] = aux;
      }
    }
  }
  return(A)
},
```

```
basis_matrix=function(u,phi,nb){
  aux = dim(u)
  ns = aux[[2]]
  V = matrix(0, nrow=ns,ncol=nb)
  for  (i in 1: nb){
    phii = function(u) {phi$value(i-1,u)}
    for (m in 1: ns){
      V[[m,i]] = phii(u[,m])
    }
  }
  return(V)
},
```

Here, `moyenne_echantillon` determines the mean of `ff` of the sample `us`. The solution is generated by the method `adaptation` below:

```
adaptation = function(psi,Xini,phi,u,ndim,nb,nitmax, prec){
  A = self$basis_sp(u,phi,nb)
  V = self$basis_matrix(u,phi,nb)
  Xsol = Xini;
  nit = 0;
  continue_iteration = TRUE
  while (continue_iteration){
    nit = nit+1
    Xnew = self$iteration_adaptation(psi,Xsol,A,V,phi,u,ndim,nb)
    d = norm(Xsol - Xnew)
    print(paste("******** d = ",d))
    Xsol = Xnew
    continue_iteration = (nit < nitmax) & (d > prec)
  }
  return(Xsol)
},
```

Remark 5.1
In many practical situations, $\boldsymbol{\Psi}(\boldsymbol{x}, u) = \boldsymbol{x} - \boldsymbol{\Phi}(\boldsymbol{x}, u)$, with $\boldsymbol{\Phi} = (\phi_1, \ldots, \phi_n)$. In such a situation, we may solve the linear systems

$$\sum_{j=1}^{k} A_{i,j} \delta x_{j,\ell}^{(r)} = \delta B_{i,\ell}^{(r)},\ 1 \leq i \leq k \text{ and } 1 \leq \ell \leq n,$$

with

$$\delta B_{i,\ell}^{(k)} = E\left(\varphi_i(U)\phi_\ell\left(\boldsymbol{P}\boldsymbol{X}^{(k)},\ U\right)\right).$$

Then,

$$x_{j,\ell}^{(r+1)} = x_{j,\ell}^{(r)} - \delta x_{j,\ell}^{(r)}, 1 \leq j \leq k \text{ and } 1 \leq \ell \leq n.$$

Example 5.5
Let us consider relaxed Newton's iterations for the resolution of the equation $x^2 - 2x + u = 0$, where u is uniformly distributed on $(0, 1)$. The equation has two solutions for each u: $x_+ = 1 + \sqrt{1-u}$ and $x_- = 1 - \sqrt{1-u}$. We have

$$P(x_- < a) = 1 - (a-1)^2, 0 < a < 1; \quad P(x_+ < a) = (a-1)^2, 1 < a < 2;$$

Newton's iterations correspond to the iteration function

$$\Psi(x,\ u) = x - \Phi(x,\ u),\ \Phi(x,\ u) = \frac{x^2 - 2x + u}{2x - 2},$$

Let us illustrate the procedure using a sample formed by 10 variates from u and the adaptation as shown in the algorithm above described, with a polynomial family and $k = 4$. The sample from u is

(continued)

Example 5.5 (continued)

```
> u
          [,1]      [,2]     [,3]      [,4]      [,5]
[1,] 0.5467892 0.3971559 0.784862 0.5158708 0.4715588
          [,6]     [,7]      [,8]      [,9]     [,10]
[1,] 0.5724601 0.978253 0.3985619 0.3755361 0.0433663
```

Starting from the initial guess `Xini` (randomly selected) below

```
> xini
          [,1]       [,2]      [,3]    [,4]      [,5]
[1,] 0.6266037 -0.7919414 -0.467521 -2.4726 0.2454318
```

The final result is `Xsol`:

```
> Xsol
           [,1]      [,2]     [,3]    [,4]    [,5]
[1,] 0.01198451 0.1547733 1.852659 -3.0259 1.90967
```

If the distribution of u is known, we can generate the CDF of the solution. For instance, knowing that the distribution of u is uniform, we obtain the CDF shown in Fig. 5.10.

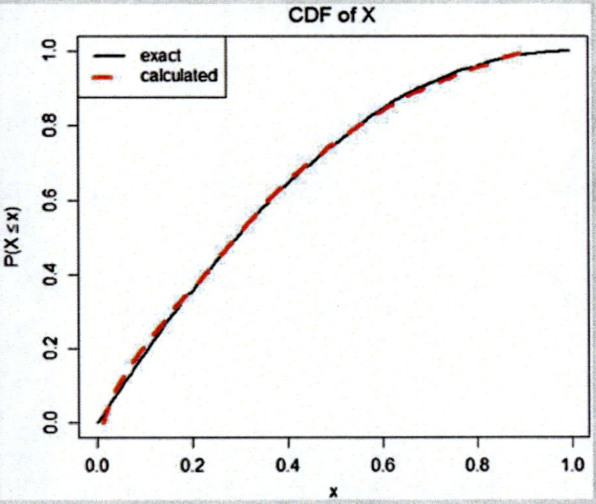

Fig. 5.10 Comparison between the exact CDF and the CDF generated by adaptation

If, in addition, we know that the support of u is (0, 1), we can generate data on this interval. In this case, the result is shown in Fig. 5.11.

(continued)

Example 5.5 (continued)

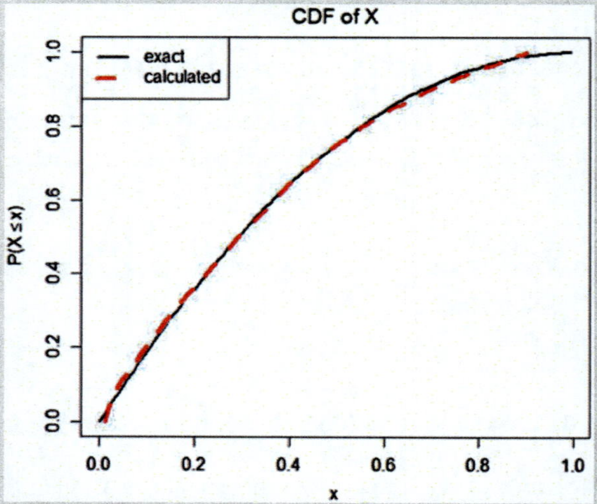

Fig. 5.11 Comparison between the exact CDF and the CDF generated by adaptation – here, the data concern u uniformly distributed on $(0, 1)$

Example 5.6

Let us consider the solution of $F(\boldsymbol{x}) = 0$ by Newton–Raphson's iterations, with

$$F(\boldsymbol{x}) = \begin{pmatrix} (1+u^2)x_1 - x_2 \\ x_1^2 + x_2^2 - (1+u^2) \end{pmatrix}, \quad DF(x) = \begin{pmatrix} 1+u^2 & -1 \\ 2x_1 & 2x_2 \end{pmatrix}.$$

The solutions are

$$x_1 = \pm\sqrt{\frac{1+u^2}{1+(1+u^2)^2}}, \quad x_2 = (1+u^2)x_1$$

(continued)

Example 5.6 (continued)

Assume that u is uniformly distributed on (1,3) and we have the sample

```
> u
         [,1]     [,2]     [,3]     [,4]     [,5]
[1,] 1.515346 2.408364 2.666862 2.556634 1.447729
         [,6]     [,7]     [,8]     [,9]    [,10]
[1,] 2.850363 1.176454 1.823339 2.729603 1.217245
        [,11]    [,12]    [,13]    [,14]    [,15]
[1,] 1.950657 2.477752 2.859946 1.323128 1.256883
        [,16]    [,17]    [,18]    [,19]    [,20]
[1,] 1.098496 1.289139 1.640641 1.233663 1.833923
```

We use as starting point (randomly chosen):

```
> Xini
           [,1]       [,2]       [,3]       [,4]
[1,] 0.05367003 -0.2541034 -1.5985338 -0.8583353
[2,] 1.04992376  0.9008442  0.7871873 -1.5041647
           [,5]      [,6]       [,7]
[1,] -0.2492248 1.1109388 -0.4067338
[2,] -0.6868230 0.7044404 -0.8625156
```

We obtain as result after 500 iterations:

```
> Xsol
          [,1]       [,2]       [,3]      [,4]
[1,] 0.6324095 -0.3775425 -0.3205309  1.023240
[2,] 1.2647823  1.7753916  0.3766401 -0.807153
          [,5]       [,6]       [,7]
[1,] -1.121169  0.6221279 -0.1439239
[2,]  1.063828 -0.7279943  0.2012620
```

The distributions of X_1 and X_2 are shown in Figs. 5.12 and 5.13.

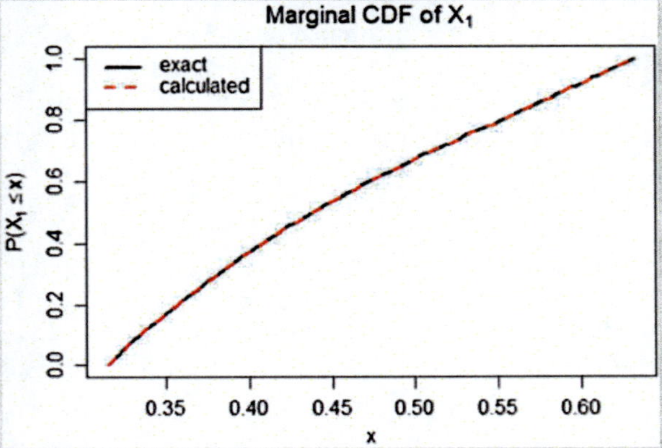

Fig. 5.12 Comparison between the exact CDF of x_1 and its CDF generated by adaptation

(continued)

Example 5.6 (continued)

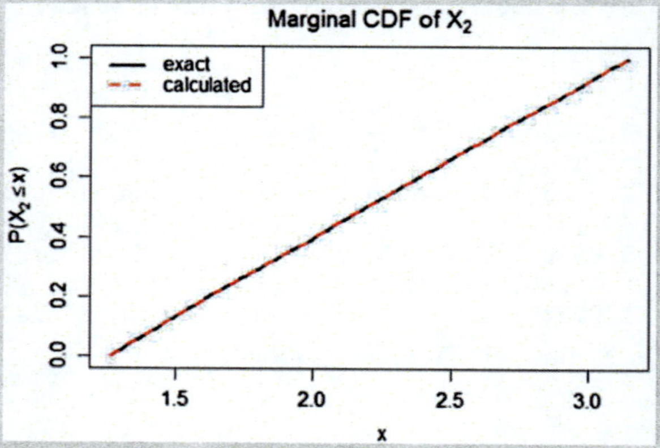

Fig. 5.13 Comparison between the exact CDF of x_2 and the CDF generated by adaptation

Example 5.7

Let us consider the solution of $F(\boldsymbol{x}) = 0$ by Newton-Raphson's iterations, with

$$F(\boldsymbol{x}) = \begin{pmatrix} x_1 - ux_2 \\ x_1^2 + x_2^2 - 1 \end{pmatrix}, \quad DF(\boldsymbol{x}) = \begin{pmatrix} 1 & -u \\ 2x_1 & 2x_2 \end{pmatrix}.$$

The solutions are

$$x_2 = \pm\sqrt{\frac{1}{1+u^2}}, \quad x_1 = ux_2$$

Assume that u is uniformly distributed on (1,3) and we have the sample

```
         [,1]     [,2]     [,3]     [,4]     [,5]
[1,] 2.477463 2.264506 1.098248 2.014669 1.462359
         [,6]     [,7]     [,8]     [,9]    [,10]
[1,] 1.168966 1.095913 2.558654 1.732113 1.722946
        [,11]    [,12]    [,13]    [,14]    [,15]
[1,] 2.179105 2.208685 1.360466 2.746472 1.258763
        [,16]    [,17]    [,18]   [,19]   [,20]
[1,] 2.430882 1.177708 2.938491 1.66744 1.29913
```

(continued)

Example 5.7 (continued)

We use as starting point (randomly chosen):

```
> xini
          [,1]       [,2]       [,3]      [,4]
[1,] -0.4866386 -0.288158 -0.4278028 0.2976413
[2,]  1.5298501 -1.210957  0.4828014 2.2833928
          [,5]      [,6]       [,7]
[1,]  0.4495766 2.240876  0.8788926
[2,] -0.5423808 1.753082 -0.6790965
> |
```

We obtain as result:

```
> xsol
          [,1]       [,2]       [,3]      [,4]
[1,] 0.7070950  0.7076596 -1.0701437 1.1412657
[2,] 0.7072085 -0.7107019  0.3964334 0.1140025
          [,5]      [,6]        [,7]
[1,] -0.8315946 0.3682475 -0.07385315
[2,] -0.4323914 0.3357569 -0.09411577
```

The distributions of X_1 and X_2 are shown in Figs. 5.14 and 5.15.

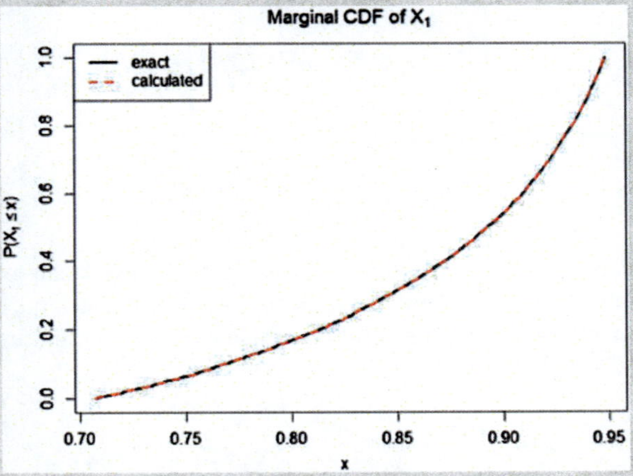

Fig. 5.14 Comparison between the exact CDF of x_1 and its CDF generated by adaptation

(continued)

Example 5.7 (continued)

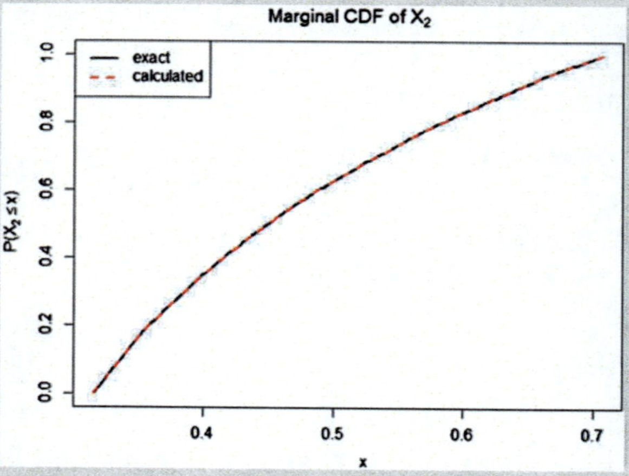

Fig. 5.15 Comparison between the exact CDF of x_2 and the CDF generated by adaptation

It is interesting to notice that practically any iterative code may be adapted by this approach. For instance, iterations from a descent method in optimization may be adapted. As an example, let us consider the iterations of the gradient descent with a fixed step $\mu > 0$.

$$\boldsymbol{X}^{(p+1)} = \boldsymbol{X}^{(p)} - \mu \nabla \boldsymbol{F}\left(\boldsymbol{X}^{(p)}, U\right) \tag{5.26}$$

These iterations correspond to

$$\Psi(\boldsymbol{x}, u) = \boldsymbol{x} - \mu \nabla \boldsymbol{F}(\boldsymbol{x}, u) \tag{5.27}$$

and may be adapted analogously to the Newton's iterations. As an example, let us consider the simple situation where $F(x, u) = (x - e^{-u})^2$, for which the exact solution is $x = e^{-u}$. Let us take u as uniformly distributed on $(-1, 1)$ and use the sample below:

```
> u
             [,1]       [,2]      [,3]      [,4]       [,5]
[1,] -0.04967899 -0.2490217 0.9093436 0.2963266 -0.8352636
           [,6]      [,7]       [,8]      [,9]     [,10]
[1,] -0.159475 -0.722855 -0.6749393 0.2547402 0.0469262
         [,11]     [,12]      [,13]      [,14]     [,15]
[1,] 0.8447855 0.3454921 -0.3945931 -0.1923467 0.4868522
         [,16]      [,17]      [,18]      [,19]     [,20]
[1,] -0.198484 -0.9670119 -0.2671101 -0.4644241 0.9028322
```

We use the polynomial family with $k = 6$ and we run 500 iterations with $\mu = 0.01$. The initial point (randomly selected) is

```
> xini
          [,1]       [,2]      [,3]       [,4]
[1,] -0.529588 -0.8569436 0.3676746 -0.6286189
           [,5]        [,6]      [,7]
[1,] -0.4558668 -0.09408194 0.9791538
```

The iterations of adaptation converge and furnish

```
> xsol
         [,1]      [,2]     [,3]      [,4]
[1,] 2.718142 -5.435973 5.430576 -3.590168
        [,5]       [,6]       [,7]
[1,] 1.71045 -0.5596861 0.09449224
```

As we see in Figs. 5.16 and 5.17, the result is close to the exact solution.

Now, let us take u as normally distributed $N(0, 1)$. We run the iterations with the same parameters: $\mu = 0.01$, and same polynomial family used in the preceding example. The sample contains 20 variates from $N(0, 1)$:

```
> u
          [,1]      [,2]      [,3]     [,4]      [,5]
[1,] -1.025688 0.1370473 0.2239263 1.096663 0.3490143
         [,6]      [,7]      [,8]      [,9]     [,10]
[1,] 1.110419 0.1811266 0.9540129 0.2440739 0.6363411
           [,11]    [,12]     [,13]     [,14]      [,15]
[1,] -0.09173458 2.012941 -1.007225 0.5771227 -0.1900536
        [,16]    [,17]     [,18]      [,19]      [,20]
[1,] 1.120953 1.426475 -1.178752 -0.6776159 -0.7424971
```

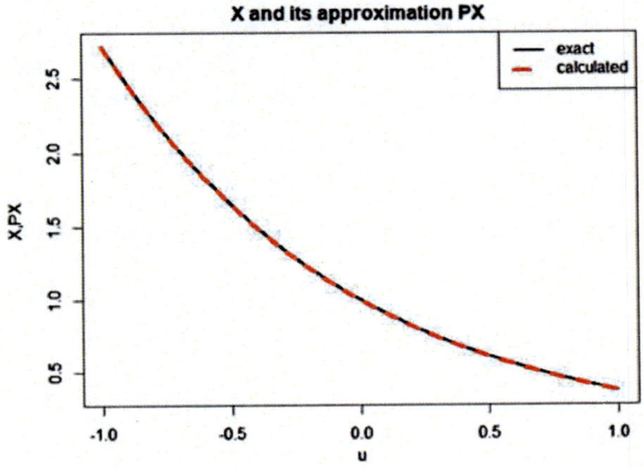

Fig. 5.16 Comparison between the exact X and its approximation PX (adaptation of gradient descent with U uniformly distributed on (−1,1)

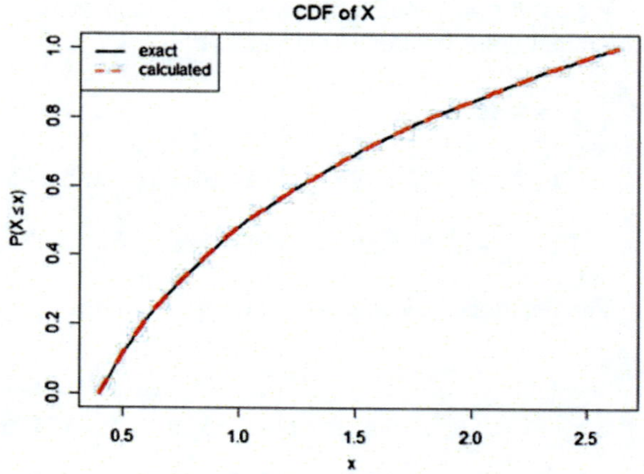

Fig. 5.17 Comparison between the exact CDF of X and the CDF of its approximation PX (adaptation of gradient descent with U uniformly distributed on (−1,1)

Adaptation iterations run 1000 iteration from a random initial point and furnish

```
> Xsol
         [,1]      [,2]     [,3]      [,4]
[1,] 2.718326 -5.436945 5.428978 -3.562407
        [,5]       [,6]       [,7]
[1,] 1.63815 -0.4874696 0.06925898
```

As we see in Figs. 5.18 and 5.19, the result is close to the exact solution.

Let us consider a final illustration of the adaptation method, to underline once again that practically any method based on recurrences can be studied by this approach: let us consider a differential equation:

$$\begin{pmatrix} x_1' \\ x_2' \end{pmatrix} = \mathbf{A}\begin{pmatrix} x_1 \\ x_2 \end{pmatrix}, \ \mathbf{A} = \begin{pmatrix} 0 & U \\ -U & 0 \end{pmatrix}, \ x_1(0) = 0, \ x_2(0) = 1, \ t \in (0, T). \tag{5.28}$$

The solution is $x_1(t) = \sin(Ut)$, $x_2(t) = \cos(Ut)$. Assume that the differential equation is discretized by Euler's method, using time steps $t_i = i\Delta t$, $0 \leq i \leq n$, $\Delta t = T/n$. Let $\boldsymbol{X}_i = (x_1(t_i), x_2(t_i))$. Then, we have

$$\boldsymbol{X}_{i+1} = \boldsymbol{\Psi}_i(\boldsymbol{X}_i, U), \tag{5.29}$$

where $\boldsymbol{\Psi}_i$ is the recurrence function associated to the method:

$$\boldsymbol{\Psi}_i(\boldsymbol{X}_i, U) = \boldsymbol{X}_i + \Delta t\, \boldsymbol{A}(U)\boldsymbol{X}_i, \tag{5.30}$$

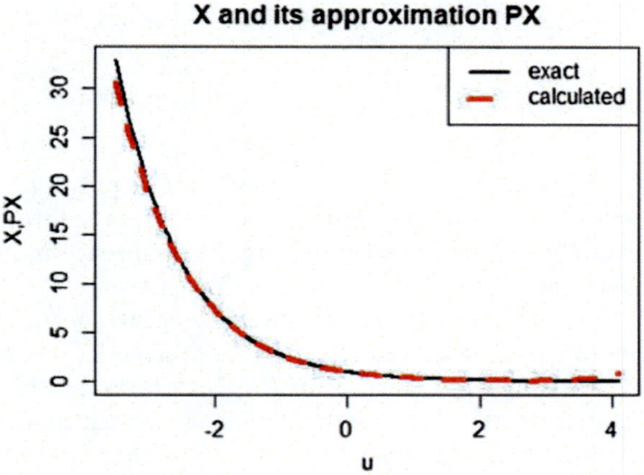

Fig. 5.18 Comparison between the exact *X* and its approximation *PX*. (Adaptation of gradient descent with *U* gaussian *N*(0,1))

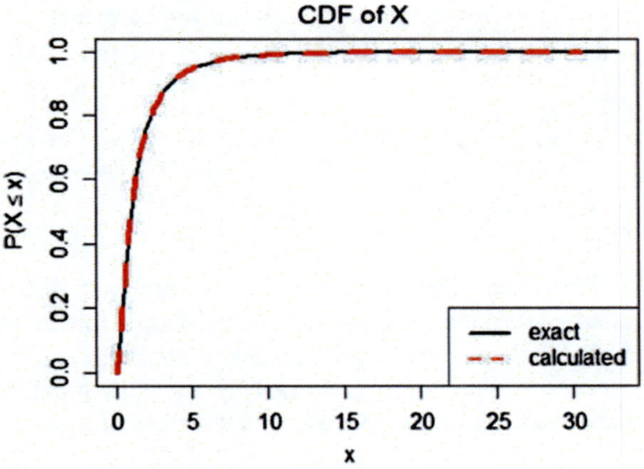

Fig. 5.19 Comparison between the exact CDF of *X* and the CDF of its approximation *PX*. (Adaptation of gradient descent with *U* gaussian *N*(0,1))

For instance, we can define the iteration function as shown at right:

```
t0 = 0
tmax = 1
nt = 1e5
dt = (tmax-t0)/nt

euler = function(X,U,F){
  aux1 = F(X,U)
  dX = dt*aux1
  v = X + dX
  return(v)
}
psi = function(X,U) {euler(X,U,F)}
```

Here, F is the right side member of the differential equation:

```
F = function(X,U){
  aux = c(U*X[[2]], -U*X[[1]])
  maux = matrix(aux, nrow = 2, ncol=1)
  return(maux)
}
```

For other Runge–Kutta methods, the expression of Ψ_n is more complex, but we do not need to know the exact expression: as previously observed, the availability of a code performing the recurrence is enough to use the adaptation (see Chap. 6).

Here, we do not look for the convergence of the iterations, but we generate a sequence of values $\boldsymbol{X}_0(U), \boldsymbol{X}_1(U), \ldots, \boldsymbol{X}_n(U)$. By the UQ approach, each $\boldsymbol{X}_i(U)$ is approximated by an expansion $\boldsymbol{PX}_i(U)$. The expansion $\boldsymbol{PX}_i(U)$ has coefficients $\boldsymbol{x}_i$ to be determined – they can be generated by adaptation of the recurrence (5.29). We can determine the values $\boldsymbol{X}_i(U)$, $1 \leq i \leq n$ using the method `progression` in class `uae`:

```
progression = function(psi,X0,nt,us,phi,nb,ndim){
  A = self$basis_sp(us,phi,nb)
  V = self$basis_matrix(u,phi,nb)
  xsol = list()
  xsol[[1]] = x0
  for (nit in (1: nt)){
    print(paste("************** nit = ", nit, "finished"))
    xsol[[nit+1]] = self$iteration_adaptation(psi,xsol[[nit]],A,V,phi,u,ndim,nb)
  }
  return(xsol)
},
```

For instance, let us assume that U is normally distributed, with a mean 2π and a standard deviation $\pi/6$. Assume that a sample formed by $ns = 20$ variates from U is available. Let us use Euler's method with different time steps and measure the error between the prevision generated by adaptation and the exact solution for all the values of U in the sample and all the values of $\boldsymbol{X}_i$, $0 \leq i \leq n$:

$$RMS = \left(\frac{1}{ns \times (n+1)} \sum_{j=1}^{ns} \sum_{i=0}^{n} \| \boldsymbol{PX}_i(U_j) - \boldsymbol{X}_i(U_j) \|^2 \right)^{\frac{1}{2}} \tag{5.31}$$

$$REL = \left(\sum_{j=1}^{ns} \sum_{i=0}^{n} \| \boldsymbol{PX}_i(U_j) - \boldsymbol{X}_i(U_j) \|^2 / \sum_{j=1}^{ns} \sum_{i=0}^{n} \| \boldsymbol{X}_i(U_j) \|^2 \right)^{\frac{1}{2}} \tag{5.32}$$

The results are shown in Table 5.11: we observe that, for a convenient time step, the adaptation furnishes results that are close to the exact ones. Figures 5.20 and 5.21 show comparisons between the components of $\boldsymbol{X}_{nt+1}(U)$ and $\boldsymbol{PX}_{nt+1}(U)$.

Table 5.11 RMS error in the adaptation of Euler's method with n time steps

n	1E2	1E3	1E4	1E5
RMS	4.8E-2	4.1E-3	4.7E-4	4.2E-5
REL	15%	1.3%	0.1%	0.01%

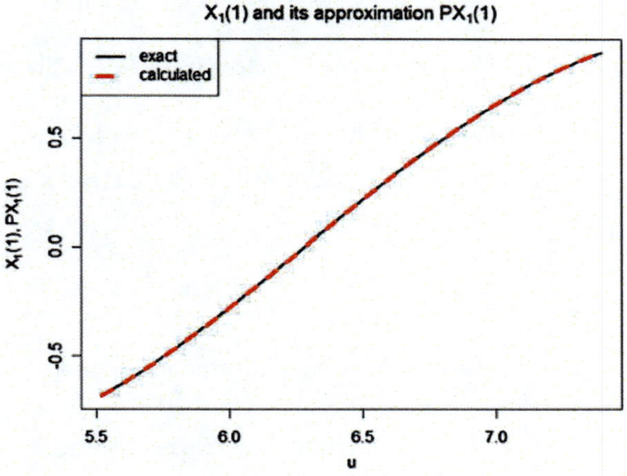

Fig. 5.20 Comparison between X_1 and its approximation PX_1 for $t = 1$. (Adaptation of Euler's method with $n = 1E4$, U gaussian $N(2\pi, \pi/6)$)

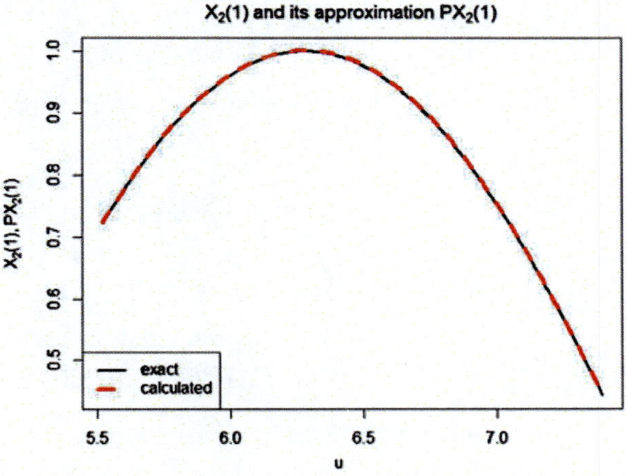

Fig. 5.21 Comparison between X_2 and its approximation PX_2 for $t = 1$. (Adaptation of Euler's method with $n = 1E4$, U gaussian $N(2\pi, \pi/6)$)

Exercises

1. Consider the adaptation of gradient descent iterations for the minimization of

$$f(\boldsymbol{x}) = \left(100 + u_1^2\right)x_1^2 + \left(50 + u_2^2\right)x_2^2 - (200 - u_1)x_1 + (50 + u_2)x_2.$$

(continued)

Assume that $u_1 \sim N(0, 0.1)$, $u_2 \sim N(0,0.2)$.

(a) Generate a sample of 10 variates from each random variable u_i.
(b) Use these values to generate a sample of 100 variates from $\boldsymbol{U} = (u_1, u_2)$.
(c) Use the sample to determine a representation of $\boldsymbol{X} = \arg\min f$ as function of $\boldsymbol{U} = (u_1, u_2)$.
(d) Determine the CDF and the PDF of $\boldsymbol{X}$.

2. Consider the adaptation of the solution by RK of the differential equations

$$\frac{d}{dt}\begin{pmatrix} x_1 \\ x_2 \end{pmatrix} = \begin{pmatrix} -u_1 & 2 \\ 4 & -u_2 \end{pmatrix}\begin{pmatrix} x_1 \\ x_2 \end{pmatrix}, \begin{pmatrix} x_1(0) \\ x_2(0) \end{pmatrix} = \begin{pmatrix} 2 \\ 5 \end{pmatrix}$$

Assume that the distributions of u_1 and u_2 are triangular: $u_1 \sim T(0, 1, 2)$ and $u_2 \sim T(0, 4, 8)$.

(a) Generate a sample of 6 variates from each random variable u_i.
(b) Use these values to generate a sample of 36 variates from $\boldsymbol{U} = (u_1, u_2)$.
(c) Use the sample to determine a representation of $\boldsymbol{X}(1) = (x_1(1), x_2(1))$ as function of $\boldsymbol{U} = (u_1, u_2)$.
(d) Determine the CDF and the PDF of $\boldsymbol{X}$.

3. Consider the adaptation of gradient descent iterations for the minimization of

$$f(\boldsymbol{x}) = (1 + u_1^2)x_1^2 + (1 + u_2^2)x_2^2 + (10 - u_1)x_1 + (15 + u_2)x_2.$$

Assume that $u_1 \sim N(0, 0.2)$, $u_2 \sim T(0, 1, 2)$.

(a) Generate a sample of 10 variates from each random variable u_i.
(b) Use these values to generate a sample of 100 variates from $\boldsymbol{U} = (u_1, u_2)$.
(c) Use the sample to determine a representation of $\boldsymbol{X} = \arg\min f$ as function of $\boldsymbol{U} = (u_1, u_2)$.
(d) Determine the CDF and the PDF of $\boldsymbol{X}$.

4. Consider the adaptation of the solution by RK of the differential equations

$$\frac{d}{dt}\begin{pmatrix} x_1 \\ x_2 \end{pmatrix} = \begin{pmatrix} 0.2 - 5u_1 & -2.5 - 5u_1 \\ 1 & 0 \end{pmatrix}\begin{pmatrix} x_1 \\ x_2 \end{pmatrix} + \begin{pmatrix} 8 \\ 3 \end{pmatrix}, \begin{pmatrix} x_1(0) \\ x_2(0) \end{pmatrix} = \begin{pmatrix} 2 \\ 5 \end{pmatrix}$$

(continued)

Assume that the distributions $u_1 \sim T(-0.1,0,0.1)$ and $u_2 \sim N(0,0.1)$.

(a) Generate a sample of 10 variates from each random variable u_i.
(b) Use these values to generate a sample of 100 variates from $\boldsymbol{U} = (u_1, u_2)$.
(c) Use the sample to determine a representation of $\boldsymbol{X}(1) = (x_1(1), x_2(1))$ as function of $\boldsymbol{U} = (u_1, u_2)$.
(d) Determine the CDF and the PDF of $\boldsymbol{X}$.
(e) Determine the CDF and the PDF of $\boldsymbol{X}$.

5. Consider the adaptation of the solution by Newton–Raphson's method of the nonlinear equations

$$F\begin{pmatrix} x_1 \\ x_2 \end{pmatrix} = \begin{pmatrix} x_1^2 + x_2^2 - u \\ x_2 - u * x_1 \end{pmatrix} = \begin{pmatrix} 0 \\ 0 \end{pmatrix}$$

Assume that $u \sim T(1,2,3)$

(a) Generate a sample of 20 variates from u.
(b) Use the sample to determine a representation of $\boldsymbol{X} = (x_1, x_2)$ as function of $\boldsymbol{U} = u$
(c) Determine the Marginal CDF and the Marginal PDF of X_1.
(d) Determine the Marginal CDF and the Marginal PDF of X_2.

6. Consider the adaptation of the solution by Newton–Raphson's method of the nonlinear equations

$$F\begin{pmatrix} x_1 \\ x_2 \end{pmatrix} = \begin{pmatrix} x_1^2 + \frac{x_2^2}{u} - 1 \\ x_2 - u * x_1 \end{pmatrix} = \begin{pmatrix} 0 \\ 0 \end{pmatrix}$$

Assume that $u \sim T(1,2,3)$

(a) Generate a sample of 40 variates from u.
(b) Use the sample to determine a representation of $\boldsymbol{X} = (x_1, x_2)$ as function of $\boldsymbol{U} = u$.
(c) Determine the Marginal CDF and the Marginal PDF of X_1.
(d) Determine the Marginal CDF and the Marginal PDF of X_2.

5.3 Iterative Evaluation of Eigenvalues

Eigenvalues and eigenvectors of matrices may be numerically determined by iterative methods. For instance, we can use Power Iterations which read as

$$X^{(p+1)} = \frac{Y^{(p)}}{\|Y^{(p)}\|}, Y^{(p)} = A(U)X^{(p)}; X^{(0)} \text{ given.} \tag{5.33}$$

For convenient $X^{(0)}$, the iterations converge to the eigenvalue of A having the largest modulus – which is said to be the dominant eigenvalue. To obtain such a convergence, the initial guess $X^{(0)}$ must have a significant projection in the direction supported by the associated eigenvector. Analogously, inverse iteration produces the eigenvalue of A having the smallest modulus:

$$X^{(p+1)} = \frac{Y^{(p)}}{\|Y^{(p)}\|}, A(U)Y^{(p)} = X^{(p)}; X^{(0)} \text{ given.} \tag{5.34}$$

Since these methods are iterative, they can be adapted as shown in the preceding section.

For instance, the code at right defines the iteration function corresponding to $A(U)$ evaluated by a function `AA(U)`.

```
puiter = function(X,AA,U){
aa = AA(U)
s = aa %*% X
v = s/norm(s,type="F")
}

psi = function(X,U) {puiter(X,matA,U)}
```

Let us consider U uniformly distributed on (0,1) and

$$A(U) = \begin{pmatrix} \cos(U) & \sin(U) \\ \sin(U) & -\cos(U) \end{pmatrix} \begin{pmatrix} 1-u & 0 \\ 0 & u \end{pmatrix} \begin{pmatrix} \cos(U) & \sin(U) \\ -\sin(U) & \cos(U) \end{pmatrix}$$

A is evaluated as follows:

```
dimX = 2

A1 = function(u){
  mat1 = matrix(c(cos(u),sin(u),-sin(u),cos(u)),
                nrow = 2, ncol = 2, byrow = TRUE)
  mat2 =   matrix(c(1-u,0,0,u),
                  nrow = 2, ncol = 2, byrow = TRUE)
  mat3 = t(mat1)
  mat = mat1 %*% mat2 %*% mat3;
  return(mat)
}
```

We consider a sample formed by 50 variates from U:

```
         [,1]      [,2]      [,3]     [,4]      [,5]      [,6]
[1,] 0.4060691 0.4661903 0.5498701 0.899938 0.5543291 0.9245256
           [,7]      [,8]     [,9]     [,10]     [,11]     [,12]
[1,] 0.01948206 0.4350821 0.432695 0.3970992 0.9723083 0.2178451
        [,13]     [,14]     [,15]     [,16]     [,17]     [,18]
[1,] 0.7370557 0.8768937 0.3043903 0.3297622 0.3555803 0.8785063
        [,19]     [,20]     [,21]     [,22]     [,23]    [,24]
[1,] 0.5441757 0.5843083 0.9439272 0.4083836 0.8521043 0.311921
        [,25]     [,26]     [,27]     [,28]     [,29]     [,30]
[1,] 0.7466468 0.2793926 0.7752472 0.4723897 0.5718907 0.6195486
        [,31]      [,32]     [,33]     [,34]     [,35]    [,36]
[1,] 0.2603154 0.08901519 0.8564545 0.3903488 0.1983744 0.264114
        [,37]     [,38]     [,39]     [,40]     [,41]     [,42]
[1,] 0.5500375 0.5290034 0.4117289 0.3702531 0.9844198 0.4206258
        [,43]     [,44]     [,45]     [,46]     [,47]      [,48]
[1,] 0.2821883 0.8731999 0.7800155 0.1478927 0.4966466 0.09749412
        [,49]     [,50]
[1,] 0.9812002 0.9385362
```

A good starting point is furnished by the eigenvectors of the mean matrix $\boldsymbol{A}$:

```
Am = A(U[[1]])
for (i in 2: length(U)){
  Am = Am + A(U[[i]])
}
Am = Am/length(U)
res = eigen(Am)
c0 = matrix(0,nrow = dimX, ncol = nb )
c0[,1] = res$vectors[,1]
```

The preceding code (algorithm of Sect. 5.2), with $k = 5$, furnishes the results shown in Figs. 5.22 and 5.23.

Fig. 5.22 Dominating eigenvalue and its approximation

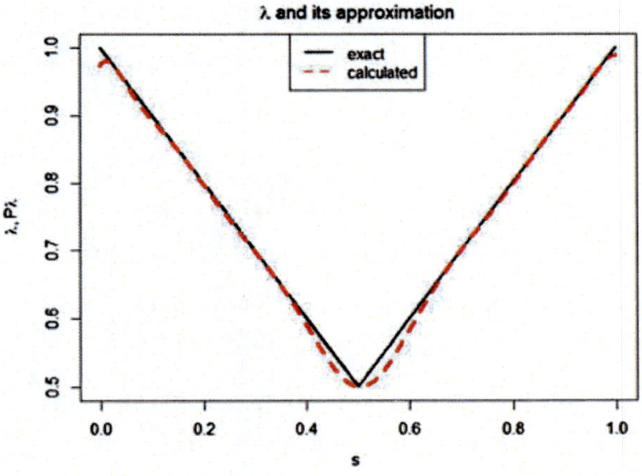

Fig. 5.23 CDF of the dominating eigenvalue and its approximation

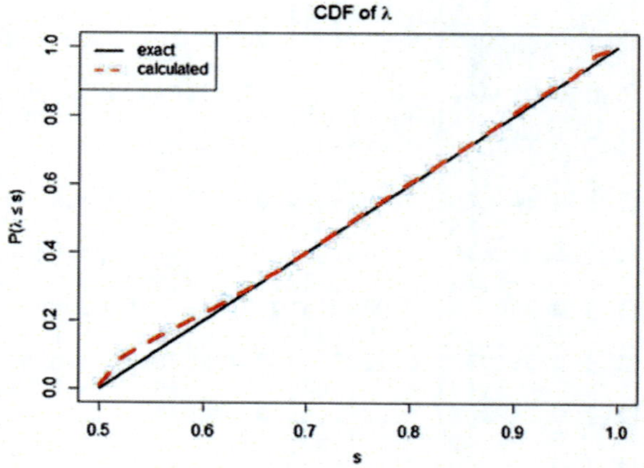

For inverse power iterations, we can use the code at right to define the iteration function. $\boldsymbol{A(U)}$ is evaluated by a function `AA(U)`.

```
inv_power_iterate = function(A,xo){
  y = solve(A , xo);
  xn = y/self$norm2(y);
  return(xn)
},
```

With $k = 5$, the preceding code furnishes the results shown in Figs. 5.24 and 5.25. As a second example, let us determine the distribution of the dominant eigenvalue of

$$\boldsymbol{A}(U) = \begin{pmatrix} \cos(U) & \sin(U) \\ \sin(U) & -\cos(U) \end{pmatrix} \begin{pmatrix} |\sin(U)| & 0 \\ 0 & |\cos(U)| \end{pmatrix} \begin{pmatrix} \cos(U) & \sin(U) \\ -\sin(U) & \cos(U) \end{pmatrix}$$

with $U \sim N(0, \pi/10)$. We consider as given a sample of 50 variates from U:

```
> U
          [,1]       [,2]       [,3]       [,4]       [,5]
[1,] -0.7470849 -0.5005908 -0.4890769 -0.4828904 -0.4643787
          [,6]       [,7]       [,8]       [,9]      [,10]
[1,] -0.4087309 -0.4007597 -0.4006056 -0.3908878 -0.3659523
         [,11]      [,12]     [,13]     [,14]     [,15]
[1,] -0.3261246 -0.2983123 -0.293623 -0.283218 -0.270673
         [,16]      [,17]      [,18]      [,19]      [,20]
[1,] -0.2541986 -0.2457912 -0.1703693 -0.1542173 -0.1540044
         [,21]      [,22]      [,23]      [,24]       [,25]
[1,] -0.1405787 -0.1323396 -0.1202506 -0.0827001 -0.07129789
          [,26]       [,27]        [,28]        [,29]
[1,] -0.03946619 -0.03142431 -0.008793515 -0.004452953
           [,30]      [,31]      [,32]      [,33]      [,34]
[1,] -0.004016822 0.01005755 0.01948478 0.05892401 0.06587436
          [,35]      [,36]    [,37]   [,38]     [,39]
[1,] 0.06780167 0.07895381 0.110299 0.14691 0.1516545
         [,40]     [,41]     [,42]    [,43]    [,44]     [,45]
[1,] 0.1521514 0.1691469 0.1747498 0.179634 0.209819 0.2148644
         [,46]     [,47]     [,48]     [,49]     [,50]
[1,] 0.2213383 0.2362803 0.3605964 0.3933454 0.8094618
> |
```

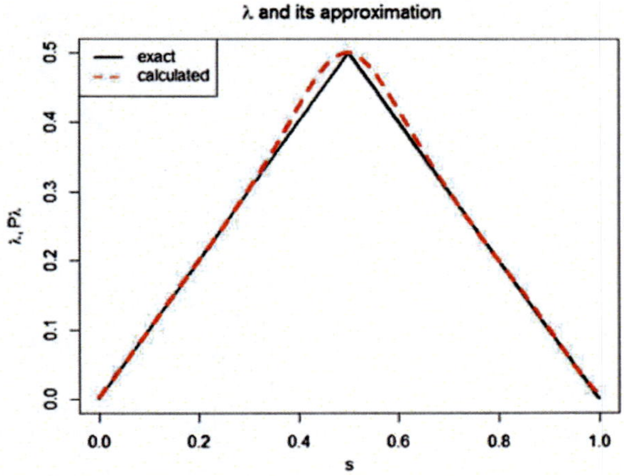

Fig. 5.24 Smallest eigenvalue and its approximation

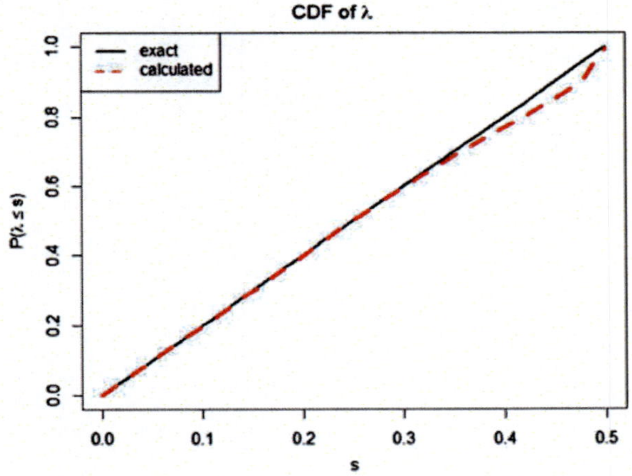

Fig. 5.25 CDF of the smallest eigenvalue and its approximation

Using $k = 5$ and a polynomial family, we got the results shown in Figs. 5.26, 5.27, and 5.28. The approximated CDF was evaluated from a large sample of 1E4 variates from $N(0, \pi/10)$, using the polynomial expansion with the coefficients determined by the adapted power iteration method. The PDF was determined by SPH derivation of the CDF. The coefficients obtained were:

```
           [,1]       [,2]      [,3]      [,4]      [,5]      [,6]
[1,] -0.4356481   8.201576 -23.84125  10.44298  24.20046 -18.76206
[2,]  0.5368746 -10.348035  33.93080 -90.59122 132.73863 -67.03419
```

Inverse Power Iterations produced the results shown in Figs. 5.29, 5.30, and 5.31.

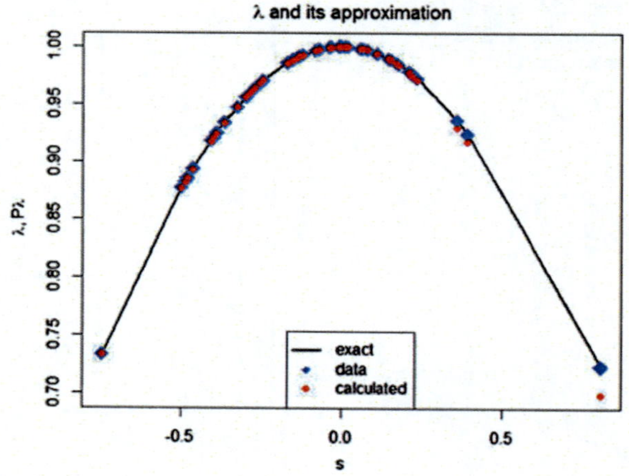

Fig. 5.26 Dominant eigenvalue and its approximation

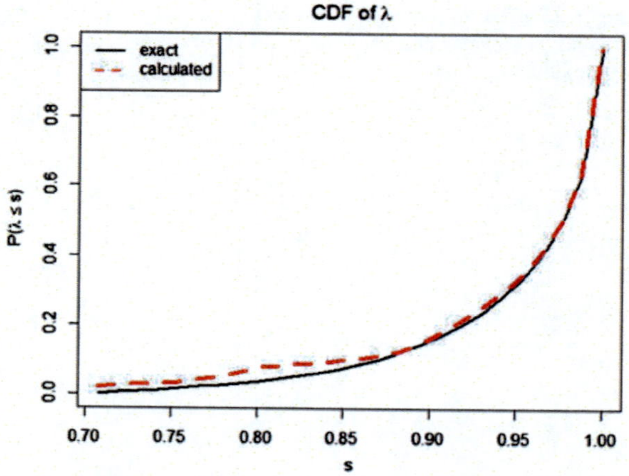

Fig. 5.27 CDF of the dominant eigenvalue and its approximation

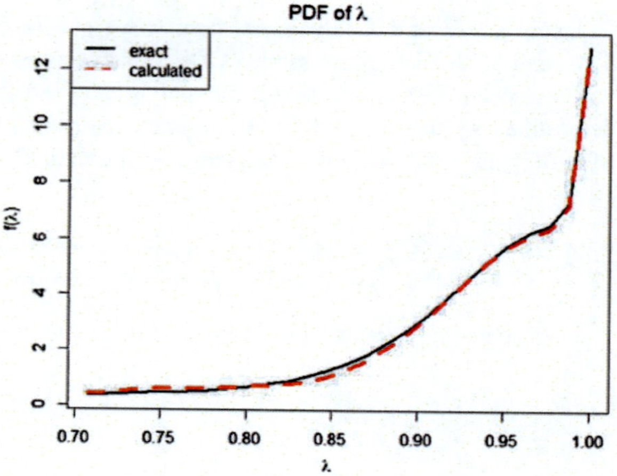

Fig. 5.28 PDF of the dominant eigenvalue and its approximation

Fig. 5.29 Smallest eigenvalue and its approximation

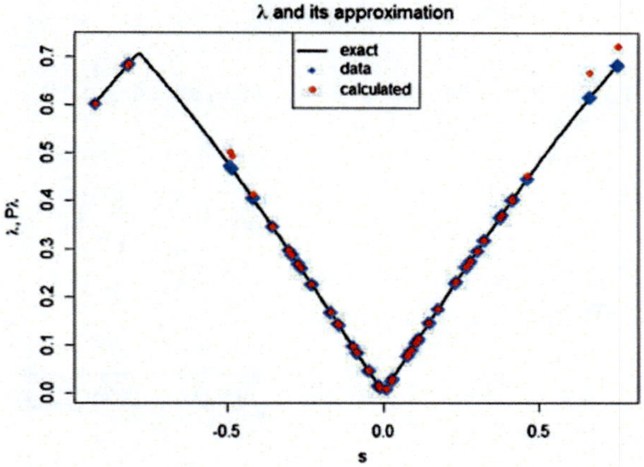

Fig. 5.30 CDF of the smallest eigenvalue and its approximation

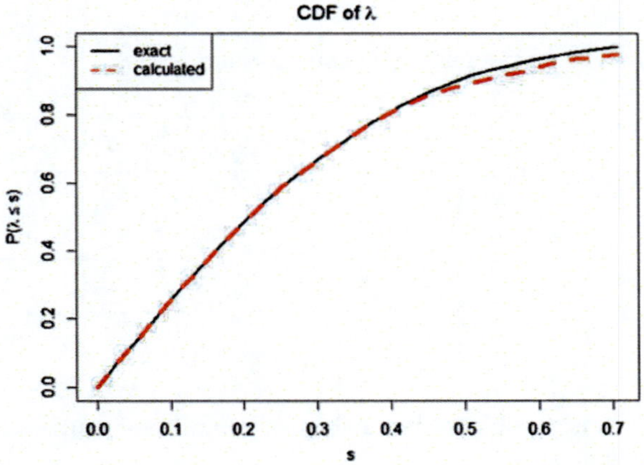

Fig. 5.31 CDF of the smallest eigenvalue and its approximation

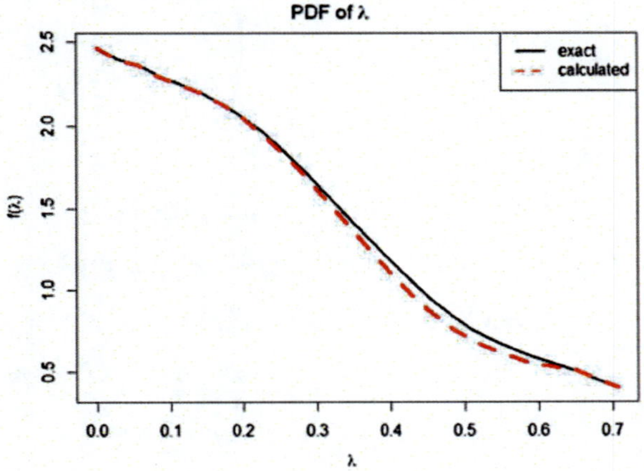

5.3.1 Very Small Matrices

If your matrix has dimension 2, you can use a closed formula for the eigenvectors and eigenvalues: let

$$A = \begin{pmatrix} X^1 & Y^1 \\ X^2 & Y^2 \end{pmatrix}.$$

Then, the eigenvalues are

$$\lambda_1 = \frac{1}{2}\left(X^1 + Y^2 - \sqrt{\left(X^1\right)^2 + 4X^2Y^1 - 2X^1Y^2 + \left(Y^2\right)^2}\right),$$

$$\lambda_2 = \frac{1}{2}\left(X^1 + Y^2 + \sqrt{\left(X^1\right)^2 + 4X^2Y^1 - 2X^1Y^2 + \left(Y^2\right)^2}\right).$$

The corresponding eigenvectors are

$$v_1 = \begin{pmatrix} -\dfrac{-X^1 + Y^2 + \sqrt{\left(X^1\right)^2 + 4X^2Y^1 - 2X^1Y^2 + \left(Y^2\right)^2}}{2X^2} \\ 1 \end{pmatrix},$$

$$v_2 = \begin{pmatrix} -\dfrac{-X^1 + Y^2 - \sqrt{\left(X^1\right)^2 + 4X^2Y^1 - 2X^1Y^2 + \left(Y^2\right)^2}}{2X^2} \\ 1 \end{pmatrix}.$$

It is possible to give analogous formulae for matrices of order 3, but they are more complex: let

$$A = \begin{pmatrix} X^1 & Y^1 & Z^1 \\ X^2 & Y^2 & Z^2 \\ X^3 & Y^3 & Z^3 \end{pmatrix}.$$

Set

$$a = -X^1 - Y^2 - Z^3,$$

$$b = -X^2Y^1 + X^1Y^2 - X^3Z^1 - Y^3Z^2 + X^1Z^3 + Y^2Z^3,$$

$$c = X^3Y^2Z^1 - X^2Y^3Z^1 - X^3Y^1Z^2 + X^1Y^3Z^2 + X^2Y^1Z^3 - X^1Y^2Z^3,$$

$$p = b - \frac{a^3}{3}, q = c + \frac{2a^3}{27} - \frac{ab}{3},$$

$$\Delta = \frac{4}{27}p^3 + q^2,\ D_+ = \frac{1}{2}\left(-q + \sqrt{\Delta}\right),\ D_- = \frac{1}{2}\left(-q - \sqrt{\Delta}\right),$$

$$\bar{z} = -\frac{1}{2} + i\frac{\sqrt{3}}{2}.$$

Then, the eigenvalues are

$$\lambda_1 = D_+^{\frac{1}{3}} + D_-^{\frac{1}{3}},\ \lambda_2 = \bar{z}D_+^{\frac{1}{3}} + \bar{z}^2 D_-^{\frac{1}{3}},\ \lambda_3 = \bar{z}^2 D_+^{\frac{1}{3}} + \bar{z}D_-^{\frac{1}{3}}.$$

The eigenvector associated with the eigenvalue λ is

$$v = \begin{pmatrix} -\lambda Z^1 + Y^2 Z^1 - Y^1 Z^2 \\ X^2 Z^1 + \lambda Z^2 - X^1 Z^2 \\ \lambda^2 - \lambda\left(X^1 + X^2\right) - X^2 Y^1 + X^1 Y^2 \end{pmatrix}.$$

This expression assumes that at least one of the lines of v is non-null. Alternative expressions exist for the situation where this assumption is not verified: for instance, we can replace ("→" means "replaced by") $X^1 \to X^1$, $Y^1 \to Z^1$, $X^2 \to X^3$, $Y^2 \to Z^3$, $Z^1 \to Y^1$, $Z^2 \to Y^3$ or $X^1 \to Y^2$, $Y^1 \to Z^2$, $X^2 \to Y^3$, $Y^2 \to Z^3$, $Z^1 \to X^2$, $Z^2 \to X^3$.

Exercises

1. Consider the matrix

$$A = \begin{pmatrix} 0.7 + u_1 & -0.5 - u_1 & -0.3 \\ -0.2 & 0.8 + u_2 & -0.3 - u_2 \\ -0.4 & -0.2 & 0.7 \end{pmatrix}$$

Assume that all the variables u_i are independent and triangularly distributed $T(0,0.2,0.5)$.

(a) Generate a sample of 10 variates from each random variable u_i.
(b) Use the values generate to generate a sample of 100 variates from $U = (u_1, u_2)$.
(c) Use the sample from to determine a representation of the dominant eigenvalue of A as function of $U = (u_1, u_2)$.
(d) Determine the Marginal CDF and PDF of the dominant eigenvalue.

2. Consider

$$A = \begin{pmatrix} u_1 & 4 \\ 1 & u_2 \end{pmatrix}$$

Assume that the distributions of u_1 and u_2 are triangular: $u_1 \sim T(1, 2, 3)$ and $u_2 \sim T(0, 5, 10)$.

(continued)

(a) Generate a sample of 10 variates from each random variable u_i.
(b) Use these values to generate a sample of 100 variates from $\boldsymbol{U} = (u_1, u_2)$.
(c) Use the sample from to determine a representation of the eigenvalues of $\boldsymbol{A}$ as function of $\boldsymbol{U} = (u_1, u_2)$.
(d) Determine the Marginal CDF and PDF of the dominant eigenvalue.

3. Consider

$$A = \begin{pmatrix} 0.2 + u_1 & u_1 \\ u_2 & -10 \end{pmatrix}$$

Assume that the distributions of u_1 and u_2 are triangular: $u_1 \sim T(1, 2, 3)$ and $u_2 \sim T(0, 1, 2)$.

(a) Generate a sample of 10 variates from each random variable u_i.
(b) Use these values to generate a sample of 100 variates from $\boldsymbol{U} = (u_1, u_2)$.
(c) Use the sample from to determine a representation of the eigenvalues of $\boldsymbol{A}$ as function of $\boldsymbol{U} = (u_1, u_2)$.
(d) Determine the Marginal CDF and PDF of the eigenvalues.

4. Consider

$$A = \begin{pmatrix} u_1 & u_2 \\ u_2 & u_1 \end{pmatrix}$$

Assume that the distributions of u_1 and u_2 are triangular: $u_1 \sim T(2, 3, 4)$ and $u_2 \sim T(0, 1, 2)$.

(a) Generate a sample of 10 variates from each random variable u_i.
(b) Use these values to generate a sample of 100 variates from $\boldsymbol{U} = (u_1, u_2)$.
(c) Use the sample from to determine a representation of the eigenvalues of $\boldsymbol{A}$ as function of $\boldsymbol{U} = (u_1, u_2)$.
(d) Determine the Marginal CDF and PDF of the eigenvalues.

5.4 The Variational Approach for Uncertain Algebraic Equations

The variational approach was introduced in Sect. 3.3. We observed that a nonlinear relation between $\boldsymbol{X}$ and $\boldsymbol{U}$ can be exploited to determine the expansion $\boldsymbol{PX}$. The method presented extends straightly to nonlinear equations

Let us consider again Eq. (5.16) involving a random variable $\boldsymbol{U}$:

$$f(\boldsymbol{X}, \boldsymbol{U}) = 0$$

Analogously to Sect. 5.1, we can consider the approximation introduced in Eq. (5.8):

$$\boldsymbol{X} \approx \boldsymbol{PX} = \sum_{j=1}^{k} \boldsymbol{x}_j \varphi_j(\boldsymbol{U}), \boldsymbol{x}_j = \left(x_{1,j}, \ \ldots, x_{n,j}\right)^t.$$

Then,

$$f(\boldsymbol{PX}(\boldsymbol{U}), \boldsymbol{U}) \approx 0,$$

so that

$$\boldsymbol{E}(\varphi_i(\boldsymbol{U})\boldsymbol{f}(\boldsymbol{PX}(\boldsymbol{U}), \ \boldsymbol{U})) \approx 0, 1 \leq i \leq k. \tag{5.35}$$

Equation (5.35) can be solved to determine the coefficients $\boldsymbol{x}_j$, $1 \leq j \leq k$. For instance, we can use `fzero`, `fminunc`, or `lsqnonlin` from `pracma` to find a solution.

To do this, we need to generate Eq. (5.35) as shown at right. `he` is a Hilbert Expansion object using basis $\{\varphi_j : 1 \leq j \leq nb\}$, `cx` a is a $n \times nb$ matrix containing the coefficients $\boldsymbol{x}_j$ (`cx[[i,j]]` $= x_{i,j}$).

```
eqs = function(cx,u,f,he,nb,ndim){
  px = matrix(NA,nrow = ndim, ncol = length(u))
  for (i in 1: ndim){
    px[i,] = he$proj(cx[i,],u)
  }
  vf = matrix(nrow = ndim,ncol = length(u))
  for (r in 1:length(u)){
    vf[,r]=f(px[,r],u[[r]])
  }
  eq = vector(length = nb*ndim)
  nn = 0
  for (i in 1:nb){
    vphi = he1$value(i-1,u)
    for (j in 1: ndim){
      nn = nn+1
      eq[[nn]]=mean(vphi*vf[j,])
    }
  }
  aux = as.matrix(eq)
  return(aux)
}
```

We must solve `eqs` $= 0$ for the unknown `cx`. However, the functions of pracma receive vectors as arguments: we need to transform a matrix into a vector and conversely: function `torow` transforms a matrix into a row vector and function `tomat` performs the inverse transformation:

```
torow = function(matcx,ndim,nb){
  aux = matrix(NA, nrow = 1, ncol = ndim*nb)
  for (i in 1: ndim){
    i1 = (i-1)*nb + 1
    i2 = i1 + nb - 1
    aux[i1:i2] = matcx[i, ]
  }
  return(aux)
}
```

```
tomat = function(vecx,ndim,nb){
  aux = matrix(NA, nrow = ndim, ncol =nb)
  for (i in 1: ndim){
    i1 = (i-1)*nb + 1
    i2 = i1 + nb - 1
    aux[i, ] = vecx[i1:i2]
  }
  return(aux)
}
```

Then, we shall solve `fvec = 0` for the unknown `vecx`, with `fvec` at right :

```
fvec = function(vecx,u,f,he,nb,ndim){
  cx = tomat(vecx,ndim,nb)
  aux = eqs(cx,u,f,he,nb)
  return(aux)
}
```

Example 5.8
Let us consider the equation $x^2 - 2x + u = 0$, where u is uniformly distributed on (, 1), solved in Example 5.5. The equation has two solutions for each u: $x_+ = 1 + \sqrt{1-u}$ and $x_- = 1 - \sqrt{1-u}$ and

$$P(x_- < a) = 1 - (a-1)^2, 0 < a < 1; P(x_+ < a) = (a-1)^2, 1 < a < 2;$$

Let us illustrate the procedure using a sample formed by 10 variates from an uniformly distributed variable u

```
u1 = c(0.110263, 0.818171, 0.218055, -0.428266, -0.521897,
       0.972777, 0.498947, -0.063814, 0.439192, 0.899228)
u1s = sort(u1)
```

We use the algorithm above described, with a polynomial family and $k = 5$:

```
k1 = 4
nb1 = k1+1
a1 = min(u1s)
b1 = max(u1s)
z1 = 1e-3
phi1 = polynom$new(a1,b1,z1)
he1 = expansion1Dmod$new(phi1)


fobj = function(vx1){fvec(vx1,u1,f1,he1,nb1,nd1)}
```

We must solve `fobj(vx) = 0`. For instance, use `lsqnonlin` with a starting point (1, 1, . . ., 1):

```
x0a = as.matrix(rep(1,nb1*nd1),nrow = 1)
library(pracma)
sol = lsqnonlin(fobj,x0a)
xpx = tomat(sol$x, nd1, nb1)
```

(continued)

Example 5.8 (continued)

The iterations converge to the coefficients of the expansion of x_+, as shown in Fig. 5.32.

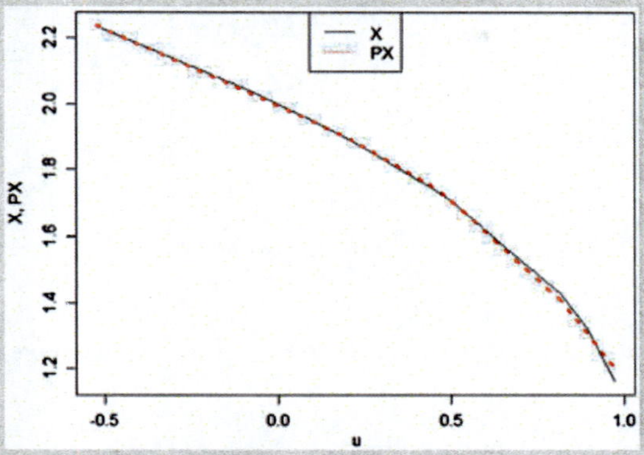

Fig. 5.32 Comparison between the expansion and the data

Since the variable is uniform, we can generate a large sample on the interval $(\min u1, \max u1)$ and use the expansion to determine the approximated CDF of X. The result is shown in Fig. 5.33.

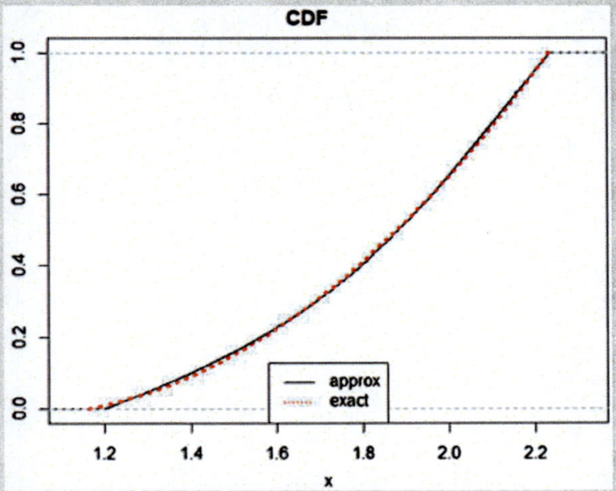

Fig. 5.33 Comparison between the exact CDF and the CDF generated by the expansion

Example 5.9
Let us consider the solution of $F(x) = 0$, with

$$F(\boldsymbol{x}) = \begin{pmatrix} (1+u^2)x_1 - x_2 \\ x_1^2 + x_2^2 - (1+u^2) \end{pmatrix}, DF(x) = \begin{pmatrix} 1+u^2 & -1 \\ 2x_1 & 2x_2 \end{pmatrix}.$$

The solutions are

$$x_1 = \pm\sqrt{\frac{1+u^2}{1+(1+u^2)^2}}, x_2 = (1+u^2)x_1$$

Assume that u is uniformly distributed on (1,3) and we have the sample

```
2.878982 2.119797 2.263830 2.080316 2.524269
2.587010 2.611016 2.770770 2.831148 2.881682
```

We use a starting point randomly chosen:

```
x0a = as.matrix(runif(nb1*nd1),ncol = 1)
library(pracma)
sol = lsqnonlin(fobj,x0a)
xpx = tomat(sol$x, nd1, nb1)
```

Examples of results are shown in Figs. 5.34 and 5.35.

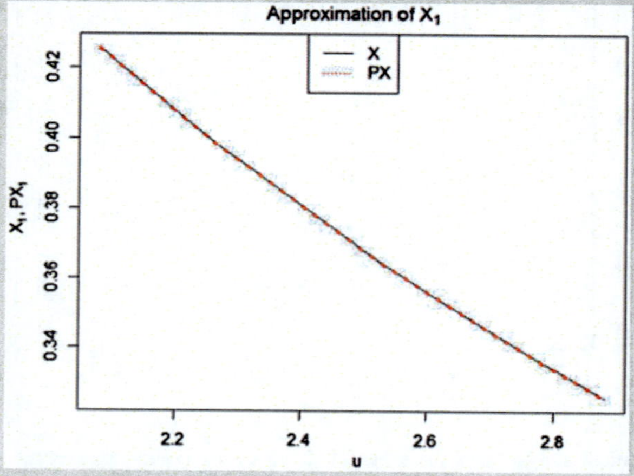

Fig. 5.34 Comparison between the expansion and the data (values of x_1)

(continued)

Example 5.9 (continued)

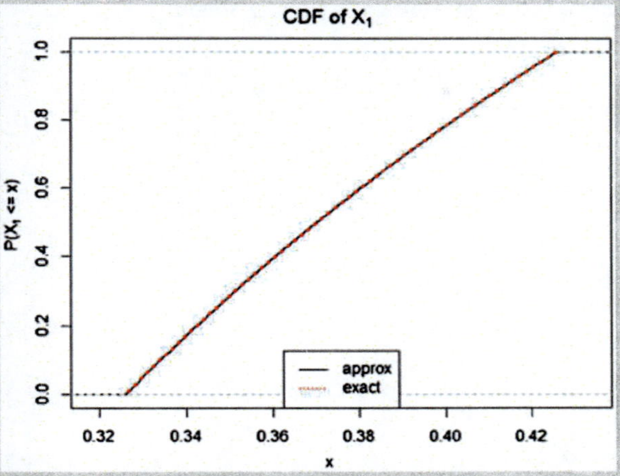

Fig. 5.35 Comparison between the exact CDF of x_1 and the CDF generated by the expansion

Example 5.10

Let us consider the solution of $F(\boldsymbol{x}) = 0$ by Newton–Raphson's iterations, with

$$F(\boldsymbol{x}) = \begin{pmatrix} x_1 - ux_2 \\ x_1^2 + x_2^2 - 1 \end{pmatrix}.$$

The solutions are

$$x_2 = \pm\sqrt{\frac{1}{1+u^2}}x_1 = ux_2$$

Assume that u is uniformly distributed on (1,3) and we have the sample

```
2.343613 2.938806 2.592832 2.757182 2.384299
2.661308 2.811384 2.040231 2.825743 2.764893
```

We use a starting point randomly chosen:

```
x0a = as.matrix(runif(nb1*nd1),ncol = 1)
library(pracma)
sol = lsqnonlin(fobj,x0a)
xpx = tomat(sol$x, nd1, nb1)
```

(continued)

Example 5.10 (continued)

Examples of results are shown in Figs. 5.36 and 5.37.

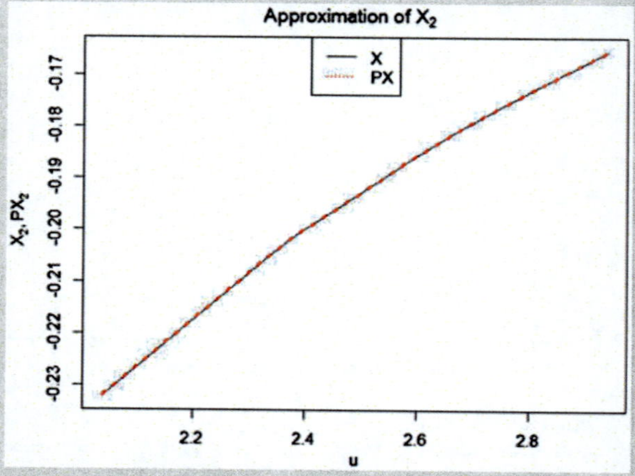

Fig. 5.36 Comparison between the expansion and the data (values of x_2)

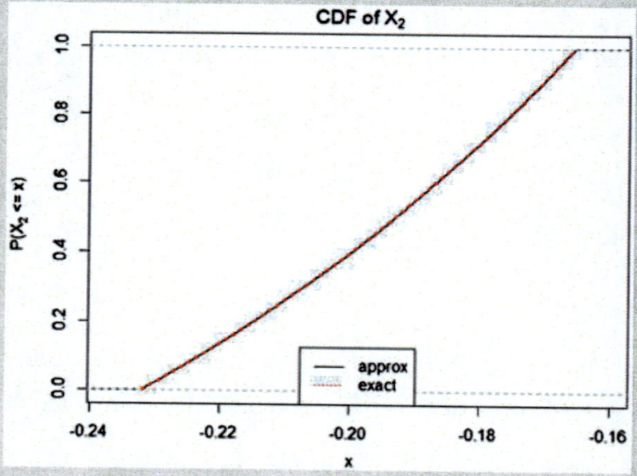

Fig. 5.37 Comparison between the exact CDF of x_2 and the CDF generated by the expansion

Exercises

1. Consider the variational solution of the nonlinear equations

$$F\begin{pmatrix} x_1 \\ x_2 \end{pmatrix} = \begin{pmatrix} x_1^2 + x_2^2 - u \\ x_2 - u * x_1 \end{pmatrix} = \begin{pmatrix} 0 \\ 0 \end{pmatrix}$$

Assume that $u \sim T(1, 2, 3)$

(e) Generate a sample of 20 variates from u.
(f) Use the sample to determine a representation of $\boldsymbol{X} = (x_1, x_2)$ as function of $\boldsymbol{U} = u$
(g) Determine the Marginal CDF and the Marginal PDF of X_1.
(h) Determine the Marginal CDF and the Marginal PDF of X_2.
(i) Consider the variational solution of the nonlinear equations

$$F\begin{pmatrix} x_1 \\ x_2 \end{pmatrix} = \begin{pmatrix} x_1^2 + \frac{x_2^2}{u} - 1 \\ x_2 - u * x_1 \end{pmatrix} = \begin{pmatrix} 0 \\ 0 \end{pmatrix}$$

Assume that $u \sim T(1, 2, 3)$

(c) Generate a sample of 40 variates from u.
(d) Use the sample to determine a representation of $\boldsymbol{X} = (x_1, x_2)$ as function of $\boldsymbol{U} = u$.
(e) Determine the Marginal CDF and the Marginal PDF of X_1.
(f) Determine the Marginal CDF and the Marginal PDF of X_2.

Chapter 6
Random Differential Equations

Abstract In this chapter, we examine methods for the determination of the probability distributions of random differential equations. We present also methods for the analysis of orbits and trajectories under uncertainty.

Differential are equations, so that the methods used for equations may be used – at least in principle.

Nevertheless, differential equations have an essential characteristic: the unknowns are fields – i.e., functions, which are elements belonging to infinitely dimensional vector spaces, contrarily to the equations manipulated in the preceding chapters, whose unknowns are vectors from $\mathbb{R}^n$, which is finite dimensional. We shall analyze the consequences of passing from finite dimension to infinite dimension in the sequel (Sect. 6.4).

As seen in Sect. 1.12, differential equations are generally solved by discretization, using packages where the classical methods of discretization are available. Notice that the discretization of a differential equations brings the unknowns to a finite dimensional space. Consequently, as shown in Sect. 5.2, we can apply the UQ approach of adaptation: the recurrence equations defining successive values generated may be adapted to generate the coefficients of the expansions of the discretized values of solution of the differential equation (see Eqs. (5.29) and (5.30)). For instance, we can use the method `progression` in class `uae`:

```
progression = function(psi,X0,nt,us,phi,nb,ndim){
  A = self$basis_sp(us,phi,nb)
  V = self$basis_matrix(u,phi,nb)
  Xsol = list()
  Xsol[[1]] = X0
  for (nit in (1: nt)){
    print(paste("*************** nit = ", nit, "finished"))
    Xsol[[nit+1]] = self$iteration_adaptation(psi,Xsol[[nit]],A,V,phi,u,ndim,nb)
  }
  return(Xsol)
}
```

Supplementary Information The online version of this chapter (https://doi.org/10.1007/978-3-031-17785-9_6) contains supplementary material, which is available to authorized users.

E. Souza de Cursi, *Uncertainty Quantification using R*, International Series in Operations Research & Management Science 335,
https://doi.org/10.1007/978-3-031-17785-9_6

When considering an iteration function `psi`, this method generates a list `Xsol` such that `Xsol[[i]]` contains the coefficients the expansion of $PX_i(U)$, for $i = 1, \ldots, nt$, where $X_i(U) = X(t_i, U)$. The initial coefficients are `X0`, corresponding to $PX_0(U)$. The Hilbert basis used is `phi` and $nb = k + 1$ is the number of coefficients for each component of PX_i. `ndim` is the dimension of X_i and `us` is a matrix having as columns the variates from U. As observed in Chap. 5, the discretization of an ordinary differential equation

$$\frac{dX}{dt} = f(t, X), X(0) = X_0$$

corresponds to iterations

$$X_{i+1} = \Psi_i\,(X_i, U), \tag{6.1}$$

which can be adapted to furnish the coefficients of PX_i, for $i = 1, \ldots, nt$. For instance, let us consider the example (5.28):

$$\begin{pmatrix} x_1' \\ x_2' \end{pmatrix} = \mathbf{A}\begin{pmatrix} x_1 \\ x_2 \end{pmatrix}, \mathbf{A} = \begin{pmatrix} 0 & U \\ -U & 0 \end{pmatrix}, x_1(0) = 0, x_2(0) = 1, \ t \in (0, T).$$

We start by defining the second member, as shown at right:

```
F = function(t,X,U){
    aux = c(U*X[[2]], -U*X[[1]])
    maux = matrix(aux, nrow = 2, ncol=1)
    return(list(maux))
}

ndim = 2
```

Then, we define a function that performs one step of a method of solution. Here, we use the method RK4.

```
RKStep = function(X,U,dt){
  tc = c(0,dt)
  f = function(t,X,lU){F(t,X,lU[[1]])}
  resu <- ode(c(X), tc, f,list(U) ,method = "rk4")
  nline = dim(resu)[[1]]
  return(matrix(resu[nline,2:3],ncol = 1))
}
```

Finally, we define the iteration function `psi`:

```
t0 = 0
tmax = 1
nt = 1e3
dt = (tmax-t0)/nt
psi = function(X,U) {RKStep(X,U,dt)}
```

Table 6.1 RMS error in the adaptation of RK4 with n time steps

n	1E2	1E3	1E4	1E5
RMS	3E-7	1E-7	3E-8	1E-8
REL	0.01%	0.001%	0.001%	0.001%

Let us consider a sample of 20 variates from $U \sim N(2\pi, \pi/6)$ and a polynomial basis with $k = 6$. The initial data $X_0 = (0, 1)^t$ correspond to

$$X0 = \begin{pmatrix} 0 & 0 & 0 & 0 & 0 & 0 & 0 & 0 \\ 1 & 0 & 0 & 0 & 0 & 0 & 0 & 0 \end{pmatrix}.$$

The method progression furnishes the results in Table 6.1 (to be compared with Table 5.11).

As a second example, let us consider the following ordinary differential equation, with $U \sim N(0, 1)$:

$$\frac{\mathrm{d}}{\mathrm{d}t}\begin{pmatrix} x_1 \\ x_2 \end{pmatrix} = \begin{pmatrix} U\sqrt{x_2} \\ 2Ux_1^2 \end{pmatrix}, x_1(0) = 1, x_2(0) = 1, \ t \in (0, T). \tag{6.2}$$

The solution is $x_1(t) = e^{Ut}, x_2(t) = e^{2Ut}$. Let us use the adaptation of the Dormand–Prince's method, with $n = 1000$, $T = 1$. In this case,

```
F = function(t,x,U){
    aux = c(U*sqrt(x[[2]]), 2*U*x[[1]]^2)
    maux = matrix(aux, nrow = 2, ncol=1)
    return(list(maux))
}

ndim = 2

RKStep = function(x,U,dt){
  tc = c(0,dt)
  f = function(t,x,lU){F(t,x,lU[[1]])}
  resu <- ode(c(x), tc, f,list(U) ,method = "ode45")
  nline = dim(resu)[[1]]
  return(matrix(resu[nline,2:3],ncol = 1))
}
```

and $\mathrm{d}t = 1E-2$, $nt = 1E2$. We use a sample of 20 variates from U. The RMS error was 5E-3, corresponding to a relative RMS error of 0.1%. An example of result is shown in Figs. 6.1 and 6.2.

In addition to adaptation, there are also specific methods for differential equations. We shall examine some of them in the next sections.

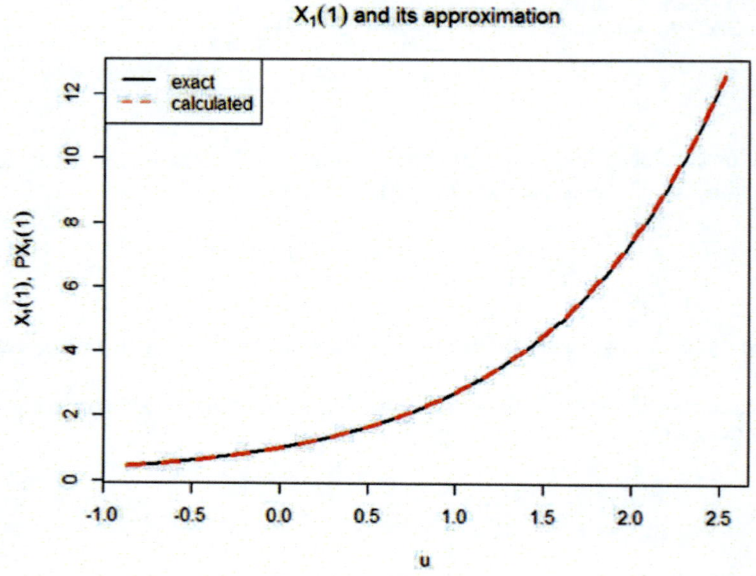

Fig. 6.1 Comparison between X_1 and its approximation PX_1 for $t = 1$. (Adaptation of Dormand–Prince's method with $n = 1E2$, U gaussian $N(0, 1)$)

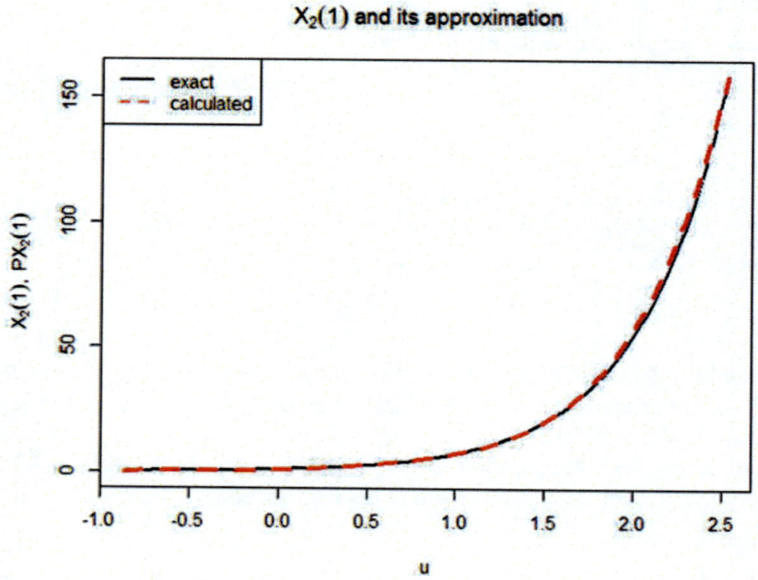

Fig. 6.2 Comparison between X_2 and its approximation PX_2 for $t = 1$. (Adaptation of Dormand–Prince's method with $n = 1E2$, U gaussian $N(0, 1)$)

6.1 Linear Differential Equations

Let us consider the ordinary differential equations (ODE):

$$\frac{dX}{dt} = A(t, U)X + B(t, U), X(0) = X_0(U). \tag{6.3}$$

Here, A is a $n \times n$ matrix, B is a $n \times 1$ vector, X_0 is a $n \times 1$ vector, X is a $n \times 1$ vector. Analogously to the approach adopted for the analysis of linear systems in Sect. 5.1, we consider an expansion analogous to Eq. (5.9):

$$X \approx PX = \sum_{j=1}^{k} x_j(t)\varphi_j(U), x_j = (x_{1,j}, \ldots, x_{n,j}) \tag{6.4}$$

Thus, we look for an expansion such that

$$\frac{d}{dt}PX = A(t, U)PX + B(t, U), \quad PX(0) = X_0(U). \tag{6.5}$$

id est,

$$\begin{gathered} \sum_{j=1}^{k} \varphi_j(U)\frac{d}{dt}x_j(t) = \sum_{j=1}^{k} \varphi_j(U)A(t, U)x_j(t) + B(t, U), \\ \sum_{j=1}^{k} x_j(0)\varphi_j(U) = X_0(U) \end{gathered} \tag{6.6}$$

Thus, for $1 \leq r \leq n$

$$\begin{gathered} \sum_{j=1}^{k} \varphi_j(U)\frac{d}{dt}x_{r,j}(t) = \sum_{j=1}^{k} \varphi_j(U)A_{rs}(t, U)x_{s,j}(t) + B_r(t, U), \\ \sum_{j=1}^{k} x_{r,j}(0)\varphi_j(U) = X_{0_r}(U) \end{gathered} \tag{6.7}$$

and

$$\begin{gathered} \sum_{j=1}^{k} \mathcal{M}_{ij}\frac{d}{dt}x_{r,j}(t) = \sum_{j=1}^{k}\sum_{s=1}^{n} \mathcal{A}_{risj}(t)x_{s,j}(t) + \mathcal{B}_{ri}(t), \\ \sum_{j=1}^{k} \mathcal{M}_{ij}x_{r,j}(0) = \mathcal{N}_{0ri} \end{gathered} \tag{6.8}$$

Here,

$$\mathcal{A}_{risj} = E(\varphi_i(U)A_{rs}(t, \ U)\varphi_j(U)), \quad \mathcal{B}_{ri} = E(\varphi_i(U)B_r(t, \ U)) \tag{6.9}$$

$$\mathcal{M}_{ij} = E(\varphi_i(U)\varphi_j(U)), \quad \mathcal{N}_{0ri} = E(\varphi_i(U)X_{0_r}(U)) \tag{6.10}$$

By setting

$$\mathcal{M}_{risj} = \mathcal{M}_{ij}\delta_{rs},$$

we have

$$\sum_{j=1}^{k}\sum_{s=1}^{n}\mathcal{M}_{risj}\frac{\mathrm{d}}{\mathrm{d}t}x_{s,j}(t) = \sum_{j=1}^{k}\sum_{s=1}^{n}\mathcal{A}_{risj}(t)x_{s,j}(t) + \mathcal{B}_{ri}(t),$$
$$\sum_{j=1}^{k}\sum_{s=1}^{n}\mathcal{M}_{risj}(t)x_{s,j}(0) = \mathcal{N}_{0ri} \tag{6.11}$$

Analogously to algebraical linear systems, we have

$$\mathbb{M}\frac{\mathrm{d}\mathbb{X}}{\mathrm{d}t} = \mathbb{A}(t)\mathbb{X} + \mathbb{B}(t), \mathbb{M}\mathbb{X}(0) = \mathbb{N}_0, \tag{6.12}$$
$$\mathbb{M}_{pq} = \mathcal{M}_{risj}, \mathbb{X}_{\mathrm{q}} = x_{s,j}, \mathbb{B}_{\mathrm{p}} = \mathcal{B}_{ri}, \mathbb{N}_{0p} = \mathcal{N}_{0ri},$$
$$p = i + (r-1)k, \quad q = j + (s-1)k.$$

These equations for a linear ODE for the $k \times n$ components of $\mathbb{X}$. Notice that – in the general situation – the matrices $\mathbb{A}$ and $\mathbb{B}$ depend on time t. If the system is autonomous, id est, if $\boldsymbol{A}$ and $\boldsymbol{B}$ are independent from t, then $\mathbb{A}$ and $\mathbb{B}$ are also independent from t. In such a situation, you can calculate $\mathbb{A}$ and $\mathbb{B}$ once. Otherwise, an evaluation at each time step is requested.

An example of code for the evaluation of matrices $\mathbb{M}$, $\mathbb{N}_0$ is given below:

```
MNvarode= function(u,ns, basis, k, n, xx0){
  nn = k * n
  mm= self$zeros(nn, nn)
  nn0= self$zeros(nn,1)
  for (m in 1 : ns){
    uu = u[[m]]
    for (r in 1 : n){
      for (i in 1 : k){
        p = i + (r - 1) * k
        aux1 = basis$value(i - 1, uu)
        nn0[[p]] = nn0[[p]] + xx0[[r]] * aux1
        for (s in 1 : n){
          for (j in 1 : k){
            q = j + (s - 1) * k
            aux2 = basis$value(j - 1, uu)
            if (r == s){
              mm[[p,q]] = mm[[p,q]] + aux1 * aux2
            }
          }
        }
      }
    }
  }
  nn0 = nn0/ns
  mm = mm/ns
  resu = list(mm,nn0)
  return(resu)
},
```

An example of code for the evaluation of matrices $\mathbb{A}(t)$, $\mathbb{B}(t)$ is the following one:

```
ABvarode= function(u, ns,basis, k, n, T,a,b){
  nn = k * n
  aa = self$zeros(nn, nn)
  bb = self$zeros(nn,1)

  for (m in 1 : ns){
    uu = u[[m]]
    Auu = a(T, uu)
    Buu = b(T, uu)
    for (r in 1 : n){
      for (i in 1 : k){
        p = i + (r - 1) * k
        aux1 = basis$value(i - 1, uu)
        bb[[p]] = bb[[p]] + Buu[[r]] * aux1
        for (s in 1 : n){
          for (j in 1 : k){
            q = j + (s - 1) * k
            aux2 = basis$value(j - 1, uu)
            aa[[p,q]] = aa[[p,q]] + aux1 * Auu[[r,s]] * aux
          }
        }
      }
    }
  }
  bb = bb/ns
  aa = aa/ns
  resu = list(aa,bb)
  return(resu)
},
```

In the autonomous case, you can use:

```
ABvarode_auto = function(u, ns, basis, k, n,a,b){
  nn = k * n
  aa = self$zeros(nn, nn)
  bb = self$zeros(1, nn)

  for (m in 1 : ns){
    uu = u[[m]]
    Auu = a( uu)
    Buu = b(uu)
    for (r in 1 : n){
      for (i in 1 : k){
        p = i + (r - 1) * k
        aux1 = basis$value(i - 1, uu)
        bb[[1,p]] = bb[[1,p]] + Buu[[r]] * aux1
        for (s in 1 : n){
          for (j in 1 : k){
            q = j + (s - 1) * k
            aux2 = basis$value(j - 1, uu)
            aa[[p,q]] = aa[[p,q]] + aux1 * Auu[[r,s]] * aux2
          }
        }
      }
    }
  }
  bb = bb/ns
  aa = aa/ns
  resu = list(aa,bb)
  return(resu)
},
```

These methods are implemented in the class `ode.R`. This class contains also a method `zeros(m,n)` to generate a $m \times n$ matrix of zeros.

Let us exemplify the use of these subprograms with the equations

$$\begin{pmatrix} x_1' \\ x_2' \end{pmatrix} = \mathbf{A}\begin{pmatrix} x_1 \\ x_2 \end{pmatrix}, \mathbf{A} = \begin{pmatrix} 0 & U \\ -U & 0 \end{pmatrix}, x_1(0) = 0, x_2(0) = 1, \ t \in (0, T).$$

The solution is $x_1(t) = \sin(Ut)$, $x_2(t) = \cos(Ut)$. U is normally distributed, with a mean π and a standard deviation $\pi/6$. We use a sample formed by $ns = 20$ variates from U:

```
> u
         [,1]     [,2]     [,3]     [,4]     [,5]
[1,] 1.775173 2.34635 2.37255 2.520298 2.662971
         [,6]     [,7]     [,8]     [,9]    [,10]
[1,] 2.717147 2.778351 2.809701 3.032209 3.173714
        [,11]    [,12]    [,13]    [,14]    [,15]
[1,] 3.212206 3.223201 3.225362 3.336063 3.353406
        [,16]    [,17]    [,18]    [,19]    [,20]
[1,] 3.355669 3.409487 3.409572 3.838237 4.011724
> |
```

Start by creating the functions that evaluate the matrices *A* and B.

```
a = function(u){
  auxa= matrix(0, nrow = 2, ncol=2)
  auxa[[1,2]] = u
  auxa[[2,1]] = -u
  return(auxa)
}

b = function(u){
  auxb= matrix(0, nrow = 2, ncol=1)
  return(auxb)
}
```

Then, define the basis to be used: here, a polynomial basis with $k = 5$.

```
umin = min(u)
umax = max(u)
zero = 1e-3|

p = polynom$new(umin,umax,zero)

k=5
nb = k+1
```

Define X_0 and determine $\mathbb{A}$, $\mathbb{B}$, $\mathbb{M}$, $\mathbb{N}_0$, $\mathbb{X}_0$. XX0 contains $\mathbb{X}_0$

```
x0 = matrix(c(0,1),ncol=1)

ode1 = ode$new()
resuAB = ode1$ABvarode_auto(u, ns, p, nb, ndim,a,b)

aa = resuAB[[1]]
bb = resuAB[[2]]

resuMN = ode1$MNvarode(u, ns, p, nb, ndim, x0)
mm = resuMN[[1]]
nn0 = resuMN[[2]]
nn = tocol(nn0)
XX0 = solve(mm, nn)
```

Create a function evaluating the right side of the equation, adapted to the package deSolve. tocol(X) transforms X into a column vector.

```
F = function(t,x,maa,mbb,mmm){
  xx = tocol(x)
  auxF = maa %*% xx + tocol(mbb)
  auxF2 = solve(mmm,auxF)
  return(list(auxF2))
}
```

Notice that the class ode proposes a function for the evaluation of the right side of the differential equation:

```
fsm_auto= function(k, n, mm, aa, bb, XX){
  nn = k * n
  yy = self$tocol(aa %*% self$tocol(XX))
  zz = self$zeros(nn,1)
  for (i in 1 : nn){
    zz[[i]] = yy[[i]] + bb[[i]]
  }
  aux = solve(mm, zz)
  return(aux)
},
```

Then, call ode in package deSolve:

```
tmax = 1
nt = 1e2

tc = seq(0,tmax,length.out = nt + 1)
f = function(t,x,lm){F(t,x,lm[[1]],lm[[2]],lm[[3]])}
lmm = list(aa,bb,mm)
resu <- ode(c(XX0), tc, f,lmm ,method = "rk4")
```

To use fsm_auto, the sequence of instructions is:

```
tmax = 1
nt = 1e2

tc = seq(0,tmax,length.out = nt + 1)

f = function(t,x,lm){list(ode1$fsm_auto(nb, ndim, mm, aa, bb, x))}
lmm = list(NA,NA)
resu <- ode(c(XX0), tc, f,lmm ,method = "rk4")
```

In both the cases, `resu` contains the coefficients $x_{s,j}(t)$. Line i of `resu` corresponds to `tc[[i]]`$=(i-1)\Delta t$, $\Delta t = tmax/nt$. `resu[[i,1]]` is the value of `tc[[i]]`; `resu[i, 2 : (nb+1)]` contains $x_{1,0}(t_i), \ldots, x_{1,k}(t_i)$ (the coefficients of the expansion of the first component x_1).; `]]`; `resu[i, (nb+2) : (2*nb+1)]` contains $x_{2,0}(t_i), \ldots, x_{2,k}(t_i)$ (the coefficients of the expansion of the second component x_2).

These coefficients were used to evaluate the solution for the values of U in the sample: for $k = 5$, the RMS error was 5E-6, corresponding to a relative error of 7E-4%. Examples of results are shown in Figs. 6.3, 6.4, 6.5, and 6.6.

As a second example, let us consider a situation where the matrices are time-dependent:

$$A(t,U) = \begin{pmatrix} 0 & 1 \\ \dfrac{U^2}{2} & -\dfrac{U}{2} \end{pmatrix}, \quad B(t,U) = \begin{pmatrix} 0 \\ \dfrac{U^2}{2}(1-tU) \end{pmatrix}, \quad X_0 = \begin{pmatrix} 1 \\ 0 \end{pmatrix}.$$

The exact solution is

$$x_1(t) = Ut + e^{-Ut}, x_2(t) = U\left(1 - e^{-Ut}\right).$$

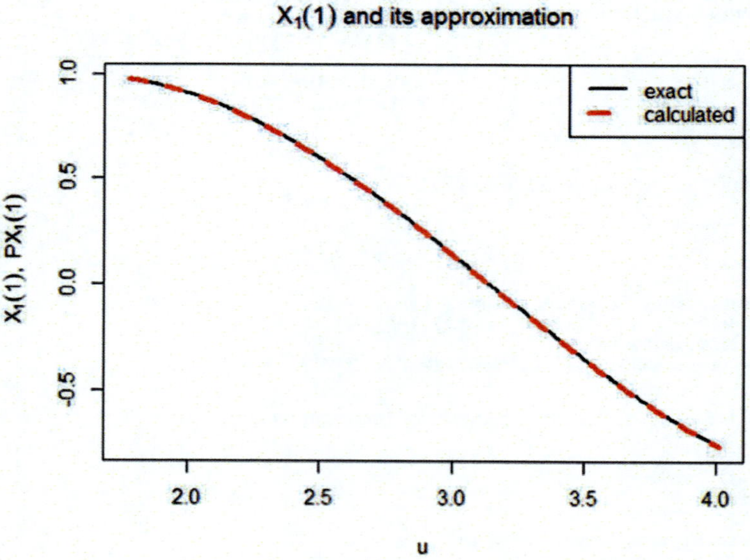

Fig. 6.3 Comparison between $X_1(1)$ and its approximation $PX_1(1)$ for $t = 1$. (Solution of Eq. (6.12) by rk4, with $n = 1E2$, U gaussian $N(\pi, \pi/6)$)

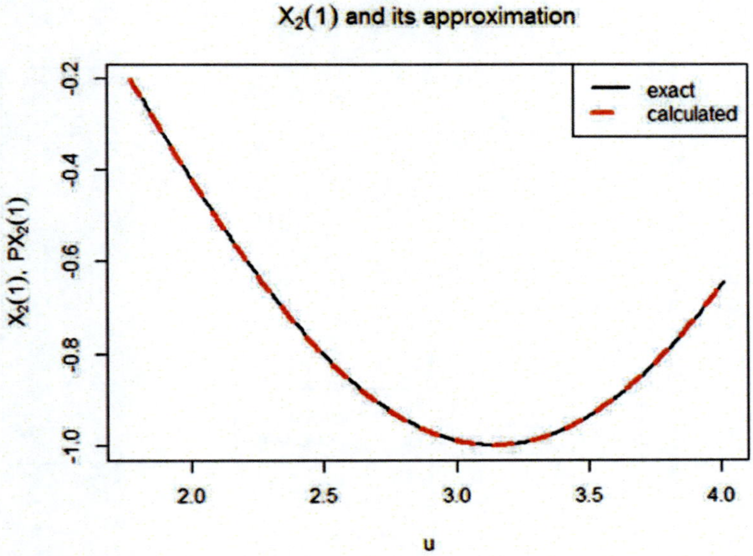

Fig. 6.4 Comparison between $X_2(1)$ and its approximation $PX_2(1)$ for $t = 1$. (Solution of Eq. (6.12) by rk4, with $n = 1E2$, U gaussian $N(\pi, \pi/6)$)

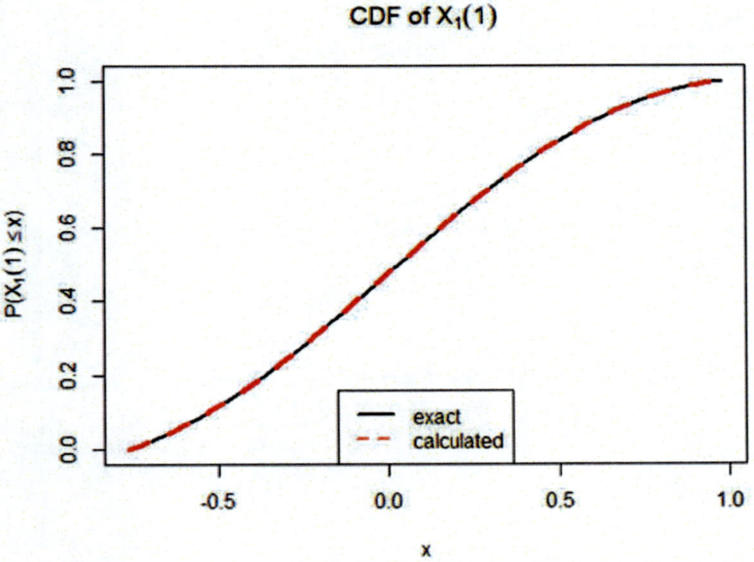

Fig. 6.5 Comparison between the CDF of $X_1(1)$ and the CDF of $PX_1(1)$ for $t = 1$. (Solution of Eq. (6.12) by rk4, with $n = 1E2$, U gaussian $N(0, 1)$)

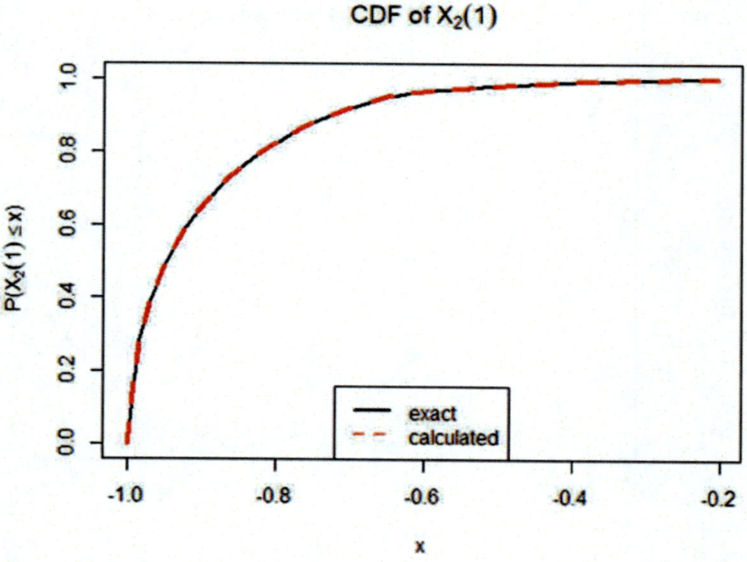

Fig. 6.6 Comparison between the CDF of $X_2(1)$ and the CDF of $PX_2(1)$ for $t = 1$. (Solution of Eq. (6.12) by rk4, with $n = 1E2$, U gaussian $N(0, 1)$)

We have

```
a = function(t,u){
  auxa= matrix(0, nrow = 2, ncol=2)
  auxa[[1,2]] = 1
  auxa[[2,1]] = 0.5*u*u
  auxa[[2,2]] = -0.5*u
  return(auxa)
}

b = function(t,u){
  auxb= matrix(0, nrow = 2, ncol=1)
  auxb[[2]] = 0.5*u*u*(1-t*u)
  return(auxb)
}
```

In this case, the function evaluating the right side of the equation, reads as shown at right:

```
F = function(t,x,maa,mbb,mmm){
  xx = tocol(x)
  auxF = maa %*% xx + tocol(mbb)
  auxF2 = solve(mmm,auxF)
  return(list(auxF2))
}
```

The class ode proposes a function for the evaluation of the right side of the differential equation in this case too:

```
fsm = function(u, ns, basis, k, n, mm, XX, T,a,b){
  resuAB = self$ABvarode(u, ns, basis, k, n, T,a,b)
  aa = resuAB[[1]]
  bb = resuAB[[2]]
  nn = k * n
  yy = self$tocol(aa %*% self$tocol(XX))
  zz = self$zeros(nn,1)
  for (i in 1 : nn){
    zz[[i]] = yy[[i]] + bb[[i]]
  }
  aux = solve(mm, zz)
  return(aux)
},
```

Now, the code to call ode in package deSolve reads as:

```
tmax = 1
nt = 1e2

tc = seq(0,tmax,length.out = nt + 1)

f = function(t,X,lm){
  resuAB = ode1$ABvarode(u, ns, p, nb, ndim,t,a,b)
  aa = resuAB[[1]]
  bb = resuAB[[2]]
  auxf = F(t,X,aa,bb,lm[[1]])
  return(auxf)
}

lmm = list(mm)
resu <- ode(c(XX0), tc, f,lmm ,method = "rk4")
```

To use fsm, the sequence of instructions is:

```
tmax = 1
nt = 1e2

tc = seq(0,tmax,length.out = nt + 1)

f = function(t,X,lm){list(ode1$fsm(u, ns, p, nb, ndim, mm, X, t,a,b))}
lmm = list(NA,NA)
resu <- ode(c(XX0), tc, f,lmm ,method = "rk4")
```

We consider U as normally distributed, with mean 0 and standard deviation 1. The sample used contains $ns = 20$ variates from U:

```
> u
          [,1]      [,2]      [,3]       [,4]       [,5]
[1,] -1.646709 -1.33608 -1.162466 -0.9290065 -0.7870194
           [,6]       [,7]      [,8]        [,9]     [,10]
[1,] -0.7373678 -0.4625434 -0.227513 -0.00864912 0.1331659
         [,11]     [,12]     [,13]     [,14]    [,15]
[1,] 0.1932613 0.1949766 0.2673599 0.3888805 1.100098
        [,16]    [,17]    [,18]    [,19]   [,20]
[1,] 1.228042 1.378303 1.604283 1.720176 1.80265
> |
```

Using $k = 5$, the RMS error was 1E-3, corresponding to a relative error of 0.1%. Figs. 6.7, 6.8, 6.9, and 6.10 exhibit examples of results.

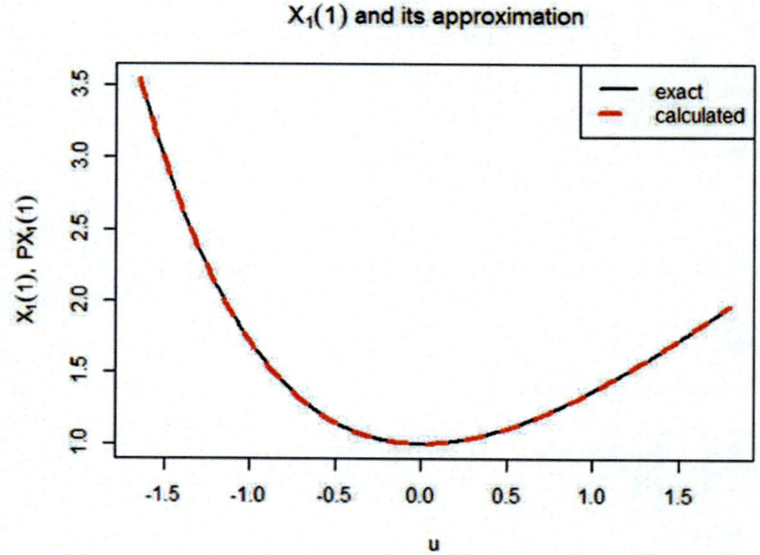

Fig. 6.7 Comparison between $X_1(1)$ and its approximation $PX_1(1)$ for $t = 1$. (Solution of Eq. (6.12) by rk4, with $n = 1E2$, U gaussian $N(0, 1)$)

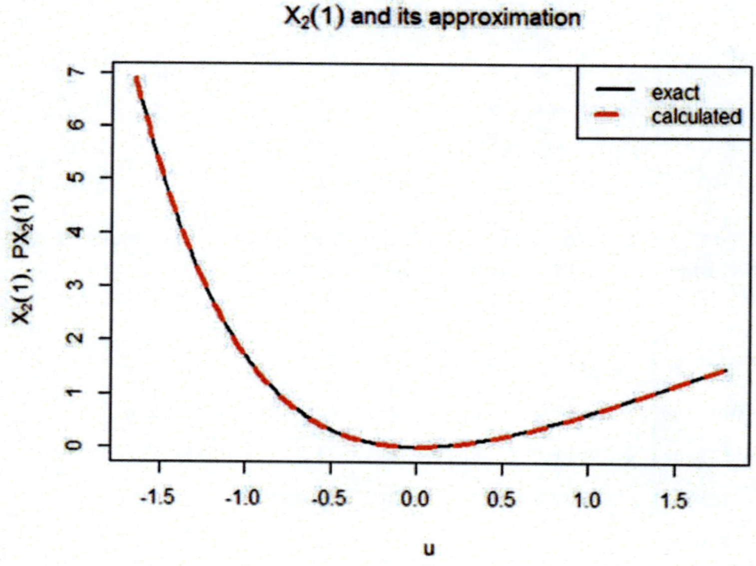

Fig. 6.8 Comparison between $X_2(1)$ and its approximation $PX_2(1)$ for $t = 1$. (Solution of Eq. (6.12) by rk4, with $n = 1E2$, U gaussian $N(0, 1)$)

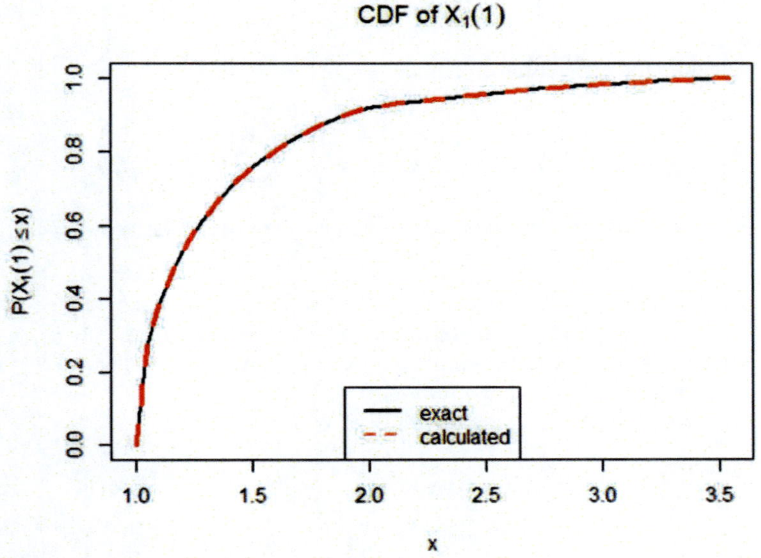

Fig. 6.9 Comparison between the CDF of $X_1(1)$ and the CDF of $PX_1(1)$ for $t = 1$. (Solution of Eq. (6.12) by rk4, with $n = 1E2$, U gaussian $N(0, 1)$)

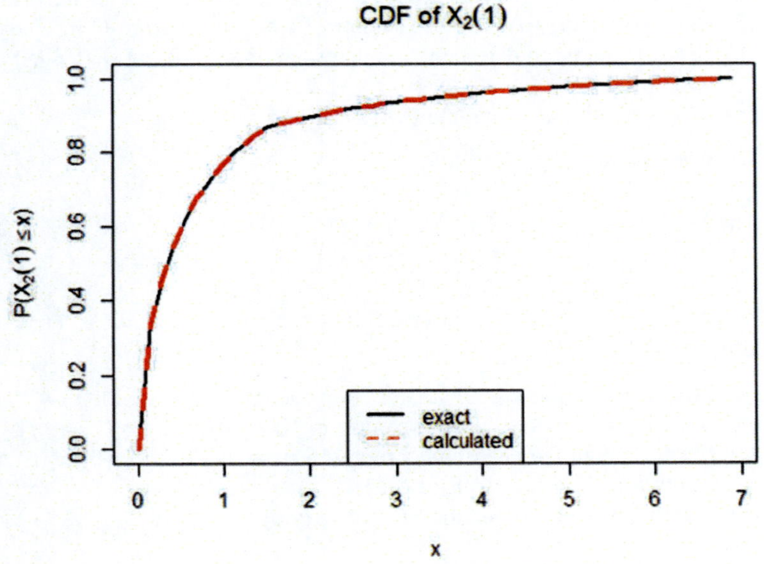

Fig. 6.10 Comparison between the CDF of $X_2(1)$ and the CDF of $PX_2(1)$ for $t = 1$. (Solution of Eq. (6.12) by rk4, with $n = 1E2$, U gaussian $N(0, 1)$)

Exercises

1. Consider the differential equations

$$\frac{\mathrm{d}}{\mathrm{d}t}\begin{pmatrix} x_1 \\ x_2 \end{pmatrix} = \begin{pmatrix} -u_1 & 2 \\ 4 & -\mathrm{u}_2 \end{pmatrix}\begin{pmatrix} x_1 \\ x_2 \end{pmatrix}, \quad \begin{pmatrix} x_1(0) \\ x_2(0) \end{pmatrix} = \begin{pmatrix} 2 \\ 5 \end{pmatrix}$$

   Assume that the distributions of u_1 and u_2 are triangular: $u_1 \sim T(0,1,2)$ and $u_2 \sim T(0,4,8)$.

   (a) Generate a sample of 6 variates from each random variable u_i.
   (b) Use these values to generate a sample of 36 variates from $\boldsymbol{U} = (u_1, u_2)$.
   (c) Use the sample to determine a representation of $\boldsymbol{X}(t) = (x_1(t), x_2(t))$ as function of $\boldsymbol{U} = (u_1, u_2)$ for $t \in (0,1)$.
   (d) Determine the CDF and the PDF of x_1 (0.5).

2. Consider

$$\frac{\mathrm{d}}{\mathrm{d}t}\begin{pmatrix} x_1 \\ x_2 \end{pmatrix} = \begin{pmatrix} 0.2 - 5u_1 & -2.5 - 5u_1 \\ 1 & 0 \end{pmatrix}\begin{pmatrix} x_1 \\ x_2 \end{pmatrix} + \begin{pmatrix} 8 \\ 3 \end{pmatrix}, \quad \begin{pmatrix} x_1(0) \\ x_2(0) \end{pmatrix} = \begin{pmatrix} 2 \\ 5 \end{pmatrix}$$

   Assume that the distributions $u_1 \sim T(-0.1, 0{,}0.1)$ and $u_2 \sim N(0, 0.1)$.

   (a) Generate a sample of 10 variates from each random variable u_i.
   (b) Use these values to generate a sample of 100 variates from $\boldsymbol{U} = (u_1, u_2)$.
   (c) Use the sample to determine a representation of $\boldsymbol{X}(t) = (x_1(t), x_2(t))$ as function of $\boldsymbol{U} = (u_1, u_2)$ for $t \in (0,1)$.
   (d) Determine the CDF and the PDF of x_1 *(0.5)*.

3. Consider the differential equation

$$\frac{\mathrm{d}}{\mathrm{d}t}\begin{pmatrix} x_1 \\ x_2 \end{pmatrix} = \begin{pmatrix} 2 + u_1 & -3 + \mathrm{u}_2 \\ 1 & 0 \end{pmatrix}\begin{pmatrix} x_1 \\ x_2 \end{pmatrix} + \begin{pmatrix} 10 + 5u_1 \\ -3 \end{pmatrix}, \quad \begin{pmatrix} x_1(0) \\ x_2(0) \end{pmatrix} = \begin{pmatrix} 2 \\ 5 \end{pmatrix}$$

   Assume that the distributions of u_1 and u_2 are triangular: $u_1 \sim T(1,2,3)$ and $u_2 \sim T(0,1,2)$.

   (a) Generate a sample of 10 variates from each random variable u_i.
   (b) Use these values to generate a sample of 100 variates from $\boldsymbol{U} = (u_1, u_2)$.
   (c) Use the sample to determine a representation of $\boldsymbol{X}(t) = (x_1(t), x_2(t))$ as function of $\boldsymbol{U} = (u_1, u_2)$ for $t \in (0,1)$.
   (d) Determine the CDF and the PDF of x_1 (0.5).

(continued)

4. Consider the differential equation

$$\frac{d}{dt}\begin{pmatrix} x_1 \\ x_2 \end{pmatrix} = \begin{pmatrix} -u_1 & -3-u_2 \\ 0.5+u_2 & -2-u_2 \end{pmatrix}\begin{pmatrix} x_1 \\ x_2 \end{pmatrix} + \begin{pmatrix} 0.2+u_1 \\ 0.1+u_2 \end{pmatrix}, \quad \begin{pmatrix} x_1(0) \\ x_2(0) \end{pmatrix} = \begin{pmatrix} 2 \\ 5 \end{pmatrix}$$

Assume that the distributions of u_1 and u_2 are triangular: $u_1 \sim T(-0.1,0,0.2)$ and $u_2 \sim T(-0.5,0,0.5)$.

(a) Generate a sample of 6 variates from each random variable u_i.
(b) Use these values to generate a sample of 36 variates from $U = (u_1, u_2)$.
(c) Use the sample to determine a representation of $X(t) = (x_1(t), x_2(t))$ as function of $U = (u_1, u_2)$ for $t \in (0, 1)$.
(d) Determine the CDF and the PDF of x_1 (0.5).

6.2 Nonlinear Differential Equations

The approach presented in the previous section extends to nonlinear ODE such as

$$\frac{d\boldsymbol{X}}{dt} = \boldsymbol{f}(t, \boldsymbol{X}, \boldsymbol{U}), \ \boldsymbol{X}(0) = \boldsymbol{X_0}(\boldsymbol{U}). \tag{6.13}$$

The procedure is analogous and leads to a formulation close to the method of adaptation, presented in Sect. 5.2: we use Eq. (6.13) to get

$$\frac{d}{dt}\boldsymbol{PX} = \boldsymbol{f}(t, \boldsymbol{PX}, \boldsymbol{U}), \ \boldsymbol{PX}(0) = \boldsymbol{X_0}(\boldsymbol{U}). \tag{6.14}$$

Thus, for $1 \leq r \leq n$

$$\begin{gathered} \sum_{j=1}^{k} \varphi_j(\boldsymbol{U}) \frac{d}{dt} x_{r,j}(t) = f_r(t, \boldsymbol{PX}, \boldsymbol{U}), \\ \sum_{j=1}^{k} x_{r,j}(0) \varphi_j(\boldsymbol{U}) = X_{0_r}(\boldsymbol{U}) \end{gathered} \tag{6.15}$$

Then,

$$\begin{gathered} \sum_{j=1}^{k} \mathcal{M}_{ij} \frac{d}{dt} x_{r,j}(t) = \mathcal{F}_{ri}(t, \boldsymbol{x}), \\ \sum_{j=1}^{k} \mathcal{M}_{ij} x_{r,j}(0) = \mathcal{N}_{0ri} \end{gathered} \tag{6.16}$$

with $\mathcal{M}_{ij}$, $\mathcal{N}_{0ri}$ previously defined (Eq. 6.10) and

$$\mathcal{F}_{ri}(t, \boldsymbol{x}) = E(\varphi_i(\boldsymbol{U}) f_r(t, \ \boldsymbol{PX}, \ \boldsymbol{U})). \tag{6.17}$$

The evaluation of $\mathcal{F}_{ri}$ is made as follows:

1. Determine $\boldsymbol{Z_m} = \boldsymbol{PX}(U_m)$, for $1 \le m \le ns$.
2. Evaluate $\boldsymbol{T_m} = f(t, \boldsymbol{Z_m}, U_m)$, for $1 \le m \le ns$. Let $\boldsymbol{T_m} = (T_{1,\ m}, \ldots, T_{n,m})$
3. Determine $V_{m,i} = \varphi_i(U_m)$, $1 \le m \le ns$, $1 \le i \le k$.
4. Determine $\mathcal{F}_{ri} \approx \frac{1}{ns} \sum_{m=1}^{ns} V_{m,i} T_{r,m}$.

Using the same map $p = i + (r - 1)k$, $q = j + (s - 1)k$ previously introduced, we have

$$\mathbb{M}\frac{\mathrm{d}\mathbb{X}}{\mathrm{d}t} = \mathbb{F}(t, \mathbb{X}), \ \mathbb{M}\mathbb{X}(0) = \mathbb{N}_0, \tag{6.18}$$

$$\mathbb{M}_{pq} = \mathcal{M}_{risj}, \mathbb{X}_q = x_{s,j}, \mathbb{F}_p = \mathcal{F}_{ri}, \mathbb{N}_{0p} = \mathcal{N}_{0ri}.$$

An example of code for the determination of $\mathbb{F}$ is given below (implemented in the class `ode.R`):

```
fvarode=function(f,u, ns, basis, k, n, mm, XX, t){
  nn = k * n
  cc = self$zeros(n,k)
  for (r in 1 : n){
    for (i in 1 : k){
      pp = i + (r - 1) * k
      cc[[r,i]] = XX[[pp]]
    }
  }
  px = self$proj(cc,u,n,k, basis)
  tabphi = self$basis_matrix(u,basis,k)
  ff = self$zeros(nn,1)
  for (m in 1 : ns){
    uu = u[[m]]
    xxxx = px[ , m]
    fuu = f(t, xxxx, uu)
    for (r in 1 : n){
      for (i in 1 : k){
        pp = i + (r - 1) * k
        aux1 = tabphi[[m,i]]
        ff[[pp]] = ff[[pp]] + fuu[[r]] * aux1
      }
    }
  }
  ff = ff/ns
  resu = solve(mm, ff)
  return(resu)
},
```

As an example, let us consider the system defined in Eq. (6.1):

$$f(t,x,u) = \begin{pmatrix} u\sqrt{x_2} \\ 2ux_1^2 \end{pmatrix}, \quad X_0 = \begin{pmatrix} 1 \\ 1 \end{pmatrix}.$$

The solution is $x_1(t) = e^{ut}$, $x_2(t) = e^{2ut}$. We consider a sample of 20 variates from N(0,1):

```
> u
           [,1]       [,2]       [,3]        [,4]     [,5]
[1,] -1.450491 0.8394019 -1.326654 -0.3824833 0.12179
         [,6]        [,7]      [,8]        [,9]     [,10]
[1,] 1.044948 0.002240034 0.1910615 -0.9913994 -1.208984
          [,11]      [,12]      [,13]      [,14]     [,15]
[1,] -1.018507 -0.5180312 0.2358686 0.2116131 -1.108532
          [,16]      [,17]     [,18]      [,19]       [,20]
[1,] -1.249894 -0.8216639 -1.676375 -0.1940404 0.009845323
```

Introduce the function $f(t, x, u)$:

```
f=function(t, x, u){
  aux1 = u * sqrt(x[[2]])
  aux2 = 2 * u * x[[1]] * x[[1]]
  aux = matrix(c(aux1,aux2),ncol=1)
  return(aux)
}
```

Define the Hilbert basis:

```
p = polynom$new(umin,umax,zero)
k=5
nb = k+1
```

Define X_0 and determine $\mathbb{M}$, $\mathbb{N}_0$, $\mathbb{X}_0$. XX0 contains $\mathbb{X}_0$.

```
x0 = matrix(c(1,1),ncol=1)
ode1 = ode$new()
resuMN = ode1$MNvarode(u, ns, p, nb, ndim, x0)
mm = resuMN[[1]]
nn0 = resuMN[[2]]
nn = tocol(nn0)
XX0 = solve(mm, nn)
```

Then solve Eq. (6.18):

```
tmax = 1
nt = 1e3
tc = seq(0,tmax,length.out = nt + 1)
fbis = function(t,X,lm){list(ode1$fvarode(f,u, ns, p, nb, ndim, mm, X, t))}
lmm = list(NA,NA)
resu <- ode(c(XX0), tc, fbis,lmm ,method = "rk4")
```

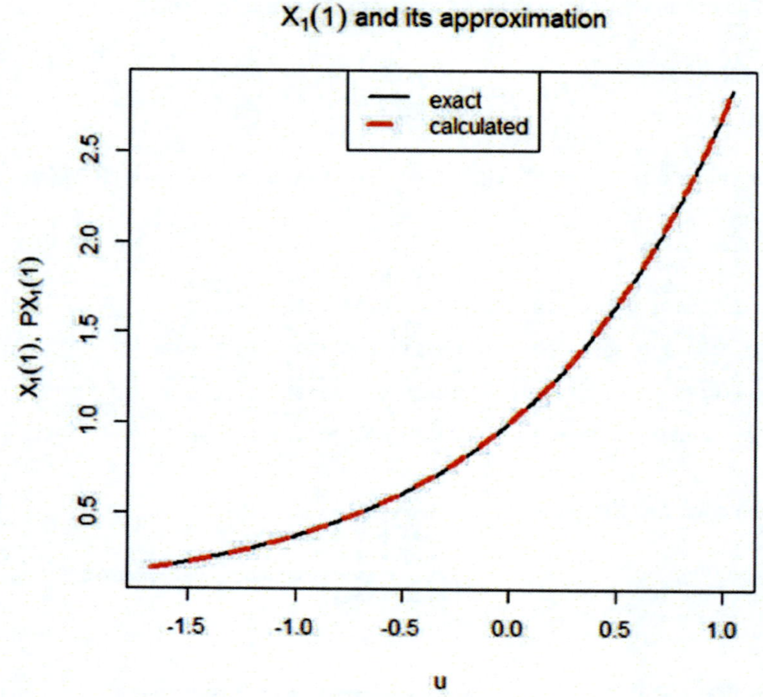

Fig. 6.11 Comparison between $X_2(1)$ and its approximation $PX_2(1)$ for $t = 1$. (Solution of Eq. (6.18) by rk4, with $n = 1E2$, U gaussian $N(0, 1)$)

With $k = 5$, the RMS error was 2E-3, corresponding to a relative error of 0.2%. Figs. 6.11, 6.12, 6.13, and 6.14 show examples of results.

Exercises

1. Consider the differential equation

$$\frac{dx}{dt} = (1 - t)x^u, \quad x(0) = 1$$

Assume that $u \sim T(1, 2, 3)$ and determine a representation of $x(t)$ as function of u for $t \in (0, 1)$. Determine the CDF and the PDF of $x(0.5)$.

(continued)

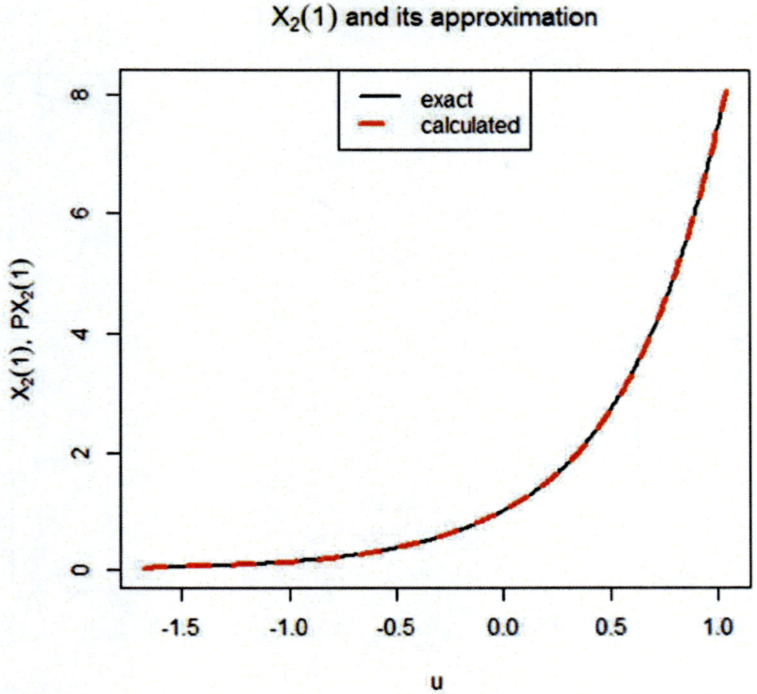

Fig. 6.12 Comparison between $X_2(1)$ and its approximation $PX_2(1)$ for $t = 1$. (Solution of Eq. (6.18) by rk4, with $n = 1E2$, U gaussian $N(0, 1)$)

2. Consider the differential equation

$$\frac{dx}{dt} = (1-t)x^{u_1}, \quad x(0) = u_2$$

Assume that $u_1 \sim T(1, 2, 3)$ and $u_2 \sim N(2, 0.1)$. Determine a representation of $x(t)$ as function of $U = (u_1, u_2)$ for $t \in (0, 1)$. Determine the CDF and the PDF of $x(0.5)$.

3. Consider the differential equation

$$\frac{d}{dt}\begin{pmatrix} x_1 \\ x_2 \end{pmatrix} = \begin{pmatrix} (2+u)x_2 \\ -(2+u)x_1 - 4x_1^2 \end{pmatrix}, \quad \begin{pmatrix} x_1(0) \\ x_2(0) \end{pmatrix} = \begin{pmatrix} 0 \\ 2 \end{pmatrix}$$

(continued)

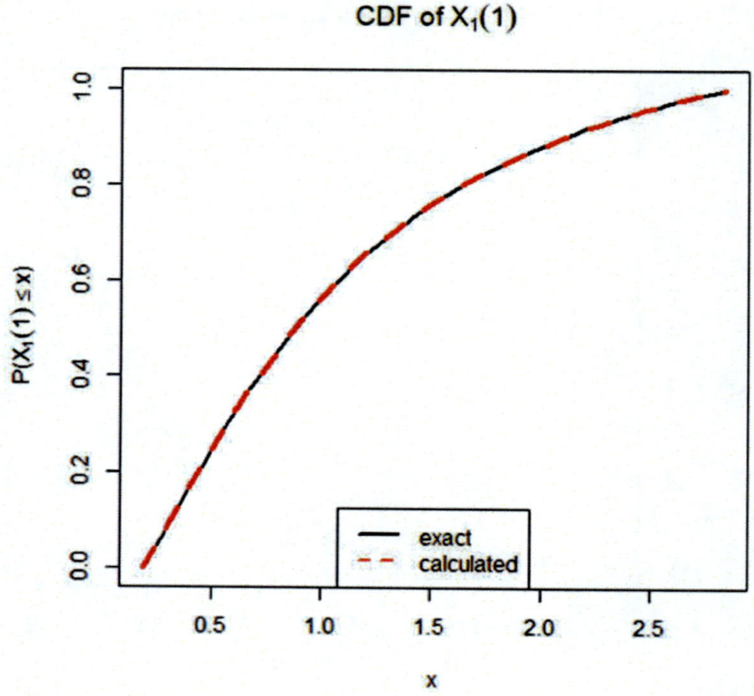

Fig. 6.13 Comparison between the CDF of $X_1(1)$ and the CDF of $PX_1(1)$ for $t = 1$. (Solution of Eq. (6.18) by rk4, with $n = 1E2$, U gaussian $N(0, 1)$)

Assume that $u \sim T(-1, 0, 1)$ and determine a representation of $X(t) = (x_1(t), x_2(t))$ as function of u for $t \in (0, 1)$. Determine the CDF and the PDF of $x_1(0.5)$.

4. Consider the differential equation

$$\frac{dx}{dt} = t^{1-u}(1 + x)^u, \quad x(0) = 1$$

Assume that $u \sim T(0,0.5,1)$ and determine a representation of $x(t)$ as function of u for $t \in (0, 1)$. Determine the CDF and the PDF of $x(0.5)$

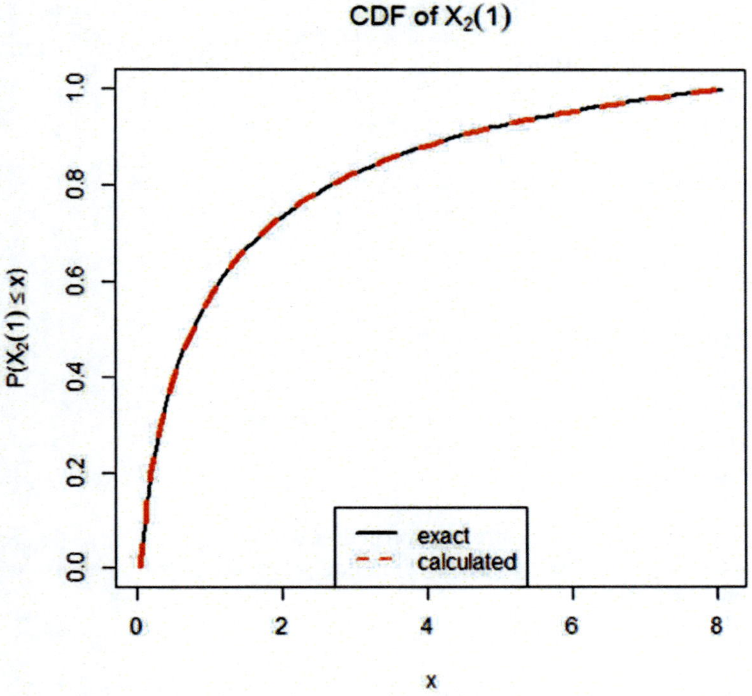

Fig. 6.14 Comparison between the CDF of $X_2(1)$ and the CDF of $PX_2(1)$ for $t = 1$. (Solution of Eq. (6.18) by rk4, with $n = 1E2$, U gaussian $N(0, 1)$)

6.3 Adaptation of ODE Solvers

In Sect. 5.2, we observed that any iterative method can be used to determine the coefficients of the expansion of the solution.

Let us illustrate how to adapt the solver `ode` from `deSolve`: recall that `ode` evaluates the solution in the times defined by the user – thus, it is possible to make a single time step using adapted parameters. For instance, let us create a function `odestep`, which performs a single step of an available method of `ode`:

```
odeStep = function(X1,U,t1,t2,method){
  tc = c(t1,t2)
  f = function(t,x,listU){F(t,x,listU[[1]])}
  resu <- ode(c(X1), tc, f,list(U), method = method)
  nline = dim(resu)[[1]]
  return(matrix(resu[nline,2:3],ncol = 1))
}
```

Here, `method` is a string containing the method to be applied, `t1` is the initial time, `t2` is the final time, `X1` is the value of the solution at time `t1`. Now, include in class `uae` two new methods:

```
progression_step = function(psi,X0,t0,t1,us,phi,nb,ndim,A,V){
  xsol = list()
  xsol[[1]] = X0
  ppsi = function(X,U){psi(X,U,t0,t1)}
  aux = self$iteration_adaptation(ppsi,X0,A,V,phi,u,ndim,nb)
  return(aux)
},
```

```
time_progression = function(psi,X0,us,phi,nb,ndim,tc){
  A = self$basis_sp(us,phi,nb)
  V = self$basis_matrix(u,phi,nb)
  xsol = list()
  xsol[[1]] = X0
  for (nt in (1: (length(tc)-1))){
    aux = self$progression_step(psi,xsol[[nt]],tc[[nt]],tc[[nt+1]],u,phi,nb,ndim,A,V)
    xsol[[nt+1]] = aux
    print(paste("*************** nt = ", nt, "finished"))
  }
  return(xsol)
},
```

These methods perform a progression in time with $\psi = \psi(X, U, t_0, t_1)$ (in Sect. 5.2, we used $\psi(X, U)$, without dependance on time). Let us illustrate their use with the differential equation

$$\frac{d\boldsymbol{x}}{dt} = \boldsymbol{f}(t, \boldsymbol{x}, u) = \begin{pmatrix} x_2 \\ \frac{U^2}{2}(1 - tU) + \frac{U^2}{2}x_1 - \frac{U x_2}{2} \end{pmatrix}, \ X_0 = \begin{pmatrix} 1 \\ 0 \end{pmatrix}.$$

We create the function that evaluates $\boldsymbol{f}(t, \boldsymbol{x}, u)$:

```
F = function(t,X,U){
  aux = c(X[[2]],0.5*((1-t*U)*(U^2) + (U^2)*X[[1]] - U*X[[2]]))
  maux = matrix(aux, nrow = 2, ncol=1)
  return(list(maux))
}
```

Then, we define the function $\psi(X, U, t_0, t_1)$

```
t0 = 0
tmax = 1
nt = 1e3
dt = (tmax-t0)/nt
tc = seq(0,1,by=dt)
psi = function(X,U,t1,t2) {odeStep(X,U,t1,t2,"ode45")}
```

We generate a sample of 20 variates from $N(0,1)$ and we define a polynomial basis:

```
ns = 20
u = matrix(sort(rnorm(ns,0,1)),nrow = 1, ncol = ns)

a = min(u)
b = max(u)

zero = 1e-3;
phi = polynom$new(a,b,zero)

nb = 5
ninc = nb + 1
```

Finally, we run the method:

```
uaet = uae$new()

x0 = matrix(0, nrow = ndim, ncol=ninc)
x0[[1,1]] = 1
x0[[2,1]] = 0

xsol = uaet$time_progression(psi,x0,u,phi,nb,ndim,tc)
```

Examples of results are shown in Fig. 6.15.

6.4 Uncertainties on Curves Connected to Differential Equations

If an ODE involves random parameters, then its solution becomes random – as seen in the preceding sections. Thus, the trajectories

$$t \rightarrow \boldsymbol{X}(t) = (x_1(t), x_2(t))$$

and the orbits

$$t \rightarrow \left(x_i(t), p_i(t) = \frac{\mathrm{d}x_i}{\mathrm{d}t}(t) \right)$$

are also random variables. For instance, let us consider the simple ODE:

$$\frac{\mathrm{d}}{\mathrm{d}t}\begin{pmatrix} x_1 \\ x_2 \end{pmatrix} = \begin{pmatrix} u_1/\sqrt{1+u_2^2} \\ u_2/\sqrt{1+u_2^2} \end{pmatrix}, x_1(0) = x_2(0) = 0,\ t \in (0,1).$$

The solution is $x_1(t) = tu_1/\sqrt{1+u_2^2}$, $x_2(t) = tu_2/\sqrt{1+u_2^2}$. If $\boldsymbol{u} = (u_1, u_2)$ is a random variable, x_1 and x_2 are also random. Thus, we may consider the analysis of the variability of these variables and look for statistics associated to them, such as, for instance, their mean and variance.

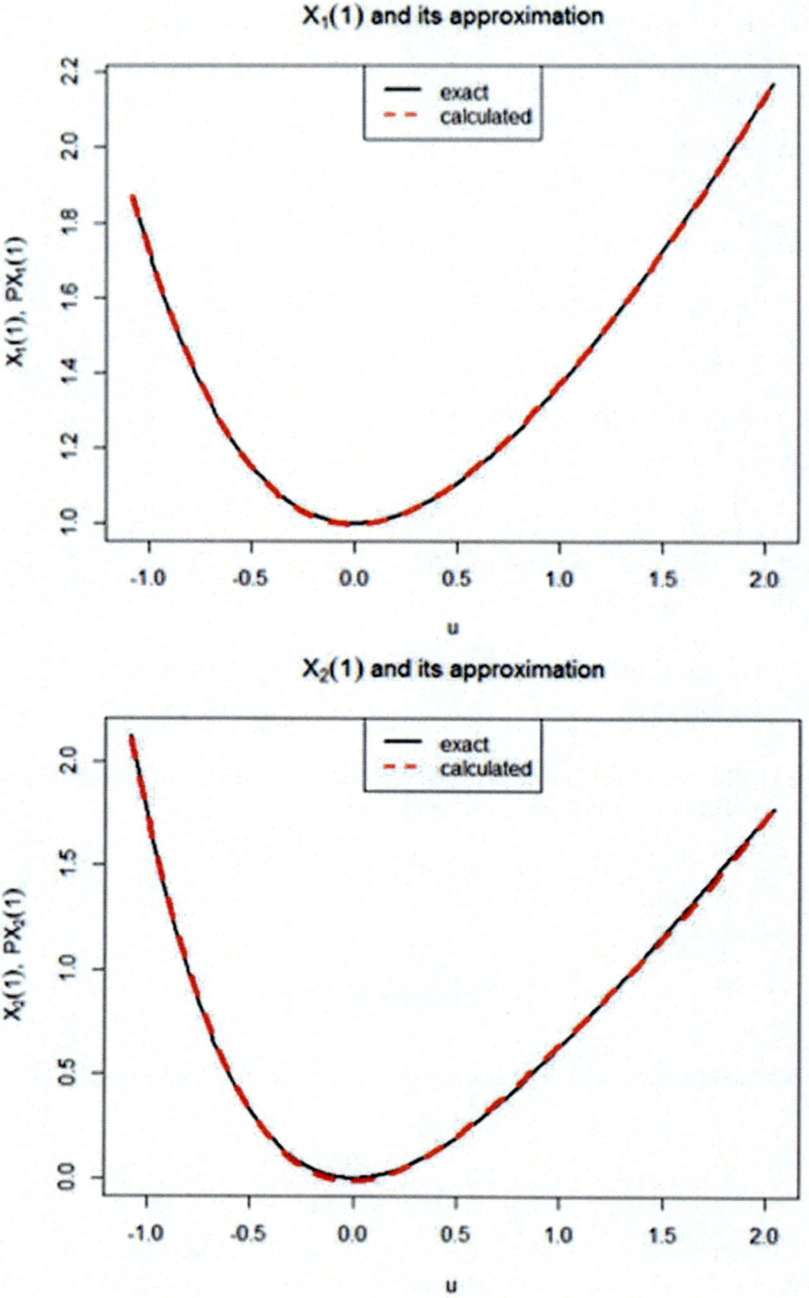

Fig. 6.15 Comparison between the prevision furnished by the expansion determined by adaptation and the exact values at $t = 1$, U gaussian $N(0, 1)$

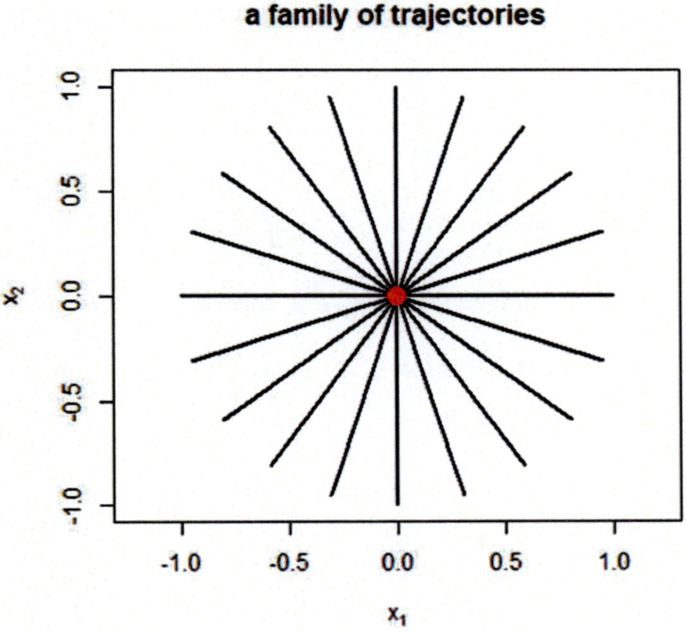

Fig. 6.16 When the mean is taken on the equations, the result is the origin, which is not representative of the family of curves: all its elements are segments of unitary length

However – as shown in Sect. 3.8 – the analysis of uncertainties in curves may request the use of approaches that are independent from the representation. For instance, let us assume that $P(u_1 = 1) = P(u_1 = -1) = 1/2$, while u_2 is uniformly distributed on $(-1, 1)$ Then, taking the mean on the equations leads to $E(x_1(t)) = 0$, $E(x_2(t)) = 0$: the resulting curve is not representative of the family, since the family is formed by segments emanating from the origin, all of length one (see Fig. 6.16). In such a situation, it appears as more adequate to determine the most representative member of the family (see Eq. (3.62)). In this case,

$$C(\boldsymbol{u}) = \left\{ \boldsymbol{X}(t,\ \boldsymbol{u}) = t\begin{pmatrix} u_1/\sqrt{1+u_2^2} \\ u_2/\sqrt{1+u_2^2} \end{pmatrix}, t \in (0,\ 1) \right\}$$

Since

$$\left(\frac{u_1}{\sqrt{1+u_2^2}}\right)^2 + \left(\frac{u_2}{\sqrt{1+u_2^2}}\right)^2 = 1,$$

we can consider

$$\sin\theta(\boldsymbol{u}) = \frac{u_1}{\sqrt{1+u_2^2}},\ \cos\theta(\boldsymbol{u}) = \frac{u_2}{\sqrt{1+u_2^2}}\ .$$

Then,

$$C(\boldsymbol{u}) = \left\{ \boldsymbol{X}(t,\boldsymbol{u}) = t\begin{pmatrix} \sin\theta(\boldsymbol{u}) \\ \cos\theta(\boldsymbol{u}) \end{pmatrix}, t \in (0,1) \right\}$$

and the Hausdorff distance between $C(\boldsymbol{u})$, $C(\boldsymbol{v})$ corresponds to $t = 1$, so that

$$dist(C(\boldsymbol{u}),\ C(\boldsymbol{v})) = \sqrt{2(\ 1 - \cos\left(\theta(\boldsymbol{u}) - \theta(\boldsymbol{v})\right)}\ ,$$

id est,

$$dist(C(\boldsymbol{u}),\ C(\boldsymbol{v})) = \left|\ \sin\left(\frac{1}{2}(\theta(\boldsymbol{u}) - \theta(\boldsymbol{v}))\right)\ \right|,$$

Thus,

$$J(C(\boldsymbol{u})) = E_{\boldsymbol{v}}(dist(C(\boldsymbol{u}),\ C(\boldsymbol{v})))$$

has a constant value for all the members of the family: any of its members may be considered as a good representative of the family.

Let us consider a second ODE

$$\frac{\mathrm{d}}{\mathrm{d}t}\begin{pmatrix} x_1 \\ x_2 \end{pmatrix} = \begin{pmatrix} 2ut \\ 2u(t-1) \end{pmatrix},\ x_1(0) = 0,\ x_2(0) = u,\ t \in (0,1).$$

The solution is $x_1(\mathrm{t}) = ut^2$, $x_2(\mathrm{t}) = u(t-1)^2$. Let us assume that u is uniformly distributed on (0,1). The evaluations of the mean trajectory by two different approaches are shown in Fig. 6.17.

In this case, $p_1(t) = 2ut$ and the orbits of the first variable are the points $(ut^2, 2ut)$. Their mean is shown in Fig. 6.18. The means taken at each time furnish

$$E(x_1(t)) = \frac{1}{2}t^2, E(x_2(t)) = \frac{1}{2}(t-1)^2, E(p_1(t)) = t, t \in (0,\ 1).$$

As a last example, let us consider the ODE

$$\frac{\mathrm{d}}{\mathrm{d}t}\begin{pmatrix} x_1 \\ x_2 \end{pmatrix} = \begin{pmatrix} -(1+|u|)x_2 \\ (1+|u|)x_1 \end{pmatrix},\ x_1(0) = 0,\ x_2(0) = u,\ t \in (0, 2\pi).$$

The solution is $x_1(\mathrm{t}) = u\cos((1 + |u|)t)$, $x_2(\mathrm{t}) = u\sin((1 + |u|)t)$. Let us assume that u is uniformly distributed on (−1,1). In this case, the means taken at each time furnish

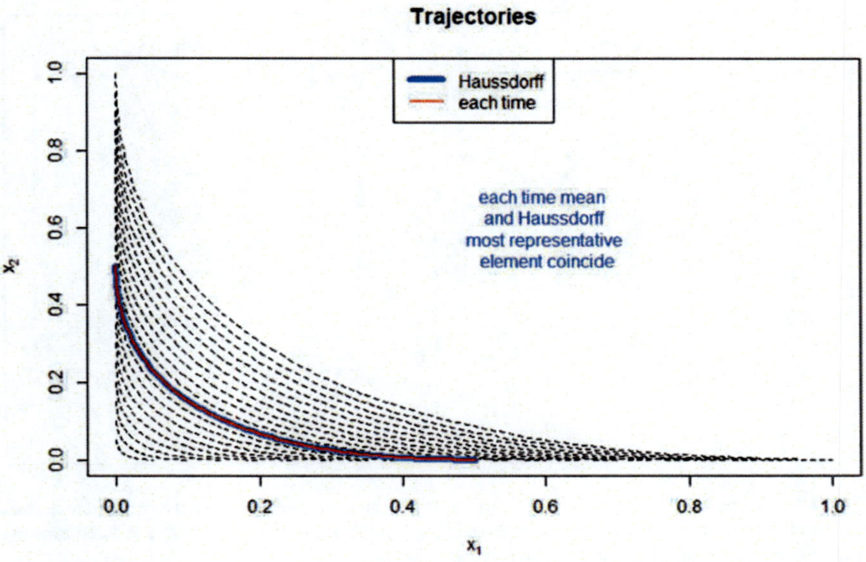

Fig. 6.17 Trajectories of the system. Here, the approach by the Hausdorff distance produces the same result as the mean $E((x_1(t), x_2(t)))$ taken at each time

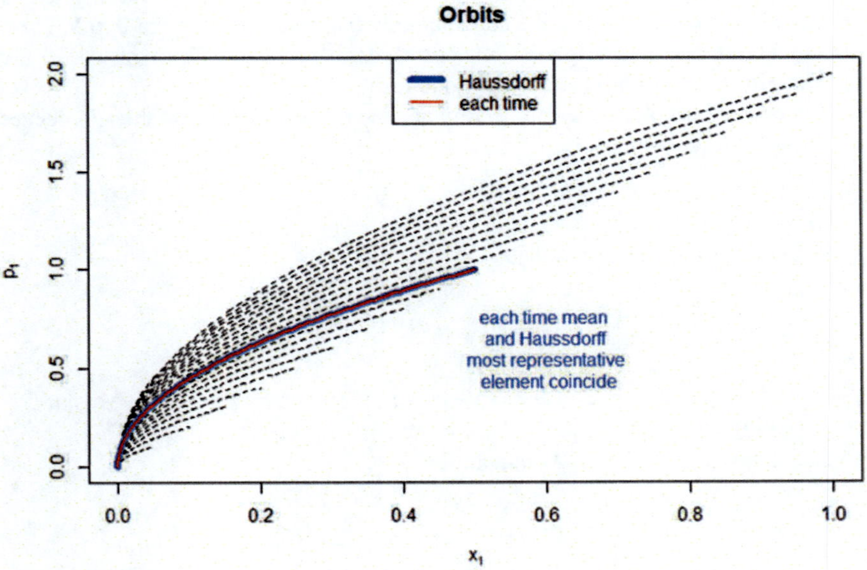

Fig. 6.18 Orbits $(x_1(t), p_1(t))$ of the system. Again, the approach by the Hausdorff distance produces the same result as the mean $E((x_1(t), p_1(t)))$ taken at each time

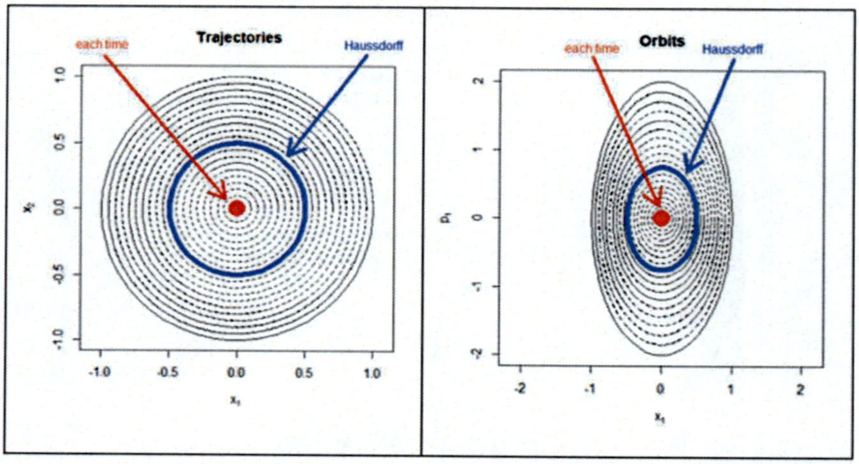

Fig. 6.19 Trajectories and orbits of the system. Here, the approach by the Hausdorff distance does not produce the same result as the mean $E((x_1(t), x_2(t)))$ taken at each time: means at fixed time are null, while Hausdorff distance produces a correct result

$$E(x_1(t)) = 0, E(x_2(t)) = 0, \ E(p_1(t)) = 0, E(p_2(t)) = 0.$$

The trajectories are concentric circles: for each u, the corresponding trajectory is the circle of radius $|u|$, centered at the origin $(0, 0)$. Considering that u and $-u$ generate the same circle of radius $|u|$, the family has radius going from 0 to 1. Thus, we expect as mean trajectory the circle corresponding to $|u| = 1/2$. In this case, only the Hausdorff approach furnishes the good result.

The evaluations of the mean trajectory and the mean orbit by both the approaches are shown in Fig. 6.19.

Exercises

1. Consider the differential equation

$$\frac{dx}{dt} = (1-t)x^u, \ \ x(0) = 1, t \in (0, 1).$$

Assume that $u \sim T(1, 2, 3)$ and determine the most representative history $(t, x(t))$ and the most representative orbit $\left(x(t), \frac{dx}{dt}(t)\right)$.

2. Consider the differential equation

$$\frac{dx}{dt} = (1-t)x^{u_1}, \ \ x(0) = u_2, t \in (0, 1).$$

Assume that $u_1 \sim T(1, 2, 3)$ and $u_2 \sim N(2, 0.1)$. Determine the most representative history $(t, x(t))$ and the most representative orbit $\left(x(t), \frac{dx}{dt}(t)\right)$.

(continued)

3. Consider the differential equation

$$\frac{d}{dt}\begin{pmatrix} x_1 \\ x_2 \end{pmatrix} = \begin{pmatrix} (2+u)x_2 \\ -(2+u)x_1 - 4x_1^2 \end{pmatrix}, \begin{pmatrix} x_1(0) \\ x_2(0) \end{pmatrix} = \begin{pmatrix} 0 \\ 2 \end{pmatrix}, t \in (0,1).$$

Assume that $u \sim T(-1,0,1)$ and determine the most representative trajectory $(x_1(t), x_2(t))$ and the most representative orbit $\left(x_1(t), \frac{dx_1}{dt}(t)\right)$.

4. Consider the differential equation

$$\frac{dx}{dt} = t^{1-u}(1+x)^u, \quad x(0)=1, \ t \in (0,1).$$

Assume that $u \sim T(0, 0.5, 1)$ and determine the most representative history $(t, x(t))$ and the most representative orbit $\left(x(t), \frac{dx}{dt}(t)\right)$.

5. Consider the differential equations

$$\frac{d}{dt}\begin{pmatrix} x_1 \\ x_2 \end{pmatrix} = \begin{pmatrix} -u_1 & 2 \\ 4 & -u_2 \end{pmatrix}\begin{pmatrix} x_1 \\ x_2 \end{pmatrix}, \begin{pmatrix} x_1(0) \\ x_2(0) \end{pmatrix} = \begin{pmatrix} 2 \\ 5 \end{pmatrix}, t \in (0,1).$$

Assume that the distributions of u_1 and u_2 are triangular: $u_1 \sim T(0,1,2)$ and $u_2 \sim T(0,4,8)$. Determine the most representative the most representative trajectory $(x_1(t), x_2(t))$ and the most representative orbit $\left(x_1(t), \frac{dx_1}{dt}(t)\right)$.

6. Consider $t \in (0,1)$ and

$$\frac{d}{dt}\begin{pmatrix} x_1 \\ x_2 \end{pmatrix} = \begin{pmatrix} 0.2-5u_1 & -2.5-5u_1 \\ 1 & 0 \end{pmatrix}\begin{pmatrix} x_1 \\ x_2 \end{pmatrix} + \begin{pmatrix} 8 \\ 3 \end{pmatrix}, \begin{pmatrix} x_1(0) \\ x_2(0) \end{pmatrix} = \begin{pmatrix} 2 \\ 5 \end{pmatrix}$$

Assume that the distributions $u_1 \sim T(-0.1, 0, 0.1)$ and $u_2 \sim N(0, 0.1)$. Determine the most representative trajectory $(x_1(t), x_2(t))$ and the most representative orbit $\left(x_1(t), \frac{dx_1}{dt}(t)\right)$.

7. Consider $t \in (0,1)$ and

$$\frac{d}{dt}\begin{pmatrix} x_1 \\ x_2 \end{pmatrix} = \begin{pmatrix} 2+u_1 & -3+u_2 \\ 1 & 0 \end{pmatrix}\begin{pmatrix} x_1 \\ x_2 \end{pmatrix} + \begin{pmatrix} 10+5u_1 \\ -3 \end{pmatrix}, \begin{pmatrix} x_1(0) \\ x_2(0) \end{pmatrix} = \begin{pmatrix} 2 \\ 5 \end{pmatrix}$$

Assume that the distributions of u_1 and u_2 are triangular: $u_1 \sim T(1,2,3)$ and $u_2 \sim T(0,1,2)$. Determine the most representative trajectory $(x_1(t), x_2(t))$ and the most representative orbit $\left(x_1(t), \frac{dx_1}{dt}(t)\right)$.

8. Consider $t \in (0,1)$ and

$$\frac{d}{dt}\begin{pmatrix} x_1 \\ x_2 \end{pmatrix} = \begin{pmatrix} -u_1 & -3-u_2 \\ 0.5+u_2 & -2-u_2 \end{pmatrix}\begin{pmatrix} x_1 \\ x_2 \end{pmatrix} + \begin{pmatrix} 0.2+u_1 \\ 0.1+u_2 \end{pmatrix}, \begin{pmatrix} x_1(0) \\ x_2(0) \end{pmatrix} = \begin{pmatrix} 2 \\ 5 \end{pmatrix}$$

Assume that the distributions of u_1 and u_2 are triangular: $u_1 \sim T(-0.1,0,0.2)$ and $u_2 \sim T(-0.5,0,0.5)$ the most representative trajectory $(x_1(t), x_2(t))$ and the most representative orbit $\left(x_1(t), \frac{dx_1}{dt}(t)\right)$.

Chapter 7
UQ in Game Theory

Abstract In this chapter, we examine mathematical games under uncertainty (for instance, in probabilities or payoffs). We examine also evolutionary dynamics associated with games and we present methods for the determination of the probability distribution of variables and the analysis of strategies. For evolutionary dynamics, we consider also trajectories and orbits.

As indicated in the title – and in the introduction of their book – Von Neumann and Morgenstern look for a new approach in Economic Sciences. In their book, they establish the basis of GT, namely the assumptions about rational behavior of the players, fixed rules, payoffs, and outcomes.

Since Von Neumann's works, GT was widely developed, including in probabilistic aspects. However, uncertainty quantification (UQ) in game theory (GT) remains rare.

In this chapter, we examine some possible uses of UQ in GT: notice that this *is not* a chapter about GT – although we recall some basic elements – but about the applications of UQ to GT – indeed, the reader can find in the literature many presentations of GT provided by specialists of the domain. Here, we focus on the specific developments concerning the use of UQ in the framework of GT.

7.1 The Language from Game Theory

It is important to recall some terms often used in the sequel. At first, the term "game" itself, which is used as an abbreviation of the complete expression "mathematical game of strategy". Von Neumann and Morgenstern consider a ***game*** as being simply a set of rules (Von Neumann & Morgenstern, 1953, p. 49). They make the difference between this abstract object and one of its concrete achievements – a ***play***. In terms of modern informatics, we may say that a game is an *object* and a play is an *instance* of a game. As said, games are defined by ***rules*** – *id est*, the laws governing the game. The rules stipulate what is allowed or forbidden, winnings and losses, the end of the

E. Souza de Cursi, *Uncertainty Quantification using R*, International Series in Operations Research & Management Science 335,
https://doi.org/10.1007/978-3-031-17785-9_7

game, the number of participants, etc. In a mathematical game of strategy, the participants must take decisions. A ***move*** is a decision took by or for a player: a ***personal move*** depends only on the decision of the player, while a ***random move*** involves some randomness. Moves may be ***simultaneous*** (when all the players reveal their decision and make their moves at the same time) or ***sequential*** (when only one player is allowed to move at a time). ***Static games*** allow only a unique simultaneous move for all the players, while ***dynamic games*** involve sequential moves or sequences of simultaneous moves. As consequence of the moves, the state of the play (and of the players) changes – the ***position*** of the play evolves from its initial state to the actual one by the moves of the players. We say that the players have a ***complete information*** when all the players know all the rules. They have a ***perfect information*** when all the players know the preceding moves and positions. The game is ***cooperative*** when negotiations between players are possible and coalitions of players may be created – otherwise, it is a ***non-cooperative game***. In a game, each player looks for the maximization of a ***utility*** function generated by the payoffs – in general, the payoffs themselves: we say that the players have a ***rational behavior***. A ***zero-sum game*** is a game where the sum of the payoffs is null: a player may have a positive payoff if and only if other players have negative payoffs corresponding to his gains. In mathematical games of strategy, it is assumed that ***there is no cheating***.

Rational players often use a ***strategy***, *id est*, predefined sequences of moves conditionally determined by the actual position of the game: if the position is P_i then the move is M_i. We say that the strategy is ***pure*** when it consists in choosing a fixed move among the possible ones. We say that the strategy is ***mixed*** when it uses different moves. In general, mixed strategies randomize among the different moves with a given probability. A player may have a ***dominant strategy***, *id est*, a strategy that maximizes his utility for any move of the other players; but it may also have a ***dominated strategy***, *id est*, a strategy that minimizes his utility for any move of the other players.

An ***equilibrium*** – sometimes called ***Nash equilibrium*** -is a set of strategies such that no player may increase his own utility by modifying only his own strategy: if all the other players keep their strategies, he must keep it too – otherwise, he will reduce his own winnings. Analogously to strategies, an equilibrium may be ***pure*** or ***mixed***, according to the use of a single fixed move or several moves with a given probability.

Games are generally represented in a ***normal form***: assume that, at some stage, k players may choose among n possible moves $\boldsymbol{M} = \{M_1, \ldots, M_n\}$: a move is represented by a $k-$ dimensional vector $\boldsymbol{m} = (m_1, \ldots, m_k) \in \boldsymbol{M}^k$, where $m_i \in \boldsymbol{M}, \forall\, i$. The result of these move is a payoff for each player, represented by a vector of payoffs $\boldsymbol{u}(\boldsymbol{m}) = (u_1(\boldsymbol{m}), \ldots, u_k(\boldsymbol{m}))$. We may represent such a game by giving the set $\boldsymbol{M}$ and the function $\boldsymbol{u}(\boldsymbol{m})$. In simple situations, the normal form of a game may be represented by a matrix: we say that the game is under **matrix form**. An alternative representation is the ***extended form*** – a tree representing the choices of the players. A second alternative form is the **multimatrix form**, where the payoffs of each player appear in a different matrix.

Uncertainties arrive to Game Theory by different ways – for instance, errors or lacks in the information, randomness involved in the moves (for instance, in mixed strategies), and uncertainty in the payoffs. A game involving some uncertainty is referred in the continuation as an ***Uncertain Game (UG).***

Examples of UG are ***Stochastic Games (SG)*** and ***Bayesian Games (BG)***. A ***SG*** is a dynamical game where the new state is randomly determined at each stage. A ***BG*** is a game where the information of the players is incomplete.

In BG, the assumptions made by one player about the characteristics of the other players are his ***beliefs***. The real set of characteristics of a player – for instance, his real payoffs – is called his ***type***. For instance, if the payoffs are uncertain – then, represented by random variables: the distribution assumed by a player about the payoff of another player is a belief. The real distribution is part of the type of the player.

In UG, it is usual introduce a special supplementary player called ***Nature***, destined to choose randomly the uncertain quantities, such as, for instance, the state of the game, the payoffs, or the type of the player.

Concerning UG, GT is mainly interested in the expected payoff of each player, *id est*, in the mean of the payoffs considered as random variables. UQ is mainly interested in the probability distributions of the payoffs – which may be used to evaluate the means but may also be used to implement other decision rules. The equivalent of the Nash equilibrium is the ***Equal Opportunity*** situation, where all the players have the same probability of obtaining a given non-negative payoff (often zero) – notice that equal opportunity does not mean that all the players have the same expected payoff: this last situation referred as ***Egalitarian Solution***. In the sequel, we examine some classical games from both the points of view.

Evolutionary Games(EG) result from the application of GT to biology and look for models describing the evolution of populations having different strategies of survival (see, for instance, Lewontin, 1961; Maynard-Smith & Price, The logic of animal conflict, 1973; Maynard-Smith, 1982). In EG, payoffs are interpreted as a measure of the adaptation – a measure of fitness to the environment. The main tool in EG is the ***replicator equation***, which describes the evolution of a population formed by groups having different strategies. Replicator equations are generated from Mathematical Games by considering the mean payoffs: groups having mean payoffs superior to the general mean are more adapted, have a higher fitness and, so, tend to increase, while groups having payoffs inferior to the general mean are less adapted, have a lower fitness, and tend to decrease. We shall examine replicator equations connected to the classical games considered. In EG, uncertainty arises by several ways: for instance, initial data are often unknown, and payoffs must be estimated.

7.2 A Simplified Odds and Evens Game

Let us consider a simplified version of the game Odds/Evens (OE): a game for two players, where both the players reveal simultaneously a number of fingers and Player 1 wins if the total number of fingers revealed is even, while Player 2 wins if the result is odd. Assume that the winner receives 1$, paid by the loser. This game is UG, since the choices of the players are uncertain.

Table 7.1 Matrix representation of the coin game

	Player 2	
Player 1	$O\ (p_2)$	$E\ (q_2 = 1 - p_2)$
$O\ (p_1)$	$(1, -1)$	$(-1, 1)$
$E\ (q_1 = 1 - p_1)$	$(-1, 1)$	$(1, -1)$

Basically, each player must decide to reveal an odd or even number of fingers: the result will be defined by the sum of the fingers revealed by each player: Player 1 wins if both the players make the same choice (odd or even) and Player 2 wins if they make different choices (one odd, the other even).

Let us denote O an odd result and E an even one. Let $C_i \in \{O, E\}$ be the choice of Player i, $i = 1, 2$. Then, the possible moves are the couples (C_1, C_2): $\boldsymbol{m} \in \{(O, O), (O, E), (E, O), (E, E)\}$. We have $\boldsymbol{u}(O, O) = (1, -1)$, $\boldsymbol{u}(O, E) = (-1, 1)$, $\boldsymbol{u}(E, O) = (-1, 1)$, $\boldsymbol{u}(E, E) = (1, -1)$. It is a zero-sum static game with random moves. Let us denote by p_i the probability of revealing an odd number of fingers for Player i, $i = 1, 2$: for a completely random strategy, $p_i = \frac{1}{2}$, but, in practice, a deviation can be observed – there is some uncertainty about p_i, which is unknown in practice. The game may be represented in normal form as shown in Table 7.1.

Let W_i be the winnings of Player i, $i = 1, 2$. From this Table, we see that:

- If Player 1 chooses O, he wins with probability p_2, so that $P(W_1 = 1 | C_1 = O) = p_2$, $P(W_1 = -1 | C_1 = O) = q_2 = 1 - p_2$.
- If Player 1 chooses E, he loses with probability p_2, so that $P(W_1 = -1 | C_1 = E) = p_2$, $P(W_1 = 1 | C_1 = E) = q_2 = 1 - p_2$.
- If Player 2 chooses O, he loses with probability p_1, so that $P(W_2 = -1 | C_2 = O) = p_1$, $P(W_2 = 1 | C_2 = O) = q_1 = 1 - p_1$.
- If player 2 chooses E, he wins with probability p_1, so that $P(W_2 = 1 | C_2 = E) = p_1$, $P(W_2 = -1 | C_2 = E) = q_1 = 1 - p_1$.

Then, we have

$$P(W_1 = 1) = P(W_1 = 1 | C_1 = O) P(C_1 = O) + P(W_1 = 1 | C_1 = E) P(C_1 = E),$$

so that

$$P(W_1 = 1) = \pi = p_1 p_2 + q_1 q_2 = (1 - p_1) + (2p_1 - 1) p_2.$$

Analogously,

$$P(W_1 = -1) = p_1 q_2 + q_1 p_2 = p_1 - (2p_1 - 1) p_2 = 1 - \pi.$$

Thus, the mean wins of Player 1 are

$$E(W_1) = P(W_1 = 1) - P(W_1 = -1) = (2p_1 - 1)(2p_2 - 1) = 2\pi - 1 = \mu.$$

and

$$E(W_1^2) = P(W_1 = 1) + P(W_1 = -1) = 1,$$

so that the variance of W_1 is

$$V(W_1) = 1 - (2\pi - 1)^2 = \beta.$$

Analogously,

$$P(W_2 = 1) = P(W_1 = -1) = 1 - \pi;$$
$$P(W_2 = -1) = P(W_1 = 1) = \pi;$$
$$E(W_2) = -E(W_1) = 1 - 2\pi.$$
$$V(W_2) = V(W_1) = 1 - (2\pi - 1)^2$$

7.2.1 GT Strategies When $\mathbf{p} = (p_1, p_2)$ *Is Known*

Let us consider Player 1: his strategy is defined by the probability p_1 of choosing O: we will examine below some possible strategies for its choice. The game may be put into a Matrix Form, as indicated in Table 7.1. The game can also be represented by a tree. From the point of view of a given player, the tree starts by his own decisions and finishes by the payoffs. For instance, the tree corresponding to the point of view of Player 1 is shown in Fig. 7.1.

Tree representation is a practical tool for the analysis of some strategies, by performing a reverse analysis on the tree. For instance, let us consider the classical ***minimax*** and ***maximin*** strategies:

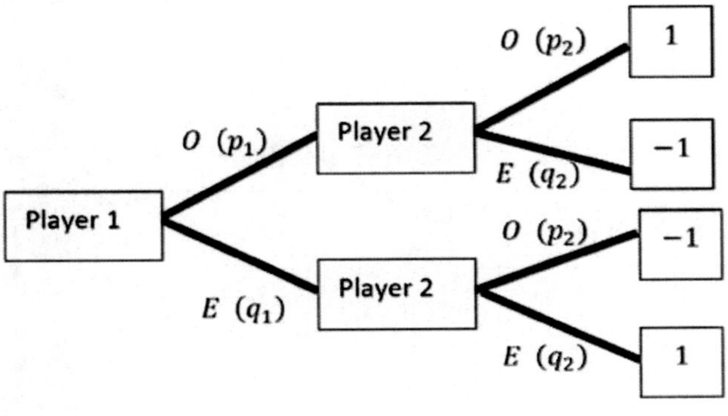

Fig. 7.1 Representation of the game Odds/Evens by a tree. Point of view of Player 1

- **Minimax**: the minimax strategy consists in minimizing the maximum loss. From the tree, we quick obtain the result by reverse analysis (see Fig. 7.2). Since the maximum loss is -1 for any decision, it is indifferent to the choice O or E.
- **Maximin**: the maximin strategy consists in maximizing the minimal payoff. Again, this strategy is indifferent to the choice O or E, since the maximum winning is 1 in each case.

Trees are also useful for the evaluation of means (Fig. 7.3) and probabilities. For instance, from Fig. 7.4, we have

$$P(W_1 = -1) = 1 - \pi, \ P(W_2 = 1) = 1 - P(W_2 = -1) = \pi.$$

Other classical strategies in Game Theory are

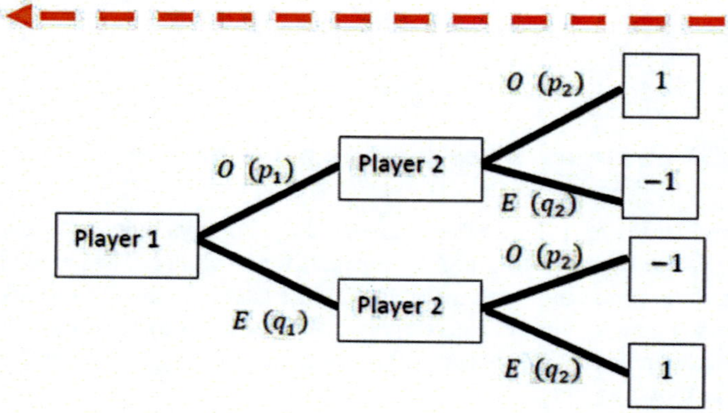

Fig. 7.2 Using the tree to find a Minimax/Maximin strategy for Player 1 in the game Odds/Evens

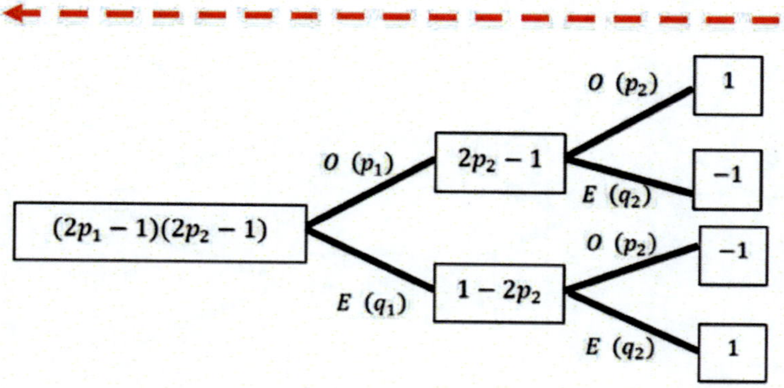

Fig. 7.3 Using the tree to evaluate the expected payoff of Player 1

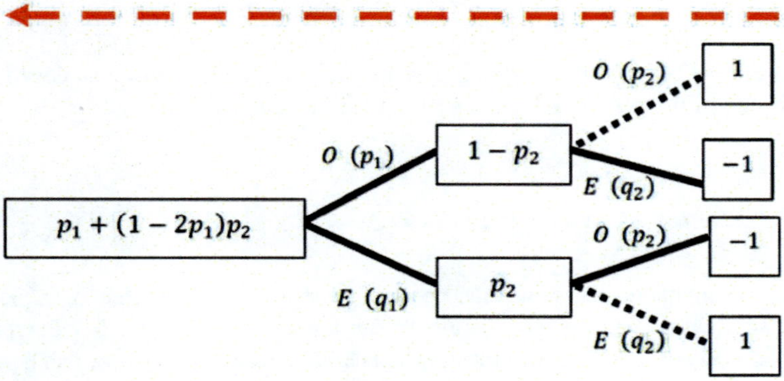

Fig. 7.4 Using the tree to evaluate $P(W_1 = -1)$

- The **fixed *O* strategy**: Player 1 always choose O – then, $p_1 = 1$. Then, $\pi = p_2$ and the expected value of his payoff is $\mu = 2p_2 - 1$, with a variance equal to $\beta = 1 - (2p_2 - 1)^2$. For Player 1, the probability of winning 1 \$ is $\pi = p_2$. This strategy is interesting if $p_2 > 1/2$: in this case, it maximizes $P(W_1 = 1)$ and $E(W_1)$.
- The **fixed *E* strategy**: Player 1 always choose E – then, $p_1 = 0$. Then, $\pi = 1 - p_2$ and the expected value of his payoff is $\mu = 1 - 2p_2$, with a variance equal to $\beta = 1 - (2p_2 - 1)^2$. For Player 1, the probability of winning 1\$ is $\pi = 1 - p_2$. Such a strategy is interesting if $p_2 < 1/2$: analogously, it maximizes $P(W_1 = 1)$ and $E(W_1)$.
- The **uniformly random strategy**: choose O or E at random, with equal probability. Then, $p_1 = \frac{1}{2}$: the expected value of his payoff is $\mu = 0$, with a variance $\beta = 1$. For Player 2, the probability of winning 1\$ is $\pi = \frac{1}{2}$.

Let us try to build a mixed strategy corresponding to a Nash's equilibrium. In Game Theory, a mixed Nash equilibrium is determined by making the players indifferent to the choice among the strategies. For instance, let us denote W_O^1, W_E^1 the expected winnings of Player 1 when ***Player 2*** (and not Player 1 himself) chooses O or E, respectively. We have

$$W_O^1 = p_1 - (1 - p_1) = 2p_1 - 1,\ W_E^1 = -p_1 + (1 - p_1) = 1 - 2p_1.$$

Player 1 is indifferent to the result when $W_O^1 = W_{\mathrm{E}}^1$, so that $p_2 = 1/2$ – this corresponds to the uniformly random strategy. In this case, the expected payoff is zero for all the strategies and all the players. We can make the same analysis for Player 2: $W_O^2 = W_E^2$ corresponds to $1 - 2p_2 = 2p_2 - 1$, *id est*, $p_2 = 1/2$ – a purely random strategy.

In this game, equal opportunity corresponds to $P(W_1 = 1) = P(W_2 = 1)$, so that $\pi = 1 - \pi \Rightarrow \pi = 1/2 \Rightarrow \mu = 0$. Analogously, an Egalitarian solution corresponds to $E(W_1) = E(W_2) \Rightarrow \mu = -\mu \Rightarrow \mu = 0 \Rightarrow \pi = 1/2$. Thus, in both the cases,

$$(2p_1 - 1)(2p_2 - 1) = \mu = 0,$$

so that $p_1 = 1/2$ or $p_2 = 1/2$: equal opportunity and Egalitarianism are achieved if and only if one of the players adopts a purely random strategy.

7.2.2 *Strategies When* p *Is Unknown*

The preceding analysis assumes the knowledge of $p = (p_1, p_2)$, what is, in practice, unrealistic, since each player does not know the strategy of the other Player. For instance, if Player 2 has not any information about the value of p_1, how can he define a strategy?

If the game is static, *id est*, if the play consists in a single move, the definition of a strategy is possible only if Player 2 has some information about p_1. Otherwise, the only option is a random choice. If the game is dynamic, *id est*, a play consists in several moves and p_1 is unknown, Player 2 may use the results of the previous moves to define a strategy.

Indeed, this game becomes BG when the value of p_1 is unknown: it is a lack of information and the value of p_1 assumed by Player 2 is his belief. As previously observed, it may be updated from the results of the preceding moves. For instance, Player 2 may use a Bayesian procedure – notice that Bayesian corrections are destined to reduce the uncertainty and are not effective to make any correction if a single value of p_1 is given: to use a Bayesian approach, it is necessary to assume a probability distribution for p_1 – the *prior*. For instance, we may assume a uniform distribution on $\left(\frac{1}{2} - \delta, \frac{1}{2} + \delta\right), \frac{1}{2} > \delta > 0$ and apply the Bayes approach to get a new distribution – the *posterior*. However, in this simple game, it is possible to define other strategies, simpler to implement. Let us examine some of them.

If the game is repeated n times, Player 2 has payoffs $W_2^1, \ldots, W_2^n$ and his total payoff is $S_2^n = W_2^1 + \ldots + W_2^n$. Notice that the possible values of S_2^n are $\{n, n - 2, \ldots, n \bmod 2, -n \bmod 2, \ldots, -n + 2, -n\}$. For instance: if $n = 2$, then the possible values are $\{2, 0, -2\}$. For $n = 3$, the possible values are $\{3, 1, -1, -3\}$. In classical GT, Player 2 looks for the maximum of the expected value $E\left(S_2^n\right)$. In the UQ approach, Player 2 appears as interested in maximizing $P\left(S_2^n \geq s\right)$ – the probability of a minimal winning s, defined by the Player – in general $s = 0$ (non-negative payoff). In UQ, the analogous of the Nash equilibrium is equal opportunity – the situation where the probabilities of winning a non-negative payoff are equal for all the players. In this simple game, it corresponds to $P\left(S_2^n \geq 0\right) = 1/2$. The equal opportunity for a single move corresponds to $\pi = 1/2$, so that $p_2 = 1/2$ – the uniformly random strategy.

If Player 2 keeps the same strategy along all the dynamic play, the distribution of S_2^n is analogous to a binomial one. Indeed, let $Y_i = \frac{1+W_2^i}{2}$: Y_i is a Bernoulli variable such that $P(Y_i = 1) = 1 - \pi$. Thus, $Z^n = Y_1 + \ldots + Y_n$ is a Binomial variable $\mathcal{B}(n, \pi)$. Since $W_2^i = 2\,Y_i - 1$, we have $S_2^n = 2Z^n - n$, so that

$$P(S_2^n = k) = P\left(Z = \frac{n+k}{2}\right) = \binom{n}{\frac{n+k}{2}} \pi^{\frac{n-k}{2}} (1-\pi)^{\frac{n+k}{2}}$$

$$E(S_2^n) = n(1-2\pi),\ V(S_2^n) = 4n\pi(1-\pi)$$

The probability of winning a non-negative payoff (at least 0$) is

$$P(S_2^n \geq 0) = P\left(Z^n \geq \frac{n}{2}\right) = 1 - P\left(Z^n < \frac{n}{2}\right)$$

Equal opportunity means that $P\left(Z^n \geq \frac{n}{2}\right) = P\left(Z^n \leq \frac{n}{2}\right)$, id est, that $n/2$ is a median of Z^n. Thus, $n(1 - \pi) = n/2$, so that $\pi = 1/2$ and, again, $p_2 = 1/2$ – as in the preceding, if one of the players adopts a uniformly random strategy, he produces equal opportunity independently of the other player.

If Player 2 does not want to offer an equal opportunity to Player 1, he may choose a different strategy. For instance, he may bet on an eventual defect in the implementation of the strategy of Player 1 – a small deviation in the real value of p_1 when compared to the value defined by the strategy of Player 1.

A first simple way to define such a strategy consists in using the previous choices of Player 1. For instance, Player 2 can choose to reveal the same result as the preceding choice of Player 1 (cyclic strategy) or the contrary of this same choice (anticyclic strategy). In this case, the game becomes a Markov chain describing the choices of the players. The chain has four states: $\{s_1 = (O, O), s2 = (O, E), s_3 = (E, O), s_4 = (E, E)\}$. For the cyclic strategy, the transition matrix is

$$M_{cyc} = \begin{pmatrix} p_1 & p_1 & 0 & 0 \\ 0 & 0 & p_1 & p_1 \\ 1-p_1 & 1-p_1 & 0 & 0 \\ 0 & 0 & 1-p_1 & 1-p_1 \end{pmatrix}.$$

For the anticyclic strategy, the transition matrix is

$$M_{anti} = \begin{pmatrix} 0 & 0 & p_1 & p_1 \\ p_1 & p_1 & 0 & 0 \\ 0 & 0 & 1-p_1 & 1-p_1 \\ 1-p_1 & 1-p_1 & 0 & 0 \end{pmatrix}.$$

These chains have stationary distributions:

$$P^s_{cyc} = \left(p_1^2, p_1(1-p_1), p_1(1-p_1), (1-p_1)^2\right),$$

$$P^s_{anti} = \left(p_1(1-p_1), p_1^2, (1-p_1)^2, p_1(1-p_1)\right),$$

In addition,

$$\forall k \geq 2 : M^k_{cyc} = \begin{pmatrix} p_1^2 & p_1^2 & p_1^2 & p_1^2 \\ p_1(1-p_1) & p_1(1-p_1) & p_1(1-p_1) & p_1(1-p_1) \\ p_1(1-p_1) & p_1(1-p_1) & p_1(1-p_1) & p_1(1-p_1) \\ (1-p_1)^2 & (1-p_1)^2 & (1-p_1)^2 & (1-p_1)^2 \end{pmatrix}.$$

$$\forall k \geq 2 : M^k_{anti} = \begin{pmatrix} p_1(1-p_1) & p_1(1-p_1) & p_1(1-p_1) & p_1(1-p_1) \\ p_1^2 & p_1^2 & p_1^2 & p_1^2 \\ (1-p_1)^2 & (1-p_1)^2 & (1-p_1)^2 & (1-p_1)^2 \\ p_1(1-p_1) & p_1(1-p_1) & p_1(1-p_1) & p_1(1-p_1) \end{pmatrix}.$$

Assume that the initial choice of Player 2 is O with probability p_2. Then, the initial state of the chain is $P^0 = (p_2p_1, (1-p_2)p_1, p_2(1-p_1), (1-p_2)(1-p_1))$. The state at step $k \geq 1$ is

$$P^k_{cyc} = M^k_{cyc}P^0 = P^s_{cyc}, P^k_{anti} = M^k_{anti}P^0 = P^s_{anti}.$$

Thus, the mean winning of Player 2 at each step is

$$\mu_{cyc} = -(1-2p_1)^2, \ \mu_{anti} = (1-2p_1)^2$$

For Player 2, the probability of winning 1$ is

$$P\left(W^2_{cyc} = 1\right) = 2p_1(1-p_1), \ P\left(W^2_{anti} = 1\right) = p_1^2 + (1-p_1)^2$$

Consequently:

- The cyclic strategy leads to $\mu_{cyc} = -(2p_1 - 1)^2 \leq 0$, and a probability of winning 1 \$ of $P\left(W^2_{cyc} = 1\right) = 2p_1(1-p_1) \leq 1/2$.
- The anticyclic choice leads to $\mu_{anti} = (2p_1 - 1)^2 \geq 0$, and a probability of winning 1 \$ of $P\left(W^2_{anti} = 1\right) = p_1^2 + (1-p_1)^2 \geq 1/2$.

Based on these results, Player 2 must choose the anticyclic strategy: his expected payoff is non-negative, and the probability of a positive payoff is at least 1/2. The cyclic strategy leads to a non-positive expected payoff and a probability of positive payoff inferior to 1/2.

Table 7.2 $P(S_2^n \geq 0)$: Anticyclic probability of winning a non-negative payoff for Player 2

	n							
p_1	10	20	30	40	50	100	1000	10,000
0.4	0.67	0.66	0.66	0.66	0.66	0.69	0.90	1.00
0.45	0.64	0.61	0.59	0.59	0.58	0.58	0.64	0.84
0.49	0.62	0.59	0.57	0.56	0.56	0.54	0.52	0.52
0.5	0.62	0.59	0.57	0.56	0.56	0.54	0.51	0.50
0.51	0.62	0.59	0.57	0.56	0.56	0.54	0.52	0.52
0.55	0.64	0.61	0.59	0.59	0.58	0.58	0.64	0.84
0.6	0.67	0.66	0.66	0.66	0.66	0.69	0.90	1.00

Table 7.3 $E(S_2^n)$: Anticyclic expected payoff for Player 2

	n							
p_1	10	20	30	40	50	100	1000	10,000
0.4	0.4	0.8	1.2	1.6	2.0	4.0	40.0	400.0
0.45	0.1	0.2	0.3	0.4	0.5	1.0	10.0	100.0
0.49	0.0	0.0	0.0	0.0	0.0	0.0	0.4	4.0
0.5	0.0	0.0	0.0	0.0	0.0	0.0	0.0	0.0
0.51	0.0	0.0	0.0	0.0	0.0	0.0	0.4	4.0
0.55	0.1	0.2	0.3	0.4	0.5	1.0	10.0	100.0
0.6	0.4	0.8	1.2	1.6	2.0	4.0	40.0	400.0

We can analyze the effects of deviation in the real value of p_1 by considering different values: recall that

$$P\left(S_{2,anti}^n = k\right) = \binom{n}{\frac{n+k}{2}} p^{\frac{n+k}{2}} (1-p)^{\frac{n-k}{2}}, p = p_1^2 + (1-p_1)^2$$

Tables 7.2 and 7.3 show, respectively, the values of $P(S_2^n \geq 0)$ and $E(S_2^n)$ for some values of p_1 and n, when Player 2 uses strategy anticyclic: they increase when p_1 deviates from ½.

A second simple strategy consists in estimating p from the previous results and applying the anticyclic strategy with the estimated value. Let O_n be the number of times where Player 1 has chosen O in the first n moves. Recall that the estimator $\widehat{p}_{1,n}$ for p_1 after n moves is $\widehat{p}_{1,n} = O_n/n$. The probability $\alpha_n = P(\widehat{p}_{1,n} < 1/2) = P(O_n < n/2)$ is given in Table 7.4. Notice that O_n is binomially distributed $\mathcal{B}(n, p_1)$.

If Player 2 uses the estimator $\widehat{p}_{1,n}$, then, at the move $n + 1$, he will choose E if $\widehat{p}_{1,n} > 1/2$, and O if $\widehat{p}_{1,n} < 1/2$. Thus,

$$P\left(W_2^{n+1} = 1\right) = \theta_n = p_1 + \alpha_n(1 - 2p_1), \; P\left(W_2^{n+1} = -1\right) = 1 - \theta_n,$$

$$E\left(W_2^{n+1}\right) = (1 - 2p_1)(2\alpha_n - 1), \; V\left(W_2^{n+1}\right) = 1 - (2p_1 - 1)^2(2\alpha_n - 1)^2.$$

Table 7.4 $\alpha_n = P\left(\widehat{p}_n < \frac{1}{2}\right)$

	n							
p_1	10	20	30	40	50	100	1000	10,000
0.4	0.63	0.76	0.82	0.87	0.90	0.97	1.00	1.00
0.45	0.50	0.59	0.64	0.68	0.72	0.82	1.00	1.00
0.49	0.40	0.45	0.47	0.49	0.50	0.54	0.73	0.98
0.5	0.38	0.41	0.43	0.44	0.44	0.46	0.49	0.50
0.51	0.35	0.38	0.38	0.39	0.39	0.38	0.25	0.02
0.55	0.26	0.25	0.23	0.21	0.20	0.13	0.00	0.00
0.6	0.17	0.13	0.10	0.07	0.06	0.02	0.00	0.00

Table 7.5 $P\left(S_2^n \geq 0\right)$ for strategy based on estimator $\widehat{p}_n$

	n							
p	10	20	30	40	50	100	1000	10,000
0.4	0.67	0.70	0.76	0.80	0.83	0.94	1.00	1.00
0.45	0.63	0.61	0.61	0.62	0.63	0.70	1.00	1.00
0.49	0.62	0.59	0.57	0.56	0.57	0.55	0.59	0.94
0.5	0.63	0.58	0.58	0.56	0.56	0.54	0.51	0.50
0.51	0.63	0.59	0.58	0.57	0.56	0.55	0.60	0.94
0.55	0.66	0.66	0.66	0.67	0.68	0.74	1.00	1.00
0.6	0.73	0.77	0.81	0.84	0.87	0.96	1.00	1.00

Table 7.6 Expected payoff for Player 2 for strategy based on the estimator $\widehat{p}_n$

	n							
p	10	20	30	40	50	100	1000	10,000
0.4	0.36	1.35	2.70	4.16	5.79	14.72	194.57	1995.30
0.45	−0.03	0.20	0.57	0.94	1.37	4.28	89.68	989.94
0.49	0.00	−0.02	−0.06	0.09	−0.11	0.17	6.21	152.96
0.5	−0.01	−0.07	−0.02	0.01	0.13	−0.03	−0.48	−0.11
0.51	0.06	0.11	0.08	0.11	0.26	0.29	7.02	155.88
0.55	0.34	0.73	1.27	1.69	2.30	5.44	91.19	991.10
0.6	0.99	2.28	3.67	5.33	7.09	16.28	196.22	1995.30

For this strategy, the exact evaluation of $P\left(S_2^n \geq 0\right)$ involves an expensive calculation when n increases. In Table 7.5, we show the values furnished by a Monte Carlo simulation with 10^4 runs. The values of $E\left(S_2^n\right)$ are shown in Table 7.6.

An alternative strategy based on the previous results consists in using statistical hypothesis testing. For example, Player 2 can test both the hypothesis $H_< : p \leq \frac{1}{2}$ and $H_> : p \geq \frac{1}{2}$. If both are accepted or both are rejected, Player 2 chooses uniformly random between O and E. If one is accepted and the other is rejected, he chooses the anticyclic strategy: either E if $H_>$ is accepted or O if $H_<$ is accepted. Notice that the decision for this kind of test is based on the number of O previously observed. Indeed, O_n is binomially distributed $\mathcal{B}(n, p)$. Thus, a decision concerning $H_<$ with a

Table 7.7 Typical values of $o_<$

	n							
α	10	20	30	40	50	100	1000	10,000
0.01	9	15	21	27	33	62	537	5116
0.05	8	14	19	25	31	58	526	5082
0.1	7	13	19	24	30	56	520	5064
0.25	6	12	17	22	27	53	511	5034

Table 7.8 Typical values of $o_>$

	n							
α	10	20	30	40	50	100	1000	10,000
0.01	1	5	9	13	17	38	463	4884
0.05	2	6	11	15	19	42	474	4918
0.1	3	7	11	16	20	44	480	4936
0.25	4	8	13	18	23	47	489	4966

significance level $\alpha_<$ may be taken by considering the maximum number $o_<$ such that $P(O_n > o_<) = \alpha_<$. If $O_n \leq o_<$, $H_<$ is accepted; otherwise, it is rejected. Analogously, $H_>$ is rejected if $O_n < o_>$, with $P(O_n < o_>) = \alpha_>$ and accepted otherwise. Typical values of $o_<$ and $o_>$ are exhibited in Tables 7.7 and 7.8. These values are furnished by the function `qbinom` of *R*.

This strategy is not as efficient as the previous ones. A Monte Carlo simulation of 10^6 plays has led to a $P\left(S_2^n \geq 0\right) \approx 0.5$ and an expected payoff of zero, for several α from 0.01 to 0.5 and values of n going from 10 to 10^4.

7.2.3 *Strategies for the Stochastic Game*

The game becomes stochastic (SG) if the probabilities change at random. For instance, imagine that the players choose a different value of p_i at each move. For instance, they can choose this value at random according to a given distribution. In this case, we say that the special player ***Nature*** chooses the probabilities and, thus, defines the types of the Players. Analogously to the preceding situation, p_i is a random variable having a distribution – which may be known or unknown. Since a draw is made at each move, Nature furnishes a different value p_i^k for each move. Then,

$$P\left(W_1^k = 1\right) = \pi^k = \left(1 - p_1^k\right) + \left(2p_1^k - 1\right)p_2^k,\ P\left(W_2^k = 1\right) = 1 - \pi^k.$$

For a given p^k and a given strategy p_2^k, the expected gain of Player 2 is

$$E\left(W_2^k | p_1 = p_1^k, p_2 = p_2^k\right) = \mu^k = 1 - 2\pi^k = \left(2p_1^k - 1\right)\left(1 - 2p_2^k\right).$$

The expected gain depends on the distribution of $p = (p_1, p_2)$ and of the strategy of the players. The equivalent of the Nash equilibrium on a single move is determined by setting $\pi^k = \frac{1}{2}$ – possible solutions are $p_1^k = \frac{1}{2}$ or $p_2^k = \frac{1}{2}$, i.e., the strategy of a random uniform choice of one of the players. Analogously, the equal opportunity and the Egalitarian solution correspond to $\pi^k = \frac{1}{2}$ and lead to the same results.

In this case, the anticyclic strategy is more difficulty to implement, since p_1^k varies with k. A solution is to use $\overline{p_1} = E(p_1)$ and take $p_2^k = 1 - \overline{p_1}$. Then,

$$\pi^k = \overline{p_1} + p_1^k - 2p_1^k \overline{p_1}, \quad E\left(W_2^k | p_1 = p_1^k\right) = \left(2p_1^k - 1\right)\left(2\overline{p_1} - 1\right).$$

Then,

$$E\left(\pi^k\right) = 2\overline{p_1}\left(1 - \overline{p_1}\right) \le \frac{1}{2}, \quad E\left(W_2^k\right) = \left(2\overline{p_1} - 1\right)^2 \ge 0$$

If the game is repeated n times, we may consider again the payoff $S_2^n = W_2^1 + \ldots + W_2^n$ of Player 2 and make an analysis similar to the preceding ones. From the preceding, we have $E\left(S_2^n\right) = n(2\overline{p} - 1)^2$: analogously to the previous situation, the expected value increases with n when $\overline{p_1} \ne \frac{1}{2}$. The distribution of S_2^n is not binomial-like, since it depends on the distribution of p_1, what makes difficulty the evaluation of $P\left(S_2^n \ge 0\right)$. To solve the difficulty, we can use a Monte Carlo simulation to estimate such a probability. The results obtained this way are shown in Table 7.9 – the calculations assume that p is uniformly distributed on $(\overline{p} - \delta, \overline{p} + \delta)$ $(\delta > 0)$.

Table 7.9 $P\left(S_n^2 \ge 0\right)$ for the anticyclic strategy in the stochastic game, evaluated by Montecarlo simulation (1e6 realizations)

$\overline{p}$ \ δ	0.01	0.05	0.1	0.15	0.2	$\overline{p}$ \ δ	0.01	0.05	0.1	0.15	0.2
n = 10						n = 100					
0.4	0.55	0.55	0.54	0.55	0.55	0.4	0.65	0.65	0.65	0.65	0.65
0.45	0.51	0.51	0.51	0.51	0.51	0.45	0.54	0.54	0.54	0.54	0.54
0.49	0.50	0.50	0.50	0.50	0.50	0.49	0.50	0.50	0.50	0.50	0.50
0.5	0.50	0.50	0.50	0.50	0.50	0.5	0.50	0.50	0.50	0.50	0.50
0.51	0.50	0.50	0.50	0.50	0.50	0.51	0.50	0.50	0.50	0.50	0.50
0.55	0.51	0.51	0.51	0.51	0.51	0.55	0.54	0.54	0.54	0.54	0.54
0.6	0.55	0.55	0.55	0.55	0.54	0.6	0.65	0.65	0.65	0.65	0.65
n = 1000						n = 10, 000					
0.4	0.89	0.89	0.89	0.89	0.89	0.4	1.00	1.00	1.00	1.00	1.00
0.45	0.62	0.62	0.62	0.62	0.62	0.45	0.84	0.84	0.84	0.84	0.84
0.49	0.50	0.50	0.51	0.50	0.51	0.49	0.52	0.52	0.52	0.52	0.52
0.5	0.50	0.50	0.50	0.50	0.50	0.5	0.50	0.50	0.50	0.50	0.50
0.51	0.50	0.50	0.51	0.50	0.50	0.51	0.52	0.52	0.52	0.52	0.52
0.55	0.62	0.62	0.62	0.62	0.62	0.55	0.84	0.84	0.84	0.84	0.84
0.6	0.89	0.89	0.89	0.89	0.89	0.6	1.00	1.00	1.00	1.00	1.00

In practice, the value of $\overline{p_1}$ may be unknown: Player 2 assumes a value which is a belief. Analogously to the preceding, Player 2 can use an estimator $\widehat{p}_{1,n} = O_n/n$ to take a decision at move $n + 1$. In this case, $\pi^k = \widehat{p}_{1,k} + p_1^k - 2p_1^k \widehat{p}_{1,k}$. If move k is independent from the past moves, we have again $E(\pi^k) = 2\overline{p_1}(1 - \overline{p_1})$ and $E(W_2^k) = (2\overline{p_1} - 1)^2$.

7.2.4 Replicator Dynamics

As observed in the introduction, Mathematical Games may be used for the analysis of population dynamics: the payoffs give the results of a competition between these populations for the resources and replicator equations describe the evolution of the population (see, for instance, Cressman & Tao, 2014; Nowak, 2006). These equations may be interpreted as generalizations of the classical Lotka–Volterra equations (Lotka, 1910, 1920; Volterra, 1926, 1928). A given game may be interpreted in terms of populations dynamics by different ways. For instance, the game represented in Table 7.1 may be interpreted as follows:

- As a first approach, we may consider the population as formed by two subpopulations which adopt the strategies O or E. Such a lecture is possible when the mean payoffs of the strategies are the same for both the players, what is the case of symmetric games. A possible refinement consists in considering that the subpopulations adopt the behavior of each player with a given probability;
- A second approach may consider the population as formed by subpopulations having the behavior of the Players: subpopulation i has the behavior of Player i, including the probabilities of each strategy;
- Finally, we may consider four subpopulations: subpopulations 1 and 2 are formed by internal subpopulations adopting strategies O or E, with the probabilities under consideration.

The replicator equation describes the evolution of a subpopulation by its access to the resources represented by the mean payoffs: the rate of variation of a subpopulation is connected to the difference between its mean payoff and the mean payoff of the whole population. According to this idea, a subpopulation which has a mean payoff larger than the mean payoff of the overall population will increase; in the inverse situation, it will decline. In other words, if the mean payoff of the overall population is $\overline{f}$ and the mean payoff of the subpopulation i is f_i, the rate of variation of the subpopulation i depends on the difference $f_i - \overline{f}$: it declines if $f_i < \overline{f}$ and increases if $f_i > \overline{f}$.

For instance, let us consider the second approach: a population has a proportion x_1 of elements adopting the behavior of Player 1 and a proportion x_2 of elements having the behavior of Player 2. We have $x_1 + x_2 = 1$, with $0 \le x_1, x_2 \le 1$ – thus, we may consider x_2 as main variable and determine $x_1 = 1 - x_2$. Using the payoffs presented in Table 7.1, the mean payoff received by an element of the subpopulation x_2 is

$$f_2 = (1 - 2p_1)(2p_2 - 1)$$

Since the game is a zero-sum game, the mean payoff of an element of the subpopulation x_1 is $f_1 = -f_2$. On the overall population, the mean is

$$\bar{f} = x_1 f_1 + x_2 f_2 = (2x_2 - 1) f_2.$$

The replicator equation reads as

$$\frac{dx_2}{dt} = x_2 \left(f_2 - \bar{f} \right) = 2x_2 (1 - x_2) f_2$$

Such an equation is a logistic equation (Verhulst, 1845, 1847). It has an explicit solution for $0 < x_2(0) < 1$:

$$x_2(t) = \frac{1}{1 + \left(\frac{1}{x_2(0)} - 1 \right) e^{-2f_2 t}}$$

We observe that x_2 grows up if $f_2 > 0$ and decreases if $f_2 < 0$. The population is stable if $f_2 = 0$. Indeed, there are 3 steady states: $x_2 = 0$, $x_2 = 1$, $f_2(x_2) = 0$. $x_2 = 0$ corresponds to the extinction of subpopulation x_2 : the whole population is formed by x_1; $x_2 = 1$ corresponds to the extinction of subpopulation x_1 : the whole population is formed by x_2; $f_2 = 0$ corresponds to the situation where $p = p_2 = \frac{1}{2}$. When explicit solutions are not available, the stability of these steady states may be analyzed by the standard methods of dynamical systems: for instance, a linearization in a neighborhood of the solution or the determination of the eigenvalues of the matrix of the linearized coefficients. Here, stability depends on the sign of f_2.

Now, let us consider a different lecture of the game: assume that the whole population is formed of individuals applying a fixed strategy O or E. Then, let us denote x_O the proportion of individuals applying the strategy O, while $x_E = 1 - x_O$ is the proportion of individuals applying the strategy E. When two individuals compete for a resource, they play as Player 1 with a given probability $\eta > 0$ and with probability $1 - \eta$ as Player 2. Then, the expected payoff of an individual adopting the strategy O is

$$f_O = \eta(x_O - x_E) + (1 - \eta)(x_E - x_O) = (2\eta - 1)(2x_O - 1).$$

Since the game is zero-sum, the expected payoff of an individual adopting the strategy T is $f_E = -f_O$. The mean payoff of the whole population is

$$\bar{f} = x_O f_O + x_E f_E = (x_O - x_E) f_O = (2\eta - 1)(2x_O - 1)^2.$$

In this case, the replicator equation reads as

$$\frac{dx_O}{dt} = x_O\left(f_O - \overline{f}\right) = 2(2\eta - 1)x_O(1 - x_O)(2x_O - 1)$$

For $\eta = ½$, the population is stable. For $\eta \neq ½$, we have three equilibrium states: $x_O = 0, x_O = 1, x_O = ½$. Stability depends on the sign of f_O.

To simulate this equation, we use a right-side member of the differential equation as follows:

```
f = function(t,x,lm){list(2*(2*lm[[1]]-1)*x*(1-x)*(2*x-1))}
```

Indeed, the growth of x_O depends upon the sign of f_O: x_O grows up if $f_O > 0$ (either $\eta > ½, x_O > ½$ or $\eta < ½, x_O < ½$) and decreases if $f_O < 0$ (either $\eta < ½, x_O > ½$ or $\eta > ½, x_O < ½$). When $\eta < \frac{1}{2}$, the system tends to the equilibrium with $x_O = ½$. For $\eta > ½$, one of the groups goes to extinction, except if there is an exact equilibrium at the start ($x_O = 1/2$). We may simulate this equation under *R*, using the methods previously shown for differential equations. For instance, we use RK4's method to generate Figs. 7.5, 7.6, 7.7, 7.8, 7.9, and 7.10.

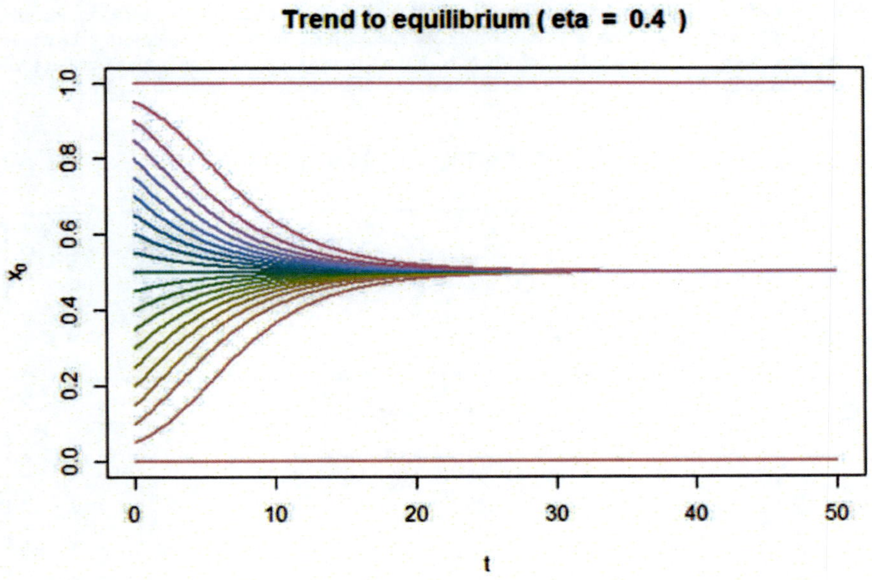

Fig. 7.5 Trend to Equilibrium when $\eta < \frac{1}{2}$: for $0 < x_O(0) < 1$, the proportion of species *O* tends to ½. $x_O = 0$ and $x_O = 1$ are equilibrium points: one species is extinct and the other forms the whole population

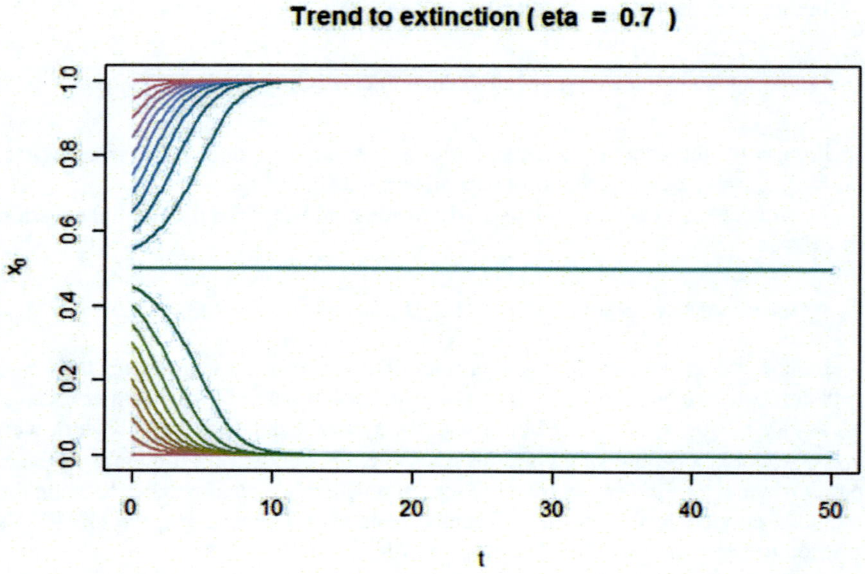

Fig. 7.6 Trend to Extinction when $\eta > \frac{1}{2}$: for $0 < x_O(0) < ½$, the proportion of species O tends to 0; for $½ < x_O(0) < 0$, the proportion of species O tends to 1, so that species E tends to extinction. $x_O = 0$ and $x_O = 1$ are equilibrium points: where one species is extinct and the other forms the whole population, $x_O = 0.5$ is an equilibrium where both the species are present, with a half of the total population each

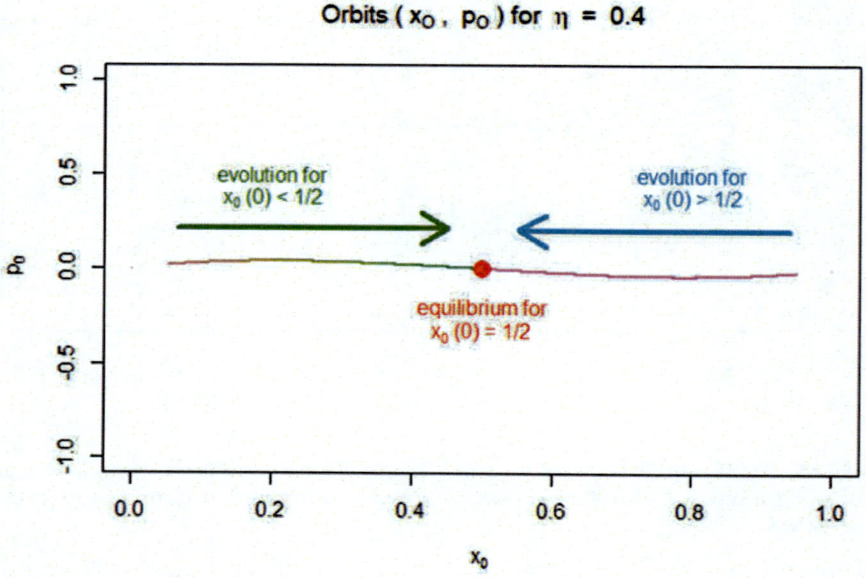

Fig. 7.7 Orbits $(x_O, p_O = \dot{x}_O)$ in the phase space for $\eta = 0.4$. Except for $x_O(0) \in \{0, 1\}$, all the curves tend to $(x_O = 0.5, p_O = 0)$, which corresponds to the equilibrium

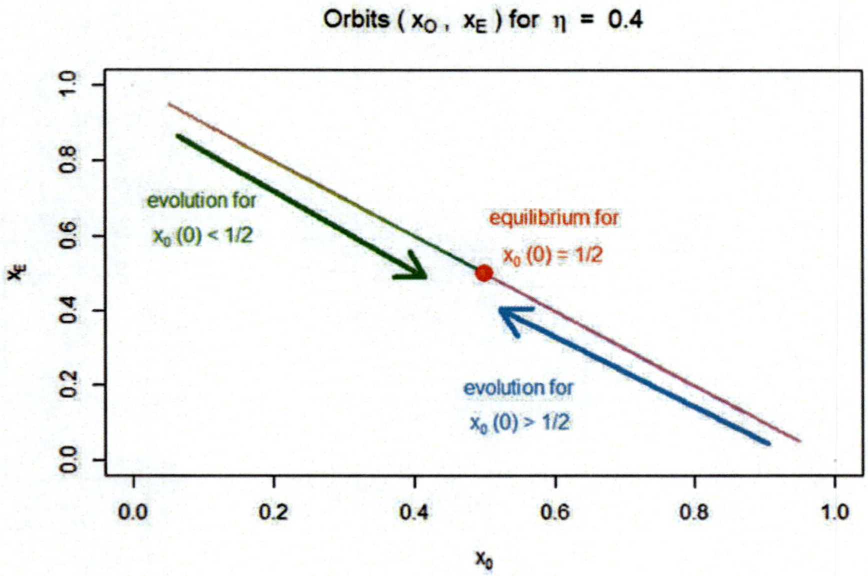

Fig. 7.8 Orbits (x_O, x_E) for $\eta = 0.4$. Except for $x_O(0) \in \{0, 1\}$, all the curves tend to $(x_O = 0.5, x_E = 0.5)$, which corresponds to the equilibrium

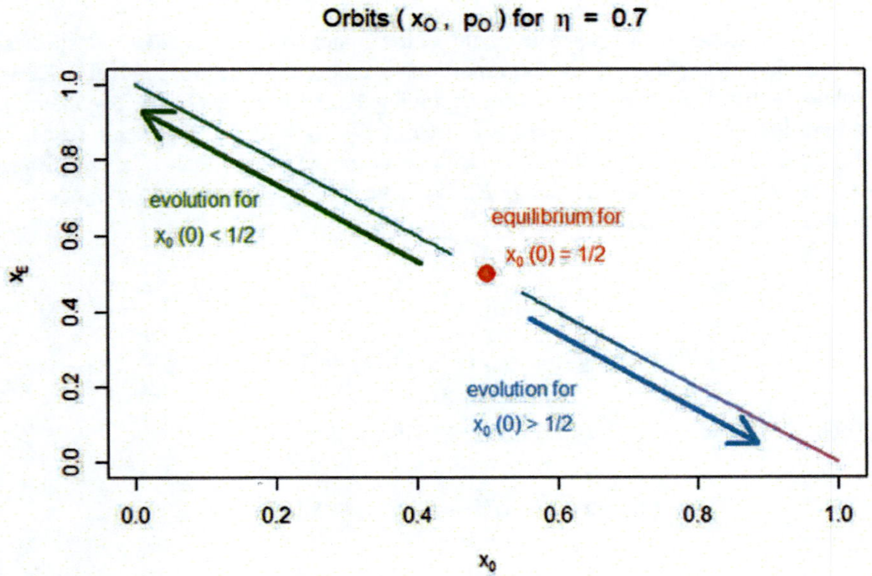

Fig. 7.9 Orbits (x_O, x_E) for $\eta = 0.7$. For $x_O(0) < 0.5$ the curves tend to $(x_O = 0, x_E = 1)$, i.e., the extinction of species O. For $x_O(0) > 0.5$ the curves tend to $(x_O = 1, x_E = 0)$, i.e., the extinction of species E

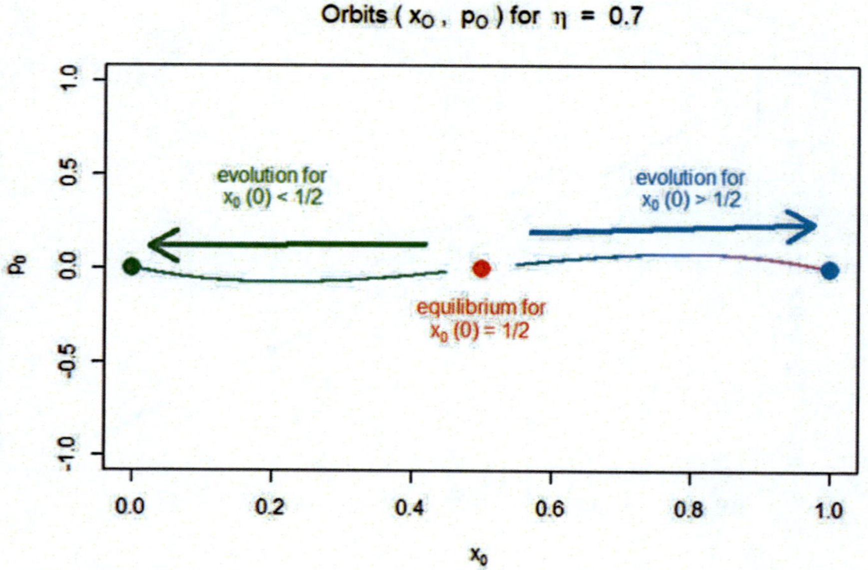

Fig. 7.10 Orbits $(x_O, p_O = \dot{x}_O)$ in the phase space for $\eta = 0.7$. For $x_O(0) < 0.5$ the curves tend to $(x_O = 0, p_O = 0)$, i.e., the extinction of species O. For $x_O(0) > 0.5$ the curves tend to $(x_O = 1, p_O = 0)$, i.e., the extinction of species E

Let us introduce other possible lecture of the game: now, we consider two species corresponding to Players 1 and 2 (proportions x_1 and $x_2 = 1 - x_1$, respectively). Each subpopulation is formed of individuals applying a fixed strategy O or E. The internal proportions are denoted x_{iO} (species i, strategy O), $x_{iE} = 1 - x_{i_o}$ (species i, strategy E). Then, the global population is divided according to proportions $x_{iO}x_i$ (species i, strategy O), $x_{iE}x_i$ (species i, strategy E). We have $x_{1_o}x_1 + x_{1_E}x_1 + x_{2_o}x_2 + x_{2_E}x_2 = 1$. In this case, the mean payoffs are

$$f_{1_o} = x_2(2x_{2_o} - 1),\ f_{1_E} = x_2(1 - 2x_{2_o}),\ \overline{f_1} = (2x_{1_o} - 1)(2x_{2_o} - 1)x_2$$

$$f_{2_o} = x_1(1 - 2x_{1_o}),\ f_{2E} = x_1(2x_{1_o} - 1),\ \overline{f_2} = (1 - 2x_{1_o})(2x_{2_o} - 1)x_1$$

Then, $\overline{f} = 0$ and the replicator dynamics leads to

$$\frac{\mathrm{d}}{\mathrm{d}t}(x_1) = x_1\overline{f_1} = x_1(1 - x_1)(2x_{1_o} - 1)(2x_{2_o} - 1),$$

$$\frac{d}{dt}(x_{1o}) = x_{1o}\left(f_{1_o} - \overline{f_1}\right) = 2x_{1_o}(1 - x_1)(1 - x_{1_o})(2x_{2_o} - 1)$$

$$\frac{d}{dt}(x_{2_o}) = x_{2o}\left(f_{2_o} - \overline{f_2}\right) = 2x_{2_o}x_1(1 - x_{2_o})(1 - 2x_{1_o})$$

In this case, the right-side member of the differential equation can be evaluated as follows:

```
f = function(t,x,lm){
  x1 = x[[1]]
  x1o = x[[2]]
  x2o = x[[3]]
  aux1 = x1*(1-x1)*(2*x1o-1)*(2*x2o-1)
  aux2 = 2*x1o*(1-x1)*(1 - x1o)*(2*x2o-1)
  aux3 = 2*x2o*x1*(1 - x2o)*(1-2*x1o)
  aux = matrix(c(aux1,aux2,aux3),ncol = 1)
  return(list(aux))
}
```

Analogous analysis may be performed in this case (see Figs. 7.11 and 7.12).

The mean evolution of the system and confidence intervals for orbits or trajectories may be obtained by using the methods presented in Chap. 3.

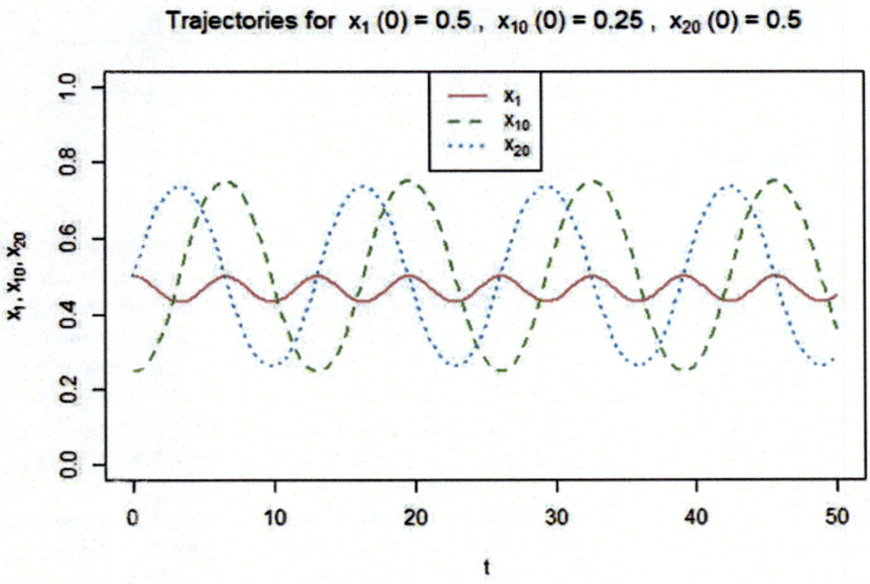

Fig. 7.11 Trajectories of some variables

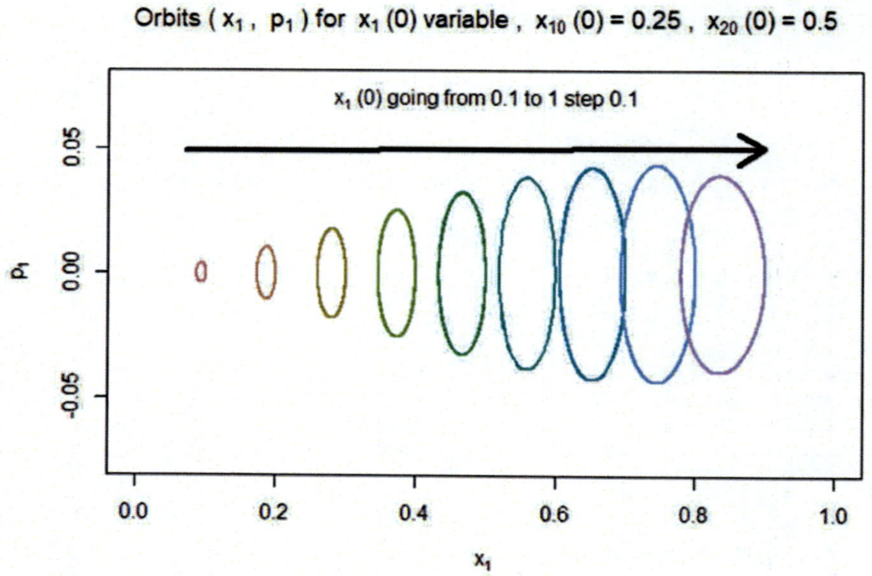

Fig. 7.12 Orbits $(x_1, p_1 = \dot{x}_1)$ for different values of $x_1(0)$

Table 7.10 The normal form of the prisoner's dilemma

	Player 2	
Player 1	C_2 (p_2)	D_2 $(q_2 = 1 - p_2)$
C_1 (p_1)	(L, L)	(V, H)
D_1 $(q_1 = 1 - p_1)$	(H, V)	(M, M)

7.3 The Prisoner's Dilemma

A classical game is prisoner's dilemma: two robbers are arrested and must choose between confess (C) or deny (D). If both confess, they will be sentenced to a light sentence of L years in prison; if one confesses and the other not, the one which has not confessed will be sentenced to a heavy sentence of H years, while the one who has confessed will be sentenced to a very light sentence of V years; if no one confess, both will have a minimal sentence M. In general, $M < V < L < H$. Each prisoner must take a decision. In practice, a prisoner has also to evaluate the probabilities of the decision taken by the other prisoner. This game may be put into a Matrix Form as in Table 7.10

If the prisoners can communicate (cooperative game), they will choose the strategy (D_1, D_2) that ensures to both the minimal sentence. If they do not communicate (non-cooperative game), the situation is more difficulty: they must evaluate the probability p_i that prisoner i confess. If a prisoner is confident in his partner, he evaluates this probability as extremely low; while if he is not so confident in his partner, he will evaluate this probably with higher value.

(D_1, D_2) is a Nash equilibrium for this game: any solitary deviation of a player results in a loss for himself – but its practical implementation cannot be made without communication and strong mutual confidence between the players. A second Nash equilibrium is (C_1, C_2): if one player confesses, the other must confess – otherwise his loss will be larger.

The existence of a mixed Nash equilibrium may be investigated as previously, by making decisions indifferent. For instance, let us consider Player 1. The mean sentence for strategy C_1 is $S_1^C = V + p_2(L - V)$. For strategy D_1, the mean sentence is $S_1^D = M + p_2(H - M)$. Indifference occurs when $S_1^C = S_1^D$, *id est*, $p_2 = p^{mix} = \frac{\Delta_{VM}}{\Delta_{VHML}}$, where

$$\Delta_{VM} = V - M > 0,\ \Delta_{HL} = H - L > 0,\ \Delta_{VHML} = \Delta_{VM} + \Delta_{HL} > 0.$$

Analogously, indifference between S_2^D and S_2^C occurs for $p_1 = p^{mix}$. If both the players follow that strategy, the mean sentence S^{mix} satisfies $V < S^{mix} < L$. Notice that the mixed strategy is not optimal. Indeed, for Player 1: $P(S_1 = H| D_1) = p_2$, $P(S_1 = M| D_1) = q_2$, $P(S_1 = L| C_1) = p_2$, $P(S_1 = V| C_1) = q_2$. So, $P(S_1 = M) = q_1 q_2$, $P(S_1 = V) = p_1 q_2$, $P(S_1 = L) = p_1 p_2$, $P(S_1 = H) = q_1 p_2$. Then, the mean sentence for Player 1 is

$$S_1 = M + p_1 \Delta_{VM} + p_2(H - M) - p_1 p_2 \Delta_{VHML}.$$

We observe that the behavior of S_1 depends on the relation between p_2 and the critical value p^{mix}: for $p_2 > p^{mix}$, S_1 decreases when p_1 increases, so that its minimum corresponds to $p_1 = 1$ (strategy C_1); for $p_2 < p^{mix}$, S_1 increases when p_1 increases, so that its minimum corresponds to $p_1 = 0$ (strategy D_1). If $p_2 = p^{mix}$, it is indifferent to choose between C_1 and D_1. Analogously, the mean sentence for Player 2 is

$$S_2 = M + p_2 \Delta_{VM} + p_1(H - M) - p_1 p_2 \Delta_{VHML}.$$

Thus, the behavior is analogous: for $p_1 > p^{mix}$, the minimum of S_2 is attained at $p_2 = 1$ (strategy C_2); for $p_1 < p^{mix}$, its minimum is attained at $p_2 = 0$ (strategy D_2). If $p_1 = p^{mix}$, it is indifferent to choose between C_2 and D_2.

Then, Player 1 may choose a strategy according to these values:

- Confess: this corresponds to $p_1 = 1$ In this case, $S_1 = S_1^C$. If $p_2 < p^{mix}$, this strategy maximizes S_1. Otherwise, it minimizes S_1.
- Deny: this corresponds to $p_1 = 0$. Then, $S_1 = S_1^D$. If $p_2 < p^{mix}$, this strategy minimizes S_1. Otherwise, it maximizes S_1.
- Apply the minimax strategy: the maximum loss is L for decision C_1 and H for decision D_1. The minimum of the maximal loss is L for decision C_1. Thus, this strategy leads to choose C_1.
- Considering as a payoff the difference between the maximal sentence and the sentence received, the maximin strategy leads to the same result: indeed, the minimal payoff is $H - L$ for C_1 and 0 for D_1. The maximum of the minimal payoff is $H - L$ for decision C_1. Thus, this strategy leads also to choose C_1.

The situation is analogous for Player 2 (the game is symmetric).

Let us adopt the point view of UQ and analyze the probabilities. Let X_1 be the sentence received by Player 1. Then

$$P(X_1 > M) = p_1 + p_2 - p_1p_2,\ P(X_1 > V) = p_2,\ P(X_1 > L) = p_2 - p_1p_2.$$

Player 1 controls his own decision and may decide the value of p_1. The value of p_2 is chosen independently by Player 2: Player 1 must estimate p_2 and take a decision based on his estimation. The decision will involve a risk of having a sentence superior to a given threshold, considered as acceptable by Player 1. Then, the strategies lead to the following results:

- Confess: this corresponds to $p_1 = 1$. If Player 1 chooses this strategy, he is looking for the sentence V or L. We have $P(X_1 > M) = p_2$. Thus, Player 1 must have a strong confidence in Player 2: if he does not estimate p_2 as small enough, he will consider the risk as excessive. Then, he may choose the betrayal: confess while expecting that Player 2 does not confess – to deceive Player 2 expecting that Player 2 will not betray.
- Deny: this corresponds to $p_1 = 0$. If Player 1 chooses this strategy, he is looking for the minimal sentence M (the other option is the maximal sentence H). Again, Player 1 must have a strong confidence in Player 2 to estimate p_2 as small enough. Otherwise, he will consider the risk as excessive and choose to confess.
- Evaluate the other as himself: the probability of C_2 is the same as C_1, what corresponds to $p_1 = p_2$. Then $P(X_1 > M) = 2p_2 - p_2^2$. Then $P(X_1 > M)$ is an increasing function of p_2 having its maximum at $p_2 = 1$. If player 1 accepts a maximal risk of α, the maximal acceptable value of p_2 is $1 - \sqrt{1-\alpha}$. For small values of α, this value is close to $\frac{\alpha}{2}$. Thus, if the Player 1 accpts a maximal risk of 0.1%, the maximal acceptable value of p_2 is about 0.05%. For a maximal risk of 0.01%, the value is about 0.005%.

As we may see, either the confidence in the partner is strong either the best strategy is to confess. If both the players make the same analysis, both will confess.

7.3.1 Replicator Dynamics

Let us consider a population formed by two species having the behavior of Players 1 (proportion x_1) and 2 (proportion x_2). Their mean payoffs f_1, f_2 are, respectively, $-S_1$, $-\ S_2$: notice that S_1, S_2 are positive. To reinterpret them as punitions, we must change their sign. The mean payoff of the whole population is

$$\overline{f} = -x_1S_1 - (1 - x_1)S_2 = -S_2 + x_1(S_2 - S_1).$$

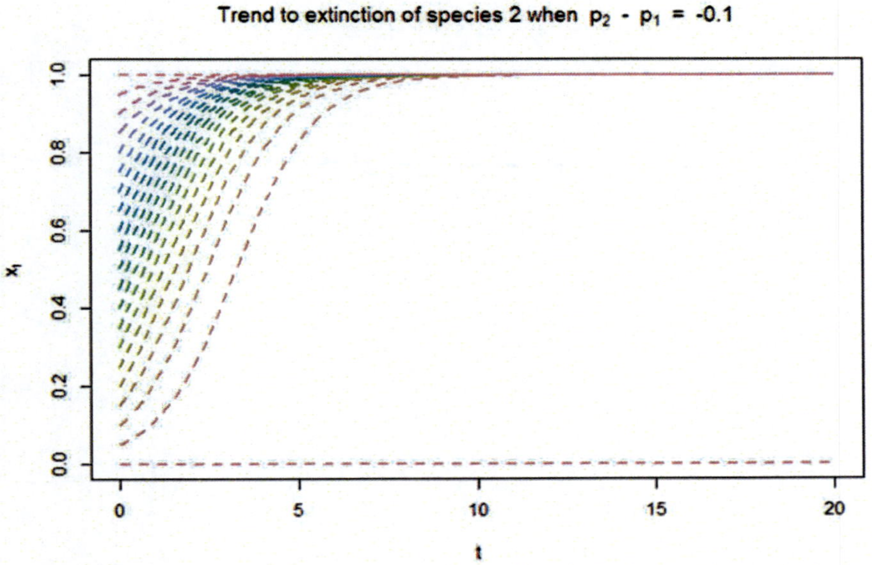

Fig. 7.13 Trend to Extinction of Species 2 when $p_2 - p_1 < 0$. It survives only if $x_1(0) = 0$, otherwise, it disappears

Thus, the replicator equation reads as

$$\frac{dx_1}{dt} = x_1(1 - x_1)(S_2 - S_1).$$

Since

$$S_2 - S_1 = (p_2 - p_1)\underbrace{(V - H)}_{<0},$$

the evolution is driven by $p_2 - p_1$: if $p_2 - p_1 < 0$, then $S_2 - S_1 > 0$ and $x_1(t) \longrightarrow 1$, while $x_2(t) \longrightarrow 0$: species 2 goes to extinction (Figs. 7.13 and 7.14). If $p_2 - p_1 > 0$, then $S_2 - S_1 < 0$ and $x_1(t) \longrightarrow 0$: species 1 goes to extinction (Figs. 7.15, 7.16 and 7.17). The populations are stable when $p_1 = p_2$. Again, C is the best strategy: $p_1 = 1$ ensures the survival of species 1. The simulations use `deSolve`. An example of function for the evaluation of the right-side member of the replicator differential equation is given below:

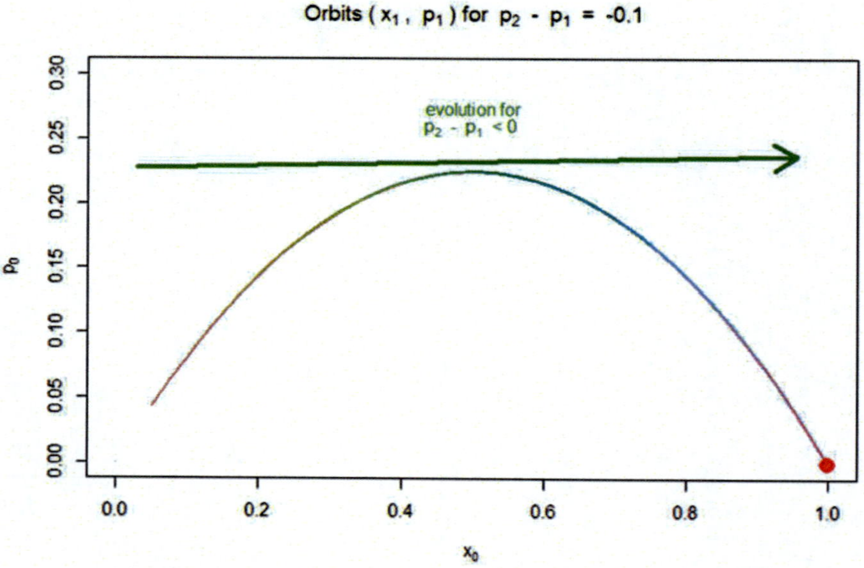

Fig. 7.14 Orbits $(x_1, p_1 = \dot{x}_1)$ for $p_2 - p_1 < 0$. They tend to the equilibrium point (1,0), corresponding to the extinction of Species 2

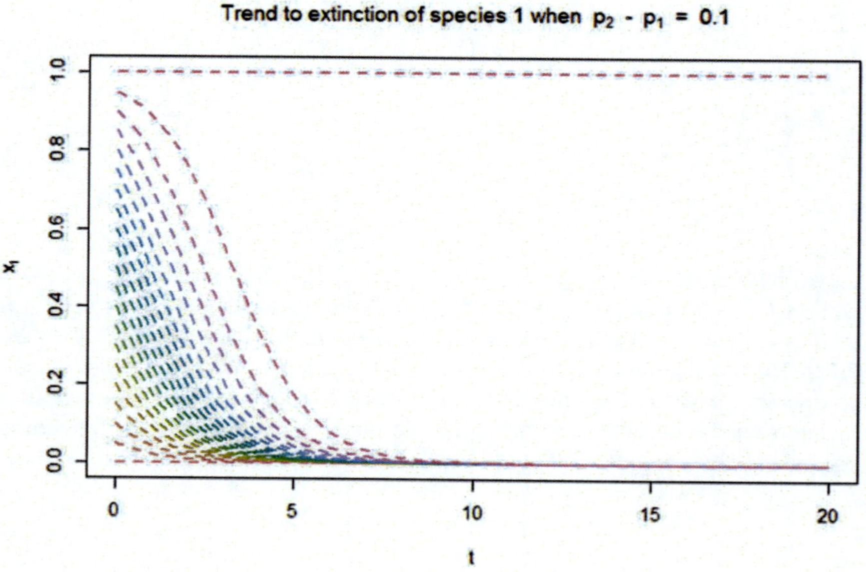

Fig. 7.15 Trend to Extinction of Species 1 when $p_2 - p_1 > 0$. It survives only if $x_2(0) = 0$, otherwise, it disappears

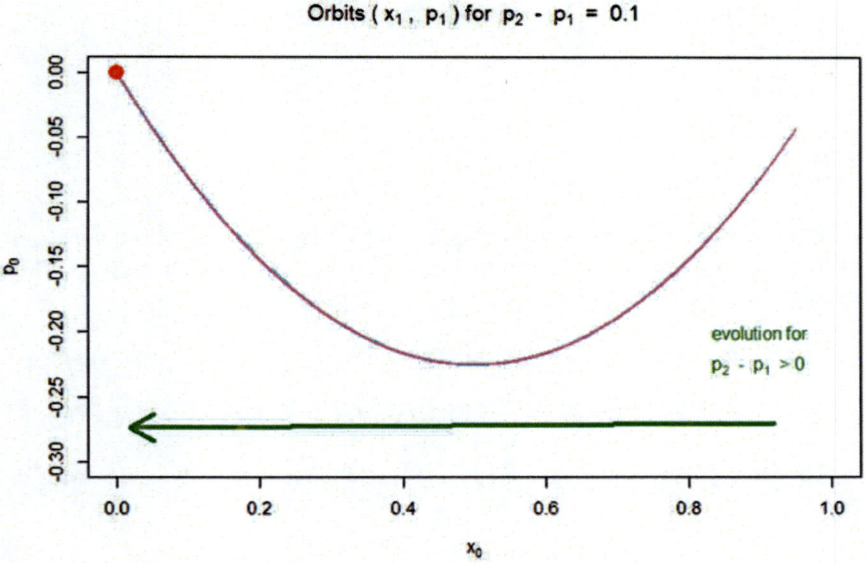

Fig. 7.16 Orbits $(x_1, p_1 = \dot{x}_1)$ for $p_2 - p_1 > 0$. They tend to the equilibrium point (0,0), corresponding to the extinction of Species 1

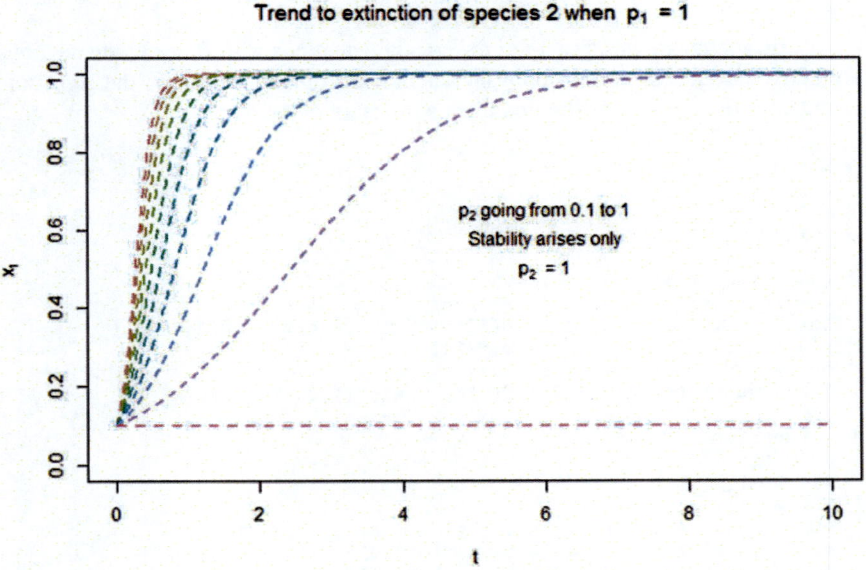

Fig. 7.17 Strategy "*C*" ensures the survival of Species 1 for any strategy of Species 2

```
f = function(t,X,lm){list(lm[[1]]*X*(1-X))}

p1 = 0.2
p2 = 0.1
dp = p2-p1
V = 1
H = 10
lmm = list((p2-p1)*(V-H))
```

If we consider subspecies applying a fixed strategy C or D, the internal proportions are x_{iC} (species i, strategy C), $x_{iD} = 1 - x_{i_D}$ (species i, strategy D). Then, analogously to the preceding situation, the mean payoffs are

$$f_{1_C} = -V - (L-V)x_{2_C}, f_{1_D} = -M - (H-M)x_{2_C}, \overline{f_1} = f_{1_D} + x_{1_C}(f_{1_C} - f_{1D})$$

$$f_{2_C} = -V - (L-V)x_{1_C}, f_{2_D} = -M - (H-M)x_{1_C}, \overline{f_2} = f_{2_D} + x_{2_C}(f_{2_C} - f_{2_D}).$$

The mean payoff is $\overline{f} = \overline{f_2} + x_1(\overline{f_1} - \overline{f_2})$ and the replicator dynamics leads to

$$\frac{\mathrm{d}}{\mathrm{d}t}(x_1) = x_1(\overline{f_1} - \overline{f}) = x_1(1-x_1)(\overline{f_1} - \overline{f_2}),$$

$$\frac{\mathrm{d}}{\mathrm{d}t}(x_{1C}) = x_{1C}(f_{1_C} - \overline{f_1}) = x_{1_C}(1-x_{1_C})(f_{1_C} - f_{1D})$$

$$\frac{\mathrm{d}}{\mathrm{d}t}(x_{2_C}) = x_{2C}(f_{2_C} - \overline{f_2}) = x_{2_C}(1-x_{2_C})(f_{2_C} - f_{2D})$$

We may simulate this system of differential equations with R, using the methods formerly introduced. An example of function for the evaluation of the right-side member of the replicator differential equation is given below:

```
M = 1
V = 2
L = 3
H = 10

f1c = function(x2c){-V - (L-V)*x2c}
f1d = function(x2c){-M - (H-M)*x2c}
f1bar = function(x1c,x2c){f1d(x2c) + x1c*(f1c(x2c)-f1d(x2c))}
f2c = function(x1c){-V - (L-V)*x1c}
f2d = function(x1c){-M - (H-M)*x1c}
f2bar = function(x1c,x2c){f2d(x2c) + x2c*(f2c(x1c)-f2d(x1c))}

f = function(t,X,lm){
  x1 = X[[1]]
  x1c = X[[2]]
  x2c = X[[3]]
  aux1 = x1*(1-x1)*(f1bar(x1c,x2c) - f2bar(x1c,x2c))
  aux2 = x1c*(1-x1c)*(f1c(x2c) - f1d(x2c))
  aux3 =  x2c*(1-x2c)*(f2c(x1c) - f2d(x1c))
  aux = matrix(c(aux1,aux2,aux3),ncol = 1)
  return(list(aux))
}
```

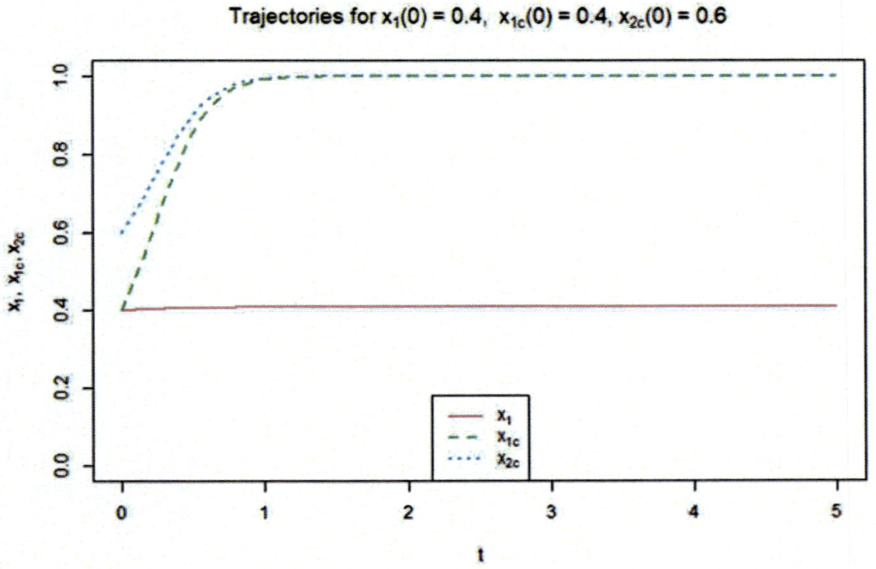

Fig. 7.18 Trajectories of x_1,x_{1_C},x_{2_C}: x_{1_C},x_{2_C} both tend to 1, so that x_{1_D},x_{2_D} both tend to zero. In addition, x_1 tends to a stable value

For instance, using `deSolve` and method `rk4`, we obtain the results in Figs. 7.18, 7.19, and 7.20. Observe that the strategy C is the best: the subpopulations applying strategy D tend to extinction in all the cases. In addition, the system evolves to a stable equilibrium between the species: the orbits $(x_1, p_1 = \dot{x}_1)$ tend to a point where $p_1 = 0$.

7.4 The Goalie's Anxiety at the Penalty Kick

Penalties in soccer interested researchers in GT since (Levitt et al., 2002). In 2003, an article (Palacios-Huerta, Professionals play minimax, 2003) analyzed penalties in soccer games using GT and a large set of data. The work of Palacios-Huerta kicked off a series of research studies about the same topic: (Garicano et al., 2005; Coloma, 2007, 2012; Palacios-Huerta & Volij, 2008; Apesteguia & Palacios-Huerta, 2010; Azar & Bar-Eli, 2011; Prakash & Garg, 2014; Palacios-Huerta, 2014; Sarkar, 2017; Boczon & Wilson, 2018; Arrondel et al., 2019). In his work, Palacios-Huerta formulates the penalty kick as a game involving the goal and the kicker. Although the paper contains many information, several studies concentrate on the choice of the side to kick for the kicker and the side to plunge for the goalkeeper. In theory, there are many options: left side, center, right side; in addition, the ball can be at ground level, in the middle or high. The paper furnishes information about the influence of the time where the penalty is kicked, about the foot of the kicker and so on. In a

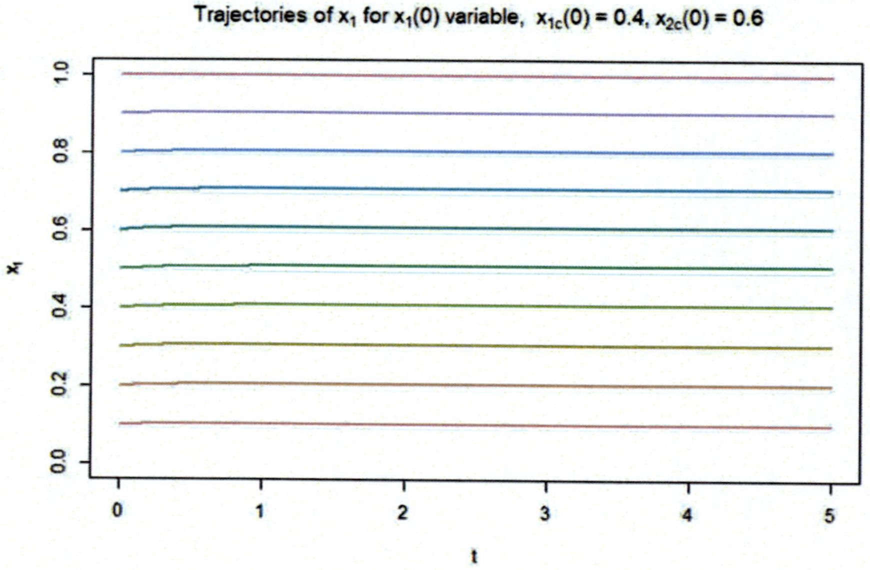

Fig. 7.19 Evolution of x_1 for different initial values. Convergence to stable values is observed in all the cases

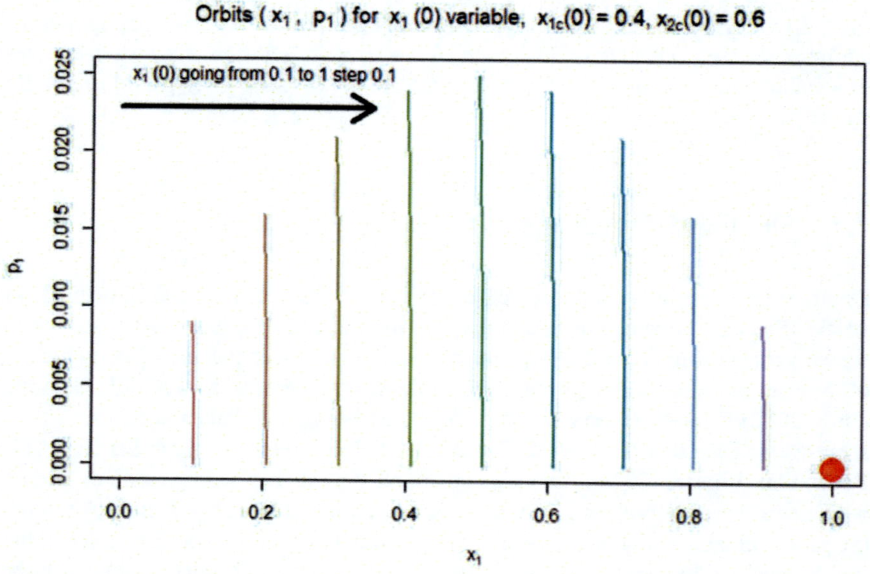

Fig. 7.20 Orbits $(x_1, p_1 = \dot{x}_1)$ for different values of $x_1(0)$. Convergence to a stable value is observed in all the cases

Table 7.11 The penalty game

	Goalkeeper G	
Kicker K	$L\ (g_L)$	$R\ (1-g_L)$
$L\ (k_L)$	$(\pi_{LL}, 1-\pi_{LL})$	$(\pi_{LR}, 1-\pi_{LR})$
$R\ (1-k_L)$	$(\pi_{RL}, 1-\pi_{RL})$	$(\pi_{RR}, 1-\pi_{RR})$

simplified model, the study is reduced to the choices L (left-hand side of the goalkeeper) and R (the right-hand side of the goalkeeper). The compilation of data about a large number of matches furnishes estimations of the probability of scoring for the kicker, which is his payoff. The payoff of the goalkeeper is the complement of the payoff of the kicker. Let us consider the kicker K as Player 1 and the goalkeeper G as the Player 2. Let g_L be the probability of the goalkeeper plunging to his left-hand side and k_L be the probability of kicking at the same left-hand side of the goalkeeper for the kicker. Thus, the game may be put into a matrix form as in Table 7.11.

Typical values for the scoring probabilities are

$$\pi_{LL} = 0.59, \pi_{RL} = 0.93, \pi_{LR} = 0.94, \ \ \pi_{RR} = 0.71.$$

Since these values result from the analysis of a sample, they are uncertain, and a margin of error must be considered. Typical values for the choices are $g_L = k_L = 0.4$ – but these values may be decision variables.

In the framework of GT, we may determine the Nash mixed strategy by considering the expected gain of each player. Let us denote by EG_{KL}, EG_{KR} the expected payoff of the Goalie when the Kicker chooses L or R, respectively. We have

$$EG_{KL} = g_L(\pi_{LR} - \pi_{LL}) + (1 - \pi_{LR}), \ EG_{KR} = (1 - \pi_{RR}) - g_L(\pi_{RL} - \pi_{RR}).$$

To make the choice indifferent, we must take $EG_{KL} = EG_{KR}$. Thus,

$$g_L = g_L^* = \frac{\pi_{LR} - \pi_{RR}}{\pi_{LR} + \pi_{RL} - \pi_{RR} - \pi_{LL}}.$$

Now, let us denote by EK_{GL}, EK_{GR} the expected payoff of the Kicker when the Goalie chooses L or R, respectively. We have

$$EK_{GL} = \pi_{RL} - k_L(\pi_{RL} - \pi_{LL}), \ EK_{GR} = \pi_{RR} + k_L(\pi_{LR} - \pi_{RR}).$$

To make the choice indifferent, we must take $EK_{GL} = EK_{GR}$. Thus,

$$k_L = k_L^* = \frac{\pi_{RL} - \pi_{RR}}{\pi_{LR} + \pi_{RL} - \pi_{RR} - \pi_{LL}}.$$

With the values given previously, we have $g_L^* \approx 0.40$, $k_L^* \approx 0.39$ – what is close to the observed values – this was the basic conclusion of (Palacios-Huerta, 2003).

Always in the framework of GT, we may analyze the minimax strategy. If the losses of a given player are evaluated as being the payoffs of the other player, the maximal losses of the Kicker are $1 - \pi_{LL}$ when he chooses L and $1 - \pi_{RR}$ when he

Table 7.12 Minimax and Maxmin strategies for the penalty kick ($\pi_{RL} \approx \pi_{LR}$)

	Minimax	Maximin
Kicker	R	
Goalie	Uniformly random	

chooses R. Thus, his minimax strategy chooses the minimum of these values, which corresponds to the largest value between π_{LL} and π_{RR}. For the given data, the choice is R. The maximin strategy for the Kicker leads to the same result: his minimal payoff are π_{LL} and π_{RR}, so that he will maximize his minimal winning by choosing the maximal value between them. Concerning the Goalie, the maximal losses are π_{LR} and π_{LR}. Let us consider that π_{RL} and π_{LR} are approximately equal: then, the maximum losses are equal for both the sides. Analogously, the minimal winnings of the Goalie are $1 - \pi_{LR}$ and $1 - \pi_{RL}$, which are approximatively equal. In this case, we obtain the result shown in Table 7.12

Assuming that the Players choose one of these strategies, the kicker chooses R and the Goalie randomly chooses between R and L. If we assume that the difference of 0.01 between π_{RL} and π_{LR} is significant, then $\pi_{RL} < \pi_{LR}$ and the minimax and maximin strategies for the Goalie are L, instead uniformly random. Notice that these choices do not correspond to a Nash Equilibrium.

Now, let us consider the point of view of UQ: the basis for decision are the probabilities. From the Kicker's point of view, the probability of scoring is S_L or S_R when he chooses to kick L or R, respectively. We have

$$S_L = \pi_{LR} - g_L(\pi_{LR} - \pi_{LL}), \quad S_R = \pi_{RR} + g_L(\pi_{RL} - \pi_{RR}).$$

The global probability of scoring is (see Fig. 7.21)

$$S = S_R + k_L(S_L - S_R).$$

If the Kicker desires to maximize this probability, so that he will choose either $k_L = 1$, if $S_R < S_L$ or $k_L = 0$, if $S_R > S_L$. When $S_L = S_R$, the Kicker chooses uniformly random between L and R. The result is logical: if the probability of scoring is larger for one of the sides, the Kicker will have the largest probability of scoring by kicking at that side. For instance, if the Goalie is known by always jump to left, $g_L = 1$, so that $S_R = \pi_{RL}$ and $S_L = \pi_{LL}$ and the decision is $k_L = 0$: the kicker strikes at right to maximize his probability of scoring. Analogously, if the Goalie is known by always jump to right, the decision will be to kick at left. Indifference occurs when $S_L = S_R$, id est,

$$g_L = g_L^*.$$

From the Goalie's point of view, the probability of not scoring is N_L or N_R when the Kicker chooses L or R, respectively. We have

$$N_L = (1 - \pi_{LR}) + k_L(\pi_{LR} - \pi_{LL}), \quad N_R = (1 - \pi_{RR}) - k_L(\pi_{RL} - \pi_{RR}).$$

The global probability of not scoring is (see Fig. 7.22)

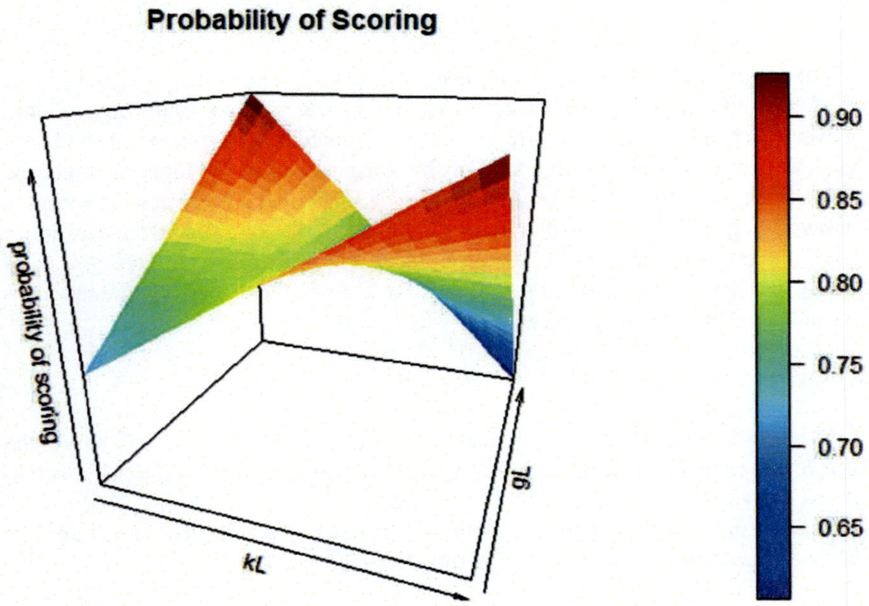

Fig. 7.21 Probability of scoring

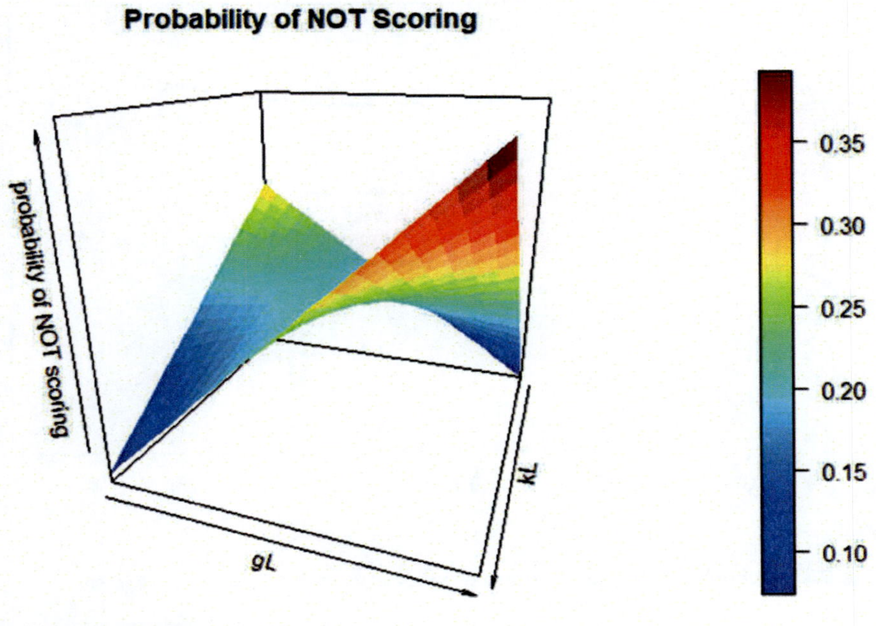

Fig. 7.22 Probability of not scoring

$$N = N_R + g_L(N_L - N_R).$$

The Goalie desires to maximize N, so that he will choose either $g_L = 0$, if $N_L < N_R$ or $g_L = 1$, if $N_L > N_R$. When $N_L = N_R$, the Goalie chooses uniformly random between L and R. Again, the result is logical: if probability of no scoring is higher at left, the Goalie will choose this side; if the same probability is higher at right, he logically chooses the right side. As an example, if the Kicker is known by always kicking at left, then $k_L = 1$, so that $N_R = 1 - \pi_{RL}$ and $S_L = 1 - \pi_{LL}$ and the decision is $g_L = 1$: the Goalie plunges at left to minimize the probability of scoring. Analogously, if the Kicker is known by always strike at right, the decision will be to jump to right. Indifference occurs when $N_L = N_R$, id est,

$$k_L = k_L^*.$$

The indifference values correspond to the Nash equilibrium: in this game, the payoffs are equal to the probabilities. For the data given, $S_L \approx 0.8$, $S_R \approx 0.8$, $S \approx 0.8$, $g_L^* \approx 0.4$, $N_L \approx 0.2$, $N_R \approx 0.2$, $N \approx 0.2$, $k_L^* \approx 0.4$.

In this game, the Kicker can use an anticyclic strategy: he may choose to strike at left with a probability $k_L = 1 - g_L$. In this case (see Fig. 7.23)

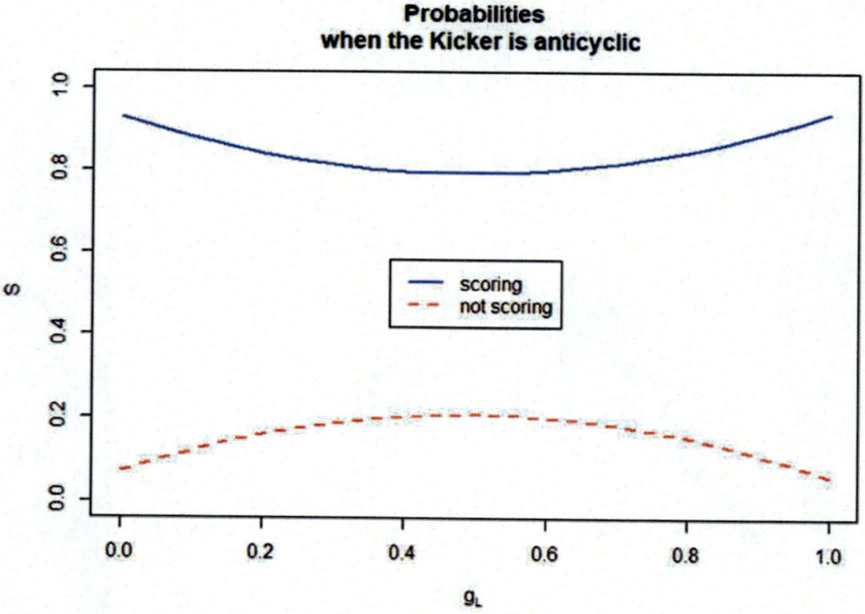

Fig. 7.23 Probability of scoring as function of g_L in the anticyclic strategy of the Kicker

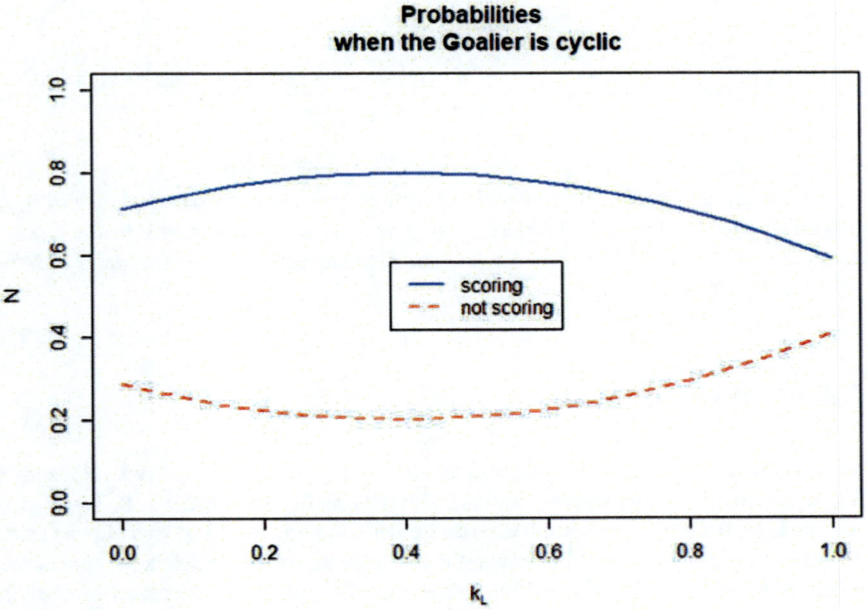

Fig. 7.24 Probability of not scoring as function of g_L in the cyclic strategy of the Goalie

$$S = S^a = \pi_{LR} + (\pi_{LL} - 2\pi_{LR} + \pi_{RR})g_L + (\pi_{LR} + \pi_{RL} - \pi_{RR} - \pi_{LL})g_l^2$$

This polynomial attains his minimal value at

$$g_L^a = \frac{2\pi_{LR} - (\pi_{LL} + \pi_{RR})}{2(\pi_{LR} + \pi_{RL} - \pi_{RR} - \pi_{LL})}$$

so that

$$S^a \geq S^a_{\min} = \frac{4\pi_{LR}\,\pi_{RL} - (\pi_{LL} + \pi_{RR})^2}{4(\pi_{LR} + \pi_{RL} - \pi_{RR} - \pi_{LL})}$$

The Goalie may use a cyclic strategy and jump to left at probability k_L. Then (see Fig. 7.24),

$$N = N^c = 1 - \pi_{RR} + k_L(2\pi_{RR} - \pi_{LR} - \pi_{RL}) + k_L^2(\pi_{LR} + \pi_{RL} - \pi_{RR} - \pi_{LL})$$

This polynomial attains his minimal value at

$$k_L^c = \frac{\pi_{LR} + \pi_{RL} - 2\pi_{RR}}{2(\pi_{LR} + \pi_{RL} - \pi_{RR} - \pi_{LL})}$$

so that

$$N^c \geq N^c_{\min} = \frac{(\pi_{LR} + \pi_{RL})^2 - 4\pi_{LL}\pi_{RR} - 4(\pi_{LR} + \pi_{RL} - \pi_{RR} - \pi_{LL})}{4(\pi_{LR} + \pi_{RL} - \pi_{RR} - \pi_{LL})}$$

For the data given, $g^a_L \approx 0.5, S^a_{\min} \approx 0.8$, $k^c_L \approx 0.4, N^c_{\min} \approx 0.2$.

In this game, equal opportunity does not exist for the considered values of π_{LL}, π_{LR}, π_{RL}, π_{RR}: indeed, Equal Opportunity gives the same probability to the Goalie and the Kicker, id est, $N = S = 0.5$, but the data correspond to a minimal S of 0.59 and a maximal N of 0.41.

7.5 Hawks and Doves

Let us consider a population of Hawks (H) and Doves (D): when a Hawk (H) faces a Dove (D), the Dove withdraws, and the Hawk wins the reward $2R > 0$. When two Doves face, the reward is equally divided: each one wins $R > 0$. If two Hawks face, they fight and, in the mean, their reward is $R - C$, where $C = \beta R > 0$ is the cost of the injuries received in the fight. This game may be put into a Matrix Form as in Table 7.13.

In this case, the mean payoffs are

$$E_{H1} = R(2 - p_{H2}(1 + \beta)); E_{D1} = (1 - p_{H2})R$$
$$E_{H2} = R(2 - p_{H1}(1 + \beta)); E_{D2} = (1 - p_{H1})R;$$

The mean playoff of Players 1 and 2 are, respectively,

$$E_1 = (1 + p_{H1} - p_{H2} - \beta p_{H1}p_{H2})R; E_2 = (1 + p_{H2} - p_{H1} - \beta p_{H1}p_{H2})R$$

The results of Player 1 are indifferent to the strategy when

$$E_{H1} = E_{D1} \Leftrightarrow 2 - p_{H2}(1 + \beta) = 1 - p_{H2};$$

The solution is

$$p_{H2} = \frac{1}{\beta}.$$

Table 7.13 The normal form of the Hawks and Doves

	Player 2	
Player 1	H (p_{H2})	D $(1 - p_{H2})$
H (p_{H1})	$((1 - \beta)R, (1 - \beta)R)$	$(2R, 0)$
D $(1 - p_{H1})$	$(0, 2R)$	(R, R)

Thus, a solution exists if and only if $\beta \geq 1$ (otherwise, $p_{H2} > 1$). – id est, a solution exists if and only if the cost of an injury is greater than the reward. The situation is analogous if we make the results of Player 2 indifferent to the strategy: indifference corresponds to $p_{H1} = 1/\beta$.

Let us analyze the replicator dynamics: let us consider a population divided into Hawks and Doves. Let us denote by x_H the proportion of Hawks and $x_D = 1 - x_H$ the proportion of Doves. Then, the mean payoff for a Hawk is

$$\overline{f_H} = x_H(1-\beta)R + 2R(1-x_H) = R(2-(1+\beta)x_H).$$

The mean payoff for a Dove is

$$\overline{f_D} = (1-x_H)R.$$

The mean payoff of the population is

$$\overline{f} = x_H\overline{f_H} + (1-x_H)\overline{f_D} = R\left(1-\beta x_H^2\right)$$

The replicator equation reads as

$$\frac{\mathrm{d}}{\mathrm{d}t}x_H = x_H\left(\overline{f_H} - \overline{f}\right) = Rx_H(1-x_H)(1-\beta x_H).$$

Then, there are three stationary solutions: $x_H \in \{0, 1, 1/\beta\}$. If $\beta < 1$, the stationary solutions are $\{0, 1\}$: the population of Doves tends to extinction. Indeed, if $x_H(0) \notin \{0, 1\}$, the population of Hawks converges to $1/\beta$ (see Figs. 7.25 and 7.26).

Thus, if the cost of an injury is less than the reward, the best strategy is "Hawk". Let us consider the population of Hawks divided into two subpopulations: Dominant Hawks (*DH*) and Hawks (*H*): when a Hawk (*H*) or a Dominant Hawk faces a Dove (*D*), the Dove withdraws, and the Hawk wins he reward $2R$. When two Doves face, the reward is equally divided: each one wins R. If two Hawks or two Dominant Hawks face, they fight and, in the mean, their reward is $R - C$, where C is the cost of the injuries received in the fight. When a Hawk faces a Dominant Hawk, the Dominant Hawk receives $2(1 - \alpha)R$ and the Hawk receives αR ($0 < \alpha < 1/2$). This game may be put into a Matrix Form as in Table 7.14.

Notice that $p_{DH1} + p_{H1} + p_{D1} = p_{DH2} + p_{H2} + p_{D2} = 1$. The mean payoffs are

$$E_{DH1} = R(2 - p_{DH2}(1+\beta) - 2\alpha p_{H2});$$
$$E_{H1} = R(2-(1+\beta)p_{H2} - 2(1-\alpha)p_{DH2});$$
$$E_{D1} = R\,(1 - p_{DH2} - p_{H2}).$$

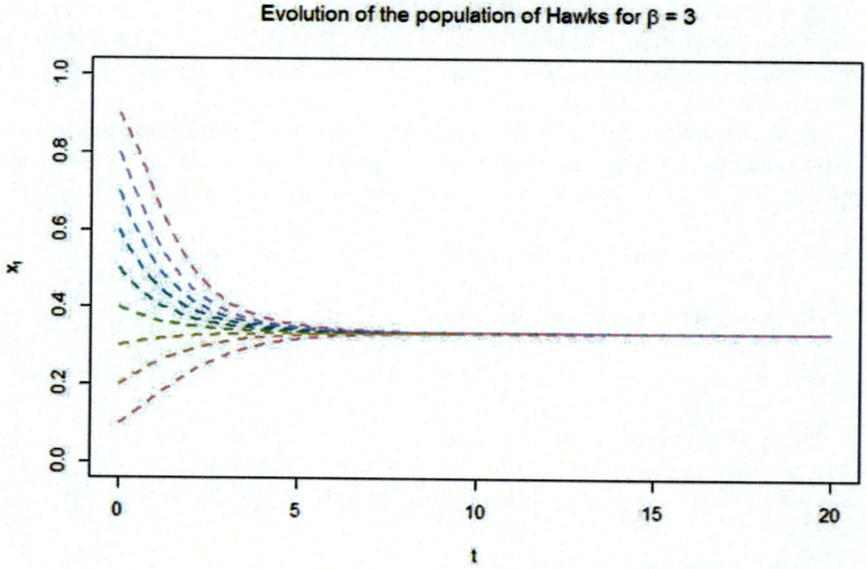

Fig. 7.25 In this example, $1/\beta \approx 0.333$, so that the fraction of "Hawks" converge to this value, for any initial value $x_H(0) \notin \{0, 1\}$

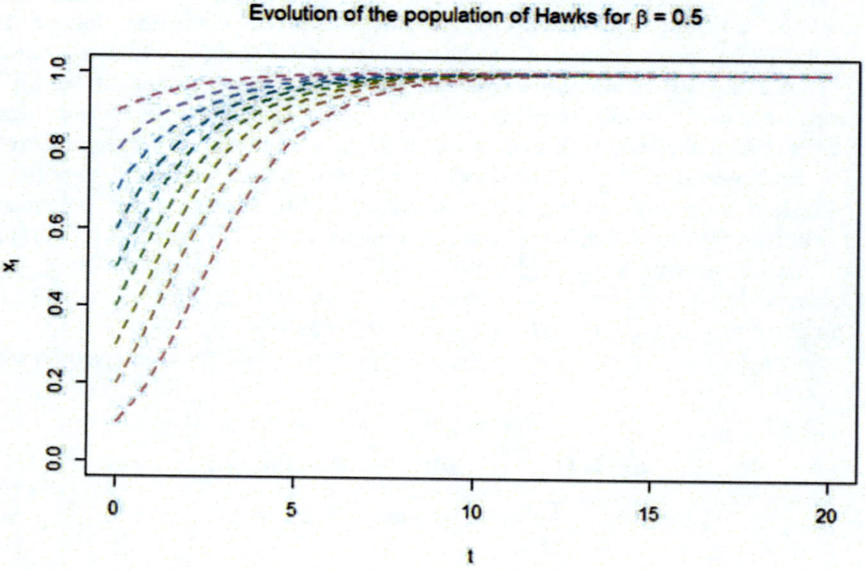

Fig. 7.26 In this example, $1/\beta = 2$, so that the fraction of "Hawks" converge to 1, for any initial value $x_H(0) \neq 0$: the species "Doves" tends to extinction

Table 7.14 The normal form of the Dominant Hawks, Hawks, and Doves

	Player 2		
Player 1	$DH\ (p_{DH2})$	$H\ (p_{H2})$	$D\ (p_{D2})$
$DH\ (p_{DH1})$	$((1-\beta)R, (1-\beta)R)$	$(2(1-\alpha)R, 2\alpha R)$	$(2R, 0)$
$H\ (p_{H1})$	$(2\alpha R, 2(1-\alpha)R)$	$((1-\beta)R, (1-\beta)R)$	$(2R, 0)$
$D\ (p_{D1})$	$(0, 2R)$	$(0, 2R)$	(R, R)

Thus, the mean payoff of Player 1 is

$$E_1 = R\Big(1 - p_{H2} + p_{H1} - p_{DH2}(1 + (1-2\alpha)p_{H1}) - \beta p_{H1}p_{H2}$$
$$+ p_{DH1}(1 + p_{H2}(1-2\alpha) - \beta p_{DH2})\Big)$$

The results of Player 1 are indifferent to the strategy when $E_{DH1} = E_{D1} = E_{H1}$, id est,

$$E_{DH1} = E_{D1} \Leftrightarrow 2 - p_{DH2}(1+\beta) - 2a p_{H2} = 1 - p_{DH2} - p_{H2};$$

$$E_{H1} = E_{D1} \Leftrightarrow 2 - (1+\beta)p_{H2} - 2(1-\alpha)p_{DH2} = 1 - p_{DH2} - p_{H2};$$

The solution is

$$p_{DH2,\mathrm{Nash}} = \frac{1-2\alpha+\beta}{(1-2\alpha)^2+\beta^2};$$
$$p_{H2,Nash} = \frac{2\alpha+\beta-1}{(1-2\alpha)^2+\beta^2}.$$

Then,

$$p_{D2,Nash} = 1 - \frac{2\beta}{a^2+(1-2\alpha)^2}$$

For these values, the mean payoff of Player 1 is

$$E_{1,Nash} = R\left(1 - \frac{2\beta}{a^2+(1-2\alpha)^2}\right) = p_{D2,Nash}R.$$

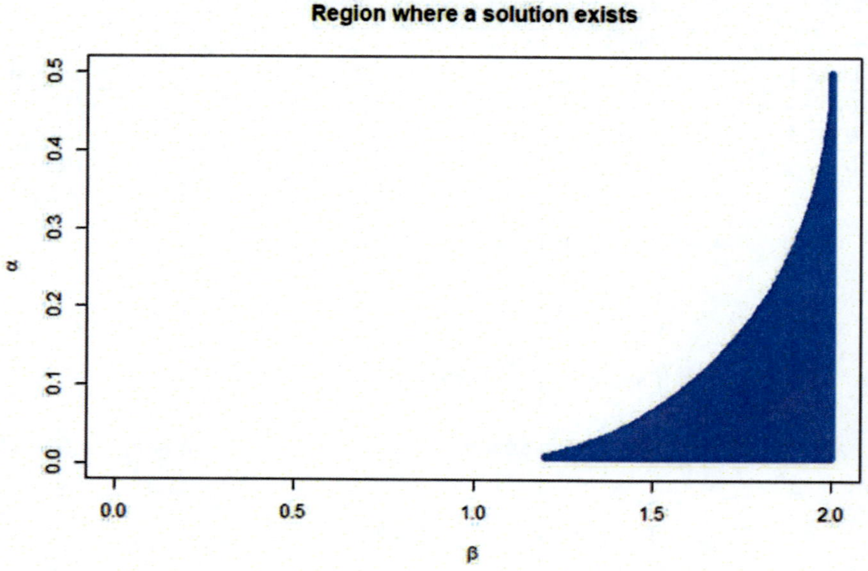

Fig. 7.27 Values $p_{DH2,\ Nash}, p_{H2,\ Nash}, p_{D2,\ Nash}$ do not exist for some pairs (β, α). The region in blue is the region where a solution exists

Solutions do not exist for some values of (β, α). Figure 7.27 shows the possible pairs. Figure 7.28 shows the corresponding values of $p_{DH2,Nash}, p_{H2,Nash}, p_{D2,Nash}$. Again, solutions with $R > C$ (id est, $a < 1$) cannot exist.

Let us examine the replicator dynamics: denoting by x_H the fraction of Hawks and x_{DH} the internal fraction of dominant Hawks, the fraction of non-dominant Hawks is $x_{NDH} = 1 - x_{DH}$ and the fraction of Doves is $x_D = 1 - x_H$. We have

$$\overline{f_{DH}} = R\ (2 + ((2\alpha - \beta - 1)x_{DH} - 2\alpha)x_H)$$
$$\overline{f_{NDH}} = R(2 + ((2\alpha + \beta - 1)x_{DH} - 1 - \beta)x_H)$$
$$\overline{f_H} = R(2 - (2\beta(x_{DH} - 1)x_{DH} + 1 + \beta)x_H)$$
$$\overline{f_D} = R(1 - x_H)$$
$$\overline{f} = R\big(1 - \beta(1 + 2(x_{DH} - 1)x_{DH})x_H^2\big).$$

Thus, the replicator equations read as

$$\frac{dx_H}{dt} = x_H\left(\overline{f_H} - \overline{f}\right) = R(x_H - 1)x_H(\beta(1 + 2(x_{DH} - 1)x_{DH})x_H - 1)$$

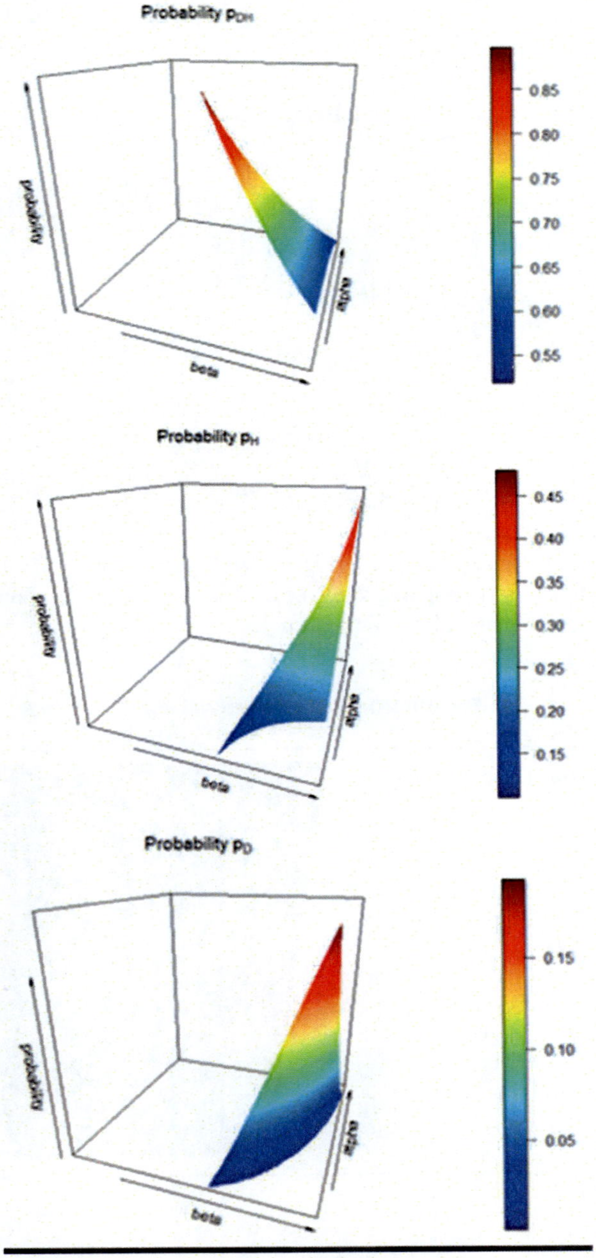

Fig. 7.28 Values $p_{DH2,\ Nash}$, $p_{H2,\ Nash}$, $p_{D2,\ Nash}$ for the pairs (a, α)

$$\frac{dx_{DH}}{dt} = x_{DH}\left(\overline{f_{DH}} - \overline{f_H}\right) = R(x_{DH} - 1)x_{DH}(2\beta x_{DH} - 1 + 2\alpha - \beta)x_H.$$

The stationary points correspond to the critical values

$$x_{DH} = \frac{1-2\alpha+\beta}{2\beta}, x_H \in \left\{0,\ 1,\ \frac{2\beta}{(1-2\alpha)^2+\beta^2}\right\};$$

$$x_{DH} \in \{0,\ 1\}, x_H \in \left\{0,\ 1,\ \frac{1}{\beta}\right\};$$

$$x_H = 0.$$

Again, the non-trivial solution

$$x_H = \frac{2\beta}{(1-2\alpha)^2+\beta^2},\ x_{DH} = \frac{1-2\alpha+\beta}{2\beta}$$

does not exist for some pairs (β, α) (Fig. 7.29). Figures 7.30, 7.31, and 7.32 show typical evolutions of the populations. Notice that Doves and non-dominant Hawks extinct, except if the cost of an injury is high.

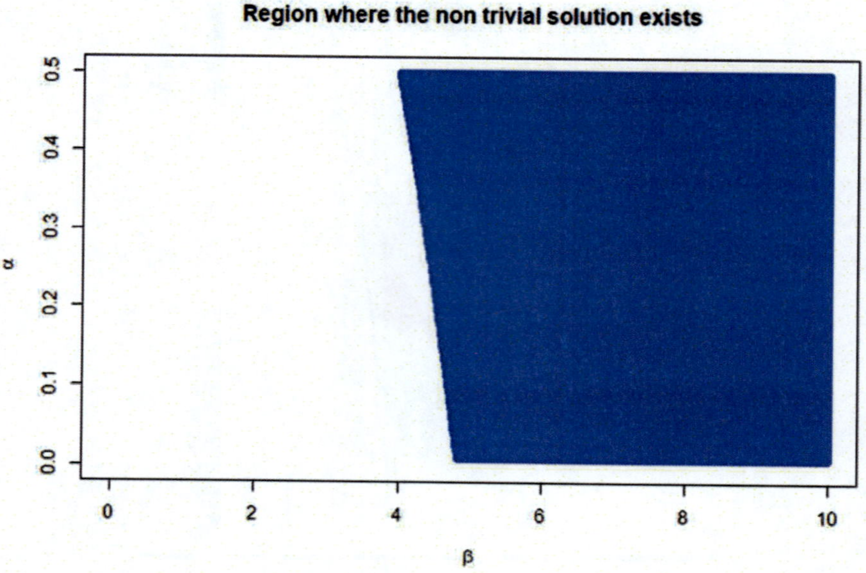

Fig. 7.29 Non-trivial stationary points do not exist for some pairs (β, α). The region in blue is the region where non-trivial stationary points do exist

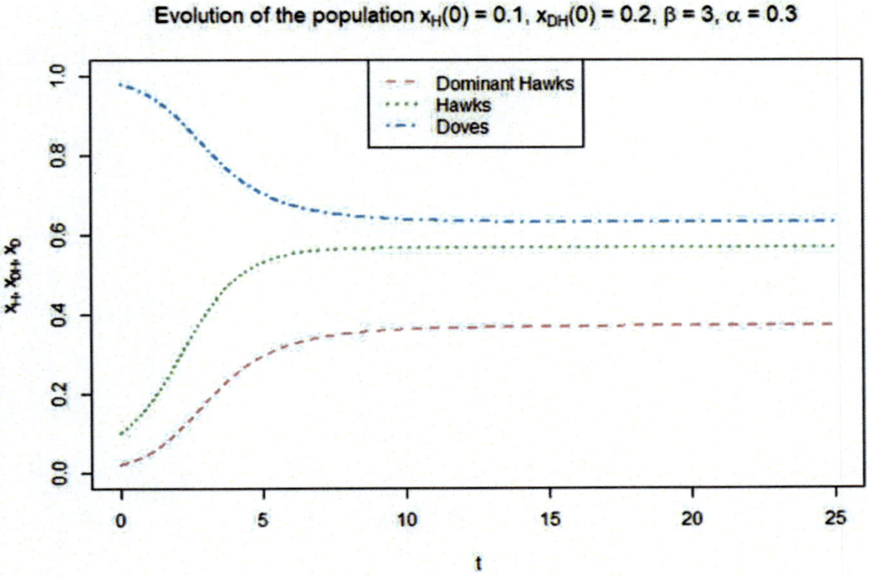

Fig. 7.30 Survival of Doves and non-dominant Hawks for a high cost for injuries

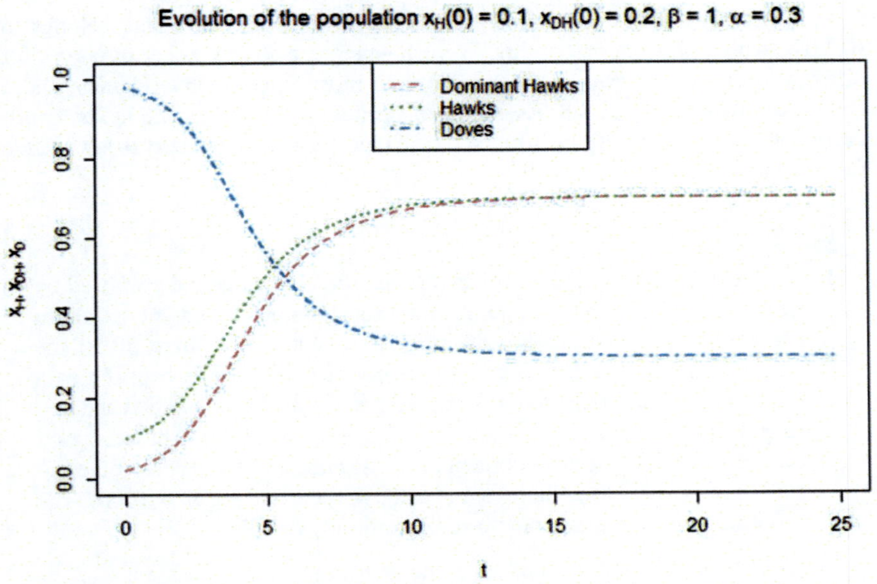

Fig. 7.31 Survival of Doves and extinction of non-dominant Hawks for a medium cost for injuries

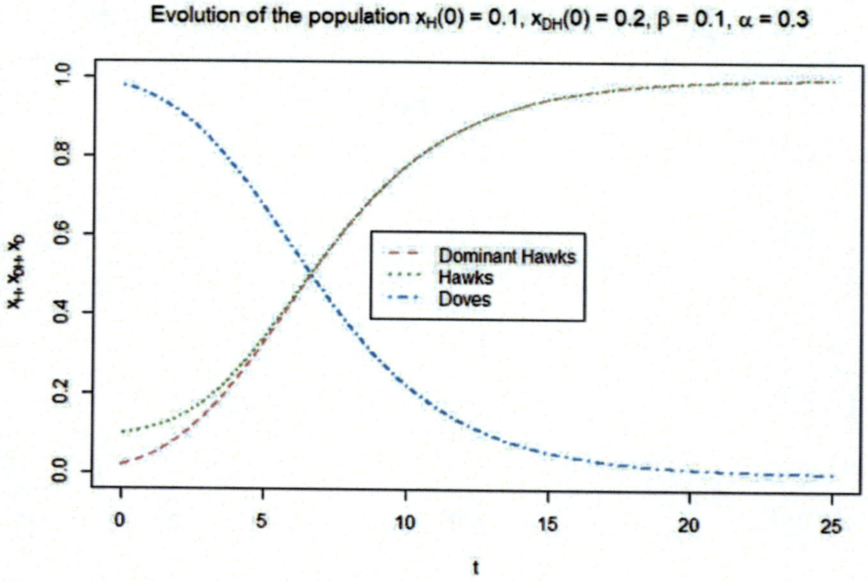

Fig. 7.32 Extinction of Doves and non-dominant Hawks for a small cost for injuries

In this model, we can get a more complex behavior. For instance, for

Assume that there is uncertainty on the value of α: the real value belongs to (0.2,0.4) with a uniform probability. We can generate a large number of curves for different α on this interval and determine their most representative element and a confidence interval, as in 3.8. Examples of results appear in Fig. 7.33. The confidence interval for a risk 20% is formed by the blue curves – black curves lay outside the confidence interval.

Exercises

1. A company must decide on the investment on a new method proposed by an inventor. There are two options: risk-based option or contract option. In risk-based option, the company and the inventor will share 50/50 the gains – if any – and the company does not pay anything, except if a gain is obtained. In the contract option, the company pay for the method and will keep 100% of the gains – if any. For the inventor, the risk-based option implies to support the development costs. Assume that these costs represent 1000 money units (but the inventor will ask for a funding $F > 1000$). The expected gain G is a random variable uniformly distributed on (a, b).

   (a) Give the normal form of the game associated to this decision.
   TIP: Two players (inventor and company) and two choices (contract, risk). The payoff is zero for both if there is no agreement. The payoffs of the inventor must consider the development costs.

(continued)

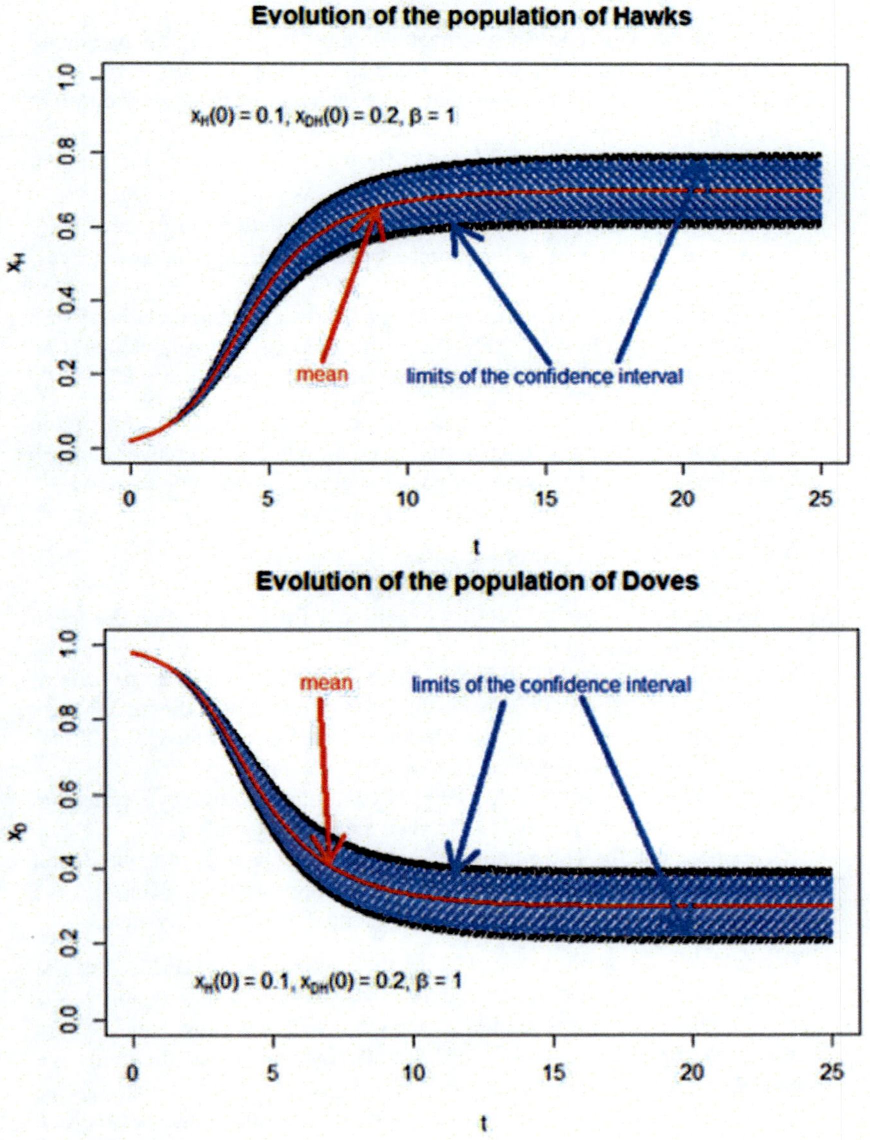

Fig. 7.33 Evolution under uncertainty on , uniformly distributed on (0.2,0.4): blue curves belong to the confidence interval. Red curve is the most representative element, taken as mean

(b) Verify that both choices (contract, contract) and (risk, risk) correspond to Nash equilibria when G and F are large enough.
(c) Find the relation between F and G for which the strategy is indifferent for the company.
(d) Find the relation between F and G for which the strategy is indifferent for the inventor.
(e) Let $F^* = \frac{b+a}{4}$. The inventor asks for a contract with $F = F^* + \delta\ (\delta > 0)$. What is the company's best choice when considering the mean payoff of the operation?
(f) The company chooses a contract but considers that the probability of a positive payoff must be 90%. What is the maximum δ acceptable? How this value is modified if the company requests a probability of 99%?

2. Two players must choose between two options. If both choose option 1, both win \$1. If both choose option 2, both loose \$2. If a player chooses option 1 and the other chooses option 2, the player choosing option 2 wins \$ α ($\alpha > 1$) and the player choosing option 1 looses \$1.

(a) Give the normal form of the game.
(b) Verify that there is no pure Nash equilibrium.
(c) Denoting by p_i the probability of option 1 for Player i, find the mean payoff of each player.
(d) Determine a mixed Nash Equilibrium. Determine the mean payoffs of the Players in this case. What is the value of α such that the mixed Nash Equilibrium corresponds to a random choice between the two options? What is the mean payoff in this case?
(e) For each Player, find the probability of not loosing money. Is there an Equal Opportunity Solution for such a minimal payoff ?
(f) Suppose that the Table find in (a) represents the payoffs of a population having a proportion x_i of individuals choosing strategy i. Determine the mean payoffs and the replicator equations. Find the steady states.

3. Two people must decide about cleaning the house. If no one makes the task, the house will be dirty and each one will have a negative utility $-d_i$ ($d\_i > 0$). If one cleans the house, the other will pay him a small reward corresponding to the utility of $r > 0$. If both clean the house, the utility is a small positive number for each one c_i ($0 < c\_i < r$). Assume that $d_1 < d_2$ and $c_1 < c_2$. Denote by p_i the probability of cleaning the house for player i,

(a) Give the normal form of the game.
(b) Has this game a pure Nash equilibrium?
(c) Determine the mean payoffs of each player.
(d) Determine a mixed Nash equilibrium. Is there a Nash equilibrium if one of the Players is little bothered by dirtiness ($d_i < c_i$)? Assume that

(continued)

$d_i > c_i$. What are the mean payoffs of each player in this case? What is the minimal value of r such that the mean payoffs are positive?

(e) Denote by (p_1^*, p_2^*) the probabilities of the mixed Nash Equilibrium found. What is the best decision for Player 1 when $p_2 < p_2^*$?
TIP: Determine the mean payoff of each strategy and choose the best.

(f) For each player, find the probability of a positive payoff. Is there a Equal Opportunity Solution for a positive payoff?

(g) For each Player, find the probability of a minimal payoff of $-r$. Is there a Equal Opportunity Solution for such a minimal payoff?

(h) Player 2 estimates that p_2 is a random variable uniformly distributed on $(p_2^* - \alpha, p_2 < p_2^* + \beta)$. Determine the probability that cleaning the house is the best strategy.

4. Two Players compete to be the first user of a resource. They must choose between facing the opponent and giving up. When facing the opponent, an injury may be inflicted, and the Player may be defected – these events are synthetized by a negative utility $-b_i$ $(b_i > 0)$ for Player i. Assume that the utility of the resource is $u_i > 0$ for the Player i. When giving up, no injury happens, but the utility is reduced to u_i $\left(0 \leq a < \frac{1}{4}\right)$.

   (a) Give the normal form of the game associated.
   TIP: Use the utilities $u_i - b_i$ when facing the opponent and to $(0, 0)$ in the case (give up, give up)

   (b) Find the conditions for (face, give up) to be a Nash equilibrium.

   (c) Find the conditions for (give up, face) to be a Nash equilibrium.

   (d) Find a mixed Nash equilibrium? What are the probabilities and the mean payoffs of each Player?

   (e) Is there an Equal Opportunity Solution with no injury for the Players?

   Assume that b_i is a random variable, uniformly distributed on $(a_i u_i, A_i u_i)$. $0 \leq a_i < A_i \leq 1$. Find the mean payoff of each strategy for each Player. Determine the mean payoff of each Player. Find the probability of facing to be a better option than giving up for each Player.

Chapter 8
Optimization Under Uncertainty

Abstract This chapter presents methods for the determination of the probability distribution of the solutions of continuous optimization problems: constrained or unconstrained, linear or nonlinear. We analyze also the distributions of Lagrange's multipliers.

In Sect. 1.13, we introduced the classical mono-objective optimization problem as

$$\boldsymbol{x}^* = \arg\min_{C} J = \arg\min \{J(\boldsymbol{y}) : \boldsymbol{y} \in C\}. \tag{8.1}$$

According to the terminology previously introduced, C is the admissible set and J is the objective function. The optimal point is $\boldsymbol{x}^*$ and the optimal value of the objective is $J(\boldsymbol{x}^*)$. In the same section, we observed that, for most of the practical applications, $J : \mathbb{R}^n \rightarrow \mathbb{R}$ and $C \subset \mathbb{R}^n$ such that

$$C = \{\boldsymbol{y} \in \mathbb{R}^n : \psi_i(\boldsymbol{y}) \leq 0,\ 1 \leq i \leq p;\ \psi_{i+p}(\boldsymbol{y}) = 0,\ 1 \leq i \leq q\}, \tag{8.2}$$

with $\psi_j : \mathbb{R}^n \rightarrow \mathbb{R},\ 1 \leq j \leq p + q$. The formulation (8.2) involves Lagrange's multipliers $\boldsymbol{\lambda} = (\lambda_1, \ldots, \lambda_{p+q})^t$.

Uncertainties may concern the objective function or constraints. If any of these elements becomes a random variable, then both $\boldsymbol{x}^*$ and $J(\boldsymbol{x}^*)$ become random variables. Optimization problems involving random variables are often referred as stochastic programming problems and the reader may find in the literature many works adopting different approaches, such as robust optimization, reliability analysis, stochastic optimization methods, etc. For instance, we may look for the solution of

$$\boldsymbol{x}^* = \arg\min \{E(J(\boldsymbol{y})) : \boldsymbol{y} \in C\}, \tag{8.3}$$

E. Souza de Cursi, *Uncertainty Quantification using R*, International Series in Operations Research & Management Science 335,
https://doi.org/10.1007/978-3-031-17785-9_8

or

$$x^* = \arg\min\{E(J(y)) : y \in C,\ V(J(y)) \le V_{\max}\}, \tag{8.4}$$

or

$$x^* = \arg\min\{V(J(y)) : y \in C,\ E(J(y)) \le J_{\max}\}. \tag{8.5}$$

We may also look for a solution such that

$$x^* = \arg\min\{E(J(y)) : \mathrm{P}(y \in C) \ge 1 - \alpha\}. \tag{8.6}$$

In this Chapter, we do not adopt these points of views. We consider the problem from a different standpoint: we look for the probability distribution of $J(x^*)$ and x^*.

8.1 Using the Methods of Representation

We can use the approaches presented in Chap. 3 to get a full characterization of the probability distributions of optimal points, Lagrange's multipliers, and optimal values.

To fix the ideas, let us consider the following example, adapted from (Kim, 1968):

$$\min\{J(y) = y_1 - 2y_2 + 3y_3,\ y \in C\},$$
$$C = \left\{y \in \mathbb{R}^3 :\ -2y_1 + y_2 + 3y_3 = u_1;\ 2y_1 + y_2 + 4y_3 = u_2;\ y_i \ge 0,\ i = 1, 2, 3\right\}$$

Assume that u_1 is uniformly distributed on $(1, 3)$, u_2 is uniformly distributed on (3, 5) and these variables are independent. If we look for a solution in the sense of (8.3), we search for the solution of

$$\min\left\{\overline{J(y)} = \overline{y_1} - 2\overline{y_2} + 3\overline{y_3},\ \overline{y} \in \overline{C}\right\},$$
$$\overline{C} = \left\{\overline{y} \in \mathbb{R}^3 :\ -2\overline{y_1} + \overline{y_2} + 3\overline{y_3} = 2; 2\overline{y_1} + \overline{y_2} + 4\overline{y_3} = 4; \overline{y_i} \ge 0,\ i = 1, 2, 3\right\}$$

The solution is $\overline{x_1} = 1/2, \overline{x_2} = 3, \overline{x_3} = 0, \overline{J(\overline{x})} = -5.5$. By this method, we do not obtain information about the distribution of $J(x^*)$.

To get the distribution of these parameters, we can use the approaches presented in Chap. 3. For instance, let us apply the collocation approach: we consider a sample from u_1 formed by the values $u_{1i} = 1 + 0.1i$, $i = 0, \ldots, 20$ and a sample from u_1 formed by the values $u_{2j} = 3 + 0.1j$, $j = 0, \ldots, 20$. We generate a sample from $U = (u_1, u_2)$ by considering the points (u_{1i}, u_{2j}). The sample is used to generate larger samples from x^* and $J(x^*)$, which are used to determine a bidimensional expansion with $k_1 = k_2 = 3$ (see Sect. 3.5). Here, we used the representation to generate a sample of $ns = 40401$ variates from x^*, using $u_{1i} = 1 + 0.01i$, $u_{2j} = 3 + 0.01j$, i, $j = 0, \ldots, 200$.

In this simple situation, we may determine the exact values:

$$x^* = \left(\frac{1}{4}(u_2 - u_1), \frac{1}{2}(u_1 + u_2), 0\right)$$

$$J(x^*) = -\frac{5}{4}u_1 - \frac{3}{4}u_2$$

A comparison between the approximated and the exact values of x^* is shown in Figs. 8.1 and 8.2. The global RMS error was 3E-13, for each component.

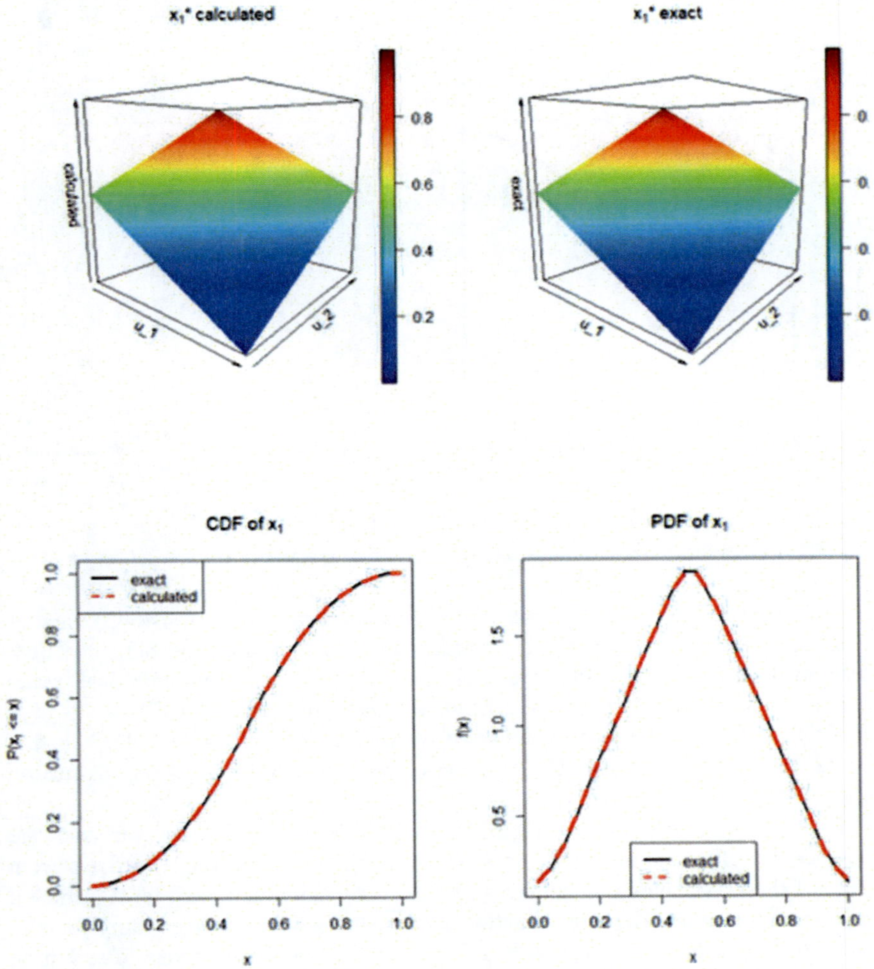

Fig. 8.1 Results for the first component x_1^* of x^* in the introductory example (*U* Uniform)

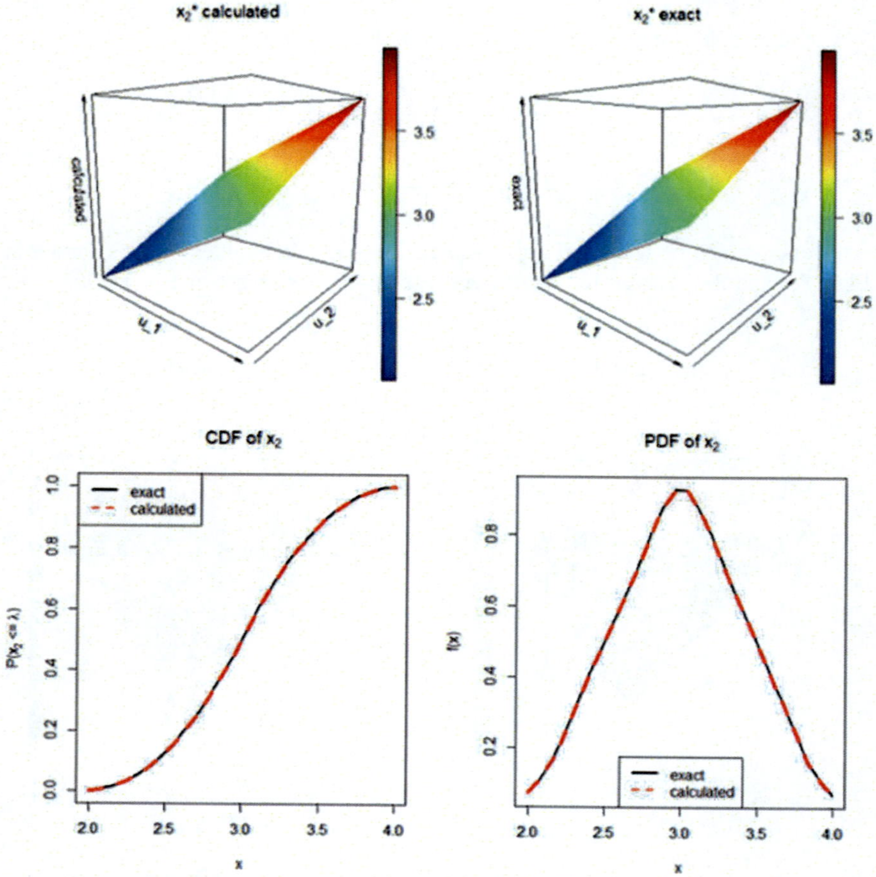

Fig. 8.2 Results for the second component x_2^* of $\boldsymbol{x}^*$ in the introductory example (*U* Uniform)

The large sample from $\boldsymbol{x}^*$ can be used to generate a sample of $ns = 40401$ variates from $J(\boldsymbol{x}^*)$, which provides the distribution of $J(\boldsymbol{x}^*)$ shown in Fig. 8.3 – the PDF was evaluated by the SPH approach (see Sect. 1.16). The empirical mean generated by such a large sample furnished the estimations $E(J(\boldsymbol{x}^*)) \approx -5.5$, $\sigma(J(\boldsymbol{x}^*)) \approx 0.85$ – notice that the exact values are $E(J(\boldsymbol{x}^*)) = -11/2 = 5.5$, $\sigma(J(\boldsymbol{x}^*)) = \sqrt{17/24} \approx 0.84$. The RMS error between $J(\boldsymbol{x}^*)$ and its approximation was 3E-12.

Now, assume that $u_1 \sim N(2, 0.25)$, $u_2 \sim N(4, 0.5)$ and these variables are independent. Into an analogous way, we consider a sample of 21 variates from each of these variables and we use again the collocation approach to determine a representation of $\boldsymbol{x}^*$. The representation was used to generate a large sample analogous to the preceding one, with 201 variates from the gaussian distribution corresponding to each component of $\boldsymbol{U}$. The CDF and the PDF of $J(\boldsymbol{x}^*)$ were determined analogously. Now, the estimations are $E(J(\boldsymbol{x}^*)) \approx -5.53$, $\sigma(J(\boldsymbol{x}^*)) \approx 0.49$, again identical up to the 5th decimal for both samples (exact and calculated).

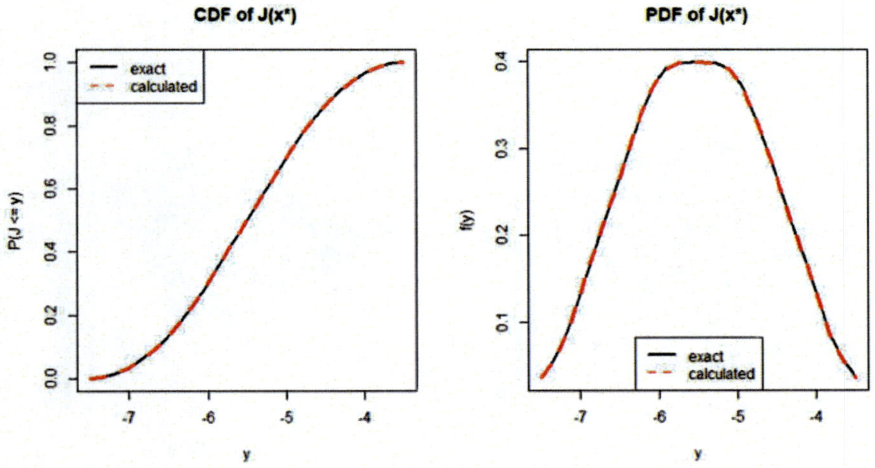

Fig. 8.3 Distribution of the optimal value $J(x^*)$ in the introductory example (U uniform)

Notice that the exact value is $\sigma(J(\boldsymbol{x}^*)) = \sqrt{61}/16 \approx 0.49$. The RMS error was 1E-4 for $J(\boldsymbol{x}^*)$ and 4E-5 for $\boldsymbol{x}^*$. Although the mean is the same for the two situations, the standard deviations and the distributions differ significantly. A comparison between the approximated and the exact values of $\boldsymbol{x}^*$ is shown in Figs. 8.4 and 8.5, while the distribution of $J(\boldsymbol{x}^*)$ appears in Fig. 8.6.

The determination of the distribution of Lagrange's multipliers can be made analogously: we can generate a sample of their values and apply one of the methods presented in Chap. 3. For this introductory and simple problem, the objective function and the restrictions are linear, with coefficients independent from $\boldsymbol{U}$, so that their gradients are constant and are not affected by the randomness introduced by $\boldsymbol{U}$. In addition, the active constraints are the same for all the solutions found, so that Lagrange's multipliers are constant. But such is not a general situation: for instance, let us modify the coefficients as follows:

$$\min\{J(\boldsymbol{y}) = (1-u_1)y_1 + (u_2-2)y_2 + (3-u_1-u_2)y_3, \boldsymbol{y} \in C\},$$
$$C = \left\{\boldsymbol{y} \in \mathbb{R}^3 : -2y_1 + y_2 + 3y_3 = u_1;\ 2y_1 + y_2 + 4y_3 = u_2;\ y_i \geq 0,\ i = 1, 2, 3\right\}$$

In this case, the gradient of the economic function J depends upon $\boldsymbol{U}$, so that the Lagrange's multipliers too. Here, the solution in the sense of (8.4) satisfies

$$\min\left\{\overline{J(\boldsymbol{y})} = -\overline{y_1} + 2\overline{y_2} - 3\overline{y_3}, \overline{\boldsymbol{y}} \in \overline{C}\right\},$$
$$\overline{C} = \left\{\overline{\boldsymbol{y}} \in \mathbb{R}^3 : -2\overline{y_1} + \overline{y_2} + 3\overline{y_3} = 2;\ 2\overline{y_1} + \overline{y_2} + 4\overline{y_3} = 4;\ \overline{y_i} \geq 0,\ i = 1, 2, 3\right\}$$

and the solution is now $\overline{x_1} = 2/7$, $\overline{x_2} = 0$, $\overline{x_3} = 6/7$, $J(\overline{\boldsymbol{x}}) = -20/7 \approx -2.86$. The exact solution is

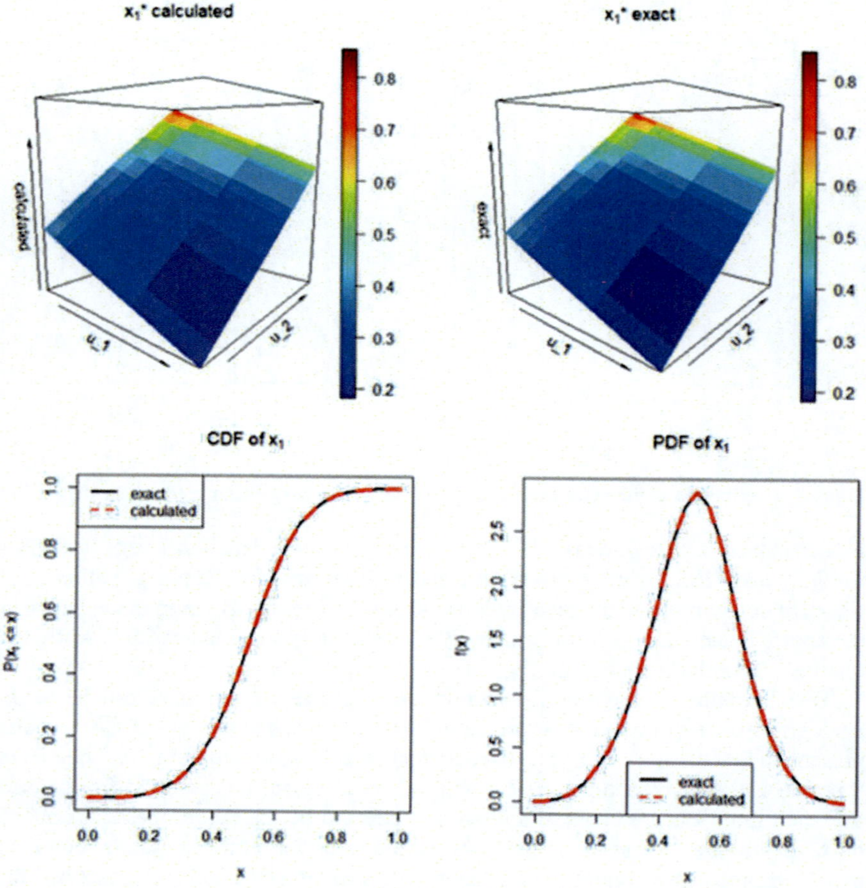

Fig. 8.4 Results for the first component x_1^* of $\boldsymbol{x}^*$ in the introductory example ($\boldsymbol{U}$ Gaussian)

$$\boldsymbol{x}^* = \left(\frac{1}{14}(-4u_1 + 3u_2), 0, \frac{1}{7}\ (u_1 + u_2)\right)$$

$$J(\boldsymbol{x}^*) = \frac{1}{14}\left(2u_1^2 + 2u_1 - 7u_1u_2 + 9u_2 - 2\ u_2^2\right)$$

We may obtain the distribution of $J(\boldsymbol{x}^*)$ by using the same samples which were considered for the preceding situation. For the uniform distribution, the results are shown in Fig. 8.7. The empirical mean generated by the large sample with 40401 variates furnished the estimations $E(J(\boldsymbol{x}^*)) \approx\ -23.75$, $\sigma(J(\boldsymbol{x}^*)) \approx 25.52$, to be compared with the exact values $-20/7 \approx -2.85714$, $\sqrt{2938/2205} \approx 1.15431$. The RMS error for the values of $J(\boldsymbol{x}^*)$ was $3E-13$, while it was of $4E$-13 for $\boldsymbol{x}^*$.

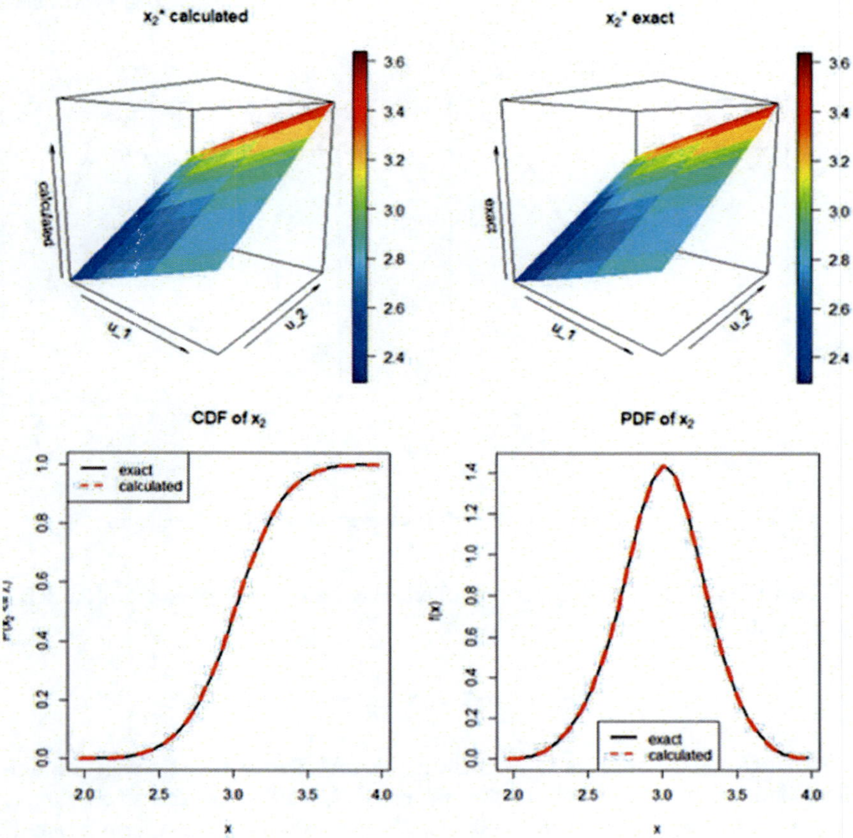

Fig. 8.5 Results for the second component x_2^* of $\boldsymbol{x}^*$ in the introductory example (*U* Gaussian)

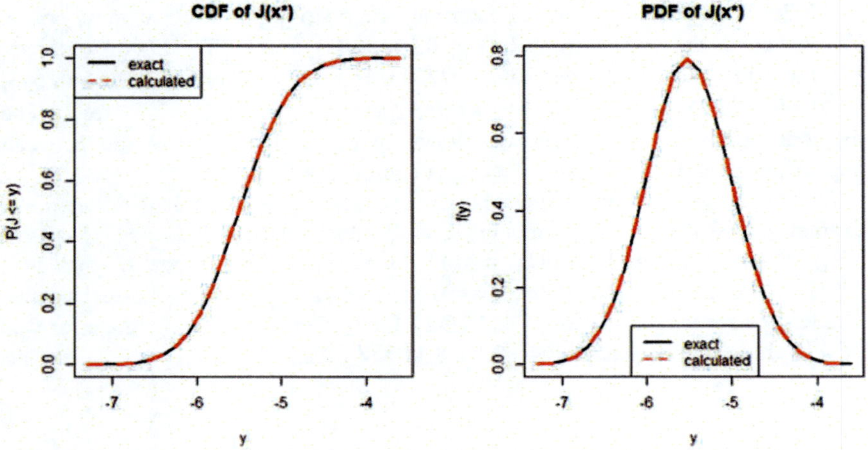

Fig. 8.6 Distribution of the optimal value $J(\boldsymbol{x}^*)$ in the introductory example (*U* Gaussian)

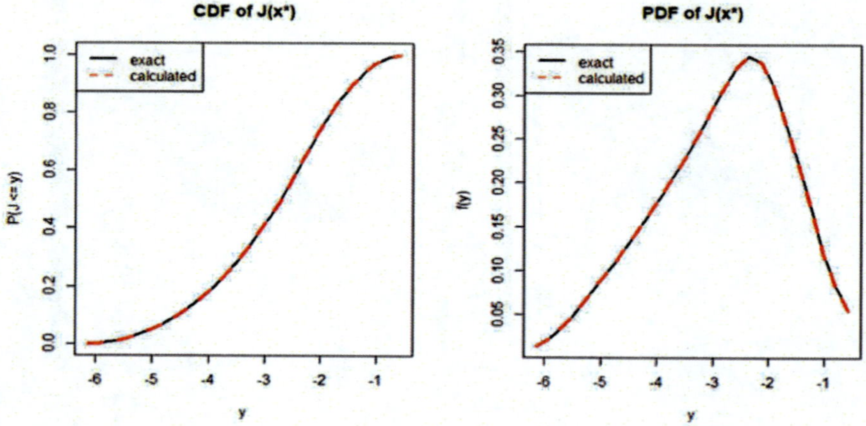

Fig. 8.7 Distribution of the optimal value $J(\boldsymbol{x}^*)$ in the example modified ($\boldsymbol{U}$ uniform)

The values of the Lagrange's multipliers may be easily derived from the KKT equations (Sect. 1.13):

$$\lambda_1 = \frac{1}{7}(-1 - u_1 + u_2), \ \lambda_2 = \frac{1}{14}(-9 + 5u_1 + 2u_2).$$

We may use the samples to estimate the distribution of the Lagrange's multipliers: indeed, we may evaluate the values of λ_1 and λ_2 on the small sample $u_{1i} = 1 + 0.1i$, $u_{2j} = 3 + 0.1j, j = 0, \ldots, 20$. The values collected form a sample from the Lagrange's multipliers, so that we may use them to find representations of these variables. Analogously to $\boldsymbol{x}^*$, we consider collocation with $k_1 = k_2 = 3$. Once the representations are determined, we use the large sample to generate new values of the Lagrange's multipliers, getting a large sample of 40401 values of each multiplier. This sample is used to determine the distribution of the shadow prices. We obtained the estimates $E(\lambda_1) \approx 0.148$, $\sigma(\lambda_1) \approx 0.081$, $E(\lambda_2) \approx 0.655$, $\sigma(\lambda_2) \approx 0.116$, close to the real values $1/7 \approx 0.143$, $\sqrt{2/147} \approx 0.117$, $9/14 \approx 0.643$, $\sqrt{29/588} \approx 0.222$ respectively. The RMS error was 8E-14 for λ_1, 5E-14 for λ_2. The distributions of the multipliers are shown in Fig. 8.8. A comparison between approximated and exact values of λ_1, λ_2 is exhibited in Fig. 8.9.

The same approach can be applied to the case where $\boldsymbol{U}$ is normally distributed. An example of distribution is shown in Fig. 8.9. We obtained the estimates $E(\lambda_1) \approx 0.15$, $\sigma(\lambda_1) \approx 0.08$, $E(\lambda_2) \approx 0.65$, $\sigma(\lambda_1) \approx 0.11$. Here, the real values are $1/7 \approx 0.143$, $\sqrt{5}/28 \approx 0.08$, $9/14 \approx 0.643$, $\sqrt{41}/56 \approx 0.11$, remarkably close to the real values. The RMS error was $2E$-13 for λ_1 and $2E$-12 for λ_2. Concerning $J(\boldsymbol{x}^*)$, the error was $2E$-11. The results appear in Figs. 8.10, 8.11 and 8.12.

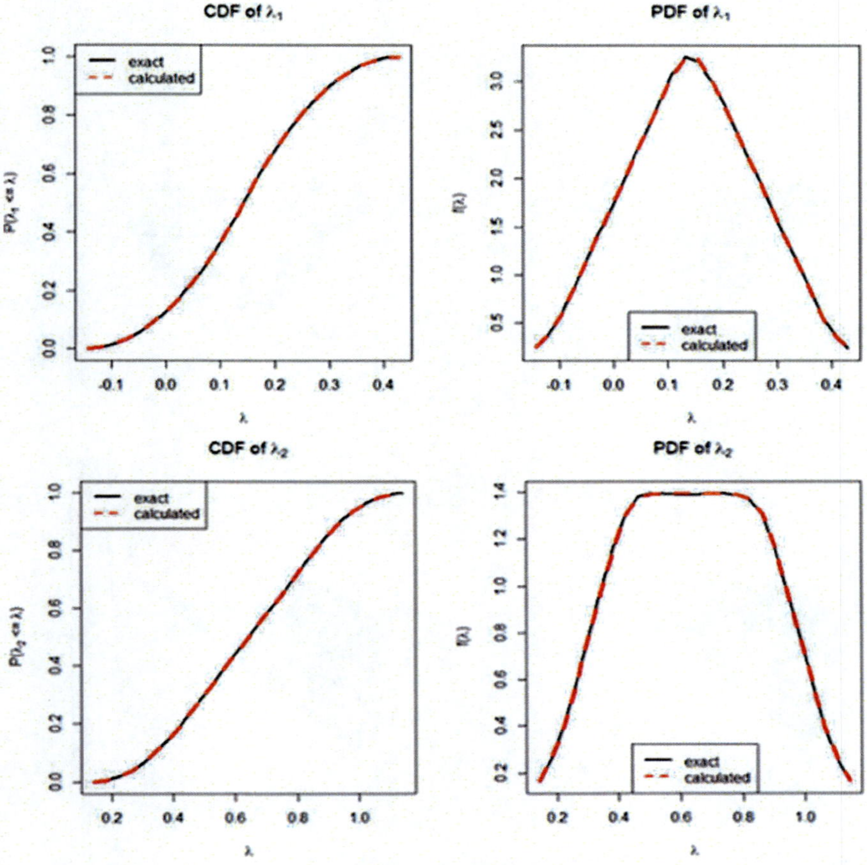

Fig. 8.8 Distributions of the Lagrange multipliers in the second example. Here, $\boldsymbol{U}$ is uniform

The same approach can be applied for nonlinear optimization. In addition, as shown in Sect. 5.2, iterative descent methods may be adapted. For instance, let us consider a regular function η and the modified Rosenbrock's function:

$$f(x) = \sum_{i=1}^{n-1} \left(100 \left(\frac{x_{i+1}}{\eta(u,\ i+1)} - \left(\frac{x_i}{\eta(u,\ i)} \right) \right)^2 + \left(\frac{x_i}{\eta(u,\ i)} - 1 \right)^2 \right). \quad (8.7)$$

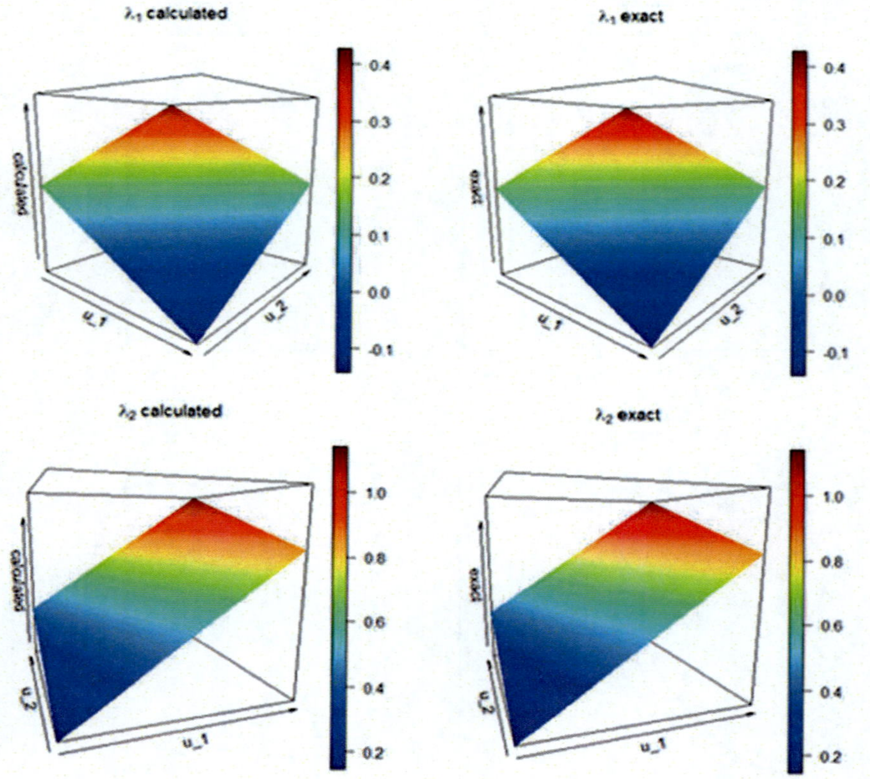

Fig. 8.9 Lagrange's multipliers in the second example (*U* Uniform)

The exact solution is $x_i^* = \eta(u, i)$, $1 \leq i \leq n$, for which $f(x^*) = 0$. Let us consider $n = 5$, $\eta(u, i) = u^2 + i$, $u \sim N(0, 1)$ and a sample of 20 variates from u given below generated by `rnorm`):

```
 [1] -0.32145143  0.17319033  0.82929878 -1.97007269  0.76815751
 [6]  0.18701457 -1.61515887 -1.03208832 -0.12796460 -1.52319995
[11]  0.34650107  1.52862311 -0.61503235  0.35338742 -0.03223157
[16] -0.04661223 -0.46591345  0.61900917  0.67721109  1.21910880
```

We use `fminunc` from `pracma` with a random initial point furnished by `runif`. `fminunc` furnishes $\mathbf{x}^*(u)$ for the elements in the sample, so that we get a sample $(u_i, \mathbf{x}^*(u_i))$ of 20 variates. This sample is used to determine the coefficients of

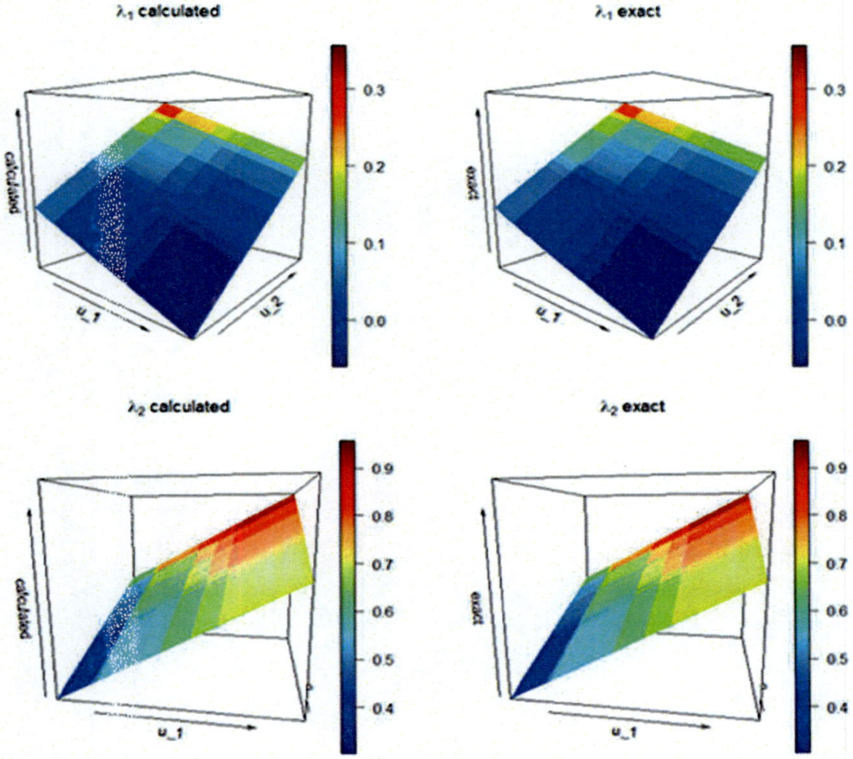

Fig. 8.10 Lagrange's multiplier λ_2 in the second example (***U*** Gaussian)

an expansion $\boldsymbol{Px}^* = \sum_{i=0}^{k} c_i \varphi_i(u)$ by collocation. Using $k = 3$, we obtain a RMS error $\|\boldsymbol{x}^* - \boldsymbol{Px}^*\| \approx 2E - 10$. The expansion is used to generate a large sample of 1E4 variates of $\boldsymbol{x}^*$: `rnorm` furnishes 1E4 variates from $\boldsymbol{u}$ and this sample furnishes 1E4 variates from $\boldsymbol{Px}^*$. Then, the RMS error is $\|\boldsymbol{x}^* - \boldsymbol{Px}^*\| \approx 3E - 9$. Examples of results are shown in Figs. 8.13 and 8.14.

Of course, this approach can also be used in constrained optimization: we can generate a sample and determine an expansion by the same way (see examples).

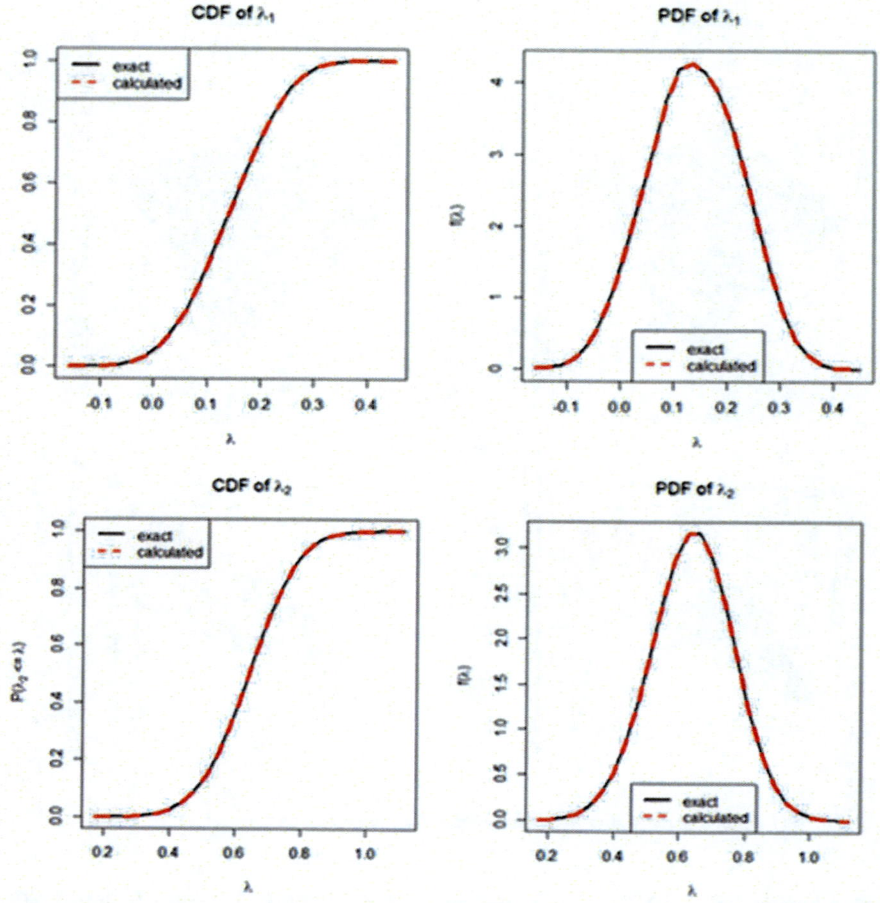

Fig. 8.11 Distributions of the Lagrange's multipliers in the second example. Here, $\boldsymbol{U}$ is Gaussian

Example 8.1

Let us consider $\boldsymbol{x} = (x_1, x_2)$ and (Cobb-Douglas function)

$$J(\boldsymbol{x}) = x_1^{\alpha} x_2^{\beta}$$

We are interested in the maximal value of $J(\boldsymbol{x})$ under the restriction

$$p_1 x_1 + p_1 x_2 = A$$

(continued)

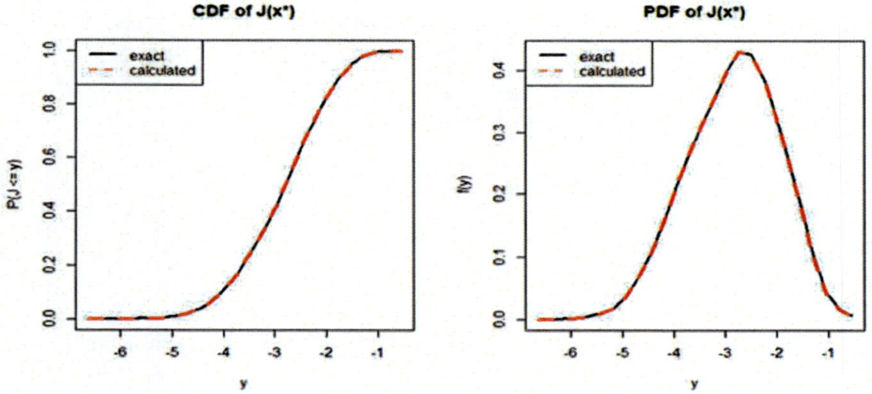

Fig. 8.12 Distribution of the optimal value $J(\boldsymbol{x}^*)$ in the example modified ($\boldsymbol{U}$ uniform)

Example 8.1 (continued)

If $p_1, p_2 > 0$, then the solution is

$$x_1^* = \frac{\alpha}{\alpha+\beta}\frac{A}{p_1}, x_2^* = \frac{\beta}{\alpha+\beta}\frac{A}{p_2}.$$

The Lagrange's multiplier associated to the restriction is

$$\lambda = -\frac{\alpha+\beta}{A}J(\boldsymbol{x}^*)$$

Let us consider uncertainties affect the exponents α, β.

In a first model, we take $\beta = 1 - \alpha$ and suppose that $\alpha \in (0.4, 0.6)$, uniformly distributed. Then

$$E(x_1^*) = \frac{1}{2}\frac{A}{p_1} = 0.5\frac{A}{p_1}, \ \sigma(x_1^*) = \frac{1}{10\sqrt{3}}\frac{A}{p_1} \approx 0.0577\frac{A}{p_1}$$

$$E(x_2^*) \approx \frac{1}{2}\frac{A}{p_2} = 0.5\frac{A}{p_2}, \ \sigma(x_2^*) \approx \frac{1}{10\sqrt{3}}\frac{A}{p_2} \approx 0.0577\frac{A}{p_2}$$

For $A = 100$, $p_1 = 2$, $p_2 = 4$, we have

$$E(x_1^*) = 25, \sigma(x_1^*) \approx 2.9, E(x_2^*) = 12.5, \sigma(x_2^*) \approx 1.4$$

(continued)

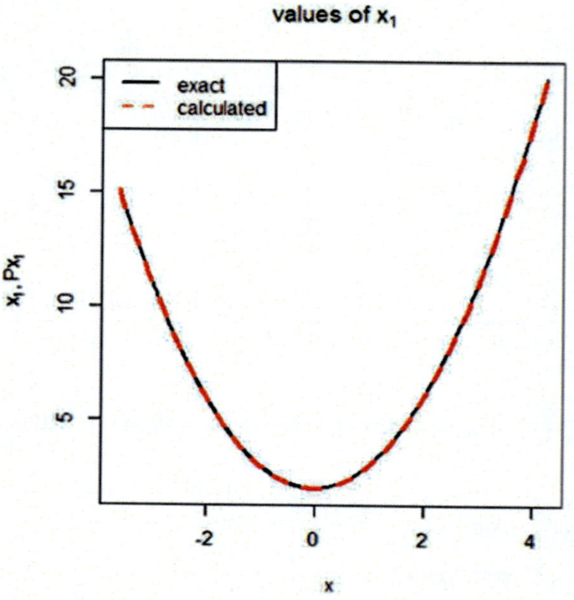

Fig. 8.13 Approximation of x_1^* on a sample of 1E4 variates from u, using the expansion determined on the sample of 20 variates from u

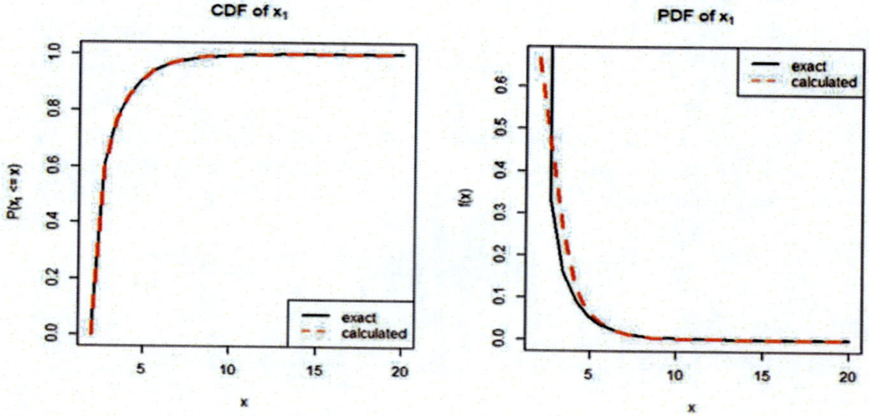

Fig. 8.14 Distribution of x_1^*, determined from on a sample of 1E4 variates from u, using the expansion determined on the sample of 20 variates from u. Since $x_1^* = 2 + u^2$, we have $x_1^* - 2 \sim \chi^2(1)$. The exact values were furnished by `pchisq` and `dchisq`

Example 8.1 (continued)

Let us apply the method presented: assume that a sample of 10 variates from α is given (Table 8.1). These values are used to determine $x^*(\alpha), J^*(\alpha) = J(x^*(\alpha))$, corresponding to the sample from α, so that we have a sample $(\alpha, x^*(\alpha), J^*(\alpha))$. Then, we can determine expansions Px^*, PJ^* by collocation.

Table 8.1 Sample from α used in the calculations

0.401836967	0.456233218	0.502981397	0.508512	0.511317
0.511581	0.521121935	0.53794	0.539049	0.596772

For $k = 5$, we obtain the RMS errors in Table 8.2.

Table 8.2 RMS errors for the first model, on the initial sample of 10 variates

x_1^*	x_2^*	λ	$J(x^*)$
2E-11	2E-11	4E-10	5E-8

Once the expansions were determined, we can generate a large sample of 1E4 variates from α and determine the distribution and statistics of x^* and J^*. We obtain the results in Table 8.3.

Table 8.3 RMS errors for the first model, on the large sample of 1E4 variates

x_1^*	x_2^*	λ	$J(x^*)$
2E-10	2E-10	6E-8	6E-6

The empirical means and standard deviations were

$$E(x_1^*) \approx 25, \sigma(x_1^*) \approx 2.9, E(x_2^*) \approx 12.5, \sigma(x_2^*) \approx 1.4$$

The distribution of $J^* = J(x^*)$ is shown in Fig. 8.15.

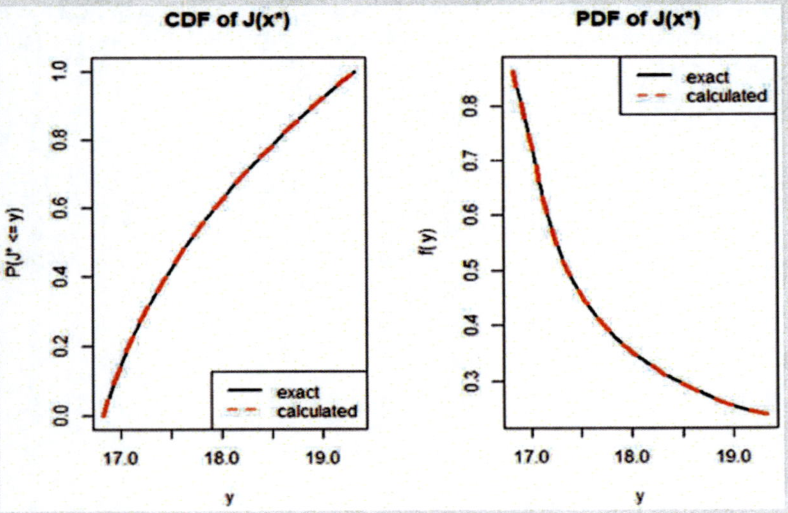

Fig. 8.15 Distribution of $J(x^*)$ with uncertain exponents $\beta = 1 - \alpha$

(continued)

Example 8.1 (continued)

Consider a second model where α and β are independent and suppose that the experts agree that $\alpha \in (0.4, 0.6)$ and $\beta \in (0.3, 0.5)$, both uniformly distributed. In this case

$$E(x_1^*) \approx 0.556 \frac{A}{p_1}, \ \sigma(x_1^*) \approx 0.0459 \frac{A}{p_1}$$

$$E(x_2^*) \approx 0.444 \frac{A}{p_2}, \ \sigma(x_2^*) \approx 0.0459 \frac{A}{p_2}$$

For $A = 100, p_1 = 2, p_2 = 4$, we have

$$E(x_1^*) \approx 27.8, \ \sigma(x_1^*) \approx 2.3$$

$$E(x_2^*) \approx 11.1, \ \sigma(x_2^*) \approx 1.2$$

Let us apply the method presented: we generate a sample of 100 values of the couple $\boldsymbol{x}^* = (x_1^*, x_2^*)$ corresponding to (α_i, β_j), from the sample of α previously introduced and the sample from β shown in Table 8.4.

Table 8.4 Sample from β used in the calculations

0.313722041	0.31488304	0.383768292	0.397799173	0.405302848
0.435980503	0.447786821	0.457281824	0.46323338	0.494181001

The expansion uses $k_1 = k_2 = 3$. The expansions determined were used to generate samples of 40421 elements from x^* and λ, by using the values $u_{1,i} = 0.2 + 0.01i$, $u_{2,j} = 0.3 + 0.01j$, $0 \le i, j \le 200$. The RMS errors between the values generated and the exact ones appear in Table 8.5. The empirical means and standard deviations were $E(x_1^*) \approx 27.8$, $\sigma(x_1^*) \approx 2.3$, $E(x_2^*) \approx 11.1$, $\sigma(x_2^*) \approx 1.2$.

Table 8.5 RMS errors for the second model

x_1^*	x_2^*	λ	$J(x^*)$
7E-4	3E-4	2E-5	1E-3

(continued)

Example 8.1 (continued)

Examples of results are shown in Figs. 8.16, 8.17, and 8.18.

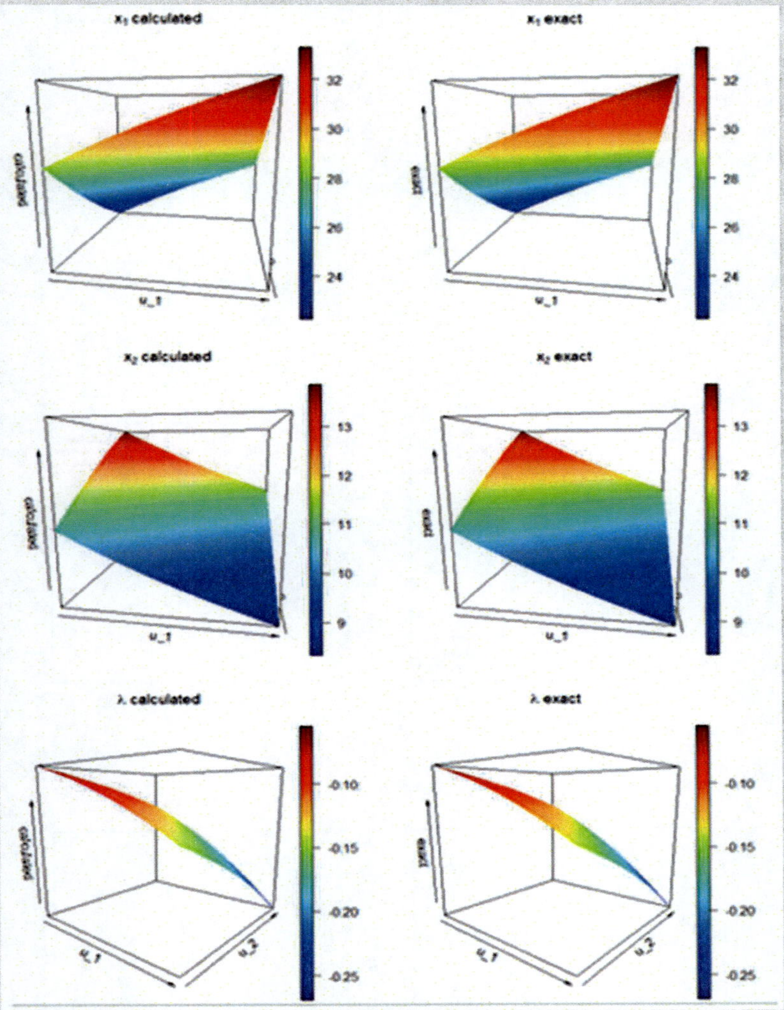

Fig. 8.16 Comparisons between calculated and exact values of x^*, λ in the second model with uncertain independent exponents α, β. At left, the exact values, at right, the calculated ones. We have $u_1 = \alpha$, $u_2 = \beta$

(continued)

Example 8.1 (continued)

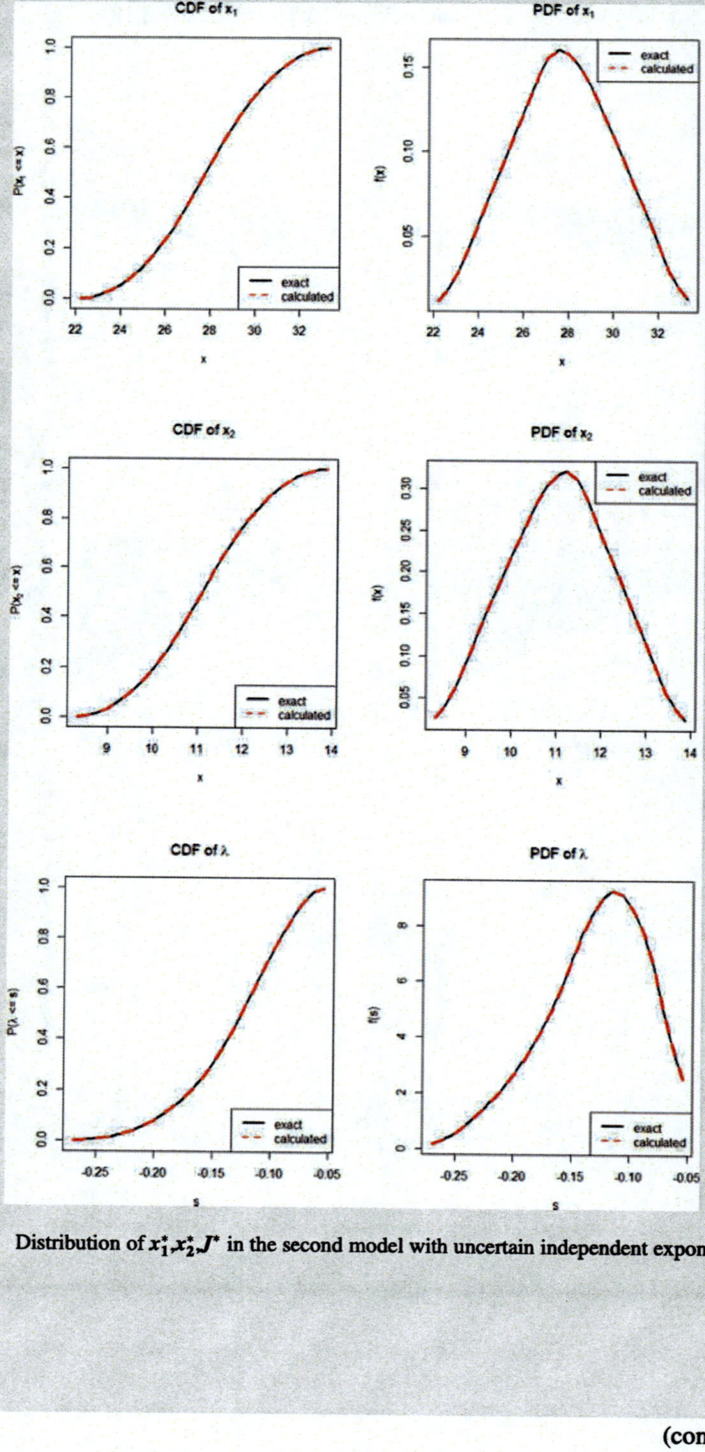

Fig. 8.17 Distribution of x_1^*, x_2^*, J^* in the second model with uncertain independent exponents α, β

(continued)

Example 8.1 (continued)

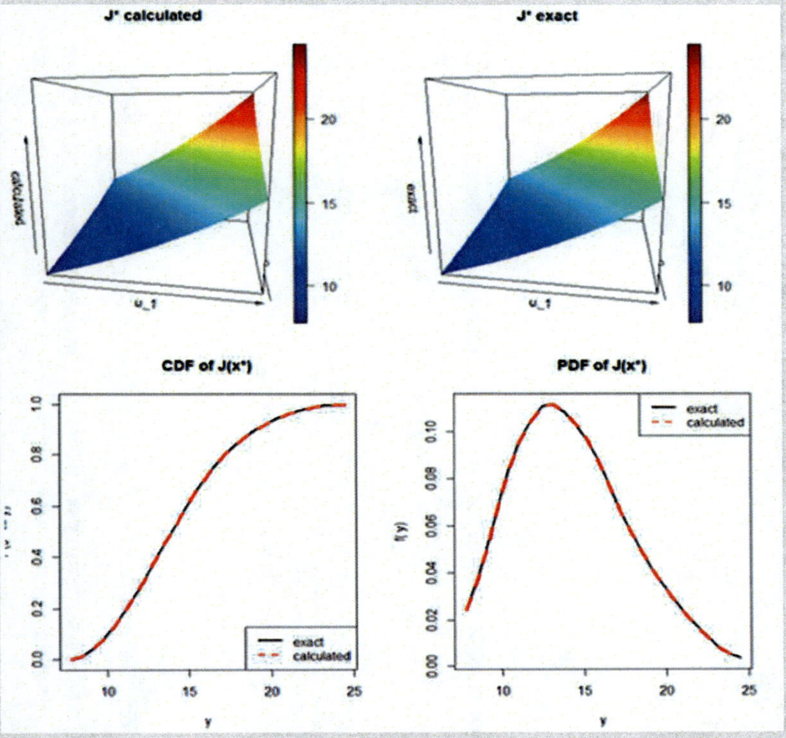

Fig. 8.18 Distribution of J^* in the second model with uncertain independent exponents α, β

Example 8.2
Let us consider the constant elasticity substitution function with two variables

$$J(\boldsymbol{x}) = \left(\alpha x_1^{\beta} + (1-\alpha) x_2^{\beta}\right)^{\frac{1}{\beta}}.$$

We are interested in the maximal value of $J(\boldsymbol{x})$ under the restriction

$$p_1 x_1 + p_1 x_2 = A$$

Consider again a model where α and β are independent, $\alpha \in (0.4, 0.6)$ and $\beta \in (0.3, 0.5)$, both uniformly distributed. We use the same sample as in the preceding example. Expansions are determined with $k_1 = k_2 = 3$. Once the expansion is found, it is used to generate a sample formed by a family of curves corresponding to na subintervals for α and nb subintervals for β, all equally spaced. The results for $na = nb = 200$ are exhibited in Figs. 8.19 and 8.20

(continued)

Example 8.2 (continued)

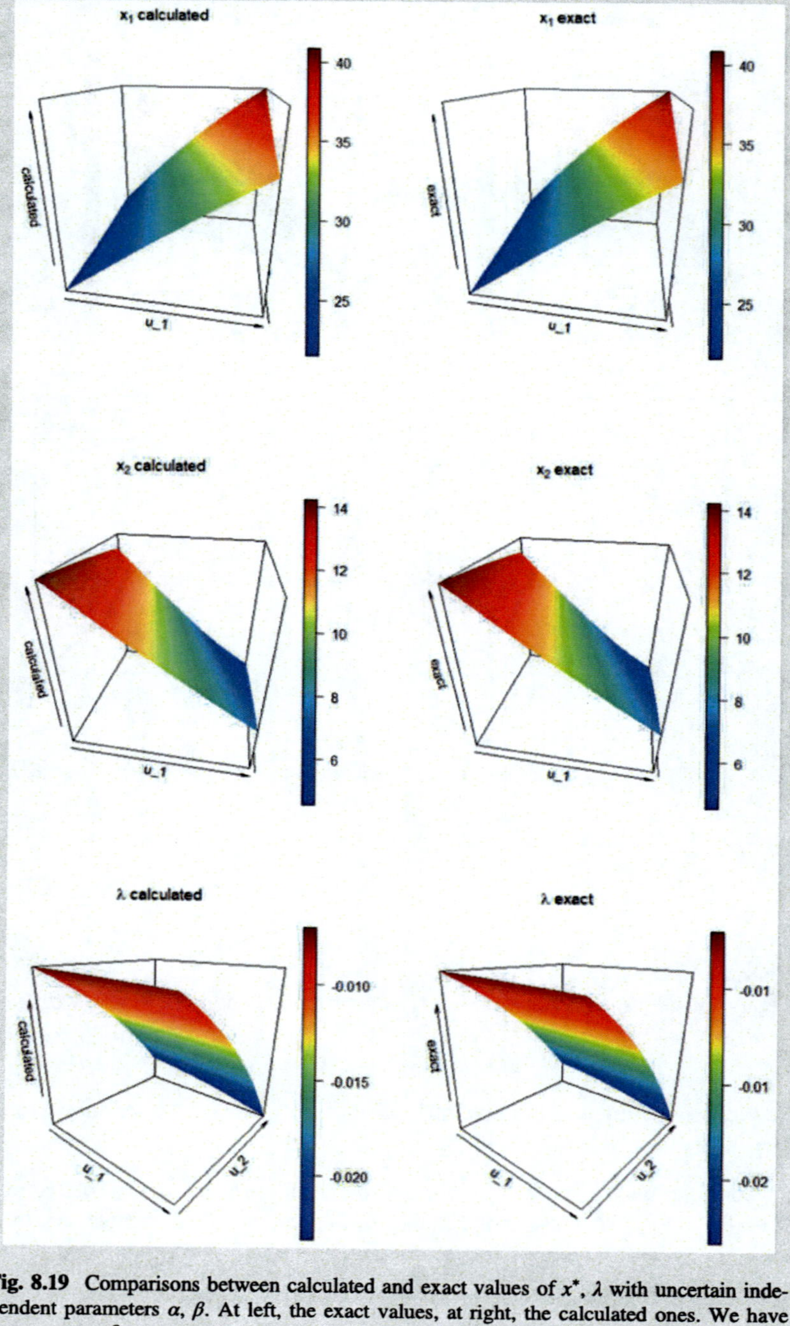

Fig. 8.19 Comparisons between calculated and exact values of x^*, λ with uncertain independent parameters α, β. At left, the exact values, at right, the calculated ones. We have $u_1 = \alpha$, $u_2 = \beta$

(continued)

Example 8.2 (continued)

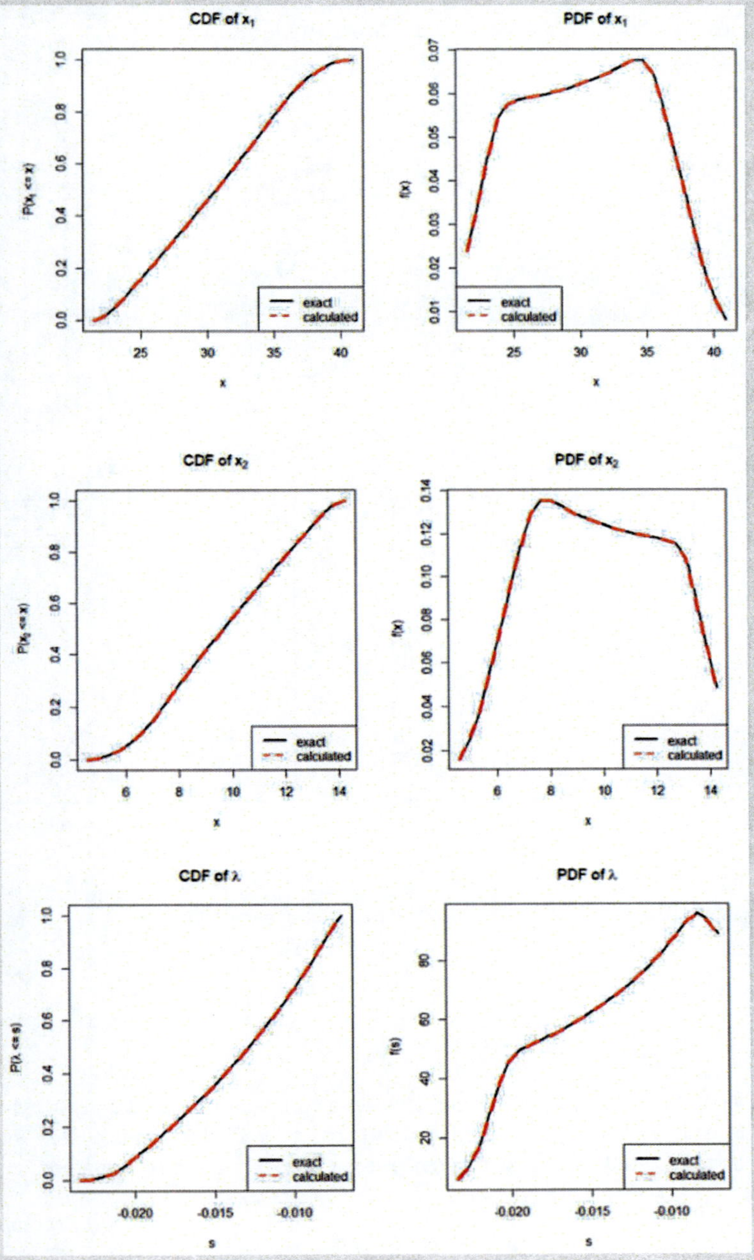

Fig. 8.20 Distribution of x_1^*, x_2^*, λ with uncertain independent parameters α, β

(continued)

Example 8.2 (continued)

The distribution of $J^* = J(\boldsymbol{x}^*)$ is shown in Fig. 8.21.

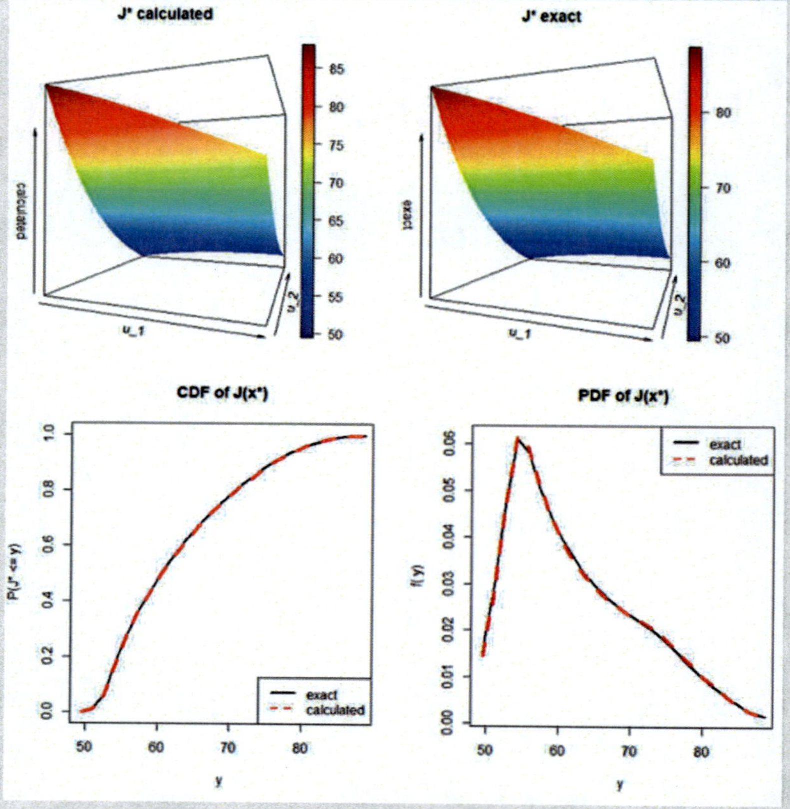

Fig. 8.21 Distribution of J^* with uncertain independent parameters α, β

Exercises

1. Red Brand Canners under uncertainty: a company produces cans of whole tomato, tomato juice, and tomato paste, sold by cases. Each case manufactured is sold at a given price, requests a mass of tomatoes, must

(continued)

respect a minimal quality score, and has a variable cost. The forecast demand is evaluated previously. The corresponding values are shown in table below:

	Whole tomato	Tomato juice	Tomato paste
Selling price	4	4.5	3.8
mass requested	$T(16,18,20)$	$T(19,20,21)$	$T(22,25,28)$
Quality Score	8	6	5
Variable cost	$T(2.4,2.5,2.6)$	$T(3,3.18,3.5)$	$T(1.8,1.95,2.1)$
Demand	Unlimited	$U(48000, 52000)$	$U(78000, 82000)$

The quality score results from the quantity et the grade of the tomatoes used in the manufacturing process: there are tomatoes grade A, which have a quality score of 9 and grade B, which have a quality score of 5. If a product is prepared using fractions x_A of grade A tomatoes and x_B grade B tomatoes, the product has a resulting score $9x_A + 5x_B$.

The company has bought the entire harvest of tomatoes of a panel of producers for a total mass of 3 million mass unities paid 180000 money units.

The harvest is formed of tomatoes grade A and tomatoes grade B – the fraction of tomatoes grade A is a variable $T(0.18,0.22,0.2)$.

The aim is to maximize the profit of the company

(a) Find a deterministic solution where each random variable is taken as its most probable value for triangular distributions and its mean value for the uniform distributions.
(b) Analyze the effects of the variability of the grades in the harvest: consider all the other variables at their most probable (triangular) or mean (uniform) values and determine the distribution of the solution.
(c) Analyze the effects of the fluctuations in the variable costs: consider all the other variables at their most probable (triangular) or mean (uniform) values and determine the distribution of the solution.
(d) Analyze the effects of the manufacturing variabilities: consider all the other variables at their most probable (triangular) or mean (uniform) values and determine the distribution of the solution.
(e) Analyze the effects of the fluctuations in the real demand: consider all the other variables at their most probable values and determine the distribution of the solution.

(continued)

(f) A producer proposes 80000 mass unities of grade A tomatoes by 6800 money unities. If the company accepts this proposition, how are modified the preceding results?

(g) The Marketing department says that an advertising campaign can modify the distribution of the demand of tomato juice to $U(70000, 740000)$. Determine the new distribution of the solution. How much money units can be invested in the advertising campaign?

2. Consider $f(\boldsymbol{x}) = x_1^{\alpha} + x_2^{1-\alpha}$. We are interested in the maximal value of $f(\boldsymbol{x})$ under the restriction

$$p_1 x_1 + p_1 x_2 = A$$

(a) Assume that $A = 100, p_1 = 2, p_2 = 4$ and $\alpha \sim T(0.4, 0.5, 0.6)$. Find the distribution of the solution.

(b) Assume that $A = 100, \alpha = 0.5, p_1 \sim T(1.5, 2, 2.5), p_2 \sim T(3, 4, 5)$. Find the distribution of the solution.

3. Consider $J(\boldsymbol{x}) = \left(\alpha x_1^{\beta} + (1-\alpha) x_2^{\beta}\right)^{\frac{1}{\beta}}$. We are interested in the maximal value of $p_1 x_1 + p_1 x_2$, under the restriction $f(\boldsymbol{x}) = 1$.

(a) Assume that $A = 100$, $p_1 = 2$, $p_2 = 4$, $\alpha \sim T(0.4, 0.5, 0.6)$, $\beta \sim T(0.3, 0.4, 0.5)$. Find the distribution of the solution.

(b) Assume that $A = 100$, $\alpha = 0.5$, $\beta = 0.4$, $p_1 \sim T(1.5, 2, 2.5)$, $p_2 \sim T(3, 4, 5)$. Find the distribution of the solution.

8.2 Using the Adaptation of a Descent Method

In Sect. 5.2, we introduced the adaptation of iterative methods. Such an approach may be applied to iterative methods of optimization. In fact, any iterative method

$$\boldsymbol{X}^{(r+1)} = \boldsymbol{\Psi}\left(\boldsymbol{X}^{(r)}, U\right)$$

can be adapted according to the approach presented in Sect. 5.2 and implemented according to the algorithm presented in the same section. In Sect. 6.3, we presented the adaptation to the use of `ode` in the package `deSolve`.

Here, we shall illustrate the approach by adapting the function fminunc of pracma: indeed, fminunc can be considered as a particular iteration function $\boldsymbol{\Psi}$ which transforms the starting point $\boldsymbol{X}^{(r)}$ into a new point $\boldsymbol{X}^{(r+1)}$. In practice, we can define $\boldsymbol{\Psi}$ as the result of fminunc at a predefined small number of iterations – recall that the number of iterations of fminunc is defined by an option. For instance, the instruction

```
fminunc(X,f,maxiter = 5)
```

fixes the number of iterations fminunc to 5. Thus, we can use fminunc as an iteration function, analogously to the gradient descent iteration function presented in Sect. 5.2.

To fix the ideas, let us consider again the optimization of the modified Rosenbrock's function given in Eq. (8.7), with $n = 5$, $k = 3$.

We start by defining the iteration function. Here, fobj evaluates Rosenbrock's function given in Eq. (8.7).

```
psi = function(x,U){
  f = function(x){fobj(x,U)}
  aux = fminunc(x,f,maxiter = 5)
  return(aux$par)
}
```

In a second step, generate the sample from U and define the Hilbert basis

```
u1 = matrix(rnorm(20),nrow = 1)
a1 = min(u1)
b1 = max(u1)
phi1 = polynom$new(a1,b1,1e-5)
k1 = 3
nb = k1+1
he1 = expansion1D$new(phi1)
```

Then, make the iterations:

```
ndim = 5
uaet = uae$new()
nitmax = 1000
prec = 1E-10
xini = matrix(1, nrow = ndim, ncol = nb)
xsol = uaet$adaptation(psi,xini,phi1,u1,ndim,nb,nitmax, prec)
```

The global RMS error was $\|\boldsymbol{x}^* - \boldsymbol{P}\boldsymbol{x}^*\| \approx 1E - 6$. Using a sample of 1E4 variates from $N(0, 1)$ furnished by rnorm and the expansion, we generated a sample of 1E4 variates from $\boldsymbol{P}\boldsymbol{x}^*$. The RMS error was $\|\boldsymbol{x}^* - \boldsymbol{P}\boldsymbol{x}^*\| \approx 2E - 6$. An example of result is given in in Fig. 8.22.

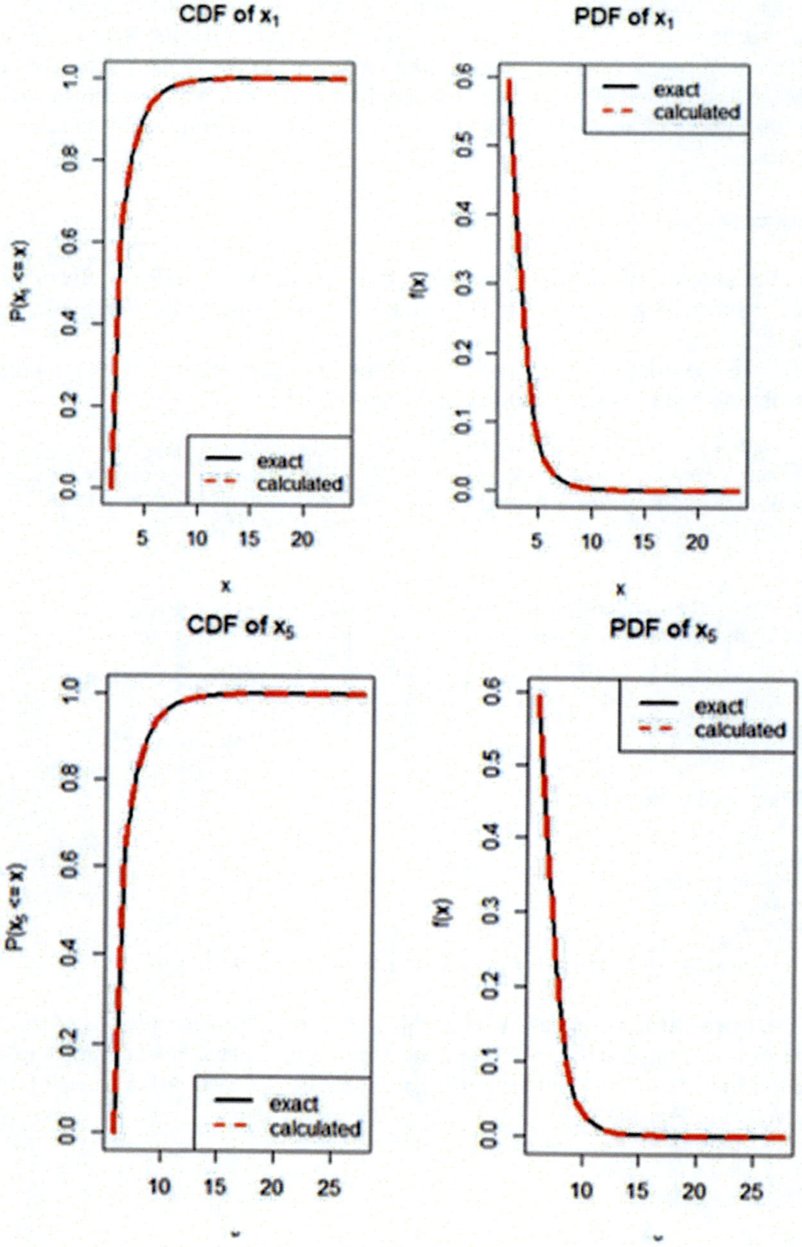

Fig. 8.22 Examples of result furnished by adaptation of `fminunc`

Exercises

1. Red Brand Canners under uncertainty: a company produces cans of whole tomato, tomato juice and tomato paste, sold by cases. Each case manufactured is sold at a given price, requests a mass of tomatoes, must respect a minimal quality score, and has a variable cost. The forecast demand is evaluated previously. The corresponding values are shown in table below:

	Whole tomato	Tomato juice	Tomato paste
Selling price	4	4.5	3.8
mass requested	$T(16,18,20)$	$T(19,20,21)$	$T(22,25,28)$
Quality Score	8	6	5
Variable cost	$T(2.4,2.52,2.6)$	$T(3,3.18,3.5)$	$T(1.8,1.95,2.1)$
Demand	Unlimited	$U(48000, 52000)$	$U(78000, 82000)$

The quality score results from the quantity et the grade of the tomatoes used in the manufacturing process: there tomatoes grade A, which have a quality score of 9 and grade B, which have a quality score of 5. If a product is prepared using fractions x_A of grade A tomatoes and x_B grade B tomatoes, the product has a resulting score $9x_A + 5x_B$.

The company has bought the entire harvest of tomatoes of a panel of producers for a total mass of 3 million mass unities paid 180000 money units.

The harvest is formed of tomatoes grade A and tomatoes grade B – the fraction of tomatoes grade A is a variable $T(0.18,0.2,0.22)$.

The aim is to maximize the profit of the company

(a) Find a deterministic solution where each random variable is taken as its most probable value for triangular distributions and its mean value for the uniform distributions.
(b) Analyze the effects of the variability of the grades in the harvest: consider all the other variables at their most probable (triangular) or mean (uniform) values and determine the distribution of the solution.
(c) Analyze the effects of the fluctuations in the variable costs: consider all the other variables at their most probable (triangular) or mean (uniform) values and determine the distribution of the solution.
(d) Analyze the effects of the manufacturing variabilities: consider all the other variables at their most probable (triangular) or mean (uniform) values and determine the distribution of the solution.
(e) Analyze the effects of the fluctuations in the real demand: consider all the other variables at their most probable values and determine the distribution of the solution.

(continued)

(f) A producer proposes 80000 mass unities of grade A tomatoes by 6800 money unities. If the company accepts this proposition, how are modified the preceding results?

(g) The Marketing department says that an advertising campaign can modify the distribution of the demand of tomato juice to $U(70000, 740000)$. Determine the new distribution of the solution. How much money units can be invested in the advertising campaign?

2. Consider $f(\boldsymbol{x}) = x_1^\alpha + x_2^{1-\alpha}$. We are interested in the maximal value of $f(\boldsymbol{x})$ under the restriction

$$p_1 x_1 + p_1 x_2 = A$$

(a) Assume that $A = 100$, $p_1 = 2$, $p_2 = 4$ and $\alpha \sim T(0.4, 0.5, 0.6)$. Find the distribution of the solution.

(b) Assume that $A = 100$, $\alpha = 0.5$, $p_1 \sim T(1.5, 2, 2.5)$, $p_2 \sim T(3, 4, 5)$. Find the distribution of the solution.

3. Consider $J(\boldsymbol{x}) = \left(\alpha x_1^\beta + (1-\alpha) x_2^\beta\right)^{\frac{1}{\beta}}$. We are interested in the maximal value of $p_1 x_1 + p_1 x_2$, under the restriction $f(\boldsymbol{x}) = 1$.

(a) Assume that $A = 100$, $p_1 = 2$, $p_2 = 4$, $\alpha \sim T(0.4, 0.5, 0.6)$, $\beta \sim T(0.3, 0.4, 0.5)$. Find the distribution of the solution.

(b) Assume that $A = 100$, $\alpha = 0.5$, $\beta = 0.4$, $p_1 \sim T(1.5, 2, 2.5)$, $p_2 \sim T(3, 4, 5)$. Find the distribution of the solution.

8.3 Combining Statistics of the Objective, the Constraints, and Expansions

Analogously to Chapter 5, Sect. 5.1, we can also use expansions to generate a new optimization problem, involving means of the objective and the constraints. For instance, we can look for the expansion

$$P\boldsymbol{x}^* = \sum_{i=1}^{k} \boldsymbol{x}_i^* \varphi_i(u), \boldsymbol{x}_i^* \in \mathbb{R}^n$$

such that $\boldsymbol{x}^* = (\boldsymbol{x}_1^*, \ldots, \boldsymbol{x}_k^*)$ verifies

$$\boldsymbol{x}^* = \arg\min \left\{ G(\boldsymbol{y}) = E\left(F\left(\sum_{i=1}^{k} \boldsymbol{y}_i \varphi_i(u),\ u \right) \right), \boldsymbol{y}_i \in \mathbb{R}^n, \boldsymbol{y} = (\boldsymbol{y}_1, \ \ldots, \ \boldsymbol{y}_k) \right\}$$

In practice, we may use a sample to estimate the mean:

$$G(y) \approx \frac{1}{ns}\sum_{n=1}^{ns} F\left(\sum_{i=1}^{k} y_i \varphi_i(u_n), u_n\right).$$

For instance, let us consider the minimization of the modified Rosenbrock's function (8.7). The minimization of G may be achieved by adapted optimization methods, such as, stochastic gradient methods (see, for instance Holdorf Lopez et al., 2011). Under R, `fminunc` can handle this minimization. For instance, let us introduce two auxiliary functions:

```
tovec = function(mat,m,n){
  aux = matrix(0,nrow = 1, ncol = m*n)
  for (i in 1:m){
    i1 = (i-1)*n+1
    i2 = i1 + n - 1
    aux[i1:i2] = mat[i,]
  }
  return(aux)
}
```

```
tomat = function(vec,m,n){
  aux = matrix(0,nrow = m, ncol = n)
  for (i in 1:m){
    i1 = (i-1)*n+1
    i2 = i1 + n - 1
    aux[i,] = vec[i1:i2]
  }
  return(aux)
}
```

`tovec` transforms a matrix into a vector containing its rows sequentially and tomat performs the inverse operation. Then, define a function `ftstat` that evaluates $G(y)$:

```
fstat = function(vec_coeff,ndim,nb,he,u,f){
  cx = tomat(vec_coeff,ndim,nb)
  px = he$proj(cx,u)
  s = 0
  for (i in 1:length(u)){
    s = s + f(px[ ,i],u[[i]])
  }
  aux = s/length(u)
  return(aux)
}
```

Here, `vec_coeff` corresponds to $\boldsymbol{y}$, `f` corresponds to F, `ndim` is the dimension of $\boldsymbol{x}$, nb $= k + 1$ is the number of coefficients by dimension, and `he` is the Hilbert expansion. Finally, define an initial point, the objective to be minimized and call `fminunc` to get a result. Finalize by setting the solution found as a matrix of coefficients

```
ndim = 5
vec0 = runif(nb*ndim)
fstatobj = function(vec_c){fstat(vec_c,ndim,nb,he1,u1,fobj)}
vecsol = fminunc(vec0,fstatobj,maxiter = 1000)
coefsol = tomat(vecsol$par,ndim,nb)
```

In this case, the RMS error was $\|\boldsymbol{x}^* - \boldsymbol{P}\boldsymbol{x}^*\| \approx 6E - 8$ on the initial sample and $\|\boldsymbol{x}^* - \boldsymbol{P}\boldsymbol{x}^*\| \approx 2E - 7$ on the large sample of 1E5 variates. Examples of results are shown in Fig. 8.23.

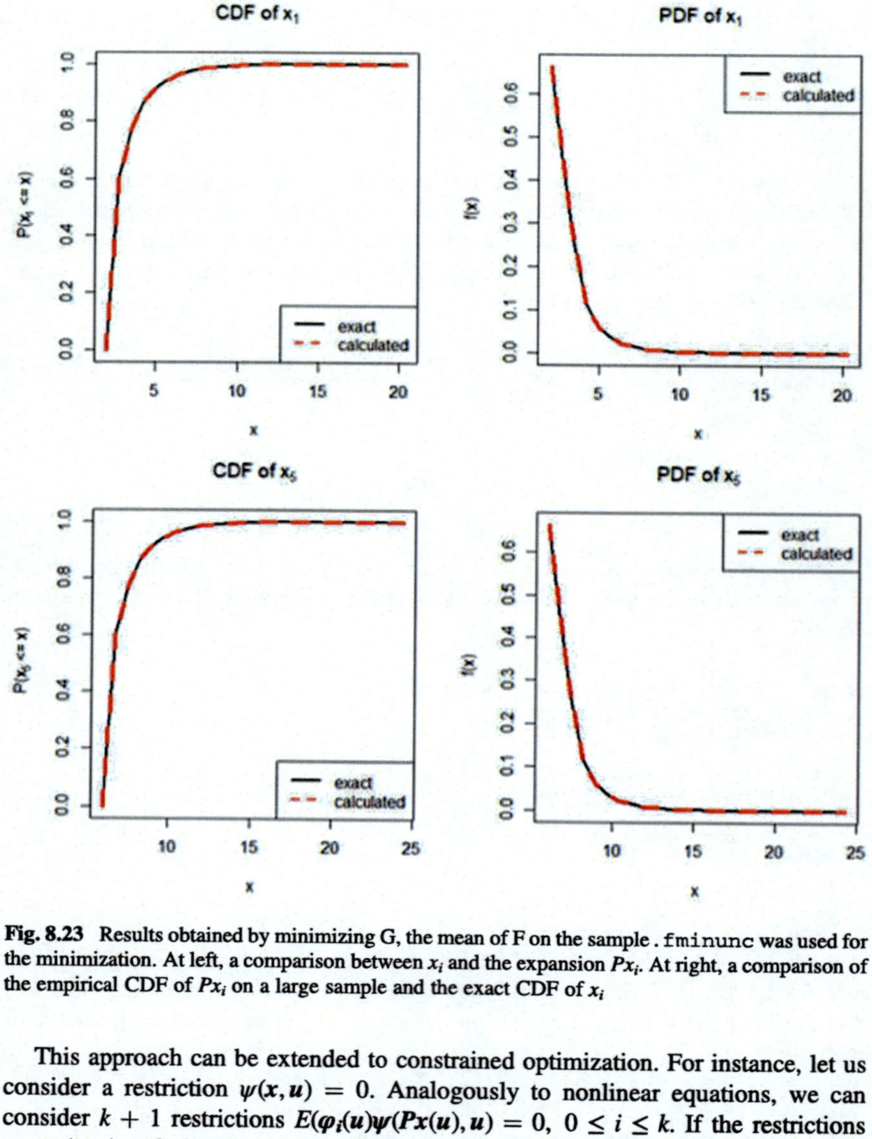

Fig. 8.23 Results obtained by minimizing G, the mean of F on the sample . `fminunc` was used for the minimization. At left, a comparison between x_i and the expansion Px_i. At right, a comparison of the empirical CDF of Px_i on a large sample and the exact CDF of x_i

This approach can be extended to constrained optimization. For instance, let us consider a restriction $\psi(\boldsymbol{x}, \boldsymbol{u}) = 0$. Analogously to nonlinear equations, we can consider $k + 1$ restrictions $E(\boldsymbol{\varphi}_i(\boldsymbol{u})\boldsymbol{\psi}(\boldsymbol{Px}(\boldsymbol{u}), \boldsymbol{u}) = 0,\ 0 \leq i \leq k$. If the restrictions are $\psi_j(\boldsymbol{x}, \boldsymbol{u}) = 0,\ 1 \leq j \leq q$, we consider $(k + 1)q$ restrictions

$$E\big(\boldsymbol{\varphi}_i(\boldsymbol{u})\boldsymbol{\psi}_j(\boldsymbol{Px}(\boldsymbol{u}),\ \boldsymbol{u})\big) = 0, 0 \leq i \leq k, 1 \leq j \leq q. \tag{8.8}$$

A vector containing these restrictions can be created by the function `nlcstat` below (it calls the function `eqs`. Function `tomat` was previously introduced in this section):

```
eqs = function(cx,u,f,he,nb,ndim,nf){
  px = matrix(NA,nrow = ndim, ncol = length(u))
  for (i in 1: ndim){
    px[i,] = he$proj(cx[i,],u)
  }
  vf = matrix(nrow = nf,ncol = length(u))
  for (r in 1:length(u)){
    vf[,r]=f(px[,r],u[[r]])
  }
  eq = vector(length = nb*nf)
  nn = 0
  for (i in 1:nb){
    vphi = he1$value(i-1,u)
    for (j in 1: nf){
      nn = nn+1
      eq[[nn]]=mean(vphi*vf[j,])
    }
  }
  aux = as.matrix(eq)
  return(aux)
}

nlcstat = function(vec_coeff,ndim,nb,he,u,nlc,ncons){
  cx = tomat(vec_coeff,ndim,nb)
  aux = eqs(cx,u,nlc,he,nb,ndim,ncons)
  return(aux)
}
```

As an example, let us consider the determination of the point on the line $x_1 + ux_2 = 3$ which is the closest to the circle $x_1^2 + x_2^2 = 1$ (see Sect. 1.13.3). Setting $\boldsymbol{x} = (x_1, x_2, x_3, x_4)$, we have

$$J(\boldsymbol{x}) = \sqrt{(x_1 - x_3)^2 + (x_2 - x_4)^2},$$

$$\varphi_1(\boldsymbol{x}) = x_1^2 + x_2^2 - 1,$$

$$\varphi_2(\boldsymbol{x}) = x_3 + ux_4 - 3,$$

Then we are looking for

$$\boldsymbol{x}^* = \arg\min \left\{ J(\boldsymbol{x}) \colon \boldsymbol{x} \in \mathbb{R}^4,\ \varphi_1(\boldsymbol{x}) = \varphi_2(\boldsymbol{x}) = 0 \right\}.$$

Let us consider u uniformly distributed on (1,2) and a sample of 10 of its variates:

```
1.014772 1.039209 1.138042 1.16512 1.177645
1.281744 1.45127 1.50283 1.741549 1.867562
```

We apply the method with a polynomial family and $k = 3$. Function `fmincon` of `pracma` is called to determine the solution:

```
u1 = c(1.014772,1.039209,1.138042,1.16512,1.177645,
       1.281744,1.45127,1.50283,1.741549,1.867562)
a1 = min(u1)
b1 = max(u1)
phi1 = polynom$new(a1,b1,1e-5)
k1 = 3
nb = k1+1
he1 = expansion1D$new(phi1)

vec0 = runif(nb*ndim)
fstatobj = function(vec_c){fstat(vec_c,ndim,nb,he1,u1,fobj,nfobj)}
ceqstat = function(vec_c){nlcstat(vec_c,ndim,nb,he1,u1,ceq,neq)}
vecsol = fmincon(vec0,fstatobj,heq = ceqstat)
coefsol = tomat(vecsol$par,ndim,nb)
```

The results obtained are shown in Figs. 8.24, 8.25, 8.26, and 8.27.

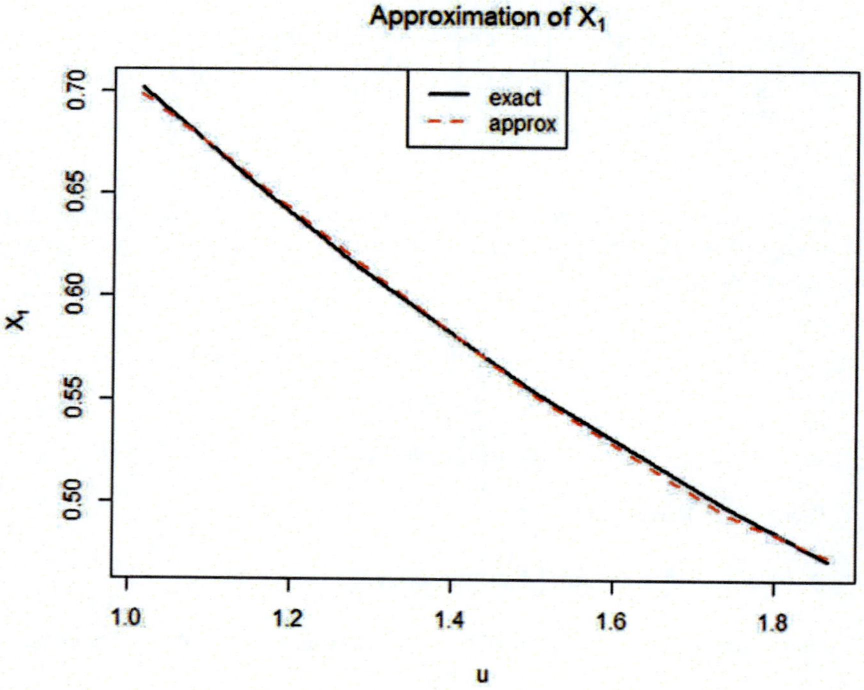

Fig. 8.24 Results obtained by minimizing G under the restrictions (8.8). `fmincon` was used for the minimization

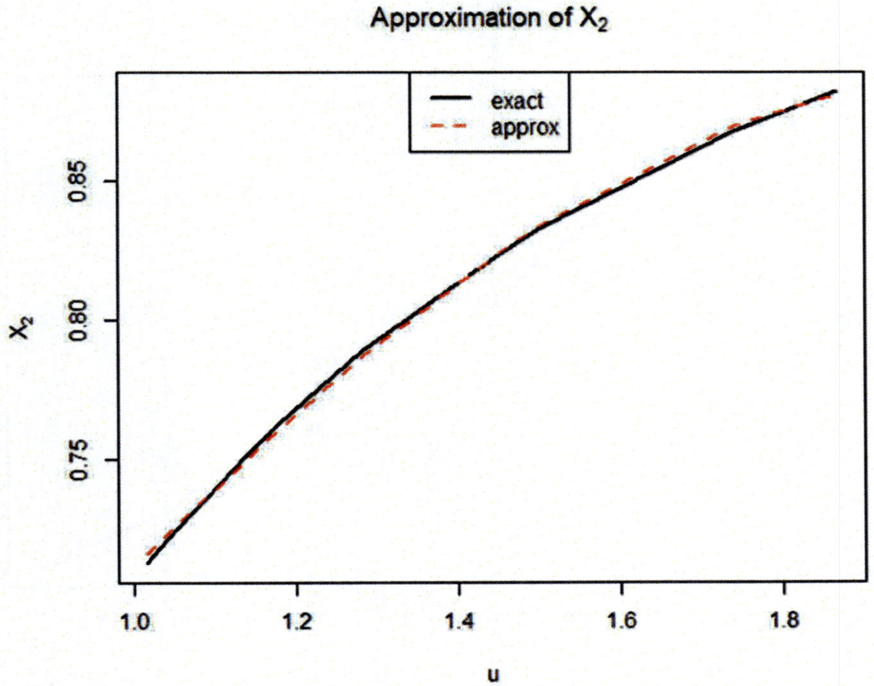

Fig. 8.25 Results obtained by minimizing G under the restrictions (8.8). `fmincon` was used for the minimization

Inequality constraints can be considered by transforming them into equality constraints. For instance, if $\boldsymbol{x} = (x_1, \ldots, x_n)$, the constraint $\psi(\boldsymbol{x}, \boldsymbol{u}) \leq 0$ can be rewritten as $\psi(\boldsymbol{x}, \boldsymbol{u}) + x_{n+1}^2 = 0$.

Alternatively, the approach introduced in Sect. 1.14.3 can be combined to the approach introduced in Sect. 5.4 to deal with inequality constraints.

Exercises

1. Use the method presented to solve the Red Brand Canners under uncertainty.
2. Consider

$$f(\boldsymbol{x}) = x_1^{\alpha} + x_2^{1-\alpha}.$$

(continued)

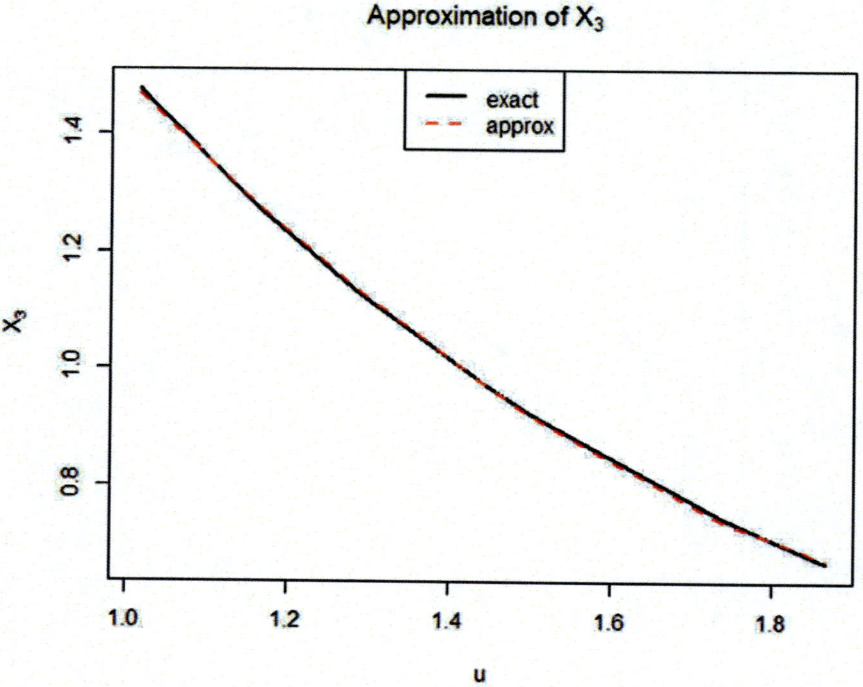

Fig. 8.26 Results obtained by minimizing G under the restrictions (8.8). `fmincon` was used for the minimization

Use the method presented to analyze the maximal value of $f(\boldsymbol{x})$ under the restriction $p_1x_1 + p_1x_2 = A$ in the following situations:

(a) Assume that $A = 100, p_1 = 2, p_2 = 4$ and $\alpha \sim T(0.4, 0.5, 0.6)$. Find the distribution of the solution.
(b) Assume that $A = 100$, $\alpha = 0.5, p_1 \sim T(1.5,2,2.5), p_2 \sim T(3, 4, 5)$. Find the distribution of the solution.

3. Consider

$$ces(\boldsymbol{x}) = \left(\alpha x_1^\beta + (1-\alpha)x_2^\beta\right)^{\frac{1}{\beta}}.$$

Use the method presented to analyze the maximal value of $p_1x_1 + p_1x_2$, under the restriction $f(\boldsymbol{x}) = 1$. Solve the following situations:

(continued)

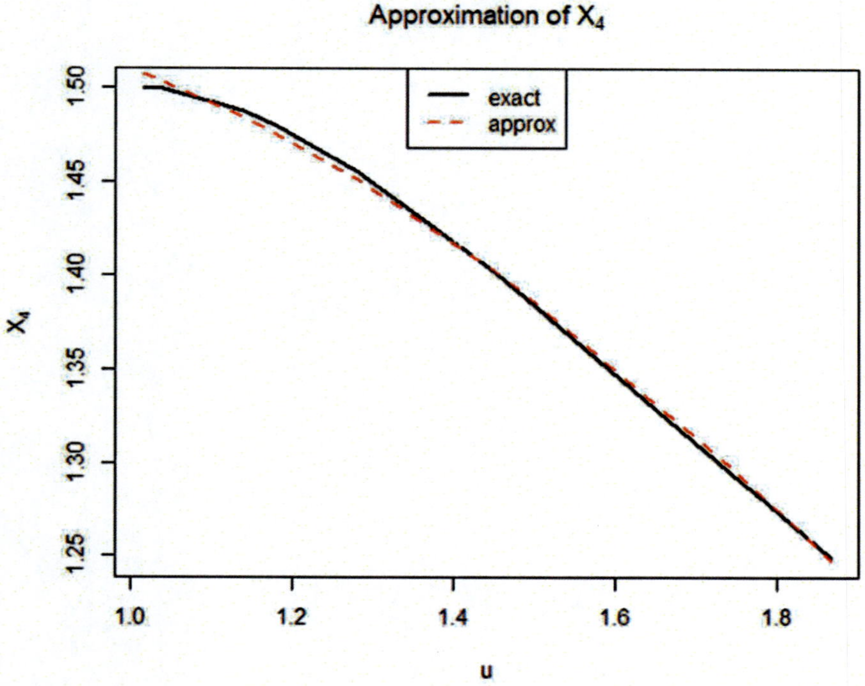

Fig. 8.27 Results obtained by minimizing G under the restrictions (8.8). `fmincon` was used for the minimization

(a) Assume that $A = 100$, $p_1 = 2$, $p_2 = 4$, $\alpha \sim T(0.4, 0.5, 0.6)$, $\beta \sim T(0.3, 0.4, 0.5)$. Find the distribution of the solution.
(b) Assume that $A = 100$, $\alpha = 0.5$, $\beta = 0.4$, $p_1 \sim T(1.5, 2, 2.5)$, $p_2 \sim T(3, 4, 5)$. Find the distribution of the solution.

Chapter 9
Reliability

Abstract In this chapter, we present some tools for reliability analysis and reliability-based design optimization, such as the notions of reliability index and its determination by different methods, transformations of Rosenblatt and nataf, FORM, SORM, and RBDO.

Indeed, civil engineers deal with various essential elements for which variability in time and space may be extremely high: external loads, values of physical parameters, soil, not to mention the future climatic events, etc. The uncertainties and risks require an adapted analysis, especially in certain situations, where the human and economic consequences of a failure can be catastrophic. More recently, other fields of engineering were brought to reliability analysis, by diverse reasons: customer satisfaction, safety, maintenance costs, possible legal liability on failures, ... Finance also adopted a culture of risk analysis connected to reliability.

Reliability analysis can be performed with the objective of studying and eventually improving an existing system or process. However, it is more efficient to include reliability since the initial studies about system's design, considering the effects of variability and uncertainties in the design – this is the modern procedure, especially when the correction of errors or the modification of already executed projects can be costly, if not impossible. This is starting point for the modern idea of reliability-based design optimization (RBDO), which consists of seeking optimal solutions that satisfy reliability requirements.

Indeed, design procedures look for the best solution in terms of cost and quality. Thus, design often makes use of optimization tools. As in any optimization procedure, the designer must define the three fundamental components of any optimization problem: the design variables, the constraints that must be respected and the objective function to be minimized. In general, the design variables are real or integers and denote the parameters for which there is some freedom of choice. The unknowns to be determined define the design. The constraints are the limitations in the freedom of choice, while the objective function gives a numerical evaluation of the design quality, providing a practical method for comparing different solutions,

E. Souza de Cursi, *Uncertainty Quantification using R*, International Series in Operations Research & Management Science 335,
https://doi.org/10.1007/978-3-031-17785-9_9

the best ones corresponding to their minima. The constraints and the objective function may involve additional parameters that are not design variables but supplied data.

If variability and uncertainty are not considered, the design variables remain deterministic: the resulting optimization problem is usually solved by a numerical optimization method, which furnishes one or several approximate solutions. The use of stochastic descent methods or stochastic modifications of descent methods does not change the deterministic character of this approach, which is usually referred as DDO (Deterministic Design Optimization).

However, in many practical situations, the framework includes variability of some parameters or stochastic phenomena, which produces a loss of robustness and reliability. In such a situation, uncertainty must be introduced in the design procedure to obtain robust and reliable solutions, able to face implementation errors and variations in the operational conditions. Different approaches were proposed to include random or uncertain aspects in the design, for instance, robust optimization and RBDO – already mentioned. RBDO looks for a control of the probabilities of some key events, such as failure: the maximal acceptable probability of a key event is introduced as a constraint, which corresponds to requesting a minimum reliability for a key event. In practice, reliability can be characterized by a reliability index, usually denoted by β and defined for systems that satisfy two basic conditions: on the one hand, the state of the system is fully defined by a vector of parameters ; on the other hand, the failure is characterized by the sign of a real variable Z – failure and safe operation correspond to opposite signs of Z – for instance, failure corresponds to $Z > 0$ and safe operation to $Z < 0$.

9.1 Limit State Curves

A situation often found in practice is the one where the failure regions are defined by inequalities or, equivalently, the limits of the failure region are defined by equations which separates regions of normal operation and regions of failure. For instance, in dimension 2, the safe and failure regions are separated by curves.

Let us denote by $\boldsymbol{x} \in \mathbb{R}^n$ the design variables of the system: the design space is $\mathbb{R}^n$. Assume that failure occurs when $Z = g(\boldsymbol{x}) > 0$ and the system operates safely when $Z = g(\boldsymbol{x}) < 0$. In such a situation, the safe region and the failure region are separated by the curve$g(\boldsymbol{x}) = 0$, which is called the *limit state curve* or *failure curve*. Then, the space of the design variables splits into two disjoint parts (Fig. 9.1):

- The failure region $\mathcal{F} = \{\boldsymbol{x} \in \mathbb{R}^n \colon g(\boldsymbol{x}) > 0\}$,
- The safe region $\mathcal{S} = \{\boldsymbol{x} \in \mathbb{R}^n \colon g(\boldsymbol{x}) \leq 0\}$

The frontier between these regions corresponds to the limit state curve

$$\mathcal{C} = \{\boldsymbol{x} \in \mathbb{R}^n \colon g(\boldsymbol{x}) = 0\}.$$

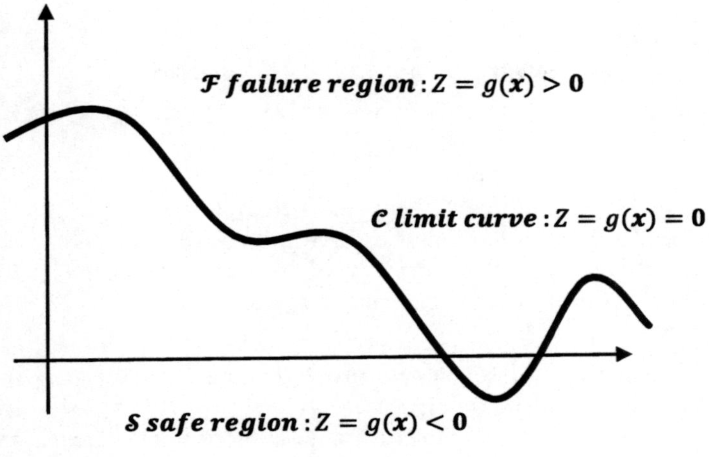

Fig. 9.1 Separation of the design space into a safe region and a failure region by a limit state curve

For the sake of convenience, we can bring $\boldsymbol{x}$ to the origin: this is equivalent to describing the points $\boldsymbol{y} \in \mathbb{R}^n$ through the auxiliary variable $\boldsymbol{u}$ such that $\boldsymbol{u} = \boldsymbol{y} - \boldsymbol{x}$. The space of the new variables $\boldsymbol{u}$ is called *centered space*. In principle, such a change of coordinates can be summarized by a simple translation. However, in a more general way, we can consider an enough regular transformation – eventually nonlinear – as, for instance, $\boldsymbol{y} = \boldsymbol{\phi}(\boldsymbol{x}, \boldsymbol{u})$, such that $\boldsymbol{\phi}(\boldsymbol{x}, \boldsymbol{0}) = 0$. Then, the value of z at a point $\boldsymbol{y}$ is $z = h(\boldsymbol{x}, \boldsymbol{u}) = g\left(\boldsymbol{\phi}(\boldsymbol{x}, \boldsymbol{u})\right) = g(\boldsymbol{y})$ and

$$\mathcal{F} = \{\boldsymbol{u} \in \mathbb{R}^n \colon h(\boldsymbol{x},\ \boldsymbol{u}) > 0\},\ \mathcal{S} = \{\boldsymbol{x} \in \mathbb{R}^n \colon h(\boldsymbol{x},\ \boldsymbol{u}) < 0\}. \tag{9.1}$$

Analogously,

$$\mathcal{C} = \{\boldsymbol{u} \in \mathbb{R}^n \colon h(\boldsymbol{x},\ \boldsymbol{u}) = 0\}. \tag{9.2}$$

Notice that the expression of h can change with $\boldsymbol{x}$. In such a general situation, the design space and the centered space may be different. In the sequel, we assume that the design space is $\mathbb{R}^n$.

Example 9.1

Let us consider the situation where $x \in \mathbb{R}^2$ and $g(x) = x_2 - x_1^2 - x_1 - 3$. Consider $x = (x_1, x_2)$, $\phi(x, u) = u + x$. We have $h(x, u) = g(u + x) = u_2 + x_2 - u_1^2 - 2u_1x_1 - x_1^2 - u_1 - x_1 - 3$, id est,

$$h(x, u) = u_2 - u_1^2 - u_1 - 2u_1x_1 + g(x).$$

In Fig. 9.2, we show the limit state curve in the design space and 4 points of interest : $x^{(1)} = (0, 0)$, $x^{(2)} = (1, 0)$, $x^{(3)} = (-1, 1)$, $x^{(4)} = (1, 2)$. In Fig. 9.3, we show the limit state curves in the centered space.

Notice that:

- In the design space, the limit state curve is the same for all the points and the position of each point is different;
- In the centered space, the limit state curve is different for each point and all the points are brought to the origin.

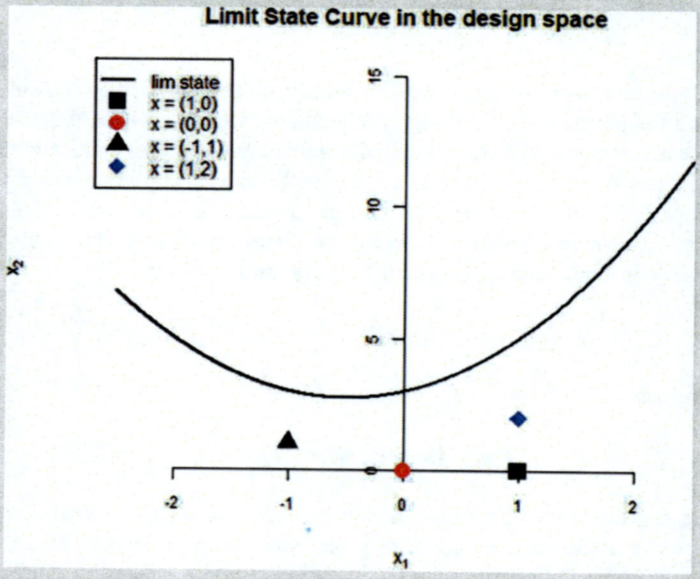

Fig. 9.2 Limit state curve in the design space: the curve is unique, and the points occupy different positions

(continued)

Example 9.1 (continued)

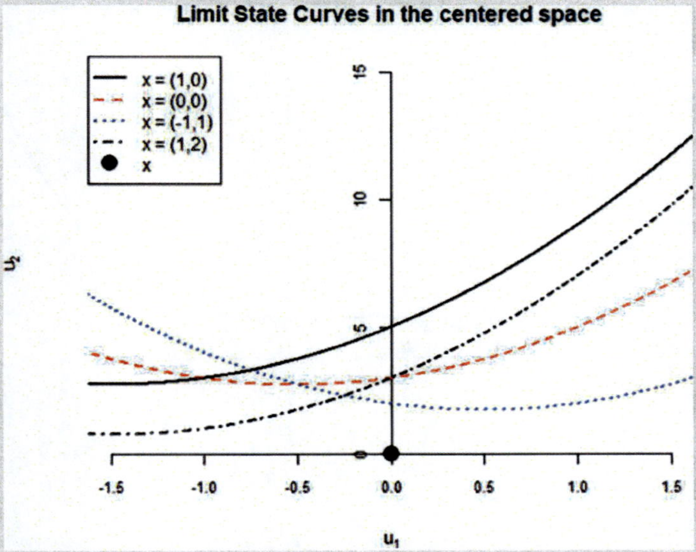

Fig. 9.3 Limit state curve in the centered space: the points of interest are brought to the origin and the curves are different for each point

Example 9.2

Let us consider the situation where $\boldsymbol{x} \in \mathbb{R}^2$ and $g(\boldsymbol{x}) = x_2 - x_1^3 - x_1^2 - 4$. Consider $\boldsymbol{x} = (x_1, x_2)$, $\phi(\boldsymbol{x}, \boldsymbol{u}) = \boldsymbol{u} + \boldsymbol{x}$. We have $h(\boldsymbol{x}, \boldsymbol{u}) = g(\boldsymbol{u} + \boldsymbol{x}) = u_2 + x_2 - u_1^2 - 2u_1x_1 - x_1^2 - u_1 - x_1 - 3$, id est,

$$h(\boldsymbol{x}, \boldsymbol{u}) = u_2 - u_1^3 - 3u_1^2x_1 - 3u_1x_1^2 - u_1^2 - 2u_1x_1 + g(\boldsymbol{x}).$$

In Fig. 9.4, we show the limit state curve in the design space and 4 points of interest : $\boldsymbol{x}^{(1)} = (0, 0)$, $\boldsymbol{x}^{(2)} = (1, 0)$, $\boldsymbol{x}^{(3)} = (-1, 1)$, $\boldsymbol{x}^{(4)} = (1, 2)$. In Fig. 9.5, we show the limit state curves in the centered space.

Again:

- In the design space, the limit state curve is the same for all the points and the position of each point is different;
- In the centered space, the limit state curve is different for each point and all the points are brought to the origin.

(continued)

Example 9.2 (continued)

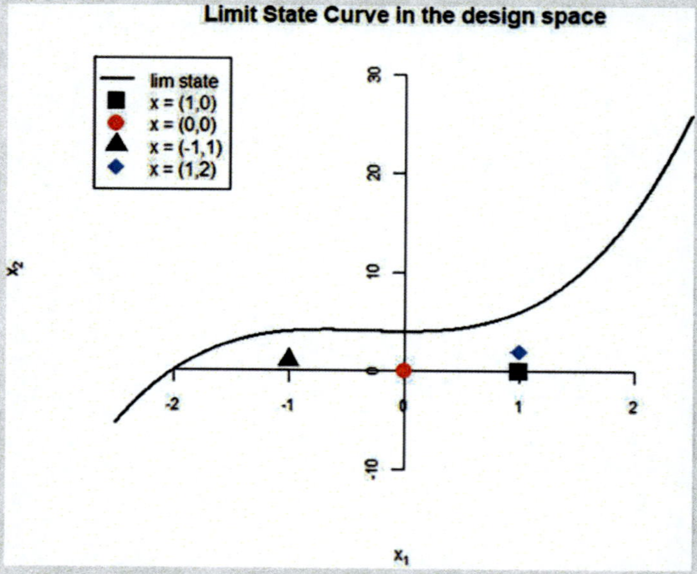

Fig. 9.4 Limit state curve in the design space: the curve is unique, and the points occupy different positions

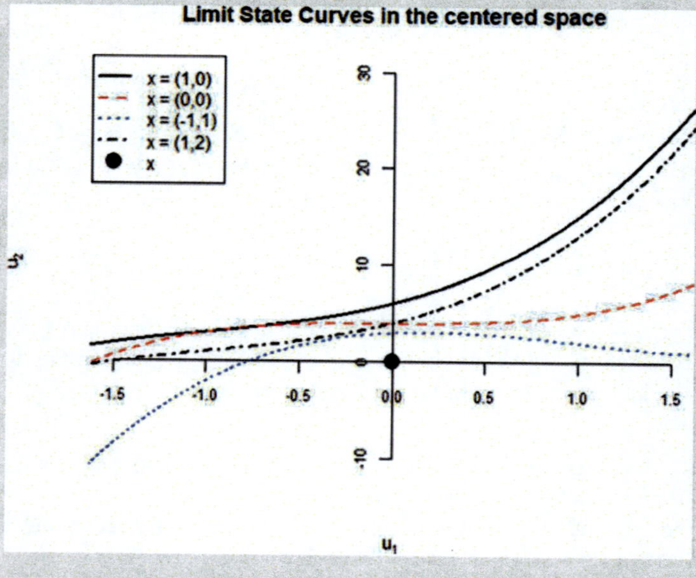

Fig. 9.5 Limit state curve in the centered space: the points of interest are brought to the origin and the curves are different for each point

Exercises

1. Let $\boldsymbol{x} \in \mathbb{R}^2$ and $g(\boldsymbol{x}) = x_1x_2 - 2$. Determine $h(\boldsymbol{x},\boldsymbol{u})$ for $\boldsymbol{x}^{(1)} = (1,0)$, $\boldsymbol{x}^{(2)} = (0,0)$, $\boldsymbol{x}^{(3)} = (1,1)$, $\boldsymbol{x}^{(4)} = (0,1)$. Draw the limit state curves $h(\boldsymbol{x}, \boldsymbol{u}) = 0$ in the centered space.
2. Let $\boldsymbol{x} \in \mathbb{R}^2$ and $g(\boldsymbol{x}) = x_1^2 + x_2^2 - 4$. Determine $h(\boldsymbol{x},\boldsymbol{u})$ for $\boldsymbol{x}^{(1)} = (1,0)$, $\boldsymbol{x}^{(2)} = (0,0)$, $\boldsymbol{x}^{(3)} = (1,1)$, $\boldsymbol{x}^{(4)} = (0,1)$. Draw the limit state curves $h(\boldsymbol{x}, \boldsymbol{u}) = 0$ in the centered space.
3. Let $\boldsymbol{x} \in \mathbb{R}^3$ and $g(\boldsymbol{x}) = x_2 + 1 - 3(2x_3 + 1)^2(3x_1 + 1)$. Determine $h(\boldsymbol{x},\boldsymbol{u})$ for $\boldsymbol{x}^{(1)} = (1,0,0)$, $\boldsymbol{x}^{(2)} = (0,0,0)$, $\boldsymbol{x}^{(3)} = (0,1,0)$, $\boldsymbol{x}^{(4)} = (0,0,1)$. Draw the limit state surfaces $h(\boldsymbol{x},\boldsymbol{u}) = 0$ in the centered space.

9.2 Design Point

In the same way that we can feel as safer as we are far from the edge of a cliff, we can consider the possible solution $\boldsymbol{x}$ as safer as it is far from the failure region – in fact, the greater the distance between $\boldsymbol{x}$ and $\mathcal{F}$, the larger can be the variations of $\boldsymbol{x}$ that will not generate a failure. Thus, the distance $\beta(\boldsymbol{x}) = d(\boldsymbol{x}, \mathcal{F})$ may be used as a measurement of the reliability of $\boldsymbol{x}$. We have

$$\beta(\boldsymbol{x}) = d(\boldsymbol{x}, \mathcal{F}) = \min\{\|\boldsymbol{y} - \boldsymbol{x}\| : \boldsymbol{y} \in \mathcal{F}\}. \tag{9.3}$$

We can recognize in this formula the definition of the orthogonal projection $\boldsymbol{x}^* = \boldsymbol{P}\boldsymbol{x}$ of $\boldsymbol{x}$ onto $\mathcal{F}$. In regular situations, we may consider the projection of $\boldsymbol{x}$ onto $\mathcal{F} \cup \mathcal{C}$:

$$\beta(\boldsymbol{x}) = \|\boldsymbol{x}^* - \boldsymbol{x}\| = \min\{\|\boldsymbol{y} - \boldsymbol{x}\| : g(\boldsymbol{y}) \geq 0\}. \tag{9.4}$$

In the centered space, we have

$$\beta(\boldsymbol{x}) = \|\boldsymbol{u}^*(\boldsymbol{x})\| = \min\{\|\boldsymbol{u}\| : h(\boldsymbol{x}, \boldsymbol{u}) \geq 0\}. \tag{9.5}$$

In the language of reliability, $\boldsymbol{x}^*$ (or $\boldsymbol{u}^*(\boldsymbol{x})$) is the *design point*, usually denoted DP. $\boldsymbol{x}^*$ (or $\boldsymbol{u}^*(\boldsymbol{x})$) is also called MPP (most probable point) or MPFP (most probable failure point). These terminologies are traditional, but their use request some precaution: we should not confuse $\boldsymbol{x}$ – the point in the design space under consideration – and $\boldsymbol{x}^*$ – the design point. Moreover, if $\boldsymbol{x}^*$ (or $\boldsymbol{u}^*(\boldsymbol{x})$) defines the failure point which is closest to $\boldsymbol{x}$, this fact does not mean that the failure will occur at that point – the evolution of the values of $\boldsymbol{x}$ may lead to another point. Finally, $\boldsymbol{x}^*$ is the most probable point only for probability distributions decreasing with distance to the failure region: although such a property seems natural, it requests that the distributions satisfy some mathematical properties. These considerations make that

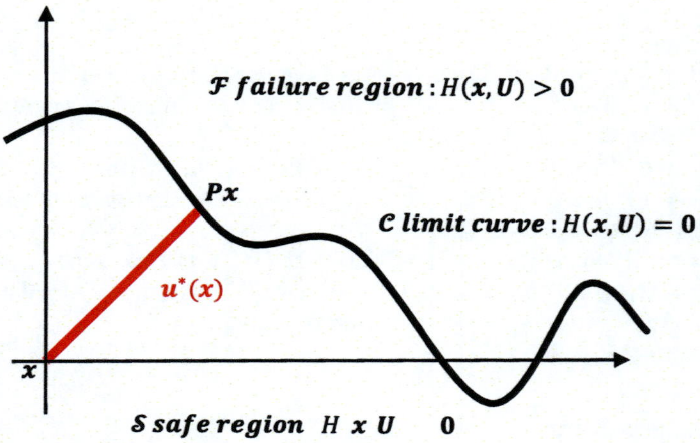

Fig. 9.6 Centered space, Hybrid space and orthogonal projection of a design point $\boldsymbol{x}$

many authors prefer to use the terminology MCP (most central point) or MCFP (most central failure point) to avoid misinterpretation. We will use the traditional terminology, and we ask the reader to keep in mind these remarks (Fig. 9.6).

The determination of $\boldsymbol{x}^*$ (or $\boldsymbol{u}^*(\boldsymbol{x})$) can be achieved by solving the optimization problem (9.5) with the optimizers available in R. For instance, we can use `fmincon` from package `pracma`:

```
ustar = function(x,g){
  hi1 = function(u){-g(x+u)}
  fobj = function(u){sqrt(sum(u^2))}
  uini = rnorm(length(x))
  sol = fmincon(uini,fobj,hin=hi1,
        maxiter=1e4,maxfeval = 1e5,tol=1E-8)
  return(sol$par)
}
```

`ustar` determines $\boldsymbol{u}^*(\boldsymbol{x})$ for a failure region $g(\boldsymbol{x}) > 0$. Then, $\beta(\boldsymbol{x}) = \|\boldsymbol{u}^*(\boldsymbol{x})\|$, $\boldsymbol{x}^* = \boldsymbol{x} + \boldsymbol{u}^*(\boldsymbol{x})$. Alternatively, we can determine $\boldsymbol{x}^*$ directly.

```
xstar = function(x,g){
  hi1 = function(y){-g(y)}
  fobj = function(y){sqrt(sum((y-x)^2))}
  yini = rnorm(length(x))
  sol = fmincon(yini,fobj,hin=hi1,
              maxiter=1e4,maxfeval = 1e5,tol=1E-8)
  return(sol$par)
}
```

`xstar` determines $\boldsymbol{x}^*$ for a failure region $g(\boldsymbol{x}) > 0$. Then, $\boldsymbol{u}^*(\boldsymbol{x}) = \boldsymbol{x}^* - \boldsymbol{x}$, $\beta(\boldsymbol{x}) = \|\boldsymbol{u}^*(\boldsymbol{x})\|$.

Example 9.3
Let us consider the situation where $\boldsymbol{x} \in \mathbb{R}^2$ and $g(\boldsymbol{x}) = x_2 - x_1^2 - x_1 - 3$.

Start by defining g :

```
g = function(x){
  aux = x[[2]] - x[[1]]^2 - x[[1]] - 3
  return(aux)
}
```

Then, give a point of interest x and call `ustar` to determine $\boldsymbol{u}^*(\boldsymbol{x})$, $\boldsymbol{x}^*$ and $\beta(\boldsymbol{x})$

```
xa = c(1,0)
ua = ustar(xa,g)
pxa = xa + ua
beta_a = sqrt(sum(ua^2))
```

We obtain the results in Table 9.1. Figures 9.7 and 9.8 show the results in the design space and in the centered space.

Table 9.1 Results funished by `ustar`

x	$u^*(x)$	$\beta(x)$	x^*
(1,0)	(–1.272838, 2.801603)	3.077189	(–0.2728376,2.8016028)
(0,0)	(–0.4232162, 2.7558958)	2.788203	(–0.4232162, 2.7558958)
(–1,1)	(0.3894887, 1.7622127)	1.804742	(–0.6105113, 2.7622127)
(2,1)	(–1, 1)	1.414214	(2.3E-9, 3)

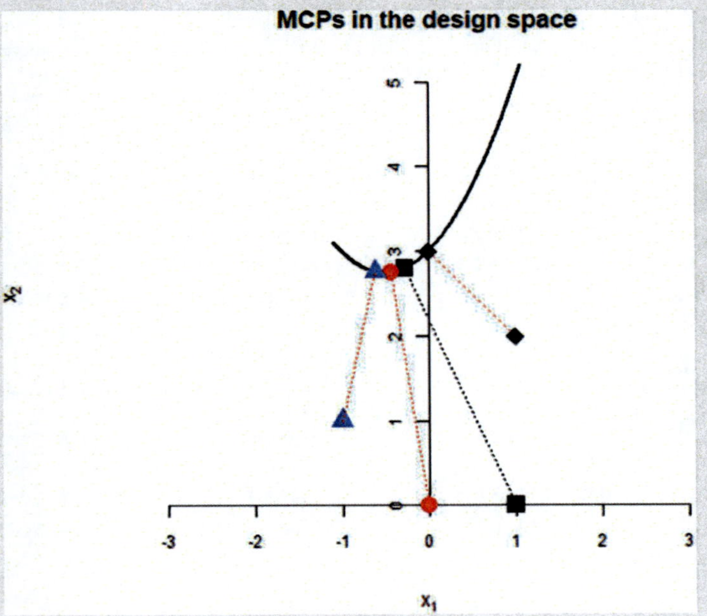

Fig. 9.7 Orthogonal projections determined by `fmincon`. The points on the failure curve are the MCPs associated to the points of interest

(continued)

Example 9.3 (continued)

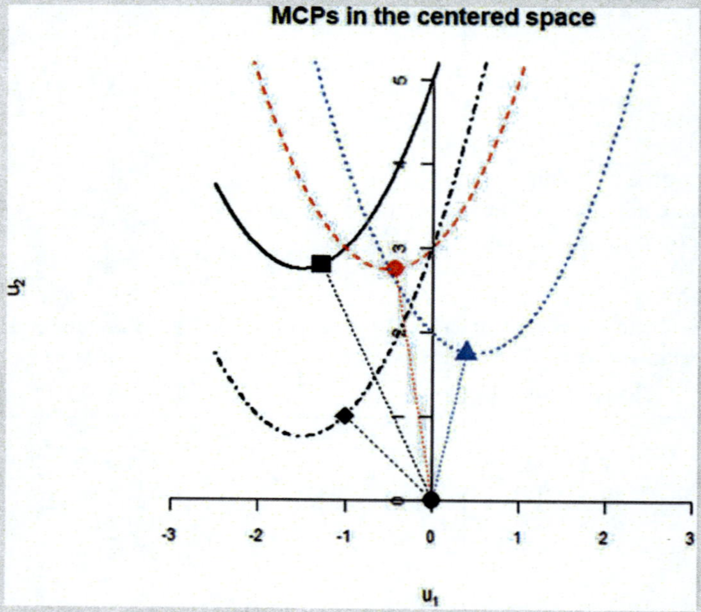

Fig. 9.8 Orthogonal projections determined by `fmincon`. The points on the failure curve are the MCPs associated to the points of interest

Example 9.4
Let us consider the situation where $\boldsymbol{x} \in \mathbb{R}^2$ and $g(\boldsymbol{x}) = x_2 - x_1^3 - x_1^2 - 4$. Running the program `xstar` furnishes the results in Table 9.2. Figures 9.9 and 9.10 show the results in the design space and in the centered space.

Table 9.2 Results funished by `xstar`

x	$u^*(x)$	$\beta(x)$	x^*
(1,0)	(–2.9492544, 0.3932201)	2.975353	(–1.9492544, 0.3932201)
(0,0)	(–1.9673004, 0.256285)	1.983924	(–1.9673004, 0.2562853)
(–1,1)	(–0.8440816, 0.1295846)	0.8539707	(–1.844082, 1.129585)
(2,1)	(–0.8351943, 2.0316372)	2.196611	(0.1648057, 4.0316372)

(continued)

Example 9.4 (continued)

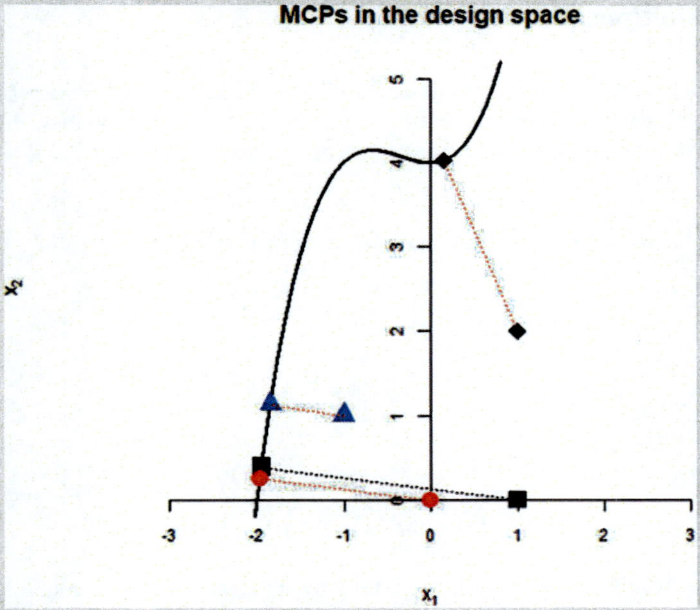

Fig. 9.9 Orthogonal projections determined by `fmincon`. The points on the failure curve are the MCPs associated to the points of interest

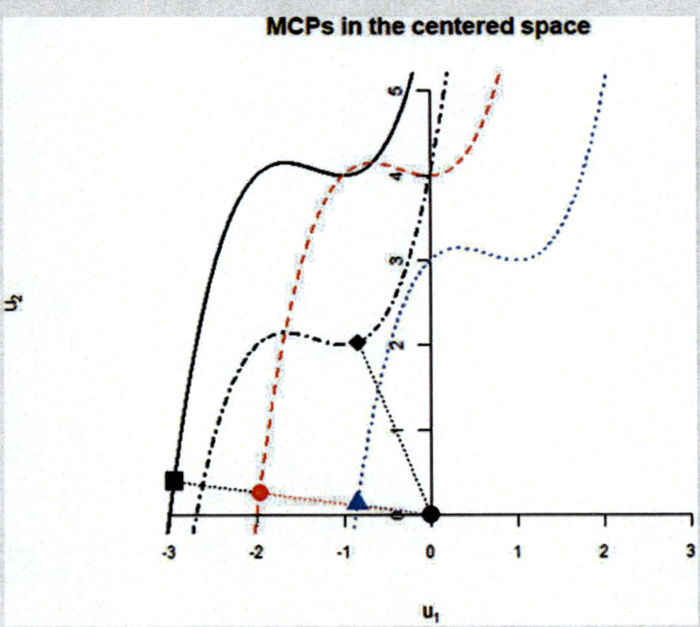

Fig. 9.10 Orthogonal projections determined by `fmincon`. The points on the failure curve are the MCPs associated to the points of interest

9.3 Multiple Failure Conditions

In practice, it can be necessary to deal with several failure conditions $g_1(\boldsymbol{x})$, $g_2(\boldsymbol{x})$, ..., $g_{ng}(\boldsymbol{x})$, so that failure occurs if $g_i(\boldsymbol{x}) > 0$ for at least one of these limit state equations. In the opposite, the safe region verifies $g_i(\boldsymbol{x}) < 0$, $\forall\, i$, $1 \le i \le ng$. Let us denote

$$\mathcal{F}_{\mathrm{i}} = \{\boldsymbol{u} \in \mathbb{R}^n : h_i(\boldsymbol{x},\ \boldsymbol{u}) > 0\},\ \mathcal{S}_{\mathrm{i}} = \{\boldsymbol{x} \in \mathbb{R}^n : h_i(\boldsymbol{x},\ \boldsymbol{u}) < 0\}.$$

Then,

$$\mathcal{F} = \bigcup_{i=1}^{ng} \mathcal{F}_{\mathrm{i}},\ \mathcal{S} = \bigcap_{i=1}^{ng} \mathcal{S}_i.$$

Thus,

$$d(\boldsymbol{x},\ \mathcal{F}) = \min\left\{d(\boldsymbol{x},\ \mathcal{F}_{\mathrm{i}}),\ 1 \le i \le ng\right\}.$$

Consequently, a simple way to extend the preceding ideas to multiple failure conditions consists in using

$$\beta(\boldsymbol{x}) = d(\boldsymbol{x},\ \mathcal{F}) = \min\left\{d(\boldsymbol{x},\ \mathcal{F}_{\mathrm{i}}),\ 1 \le i \le ng\right\}.$$

Let

$$\beta_i(\boldsymbol{x}) = d(\boldsymbol{x},\ \mathcal{F}_{\mathrm{i}}) = \min\left\{\|\boldsymbol{y} - \boldsymbol{x}\| : \boldsymbol{y} \in \mathcal{F}_{\mathrm{i}}\right\}.$$

Then

$$\beta(\boldsymbol{x}) = \min\left\{\beta_i(\boldsymbol{x}),\ 1 \le i \le ng\right\}.$$

We have also

$$\beta_i(\boldsymbol{x}) = \|\boldsymbol{u}_i^*\| = \min\left\{\|\boldsymbol{u}\| : h_i(\boldsymbol{x},\ \boldsymbol{u}) \ge 0\right\}$$

and

$$\beta(\boldsymbol{x}) = \|\boldsymbol{u}^*\| = \min\left\{\|\boldsymbol{u}_i^*\|,\ 1 \le i \le ng\right\}.$$

Thus, a simple way to extend the preceding ideas consists in determining individual design points for each limit state curve and taking the one corresponding to the minimal norm. For instance, we can use the functions below:

```
findsolm = function(x,listls){
  us = list()
  betta = numeric()
  for (i in 1: length(listls)){
    g = listls[[i]][[1]]
    gradg = listls[[i]][[2]]
    sol = findsol(x,g,gradg)
    betta[[i]] = sol$beta
    us[[i]] = sol$ustar
  }
  ind = which.min(betta)
  iii = ind[[1]]
  aux = list(us[[iii]],betta[[iii]])
  names(aux) = c("ustar","beta")
  return(aux)
}
```

```
findum = function(x,listls){
  us = list()
  betta = numeric()
  for (i in 1: length(listls)){
    g = listls[[i]][[1]]
    gradg = listls[[i]][[2]]
    sol = findu(x,g,gradg)
    betta[[i]] = sol$beta
    us[[i]] = sol$ustar
  }
  ind = which.min(betta)
  iii = ind[[1]]
  aux = list(us[[iii]],betta[[iii]])
  names(aux) = c("ustar","beta")
  return(aux)
}
```

These functions receive as arguments the point of interest x and a list, whose elements are lists of two elements: a limit state equations and its gradient (see examples)

Example 9.5
Let us consider the situation where $\boldsymbol{x} \in \mathbb{R}^2$ and we have three failure conditions, corresponding to $g_1(\boldsymbol{x}) = x_2 - x_1^2 - x_1 - 3$, $g_2(\boldsymbol{x}) = x_1 - x_2^2 - x_2 - 2$, $g_3(\boldsymbol{x}) = x_1 + x_1x_2 - 1$. In this case,

```
listg1 = list(g1,gradg1)
listg2 = list(g2,gradg2)
listg3 = list(g3,gradg3)
listg = list(listg1,listg2,listg3)
```

Each pair $(g_i, \boldsymbol{gradg_i})$ is formed by a limit state equation and its gradient. For instance,

```
gradg1 = function(x){
     aux1 = -2*x[[1]] - 1
     aux2 = 1
     aux = matrix(c(aux1,aux2),nrow = 2, ncol = 1)
     return(aux)
}
```

We determine the values of $\beta_i(\boldsymbol{x})$ for the points $\boldsymbol{x} \in \{(-2,2), (-1,0), (1,-1)\}$. The results appear in Table 9.3. Figure 9.11 shows the results in the design space.

Table 9.3 Results funished by `xstar`

x	$\beta_1(x)$	$\beta_2(x)$	$\beta_3(x)$	$\beta(x) = \min \beta_i(x)$
(–2,2)	1.414214	4.425427	2.321008	1.414214
(–1,0)	2.788203	2.788203	1.732051	1.732051
(1,–1)	4.006089	0.844989	0.8182296	0.8182296

(continued)

Example 9.5 (continued)

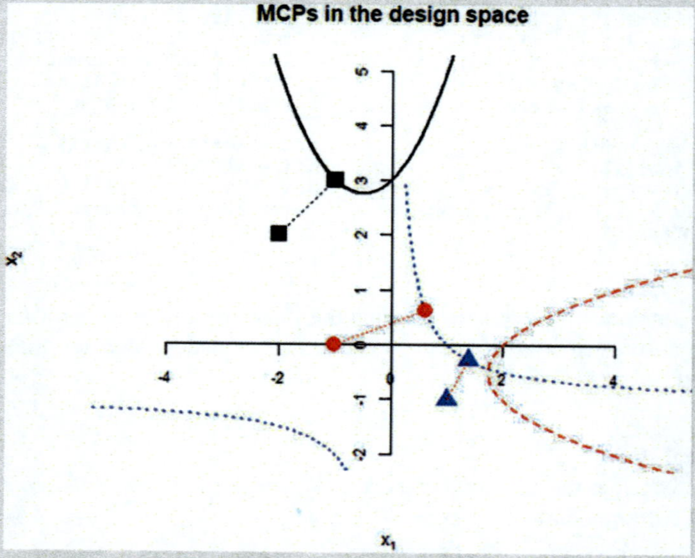

Fig. 9.11 Orthogonal projections determined by `fmincon`. The points on the failure curve are the MCPs associated to the points of interest

We can use the programs previously introduced. For instance:

```
> xa = c(-2,2)
> sola = findum(xa,listg)
> ua = sola$ustar
> beta_a = sola$beta
> pxa = xa + ua
> ua
[1] 1 1
> beta_a
[1] 1.414214
> pxa
[1] -1  3
> |
```

Example 9.6
Let us consider the situation where $\boldsymbol{x} \in \mathbb{R}^2$ and we have three failure conditions, corresponding to $g_1(\boldsymbol{x}) = x_2 - x_1^2 - x_1 - 3$, $g_2(\boldsymbol{x}) = x_1 - x_2^2 - x_2 - 2$, $g_3(\boldsymbol{x}) = x_1 + x_2 + x_1^2 + x_2^2 - 15$. We determine the values of $\beta_i(\boldsymbol{x})$ for the points $\boldsymbol{x} \in \{(-2,2), (-1,0), (1,-1)\}$. The results appear in Table 9.4. Figure 9.12 shows the results in the design space.

Table 9.4 Results funished by `xstar`

x	$\beta_1(x)$	$\beta_2(x)$	$\beta_3(x)$	$\beta(x) = \min \beta_i(x)$
(–2,2)	1.414214	4.425427	1.021528	1.021528
(–1,0)	2.788203	2.788203	3.229897	2.788203
(1,–1)	4.006089	0.844989	2.35586	0.844989

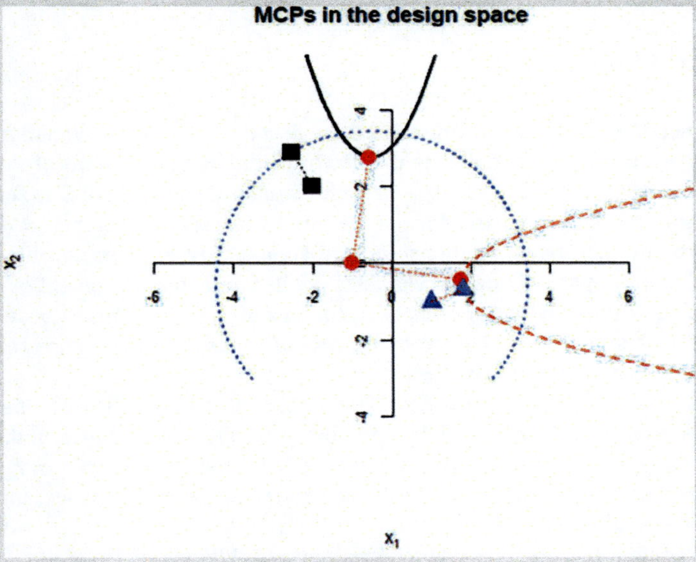

Fig 9.12 Orthogonal projections determined by `fmincon`. The points on the failure curve are the MCPs associated to the points of interest. Notice that there are two solutions for (–1,0)

Exercises

1. Let $\boldsymbol{x} \in \mathbb{R}^2$ and $g(\boldsymbol{x}) = x_2 - x_1^2 - x_1 - 3$. Determine the design point for $\boldsymbol{x}^{(1)} = (2, 2)$, $\boldsymbol{x}^{(3)} = (1, 1)$, $\boldsymbol{x}^{(4)} = (0, 1)$.
2. Let $\boldsymbol{x} \in \mathbb{R}^2$ and $g(\boldsymbol{x}) = x_2 - x_1^3 - x_1^2 - 4$. Determine the design point for $\boldsymbol{x}^{(1)} = (1, 0)$, $\boldsymbol{x}^{(3)} = (1, 1)$, $\boldsymbol{x}^{(4)} = (0, 1)$.
3. Let $\boldsymbol{x} \in \mathbb{R}^2$ and $g(\boldsymbol{x}) = x_1 x_2 - 2$. Determine the design point for $\boldsymbol{x}^{(1)} = (1, 0)$, $\boldsymbol{x}^{(2)} = (0, 0)$, $\boldsymbol{x}^{(3)} = (1, 1)$, $\boldsymbol{x}^{(4)} = (0, 1)$.
4. Let $\boldsymbol{x} \in \mathbb{R}^2$ and $g(\boldsymbol{x}) = x_1^2 + x_2^2 - 4$. Determine the design point for $\boldsymbol{x}^{(1)} = (1, 0)$, $\boldsymbol{x}^{(2)} = (0, 0)$, $\boldsymbol{x}^{(3)} = (1, 1)$, $\boldsymbol{x}^{(4)} = (0, 1)$.
5. Let $\boldsymbol{x} \in \mathbb{R}^3$ and $g(\boldsymbol{x}) = x_2 + 1 - 3(2x_3 + 1)^2(3x_1 + 1)$. Determine the design point for $\boldsymbol{x}^{(1)} = (1, 0, 0)$, $\boldsymbol{x}^{(2)} = (0, 0, 0)$, $\boldsymbol{x}^{(3)} = (0, 1, 0)$, $\boldsymbol{x}^{(4)} = (0, 0, 1)$.
6. Let $\boldsymbol{x} \in \mathbb{R}^2$ and $g_1(\boldsymbol{x}) = x_1 - x_2^2 - x_2 - 2$, $g_2(\boldsymbol{x}) = x_2 - x_1^3 - x_1^2 - 4$, $g_2(\boldsymbol{x}) = x_1 + x_2 + x_1^2 + x_2^2 - 6$. Determine the design point for $\boldsymbol{x}^{(1)} = (1, 0)$, $\boldsymbol{x}^{(2)} = (0, 0)$, $\boldsymbol{x}^{(3)} = (1, 1)$, $\boldsymbol{x}^{(4)} = (0, 1)$.

9.4 Reliability Analysis

When uncertainties must be considered, $\boldsymbol{x}$ becomes a random variable $\boldsymbol{X}$. For example, we can consider that $\boldsymbol{x}$ is the vector of the nominal values of the design variables and $\boldsymbol{U}$ and the uncertainty – in such a situation, the mean of $\boldsymbol{U}$ is zero. As in the deterministic case, we can consider the more general case where $\boldsymbol{X} = \phi(\boldsymbol{x}, \boldsymbol{U})$ – with a possibly non-linear relationship. In the context of uncertainty, the centered space becomes a space of random variables, while $\boldsymbol{x}$ remains an element of $\mathbb{R}^n$ – the pair $(\boldsymbol{x}, \boldsymbol{U})$ is formed by two objects of distinct natures: in the context of reliability, we say that this pair is an element of the *hybrid space*, that is, a space formed by vectors of $\mathbb{R}^n$ and random variables.

In this case, $Z = H(\boldsymbol{x}, \boldsymbol{U}) = g\,(\phi(\boldsymbol{x}, \boldsymbol{U}))$ – notice that H is defined on the hybrid space. Z is a random variable and the probability of failure at a point $\boldsymbol{x}$ on the design space is $F = P\,(Z > 0)$. Similarly, the reliability of the same point is $R = P\,(Z \leq 0)$: the determination of F and R involves calculations in the hybrid space, that is, the space of the random variables $\boldsymbol{U}$.

In the context of reliability, the terminology *reliability analysis of the point* $\boldsymbol{x}$ of the design space refers to the numerical estimation of the probability of failure associated with $\boldsymbol{x}$. Thus, reliability analysis consists of the numerical determination of $F = P(Z > 0)$ – of course, it is equivalent to determine the reliability $R = P(Z \leq 0)$. Let f_X denotes the PDF of X: we have

$$F = P(Z > 0) = \int_{\mathcal{F}} f_X(\boldsymbol{x})d\boldsymbol{x},\ R = P(Z \leq 0) = \int_{\mathcal{SUC}} f_X(\boldsymbol{x})d\boldsymbol{x}. \tag{9.6}$$

As shown by Eq. (9.6), reliability analysis requests the knowledge of f_X – id est, information about the joint distribution of the components of $\boldsymbol{U}$ (or – equivalently – of $\boldsymbol{X}$) is necessary to carry the evaluation of F or R.

In practice, these values can rarely be determined analytically and must be evaluated using numerical approximations – generally Monte Carlo simulation. An additional difficulty arises from the fact that the values of the probabilities of failure are generally small: in general, $F \lll 1$. This difficulty makes that, in practice, it is more convenient to use a reliability index to estimate the probability of failure.

Example 9.7

Let us consider the situation where $\boldsymbol{x} \in \mathbb{R}^2$ and $g(\boldsymbol{x}) = x_2 - x_1^2 - x_1 - 3$. Consider $\boldsymbol{x} = (x_1, x_2)$, $\phi(\boldsymbol{x}, \boldsymbol{u}) = \boldsymbol{u} + \boldsymbol{x}$. We have $h(\boldsymbol{x}, \boldsymbol{u}) = g(\boldsymbol{u} + \boldsymbol{x}) = u_2 + x_2 - u_1^2 - 2u_1x_1 - x_1^2 - u_1 - x_1 - 3$, id est,

$$h(\boldsymbol{x}, \boldsymbol{u}) = u_2 - u_1^2 - u_1 - 2u_1x_1 + g(\boldsymbol{x}).$$

The probability of failure at the point $\boldsymbol{x}$ is

$$F = P(h(\boldsymbol{x}, \ \boldsymbol{u}) > 0).$$

Assume that u_1, u_2 are independent random variables having the same distribution $N(0, 1)$. We can evaluate the probability of failure by a Monte Carlo method. For instance, let us generate two samples of ns variates for u_1 and u_2. The, we consider the sample of ns^2 elements formed by $\boldsymbol{u}_{ij} = (u_{1,i}, u_{2,j})$, $1 \leq i, j \leq ns$. We count the number n_+ of pairs in the sample such that $h(\boldsymbol{x}^{(2)}, \boldsymbol{u}_{ij}) > 0$ and we evaluate the probability of failure as $F \approx n_+/ns^2$. Since the result depends on the sample used, we can consider nr realizations of F, denote $F_1, \ldots, F_{nr}$, and evaluate $F \approx \overline{F_{nr}}$, the empirical mean of the values obtained. Typical results are shown in Tables 9.5 and 9.6.

Table 9.5 Monte Carlo Evaluation of $\boldsymbol{F}$: values of $\overline{F_{nr}} \pm s_n(F_{nr})$ with nr = 10

ns	100	1000	10000
(1,0)	4.6E-4 ± 7.8E-4	4.9E-4 ± 2.3E-4	4.0E-4 ± 8.2E-5
(0,0)	1.5E-3 ± 2.6E-3	7.5E-4 ± 6.6E-4	9.0E-4 ± 2.2E-4
(–1,1)	1.3E-3 ± 7.7E-3	1.4E-2 ± 2.5E-3	1.5E-2 ± 7.4E-4
(1,2)	5.7E-2 ±1.4E-2	4.4E-2 ± 3.6E-3	4.5E-2 ± 1.0E-3

(continued)

Example 9.7 (continued)

Table 9.6 Monte Carlo Evaluation of F: values of $\bar{F}_{nr} \pm s_n(F_{nr})$ with nr = 100

ns	100	1000	10000
(1, 0)	4.0E-4 ± 8.3E-4	3.7E-4 ± 2.5E-4	4.2E-4 ± 8.0E-5
(0,0)	7.3E-4 ± 1.5E-3	1.1E-3 ± 6.7E-4	9.9E-4 ± 1.7E-4
(−1,1)	1.6E-2 ± 8.7E-3	1.4E-2 ± 2.8E-3	1.5E-2 ± 8.1E-4
(1,2)	4.4E-2 ±1.2E-2	4.6E-2 ± 3.5E-3	4.6E-2 ± 1.1E-3

The main difficulty in this evaluation is the need of the use of exceptionally large samples, due to the smallness of F. Special sampling methods can be used to diminish the computational cost. For instance, importance sampling can be used. (Papaioannou et al., 2016; Li & Wu, 2007).

Exercises

1. Let $\boldsymbol{x} \in \mathbb{R}^2$ and $g(\boldsymbol{x}) = x_2 - x_1^2 - x_1 - 3$. Assuming normality $N(0, 1)$, generate two samples of 1000 variates of u_1, u_2 and use it to evaluate the reliability of each point: $\boldsymbol{x}^{(1)} = (-1, 0)$, $\boldsymbol{x}^{(2)} = (1, 1)$, $\boldsymbol{x}^{(3)} = (0, 1)$.
2. Let $\boldsymbol{x} \in \mathbb{R}^2$ and $g(\boldsymbol{x}) = x_2 - x_1^3 - x_1^2 - 4$. Assuming normality $N(0, 1)$, generate two samples of 1000 variates of u_1, u_2 and use it to evaluate the reliability of each point: $\boldsymbol{x}^{(1)} = (1, 0)$, $\boldsymbol{x}^{(2)} = (1, 1)$, $\boldsymbol{x}^{(\cdot)} = (0, 1)$.
3. Let $\boldsymbol{x} \in \mathbb{R}^2$ and $g(\boldsymbol{x}) = x_1 x_2 - 2$. Assuming normality $N(0, 1)$, generate two samples of 1000 variates of u_1, u_2 and use it to evaluate the reliability of each point: $\boldsymbol{x}^{(1)} = (1, 0)$, $\boldsymbol{x}^{(2)} = (0, 0)$, $\boldsymbol{x}^{(3)} = (1, 1)$, $\boldsymbol{x}^{(4)} = (0, 1)$.
4. Let $\boldsymbol{x} \in \mathbb{R}^2$ and $g(\boldsymbol{x}) = x_1^2 + x_2^2 - 4$. Assuming normality $N(0, 1)$, generate two samples of 1000 variates of u_1, u_2 and use it to evaluate the reliability of each point: $\boldsymbol{x}^{(1)} = (1, 0)$, $\boldsymbol{x}^{(2)} = (0, 0)$, $\boldsymbol{x}^{(3)} = (1, 1)$, $\boldsymbol{x}^{(4)} = (0, 1)$.
5. Let $\boldsymbol{x} \in \mathbb{R}^3$ and $g(\boldsymbol{x}) = x_2 + 1 - 3(2x_3 + 1)^2(3x_1 + 1)$. Assuming normality $N(0, 1)$, generate three samples of 100 variates of u_1, u_2, u_3 and use it to evaluate the reliability of each point: $\boldsymbol{x}^{(1)} = (1, 0, 0)$, $\boldsymbol{x}^{(2)} = (0, 0, 0)$, $\boldsymbol{x}^{(3)} = (0, 1, 0)$, $\boldsymbol{x}^{(4)} = (0, 0, 1)$.

9.5 Hasofer-Lind Reliability Index

Among the existing reliability indexes, one of the most popular is the Hasofer-Lind reliability index – the distance between a point in the design space and the failure region, calculated through the design point – the MPP or MPFP.

9.5.1 The General Situation

The Hasofer-Lind reliability index of a design point $\boldsymbol{x}$ is $\beta(\boldsymbol{x})$ given by

$$\beta(\boldsymbol{x}) = \|\boldsymbol{u}^*(\boldsymbol{x})\| = \min\{\|\boldsymbol{u}\|: H(\boldsymbol{x}, \boldsymbol{u}) \geq 0\}. \tag{9.7}$$

For $x \in C$, an alternative form is

$$\beta(\boldsymbol{x}) = \|\boldsymbol{u}^*(\boldsymbol{x})\| = \min\{\|\boldsymbol{u}\|: H(\boldsymbol{x}, \boldsymbol{u}) = 0\}. \tag{9.8}$$

Three interesting properties of the Hasofer-Lind reliability index are (see Fig. 9.13)

- The collinearity between the design point $\boldsymbol{u}^* = \boldsymbol{u}^*(\boldsymbol{x})$ and the normal $\boldsymbol{n}^*$ to the limit state curve at the point $\boldsymbol{u}^*$: $\boldsymbol{u}^*(\boldsymbol{x}) = \beta(\boldsymbol{x})\boldsymbol{n}^*$;
- an upper bound for the failure probability F: $F \leq P(\|\boldsymbol{u}\| \geq \beta(\boldsymbol{x}))$;
- a nonlinear system of equations defining $\boldsymbol{u}^*$ and β:

$$\boldsymbol{u}^*\|\nabla_{\boldsymbol{u}}\boldsymbol{H}(\boldsymbol{x}, \boldsymbol{u}^*)\| = \beta\nabla_{\boldsymbol{u}}\boldsymbol{H}(\boldsymbol{x}, \boldsymbol{u}^*), H(\boldsymbol{x}, \boldsymbol{u}^*) = 0 \tag{9.9}$$

Let us examine the first property: if $\beta(\boldsymbol{x}) > 0$ and H is a differentiable function, then $\boldsymbol{u}^*(\boldsymbol{x})$ is collinear to the unit normal to the limit state curve at $\boldsymbol{u}^*(\boldsymbol{x})$, id est,

$$\boldsymbol{u}^*(\boldsymbol{x}) = \beta(\boldsymbol{x})\boldsymbol{n}^*, \boldsymbol{n}^* = \frac{\boldsymbol{d}^*}{\|\boldsymbol{d}^*\|}, \boldsymbol{d}^* = \nabla_{\boldsymbol{u}}\boldsymbol{H}(\boldsymbol{x}, \boldsymbol{u}^*(\boldsymbol{x})),$$

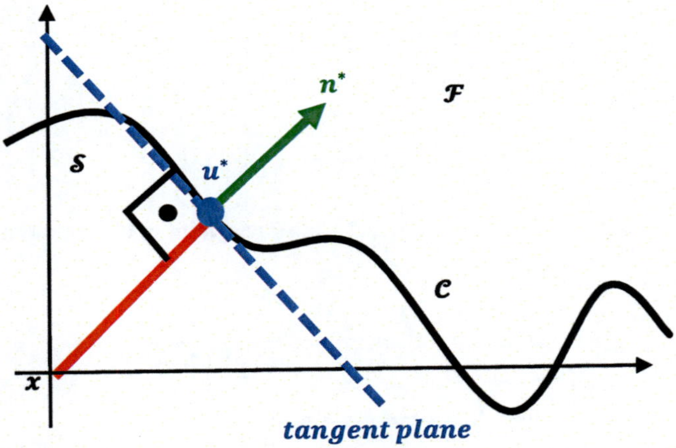

Fig. 9.13 Collinearity between the design point and the normal

$$\nabla_u H(x, u) = \left(\frac{\partial H}{\partial u_1}(x, u), \ldots, \frac{\partial H}{\partial u_n}(x, u)\right).$$

Indeed, let λ be the Lagrange's multiplier associated to the condition :$H(x, u) \geq 0$. Then, the KKT equations show that (see Sect. 1.13)

$$2u^* - \lambda d^* = 0, \lambda \geq 0, H(x, u^*) \geq 0, \lambda H(x, u^*) = 0.$$

Thus, on the one hand, $\lambda > 0$ (otherwise, $u^* = 0 \Rightarrow \beta(x) = 0$) and, on the other hand,

$$u^* = \frac{\lambda}{2} d^* \Rightarrow \beta(x) = \|u^*\| = \frac{\lambda}{2}\|d^*\|.$$

Consequently,

$$u^* = \beta(x) \frac{d^*}{\|d^*\|} = \beta(x) n^*.$$

The second useful property of the Hasofer-Lind reliability index is an upper bound of the probability of failure: let $A(r)$ be the exterior of the open ball of radius r:

$$A(r) = \{u: \|u\| \geq r\}.$$

We have

$$r_1 > r_2 \Rightarrow A(r_1) \subset A(r_2) \Rightarrow P(A(r_1)) \leq P(A(r_2)).$$

Thus, $r \longrightarrow P(A(r))$ is decreasing. Let us denote $A^*(x) = A(\beta(x))$. We have

$$u \in \mathcal{F} \Rightarrow \|u\| \geq \|u^*\| = \beta(x) \Rightarrow u \in A^*(x).$$

Consequently,

$$\mathcal{F} \subset A^*(x) \Rightarrow F = P(\mathcal{F}) \leq P(A^*(x)).$$

The third property yields from the fact that u^* is a point of the common boundary of $\mathcal{S}$ and $\mathcal{F}$, which is $\mathcal{C}$. Thus,

$$H(x, u^*) = 0 \text{ and } u^* = \beta \frac{\nabla_u H(x, u^*)}{\|\nabla_u H(x, u^*)\|}. \tag{9.10}$$

If $\phi(x, u) = u + x$, then $H(x, u) = g(x + u)$, and

$$\nabla_{\mathbf{u}} H(\boldsymbol{x}, \boldsymbol{u}) = \nabla_x g(\boldsymbol{x} + \boldsymbol{u})$$

In this case, Eqs. (9.10) can be rewritten as:

$$g(\boldsymbol{x} + \boldsymbol{u}^*) = 0 \text{ and } \boldsymbol{u}^* = \beta \frac{\nabla_x g(\boldsymbol{x} + \boldsymbol{u}^*)}{\|\nabla_x g(\boldsymbol{x} + \boldsymbol{u}^*)\|}. \tag{9.11}$$

These equations can be solved by R (see examples):

Start by creating a function that determines $\frac{\nabla_x g(\boldsymbol{x}+\boldsymbol{u})}{\|\nabla_x g(\boldsymbol{x}+\mathbf{u})\|}$

```
nxg = function(x,gradg){
  dimx = length(x)
  aux1 = gradg(x)
  aux2 = sqrt(sum(aux1^2))
  aux3 = aux1/aux2
  aux = matrix(aux3, nrow = dimx, ncol = 1)
  return(aux)
}
```

The, create a function that evaluates the Eqs. (9.11)

```
eqs = function(unks,x,g,gradg){
  dimx = length(x)
  u = matrix(unks[1:dimx], nrow = dimx, ncol = 1)
  betta = unks[[dimx+1]]
  px = x+u
  nx = nxg(px,gradg)
  aux = matrix(nrow = (dimx+1), ncol=1)
  aux[[1]] = g(px)
  aux[2:(dimx+1)] = u - betta*nx
  return(aux)
}
```

To determine the solution, we must solve `eqs` = `0` for the unknowns `unks`. This can be achieved by the methods presented in Sect. 1.14.2. For instance, we can use the function `lsqnonlin` of package `pracma` as follows (see examples):

```
findsol = function(x,g,gradg){
  nunk = length(x) + 1
  fx = function(unks){eqs(unks,x,g,gradg)}
  unkini = matrix(randn(nunk),nrow = nunk, ncol = 1)
  solx = lsqnonlin(fx,unkini)
  us= solx$x[1:(nunk-1)]
  betta = solx$x[[nunk]]
  aux = list(us,betta)
  names(aux) = c("ustar","beta")
  return(aux)
}
```

Notice that you can reduce the number of unknowns by eliminating β:

In this case, the equations to solve are evaluated as :

```
eqsu = function(unks,x,g,gradg){
  dimx = length(x)
  u = matrix(unks, nrow = dimx, ncol = 1)
  betta = unks[[dimx]]
  px = x+u
  nx = nxg(px,gradg)
  aux = matrix(nrow = (dimx+1), ncol=1)
  aux[[1]] = g(px)
  aux[2:(dimx+1)] = u - sqrt(sum(u^2))*nx
  return(aux)
}
```

They are solved by the function findu:

```
findu = function(x,g,gradg){
  nunk = length(x)
  fx = function(unks){eqsu(unks,x,g,gradg)}
  unkini = matrix(randn(nunk),nrow = nunk, ncol = 1)
  solx = lsqnonlin(fx,unkini)
  us= solx$x
  betta = sqrt(sum(us^2))
  aux = list(us,betta)
  names(aux) = c("ustar","beta")
  return(aux)
}
```

Example 9.8

Let us consider the situation where $\boldsymbol{x} \in \mathbb{R}^2$, $g(\boldsymbol{x}) = x_2 - x_1^2 - x_1 - 3$, $\phi(\boldsymbol{x}, \boldsymbol{u}) = \boldsymbol{u} + \boldsymbol{x}$. Then,

$$\nabla_{\boldsymbol{u}} H(\boldsymbol{x}, \boldsymbol{u}) = \nabla_{\boldsymbol{x}} g(\boldsymbol{x} + \boldsymbol{u}) = (-2(x_1 + u_1) - 1, 1),$$

Indeed, ,$h(\boldsymbol{x}, \boldsymbol{u}) = u_2 - u_1^2 - u_1 - 2u_1x_1 + g(\boldsymbol{x})$ and the equality above holds. Then, $\boldsymbol{u}^*$ is the solution of Eqs. (9.11).

We can look for the solution of these equations using `findsol` defined above. Initially, we define `g` and `gradg`:

```
g = function(x){
  aux = x[[2]] - x[[1]]^2 - x[[1]] - 3
  return(aux)
}

gradg = function(x){
    aux1 = -2*x[[1]] - 1
    aux2 = 1
    aux = matrix(c(aux1,aux2),nrow = 2, ncol = 1)
    return(aux)
}
```

(continued)

Example 9.8 (continued)
Then, we call `findsol`:

```
> xa = c(1,0)
> sola = findsol(xa,g,gradg)
> sola$ustar
[1] -1.272838  2.801603
> sola$beta
[1] 3.077189
```

In Table 9.4, we found the same values. We can use this method to evaluate the values of $\boldsymbol{x}^*$, $\boldsymbol{u}^*$, $\beta(\boldsymbol{x})$ for the same points evaluated in Table 9.4: we obtain the same results. Figures 9.14 and 9.15 show the results.

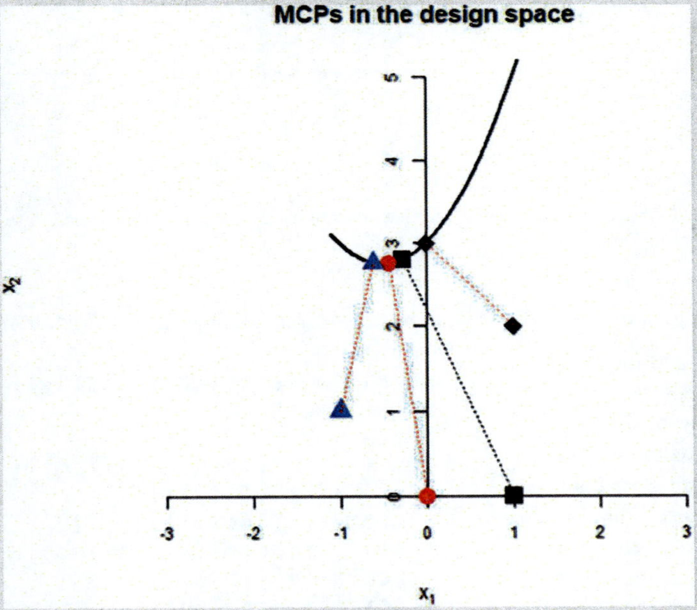

Fig. 9.14 Orthogonal projections determined by solving Eqs. (9.11) with `findsol`. The results are the same as in Table 9.4

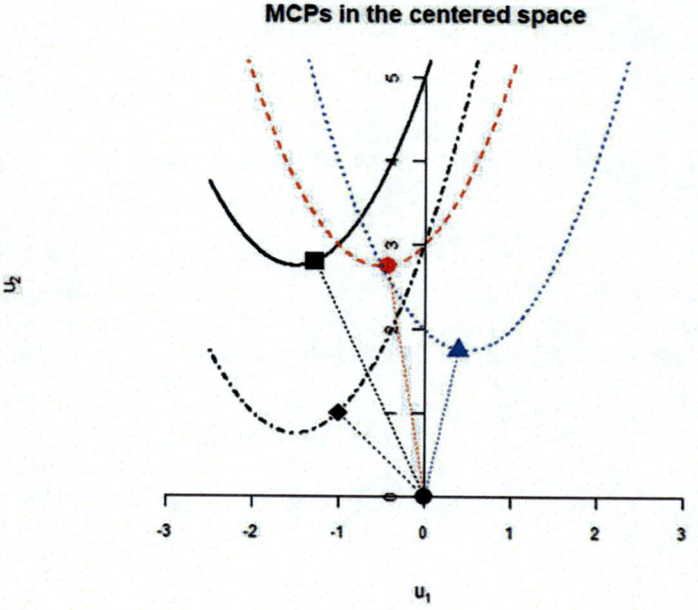

Fig. 9.15 Orthogonal projections determined by solving Eqs. (9.11) with `findsol`. The results are the same as in Table 9.4

Multiple failure conditions can be considered, analogously to Sect. 9.3:

Define a function that determines the value of $\beta_i(\boldsymbol{x})$ for a list lists of Limit States Equations. Each limit state equation is a list itself, having the limit state equation at position 1 and its gradient at position 2. The function returns the minimal β and the corresponding $\boldsymbol{u}^*$.

```
findsolm = function(x,listls){
  us = list()
  betta = numeric()
  for (i in 1: length(listg)){
    g = listls[[i]][[1]]
    gradg = listls[[i]][[2]]
    sol = findsol(x,g,gradg)
    betta[[i]] = sol$beta
    us[[i]] = sol$ustar
  }
  ind = which.min(betta)
  iii = ind[[1]]
  aux = list(us[[iii]],betta[[iii]])
  names(aux) = c("ustar","beta")
  return(aux)
}
```

Example 9.9
Let us consider the situation where $\mathbf{x} \in \mathbb{R}^2$ and we have three failure conditions, corresponding to $g_1(\boldsymbol{x}) = x_2 - x_1^2 - x_1 - 3$, $g_2(\boldsymbol{x}) = x_1 - x_2^2 - x_2 - 2$, $g_3(\boldsymbol{x}) = x_1 + x_2 + x_1^2 + x_2^2 - 6$.

We define the list of limit states:

```
listg1 = list(g1,gradg1)
listg2 = list(g2,gradg2)
listg3 = list(g3,gradg3)
listg = list(listg1,listg2,listg3)
```

Then, we call findum:

```
xa = c(-2,2)
sola = findum(xa,listg)
ua = sola$ustar
beta_a = sola$beta
xas = xa + ua
```

We obtain:

```
[1] 1 1
> beta_a
[1] 1.414214
> xas
[1] -1  3
```

In Table 9.3, we found the same values for β. We can use this method to evaluate the values of $\boldsymbol{x}^*$, $\boldsymbol{u}^*$, $\beta(\boldsymbol{x})$ for the same points evaluated in Table 9.4: we obtain analogous results (notice that the design point for (–1,0) is different, but the value of β is identical). Figure 9.16 shows the results obtained.

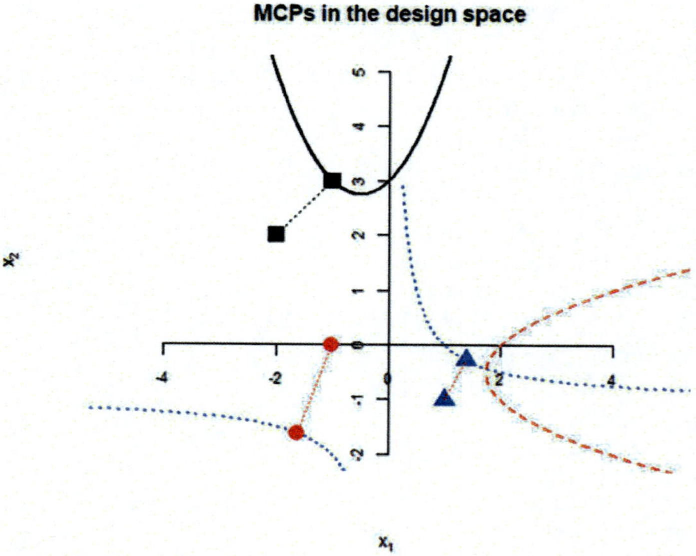

Fig. 9.16 Orthogonal projections determined by solving Eqs. (9.11) with `findsol`. The results are the same as in Table 9.4

Exercises

1. Let $\boldsymbol{x} \in \mathbb{R}^2$ and $g(\boldsymbol{x}) = x_2 - x_1^2 - x_1 - 3$. Determine $\boldsymbol{u}^*$ for $\boldsymbol{x}^{(1)} = (2, 2)$, $\boldsymbol{x}^{(2)} = (1, 1)$, $\boldsymbol{x}^{(3)} = (0, 1)$.
2. Let $\boldsymbol{x} \in \mathbb{R}^2$ and $g(\boldsymbol{x}) = x_2 - x_1^3 - x_1^2 - 4$. Determine $\boldsymbol{u}^*$ for $\boldsymbol{x}^{(1)} = (1, 0)$, $\boldsymbol{x}^{(2)} = (1, 1)$, $\boldsymbol{x}^{(3)} = (0, 1)$.
3. Let $\boldsymbol{x} \in \mathbb{R}^2$ and $g(\boldsymbol{x}) = x_1 x_2 - 2$. Determine $\boldsymbol{u}^*$ for $\boldsymbol{x}^{(1)} = (1, 0)$, $\boldsymbol{x}^{(2)} = (0, 0)$, $\boldsymbol{x}^{(3)} = (1, 1)$, $\boldsymbol{x}^{(4)} = (0, 1)$.
4. Let $\boldsymbol{x} \in \mathbb{R}^2$ and $g(\boldsymbol{x}) = x_1^2 + x_2^2 - 4$. Determine $\boldsymbol{u}^*$ for $\boldsymbol{x}^{(1)} = (1, 0)$, $\boldsymbol{x}^{(2)} = (0, 0)$, $\boldsymbol{x}^{(3)} = (1, 1)$, $\boldsymbol{x}^{(4)} = (0, 1)$.
5. Let $\boldsymbol{x} \in \mathbb{R}^3$ and $g(\boldsymbol{x}) = x_2 + 1 - 3(2x_3 + 1)^2(3x_1 + 1)$. Determine $\boldsymbol{u}^*$ for $\boldsymbol{x}^{(1)} = (1, 0, 0)$, $\boldsymbol{x}^{(2)} = (0, 0, 0)$, $\boldsymbol{x}^{(3)} = (0, 1, 0)$, $\boldsymbol{x}^{(4)} = (0, 0, 1)$.

9.5.2 *The Case of Affine Limit State Equations*

Let us consider the situation where $g(\boldsymbol{x}) = \boldsymbol{a}^t\boldsymbol{x} + b$, $\boldsymbol{a} \in \mathbb{R}^n$, $b \in \mathbb{R}$. In this case,

$$h(\boldsymbol{x}, \boldsymbol{u}) = g(\boldsymbol{x} + \boldsymbol{u}) = \boldsymbol{a}^t(\boldsymbol{x} + \boldsymbol{u}) + \boldsymbol{b} = \boldsymbol{a}^t\boldsymbol{u} + g(\boldsymbol{x}).$$

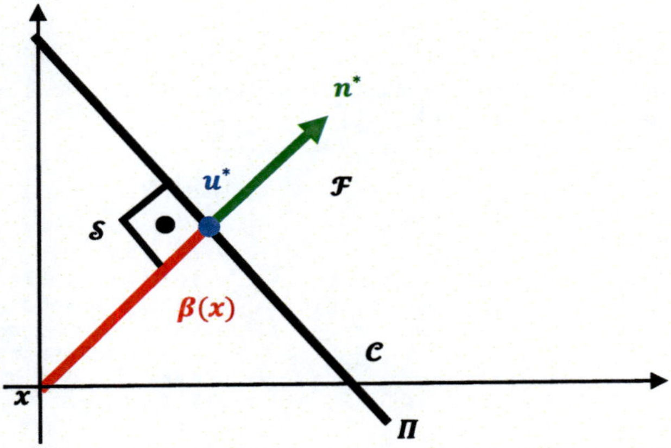

Fig. 9.17 If the limit state is a hyperplane Π, then β is the distance from Π to the origin

Thus, $H(\boldsymbol{x}, \boldsymbol{U}) = \boldsymbol{a}^t\boldsymbol{U} + g(\boldsymbol{x})$: we have $g(\boldsymbol{x}) < 0$ when $\boldsymbol{x} \in S$. In such a case, C is a hyperplane Π of equation $H(\boldsymbol{x}, \boldsymbol{U}) = 0$. (Fig. 9.17)

Let $\boldsymbol{n}^*$ be the unit normal pointing outwards the safe region: if $\boldsymbol{a} \neq \boldsymbol{0}$, then $\boldsymbol{n}^* = \boldsymbol{a}/\|\boldsymbol{a}\|$. Then,

$$H(\boldsymbol{x}, \boldsymbol{u}^*) = \boldsymbol{a}^t\boldsymbol{u}^* + g(\boldsymbol{x}) = \beta(\boldsymbol{x})\frac{\boldsymbol{a}^t\boldsymbol{a}}{\|\boldsymbol{a}\|} + g(\boldsymbol{x}) = 0.$$

Thus,

$$\beta(\boldsymbol{x}) = -\frac{g(\boldsymbol{x})}{\|\boldsymbol{a}\|}, \boldsymbol{u}^* = -\frac{g(\boldsymbol{x})}{\|\boldsymbol{a}\|^2}\boldsymbol{a}. \tag{9.12}$$

Notice that $\beta(\boldsymbol{x}) > 0$ for $\boldsymbol{x} \in \mathcal{S}$. Eq. (9.12) can be solved by R as if shown at right. Function `solaff` returns a list containing $\beta(\boldsymbol{x})$, $\boldsymbol{u}^*$, $\boldsymbol{x}^*$(see example).

```
solaff = function(x,a,b){
  na = sqrt(sum(a^2))
  nnn = a/na
  gx = sum(a*x) + b
  betta = -gx/na
  u = betta*nnn
  px = x + u
  aux = list(betta,u,px)
  names(aux) = c("beta","ustar","xstar")
  return(aux)
}
```

Example 9.10

Let us consider the situation where $\boldsymbol{x} \in \mathbb{R}^2$, $g(\boldsymbol{x}) = x_2 + x_1 - 3$. Then, $\boldsymbol{a} = (1, 1)^t$, $b = -3$, $\|\boldsymbol{a}\| = \sqrt{2}$.

Let us consider the point $\boldsymbol{x} = (1, 0)$. Then, $g(\boldsymbol{x}) = -2$. Thus,
$\beta(\boldsymbol{x}) = \sqrt{2}$,
$\boldsymbol{u}^*(\boldsymbol{x}) = (1, 1)^t$,
$\boldsymbol{x}^* = (2, 1)^t$.

Using R:

```
> xa = c(1,0)
> sola = solaff(xa,a,b)
> beta_a = sola$beta
> ua = sola$ustar
> pxa = sola$xstar
> beta_a
[1] 1.414214
> ua
[1] 1 1
> pxa
[1] 2 1
```

Now, consider the point $\boldsymbol{x} = (0, 0)$. Then, $g(\boldsymbol{x}) = -3$. Thus,
$\beta(\boldsymbol{x}) = \frac{3}{\sqrt{2}}$,
$\boldsymbol{u}^*(\boldsymbol{x}) = \left(\frac{3}{2}, \frac{3}{2}\right)^t$,
$\boldsymbol{x}^* = \left(\frac{3}{2}, \frac{3}{2}\right)^t$.

Using R:

```
> xb = c(0,0)
> solb = solaff(xb,a,b)
> beta_b = solb$beta
> ub = solb$ustar
> pxb = solb$xstar
> beta_b
[1] 2.12132
> ub
[1] 1.5 1.5
> pxb
[1] 1.5 1.5
```

Let consider the point $\boldsymbol{x} = (0, 1)$. Again, $g(\boldsymbol{x}) = -2$. Thus,
$\beta(\boldsymbol{x}) = \sqrt{2}$,
$\boldsymbol{u}^*(\boldsymbol{x}) = (1, 1)^t$,
$\boldsymbol{x}^* = (1, 2)^t$.

Using R:

```
> xc = c(0,1)
> solc = solaff(xc,a,b)
> beta_c = solc$beta
> uc = solc$ustar
> pxc = solc$xstar
> beta_c
[1] 1.414214
> uc
[1] 1 1
> pxc
[1] 1 2
```

(continued)

Example 9.10 (continued)

Let consider the point $\boldsymbol{x} = (1, -1)$. Here, $g(\boldsymbol{x}) = -3$. Thus,
$\beta(\boldsymbol{x}) = \frac{3}{\sqrt{2}}$,
$\boldsymbol{u}^*(\boldsymbol{x}) = \left(\frac{3}{2}, \frac{3}{2}\right)^t$,
$\boldsymbol{x}^* = \left(\frac{5}{2}, \frac{1}{2}\right)^t$.

Using R:

```
> xd = c(1,-1)
> sold = solaff(xd,a,b)
> beta_d = sold$beta
> ud = sold$ustar
> pxd = sold$xstar
> beta_d
[1] 2.12132
> ud
[1] 1.5 1.5
> pxd
[1] 2.5 0.5
```

Figure 9.16 shows the results obtained (Fig. 9.18).

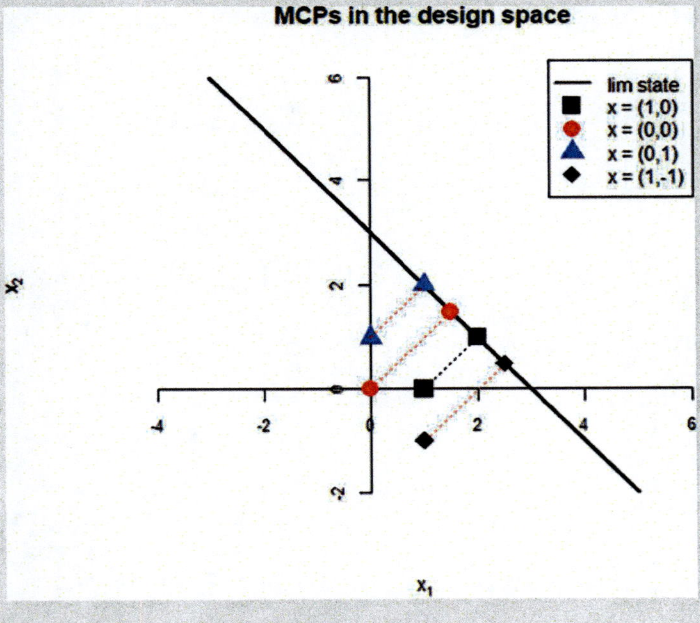

Fig. 9.18 Orthogonal projections determined by Eq. (9.12)

Analogously to Sect. 9.3, multiple linear failure conditions can be considered by taking the minimal β. For instance, we can use the function at right. It receives a list `listls` formed by lists, whose elements are a_i, b_i (see example)

```
solaffm = function(x,listls){
  us = list()
  xs = list()
  betta = numeric()
  for (i in 1: length(listls)){
    a = listls[[i]][[1]]
    b = listls[[i]][[2]]
    sol = solaff(x,a,b)
    betta[[i]] = sol$beta
    us[[i]] = sol$ustar
    xs[[i]] = sol$xstar
  }
  ind = which.min(betta)
  iii = ind[[1]]
  aux = list(betta[[iii]],us[[iii]],xs[[iii]])
  names(aux) = c("beta","ustar","xstar")
  return(aux)
}
```

Example 9.11

Let us consider the situation where $\boldsymbol{x} \in \mathbb{R}^2$, $g_1(\boldsymbol{x}) = x_2 + x_1 - 3$, $g_2(\boldsymbol{x}) = x_2 - x_1 - 1$, $g_3(\boldsymbol{x}) = -x_2 - 4$. Here, $\boldsymbol{a_1} = (1, 1)^t$, $b_1 = -3$, $\boldsymbol{a_2} = (-1, 1)^t$, $b_2 = -1$, $\boldsymbol{a_3} = (0, -1)^t$, $b_3 = -4$, $\|\boldsymbol{a_1}\| = \sqrt{2}$, $\|\boldsymbol{a_2}\| = \sqrt{2}$, $\|\boldsymbol{a_3}\| = 1$.

Let us consider the point $\boldsymbol{x} = (0, 0)$: $g_1(\boldsymbol{x}) = -3$, $g_2(\boldsymbol{x}) = -1$, $g_3(\boldsymbol{x}) = -4$. Thus,

$$\beta_1(\boldsymbol{x}) = \frac{3}{\sqrt{2}}, \beta_2(\boldsymbol{x}) = \frac{1}{\sqrt{2}}, \beta_3(\boldsymbol{x}) = 4.$$

Thus, $\beta(\boldsymbol{x}) = \beta_2(\boldsymbol{x})$ and

$$\boldsymbol{u}^*(\boldsymbol{x}) = \left(-\frac{1}{2}, \frac{1}{2}\right)^t, \boldsymbol{x}^* = \left(-\frac{1}{2}, \frac{1}{2}\right)^t.$$

To use R, we call the function `solaffm`:

```
a1 = c(1,1)
b1 = -3
a2 = c(-1,1)
b2 = -1
a3 = c(0,-1)
b3 = -4
ls1 = list(a1,b1)
ls2 = list(a2,b2)
```

```
ls3 = list(a3,b3)
lls = list(ls1,ls2,ls3)

xa = c(0,0)
sola = solaffm(xa,lls)
beta_a = sola$beta
ua = sola$ustar
pxa = sola$xstar
```

We obtain the results at right:

```
> beta_a
[1] 0.7071068
> ua
[1] -0.5  0.5
> pxa
[1] -0.5  0.5
```

We consider the points (2,–2), (1,0), (–2,–2). Figure 9.16 shows the results obtained (Fig. 9.19).

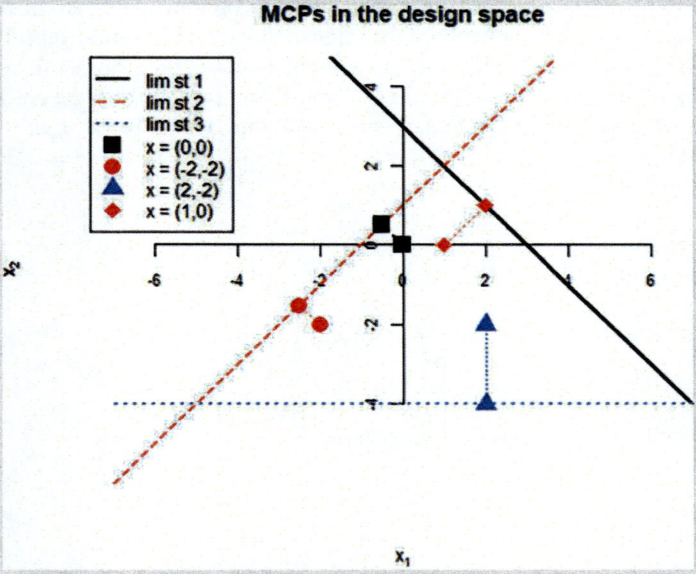

Fig. 9.19 Orthogonal projections determined by solving Eqs. (9.11) with `findsol`. The results are the same as in Table 9.4

Exercises

1. Let $\boldsymbol{x} \in \mathbb{R}^3$ and $g(\boldsymbol{x}) = x_2 - 2x_1 - x_3 - 3$. Determine $\boldsymbol{u}^*$ for $\boldsymbol{x}^{(1)} = (1, 0, 0)$, $\boldsymbol{x}^{(2)} = (0, 0, 0)$, $\boldsymbol{x}^{(3)} = (1, 1, 1)$, $\boldsymbol{x}^{(4)} = (0, 1, 0)$.
2. Let $\boldsymbol{x} \in \mathbb{R}^3$ and $g(\boldsymbol{x}) = x_3 - x_1 - x_2 - 4$. Determine $\boldsymbol{u}^*$ for $\boldsymbol{x}^{(1)} = (1, 0, 0)$, $\boldsymbol{x}^{(2)} = (0, 0, 0)$, $\boldsymbol{x}^{(3)} = (1, 1, 1)$, $\boldsymbol{x}^{(4)} = (0, 1, 0)$.

(continued)

3. Let $x \in \mathbb{R}^4$ and $g(x) = x_3 + 2x_4 - 3x_1 - x_2 - 4$. Determine u^* for $x^{(1)} = (1,0,0,1)$, $x^{(2)} = (0,0,0,0)$, $x^{(3)} = (1,0,1,0)$, $x^{(4)} = (0,1,0,1)$.
4. Let $x \in \mathbb{R}^4$ and $g(x) = x_3 + x_4 - x_1 - x_2 - 4$. Determine u^* for $x^{(1)} = (1,0,0,1)$, $x^{(2)} = (0,0,0,0)$, $x^{(3)} = (1,0,1,0)$, $x^{(4)} = (0,1,0,1)$.

9.5.3 Convex Failure Regions

This property extends to the situation where the failure region is convex: $\beta(x)$ corresponds to the distance between the origin and the failure region yet (Fig. 9.20).

Recalling that a regular convex region is the intersection of all the hyperplanes containing it (see, for instance, (Souza de Cursi & Sampaio, 2010)), the convexity of the failure region can be exploited to characterize the reliability index as

$$\beta(x) = \sup\left\{ -\frac{a^t x + b}{\|a\|} : a,\ b; a^t y + b \geq 0,\ \forall y \in \mathcal{F} \right\}, \tag{9.13}$$

or, equivalently,

$$\beta(x) = \sup\left\{ -\frac{\nabla g(y)^t (x - y)}{\|\nabla g(y)\|} : y \in C \right\}. \tag{9.14}$$

Equation (9.14) can be solved as follows:

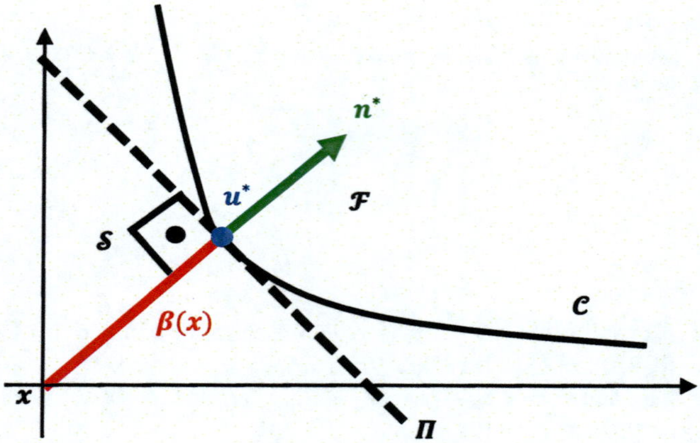

Fig. 9.20 If the failure region is convex, then β is the distance from $\mathcal{F}$ to the origin, which coincides with the maximal distance between the origin and a hyperplane containing $\mathcal{F}$

Create the function to be minimized. Here, `fhl (y)` evaluates $\frac{\nabla g(y)^t(x-y)}{\|\nabla g(y)\|}$. `nxg` was defined in Sect. 9.5.1.

```
fhl = function(y,x,g,gradg){
  dimx = length(x)
  ny = nxg(y,gradg)
  aux1 = matrix(x-y,nrow = dimx, ncol = 1)
  aux = sum(ny*aux1)
  return(aux)
}
```

Create the constraints. Here, `clh (y)` evaluates (y) , which must be equal to zero.

```
chl = function(y,g){
  aux = g(y)
  return(aux)
}
```

Create a function that determines the minimum of `fhl` for given `x`, `g`, `gradg`. Here, we call `fmincon` of `pracma`. `solconv` returns the solution: β, u^*,x^*.

```
solconv = function(x,g,gradg){
  dimx = length(x)
  yini = runif(dimx)
  fobj = function(y){fhl(y,x,g,gradg)}
  he = function(y){chl(y,g)}
  sol = fmincon(yini, fobj,heq = he)
  betta = -sol$value
  px = sol$par
  u = px - x
  aux = list(betta,u,px)
  names(aux) = c("beta","ustar","xstar")
  return(aux)
}
```

Example 9.12
Let us consider the situation where $x \in \mathbb{R}^2$, $g(x) = x_2 - x_1^2 - x_1 - 3$.

Let us consider the point $x = (0,0)$. We run the code at right. The results are identical to those in Table 9.1.

```
> xa = c(0,0)
> sola = solconv(xa,g,gradg)
> beta_a = sola$beta
> ua = sola$ustar
> pxa = sola$xstar
> beta_a
[1] 2.788203
> ua
[1] -0.4232162  2.7558958
> pxa
[1] -0.4232162  2.7558958
```

Now, consider the point $\boldsymbol{x} = (-1, 1)$. The results are identical to those in Table 9.1.

```
> xb = c(-1,1)
> solb = solconv(xb,g,gradg)
> beta_b = solb$beta
> ub = solb$ustar
> pxb = solb$xstar
> beta_b
[1] 1.804742
> ub
[1] 0.3894888 1.7622127
> pxb
[1] -0.6105112  2.7622127
```

Figure 9.16 shows the results obtained for the points considered in Sect. 9.2: the results are identical (Fig. 9.21).

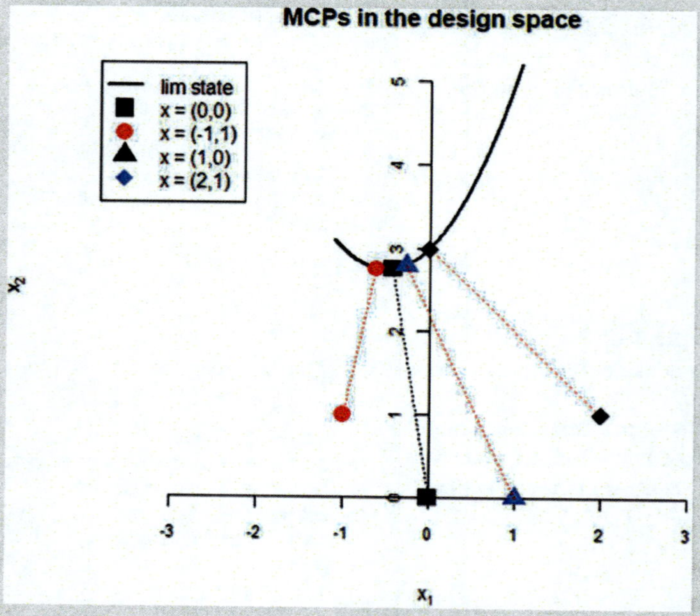

Fig. 9.21 Orthogonal projections determined by `solconv`

Exercises

1. Let $\boldsymbol{x} \in \mathbb{R}^2$ and $g(\boldsymbol{x}) = x_2 - x_1^3 - x_1^2 - 4$. Use the method of this section to determine $\beta(\boldsymbol{x})$ and $\boldsymbol{x}^*$ for $\boldsymbol{x}^{(1)} = (1,0)$, $\boldsymbol{x}^{(2)} = (0,0)$, $\boldsymbol{x}^{(3)} = (1,1)$, $\boldsymbol{x}^{(4)} = (0,1)$.

(continued)

2. Let $x \in \mathbb{R}^2$ and $g(x) = x_1x_2 - 2$. Use the method of this section to determine $\beta(x)$ and x^* for $x^{(1)} = (1,0)$, $x^{(2)} = (0,0)$, $x^{(3)} = (1,1)$, $x^{(4)} = (0,1)$.
3. Let $x \in \mathbb{R}^2$ and $g(x) = x_1^2 + x_2^2 - 4$. Use the method of this section to determine $\beta(x)$ and x^* for $x^{(1)} = (1,0)$, $x^{(2)} = (0,0)$, $x^{(3)} = (1,1)$, $x^{(4)} = (0,1)$.
4. Let $x \in \mathbb{R}^3$ and $g(x) = x_2 + 1 - 3(2x_3 + 1)^2(3x_1 + 1)$. Use the method of this section to determine $\beta(x)$ and x^* for $x^{(1)} = (1,0,0)$, $x^{(2)} = (0,0,0)$, $x^{(3)} = (0,1,0)$, $x^{(4)} = (0,0,1)$.

9.6 Using the Reliability Index to Estimate the Probability of Failure

When the random variables $\boldsymbol{U}$ form a gaussian vector, the reliability index may be used to estimate the probability of failure. In general, the determination assumes that the distribution of $\boldsymbol{U}$ is $N(\boldsymbol{0}, \boldsymbol{Id})$, id est, the components of $\boldsymbol{U}$ are independent and have the same distribution $N(0,1)$. For a general Gaussian vector having a mean $\boldsymbol{m}$ and a covariance matrix $\boldsymbol{C}$, id est, a vector $\boldsymbol{U}$ having the distribution $N(\boldsymbol{m}, \boldsymbol{C})$, a transformation $\boldsymbol{U} = \boldsymbol{T}(\boldsymbol{W})$ is used to bring the vector $\boldsymbol{U}$ to a $N(\boldsymbol{0}, \boldsymbol{Id})$ vector $\boldsymbol{W}$.

If $\boldsymbol{U} = (u_1, \ldots, u_n)$ is a gaussian vector, id est, if its components are independent, then $\boldsymbol{C}$ is a diagonal matrix such that $C_{ii} = \sigma_i^2$. Then, each u_i is an independent gaussian variable having as distribution $N(m_i, \sigma_i)$. In such a situation, the transformation reads as

$$w_i = \frac{u_i - m_i}{\sigma_i} \Leftrightarrow u_i = m_i + \sigma_i w_i. \tag{9.15}$$

In the context of Reliability, this transformation is referred as the *transformation of Hasofer-Lind*.

If $\boldsymbol{C}$ is not a diagonal matrix, then start by finding a decomposition $\boldsymbol{C} = \boldsymbol{P}\boldsymbol{D}\boldsymbol{P}^t$, where $\boldsymbol{D}$ is a diagonal matrix and $\boldsymbol{P}^t\boldsymbol{P} = \boldsymbol{P}\boldsymbol{P}^t = \boldsymbol{Id}$. Such a decomposition can be made by determining the eigenvalues and eigenvectors of $\boldsymbol{C}$: $\boldsymbol{D}$ is a diagonal matrix containing the eigenvalues of $\boldsymbol{C}$ as diagonal and $\boldsymbol{P}^t$ is formed by the orthonormalized eigenvectors of $\boldsymbol{C}$, id est,

$$\boldsymbol{P} = \begin{pmatrix} \boldsymbol{P}_1 \\ \ldots \\ \boldsymbol{P}_m \end{pmatrix}, \ \boldsymbol{P}_i\boldsymbol{P}_j^t = \begin{cases} 0, & \text{if } i \neq j \\ 1, & \text{if } i = j \end{cases}, \ \boldsymbol{C}\boldsymbol{P}_i^t = d_{ii}\boldsymbol{P}_i^t$$

Under R, the command `eigen` furnishes both the values of the eigenvalues and of the eigenvectors, but they can be non-orthonormalized: it is necessary to call a Gram-Schmidt subroutine to get $\boldsymbol{P}$ – such a subroutine is available in packages `pracma` (`gramSchmidt`, with a lowercase `g`) and `matlib` (GramSchmidt, with an uppercase `G`).

Notice that all the eigenvalues of $\boldsymbol{C}$ are non-negative, since $\boldsymbol{C}$ is semidefinite: $\forall \boldsymbol{a} \in \mathbb{R}^n$: taking $v = \boldsymbol{a}^t(\boldsymbol{u} - \boldsymbol{m})$, we have $\boldsymbol{a}^t\boldsymbol{C}\boldsymbol{a} = E(vv^t) \geq 0$. Assume that the eigenvalues are ranged in a decreasing order: $d_{11} \geq d_{22} \geq \ldots \geq d_{nn} \geq 0$. Let $m = \max \{i: d_{ii} > 0\}$. Let $\boldsymbol{A}$ be the matrix formed by the first m columns of $\boldsymbol{P}$ multiplied by the square root $\sqrt{d_{ii}}$, id est,

$$\boldsymbol{A} = (\boldsymbol{A}_1, \ \ldots, \boldsymbol{A}_m), \boldsymbol{A}_i = \sqrt{d_{ii}}\boldsymbol{P}_i.$$

Let $\boldsymbol{w} = (w_1, \ldots, w_m)$ be a Gaussian vector having as distribution $N(\boldsymbol{0}, \boldsymbol{Id})$. Then, $\boldsymbol{U} = \boldsymbol{A}\boldsymbol{W} + \boldsymbol{m}$ has as distribution $(\boldsymbol{m}, \boldsymbol{C})$: indeed, $E(\boldsymbol{U}) = \boldsymbol{A}E(\boldsymbol{W}) + \boldsymbol{m} = \boldsymbol{m}$ and

$$E\left((\boldsymbol{U} - \boldsymbol{m})(\boldsymbol{U} - \boldsymbol{m})^t\right) = \boldsymbol{A}E(\boldsymbol{W}\boldsymbol{W}^t)\boldsymbol{A}^t = \boldsymbol{A}\boldsymbol{A}^t = \boldsymbol{P}\boldsymbol{D}\boldsymbol{P}^t = \boldsymbol{C}$$

As an alternative, we can determine a decomposition $\boldsymbol{C} = \boldsymbol{A}\ \boldsymbol{A}^t$ – this can be achieved by Cholesky's factorization of $\boldsymbol{C}$: under R, the command `chol` furnishes the Cholesky factorization of a real symmetric positive-definite square matrix. Whichever the case or the method used, the transformation is affine

$$\boldsymbol{U} = \boldsymbol{T}\boldsymbol{W} + \boldsymbol{m},$$

so that the limit state curve can be written as

$$G(\boldsymbol{x}, \boldsymbol{W}) = H(\boldsymbol{x}, \boldsymbol{T}\boldsymbol{W} + \boldsymbol{m}).$$

Consequently, the analysis can be brought to the case where the vector of random variables is formed by independent gaussian variables of same distribution $N(0, 1)$. In the sequel of this subsection, we assume that $\boldsymbol{U}$ satisfies such a condition.

Example 9.13
Let us consider two independent random variables normally distributed: $u_1 \sim N(0, 1)$, $u_2 \sim N(0, 1)$. Let $X_1 = u_1 + u_2$, $X_2 = 3u_1 + 2u_2$, $X_3 = 5u_1 - 4\ u_2$. Let $X = (X_1, X_2, X_3)$. The covariance matrix of X is

$$\boldsymbol{C} = \begin{pmatrix} 2 & 5 & 1 \\ 5 & 13 & 7 \\ 1 & 7 & 41 \end{pmatrix}$$

Let us use `pracma` and `matlib` to get the matrix A: with the code at right, `A1` is the result furnished by `pracma`, `A2` is the result furnished by `matlib`. We verify that $AA^t = C$:

```
> print(A1 %*% t(A1))
     [,1] [,2] [,3]
[1,]    2    5    1
[2,]    5   13    7
[3,]    1    7   41
```

```
> print(A2 %*% t(A2))
     [,1] [,2] [,3]
[1,]    2    5    1
[2,]    5   13    7
[3,]    1    7   41
```

```
matC = matrix(c(2,5,1,
                5,13, 7,
                1,7,41),
              nrow = 3, ncol=3,
              byrow = TRUE)

tol = 1e-10
r = eigen(matC)
vp = r$values
ind = which(vp > tol
V = r$vectors[,ind]
vpp = r$values[ind]
D = diag(sqrt(vpp))
aux = gramSchmidt(V)
P1 = aux$Q
A1 = P1 %*% D
print(A1 %*% t(A1))

P2 = GramSchmidt(V)
A2 = P2 %*% D
print(A2 %*% t(A2))
```

Now let us use Cholesky's decomposition:

```
> U3 = chol(matC)
> A3 = t(U3)
> print(A3 %*% t(A3))
     [,1] [,2] [,3]
[1,]    2    5    1
[2,]    5   13    7
[3,]    1    7   41
```

Exercises

1. Consider a vector of 3 gaussian variables (X_1, X_2, X_3) having as covariance matrix

$$C = \begin{pmatrix} 5 & 7 & 1 \\ 7 & 8 & 3 \\ 1 & 3 & 10 \end{pmatrix}$$

What is the minimal number of independent $N(0,1)$ variables requested to describe this distribution?

2. Consider a vector of 4 gaussian variables (X_1, X_2, X_3, X_4) having as covariance matrix

$$C = \begin{pmatrix} 19 & 19 & 9 & -4 \\ 19 & 29 & -1 & 2 \\ 9 & -1 & 19 & -18 \\ -4 & 2 & -18 & 26 \end{pmatrix}$$

What is the minimal number of independent $N(0,1)$ variables requested to describe this distribution?

3. Consider a vector of 4 gaussian variables (X_1, X_2, X_3, X_4) having as covariance matrix

$$C = \begin{pmatrix} 25 & -2 & 7 & 9 \\ -2 & 5 & -1 & 5 \\ 7 & -1 & 2 & 2 \\ 9 & 5 & 2 & 10 \end{pmatrix}$$

What is the minimal number of independent $N(0,1)$ variables requested to describe this distribution?

4. Consider a vector of 5 gaussian variables $(X_1, X_2, X_3, X_4, X_5)$ having as covariance matrix

$$C = \begin{pmatrix} 34 & 11 & 28 & -12 & 7 \\ 11 & 18 & 19 & 0 & -16 \\ 28 & 19 & 33 & -3 & -8 \\ -12 & 0 & -3 & 11 & -9 \\ 7 & -16 & -8 & -9 & 25 \end{pmatrix}$$

What is the minimal number of independent $N(0,1)$ variables requested to describe this distribution?

(continued)

5. Consider a vector of 5 gaussian variables $(X_1, X_2, X_3, X_4, X_5)$ having as covariance matrix

$$C = \begin{pmatrix} 5 & 2 & 4 & 6 & -8 \\ 2 & 17 & -2 & -12 & 4 \\ 4 & -2 & 4 & 8 & -8 \\ 6 & -12 & 8 & 20 & -16 \\ -8 & 4 & -8 & -16 & 16 \end{pmatrix}$$

What is the minimal number of independent $N(0, 1)$ variables requested to describe this distribution?

9.6.1 The Case of Affine Limit State Equations

In this section, we examine the situation where $H(\boldsymbol{x}, \boldsymbol{U}) = \boldsymbol{a}^t\boldsymbol{U} + b(\boldsymbol{x})$. As indicated above, we assume that $\boldsymbol{U}$ is a centered Gaussian vector $N(\boldsymbol{0}, \boldsymbol{Id})$, i.e., all its components are independent and have the same distribution $N(0, 1)$ – otherwise the transformation previously introduced must be used to bring the analysis to such a situation.

Under these assumptions, $Z = H(\boldsymbol{x}, \boldsymbol{U})$ is a Gaussian variable such that $E(Z) = b(\boldsymbol{x})$ and $V(Z) = \boldsymbol{a}^t\boldsymbol{a} = \|\boldsymbol{a}\|^2$. Thus,

$$S = \frac{Z - b(\boldsymbol{x})}{\|\boldsymbol{a}\|} \sim N(0, 1).$$

Consequently,

$$F = P(\mathcal{F}) = P(Z > 0) = P\left(S > -\frac{b(\boldsymbol{x})}{\|\boldsymbol{a}\|}\right) = P(S > \beta(\boldsymbol{x}))$$

Let Φ be the CDF associated to $N(0, 1)$. We have

$$F = 1 - \Phi(\beta(\boldsymbol{x})) = \Phi(-\beta(\boldsymbol{x})). \tag{9.16}$$

The value of F may be furnished by `pnorm(`$\beta(x)$`, lower.tail = FALSE)` or `pnorm(`$-\beta(x)$`)`. Typical values are:

```
> pnorm(1:12,lower.tail=FALSE)
 [1] 1.586553e-01 2.275013e-02 1.349898e-03 3.167124e-05
 [5] 2.866516e-07 9.865876e-10 1.279813e-12 6.220961e-16
 [9] 1.128588e-19 7.619853e-24 1.910660e-28 1.776482e-33
> pnorm(-(1:12))
 [1] 1.586553e-01 2.275013e-02 1.349898e-03 3.167124e-05
 [5] 2.866516e-07 9.865876e-10 1.279813e-12 6.220961e-16
 [9] 1.128588e-19 7.619853e-24 1.910660e-28 1.776482e-33
```

We see that $\beta = 4$ corresponds to $F \approx 3E - 5$, while $\beta = 8$ corresponds to $F \approx 6E - 16$ and $\beta = 12$ to $F \approx 2E - 33$.

Example 9.14
Let us consider the situation where the failure corresponds to the difference between two variables x_1 and x_2 : failure occurs when $x_1 > x_2$. Then, denoting $\boldsymbol{x} = (x_1, x_2)$:

$$g(\boldsymbol{x}) = x_1 - x_2.$$

If $X_1 = x_1 + \sigma_1 u_1$, $X_2 = x_2 + \sigma_2 u_2$, with $u_i \sim N(0, 1)$, $i = 1, 2$. Then, with $\boldsymbol{u} = (u_1, u_2)$:

$$H(\boldsymbol{x}, \boldsymbol{u}) = x_1 - x_2 + \sigma_1 u_1 - \sigma_2 u_2.$$

In this case, $\boldsymbol{a} = (\sigma_1, -\sigma_2)^t$. Thus,

$$\beta(\boldsymbol{x}) = -\frac{x_1 - x_2}{\sqrt{\sigma_1^2 + \sigma_2^2}} = \frac{x_2 - x_1}{\sqrt{\sigma_1^2 + \sigma_2^2}}.$$

The probability of failure is

$$F = \Phi\left(-\frac{x_2 - x_1}{\sqrt{\sigma_1^2 + \sigma_2^2}}\right) = \Phi\left(\frac{x_1 - x_2}{\sqrt{\sigma_1^2 + \sigma_2^2}}\right).$$

Exercises
1. Let $\boldsymbol{x} \in \mathbb{R}^2$ and $g(\boldsymbol{x}) = x_2 - 3x_1 - 3$. Let $X_i = x_i + \sigma_i u_i$ with $u_i \sim N(0, 1)$, $i = 1, 2$. Determine the probability of failure for the points (0,0), (1,0), (0,1), when $\sigma_1 = 1$, $\sigma_2 = 2$.

(continued)

2. Let $\boldsymbol{x} \in \mathbb{R}^3$ and $g(\boldsymbol{x}) = x_3 - 2x_1 - x_2 - 2$. Let $X_i = x_i + \sigma_i u_i$ with $u_i \sim N(0, 1)$, $i = 1, 2, 3$. Determine the probability of failure for the points (0,0,0), (1,0,0), (0,1,0), (0,0,1) when $\sigma_1 = 1$, $\sigma_2 = 2$, $\sigma_3 = 1$.
3. Let $\boldsymbol{x} \in \mathbb{R}^3$ and $g(\boldsymbol{x}) = x_2 - 2x_3 - x_1 - 2$. Let $X_i = x_i + \sigma_i u_i$ with $u_i \sim N(0, 1)$, $i = 1, 2, 3$. Determine the probability of failure for the points (0,0,0), (1,0,0), (0,1,0), (0,0,1) when $\sigma_1 = 1$, $\sigma_2 = 2$, $\sigma_3 = 1$.
4. Let $\boldsymbol{x} \in \mathbb{R}^4$ and $g(\boldsymbol{x}) = x_4 + x_3 - 2x_2 - x_1 - 2$. Let $X_i = x_i + \sigma_i u_i$ with $u_i \sim N(0, 1)$, $i = 1, 2, 3, 4$. Determine the probability of failure for the points (0,0,0,0), (1,0,0,0), (0,1,0,0), (0,0,1,0) and (0,0,0,1) when $\sigma_1 = 1$, $\sigma_2 = 2$, $\sigma_3 = 1$, $\sigma_4 = 2$.
5. Let $\boldsymbol{x} \in \mathbb{R}^4$ and $g(\boldsymbol{x}) = x_4 + 2x_3 - 3x_2 - 2x_1 - 3$. Let $X_i = x_i + \sigma_i u_i$ with $u_i \sim N(0, 1)$, $i = 1, 2, 3, 4$. Determine the probability of failure for the points (0,0,0,0), (1,0,0,0), (0,1,0,0), (0,0,1,0) and (0,0,0,1) when $\sigma_1 = 1$, $\sigma_2 = 2$, $\sigma_3 = 1$, $\sigma_4 = 2$.

9.6.2 The Case of a Convex Failure Region

Assume that the region $\mathcal{F}$ is convex. In this case (see, for instance, (Souza de Cursi & Sampaio, 2010):

$$\boldsymbol{u} \in \mathcal{F} \Leftrightarrow -(\boldsymbol{u}^*)^t(\boldsymbol{u} - \boldsymbol{u}^*) \leq 0. \tag{9.17}$$

Considering that $\boldsymbol{u}^* = \beta \boldsymbol{n}^*$, we have

$$\boldsymbol{u} \in \mathcal{F} \Leftrightarrow (\boldsymbol{n}^*)^t(\boldsymbol{u} - \boldsymbol{u}^*) \geq 0. \tag{9.18}$$

Let Π^* be the hyperplane having as equation

$$\Pi^*: p^*(\boldsymbol{u}) = (\boldsymbol{n}^*)^t \boldsymbol{u} - (\boldsymbol{n}^*)^t \boldsymbol{u}^* = 0. \tag{9.19}$$

Π^* splits the space in two half spaces, according to the sign of $p^*(\boldsymbol{u})$. We have

$$\boldsymbol{u} \in \mathcal{F} \Rightarrow p^*(\boldsymbol{u}) \geq 0 \Rightarrow F = P(\mathcal{F}) \leq P(p^*(\boldsymbol{u}) \geq 0).$$

$P(p^*(\boldsymbol{u}) \geq 0)$ is the probability of failure associated to the limit state curve defined by Π^*, for which $\boldsymbol{a} = \boldsymbol{n}^*$, $g(\boldsymbol{x}) = -(\boldsymbol{n}^*)^t \boldsymbol{x}$. Thus,

$$\beta(x) = \|\boldsymbol{u}^*\| = (\boldsymbol{n}^*)^t \boldsymbol{u}^*. \tag{9.20}$$

Assume yet that $\boldsymbol{U}$ is a centered Gaussian vector $N(\mathbf{0}, \boldsymbol{Id})$, i.e., all its components are independent and have the same distribution $N(0,1)$. Then, denoting by Φ the CDF of $N(0,1)$:

$$F = P(\mathcal{F}) \leq \Phi(-\beta(\boldsymbol{x})). \tag{9.21}$$

Thus, pnorm($\beta(x)$, lower.tail = FALSE) or pnorm($-\beta(x)$) can be used to furnish an upper bound to the probability of failure.

Example 9.15

Let us consider the situation where $\boldsymbol{x} = (x_1, x_2)$ and the failure occurs when $x_1 x_2 > 1$.

$$g(\boldsymbol{x}) = x_1 x_2 - 1, \ \mathcal{F} = \{\boldsymbol{x}: x_1, x_2 \geq 0, \ g(\boldsymbol{x}) > 0\}.$$

In this situation, $\mathcal{F}$ is convex. Let $X_1 = x_1 + \sigma_1 u_1$, $X_2 = x_2 + \sigma_2 u_2$, with $u_i \sim N(0,1)$, $i = 1, 2$. Then, with $\boldsymbol{u} = (u_1, u_2)$:

$$H(\boldsymbol{x}, \boldsymbol{u}) = x_1 x_2 - 1 + \sigma_1 \sigma_2 u_1 u_2 + \sigma_1 u_1 x_2 + \sigma_2 u_2 x_1.$$

Consider the point $\boldsymbol{x} = (0,0)^t$. In this case, $H(\boldsymbol{x}, \boldsymbol{u}) = -1 + \sigma_1 \sigma_2 u_1 u_2$. Then

$$u^* = \left(\frac{a}{\sigma_1}, \frac{a}{\sigma_2}\right), \ a = \frac{1}{(\sigma_1^2 + \sigma_2^2)^{\frac{1}{4}}}$$

and

$$\beta = \sqrt{\frac{1}{\sigma_1^2} + \frac{1}{\sigma_2^2}}.$$

Assume that $\sigma_1 = \sigma_2 = \sigma$. Then $\beta = \sqrt{2}/\sigma$ and we have

$$F \leq \Phi\left(-\frac{\sqrt{2}}{\sigma}\right).$$

Let us consider $\sigma = 1, 1/2, 1/3, \ldots, 1/12$. We have

```
> sig= 1/(1:12)
> pnorm(-sqrt(2)/sig)
 [1] 7.864960e-02 2.338867e-03 1.104525e-05 7.708629e-09
 [5] 7.687299e-13 1.075987e-17 2.091913e-23 5.612149e-30
 [9] 2.068516e-37 1.044244e-45 7.204331e-55 6.781306e-65
>
```

(continued)

Example 9.15 (continued)

The exact value of the probability of failure can be obtained by integration. Indeed, let us denote by Φ_σ the CDF of x_i, $i = 1, 2$: $\Phi_\sigma(s) = \Phi(s/\sigma)$, where Φ is the CDF of $N(0, 1)$. Let $\phi_\sigma = \Phi'_\sigma$ be the associated PDF: $\phi_\sigma(s) = \phi(s/\sigma)/\sigma$, where $\phi = \Phi'$ is the PDF of $N(0, 1)$,

$$\phi(s) = \frac{1}{\sigma\sqrt{2\pi}} \exp\left(-\frac{1}{2}\left(\frac{s}{\sigma}\right)^2\right).$$

Then

$$F = \int_0^{+\infty} dx_1 \phi_\sigma(x_1) \int_{\frac{1}{x_1}}^{+\infty} \phi_\sigma(x_2) dx_2 = \int_0^{+\infty} \left(1 - \Phi_\sigma\left(\frac{1}{x_1}\right)\right) \phi_\sigma(x_1)\, dx_1.$$

This integral can be evaluated numerically. For instance, let us use R and its command `integral` :

```
f = function(x,sig){
  aux1 = dnorm(x,0,sig)
  aux2 = pnorm(1/x,0,sig,lower.tail=FALSE)
  aux = aux1*aux2
  return(aux)
}

resu = numeric()
for (s in (1:12) ){
  sig = 1/s
  ff = function(x){f(x,sig)}
  fff = Vectorize(ff)
  resu[[s]] = integrate(fff,0,Inf)$value
}
```

We obtain

```
> resu
 [1] 5.224842e-02 1.614906e-03 7.716444e-06 5.411758e-09
 [5] 5.410518e-13 7.585222e-18 1.474067e-23 3.951324e-30
 [9] 1.458919e-37 7.400951e-46 5.136661e-55 4.856455e-65
> |
```

The results present a good accordance with the upper bound.

Exercises

1. Let $x \in \mathbb{R}^2$ and $g(x) = x_2 - x_1^2 - x_1 - 3$. Let $X_i = x_i + \sigma_i u_i$ with $u_i \sim N(0, 1)$, $i = 1, 2$. Determine an upper bound for the probability of failure for the points (0,0), (1,0), (0,1), when $\sigma_1 = 1, \sigma_2 = 2$.
2. Let $x \in \mathbb{R}^2$ and $g(x) = x_2 - x_1^3 - x_1^2 - 4$. Let $X_i = x_i + \sigma_i u_i$ with $u_i \sim N(0, 1)$, $i = 1, 2$. Determine an upper bound for the probability of failure for the points (0,0), (1,0), (0,1), when $\sigma_1 = 1, \sigma_2 = 2$.
3. Let $x \in \mathbb{R}^2$ and $g(x) = x_1 x_2 - 2$. Let $X_i = x_i + \sigma_i u_i$ with $u_i \sim N(0, 1)$, $i = 1$, 2. Determine an upper bound for the probability of failure for the points (0,0), (1,0), (0,1), when $\sigma_1 = 1, \sigma_2 = 2$.
4. Let $x \in \mathbb{R}^4$ and $g(x) = x_1 x_2 x_3 x_4 - 2$. Let $X_i = x_i + \sigma_i u_i$ with $u_i \sim N(0, 1)$, $i = 1, 2, 3, 4$. Determine an upper bound for the probability of failure for the points (0,0,0,0), (1,0,0,0), (0,1,0,0), (0,0,1,0) and (0,0,0,1) when $\sigma_1 = 1$, $\sigma_2 = 2, \sigma_3 = 1, \sigma_4 = 2$.
5. Let $x \in \mathbb{R}^4$ and $g(x) = x_4 - x_1 x_2 x_3 - 3$. Let $X_i = x_i + \sigma_i u_i$ with $u_i \sim N(0, 1)$, $i = 1, 2, 3, 4$. Determine an upper bound for the probability of failure for the points (0,0,0,0), (1,0,0,0), (0,1,0,0), (0,0,1,0) and (0,0,0,1) when $\sigma_1 = 1$, $\sigma_2 = 2, \sigma_3 = 1, \sigma_4 = 2$.

9.6.3 General Failure Regions

Equations (9.16) and (9.21) furnish the most popular estimations of the probability failure using the reliability index. They may be used in the situations where the failure region $\mathcal{F}$ is a subset of the non-negative half space Π_+^* defined by the hyperplane Π^* given by (9.19), which corresponds to the tangent hyperplane to the failure region at the point $\boldsymbol{u}^*$. Notice that condition (9.18) is sufficient to guarantee that $\mathcal{F} \subset \mathcal{F} \subset \Pi_+^*$.

If $\mathcal{F} \not\subset \Pi_+^*$ (for instance, in the situation of Fig. 9.22), we need to use a different approach. Recall that we continue under the assumption that $\boldsymbol{U}$ is a centered Gaussian vector of covariance $\boldsymbol{Id}$ (identity matrix), i.e., all its components are $N(0, 1)$ independent variables – otherwise, transformations must be used (see Sect. 9.7). Thus, the probability density of $\boldsymbol{U}$ is

$$\phi_U(\boldsymbol{u}) = \prod_{i=1}^{n} \phi(u_i) = \frac{1}{\sqrt{2^n \pi^n}} \exp\left(-\frac{1}{2}\|\boldsymbol{u}\|^2\right) \tag{9.22}$$

Therefore, the probability density depends only on the distance to the origin $r = \|\boldsymbol{u}\|$ and decreases with r. We will use this property to establish estimates of $F = P(\mathcal{F})$.

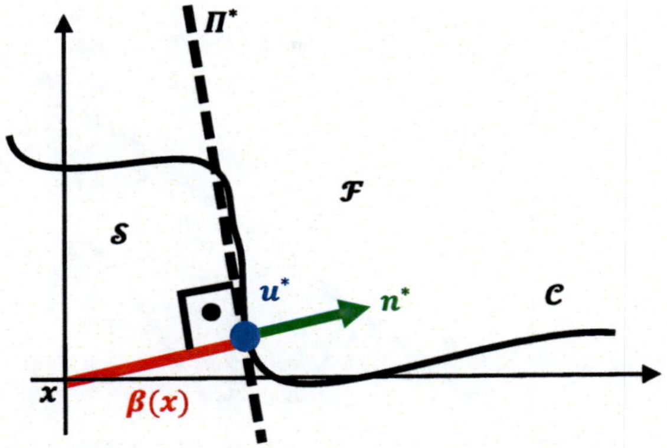

Fig. 9.22 If $\mathcal{F} \not\subset \Pi_+^*$, then a different approach must be used

Let us consider

$$\varepsilon = \{e \in \mathbb{R}^n: \|e\| = 1\},\ \mathcal{M}(e) = \{\lambda e: \lambda \geq 0\}. \tag{9.23}$$

Then

$$0 \in \mathcal{M}(e), \forall e \in \varepsilon,\ x \neq 0 \Rightarrow x \in \mathcal{M}\left(\frac{x}{\|x\|}\right),$$

so that

$$\mathbb{R}^n \subset \bigcup_{e \in \varepsilon} \mathcal{M}(e) \subset \mathbb{R}^n \Rightarrow \mathbb{R}^n = \bigcup_{e \in \varepsilon} \mathcal{M}(e) \tag{9.24}$$

Let us consider (see Fig. 9.23)

$$\mathcal{F}(e) = \mathcal{F} \cap \mathcal{M}(e),\ \Lambda(e) = \{\lambda \in \mathbb{R}: \lambda \geq 0,\ \lambda e \in \mathcal{F}\}.$$

We have

$$\mathcal{F}(e) = \{\lambda e: \lambda \geq 0,\ H(x,\ \lambda e) \geq 0\},\ \Lambda(e) = \{\lambda: \lambda \geq 0,\ H(x,\ \lambda e) \geq 0\}$$

In addition,

$$\mathcal{F} = \bigcup_{e \in \varepsilon} \mathcal{F}(e).$$

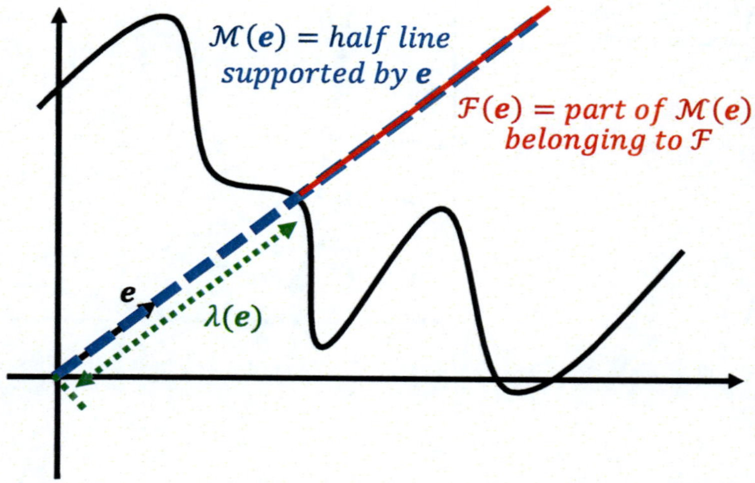

Fig. 9.23 $\mathcal{M}(e)$ is the half line oriented by the vector e. $\mathcal{F}(e)$ is the part of the half line belonging to the failure region $\mathcal{F}$. $\lambda(e)$ is the distance between the first point of $\mathcal{F}(e)$ and the origin of the centered space

Let

$$\lambda(e) = \begin{cases} -\infty, \ if \ \Lambda(e) = \varnothing, \\ \inf \Lambda(e), \ if \ \Lambda(e) \neq \varnothing. \end{cases}$$

Notice that the inf exists for $\Lambda(e) \neq \varnothing$, since $\lambda \geq 0, \forall \lambda \in \Lambda(e)$. If H is a continuous function, then $\lambda(e)$ is the smallest non-negative solution of the equation $H(x, \lambda e) = 0$. Thus,

$$\lambda \geq 0 \ and \ H(x, \lambda e) \geq 0 \Rightarrow \lambda \geq \lambda(e).$$

Let $\mathcal{L}(e) = \{\lambda e : \lambda \geq \lambda(e)\}$. We have

$$\forall e \in \varepsilon: \mathcal{F}(e) \subset \mathcal{L}(e) \Rightarrow \mathcal{F} \subset \bigcup_{e \in \varepsilon} \mathcal{L}(e)$$

and

$$F = P(\mathcal{F}) \leq P\left(\bigcup_{e \in \varepsilon} \mathcal{L}(e) \right). \quad (9.25)$$

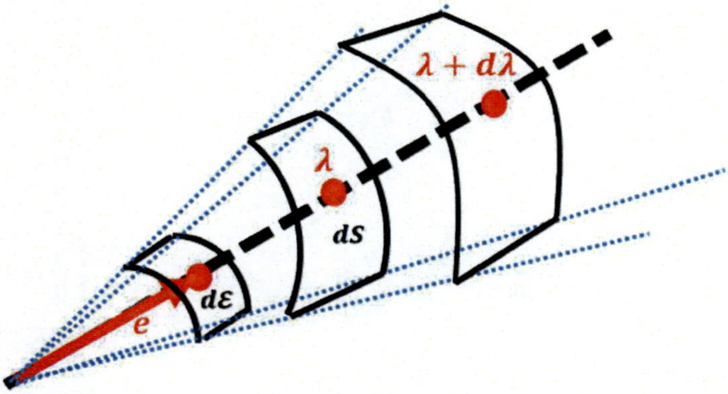

Fig. 9.24 Elements of volume used in the integration

This last value can be determined by integration: we have

$$P\left(\bigcup_{e\in\varepsilon}\mathcal{L}(e)\right) = \int_{u\in\bigcup_{e\in\varepsilon}\mathcal{L}(e)} \phi_U(u)du$$

The element of volume is

$$du = dSd\lambda,$$

where dS is the element of surface for a sphere of radius λ (see Fig. 9.24). The surface of a hyperball of radius r is

$$surf(r) = meas(\|u\| = r) = \frac{2\pi^{\frac{n}{2}}r^{n-1}}{\Gamma\left(\frac{n}{2}\right)}$$

In addition,

$$\frac{dS}{surf(\lambda)} = \frac{d\varepsilon}{surf(1)} \Rightarrow dS = \frac{surf(\lambda)}{surf(1)}d\varepsilon$$

and we have

$$P\left(\bigcup_{e\in\varepsilon}\mathcal{L}(e)\right) = \frac{1}{surf(1)}\int_{e\in\varepsilon} d\varepsilon \int_{\lambda(e)}^{+\infty} surf(\lambda)\phi_U(u)du.$$

Thus, using that

$$\phi_U(\lambda e) = \frac{1}{\sqrt{2^n \pi^n}} \exp\left(-\frac{\lambda^2}{2}\right),$$

we have

$$P\left(\bigcup_{e \in \varepsilon} \mathcal{L}(e)\right) = \frac{1}{surf(1)} \frac{1}{2\pi^{\frac{n}{2}}} \int_{e \in \varepsilon} d\varepsilon \int_{\lambda(e)}^{+\infty} surf(\lambda) \exp\left(-\frac{\lambda^2}{2}\right) d\lambda.$$

Noticing that

$$\int_{\lambda(e)}^{+\infty} surf(\lambda) \exp\left(-\frac{\lambda^2}{2}\right) du = \frac{2\pi^{\frac{n}{2}}}{\Gamma\left(\frac{n}{2}\right)} \int_{\lambda(e)}^{+\infty} \lambda^{n-1} \exp\left(-\frac{\lambda^2}{2}\right) d\lambda,$$

We have

$$\int_{\lambda(e)}^{+\infty} surf(\lambda) \exp\left(-\frac{\lambda^2}{2}\right) du = \frac{2^{\frac{n}{2}}\pi^{\frac{n}{2}}}{\Gamma\left(\frac{n}{2}\right)} \int_{\frac{\lambda(e)^2}{2}}^{+\infty} t^{\frac{n}{2}-1} e^{-t} dt$$

and

$$\int_{\lambda(e)}^{+\infty} surf(\lambda) \exp\left(-\frac{\lambda^2}{2}\right) du = \frac{2^{\frac{n}{2}}\pi^{\frac{n}{2}}}{\Gamma\left(\frac{n}{2}\right)} \Gamma\left(\frac{n}{2}, \frac{\lambda(e)^2}{2}\right),$$

where

$$\Gamma(\mathrm{s}, \mathrm{b}) = \int_{\mathrm{b}}^{+\infty} t^{\mathrm{s}-1} e^{-t} dt$$

is the incomplete Gamma function. Thus,

$$P\left(\bigcup_{e \in \varepsilon} \mathcal{L}(e)\right) = \frac{1}{surf(1)} \int_{\varepsilon} \frac{\Gamma\left(\frac{n}{2}, \frac{\lambda(e)^2}{2}\right)}{\Gamma\left(\frac{n}{2}\right)} d\varepsilon = \mathrm{E}\left(\frac{\Gamma\left(\frac{n}{2}, \frac{\lambda(e)^2}{2}\right)}{\Gamma\left(\frac{n}{2}\right)}\right).$$

and we have

$$F \leq \mathrm{E}\left(\frac{\Gamma\left(\frac{n}{2}, \frac{\lambda(\boldsymbol{e})^2}{2}\right)}{\Gamma\left(\frac{n}{2}\right)}\right). \tag{9.26}$$

Since (see, for instance, (Souza de Cursi, 1992))

$$P\left(\chi^2(n) \geq \lambda(\boldsymbol{e})^2\right) = \frac{\Gamma\left(\frac{n}{2}, \frac{\lambda(\boldsymbol{e})^2}{2}\right)}{\Gamma\left(\frac{n}{2}\right)},$$

we have

$$F \leq E\left(P\left(\chi^2(n) \geq \lambda(\boldsymbol{e})^2\right)\right). \tag{9.27}$$

These values may be estimated by integration or Monte Carlo methods. For instance, let us describe ε using hyperspherical coordinates $\boldsymbol{\theta} = (\theta_1, \ldots, \theta_{n-1})$: let $e(\boldsymbol{\theta}) = (e_1(\boldsymbol{\theta}), \ldots, e_n(\boldsymbol{\theta}))$ be given by

$$e_1(\boldsymbol{\theta}) = \prod_{j=1}^{n-1} \sin(\theta_j),$$

$$e_i(\boldsymbol{\theta}) = \cos(\theta_{n-i+1}) \prod_{j=1}^{i-1} \sin(\theta_j), 2 \leq i \leq n-1,$$

$$e_n(\boldsymbol{\theta}) = \cos(\theta_1).$$

Then,

$$\varepsilon = \left\{ e(\boldsymbol{\theta}) : \boldsymbol{\theta} \in (0, \pi)^{n-2} \times (0, 2\pi) \right\}.$$

The Jacobian associated to these coordinates is

$$J(\boldsymbol{\theta}) = (-1)^{[n/2]} \prod_{j=1}^{n-2} \sin^{\mathrm{j}}(\theta_{n-j-1}) = (-1)^{[n/2]} \prod_{j=1}^{n-2} \sin^{\mathrm{n-j-1}}(\theta_j).$$

Thus,

$$\int_{\varepsilon} P\left(\chi^2(n) \geq \lambda(\boldsymbol{e})^2\right) d\varepsilon = \int_{\boldsymbol{\theta} \in (0, \pi)^{n-2} \times (0, 2\pi)} \frac{\Gamma\left(\frac{n}{2}, \frac{\lambda(e(\boldsymbol{\theta}))^2}{2}\right)}{\Gamma\left(\frac{n}{2}\right)} J(\boldsymbol{\theta}) d\boldsymbol{\theta}. \tag{9.28}$$

The evaluation of the estimate in Eq. (9.28) can be performed by considering a discretization of ε. For instance, in dimension 2, we can consider a given number of steps nsp and a step $\Delta\theta = 2\pi/nsp$. Then, we generate the angles $\theta_i = i\Delta\theta$, $i = 0, \ldots, nsp + 1$ and $\boldsymbol{e}(\theta_i) = (\cos(\theta_i), \sin(\theta_i))$. We obtain the discretized values $F_i = P\left(\chi^2(n) \geq \lambda_i^2\right)$, $\lambda_i = \lambda(\boldsymbol{e}(\theta_i))$ which can be used for a numerical integration. For instance, the trapezoidal rule reads as

$$E\left(P\left(\chi^2(n) \geq \lambda(\boldsymbol{e})^2\right)\right) \approx \frac{\Delta\theta}{2\pi}\left(\frac{F_0 + F_{nsp}}{2} + \sum_{j=1}^{nsp-1} F_j\right). \tag{9.29}$$

Example 9.16

Let us consider the situation where $\boldsymbol{x} = (0, 0)$ and the failure occurs when $g(\boldsymbol{x}) = x_2 - x_1^2 - x_1 - 3 > 0$. We have $g(\boldsymbol{x} + \lambda\boldsymbol{e}(\theta_i)) = a\lambda^2 + b\lambda + c$, with $a = -\cos^2(\theta_i)$, $b = (\sin(\theta_i) - \cos(\theta_i))$, $c = -3$. λ_i is a positive root the equation $a\lambda^2 + b\lambda + c = 0$, if such a root exists. If not any positive solution exists, then $F_i = 0$, what corresponds to $\lambda(\theta_i) = +\infty$.

- Let $\theta_i = \pi/2$: $a = 0$, $b = 1$ and $\lambda_i = 3$.
- Let $\theta_i = 3\pi/2$: $a = 0$, $b = -1$ and $\lambda_i = -3$. Thus, there is no positive solution. In this case, we take $F_i = 0$, what corresponds to $\lambda(\theta_i) = +\infty$.
- Let $\theta_i \neq \pi/2$ and $\theta_i \neq 3\pi/2$: the equation has degree 2. Let $\Delta = b^2 - 4ac$. Real roots exist only for $\Delta \geq 0$. In this case, the roots are

$$\lambda_\pm = \frac{-b \pm \sqrt{\Delta}}{2a}.$$

λ_i corresponds to the smaller positive root. Again, if no positive root exists, we take $F_i = 0$.

We use Eq. (9.29) with different values of nsp to evaluate F (this probability was evaluated in Example 9.7 by a simple Monte Carlo method). The results for $nsp = 10^1, 10^2, \ldots, 10^{10}$ are the following:

```
resu
[1] 0.0007762962 0.0107546957 0.0019562500 0.0010762316 0.0009882312
[6] 0.0009794312 0.0009787468 0.0009787566 0.0009787654 0.0009787647
```

The results are in accordance with the Monte Carlo simulation.

Example 9.17
Let us consider the situation where $\boldsymbol{x} = (x_1, x_2)$ and the failure occurs when $h(\boldsymbol{x}, \boldsymbol{u}) = (x_1 + u_1)^2 + (x_2 + u_2)^2 > \alpha^2 (\alpha > 0)$. Let $\boldsymbol{x} = (0, 0)^t$: then, $\boldsymbol{u}^*$ is not unique since any point $\boldsymbol{u}^*$ such that $\|\boldsymbol{u}^*\| = \alpha$ is a design point. Then, $\beta(\boldsymbol{x}) = \alpha$ and the probability of failure is

$$F = \frac{1}{2\pi} \int_{\|\boldsymbol{u}\| \geq \alpha} \exp\left(-\frac{1}{2}\|\boldsymbol{u}\|^2\right) d\boldsymbol{u},$$

id est,

$$F = \frac{1}{2\pi} \int_{\alpha}^{+\infty} 2\pi r \exp\left(-\frac{1}{2} r^2\right) dr = \exp\left(-\frac{\alpha^2}{2}\right).$$

In this case, $h(\boldsymbol{x}, \lambda \boldsymbol{e}) = \lambda^2$, so that $\lambda(\boldsymbol{e}(\theta)) = \alpha, \forall\, \theta$. Thus, from Eq. (9.27):

$$F \leq E\left(P\left(\chi^2(2) \geq \alpha^2\right)\right) = P\left(\chi^2(2) \geq \alpha^2\right)$$

We can compare these values using the program below:

```
ndim = 2
resu1 = numeric()
resu2 = numeric()

for (i in 1:10){
  resu1[[i]] = pchisq(i^2,ndim,lower.tail = FALSE)
  resu2[[i]] = exp(-i^2/2)
}
```

We obtain:

```
> resu1
 [1] 6.065307e-01 1.353353e-01 1.110900e-02 3.354626e-04 3.726653e-06
 [6] 1.522998e-08 2.289735e-11 1.266417e-14 2.576757e-18 1.928750e-22
> resu2
 [1] 6.065307e-01 1.353353e-01 1.110900e-02 3.354626e-04 3.726653e-06
 [6] 1.522998e-08 2.289735e-11 1.266417e-14 2.576757e-18 1.928750e-22
> |
```

Example 9.18

Let us consider the situation where $\boldsymbol{x} = (x_1, x_2)$ and the failure occurs when $x_1 x_2 > 1$, x_1, $x_2 > 0$. Let $X_1 = x_1 + \sigma_1 u_1$, $X_2 = x_2 + \sigma_2 u_2$, with $u_i \sim N(0, 1)$, $i = 1, 2$. Let us consider $\boldsymbol{x} = (0, 0)^t$. In this case, $H(\boldsymbol{x}, \boldsymbol{u}) = -1 + \sigma_1 \sigma_2 u_1 u_2$ (see Example 9.15). Then, $h(\boldsymbol{x}, \lambda \boldsymbol{e}(\boldsymbol{\theta})) = \lambda^2 \sigma_1 \sigma_2 \sin\theta \cos\theta - 1$, so that

$$\lambda(\boldsymbol{e}(\boldsymbol{\theta})) = \begin{cases} \sqrt{\dfrac{2}{\sigma_1 \sigma_2 \sin(2\theta)}}, & \text{if } \theta \in \left(0,\ \dfrac{\pi}{2}\right) \\ +\infty, & \text{otherwise} \end{cases}.$$

We can evaluate F as follows:

```
ndim = 2
resu = numeric()
nsp = 10^5
teta = seq(0,2*pi,length.out = (nsp+1))
hteta = 2*pi/nsp
for (n in (1:12)){
  s = 0
  sig = 1/n
  for (j in 2:nsp){
    ang = teta[[j]]
    if (ang < 0.5*pi){
      lbd2 = 2/(sig*sig*sin(2*ang))
      fs = pchisq(lbd2,ndim,lower.tail=FALSE)
      s = s + fs
    }
  }
  resu[[n]] = hteta*s/(2*pi)
}
```

We obtain:

```
resu
[1] 5.224842e-02 1.614906e-03 7.716464e-06 5.411826e-09 5.410101e-13 7.583027e-18
[7] 1.475546e-23 3.960819e-30 1.460447e-37 7.374829e-46 5.089035e-55 4.790989e-65
```

Now assume that x_1, x_2 can take negative signs. In this case,

$$\lambda(\boldsymbol{e}(\boldsymbol{\theta})) = \begin{cases} \sqrt{\dfrac{2}{\sigma_1 \sigma_2 \sin(2\theta)}}, & \text{if } \theta \in \cup\left(0,\ \dfrac{\pi}{2}\right) \cup \left(\pi,\ \dfrac{3\pi}{2}\right) \\ +\infty, & \text{otherwise} \end{cases}.$$

In this case,

```
> resu2
 [1] 1.044968e-01 3.229813e-03 1.543293e-05 1.082365e-08 1.082020e-12 1.516605e-17
 [7] 2.951093e-23 7.921639e-30 2.920893e-37 1.474966e-45 1.017807e-54 9.581979e-65
```

As expected, the probability of failure doubles.

Exercises

1. Let $\boldsymbol{x} \in \mathbb{R}^2$ and $g(\boldsymbol{x}) = x_2 - x_1^2 - x_1 - 3$. Let $X_i = x_i + u_i$ with $u_i \sim N(0, 1)$, $i = 1, 2$. Determine an upper bound for the probability of failure at the points (1,0), (0,1), (1,1).
2. Let $\boldsymbol{x} \in \mathbb{R}^2$ and $g(\boldsymbol{x}) = x_2 - x_1^3 - x_1^2 - 4$. Let $X_i = x_i + u_i$ with $u_i \sim N(0, 1)$, $i = 1, 2$. Determine an upper bound for the probability of failure at the points (0,0), (1,0), (0,1).
3. Let $\boldsymbol{x} \in \mathbb{R}^2$ and $g(\boldsymbol{x}) = x_1 x_2 - 2$. Let $X_i = x_i + u_i$ with $u_i \sim N(0, 1)$, $i = 1$, 2. Determine an upper bound for the probability of failure at the points (0,0), (1,0), (0,1).

9.7 The Transformations of Rosenblatt and Nataf

The preceding analysis was limited to Gaussian variables $\boldsymbol{U}$. When $\boldsymbol{U}$ is not gaussian, it is necessary to use more complex transformations, such as the ***Rosenblatt Transformation*** (Rosenblatt, 1952), which is defined as follows: let F_1 be the marginal CDF of U_1 and, for $i > 1$, F_i be the CDF of U_i conditional to $(U_1, \ldots, U_{i-1})$:

$$F_i(u_i | U_1 = u_1, \ \ldots, \ U_{i-1} = u_{i-1}) = P(U_i < u_i | U_1 = u_1, \ \ldots, \ U_{i-1} = u_{i-1})$$

Let us consider the CDF Φ of a Gaussian variable $N(0, 1)$ and introduce new variables $\boldsymbol{W} = (W_1, \ldots, W_n)$ such that $P(W_i < w_i) = \Phi(w_i)$ with

$$\begin{gathered} w_1 = \Phi^{-1}(F_1(u_1)), \\ w_2 = \Phi^{-1}(F_2(u_2 | U_1 = u_1)), \\ w_3 = \Phi^{-1}(F_3(u_3 | U_1 = u_1, \ U_2 = u_2)), \\ \vdots \\ w_n = \Phi^{-1}(F_n(u_n | U_1 = u_1, \ U_2 = u_2, \ \ldots, \ U_{n-1} = u_{n-1})) \end{gathered}$$

Let Φ_i be the CDF of W_i conditional to $(W_1, \ldots, W_{i-1})$: we have

$$\Phi_i(w_i | W_1 = w_1, \ \ldots, \ W_{i-1} = w_{i-1}) = F_i(u_i | U_1 = u_1, \ \ldots, \ U_{i-1} = u_{i-1}) = \Phi(w_i),$$

so that $\boldsymbol{W}$ is a gaussian vector of independent variables. Notice that Rosenblatt transformation may generate non-gaussian variables if $\boldsymbol{U}$ contains dependent variables (see example below). Under R, packages `rvinecopulib` and `HMMcopula` propose function to make Rosenblatt's transformations.

Example 9.19
Let us consider the situation where $\boldsymbol{U} = (U_1, f(U_1))$, where f is a regular bijective function and U_1 is uniformly distributed on $(-1, 1)$. Here,

$$F_1(u_1) = \frac{1+u_1}{2} \Rightarrow w_1 = \Phi^{-1}\left(\frac{1+u_1}{2}\right).$$

In addition,

$$F_2(u_2|U_1 = u_1) = 0, \textit{ if } u_2 \leq f(U_1); F_2(u_2|U_1 = u_1) = 1, \textit{ otherwise}.$$

Thus,

$$w_2 = \Phi^{-1}(0) = -\infty, \textit{ if } u_2 \leq f(U_1); w_2 = \Phi^{-1}(1) = +\infty, \textit{ otherwise}.$$

In this situation, the variable U_2 can be eliminated and the description of the uncertainties can be made by using only U_1 and w_1: we have

$$\boldsymbol{U} = (2\Phi(w_1) - 1, f(2\Phi(w_1) - 1)).$$

The use of the Rosenblatt transformation requests the complete knowledge of the distribution of the vector $\boldsymbol{U}$, so that its use may result difficulty in practice – in general, only the marginal laws are known. An alternative is to use an approximation: the *Nataf transformation* (Nataf, 1962), which uses only the marginal distributions of the variables. The Nataf transformation generates a vector $\boldsymbol{W}$ of independent Gaussian variables in two steps: first, we generate a vector $\boldsymbol{V} = (v_1, \ldots, v_n)$ by using the marginal laws:

$$v_i = \Phi^{-1}(F_i(u_i)), F_i(u_i) = P(U_i < u_i).$$

In the second step, we look for a vector $\boldsymbol{W} = (W_1, \ldots, W_m)$, $m \leq n$ formed by independent Gaussian variables of same distribution $N(0, 1)$, such that $\boldsymbol{V} = \boldsymbol{MW}$, with $\boldsymbol{M}$ a $n \times m$ matrix, which may be determined by the methods presented in Chapter 2, Sect. 2.13 and Sect. 9.6 above. For instance, by solving the equation

$$cov\,(\boldsymbol{V}, \boldsymbol{V}) = \boldsymbol{MM}^t.$$

As previously indicated (Sect. 9.6), $\boldsymbol{M}$ can be determined by Cholesky's decomposition. In practice, it may be useful to avoid the evaluation of $cov\,(\boldsymbol{V}, \boldsymbol{V})$, using an approximation: for instance, $cov(\boldsymbol{V}, \boldsymbol{V}) \approx cov\,(\boldsymbol{U}, \boldsymbol{U})$. As an alternative, the procedure of factorization can be applied to the correlation matrix $\boldsymbol{R_V} = \rho(\boldsymbol{V}, \boldsymbol{V})$, which may be approximated into an analogous way: $\boldsymbol{R_V} \approx \boldsymbol{R_U} = \rho(\boldsymbol{U}, \boldsymbol{U})$.

Under R, the package `mistral` proposes functions to make Nataf's transformations.

Example 9.20
Let us consider again the situation where $\boldsymbol{U} = (U_1, f(U_1))$, where f is a regular bijective function and U_1 is uniformly distributed on $(-1, 1)$. Then,

$$F_1(u_1) = \frac{1+u_1}{2},\ F_2(u_2) = \frac{1+f^{-1}(u_2)}{2}.$$

Thus,

$$v_1 = \Phi^{-1}\left(\frac{1+u_1}{2}\right),\ v_2 = \Phi^{-1}\left(\frac{1+f^{-1}(u_2)}{2}\right).$$

Let $r = \rho(U_1, U_2)$: the correlation matrix is

$$\boldsymbol{R_U} = \begin{pmatrix} 1 & r \\ r & 1 \end{pmatrix}$$

and

$$\boldsymbol{M} = \begin{pmatrix} 1 & 0 \\ r & \sqrt{1-r^2} \end{pmatrix} \Rightarrow \boldsymbol{MM}^t = \boldsymbol{R_U}.$$

Thus, using the approximation $\boldsymbol{R_V} \approx \boldsymbol{R_U}$, we have

$$u_1 = 2\Phi(w_1) - 1,\ u_2 = f\left(2\Phi\left(rw_1 + w_2\sqrt{1-r^2}\right) - 1\right).$$

One of the effects of the transformations of Rosenblatt and Nataf is the modification of the geometry of the failure region and of the limit state curve. For instance, let us consider the situation where $n = 2$ and the boundary curve is linear: $g(\boldsymbol{x}) = x_1 - x_2$. Consider the point $\boldsymbol{x} = (3, 4)$ and $X_i = x_i + u_i$, where u_1, u_2 are independent variables uniformly distributed on $(-1, 1)$. The associated Rosenblatt transformation modifies the geometry as shown in Fig. 9.25. Notice that the original failure region is convex.

In this same situation, let us consider the case where X_1 is log-normal distributed with mean x_1: we can use $u_1 = (\log(X_1) - \mu_1)/\sigma_1$, id est, $X_1 = \exp(\mu_1 + \sigma_1 u_1)$. Since $u_1 = 0$ for $X_1 = x_1$(centered space), we have $\mu_1 = \log(x_1)$. The Rosenblatt transformation is illustrated in Fig. 9.26, where we consider $\sigma_1 = 1$ – the other data remain identical. In this case, the failure region becomes non-convex.

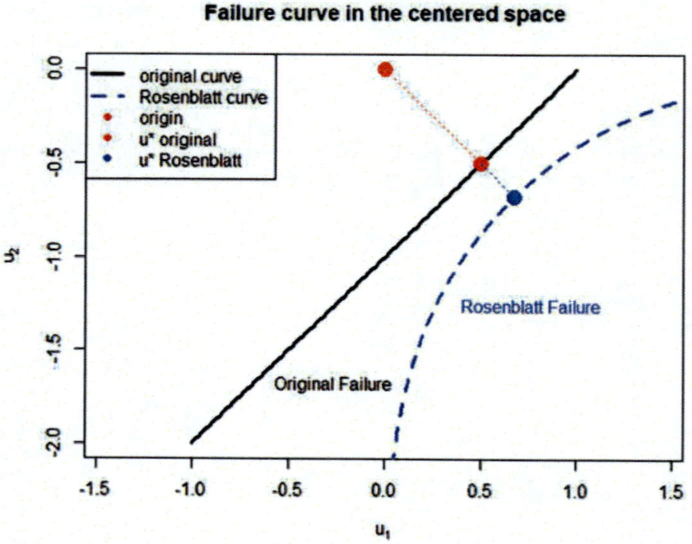

Fig. 9.25 Example of modification of the geometry by the Rosenblatt Transformation. The failure region is convex

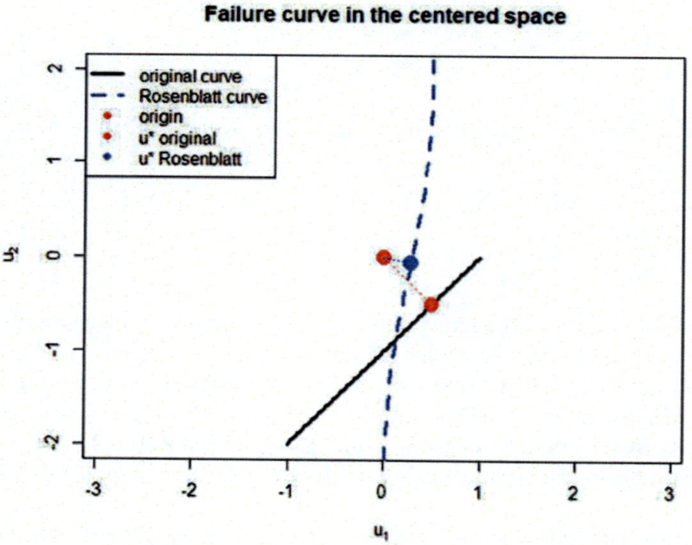

Fig. 9.26 Example of modification of the geometry by the Rosenblatt Transformation. The failure region is convex

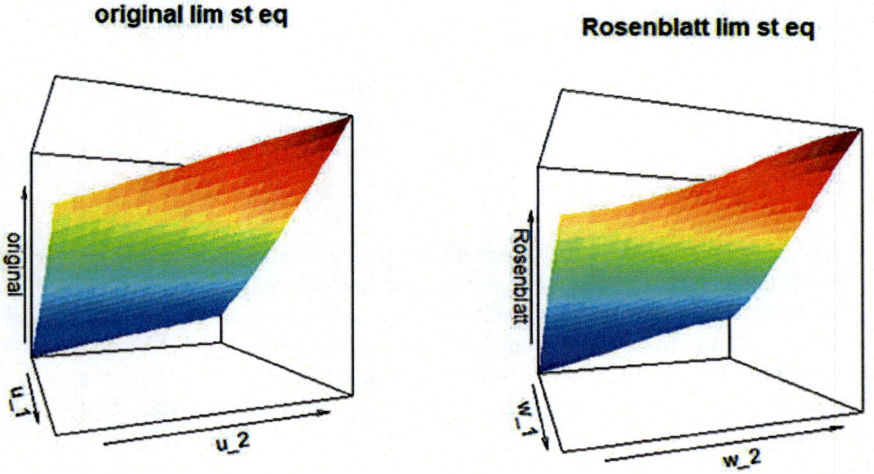

Fig. 9.27 A third example of modification of the geometry by the Rosenblatt Transformation. The failure region becomes non-convex

As a last example, let us consider the situation where X_1, X_2, X_3 are uniformly distributed and $g(\boldsymbol{x}) = x_3 - x_1^2 x_2$, with $X_1 \sim U(3, 5)$, $X_2 \sim U(250{,}350)$, $X_3 \sim U(2E3, 9E3)$. The Rosenblatt transformation is illustrated in Fig. 9.27. In this case, the failure region becomes non-convex.

Exercises

1. Let $\boldsymbol{x} \in \mathbb{R}^2$ and $g(\boldsymbol{x}) = x_2 - x_1^2$. Let $X_i = x_i + u_i$ with u_i lognormally distributed with $\sigma_i = 1$, $i = 1, 2$. Plot the Rosenblatt transformed limit curve.
2. Let $\boldsymbol{x} \in \mathbb{R}^2$ and $g(\boldsymbol{x}) = x_2 - x_1^3 - x_1^2 - 4$. Let $X_i = x_i + u_i$ with u_i lognormally distributed with $\sigma_i = 1$, $i = 1, 2$. Plot the Rosenblatt transformed limit curve
3. Let $\boldsymbol{x} \in \mathbb{R}^2$ and $g(\boldsymbol{x}) = x_1 x_2 - 2$. Let $X_i = x_i + u_i$ with u_i lognormally distributed with $\sigma_i = 1$, $i = 1, 2$. Plot the Rosenblatt transformed limit curve.
4. Let $\boldsymbol{x} \in \mathbb{R}^2$ and $g(\boldsymbol{x}) = x_2 - x_1^2$. Let $X_i = x_i + u_i$ with u_i uniformly distributed on (–1,1) ,$i = 1, 2$. Plot the Rosenblatt transformed limit curve.
5. Let $\boldsymbol{x} \in \mathbb{R}^2$ and $g(\boldsymbol{x}) = x_2 - x_1^3 - x_1^2 - 4$. Let $X_i = x_i + u_i$ with u_i uniformly distributed on (–1,1) ,$i = 1, 2$. Plot the Rosenblatt transformed limit curve.
6. Let $\boldsymbol{x} \in \mathbb{R}^2$ and $g(\boldsymbol{x}) = x_1 x_2 - 2$. Let $X_i = x_i + u_i$ with u_i uniformly distributed on (–1,1) ,$i = 1, 2$. Plot the Rosenblatt transformed limit curve.
7. Let $\boldsymbol{x} \in \mathbb{R}^3$ and $g(\boldsymbol{x}) = x_3 - x_1^2 x_2$. Let $X_i = x_i + u_i$ with u_i lognormally distributed with $\sigma_i = 1$, $i = 1, 2, 3$. Plot the Rosenblatt transformed limit curve.

9.8 FORM and SORM

As shown in Sect. 9.5, the reliability index of Hasofer-Lind corresponds to the value of an optimization problem:

$$\beta(\boldsymbol{x}) = \|\boldsymbol{u}^*(\boldsymbol{x})\|, \ \boldsymbol{u}^*(\boldsymbol{x}) = \arg\min \left\{ \|\boldsymbol{u}\|^2 : h(\boldsymbol{x}, \ \boldsymbol{u}) \geq 0 \right\}.$$

The solution of the optimization problem above provides $\boldsymbol{u}^*$, which furnishes the value of the reliability index. We presented methods of solution in Sect. 9.5, but the reader will find in the literature about optimization many methods, algorithms, and software for its numerical solution – eventually, global optimization methods must be considered, if the safe or the failure region is non-convex. As shown in the preceding, we may also determine its value by solving a system of algebraic equations:

$$\boldsymbol{u}^* \|\nabla_{\boldsymbol{u}} h(\boldsymbol{x}, \ \boldsymbol{u}^*)\| = \beta \nabla_{\boldsymbol{u}} h(\boldsymbol{x}, \boldsymbol{u}^*), \ \ h(\boldsymbol{x}, \boldsymbol{u}^*) = 0.$$

Other than these standard methods, there are methods specific to reliability, which are examined below.

9.8.1 *First-Order Reliability Method (FORM)*

The first – and extremely popular method – is the first order reliability method (FORM), which consists in solving a sequence of problems with affine limit state curves: at each step, the limit state curve is approximated by a hyperplane and we use the expression given in Eq. (9.12) to get an approximated value of β (see Fig. 9.28). FORM can be interpreted as an adaptation of the sequential quadratic programming procedure (SQP) to reliability analysis. Assuming that all the variables are independent and $N(0, 1)$ – otherwise, transformations must be considered – the algorithm of the method reads as

Algorithm of the Method FORM
Input: the point $\boldsymbol{x}$, starting point $\boldsymbol{u}^{(0)} \in \mathcal{S}$, stopping conditions (for instance, maximal iteration number, minimal step at each iteration). Programs evaluating $h(\boldsymbol{x}, \boldsymbol{u})$ and $\nabla_{\boldsymbol{u}} h(\boldsymbol{x}, \boldsymbol{u})$, a program for the correction of a tentative design point $\boldsymbol{u}$ by determining a point $\overline{\boldsymbol{u}}(\boldsymbol{u})$ such that $|h(\boldsymbol{x}, \overline{\boldsymbol{u}}(\boldsymbol{u}))| < tol$, where tol is a predefined tolerance.

(continued)

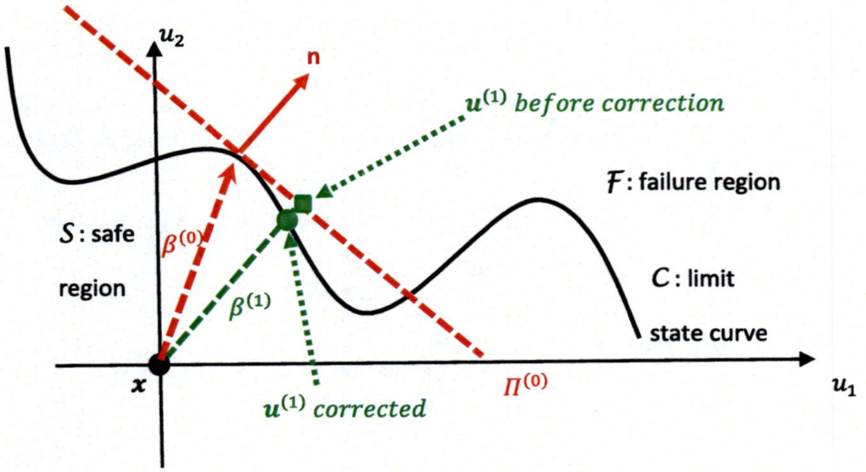

Fig. 9.28 Iterative determination of the reliability index by FORM

Output: estimations of $\boldsymbol{u}^*(\boldsymbol{x})$ and $\beta(\boldsymbol{x})$.

Initialization: If the initial point does not verify $|h(\boldsymbol{x}, \boldsymbol{u}^{(0)})| < tol$, then correct it: $\boldsymbol{u}^{(0)} \longleftarrow \overline{\boldsymbol{u}}(\boldsymbol{u}^{(0)})$. Initialize $\beta^{(0)} = \|\boldsymbol{u}^{(0)}\|$. Set the iteration number to zero : $i \longleftarrow 0$.

1. Approximate the limit state curve by a hyperplane $\Pi^{(i)}: p^{(i)}(\boldsymbol{u}) = (\boldsymbol{a}^{(i)})^t\boldsymbol{u} + b^{(i)}(\boldsymbol{x})$. For instance, use a Taylor-MacLaurin series: $\boldsymbol{a}^{(i)} = \nabla_{\boldsymbol{u}} h(\boldsymbol{x}^{(i)}, \boldsymbol{u}^{(i)})$, $b^{(i)}(\boldsymbol{x}) = h(\boldsymbol{x}, \boldsymbol{u}^{(i)}) - (\boldsymbol{a}^{(i)})^t\boldsymbol{u}^{(i)}$.
2. Set $\boldsymbol{n}^{(i)} = \boldsymbol{a}^{(i)}/\|\boldsymbol{a}^{(i)}\|$, $\beta^{(i+1)} = -b^{(i)}(\mathrm{x})/\|\boldsymbol{a}^{(i)}\|$, $\boldsymbol{u}^{(i+1)} = \beta^{(i+1)}\boldsymbol{n}^{(i)}$.
3. If $|h(\boldsymbol{x}, \boldsymbol{u}^{(i+1)})| \geq tol$ then $\boldsymbol{u}^{(i+1)} \longleftarrow \overline{\boldsymbol{u}}(\boldsymbol{u}^{(i+1)})$.
4. Set $\beta^{(i+1)} = \|\boldsymbol{u}^{(i+1)}\|$.
5. Increment the iteration number $i \longleftarrow i + 1$.
6. Test for the stopping conditions. If the iterations continue, then go to 1, else estimate $\beta(\boldsymbol{x}) \approx \beta^{(i)}$, $\boldsymbol{u}(\boldsymbol{x}) \approx \boldsymbol{u}^{(i)}$.

To make this algorithm to work correctly, it is necessary to choose the starting point in the safe region. We must also ensure that the generated sequence of points maintains $b^{(i)}(\boldsymbol{x}) < 0$ during the iterations – notice that $\boldsymbol{0} \in \mathcal{S}$, so that $\boldsymbol{u}^{(i)}$ generally satisfies the condition $(\boldsymbol{a}^{(i)})^t\boldsymbol{u}^{(i)} < 0$. When the candidate point $\boldsymbol{u}^{(i+1)}$ does not belong to the safe region $\mathcal{S}$, it is necessary to perform the correction $\overline{\boldsymbol{u}}(\boldsymbol{u}^{(i+1)})$, to bring it to the safe region (with a small enough tolerance *tol*). For this purpose, optimization methods can be used, such as, for example, those presented in (Souza de Cursi et al., 2004).

In general, FORM converges quickly. The reader will find in the literature many works on FORM, with implementations available on the net. Under R, package `mistral` proposes a pre-implemented FORM.

Below, we illustrate this method with a simple implementation as follows:

```
FORM = function(x,ut,h,gradh,nitmax,tol,omr){
  aux = h(x, ut)
  if (abs(aux) > tol) {
    uac = correction(x,ut,h)
  }else{uac = ut}
  xac = x + uac
  betta = sqrt(sum(uac^2))
  prec = 1 + tol
  nit = 0
  continue = TRUE
  while (continue) {
    a = gradh(x, uac)
    b = h(x, uac) - sum(a*uac)
    betta = -b / sqrt(sum(a^2))
    un = betta*a / sqrt(sum(a^2))
    aux = h(x, un)
    if (abs(aux) > tol) {
      un = correction(x,un,h)
    }
    xn = x + un
    betta = sqrt(sum(un^2))
    nit = nit + 1
    prec = sqrt(sum((un-uac)^2))/abs(omr)
    uac = omr*un + (1-omr)*uac
    continue = (nit <nitmax) & (prec > tol)
  }
  aux = list(betta,un,xn,nit,prec)
  names(aux) = c("beta","ustar","xstar"
                ,"nit","prec")
  return(aux)
}
```

Notice that this is not an optimal implementation and that you can find on Internet sophisticated software implementing FORM, but we focus on other aspects, more pedagogical and coherent with the spirit of this book.

Example 9.21
Let us consider the case where $x = (0,0)$ and $g(x) = x_2 - x_1^2 - x_1 - 3$. This problem was previously solved (Example 9.30). We run the program as follows:

```
g = function(x){
  x[[2]] - x[[1]]^2 - x[[1]] - 3
}

gradg = function(x){
  aux1 = -2*x[[1]] - 1
  aux2 = 1
  return(c(aux1,aux2))
}

h = function(x,u){g(x+u)}

gradh = function(x,u){gradg(x+u)}

xa = c(0,0)
uini = runif(2)
nitmax = 1000
tol = 1e-6
omr = 0.1
sol = FORM(xa,uini,h,gradh,nitmax,tol,omr)
```

The result is

```
> sol$beta
[1] 2.788203
> sol$ustar
[1] -0.4232162  2.7558957
> sol$xstar
[1] -0.4232162  2.7558957
> sol$nit
[1] 121
> sol$prec
[1] 9.089252e-07
> |
```

In Example 9.30, we got $u^* = (-0.4232162, 2.7558958)$, $x^* = (-0.4232162, 2.7558958)$, $\beta = 2.788203$.

Example 9.22
Multiple failure conditions can be treated by the simplified previously introduced in Sect. 9.3 and exploited in the sequel. To do this, we need to run FORM for each failure equation.

Let us illustrate the procedure by using the same example considered in the preceding: $\mathbf{x} \in \mathbb{R}^2$ and three failure conditions, corresponding to $g_1(\boldsymbol{x}) = x_2 - x_1^2 - x_1 - 3$, $g_2(\boldsymbol{x}) = x_1 - x_2^2 - x_2 - 2$, $g_3(\boldsymbol{x}) = x_1 + x_1x_2 - 1$.

We determine three solutions `sol1`, `sol2`, `sol3`, corresponding top each g_i, i=1,2,3. We obtain:

```
> sol1$beta
[1] 2.788203
> sol2$beta
[1] 2.788203
> sol3$beta
[1] 1.804913
```

Consequently, $\beta = 1.804913$. For the point (1,–1), we obtain $\beta = 0.8946934$.

Exercises

1. Let $\boldsymbol{x} \in \mathbb{R}^2$ and $g(\boldsymbol{x}) = x_2 - x_1^2 - x_1 - 3$. Let $X_i = x_i + u_i$ with $u_i \sim N(0, 1)$, $i = 1, 2$. Use FORM to evaluate the reliability index of the points (1,0), (0,1).
2. Let $\boldsymbol{x} \in \mathbb{R}^2$ and $g(\boldsymbol{x}) = x_2 - x_1^3 - x_1^2 - 4$. Let $X_i = x_i + u_i$ with $u_i \sim N(0, 1)$, $i = 1, 2$. Use FORM to evaluate the reliability index of the points (0,0), (1,0), (0,1).
3. Let $\boldsymbol{x} \in \mathbb{R}^2$ and $g(\boldsymbol{x}) = x_1x_2 - 2$. Let $X_i = x_i + u_i$ with $u_i \sim N(0, 1)$, $i = 1, 2$. Use FORM to evaluate the reliability index of the points (0,0), (1,0), (0,1).
4. Let $\boldsymbol{x} \in \mathbb{R}^4$ and $g(\boldsymbol{x}) = x_1x_2x_3x_4 - 2$. Let $X_i = x_i + u_i$ with $u_i \sim N(0, 1)$, $i = 1, 2, 3, 4$. Use FORM to evaluate the reliability index of the points (0,0,0,0), (1,0,0,0), (0,1,0,0), (0,0,1,0) and (0,0,0,1).
5. Let $\boldsymbol{x} \in \mathbb{R}^4$ and $g(\boldsymbol{x}) = x_4 - x_1x_2x_3 - 3$. Let $X_i = x_i + u_i$ with $u_i \sim N(0, 1)$, $i = 1, 2, 3, 4$. Use FORM to evaluate the reliability index of the points (0,0,0,0), (1,0,0,0), (0,1,0,0), (0,0,1,0) and (0,0,0,1).

9.8.2 *Second Order Reliability Method (SORM)*

The second – and less popular than FORM – is the second order reliability method (SORM), which is a variant from FORM, where a quadratic approximation is used instead the affine one. In SORM, the limit state curve is approximated by a quadratic equation and we must solve at each step an optimization problem involving a quadratic objective function and quadratic constraints – such a problem is said a quadratically constrained quadratic problem (QCQP). Analogously to FORM, the reader will find many works and implementations of SORM. We give below a simple implementation intended to a pedagogical aim.

The algorithm of the method reads as

Algorithm of the Method SORM

Input: the point $\boldsymbol{x}$, starting point $\boldsymbol{u}^{(0)} \in \mathcal{S}$, stopping conditions (for instance, maximal iteration number, minimal step at each iteration). Programs evaluating $h(\boldsymbol{x}, \boldsymbol{u})$, $\nabla_{\boldsymbol{u}} h(\boldsymbol{x}, \boldsymbol{u})$ and the Hessian $\nabla^2_{\boldsymbol{u}} h(\boldsymbol{x}, \boldsymbol{u})$, a program for the correction of a tentative design point $\boldsymbol{u}$ by determining a point $\overline{\boldsymbol{u}}(\boldsymbol{u})$ such that $|h(\boldsymbol{x}, \overline{\boldsymbol{u}}(\boldsymbol{u}))| < tol$, where tol is a predefined tolerance.

Output: estimations of $\boldsymbol{u}^*(\boldsymbol{x})$ and $\beta(\boldsymbol{x})$.

Initialization: If the initial point does not verify $|h(\boldsymbol{x}, \boldsymbol{u}^{(0)})| < tol$, then correct it: $\boldsymbol{u}^{(0)} \longleftarrow \overline{\boldsymbol{u}}(\boldsymbol{u}^{(0)})$. Initialize $\beta^{(0)} = \|\boldsymbol{u}^{(0)}\|$. Set the iteration number to zero : $i \longleftarrow 0$.

1. Approximate the limit state curve by a quadratic form $Q^{(i)}$: $q^{(i)}(\boldsymbol{u}) = \boldsymbol{u}^t \boldsymbol{A}^{(i)} \boldsymbol{u} / 2 + \boldsymbol{u}^t \boldsymbol{B}^{(i)} + C^{(i)}(\boldsymbol{x})$. For instance, use a Taylor-MacLaurin series: $\boldsymbol{A}^{(i)} = \nabla^2_{\boldsymbol{u}} h(\boldsymbol{x}^{(i)}, \boldsymbol{u}^{(i)})$, $\boldsymbol{B}^{(i)} = \nabla_{\boldsymbol{u}} h(\boldsymbol{x}^{(i)}, \boldsymbol{u}^{(i)}) - \boldsymbol{A}^{(i)} \boldsymbol{u}^{(i)}$, $C^{(i)}(\boldsymbol{x}) = h(\boldsymbol{x}, \boldsymbol{u}^{(i)}) - (\boldsymbol{B}^{(i)})^t \boldsymbol{u}^{(i)} + (\boldsymbol{u}^{(i)})^t \boldsymbol{A}^{(i)} (\boldsymbol{u}^{(i)})^t / 2$.
2. Find the solution of the QCQP: $\boldsymbol{u}^{(i+1)} = \arg\min \{\boldsymbol{u}^t \boldsymbol{u}: q^{(i)}(\boldsymbol{u}) \geq 0\}$
3. If $|h(\boldsymbol{x}, \boldsymbol{u}^{(i+1)})| \geq tol$ then $\boldsymbol{u}^{(i+1)} \longleftarrow \overline{\boldsymbol{u}}(\boldsymbol{u}^{(i+1)})$.
4. Set $\beta^{(i+1)} = \|\boldsymbol{u}^{(i+1)}\|$.
5. Increment the iteration number $i \longleftarrow i + 1$.
6. Test for the stopping conditions. If the iterations continue, then go to 1, else estimate $\beta(\boldsymbol{x}) \approx \beta^{(i)}$, $\boldsymbol{u}(\boldsymbol{x}) \approx \boldsymbol{u}^{(i)}$.

Th implementation used is the following:

```
SORM = function(x,ut,h,gradh,d2h,nitmax,tol,omr){
  aux = h(x, ut)
  if (abs(aux) > tol) {
    uac = correction(x,ut,h)
  }else{uac = ut}

  xac = x + uac
  betta = sqrt(sum(uac^2))
  prec = 1 + tol
  nit = 0
  continue = TRUE
  while (continue) {
    a = d2h(x, uac)
    b = gradh(x, uac) - a %*% uac
    y = a %*% uac
    ccc = h(x, uac) - sum(b*uac) + sum(uac* y) / 2
    he = function(u){t(u) %*% a %*% u + sum(b*u) + ccc}
    fobj = function(u){sqrt(sum(u^2))}
    sol = fmincon(uac,fobj,heq=he)
    un = sol$par
    aux = h(x, un)
    if (abs(aux) > tol) {
      un = correction(x,un,h)
    }
    xn = x + un
    betta = sqrt(sum(un^2))
    nit = nit + 1
    prec = sqrt(sum((un-uac)^2))/abs(omr)
    uac = omr*un + (1-omr)*uac
    continue = (nit <nitmax) & (prec > tol)
  }
  aux = list(betta,un,xn,nit,prec)
  names(aux) = c("beta","ustar","xstar"
                 ,"nit","prec")
  return(aux)
}
```

Example 9.23

Let us consider the case where $x = (0, 0)$ and $g(x) = x_2 - x_1^2 - x_1 - 3$. This problem was previously solved (Example 9.30). In this case,

```
d2g = function(x){
  aux = matrix(0,nrow = 2, ncol = 2)
  aux[[1,1]] = -2
  return(aux)
}

h = function(x,u){g(x+u)}

gradh = function(x,u){gradg(x+u)}

d2h = function(x,u){d2g(x+u)}
```

(continued)

Example 9.23 (continued)
The result is

```
> sol$beta
[1] 2.857062
> sol$ustar
[1] -0.1817633  2.8512746
> sol$xstar
[1] -0.1817633  2.8512746
> sol$nit
[1] 126
> sol$prec
[1] 9.334006e-07
```

Thus, the estimation is $\beta = 2.857062$.

Example 9.24
Multiple failure conditions can be treated by the simplified approach previously presented. Analogously to FORM, we need to run form with different parameters. Let us illustrate the procedure by using the same example considered in the preceding: $\mathbf{x} \in \mathbb{R}^2$ and three failure conditions, corresponding to $g_1(\boldsymbol{x}) = x_2 - x_1^2 - x_1 - 3$, $g_2(\boldsymbol{x}) = x_1 - x_2^2 - x_2 - 2$, $g_3(\boldsymbol{x}) = x_1 + x_1x_2 - 1$.

We determine three solutions `sol1`, `sol2`, `sol3`, corresponding top each g_i, i=1,2,3. We obtain:

```
> sol1$beta
[1] 2.857062
> sol2$beta
[1] 2.798287
> sol3$beta
[1] 1.880304
```

Consequently, the estimation is $\beta = 1.80304$. For the point (1,–1), we obtain $\beta = 1.034464$.

Exercises

1. Let $\boldsymbol{x} \in \mathbb{R}^2$ and $g(\boldsymbol{x}) = x_2 - x_1^2 - x_1 - 3$. Let $X_i = x_i + u_i$ with $u_i \sim N(0, 1)$, $i = 1, 2$. Use SORM to evaluate the reliability index of the points (1,0), (0,1).
2. Let $\boldsymbol{x} \in \mathbb{R}^2$ and $g(\boldsymbol{x}) = x_2 - x_1^3 - x_1^2 - 4$. Let $X_i = x_i + u_i$ with $u_i \sim N(0, 1)$, $i = 1, 2$. Use SORM to evaluate the reliability index of the points (0,0), (1,0), (0,1).

(continued)

3. Let $\boldsymbol{x} \in \mathbb{R}^2$ and $g(\boldsymbol{x}) = x_1x_2 - 2$. Let $X_i = x_i + u_i$ with $u_i \sim N(0, 1)$, $i = 1$, 2. Use SORM to evaluate the reliability index of the points (0,0), (1,0), (0,1).
4. Let $\boldsymbol{x} \in \mathbb{R}^4$ and $g(\boldsymbol{x}) = x_1x_2x_3x_4 - 2$. Let $X_i = x_i + u_i$ with $u_i \sim N(0, 1)$, $i = 1$, 2, 3, 4. Use SORM to evaluate the reliability index of the points (0,0,0,0), (1,0,0,0), (0,1,0,0), (0,0,1,0) and (0,0,0,1).
5. Let $\boldsymbol{x} \in \mathbb{R}^4$ and $g(\boldsymbol{x}) = x_4 - x_1x_2x_3 - 3$. Let $X_i = x_i + u_i$ with $u_i \sim N(0, 1)$, $i = 1, 2, 3, 4$. Use SORM to evaluate the reliability index of the points (0,0,0,0), (1,0,0,0), (0,1,0,0), (0,0,1,0) and (0,0,0,1).

9.9 Reliability Based Design Optimization

As we have already observed, optimization problems may involve uncertainty. For instance, when considering

$$\boldsymbol{x}^* = \arg\min\{f(\boldsymbol{x}) : \boldsymbol{x} \in C \subset \mathbb{R}^n\} \tag{9.30}$$

both the objective f and the admissible set C may contain uncertainty, arising from different variability sources.

Thus, variability and uncertainties should be considered in the optimization procedure to generate a robust result. Indeed,

- The effects of variations of the parameters involved in the objective and constraints must be analyzed, to prevent the situation where small variations of some parameters can cause significant variations of the result (loss of robustness)
- The impacts of implementation errors of the solution found must be considered, to avoid situations where small errors in the implementation produce substantial variations in the real result (variability and uncertainty)
- the effects of fluctuations of the external environment must be evaluated to ensure minimal safety conditions (unreliability).

To avoid these difficulties, it is necessary to introduce uncertainties in the design process. A possible aim is to get a complete characterization of the variability of the solutions: as shown in the preceding, UQ furnishes tools for such a characterization.

However, in the case where we are interested in the determination of robust solutions without the full determination of the variability of the solution, alternatives exist in the literature.

For instance, we can look for solutions having a small sensitivity to the variations of some parameters (*Robust Optimization*): in such a situation, we must introduce an index of robustness $\beta(\boldsymbol{x})$, analogous to the reliability index, which gives to each possible solution $\boldsymbol{x}$ a value measuring the robustness. Assuming that robustness increases with β, we may either look for the solution that minimizes the objective on the admissible set and ensures a minimal safety:

$$x^* = \arg\min \{f(x) : x \in C \subset \mathbb{R}^n, \beta(x) \geq \beta_{min}\} \quad (9.31)$$

or determine the solution that ensures the maximal safety for a maximal value of the objective:

$$x^* = \arg\min \{\beta(x) : x \in C \subset \mathbb{R}^n, f(x) \leq f_{max}\}. \quad (9.32)$$

In these formulations, the robustness index may be replaced by the reliability index: in this case, we refer to these optimization problems as being *Reliability Based Design Optimization* (RBDO) problems. More generally, we can look for the solution that controls the probability of failure:

$$x^* = \arg\min \{f(x) : x \in C \subset \mathbb{R}^n, F = P(\mathcal{F}) \leq p_{max}\} \quad (9.33)$$

or

$$x^* = \arg\min \{F = P(\mathcal{F}) : x \in C \subset \mathbb{R}^n, f(x) \leq f_{max}\}. \quad (9.34)$$

9.9.1 *The Bilevel or Double Loop Approach for a Desired β*

The formulation (9.31) leads to a bi-level optimization problem, since it deals with nested optimization problems: on the one hand, the determination of $\beta(x)$ and, on the other hand, the minimization of $f(x)$. Indeed, typical optimization iterations start from an initial guess $x^{(0)}$ to generate a sequence of points $x^{(1)}, x^{(2)}, \ldots$. At each step, the generation of $x^{(i+1)}$ requests the evaluation of $\beta(x)$ for the candidate points $\beta(x)$ – so, it requests the solution of an internal optimization problem. A typical algorithm is given below:

Bi-Level (or Double Loop) Algorithm for RBDO (Desired β)
Input: the starting point $x^{(0)}$, stopping conditions (for instance, maximal iteration number, minimal step at each iteration), a program determining $u^*(x)$, a program for the correction of x for a given u^*.
Output: estimations of x^*, $u^*(x^*)$ and $\beta(x^*)$.
Initialization: Set the iteration number to zero: $k \longleftarrow 0$.

1. **Inferior level (interior loop)**: find $(u)^k = u^*(x^{(k)})$. Set $\beta^{(k)} = \|(u^*)^k\|$.
2. **Superior level (external loop)**: Determine $x^{(k+1)}$ as a correction of $x^{(k)}$ using $(u)^k$.

(continued)

3. Increment the iteration number $k \longleftarrow k + 1$.
4. Test for the stopping conditions. If the iterations continue, then go to 1, else estimate $\boldsymbol{x}^* \approx \boldsymbol{x}^{(k)}$, $\boldsymbol{u}^*(\boldsymbol{x}^*) \approx \boldsymbol{u}^*(\boldsymbol{x}^{(k)})$, $\beta(\boldsymbol{x}^*) \approx \|\boldsymbol{u}^*(\boldsymbol{x}^{(k)})\|$.

Examples of methods for the correction of $\boldsymbol{x}^{(k)}$ can be found in the literature. We presented some methods in Sects. 9.2 and 9.8.1 (see also (Souza de Cursi et al., 2004)). The determination of $(\boldsymbol{u})^k$ is usually called the *reliability level*, while the determination of $\boldsymbol{x}^{(k+1)}$ is called the *optimization level*. The numerical solution by iterative methods may request a large number of calls to each of these levels. Thus, simplification methods have been proposed in the literature –, for example, the use of FORM at the reliability level.

A popular approach consists in generating $\boldsymbol{x}^{(k+1)}$ by determining successively: at the internal loop,

$$\boldsymbol{u}^{(k)} = \arg\min\left\{\|\boldsymbol{u}\|: h\left(\boldsymbol{x}^{(k)}, \boldsymbol{u}\right) \geq 0 \text{ } and \text{ } \|\boldsymbol{u}\| \geq \beta_{min}\right\}$$

and, at the external loop (correction of $\boldsymbol{x}^{(k)}$):

$$\boldsymbol{x}^{(k+1)} = \arg\min\left\{f(x): h\left(\boldsymbol{x}, \boldsymbol{u}^{(k)}\right) \leq 0\right\}.$$

The main difficulty in the implementation of the double loop is the fact that there are two problems of optimization intertwined: the reliability analysis and the optimization of the objective function. Nested calls to an optimizer are a difficulty task. The best solution consists in the use an optimizer only for one of these optimization problems and the use of another method to the other problem. For instance, reliability analysis can be performed by FORM.

Example 9.25

Let us consider the simple objective function $f(\boldsymbol{x}) = x_1^2 + (x_2 - 1)^2$ and the failure region defined by $g(\boldsymbol{x}) = x_1 + x_2 - 3$. We look for a RBDO solution with $\beta(\boldsymbol{x}) \geq \beta_{min}$. The global minimum of f is $\boldsymbol{x}_g^* = (0, 1)$, for which $\beta\left(\boldsymbol{x}_g^*\right) = \sqrt{2}$. For $\beta_{min} \geq 2$, $\boldsymbol{x}_g^*$ is not a solution of the RBDO problem.

Define the parameters, the objective function, the limit state equation and the reliability index: since the limit state curve is affine, we can evaluate the reliability index by Eq. (9.12).

```
a = c(1,1)
b = -3
betamin = 5
na = sqrt(sum(a^2))
fobj = function(x){x[[1]]^2 + (x[[2]]-1)^2}
g = function(x){sum(a*x) + b}
hlind = function(x){-g(x)/na}
```

(continued)

Example 9.25 (continued)

Minimize the objective function under the restrictions: the solution must belong to the safe region ($g(\boldsymbol{x}) \le 0$) and $\beta(\boldsymbol{x}) \ge \beta_{min}$.

```
library(pracma)
h = function(x,betamin){
  aux1 = g(x)
  aux2 = betamin - hlind(x)
  aux = c(aux1,aux2)
  return(aux)
}

hi = function(x){h(x,betamin)}
sol = fmincon(c(0,0),fobj,hin = hi)
```

We obtain the solution

```
> sol$par
[1] -2.535534 -1.535534
```

Make a loop to find solutions for several values of β_{min}

```
xini = c(0,0)
for (betamin in seq(1,5,by=0.5)){
  hi = function(x){h(x,betamin)}
  sol = fmincon(xini,fobj,hin = hi)
  xini = sol$par
  print(paste("beta_min = ",betamin,
        ": x* = (", sol$par[[1]],", ",
    sol$par[[2]]," )"))
}
```

We obtain:

```
[1] "beta_min =  1 : x* = ( -7.48505621972409e-09 ,  0.99999999243047  )"
[1] "beta_min =  1.5 : x* = ( -0.0606601717638506 ,  0.939339828204208  )"
[1] "beta_min =  2 : x* = ( -0.414213551239461 ,  0.585786426493271  )"
[1] "beta_min =  2.5 : x* = ( -0.76776691937133 ,  0.232233013438594  )"
[1] "beta_min =  3 : x* = ( -1.12132031863622 ,  -0.12132036848307  )"
[1] "beta_min =  3.5 : x* = ( -1.47487370868293 ,  -0.474873759622904  )"
[1] "beta_min =  4 : x* = ( -1.82842711660503 ,  -0.828427132887355  )"
[1] "beta_min =  4.5 : x* = ( -2.18198049618097 ,  -1.18198053449796  )"
[1] "beta_min =  5 : x* = ( -2.53553387088458 ,  -1.53553394098089  )"
```

Try it in the reverse order – just change `seq(1,5,by=05)` into `seq(5,1,by=-0.5)` – the results are close to the preceding:

```
[1] "beta_min =  5 : x* = ( -2.53553391321926 ,  -1.53553389864622  )"
[1] "beta_min =  4.5 : x* = ( -2.18198041482455 ,  -1.18198061585437  )"
[1] "beta_min =  4 : x* = ( -1.82842699299184 ,  -0.828427256500535  )"
[1] "beta_min =  3.5 : x* = ( -1.47487360050249 ,  -0.474873867803344  )"
[1] "beta_min =  3 : x* = ( -1.12132011615258 ,  -0.121320570966679  )"
[1] "beta_min =  2.5 : x* = ( -0.767766683088464 ,  0.232232777155747  )"
[1] "beta_min =  2 : x* = ( -0.414213320325209 ,  0.585786195579028  )"
[1] "beta_min =  1.5 : x* = ( -0.0606599319460754 ,  0.939339588386442  )"
[1] "beta_min =  1 : x* = ( 7.47712811111079e-09 ,  0.999999992539611  )"
```

Example 9.26

Let us consider again the simple objective function $f(x) = x_1^2 + (x_2 - 1)^2$ and the failure region defined by $g(x) = x_2 - x_1^2 - x_1 - 3$. We look for a RBDO solution with $\beta(x) \geq \beta_{min}$. In this case, the evaluation of $\beta(x)$ must be made by a different method. For instance, we can use Eq. (9.11) and the function `findsol` introduced in Sect. 9.5.1. In this case, `findsol` furnishes $\beta\left(x_g^*\right) \approx 1.8$ for $x_g^* = (0, 1)$.

To do this, we modify the function `hlind` as shown at right. Notice that we need to define `gradg`, which evaluates ∇g

```
hlind = function(x){
  aux = findsol(x,g,gradg)
  return(aux$beta)
}
```

Minimizing the objective function under the same restrictions as in the preceding example furnishes the results at right, for $\beta_{min} = 5$.

```
> sol$par
[1]  0.457406 -2.170733
```

Here,

```
"beta_min =  1 : x* = ( -7.45058059692383e-09 ,  0.999999992549419 )"
"beta_min =  1.5 : x* = ( 7.45058059692383e-09 ,  0.999999992549419 )"
"beta_min =  2 : x* = ( 0.019648585187902 ,  0.804479775374156 )"
"beta_min =  2.5 : x* = ( 0.181773370441229 ,  0.328317872243523 )"
"beta_min =  3 : x* = ( 0.0672753480366363 ,  -0.20827131591478 )"
"beta_min =  3.5 : x* = ( 0.167775596772007 ,  -0.693915707079744 )"
"beta_min =  4 : x* = ( 0.787898895465687 ,  -1.06223651337879 )"
"beta_min =  4.5 : x* = ( 0.639960564365806 ,  -1.61849675948469 )"
"beta_min =  5 : x* = ( 0.644054350954818 ,  -2.12982143355828 )"
```

In the reverse order:

```
"beta_min =  5 : x* = ( 1.24592507033111 ,  -1.96875557635976 )"
"beta_min =  4.5 : x* = ( 0.780980508965978 ,  -1.583429789725 )"
"beta_min =  4 : x* = ( 0.519953775949149 ,  -1.14029010631964 )"
"beta_min =  3.5 : x* = ( 0.28687430126925 ,  -0.671938781515596 )"
"beta_min =  3 : x* = ( -0.234345459553115 ,  -0.239907317570673 )"
"beta_min =  2.5 : x* = ( 0.287631041864963 ,  0.354913375450224 )"
"beta_min =  2 : x* = ( 0.00575958604182898 ,  0.801582494352806 )"
"beta_min =  1.5 : x* = ( 8.93057004822811e-09 ,  0.999999992390904 )"
"beta_min =  1 : x* = ( -7.45058059692383e-09 ,  0.999999992549419 )"
```

Example 9.27

Let us consider again the simple objective function $f(x) = x_1^2 + (x_2 - 1)^2$ and the failure region defined by $g(x) = x_2 - x_1^3 - x_1^2 - 4$. In this case, the global minimum $\boldsymbol{x}_g^* = (0, 1)$ is admissible and $\beta\left(\boldsymbol{x}_g^*\right) \approx 4.5$ (value generated by `findsol`). Indeed, all the results are close to $\boldsymbol{x}_g^*$:

```
[1] "beta_min = 1 : x* = ( 1.24753914149226e-08 , 1.000000004505 )"
[1] "beta_min = 1.5 : x* = ( -7.45058063048871e-09 , 0.999999992549419 )"
[1] "beta_min = 2 : x* = ( 7.45058059692383e-09 , 0.999999992549419 )"
[1] "beta_min = 2.5 : x* = ( 7.45058059692383e-09 , 0.999999992549419 )"
[1] "beta_min = 3 : x* = ( -7.45058059692383e-09 , 0.999999992549419 )"
[1] "beta_min = 3.5 : x* = ( 7.45058059692383e-09 , 0.999999992549419 )"
[1] "beta_min = 4 : x* = ( 3.27358568740462e-09 , 0.9999999254942 )"
[1] "beta_min = 4.5 : x* = ( 3.27358568740462e-09 , 0.9999999254942 )"
```

In the reverse order, the results are analogous:

```
[1] "beta_min = 4.5 : x* = ( 0.0107464119329971 , 1.05146789677922 )"
[1] "beta_min = 4 : x* = ( -7.83410300519982e-09 , 0.999999992227175 )"
[1] "beta_min = 3.5 : x* = ( 7.45058059692382e-09 , 0.999999992549419 )"
[1] "beta_min = 3 : x* = ( 7.45058059692382e-09 , 0.999999992549419 )"
[1] "beta_min = 2.5 : x* = ( 7.45058059692382e-09 , 0.999999992549419 )"
[1] "beta_min = 2 : x* = ( 7.45058059692382e-09 , 0.999999992549419 )"
[1] "beta_min = 1.5 : x* = ( 7.45058059692382e-09 , 0.999999992549419 )"
[1] "beta_min = 1 : x* = ( 7.45058059692382e-09 , 0.999999992549419 )"
```

Exercises

1. Let $\boldsymbol{x} \in \mathbb{R}^2$, $f(x) = x_1^2 + x_2^2$ and $g(\boldsymbol{x}) = x_1 x_2 - 1$. Let $X_i = x_i + u_i$ with $u_i \sim N(0, 1)$, $i = 1, 2$. Find a RBDO solution with $\beta = 3, 4, 5, 6$.
2. Let $\boldsymbol{x} \in \mathbb{R}^2$, $f(x) = x_1^2 + x_2^2$ and $g(\boldsymbol{x}) = x_1 x_2^2 - 1$. Let $X_i = x_i + u_i$ with $u_i \sim N(0, 1)$, $i = 1, 2$. Find a RBDO solution with $\beta = 3, 4, 5, 6$.
3. Let $\boldsymbol{x} \in \mathbb{R}^2$, $f(x) = (x_1 + x_2 - 6)^2 + (x_1 - x_2 - 10)^2$ and $g(\boldsymbol{x}) = x_1 - 3x_2^2 + 2$. Let $X_i = x_i + u_i$ with $u_i \sim N(0, 1)$, $i = 1, 2$. Find a RBDO solution with $\beta = 3, 4, 5, 6$.
4. Let $\boldsymbol{x} \in \mathbb{R}^2$, $f(x) = x_1^2 + x_2^2$ and $g(\boldsymbol{x}) = x_1^2 + 8x_2 - 2$. Let $X_i = x_i + u_i$ with $u_i \sim N(0, 1)$, $i = 1, 2$. Find a RBDO solution with $\beta = 3, 4, 5, 6$.
5. Let $\boldsymbol{x} \in \mathbb{R}^2$, $f(x) = (x_1 + x_2 - 6)^2 + (x_1 - x_2 - 10)^2$ and $g(\boldsymbol{x}) = x_1 + x_2^3 - 2$. Let $X_i = x_i + u_i$ with $u_i \sim N(0, 1)$, $i = 1, 2$. Find a RBDO solution with $\beta = 3, 4, 5, 6$.

9.9.2 The Bilevel or Double Loop Approach for a Desired Objective

As observed, an alternative formulation is furnished by (9.34): we can exchange the rules of β and f, to look for the maximal β for a prescribed maximal value of f. In this case, we run the SOLVER to maximize β under some restrictions.

Example 9.28

Let us consider the simple objective function $f(x) = x_1^2 + (x_2 - 1)^2$ and the failure region defined by $g(x) = x_1 + x_2 - 3$. We look for a RBDO solution that maximizes $\beta(x)$ with the maximal value $f(x) \leq f_{max}$.

Define the parameters, the objective function, the limit state equation and the reliability index: as previously, the limit state curve is affine, we can evaluate the reliability index by Eq. (9.12).

```
a = c(1,1)
na = sqrt(sum(a^2))
b = -3
fmax = 3
f = function(x){x[[1]]^2 + (x[[2]]-1)^2}
g = function(x){sum(a*x) + b}
hlind = function(x){-g(x)/na}
fobj = function(x) {-hlind(x)}
```

Minimize the objective function under the restrictions: the solution must belong to the safe region ($g(\boldsymbol{x}) \leq 0$) and $f(\boldsymbol{x}) \leq f_{max}$.

```
library(pracma)
h = function(x,fmax){
  aux1 = g(x)
  aux2 = f(x) - fmax
  aux = c(aux1,aux2)
  return(aux)
}
hi = function(x){h(x,fmax)}
sol = fmincon(c(0,0),fobj,hin = hi)
```

Make a loop to get several results for different values of f_{max}.

```
xini = c(0,0)
for (fmax in seq(10,1,by=-1)){
  hi = function(x){h(x,fmax)}
  sol = fmincon(xini,fobj,hin = hi)
  xini = sol$par
  print(paste("fmax = ",fmax,
            " : beta = ",-sol$value))
}
```

We obtain:

```
[1] "fmax =  10  : beta =  4.57649122254147"
[1] "fmax =  9  : beta =  4.41421356237451"
[1] "fmax =  8  : beta =  4.24264068712256"
[1] "fmax =  7  : beta =  4.05996487344589"
[1] "fmax =  6  : beta =  3.8637033051801"
[1] "fmax =  5  : beta =  3.65028153995589"
[1] "fmax =  4  : beta =  3.41421356274558"
[1] "fmax =  3  : beta =  3.14626437238844"
[1] "fmax =  2  : beta =  2.82842715542529"
[1] "fmax =  1  : beta =  2.41421356237422"
```

In the reverse order, the results are analogous:

```
[1] "fmax =  1  : beta =  2.41421356237313"
[1] "fmax =  2  : beta =  2.82842712474779"
[1] "fmax =  3  : beta =  3.14626436994197"
[1] "fmax =  4  : beta =  3.41421356519868"
[1] "fmax =  5  : beta =  3.65028154028874"
[1] "fmax =  6  : beta =  3.86370330524722"
[1] "fmax =  7  : beta =  4.05996487346332"
[1] "fmax =  8  : beta =  4.24264068712791"
[1] "fmax =  9  : beta =  4.41421356237644"
[1] "fmax =  10  : beta =  4.57649122254296"
```

Example 9.29

Let us consider again the simple objective function $f(x)=x_1^2+(x_2-1)^2$ and the failure region defined by $g(x)=x_2-x_1^2-x_1-3$. We look for a RBDO solution that maximizes $\beta(x)$ with the maximal value $f(x)\le f_{max}$. As previously, we use Eq. (9.11) and the function `findsol` introduced in Sect. 9.5.1.

```
[1] "fmax =  10  : beta =  4.95105262774614"
[1] "fmax =  9  : beta =  4.74402258136926"
[1] "fmax =  8  : beta =  4.57984352138434"
[1] "fmax =  7  : beta =  4.39475020139828"
[1] "fmax =  6  : beta =  4.2464560163457"
[1] "fmax =  5  : beta =  4.03663354914799"
[1] "fmax =  4  : beta =  3.79928484221252"
[1] "fmax =  3  : beta =  3.53439684063723"
[1] "fmax =  2  : beta =  3.21868715962584"
[1] "fmax =  1  : beta =  2.80396857763235"
```

(continued)

Example 9.29 (continued)

In the reverse order:

```
[1] "fmax =  1   : beta =  2.79794285705977"
[1] "fmax =  2   : beta =  3.20356695793978"
[1] "fmax =  3   : beta =  3.52987841391897"
[1] "fmax =  4   : beta =  3.79896917669451"
[1] "fmax =  5   : beta =  4.03539027260714"
[1] "fmax =  6   : beta =  4.2540155090842"
[1] "fmax =  7   : beta =  4.44987840736915"
[1] "fmax =  8   : beta =  4.6268504115081"
[1] "fmax =  9   : beta =  4.80473719618067"
[1] "fmax =  10  : beta =  4.96039897953989"
```

Example 9.30

Let us consider again the simple objective function $f(x) = x_1^2 + (x_2 - 1)^2$ and the failure region defined by $g(x) = x_2 - x_1^3 - x_1^2 - 4$. Now:

```
[1] "fmax =  10  : beta =  7.12631645151213"
[1] "fmax =  9   : beta =  7.4303589720892"
[1] "fmax =  8   : beta =  6.44542947728034"
[1] "fmax =  7   : beta =  6.83165270320154"
[1] "fmax =  6   : beta =  5.20534635620564"
[1] "fmax =  5   : beta =  6.54724761494352"
[1] "fmax =  4   : beta =  6.15280535355559"
[1] "fmax =  3   : beta =  4.87628191955161"
[1] "fmax =  2   : beta =  5.85565538137503"
[1] "fmax =  1   : beta =  3.62794725650564"
```

In the reverse order:

```
[1] "fmax =  1   : beta =  4.43254575609422"
[1] "fmax =  2   : beta =  4.40750729965579"
[1] "fmax =  3   : beta =  6.1428896494357"
[1] "fmax =  4   : beta =  6.47136423742456"
[1] "fmax =  5   : beta =  6.54530039449001"
[1] "fmax =  6   : beta =  6.64545442788239"
[1] "fmax =  7   : beta =  6.65591450991448"
[1] "fmax =  8   : beta =  6.72208222991462"
[1] "fmax =  9   : beta =  7.04355453342922"
[1] "fmax =  10  : beta =  6.81789810566949"
```

Notice that there are some inconsistencies in the results – this is due to the difficulty in the optimization, which involves non-convexities.

Exercises

1. Let $\boldsymbol{x} \in \mathbb{R}^2$, $f(x) = x_1^2 + x_2^2$ and $g(\boldsymbol{x}) = x_1x_2 - 1$. Let $X_i = x_i + u_i$ with $u_i \sim N(0, 1)$, $i = 1, 2$. Find a solution that maximizes β for $f(x) \le 2$, $f(x) \le 3$, $f(x) \le 4$, $f(x) \le 5$.
2. Let $\boldsymbol{x} \in \mathbb{R}^2$, $f(x) = x_1^2 + x_2^2$ and $g(\boldsymbol{x}) = x_1x_2^2 - 1$. Let $X_i = x_i + u_i$ with $u_i \sim N(0, 1)$, $i = 1, 2$. Find a solution that maximizes β for $f(x) \le 2$, $f(x) \le 3$, $f(x) \le 4$, $f(x) \le 5$.
3. Let $\boldsymbol{x} \in \mathbb{R}^2$, $f(x) = (x_1 + x_2 - 6)^2 + (x_1 - x_2 - 10)^2$ and $g(\boldsymbol{x}) = x_1 - 3x_2^2 + 2$. Find a solution that maximizes β for $f(x) \le 2$, $f(x) \le 3$, $f(x) \le 4$, $f(x) \le 5$.
4. Let $\boldsymbol{x} \in \mathbb{R}^2$, $f(x) = x_1^2 + x_2^2$ and $g(\boldsymbol{x}) = x_1^2 + 8x_2 - 2$. Let $X_i = x_i + u_i$ with $u_i \sim N(0, 1)$, $i = 1, 2$. Find a solution that maximizes β for $f(x) \le 2$, $f(x) \le 3$, $f(x) \le 4$, $f(x) \le 5$.
5. Let $\boldsymbol{x} \in \mathbb{R}^2$, $f(x) = (x_1 + x_2 - 6)^2 + (x_1 - x_2 - 10)^2$ and $g(\boldsymbol{x}) = x_1 + x_2^3 - 2$. Find a solution that maximizes β for $f(x) \le 2$, $f(x) \le 3$, $f(x) \le 4$, $f(x) \le 5$.

Bibliography

Akaike, H. (1974, December). A new look at the statistical model identification. *IEEE Transactions on Automatic Control, 19*(6), 716–723. https://doi.org/10.1109/TAC.1974.1100705

Apesteguia, J., & Palacios-Huerta, I. (2010, December). Psychological pressure in competitive environments: Evidence from a randomized natural experiment. *American Economic Review, 100*(5), 2548–2564.

Arrondel, L., Duhautois, R., & Laslier, J.-F. (2019). Decision under psychological pressure: The shooter's anxiety at the penalty kick. *Journal of Economic Psychology, 70*(C), 22–35.

Arrous, J. (2000). *Algèbre linéaire et Science économique, un cas exemplaire*. Retrieved 1 13, 2021, from Research Gate page of Jean Arrous: https://www.researchgate.net/publication/242301525_ALGEBRE_LINEAIRE_ET_SCIENCE_ECONOMIQUE_UN_CAS_EXEMPLAIRE

Autrique, L., & Souza de Cursi, E. (1997). On stochastic modification for global optimization problems: An efficient implementation for the control of the vulcanization process. *International Journal of Control, 67*(1), 1–22.

Azar, O. H., & Bar-Eli, M. (2011). Do soccer players play the mixed-strategy Nash equilibrium? *Applied Economics, 43*(25), 3591–3601. https://doi.org/10.1080/00036841003670747

Bachelier, L. (1900). Théorie de la spéculation. *Annales scientifiques de l'École Normale Supérieure, série, 3*(17), 21–86. https://doi.org/10.24033/asens.476

Bachelier, L. (1901). Théorie mathématique du jeu. *Annales scientifiques de l'E.N.S., 18*(3eme série), pp. 143–209.

Bassi, M., Souza de Cursi, E., Pagnacco, E., & Ellaia, R. (2018). Statistics of the Pareto front in Multi-objective Optimization under Uncertainties. *Latin American Journal of Solids and Structures, 15*(11), e130. https://doi.org/10.1590/1679-78255018

Black, F., & Scholes, M. (1973). The pricing of options and corporate liabilities. *Journal of Political Economy, 81*(3), 637–654. https://doi.org/10.1086/260062

Blankenship, G., & Baras, J. (1981). Accurate evaluation of stochastic Wiener integrals with applications to scattering in random media and to nonlinear filtering. *SIAM Journal of Applied Mathematics, 41*(3), 518–552. https://doi.org/10.1137/0141043

Boczon, M., & Wilson, A. J. (2018). *Goals, constraints, and public assignment: A field study of the UEFA champions league*. University of Pittsburgh. Retrieved June 2, 2020, from https://www.econ.pitt.edu/sites/default/files/working_papers/Working%20Paper.18.16.pdf

Booth, T. (1982). Regional Monte Carlo Solution of Elliptic Partial Differential Equations. *Journal of Computational Physics*, 281–290. https://doi.org/10.1016/0021-9991(82)90079-1

E. Souza de Cursi, *Uncertainty Quantification using R*, International Series in Operations Research & Management Science 335,
https://doi.org/10.1007/978-3-031-17785-9

Borel, E. (1921, 7 11). La théorie du jeu et les équations intégrales à noyau symétrique. *Comptes rendus hebdomadaires des séances de l'Académie des Sciences de Paris, 173*, 1304–1308.

Bouhadi, M., Ellaia, R., & Souza de Cursi, E. (2004a). Global optimization under nonlinear restrictions by using stochastic perturbations of the projected gradient. In C. A. Floudas & P. N. Pardalos (Eds.), *Frontiers in global optimization* (pp. 541–561). Kluwer Academic Press. https://doi.org/10.1007/978-1-4613-0251-3_29

Bouhadi, M., Ellaia, R., & Souza de Cursi, E. (2004b). Stochastic perturbation methods for affine restrictions. In C. A. Floudas & P. Pardalos (Eds.), *Advances in convex analysis and global optimization* (pp. 487–499). Kluwer Academic Press.

Boukhetala, K., & Guidoum, A. (2011). *Sim.DiffProc: A package for simulation of diffusion processes in R.* HAL. Retrieved 1 04, 2021, from https://hal.archives-ouvertes.fr/hal-00629841/en/

Box, G. E. (1958). A note on the generation of random normal deviates. *Annals of Mathematical Statistics, 29*(2), 610–611.

Brown, R. (1828). A brief account of microscopical observations made in the months of June, July, and August, 1827, on the particles contained in the pollen of plants; and on the general existence of active molecules in organic and inorganic bodies. *The Philosophical Magazine, 4*(21), 161–173. https://doi.org/10.1080/14786442808674769

Chen, F. (2008, December 18). *Random music generator.* Retrieved from OpenStax CNX: http://cnx.org/contents/62219fba-96df-418a-b9aa-14fddd7c30fa@2

Choi, B. (1992). *ARMA model identification*. Springer. https://doi.org/10.1007/978-1-4613-9745-8

Choleski, A.-L. (1910). *Sur la résolution numérique des systèmes d'équations linéaires.* manuscript.

Chorin, A. J. (1973). Accurate evaluation of Wiener integrals. *Mathematics of Computation, 27*(121), 1–15. https://doi.org/10.2307/2005242

Coloma, G. (2007). Penalty kicks in soccer: An alternative methodology for testing mixed-strategy equilibria. *Journal of Sports Economics, 8*(5), 530–545.

Coloma, G. (2012). *The penalty-kick game under incomplete information*. Universidad del CEMA.

Cournot, A. (1838). *Recherches Sur les principes mathématiques de la théorie des richesses*. Hachette.

Converse, P. D. (1949). New Laws of retail gravitation. *Journal of Marketing, 14*, 379–384. https://doi.org/10.1177/002224295001400303

Cowles, A. (1933). Can stock market forecasters forecast? *Econometrica, 1*(3), 309–324. https://doi.org/10.2307/1907042

Cox, J., Ingersoll, J., & Ross, S. (1985). A theory of the term structure of interest rates. *Econometrica, 53*(2), 385–407. https://doi.org/10.2307/1911242

Clark, C. (1951). Urban population densities. *Journal of the Royal Statistical Society: Series A (General), 114*(4), 490–496. https://doi.org/10.2307/2981088

Cramer, G. (1750). *Introduction à l'Analyse des Lignes Courbes Algébriques*. https://doi.org/10.3931/e-rara-4048

Cressman, R., & Tao, Y. (2014). The replicator equation and other game dynamics. *Proceedings of the National Academy of Sciences of the United States of America, 111*(Supplement 3), 10810–10817. https://doi.org/10.1073/pnas.1400823111

Cui, T., & Li, S. (2020). *Space fault tree theory and system reliability analysis*. EDP Sciences.

Dautray, R., & Lions, J.-L. (2012). *Mathematical analysis and numerical methods for science and technology* (Vol. 3). Springer.

Dickey, D. A., & Fuller, W. A. (1979). Distribution of the estimators for autoregressive time series with a unit root. *Journal of the American Statistical Association, 366a*, 427–431. https://doi.org/10.1080/01621459.1979.10482531

Einstein, A. (1905). Über die von der molekularkinetischen Theorie der Wärme geforderte Bewegung von in ruhenden Flüssigkeiten suspendierten Teilchen. *Annalen der Physik, 17*, 549–560.

El Mouatasim, A., Ellaia, R., & Souza de Cursi, E. (2006). Random perturbation of the variable metric method for unconstrained nonsmooth nonconvex optimization. *Applied Mathematics and Computer Science, 16*(4), 463–474.

El Mouatasim, A., Ellaia, R., Souza de Cursi, E. (2011). Random perturbation of the projected variable metric method for nonsmooth nonconvex optimization problems with linear constraints. *International Journal of Applied Mathematics and Computer Science*: 317–329. http://eudml.org/doc/208050

enseignements. (2020, 12 13). Retrieved from page personnelle de V. *Lemaire*. https://www.lpsm.paris/pageperso/lemaire/docs/stats-ENPC/series-chrono.pdf

Elepov, B. S., & Mikhailov, G. A. (1969). Solution of the Dirichlet problem by a model of "walks on Spheres". *USSR Computational Mathematics and Mathematical Physics,* 9(3), 194–204. https://doi.org/10.1016/0041-5553(69)90070-6

Ericson, C. A. (2011). *Fault tree analysis primer*. CreateSpace Incorporated.

Fisk, D. L. (1963). *Quasi-martingales and stochastic integrals*. Michigan State University, Department of Statistics. Office of Naval Research. Retrieved 12 27, 2020, from https://apps.dtic.mil/dtic/tr/fulltext/u2/414838.pdf

Föllmer, H. (1981). Calcul d'Ito sans probabilités. *Séminaire de probabilités (Strasbourg), 15*, 143–150. Retrieved from http://www.numdam.org/item?id=SPS_1981__15__143_0

Furness, K. P. (1965). Time function iteration. *Traffic Engineering and Control,* 458–460.

Garicano, L., Palacios-Huerta, I., & Prendergast, C. (2005). Favoritism under social pressure. *The Review of Economics and Statistics, 87*(2), 208–216.

Gauss, C. F. (1811). Disquisitio de elementis ellipticis Palladis ex oppositionibus annorum 1803, 1804, 1805, 1807, 1808, 1809. *Commentationes societatis regiæ scientiarum Gottingensis recentiores - commentationes mathematicae*, 1–26.

Graham, C., & Talay, D. (2013). *Stochastic simulation and Monte Carlo methods*. Springer. https://doi.org/10.1007/978-3-642-39363-1

Gonçalves, M. B., & Souza de Cursi, E. (2001). Parameter estimation in a trip distribution model by random. *Transportation Research part B, 35*(2), 137–161. https://doi.org/10.1016/S0191-2615(99)00043-0

Hannan, E. J., & Quinn, B. G. (1979). The determination of the order of an autoregression. *Journal of the Royal Statistical Society Series B (Methodological), 41*(2), 190–195. Retrieved from http://www.jstor.org/stable/2985032

Hansen, W. G. (1959). How accessibility shapes Land Use. *Journal of the American Institute of Planners, 25*(2), 73–76. https://doi.org/10.1080/01944365908978307

Holdorf Lopez, R., Souza de Cursi, J. E., & Lemosse, D. (2011). Approximating the probability density function of the optimal point of an optimization problem. *Engineering Optimization, 43*(3), 281–303. https://doi.org/10.1080/0305215X.2010.489607

Hoyt, H. (1939). *The Structure and Growth of Residential Neighborhoods in American Cities*. Washington: U.S. Government Printing Office.

Huff, D. L. (1964). Defining and estimating a trading area. *Journal of Marketing, 28*(3), 34–38. https://doi.org/10.2307/1249154

Hwang, C.-O., & Mascagni, M. (2001). Efficient modified "walk on spheres" algorithm for the linearized Poisson-Bolzmann. *Applied Physics Letters* 78, 787.https://doi.org/10.1063/1.1345817

Hwang, C.-O., Mascagni, M., & Given, J. A. (2003). A Feynman–Kac path-integral implementation for Poisson's equation using an h-conditioned Green's function. *Mathematics and Computers in Simulation, 62*(3-6), 347–355. https://doi.org/10.1016/S0378-4754(02)00224-0

Itô, K. (1944). Stochastic integral. *Proceedings of the Imperial Academy, 20*(8), 519–524. https://doi.org/10.3792/pia/1195572786

Itô, K. (1946). On a stochastic integral equation. *Proceedings. Japan Academy, 22*(2), 32–35. https://doi.org/10.3792/pja/1195572371

Itô, K. (1950). Stochastic differential equations in a differentiable manifold. *Nagoya Mathematical Journal, 1*, 35–47. https://doi.org/10.1017/S0027763000022819

Itô, K. (1951). On a formula concerning stochastic differentials. *Nagoya Mathematical Journal, 3*, 55–65. Retrieved from https://projecteuclid.org/euclid.nmj/1118799221

Jafari, M. A., & Abbasian, S. (2017). The moments for solution of the Cox-Ingersoll-Ross interest rate model. *Journal of Finance and Economics, 5*(1), 34–37. https://doi.org/10.12691/jfe-5-1-4

Iacus, S. M. (2008). *Simulation and Inference for Stochastic Differential Equations*. New York, NY, USA: Springer. https://doi.org/10.1007/978-0-387-75839-8

Jordan, W. (1962). *Handbook of geodesy* (Vol. 1). (M. W. Carta, Trans.)

Isard, W. (1956). *Location and Space-economy; a General Theory Relating to Industrial Location, Market Areas, Land Use, Trade, and Urban Structure*. Cambridge: MIT Press and Wiley.

Kharroubi, I., & Pham, H. (2015). Feynman–Kac representation for Hamilton–Jacobi–Bellman IPDE. *The Annals of Probability, 43*(4), 1823–1865. https://doi.org/10.1214/14-AOP920

Kim, N. J. (1968). *Linear programming with random requirements*. Utah Water Research Laboratory. Retrieved February 7, 2021, from https://digitalcommons.usu.edu/water_rep/272?utm_source=digitalcommons.usu.edu%2Fwater_rep%2F272&utm_medium=PDF&utm_campaign=PDFCoverPages

Kloeden, P., & Platen, E. (1989). A survey of numerical methods for stochastic differential equations. *Stochastic Hydrology and Hydraulics, 3*, 155–178. https://doi.org/10.1007/BF01543857

Kolmogorov, A. N. (1950). *Foundations of the theory of probability*. Chelsea Publishing Company.

Kress, R. (1998). *Numerical analysis*. Springer. https://doi.org/10.1007/978-1-4612-0599-9

Kuhn, H. W. (1957). Extensive games and the problem and information. In H. W. Kuhn & A. W. Tucker (Eds.), *Contributions to the theory of games, annals of mathematical studies* (pp. 193–216). Princeton University Press.

Kwiatkowski, D., Phillips, P. C., Schmidt, P., & Shin, Y. (1992). Testing the null hypothesis of stationarity against the alternative of a unit root: How sure are we that economic time series have a unit root? *Journal of Econometrics, 54*(1–3), 159–178. https://doi.org/10.1016/0304-4076(92)90104-Y

Langevin, P. (1908). Sur la théorie du mouvement brownien. *Comptes-Rendus de l'Académie des Sciences, 146*, 530–532.

Leontieff, W. W. (1936). Quantitative input and output relations in the economic systems of the United States. *The Review of Economics and Statistics, 18*(3), 105–125. https://doi.org/10.2307/1927837

Leontieff, W. W. (1937). Interrelation of prices, output, savings, and investment. *The Review of Economics and Statistics*, (3), 109–132. https://doi.org/10.2307/1927343

Leontieff, W. W. (1986). *Input-output economics*. Oxford University Press.

Levitt, S., Chiappori, P., & Groseclose, T. (2002). Testing mixed-strategy equilibria when players are heterogeneous: The case of penalty kicks in soccer. *American Economic Review, 92*, 1138–1151. https://doi.org/10.1257/00028280260344678

Lévy, P. (1922, November 13). Sur la détermination des lois de probabilité par leurs fonctions caractéristiques. *Comptes Rendus de l'Académie des Sciences de Paris*, pp. 854–856.

Lewontin, R. C. (1961, July). Evolution and the theory of games. *Journal of Theoretical Biology, 1*(3), 382–403.

Leybourne, S. J., & McCabe, B. P. (1994). A consistent test for a unit root. *Journal of Business and Economic Statistics, 12*(2), 157–166. https://doi.org/10.1080/07350015.1994.10510004

Li, F., & Wu, T. (2007). An importance sampling based approach for reliability analysis. *2007 IEEE international conference on automation science and engineering* (pp. 956–961). IEEE. https://doi.org/10.1109/COASE.2007.4341815.

Limnios, N. (2007). *Fault trees*. ISTE/Wiley.

Lotka, A. J. (1910, March 1). Contribution to the theory of periodic reactions. *The Journal of Physical Chemistry, 14*(3), 271–274. https://doi.org/10.1021/j150111a004

Lotka, A. J. (1920, July 1). Analytical note on certain rhythmic relations in organic systems. *Proceedings of the National Academy of Sciences of the USA, 6*(7), 410–415. https://doi.org/10.1073/pnas.6.7.410

Maynard-Smith, J. (1982). *Evolution and the theory of games*. Cambridge University Press.

Maynard-Smith, J., & Price, G. R. (1973, November 2). The logic of animal conflict. *Nature, 246*, 15–18.

Merton, R. C. (1973). Theory of rational option pricing. *The Bell Journal of Economics and Management Science, 4*(1), 141–183. https://doi.org/10.2307/3003143

Milstein, G. N. (1973). Approximate integration of stochastic differential equations. *Theory of Probability and Its Applications, 19*(3), 557–562. https://doi.org/10.1137/1119062

Milstein, G., & Tretyakov, M. (2004). *Stochastic Numerics for mathematical physics*. Springer. https://doi.org/10.1007/978-3-662-10063-9

Milstein, G., & Tretyakov, M. (2012). Solving the Dirichlet problem for Navier–stokes equations by probabilistic approach. *BIT Numerical Mathematics, 52*, 141–153. https://doi.org/10.1007/s10543-011-0347-z

Milstein, G., & Tretyakov, M. (2013). Probabilistic methods for the incompressible Navier–stokes equations with space periodic conditions. *Adv. in Appl. Probab., 45*(3), 742–772. https://doi.org/10.1239/aap/1377868537

Milstein, G., & Tretyakov, M. (2020). Mean-Square approximation of Navier-stokes equations with additive noise in vorticity-velocity formulation. *Numerical Mathematics: Theory, Methods and Applications, 14*(1), 1–30. https://doi.org/10.4208/nmtma.OA-2020-0034

Morillon, J.-P. (1997). Numerical solutions of linear mixed boundary value problems using stochastic representations. *International Journal for Numerical Methods in Engineering, 40*, 387–405. https://doi.org/10.1002/(SICI)1097-0207(19970215)40:3%3C387::AID-NME69%3E3.0.CO;2-D

Muller, M. E. (1958). An inverse method for the generation of random Normal deviates on large-scale computers. *Mathematical Tables and Aids to Computation, 12*, 167–174.

Muller, M. E. (1959). A comparison of methods for generating Normal deviates on digital computers. *Journal of the Association for Computing Machinery, 6*, 376–383.

Muller, M. E. (n.d.). *Generation of normal deviates* (Technical report no. 13. Statistical techniques research group). Princeton University Press.

Nash, J. F., Jr. (1950a). Equilibrium points in n-person games. *Proceedings of the National Academy of Sciences of the United States of America, 36*(1), 48–49.

Nash, J. F., Jr. (1950b, April). The bargaining problem. *Econometrica, 18*(2), 155–162.

Nataf, A. (1962). Détermination des distributions de probabilités dont les marges sont données. *Comptes rendus de l'Académie des Sciences Paris, 225*, 42–43.

Nowak, M. (2006). *Evolutionary dynamics: Exploring the equations of life*. Harvard University Press.

Organisation for Economic Co-operation and Development. (2021, 1 13). *Input-Output Tables (IOTs)*. Retrieved from OECD Homepage: https://www.oecd.org/sti/ind/input-outputtables.htm

Palacios-Huerta, I. (2003). Professionals play minimax. *Review of Economic Studies, 70*, 395–415.

Palacios-Huerta, I. (2014). *Beautiful game theory: How soccer can help economics*. Princeton University Press.

Palacios-Huerta, I., & Volij, O. (2008, January). Experientia Docet: Professionals play minimax in laboratory experiments. *Econometrica, 76*(1), 71–115.

Paley, R., Wiener, N., & Zygmund, A. (1933). Notes on random functions. *Mathematische Zeitschrift, 37*, 647–668. https://doi.org/10.1007/BF01474606

Papaioannou, I., Papadimitriou, C., & Straub, D. (2016). Sequential importance sampling for structural reliability analysis. *Structural Safety, 62*, 66–75. https://doi.org/10.1016/j.strusafe.2016.06.002

Pardoux, E., & Talay, D. (1985). Discretization and simulation of stochastic differential equations. *Acta Applicandae Mathematicae, 3*, 23–47. https://doi.org/10.1007/BF01438265

Pearson, K. (1905). The problem of random walk. *Nature, 294*, 342.

Phillips, P. C., & Perron, P. (1988, June). Testing for a unit root in time series regression. *Biometrika, 75*(2), 335–346. https://doi.org/10.1093/biomet/75.2.335

Pogu, M., & Souza de Cursi, E. (1994). Global optimization by random perturbation of the gradient method with a fixed parameter. *Journal of Global Optimization, 5*(2), 159–180.

Pourahmadi, M. (2001). *Foundations of time series analysis and prediction theory*. Wiley.

Prakash, P., & Garg, R. (2014, September). Preferred side of the penalty kick. *International Journal of Science, Technology and Management, 3*(9), 7–12.

Ravenstein, E. G. (1885, June). The Laws of Migration. *Journal of the Statistical Society, 48*(2), 167–227. https://doi.org/10.2307/2979181

Regnault, J. (1863). *Calcul des chances et philosophie de la bourse*. Castel, Mallet-Bachelier.

Reilly, W. J. (1929). *Methods for the Study of Retail Relationships.* Austin: University of Texas at Austin. https://doi.org/10.15781/T2XD0RC8G

Reilly, W. J. (1931). *The law of retail gravitation.* New York: Pilsbury Press.

Robert, C. P. (2006). *Le choix bayésien.* Springer.

Romeo, G. (2020). *Elements of mathematical economics with EXCEL.* Academic Press. https://doi.org/10.1016/C2018-0-02476-7

Rosenblatt, M. (1952). Remarks on a multivariate transformation. *Annals of Mathematical Statistics, 23*(3), 470–472. https://doi.org/10.1214/aoms/1177729394

Sarkar, A. (2017). *The Gambler's fallacy and hot outcome: Cognitive biases or adaptive thinking for Goalkeepers' decisions on dive direction during penalty shootouts*. Master thesis. Bowling Green State University.

Schwarz, G. (1978). Estimating the dimension of a model. *The Annals of Statistics, 6*(2), 461–464. https://doi.org/10.1214/aos/1176344136

Shapley, L. S. (1953). Stochastic games. *Proceedings of the National Academy of Sciences of the United States of America, 39*(10), 1095–1100.

Sinclair, B. (2005, June 9). *Machine repair model.* Retrieved from OpenStax CNX: http://cnx.org/contents/56f1bed0-bd34-4c28-a2ec-4a3f9ded8e18@3

Soetaert, K., Cash, J., & Mazzia, F. (2012). *Solving differential equations in R*. Springer. https://doi.org/10.1007/978-3-642-28070-2

Souza de Cursi, E. (1992). *Introduction aux Probabilités et Statistiques*. Ecole Centrale Nantes.

Souza de Cursi, E. (1994). Numerical methods for linear boundary value problems based on Feynman-Kac representations. *Mathematics and Computers in Simulation, 36*(1), 1–16.

Souza de Cursi, E. (2015). *Variational methods for engineers with Matlab*. ISTE/Wiley.

Souza de Cursi, E. (2021). Uncertainty quantification in game theory. *Chaos, Solitons & Fractals, Elsevier, 143*, 110558. https://doi.org/10.1016/j.chaos.2020.110558

Souza de Cursi, E., & Sampaio, R. (2010). Modelling and convexity.. ISTE-Wiley.

Souza de Cursi, E., & Sampaio, R. (2015). *Uncertainty quantification and stochastic modelling with Matlab*. ISTE/Elsevier.

Souza de Cursi, J. E., Ellaia, R., & Bouhadi, M. (2004). Global optimization under nonlinear restrictions by using stochastic perturbations of the projected gradient. In C. A. Floudas & P. Pardalos (Eds.), *Frontiers in global optimization* (pp. 541–561). Springer.

Stewart, J. Q. (1947). Empirical mathematical rules concerning the distribution and equilibrium of population. *Geographical Review, 37*(3), 461–485. https://doi.org/10.2307/211132

Stewart, J. Q. (1948). Demographic Gravitation: Evidence and Application. Sociometry, XI, 31–58. https://doi.org/10.2307/2785468

Stouffer, S. A. (1940). Intervening opportunities: A theory relating to mobility and distance. *American Sociological Review, 5*(6), 845–867. https://doi.org/10.2307/2084520

Stratonovich, R. L. (1966). A new representation for stochastic integrals and equations. *SIAM Journal on Control and Optimization, 4*(2), 362–371. https://doi.org/10.1137/0304028

Talay, D. (2015). Simulation of stochastic differential equations. In B. Engquist (Ed.), *Encyclopedia of applied and computational mathematics*. Springer. https://doi.org/10.1007/978-3-540-70529-1_346

Tobler, W. (1976). Spatial interaction patterns. *Journal of Environmental Systems, 6*(4), 271–301. https://doi.org/10.2190/VAKC-3GRF-3XUG-WY4W

Vasicek, O. (1977). An equilibrium characterization of the term structure. *Journal of Financial Economics, 5*(2), 177–188. https://doi.org/10.1016/0304-405X(77)90016-2

Verhulst, P. (1845). Recherches mathématiques sur la loi d'accroissement de la population. *Nouveaux mémoires de l'Académie Royale des Sciences et Belles-Lettres de Bruxelles, 18*, 14–54. Retrieved August 15, 2020, from http://eudml.org/doc/182533

Verhulst, P. (1847). Deuxième Mémoire sur la Loi d'Accroissement de la Population. *Mémoires de l'Académie Royale des Sciences, des Lettres et des Beaux-Arts de Belgique, 20*, 1–32. Retrieved August 20, 2020, from http://eudml.org/doc/178976

Volterra, V. (1926). *Variazioni e fluttuazioni del numero d'individui in specie animali conviventi* (Vol. 2). (http://www.liberliber.it, Ed.) Memoria della Reale Accademia Nazionale dei Lincei. Ser. VI. Retrieved August 12, 2020, from https://www.europeana.eu/pt/item/2022117/urn_axmedis_00000_obj_bd05ae74_d168_4c92_9a65_4f461377f7bd

Volterra, V. (1928, April). Variations and fluctuations of the number of individuals in animal species living together. *ICES Journal of Marine Science, 3*(1), 3–51. https://doi.org/10.1093/icesjms/3.1.3

Von Neumann, J. (1928). Zur Theorie der Gesellschaftsspiele. *Mathematische Annalen, 100*(1), 295–320.

Von Neumann, J., & Morgenstern, O. (1944). *Theory of games and economic behavior*. Princeton University Publishing.

Von Neumann, J., & Morgenstern, O. (1953). *Theory of games and economic behavior*. Princeton University Press.

Wilson, A. G. (1967). A Statistical theory of spatial distribution models. *Transportation Research*, 1(3), 253–269. https://doi.org/10.1016/0041-1647(67)90035-4

Wilson, A. G. (1970). *Entropy in Urban and Regional Modelling*. London: Pion. https://doi.org/10.4324/9780203142608

Wilson, A. G. (2010). Entropy in Urban and Regional Modelling: Retrospect and Prospect. *Geographical Analysis, 42*(4), 364–394. https://doi.org/10.1111/j.1538-4632.2010.00799.x

Zermelo, E. (1912). Über eine Anwendung der Mengenlehre auf die Theorie des Schachspiels. *Proceedings of the fifth international congress of mathematicians* (pp. 501–504). Cambridge University Press.

Zhou, Y., & Cai, W. (2016). Numerical solution of the Robin problem of Laplace equations with a Feynman-Kac formula and reflecting Brownian motions. *Journal of Scientific Computing, 69*(1), 107. https://doi.org/10.1007/s10915-016-0184-y

Zidani, H. (2013). *Représentation de solution en optimisation continue, multi-objectif et applications*. INSA/EMI.

Zipf, G. K. (1946). The P1 P2/D Hypothesis: On the intercity movement of persons. *American Sociological Review, 11*(6), 677–686. https://doi.org/10.2307/2087063

Zipf, G. K. (1949). *Human behavior and the principle of least effort: An introduction to human eoclogy*. Cambridge, MA: Addison-Wesley.

Index

E. Souza de Cursi, *Uncertainty Quantification using R*, International Series in Operations Research & Management Science 335,
https://doi.org/10.1007/978-3-031-17785-9